MW01251740

Compulsive Eating Behavior and Food Addiction

Compulsive Eating Behavior and Food Addiction

Emerging Pathological Constructs

Edited by

Pietro Cottone

Boston University School of Medicine, Boston, MA, USA

Valentina Sabino

Boston University School of Medicine, Boston, MA, USA

Catherine F. Moore

Boston University School of Medicine, Boston, MA, USA

George F. Koob

National Institute on Alcohol Abuse and Alcoholism, National Institutes of Health, Bethesda, MD, USA

Academic Press is an imprint of Elsevier
125 London Wall, London EC2Y 5AS, United Kingdom
525 B Street, Suite 1650, San Diego, CA 92101, United States
50 Hampshire Street, 5th Floor, Cambridge, MA 02139, United States
The Boulevard, Langford Lane, Kidlington, Oxford OX5 1GB, United Kingdom

Copyright © 2019 Elsevier Inc. All rights reserved.

No part of this publication may be reproduced or transmitted in any form or by any means, electronic or mechanical, including photocopying, recording, or any information storage and retrieval system, without permission in writing from the publisher. Details on how to seek permission, further information about the Publisher's permissions policies and our arrangements with organizations such as the Copyright Clearance Center and the Copyright Licensing Agency, can be found at our website: www.elsevier.com/permissions.

This book and the individual contributions contained in it are protected under copyright by the Publisher (other than as may be noted herein).

Notices

Knowledge and best practice in this field are constantly changing. As new research and experience broaden our understanding, changes in research methods, professional practices, or medical treatment may become necessary.

Practitioners and researchers must always rely on their own experience and knowledge in evaluating and using any information, methods, compounds, or experiments described herein. In using such information or methods they should be mindful of their own safety and the safety of others, including parties for whom they have a professional responsibility.

To the fullest extent of the law, neither the Publisher nor the authors, contributors, or editors, assume any liability for any injury and/or damage to persons or property as a matter of products liability, negligence or otherwise, or from any use or operation of any methods, products, instructions, or ideas contained in the material herein.

Library of Congress Cataloging-in-Publication Data
A catalog record for this book is available from the Library of Congress

British Library Cataloguing-in-Publication Data
A catalogue record for this book is available from the British Library

ISBN: 978-0-12-816207-1

For information on all Academic Press publications visit our website at https://www.elsevier.com/books-and-journals

Publisher: Nikki Levy
Acquisition Editor: Joslyn Chaiprasert-Paguio
Editorial Project Manager: Sandra Harron
Production Project Manager: Paul Prasad Chandramohan
Cover Designer: Christian Bilbow

Typeset by TNQ Technologies

Contents

Contributors

Iris M. Balodis
Peter Boris Centre for Addictions Research, Department of Psychiatry and Behavioural Neurosciences, McMaster University, St. Joseph's Healthcare Hamilton, Hamilton, ON, Canada

Revi Bonder
York University, Toronto, ON, Canada

Benjamin Boutrel
Center for Psychiatric Neuroscience, Department of Psychiatry, Lausanne University Hospital, Switzerland; Division of Adolescent and Child Psychiatry, Department of Psychiatry, Lausanne University Hospital, University of Lausanne, Switzerland

Timothy D. Brewerton
Department of Psychiatry and Behavioral Sciences, Medical University of South Carolina, Charleston, SC, United States

Jonathan E. Cheng
Laboratory of Addictive Disorders, Departments of Pharmacology and Psychiatry, Boston University School of Medicine, Boston, MA, United States

Pietro Cottone
Laboratory of Addictive Disorders, Departments of Pharmacology and Psychiatry, Boston University School of Medicine, Boston, MA, United States

Caroline Davis
York University, Toronto, ON, Canada

Fernando Fernández-Aranda
Department of Psychiatry, Bellvitge University Hospital-IDIBELL, Barcelona, Spain; Ciber Fisiopatología Obesidad y Nutrición (CIBERObn), Instituto de Salud Carlos III, Madrid, Spain; Department of Clinical Sciences, School of Medicine, University of Barcelona, Barcelona, Spain

Ashley N. Gearhardt
Department of Psychology, University of Michigan, Ann Arbor, MI, United States

Kirstie M. Herb
Eastern Michigan University, Department of Psychology, Ypsilanti, MI, United States

Susana Jiménez-Murcia
Department of Psychiatry, Bellvitge University Hospital-IDIBELL, Barcelona, Spain; Ciber Fisiopatología Obesidad y Nutrición (CIBERObn), Instituto de Salud Carlos III, Madrid, Spain; Department of Clinical Sciences, School of Medicine, University of Barcelona, Barcelona, Spain

Paul J. Kenny
Nash Family Department of Neuroscience, Icahn School of Medicine at Mount Sinai, New York, NY, United States

George F. Koob
National Institute on Alcohol Abuse and Alcoholism, National Institutes of Health, Bethesda, MD, United States; Neurobiology of Addiction Section, Intramural Research Program, National Institute on Drug Abuse, Baltimore, MD, United States

James MacKillop
Peter Boris Centre for Addictions Research, McMaster University/St. Joseph's Healthcare Hamilton, Hamilton, ON, United States

Gemma Mestre-Bach
Department of Psychiatry, Bellvitge University Hospital-IDIBELL, Barcelona, Spain; Ciber Fisiopatología Obesidad y Nutrición (CIBERObn), Instituto de Salud Carlos III, Madrid, Spain

Adrian Meule
Department of Psychology, University of Salzburg, Salzburg, Austria

Catherine F. Moore
Laboratory of Addictive Disorders, Departments of Pharmacology and Psychiatry, Boston University School of Medicine, Boston, MA, United States; Graduate Program for Neuroscience, Boston University School of Medicine, Boston, MA, United States

Cara M. Murphy
Center for Alcohol and Addiction Studies, Brown University, Providence, RI, United States

Katherine R. Naish
Peter Boris Centre for Addictions Research, Department of Psychiatry and Behavioural Neurosciences, McMaster University, St. Joseph's Healthcare Hamilton, Hamilton, ON, Canada

Marc N. Potenza
Department of Psychiatry, Yale School of Medicine, New Haven, CT, United States; Connecticut Council on Problem Gambling, Wethersfield, CT, United States; Connecticut Mental Health Center, New Haven, CT, United States; Department of Neuroscience and Child Study Center, Yale School of Medicine, New Haven, CT, United States

Clara Rossetti
Center for Psychiatric Neuroscience, Department of Psychiatry, Lausanne University Hospital, Switzerland; Division of Adolescent and Child Psychiatry, Department of Psychiatry, Lausanne University Hospital, University of Lausanne, Switzerland

Valentina Sabino
Laboratory of Addictive Disorders, Departments of Pharmacology and Psychiatry, Boston University School of Medicine, Boston, MA, United States

Karen K. Saules
Eastern Michigan University, Department of Psychology, Ypsilanti, MI, United States

Emma T. Schiestl
Department of Psychology, University of Michigan, Ann Arbor, MI, United States

Erica M. Schulte
Department of Psychology, University of Michigan, Ann Arbor, MI, United States

Eric Stice
Oregon Research Institute, Eugene, Oregon, United States

Sonja Yokum
Oregon Research Institute, Eugene, Oregon, United States

Eric P. Zorrilla
Department of Neuroscience, The Scripps Research Institute, La Jolla, CA, United States

Preface

With the sharp increase in rates of obesity and eating disorders in Western countries, a focus on the potential addicting properties of food has become a point of emphasis for researchers attempting to explain behaviors and neurobiological processes that may contribute to this growing epidemic. Drawing from analogous concepts in the addiction literature, compulsive eating behavior has emerged as a transdiagnostic construct, consisting of a pathological form of feeding that phenotypically, neurobiologically, and conceptually resembles compulsive-like behavior associated with both substance/alcohol-use disorders and behavioral addictions.

Recently, the scientific community has begun to embrace and evaluate the concept of addictive and compulsive eating behavior. A Web of Science search reveals a persistent, steady increase in compulsive eating research over time, coupled with a recent explosion of "food addiction" studies following the creation and validation of new diagnostic tools in 2009 (Fig. 1). While the scientific discussion on food addiction and compulsive eating behavior is in its nascent stage and the concepts are still somewhat controversial, this research holds enormous potential for improving treatment and prevention strategies for millions of people.

The book begins with "*A History of Food Addiction*" to place this concept and the current state of research into a historical context. Furthermore, the term and diagnosis of "food addiction" is explained in great detail by the researchers key to its development in "*Food Addiction Prevalence, Development, and Validation of*

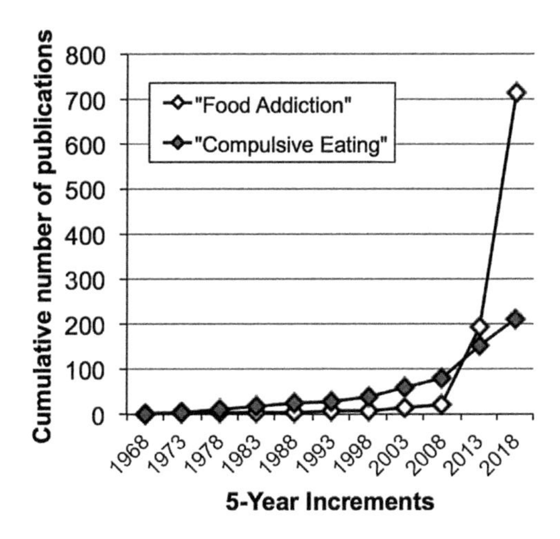

FIGURE 1 Number of scientific publications on food addiction and compulsive eating in recent decades.

Values were obtained by a Web of Science search for each 5-year span using the search terms "food addiction" and "compulsive eating."

Diagnostic Tools." Following this, we, the editors, wrote a chapter detailing on what have been identified as the elements of compulsive eating behavior in "*Dissecting Compulsive Eating Behavior into Three Elements*." This chapter describes the elements of (1) *habitual overeating*, (2) *overeating to relieve a negative emotional state,* and (3) *overeating despite negative consequences* and outlines their conception from the field of drug addiction after many behavioral and neurobiological overlaps were observed. A more in-depth breakdown of each element follows in Chapters 4–8. Chapter 4, "*Habitual Overeating,*" overviews the ways in which eating behavior can become inflexible and rote in compulsive eating. "*Reward Deficits in Compulsive Eating*" and "*The Dark Side of Compulsive Eating and Food Addiction*" describe the dual processes that make up the element of overeating to relieve a negative emotional state. In "*Food Addiction and Self-Regulation,*" the inhibitory control processes that underlie overeating despite negative consequences are discussed at length.

Other chapters serve to illustrate the overlaps among the elements of compulsive eating behavior and between these elements and other biological mechanisms. "*Reward Processing in Food Addiction and Overeating*" investigates the intersection of striatal reward processes with prefronto-cortical control circuits. In "*Interactions of Hedonic and Homeostatic Systems in Compulsive Eating*," the elements of compulsive eating are discussed in the context of highly relevant homeostatic feeding mechanisms.

For updates into specific technical fields of research into compulsive eating behavior, chapters on genetics ("*Genetics and Epigenetics of Food Addiction*"), neuroimaging ("*Neuroimaging of Compulsive Disorders: Similarities of Food Addiction with Drug Addiction*"), and animal models ("*Modeling and Testing Compulsive Eating Behavior in Animals*") were included.

We have also included a chapter to address the highly relevant topic of "*Sex and Gender Differences in Compulsive Overeating*." This chapter not only does an excellent and thorough job detailing the current evidence in this area but also highlights an area of inquiry with much left to understand in terms of biological mechanistic sex and gender differences in compulsive eating.

Furthermore, as the concept of food addiction and the consideration of forms of pathological overeating as addictive behaviors have been fraught with debate, we have a chapter devoted to "*Addressing Controversies Surrounding Food Addiction*." While we have a clear bias as editors of this book, this chapter serves to clarify some of the most prominent arguments that continue to fuel discussion on this topic. We hope to continue the discourse, while also using this book to highlight the undeniable breakthroughs in knowledge and mechanisms that have come about from the food addiction concept.

In the final chapter, "*Food Addiction and Its Associations to Trauma, Severity of Illness, and Comorbidity*," the utility of, and implications for, studying food addiction within the context of overall mental and physical health is discussed.

Each chapter stands on its own, and together all the chapters form a comprehensive picture of what drives compulsive eating behavior, how the prevalence compulsive eating, and how future therapeutic strategies may look. We have intended that this book be a bridge between preclinical and clinical researchers and drives further excitement in this rich and continually developing field.

Some common definitions of terms used throughout the book:

1. *Binge eating*: Eating within a 2-hour period of time an amount of food larger than what most people would eat in a similar period of time under similar circumstances and a sense of lack of control over eating during the episode.
2. *Binge eating disorder*: Recurrent (i.e., >1x weekly for 3 months, on average) binge eating as defined above, coupled with marked distress regarding binge eating and three or more cognitive symptoms (e.g., eating alone out of embarrassment, feeling disgusted/guilty). Importantly, no compensatory behaviors (e.g., purging) are associated with binge eating.
3. *Compulsivity*: Repetitive behaviors in the face of adverse consequences as well as repetitive behaviors that are inappropriate to a particular situation. Compulsivity has historical roots in the symptoms related to obsessive compulsive disorder, impulse control disorders, and substance-use disorders and may involve engagement in compulsive behaviors to prevent or relieve distress, anxiety, or stress.
4. *Compulsive eating*: Broadly defined as an irresistible, uncontrollable urge to overeat despite efforts to control this behavior. Compulsive eating behavior manifests as one or more of its constituent elements: habitual overeating, overeating to alleviate a negative emotional state, or overeating despite negative consequences.
5. *Food addiction*: Eating-related problems assessed by a recently created psychometric measurement tool: the Yale Food Addiction Scale (YFAS). This scale was originally modified in 2009 from the substance-dependence criteria described by the Diagnostic and Statistical Manual (DSM, fourth ed.) and subsequently updated to reflect changes to the substance-use disorder diagnosis in the DSM-5. A diagnosis of food addiction is given when a patient displays clinically significant impairment or distress and meets criteria, such as eating much more than intended and experiencing problems in ability to function because of food. In the updated YFAS, a severity score is calculated based on number of symptoms endorsed (2−3 = mild, 4−5 = moderate, 6+ = severe). Importantly, this emerging but not fully established condition is different from the already well-recognized feeding-related pathologies, and further validation is necessary.
6. *Overeating*: Consuming an excessive amount of food relative to energy expended.
7. *Overweight/obesity*: A body mass index ≥25 and ≥30 is considered overweight and obese, respectively, as defined by guidelines set forth by the World Health Organization. Overweight/obesity is neither necessary nor sufficient to characterize compulsive eating.

8. *Dopamine*: A key neurotransmitter in the basal ganglia, long implicated not only in Parkinson's disease but also in incentive salience. In driving, incentive salience dopamine conveys motivational properties to previously neutral stimuli perpetuating cue and context reward seeking. Dopamine is not a reward neurotransmitter per se. In fact, the midbrain dopamine system neurons decrease firing to repeated presentation of predicted rewards, but it is reactivated by unpredictable rewards. It is critical for the rewarding properties of psychostimulant drugs and through its incentive motivational actions promotes reward seeking in general.
9. *Addiction*: A chronically relapsing disorder characterized by compulsive drug seeking, loss of control over drug intake, and emergence of a negative emotional state when the drug is removed.
10. *Brain reward system*: The medial forebrain bundle and its connections historically forms the brain reward system. It supports brain stimulation reward with the lowest currents of all structures in the brain. It is composed of not only ascending monoamine pathways but also prominent descending pathways from the basal forebrain to the midbrain and brainstem. The mesocorticolimbic dopamine pathway projects from the ventral tegmental area and parts of the substantia nigra to the ventral striatum and prefrontal cortex. It is not the reward system per se but contributes to incentive salience (as described above).

CHAPTER

1 A history of "food addiction"

Adrian Meule

Department of Psychology, University of Salzburg, Salzburg, Austria

A chocolate inebriate has appeared. His addiction has been for three years, and his general health is much impaired, principally the digestion. His only thought night and day is how to get chocolate.

The Quarterly Journal of Inebriety, Volume 12, Issue 4, October 1890 (p. 392)

Introduction

Concepts of diseases and mental disorders are not set in stone. References to *drink madness* can be found in ancient civilizations and terms such as *drunkenness*, *intemperance*, *inebriety*, *dipsomania*, or *alcoholism* were used in the 18th and 19th centuries to describe substance-related addictive disorders (White, 2000). While the fourth version of the Diagnostic and Statistical Manual of Mental Disorders (DSM) distinguished between *substance abuse* and *substance dependence* (American Psychiatric Association, 1994), this distinction has been repealed in its fifth revision. The DSM-5 now lists several *substance use disorders* and, for the first time, a non–substance-related addiction: *gambling disorder* (American Psychiatric Association, 2013).

Similar dynamics can be found in the field of eating disorders. Anorexia nervosa was the first eating disorder included in DSM-I in 1952 and appeared along pica and rumination in DSM-II in 1968 (Dell'Osso et al., 2016). Bulimia nervosa was added to the DSM-III in 1980. The DSM-IV yet again involved some slight changes in the categorization of eating disorders and now—in addition to changes made to the diagnostic criteria for anorexia and bulimia nervosa—the DSM-5 lists pica, rumination disorder, avoidant/restrictive food intake disorder, anorexia nervosa, bulimia nervosa, binge eating disorder, and other specified eating disorders (e.g., night eating syndrome).

In the light of high prevalence rates of obesity in the past decades, there is an increased interest if certain foods may have an addiction potential and if obese individuals—or at least a subgroup of them—can be considered "food-addicted." In fact, it seems widely accepted that "food addiction" is a relatively new idea that was conceived in the past 20 years to explain the rising obesity prevalence (Davis, Edge, & Gold, 2014; Yau, Gottlieb, Krasna, & Potenza, 2014). Yet, is this

Compulsive Eating Behavior and Food Addiction. https://doi.org/10.1016/B978-0-12-816207-1.00001-9
Copyright © 2019 Elsevier Inc. All rights reserved.

alleged "new disorder" really a new concept in an attempt to explain why nowadays so many people are obese? This chapter will demonstrate that the concept of "food addiction" actually has a long history and did not arise from the obesity pandemic.

References to addiction in relation to food in the 19th century

In the scientific literature, references to addiction in relation to food have been made as early as the late 19th century. In the first journal of addiction medicine—the *Journal of Inebriety* (1876–1914)—food was routinely mentioned (Davis & Carter, 2014; Weiner & White, 2007). When describing "diseased cravings," for example, Clouston (1890) referred to the stimulating effects of, craving for, and dependence on both food and alcohol (Table 1.1). Similarly, Crothers (1890a) cautions against some stimulating foods when describing how diseases in children with "alcoholic ancestors" should be treated (Table 1.1). Finally, a case of a "chocolate inebriate" is mentioned in the journal (Crothers, 1890b), describing his persistent craving for and preoccupation with chocolate as an addiction (Table 1.1).

A description of eating disorders in 1932

Mosche Wulff was a Soviet-Israeli physician and psychoanalyst who lived from 1878 to 1971. In 1932, he published an article in German in the International Journal of Psychoanalysis (Fig. 1.1), in which he describes case studies of five of his patients (Wulff, 1932). I refer interested readers to an article by Stunkard (1990) that provides a short biographical note on Moshe Wulff along with an English translation of some excerpts of his article. In a nutshell, Wulff's case studies include the description of binge eating, including precedent food craving and subsequent feelings of guilt as well as aspects of emotional eating (eating more in response to negative affect, eating less when in a positive mood) and restrained eating (periods of restriction between eating binges). Importantly, he calls the symptomatology of all five cases "eating addiction" (German: *Esssucht*) throughout the article and provides an explanation for using this term at the end (Table 1.1).

"Food addiction" in the 1950s

Following up on Wulff's observations, Hamburger (1951) noted the apparent parallels between recurrent binge eating episodes and gambling or drinking: "it is this eating pattern that most readily invites the label 'addictive'" (Table 1.1). The American physician Theron Randolph (1906–95) first used the term "food addiction" in the scientific literature in 1956 (Table 1.1). In contrast to modern views that associate addiction with the consumption of highly processed foods (Ifland et al., 2015;

Table 1.1 Some references and quotes demonstrating the long history of the "food addiction" idea.

References	Quote
Clouston (1890)	"It is a fact that some foods are more stimulating to the brain cortex than others, e.g., strong beef-tea than milk, flesh than bread. [...] If from childhood upwards the possessor of such a brain has depended on stimulating diet and drink for its restoration when exhausted, there is an intense and irresistible craving set up for such food and drink stimulants whenever there is fatigue. Such a brain has developed an affinity for them, and for such alone. Milk and farinaceous diet often become repugnant, and when taken do not satisfy the brain craving. Its owner becomes physiologically a flesh-eater and an alcohol-drinker." (p. 207)
Crothers (1890a)	"There is a special affinity for all nerve stimulants by those higher brain centers. Their use constantly interferes with the natural development of brain energy from food. Thus, alcohol, tea, coffee, and other substances have a peculiar delusive effect. [...] The diet should not include meats of any kind, because of their stimulating character; while meats contain much food force, they act as stimulants to a brain already over stimulated and exhausted, and increase the peril of nervous disease. The pathological tendency of all these cases is to become alcohol-takers and meat-eaters, hence the diet should always be non-stimulating and farinaceous, and should be carried out with military regularity." (p. 285)
Crothers (1890b)[a]	"A chocolate inebriate has appeared. His addiction has been for three years, and his general health is much impaired, principally the digestion. His only thought night and day is how to get chocolate." (p. 392)
Wulff (1932)[b]	"I have used the term "eating addiction" above without justifying why I deem it important to call it an "addiction" and not, for example, a compulsion. I believe that the nature of this compulsive eating can be best characterized by the term *addiction*. How do addiction and compulsion differ from each other except regarding the different manners through which they are experienced? [...] Another characteristic of a compulsion is the fact that its suppression produces anxiety while suppression of a compulsive, addiction-related urge increases the tension of the addictive desire (if withdrawal symptoms do not complicate the picture)—just as it was observed in the cases described here with regard to eating." (p. 299)
Hamburger (1951)	"A number of authors have described people who show extreme preoccupation with food and weight, who episodically consume enormous amounts of food, in short periods of time in an 'orgiastic' manner (episodes varying in frequency from more than once a day to once every few weeks), and who experience guilt, shame, depression and self-condemnation following 'binges.' The parallel with apparently 'compulsive' patterns of gambling or drinking is immediately striking. Indeed it is this eating pattern that most readily invites the label 'addictive'." (p. 487)
Randolph (1956)	"Food addiction—a specific adaptation to one or more regularly consumed foods to which a person is highly sensitive—produces a

Continued

Table 1.1 Some references and quotes demonstrating the long history of the "food addiction" idea.—*cont'd*

References	Quote
	common pattern of symptoms descriptively similar to those of other addictive processes." (p. 221)
Hinkle et al. (1959)	"One of the most common and difficult problems we face is that of food addiction, both in the genesis of diabetes and in its treatment. Are there physiological factors involved in this mechanism or is it all psychological? What is its relation to alcohol addiction and addiction to narcotics? [...] That is a good question, because these terms "food addiction" and "compulsive eating" are widely used and widely misunderstood." (p. 377)
Bell (1960)	"Social custom and occupational contact, as well as medical treatment and physiological need, can be responsible for the introduction of a person to the chemicals involved in addiction. Food addiction is the only one in which the chemicals and the person come together initially out of physiological necessity." (p. 1348)
Bell (1965)	"It is important for the physician to explain that an uncontrollable need for a drug or for alcohol is not planned; that initially these agents were used to produce temporary improvement in wellbeing; and that addiction is a very common type of human disability. It is helpful to compare tobacco and food addiction to alcohol and drug addiction, and to remove as much of the guilt and shame as possible at the first interview. The physician has a good chance of initiating motivation if at the end of the first interview the patient feels that he does not need to down-grade himself in order to accept his illness." (p. 230)

[a] *This quote is from an editorial for which authorship was not specified and, thus, the editor (T.D. Crothers) is indicated as author here.*
[b] *This article is in German and the quotes have been translated by the author of this chapter. An English translation of some excerpts of this article can be found in Stunkard (1990).*

Schulte, Avena, & Gearhardt, 2015), however, he noted that "most often involved are corn, wheat, coffee, milk, eggs, potatoes, and other frequently eaten foods" (Randolph, 1956, p. 221). Although "food addiction" did not appear in other scientific articles around this time, famous psychiatrist Albert J. Stunkard (1922–2014) noted during a panel discussion in 1959 that the term "food addiction" was widely used back then (Table 1.1; Hinkle, Knowles, Fischer, & Stunkard, 1959).

Varying themes in the second half of the 20th century

In 1960, Overeaters Anonymous was founded. This self-help organization is based on the 12-step program of Alcoholics Anonymous and, accordingly, uses an addiction framework for overeating. For example, in contrast to cognitive behavioral therapy, which emphasizes flexible food choices with no forbidden foods (Wilson, 2010), Overeaters Anonymous advocates abstinence from certain foods (Russell-Mayhew, von Ranson, & Masson, 2010). Yet, the term "food addiction"

Internationale Zeitschrift
für Psychoanalyse

Herausgegeben von Sigm. Freud

XVIII. Band 1932 Heft 3

Über einen interessanten oralen Symptomenkomplex und seine Beziehung zur Sucht

Vortrag in der Deutschen Psychoanalytischen Gesellschaft, am 12. April 1932

Von
M. Wulff
Berlin

FIGURE 1.1

Excerpt from the title page of an article by Mosche Wulff. It reads "International Journal of Psychoanalysis; Edited by Sigmund Freud; Volume 18; 1932; Issue 3; On an interesting oral symptom complex and its relationship to addiction; Lecture at the German Psychoanalytical Society, April 12th, 1932; By Mosche Wulff; Berlin."

was only occasionally mentioned in scientific articles in the 1960s and 1970s, primarily in the context of obesity (Table 1.1; Bell, 1960, 1965; Clemis, Shambaugh, & Derlacki, 1966; Swanson & Dinello, 1970; Thorner, 1970).

Notably, however, some cases of bulimia nervosa or binge/purge-subtype anorexia nervosa were described as an addiction in these decades as well (Vandereycken, 1994). For example, Ziolko (1966) presents a case of "hyperorexia," which he denotes—similar to Wulff (1932)—as "eating addiction" (i.e., *Esssucht* in German). In a report about an expert group discussion about overeating and vomiting, Garrow (1976) notes that "one group of subjects with chronic anorexia nervosa exemplify many aspects of addiction; they habitually/constantly ingest and vomit food in large quantities" (p. 407).

In the 1980s, the excessive food restriction displayed by individuals with anorexia nervosa was mentioned for the first time in the context of addiction (Scott, 1983). Similarly, Szmukler and Tantam (1984) described anorexia nervosa as an addiction—what they called *starvation dependence.* For example, they note that "patients with anorexia nervosa are dependent on the psychological and possibly physiological effects of starvation. Increased weight loss results from tolerance to starvation necessitating greater restriction of food to obtain the desired effect, and the later development of unpleasant 'withdrawal' symptoms on eating." (p. 309). Finally, Marrazzi et al. (Marrazzi et al., 1990; Marrazzi & Luby, 1986) compared anorexic phenomenology with addictive states in their auto-addiction opioid model of chronic anorexia nervosa.

Another approach stemming from an addiction perspective on eating was the examination of addictive personality in individuals with anorexia nervosa, bulimia nervosa, or obesity (Davis & Claridge, 1998; Feldman & Eysenck, 1986; Kayloe, 1993; Leon, Eckert, Teed, & Buchwald, 1979). Several studies compared whether individuals with anorexia nervosa, bulimia nervosa, or obesity scored higher than healthy controls and similar to individuals with tobacco use, alcohol use, or gambling disorder on certain addiction personality questionnaires (de Silva & Eysenck, 1987; Hatsukami, Owen, Pyle, & Mitchell, 1982; Kagan & Albertson, 1986; Leon, Kolotkin, & Korgeski, 1979).

In the 1990s, a particular interest emerged on addiction-like consumption of chocolate. Characteristics of chocolate such as its macronutrient composition, sensory properties, and ingredients such as caffeine and theobromine were discussed as contributors to its addictive-like nature (Bruinsma & Taren, 1999; Patterson, 1993; Rozin, Levine, & Stoess, 1991). Some studies investigated self-identified "chocolate addicts" (Hetherington & Macdiarmid, 1993; Macdiarmid & Hetherington, 1995; Tuomisto et al., 1999) or examined associations between "chocolate addiction" and other addictive behaviors (Greenberg, Lewis, & Dodd, 1999; Rozin & Stoess, 1993).

Besides these themes, a variety of different topics were covered in one or few single articles in the 1980s and 1990s. These include discussions of the role of endorphins in terms of an addictive response in obesity (Gold & Sternbach, 1984; Wise, 1981), substance abuse as a metaphor in the treatment of bulimia nervosa (Slive & Young, 1986), a "foodaholics" group treatment program (Stoltz, 1984), and some unusual case studies of addiction-like carrot consumption (Kaplan, 1996; Černý; Černý, 1992). Finally, the first critical reviews were published, which scrutinized adopting an addiction framework in the treatment of eating disorders (Bemis, 1985) and questioned the overall "food addiction" approach based on conceptual and physiological considerations (Rogers & Smit, 2000; Vandereycken, 1990; Wilson, 1991, 2000).

Increased popularity in the 21st century

Increased interest in "food addiction" in the early 2000s was largely driven by brain imaging studies in humans—particularly in individuals with obesity or binge eating disorder (Volkow, Wang, Fowler, & Telang, 2008)—and by animal models of addiction-like sugar intake (Avena, Rada, & Hoebel, 2008). Besides these lines of research, numerous review articles were published that discussed behavioral, cognitive, and neural parallels between obesity or binge eating disorder and substance dependence and examined whether the diagnostic criteria for substance dependence can be applied to food and eating (e.g., Barry, Clarke, & Petry, 2009; Corsica & Pelchat, 2010; Davis & Carter, 2009; Gearhardt, Corbin, & Brownell, 2009a; Ifland et al., 2009; Pelchat, 2009; Thornley, McRobbie, Eyles, Walker, & Simmons, 2008).

Correspondingly, several approaches were developed to measure addiction-like eating in humans based on translating substance dependence criteria to food and eating (Meule, 2011). For example, Cassin and von Ranson (2007) replaced references to *substance* by *binge eating* in the substance dependence module of the structured clinical interview for DSM-IV axis I disorders to "diagnose" addiction-like eating in individuals with binge eating disorder. Relatedly, Gearhardt, Corbin, and Brownell (2009b) developed the Yale Food Addiction Scale by adapting DSM-IV substance dependence criteria to food and eating. Scoring of this self-report questionnaire allows for a dichotomous classification of the presence or absence of "food addiction." It may be because of this uniqueness that the scale turned out to be widely used in the years that followed (Meule & Gearhardt, 2014).

Current developments

In 2013, gambling disorder was the first behavioral addiction that was added as an addictive disorder in addition to substance use disorders in DSM-5. Reflecting this nosological change, researchers have proposed that framing addiction-like eating as a behavioral addiction may be more appropriate than framing it as a substance-related disorder (Hebebrand et al., 2014). This approach has intuitive appeal and, at first glance, seems to resolve some controversies that are inherent in the substance-based "food addiction" approach. Yet, the "eating addiction" approach may create more problems than it solves. For example, efforts have been made to develop self-report measures for capturing "eating addiction" (Ruddock, Christiansen, Halford, & Hardman, 2017). Yet, "eating addiction" may be in fact even harder to distinguish than "food addiction" from existing concepts such as binge eating–related disorders (Schulte, Potenza, & Gearhardt, 2018; Vainik & Meule, 2018).

The current state of affairs can be broken down into three different views:

(1) certain foods are regarded as addictive substance(s), and, thus, so-called "food addiction" represents a substance-related addictive disorder (Ifland et al., 2015; Schulte, Potenza, & Gearhardt, 2017),

(2) it is not possible to identify a specific substance in foods that is addictive (similar to nicotine in tobacco, ethanol in alcoholic beverages, tetrahydrocannabinol in cannabis, etc.), and, thus, so-called "eating addiction" represents a behavioral addictive disorder (Hebebrand et al., 2014; Ruddock et al., 2017),

(3) neither "food addiction" nor "eating addiction" represent valid concepts or—even if they are—they are not necessary (Finlayson, 2017; Rogers, 2017; Ziauddeen & Fletcher, 2013).

While most writings on this topic clearly take up one of these three positions, it has also been argued that the addiction perspective on eating requires a more nuanced view (Fletcher & Kenny, 2018). For example, Lacroix, Tavares, and von Ranson (2018) emphasize that alternative conceptualizations of addictive-like eating may be overlooked when the discussion is framed as a dichotomous debate between food and eating addiction models. Such alternative views include, for example, considering compulsivity as a transdiagnostic construct in both addiction and pathological overeating (Moore, Sabino, Koob, & Cottone, 2017).

Conclusions

"Food addiction" is not a new idea that emerged in the 21st century because of the obesity pandemic. Instead, researchers have discussed for many decades whether humans can be addicted to certain foods and whether certain eating behaviors represent an addictive behavior. The history of "food addiction" research involves different perspectives, which range from mentioning food in the context of addiction in the late 19th century, describing binge eating as "eating addiction" in the 1930s, establishing the term "food addiction" in the 1950s, acknowledging the addiction-like character of binge eating in individuals with bulimia and binge/purge-subtype anorexia nervosa in the 1960 and 1970s to characterizing the self-starvation of individuals with anorexia nervosa as an addiction in the 1980s, and many more. Thus, research on "food addiction" encompasses a long history with dynamically changing but recurring themes. These include the types of food involved (e.g., chocolate and other foods), discussions about the appropriateness of a "food addiction" versus "eating addiction" rationale, and which type of individuals are involved (e.g., individuals with anorexia nervosa, bulimia nervosa, binge eating disorder, and/or obesity).

In spite of its long history, the "food addiction" versus "eating addiction" versus "not-an-addiction" discussion has developed to a lively debate in recent years. To move the field forward, researchers need to generate—and agree upon—testable predictions, which may include neural mechanisms (Fletcher & Kenny, 2018) or whether the construct of addictive-like eating holds incremental clinical utility over and above existing eating disorder diagnoses (Lacroix et al., 2018). Furthermore, providing an addiction framework in the prevention and treatment of eating disorders and obesity will likely be helpful in some instances but may be unnecessary or even counterproductive in others (Meule, 2019). Therefore, future studies need to systematically examine under which circumstances and for whom an addiction perspective on eating is beneficial for normalizing food intake and reduce eating-related distress.

References

American Psychiatric Association. (1994). *Diagnostic and Statistical Manual of Mental Disorders* (4th ed.). Washington, DC: American Psychiatric Association.

American Psychiatric Association. (2013). *Diagnostic and Statistical Manual of Mental Disorders* (5th ed.). Washington, DC: American Psychiatric Association.

Avena, N. M., Rada, P., & Hoebel, B. G. (2008). Evidence for sugar addiction: Behavioral and neurochemical effects of intermittent, excessive sugar intake. *Neuroscience & Biobehavioral Reviews, 32*, 20–39. https://doi.org/10.1016/j.neubiorev.2007.04.019.

Barry, D., Clarke, M., & Petry, N. M. (2009). Obesity and its relationship to addictions: Is overeating a form of addictive behavior? *American Journal on Addictions, 18*, 439–451. https://doi.org/10.3109/10550490903205579.

Bell, R. G. (1960). A method of clinical orientation to alcohol addiction. *Canadian Medical Association Journal, 83*, 1346–1352.

Bell, R. G. (1965). Defensive thinking in alcohol addicts. *Canadian Medical Association Journal, 92*, 228–231.

Bemis, K. M. (1985). "Abstinence" and "nonabstinence" models for the treatment of bulimia. *International Journal of Eating Disorders, 4*, 407–437. https://doi.org/10.1002/1098-108X(198511)4:4<407.

Bruinsma, K., & Taren, D. L. (1999). Chocolate: Food or drug? *Journal of the American Dietetic Association, 99*, 1249–1256. https://doi.org/10.1016/S0002-8223(99)00307-7.

Cassin, S. E., & von Ranson, K. M. (2007). Is binge eating experienced as an addiction? *Appetite, 49*, 687–690. https://doi.org/10.1016/j.appet.2007.06.012.

Černý, L., & Černý, K. (1992). Can carrots be addictive? An extraordinary form of drug dependence. *British Journal of Addiction, 87*, 1195–1197. https://doi.org/10.1111/j.1360-0443.1992.tb02007.x.

Clemis, J. D., Shambaugh, G. E., Jr., & Derlacki, E. L. (1966). Withdrawal reactions in chronic food addiction as related to chronic secretory otitis media. *Annals of Otology, Rhinology & Laryngology, 75*, 793–797. https://doi.org/10.1177/000348946607500319.

Clouston, T. S. (1890). Diseased cravings and paralyzed control: Dipsomania; morphinomania; chloralism; cocainism. *Journal of Inebriety, 12*, 203–245.

Corsica, J. A., & Pelchat, M. L. (2010). Food addiction: True or false? *Current Opinion in Gastroenterology, 26*, 165–169. https://doi.org/10.1097/MOG.0b013e328336528d.

Crothers, T. D. (1890a). Alcoholic heredity in diseases of children. *Journal of Inebriety, 12*, 281–286.

Crothers, T. D. (1890b). Editorial: Legal treatment of inebriates. *Journal of Inebriety, 12*, 386–392.

Davis, C., & Carter, J. C. (2009). Compulsive overeating as an addiction disorder. A review of theory and evidence. *Appetite, 53*, 1–8. https://doi.org/10.1016/j.appet.2009.05.018.

Davis, C., & Carter, J. C. (2014). If certain foods are addictive, how might this change the treatment of compulsive overeating and obesity? *Current Addiction Reports, 1*, 89–95. https://doi.org/10.1007/s40429-014-0013-z.

Davis, C., & Claridge, G. (1998). The eating disorders as addiction: A psychobiological perspective. *Addictive Behaviors, 23*, 463–475. https://doi.org/10.1016/S0306-4603(98)00009-4.

Davis, A. A., Edge, P. J., & Gold, M. S. (2014). New directions in the pharmacological treatment of food addiction, overeating, and obesity. In K. P. Rosenberg, & L. C. Feder (Eds.), *Behavioral Addictions* (pp. 185–213). London, UK: Academic Press.

de Silva, P., & Eysenck, S. (1987). Personality and addictiveness in anorexic and bulimic patients. *Personality and Individual Differences, 8*, 749–751. https://doi.org/10.1016/0191-8869(87)90077-8.

Dell'Osso, L., Abelli, M., Carpita, B., Pini, S., Castellini, G., Carmassi, C., & Ricca, V. (2016). Historical evolution of the concept of anorexia nervosa and relationships with orthorexia nervosa, autism, and obsessive–compulsive spectrum. *Neuropsychiatric Disease and Treatment, 12*, 1651–1660. https://doi.org/10.2147/NDT.S108912.

Feldman, J., & Eysenck, S. (1986). Addictive personality traits in bulimic patients. *Personality and Individual Differences, 7*, 923–926. https://doi.org/10.1016/0191-8869(86)90097-8.

Finlayson, G. (2017). Food addiction and obesity: Unnecessary medicalization of hedonic overeating. *Nature Reviews Endocrinology, 13*, 493–498. https://doi.org/10.1038/nrendo.2017.61.

Fletcher, P. C., & Kenny, P. J. (2018). Food addiction: A valid concept? *Neuropsychopharmacology, 43*, 2506–2513. https://doi.org/10.1038/s41386-018-0203-9.

Garrow, J. S. (1976). Pathology of eating group report. In T. Silverstone (Ed.), *Appetite and Food Intake* (pp. 405–416). Berlin: Abakon.

Gearhardt, A. N., Corbin, W. R., & Brownell, K. D. (2009a). Food addiction: An examination of the diagnostic criteria for dependence. *Journal of Addiction Medicine, 3*, 1–7. https://doi.org/10.1097/ADM.0b013e318193c993.

Gearhardt, A. N., Corbin, W. R., & Brownell, K. D. (2009b). Preliminary validation of the Yale Food Addiction Scale. *Appetite, 52*, 430–436. https://doi.org/10.1016/j.appet.2008.12.003.

Gold, M. S., & Sternbach, H. A. (1984). Endorphins in obesity and in the regulation of appetite and weight. *Integrative Psychiatry, 2*, 203–207.

Greenberg, J. L., Lewis, S. E., & Dodd, D. K. (1999). Overlapping addictions and self-esteem among college men and women. *Addictive Behaviors, 24*, 565–571. https://doi.org/10.1016/S0306-4603(98)00080-X.

Hamburger, W. W. (1951). Emotional aspects of obesity. *Medical Clinics of North America, 35*, 483–499.

Hatsukami, D., Owen, P., Pyle, R., & Mitchell, J. (1982). Similarities and differences on the MMPI between women with bulimia and women with alcohol or drug abuse problems. *Addictive Behaviors, 7*, 435–439. https://doi.org/10.1016/0306-4603(82)90015-6.

Hebebrand, J., Albayrak, Ö., Adan, R., Antel, J., Dieguez, C., de Jong, J., ... Murphy, M. (2014). "Eating addiction", rather than "food addiction", better captures addictive-like eating behavior. *Neuroscience & Biobehavioral Reviews, 47*, 295–306. https://doi.org/10.1016/j.neubiorev.2014.08.016.

Hetherington, M. M., & Macdiarmid, J. I. (1993). "Chocolate addiction": A preliminary study of its description and its relationship to problem eating. *Appetite, 21*, 233–246. https://doi.org/10.1006/appe.1993.1042.

Hinkle, L. E., Knowles, H. C., Fischer, A., & Stunkard, A. J. (1959). Role of environment and personality in management of the difficult patient with diabetes mellitus: Panel discussion. *Diabetes, 8*, 371–378. https://doi.org/10.2337/diab.8.5.371.

Ifland, J., Preuss, H. G., Marcus, M. T., Rourk, K. M., Taylor, W. C., Burau, K., ... Manso, G. (2009). Refined food addiction: A classic substance use disorder. *Medical Hypotheses, 72*, 518–526. https://doi.org/10.1016/j.mehy.2008.11.035.

Ifland, J., Preuss, H. G., Marcus, M. T., Rourke, K. M., Taylor, W., & Wright, H. T. (2015). Clearing the confusion around processed food addiction. *Journal of the American College of Nutrition, 34*, 240–243. https://doi.org/10.1080/07315724.2015.1022466.

Kagan, D. M., & Albertson, L. M. (1986). Scores on MacAndrew Factors - bulimics and other addictive populations. *International Journal of Eating Disorders, 5*, 1095–1101. https://doi.org/10.1002/1098-108X(198609)5:6.

Kaplan, R. (1996). Carrot addiction. *Australian and New Zealand Journal of Psychiatry, 30*, 698–700. https://doi.org/10.3109/00048679609062670.

Kayloe, J. C. (1993). Food addiction. *Psychotherapy, 30*, 269–275. https://doi.org/10.1037/0033-3204.30.2.269.

Lacroix, E., Tavares, H., & von Ranson, K. M. (2018). Moving beyond the “eating addiction” versus “food addiction” debate: Comment on Schulte et al. (2017). *Appetite, 130*, 286–292. https://doi.org/10.1016/j.appet.2018.06.025.

Leon, G. R., Eckert, E. D., Teed, D., & Buchwald, H. (1979a). Changes in body image and other psychological factors after intestinal bypass surgery for massive obesity. *Journal of Behavioral Medicine, 2*, 39–55. https://doi.org/10.1007/BF00846562.

Leon, G. R., Kolotkin, R., & Korgeski, G. (1979b). MacAndrew Addiction Scale and other MMPI characteristics associated with obesity, anorexia and smoking behavior. *Addictive Behaviors, 4*, 401–407. https://doi.org/10.1016/0306-4603(79)90011-X.

Macdiarmid, J. I., & Hetherington, M. M. (1995). Mood modulation by food: An exploration of affect and cravings in ’chocolate addicts’. *British Journal of Clinical Psychology, 34*, 129–138. https://doi.org/10.1111/j.2044-8260.1995.tb01445.x.

Marrazzi, M. A., & Luby, E. D. (1986). An auto-addiction opioid model of chronic anorexia nervosa. *International Journal of Eating Disorders, 5*, 191–208. https://doi.org/10.1002/1098-108X(198602)5:2<191.

Marrazzi, M. A., Mullingsbritton, J., Stack, L., Powers, R. J., Lawhorn, J., Graham, V., … Gunter, S. (1990). Atypical endogenous opioid systems in mice in relation to an auto-addiction opioid model of anorexia nervosa. *Life Sciences, 47*, 1427–1435. https://doi.org/10.1016/0024-3205(90)90521-R.

Meule, A., & Gearhardt, A. N. (2014). Five years of the Yale Food Addiction Scale: Taking stock and moving forward. *Current Addiction Reports, 1*, 193–205. https://doi.org/10.1007/s40429-014-0021-z.

Meule, A. (2011). How prevalent is "food addiction"? *Frontiers in Psychiatry, 2*(61), 1–4. https://doi.org/10.3389/fpsyt.2011.00061.

Meule, A. (2019). A critical examination of the practical implications derived from the food addiction concept. *Current Obesity Reports, 8*, 11–17. https://doi.org/10.1007/s13679-019-0326-2.

Moore, C. F., Sabino, V., Koob, G. F., & Cottone, P. (2017). Pathological overeating: Emerging evidence for a compulsivity construct. *Neuropsychopharmacology, 42*, 1375–1389. https://doi.org/10.1038/npp.2016.269.

Patterson, R. (1993). Recovery from this addiction was sweet indeed. *Canadian Medical Association Journal, 148*, 1028–1032.

Pelchat, M. L. (2009). Food addiction in humans. *Journal of Nutrition, 139*, 620–622. https://doi.org/10.3945/jn.108.097816.

Randolph, T. G. (1956). The descriptive features of food addiction: Addictive eating and drinking. *Quarterly Journal of Studies on Alcohol, 17*, 198–224.

Rogers, P. J., & Smit, H. J. (2000). Food craving and food "addiction": A critical review of the evidence from a biopsychosocial perspective. *Pharmacology, Biochemistry and Behavior, 66*, 3–14. https://doi.org/10.1016/S0091-3057(00)00197-0.

Rogers, P. J. (2017). Food and drug addictions: Similarities and differences. *Pharmacology, Biochemistry and Behavior, 153*, 182–190. https://doi.org/10.1016/j.pbb.2017.01.001.

Rozin, P., & Stoess, C. (1993). Is there a general tendency to become addicted? *Addictive Behaviors, 18*, 81–87. https://doi.org/10.1016/0306-4603(93)90011-W.

Rozin, P., Levine, E., & Stoess, C. (1991). Chocolate craving and liking. *Appetite, 17*, 199–212. https://doi.org/10.1016/0195-6663(91)90022-K.

Ruddock, H., Christiansen, P., Halford, J., & Hardman, C. (2017). The development and validation of the Addiction-Like Eating Behaviour Scale. *International Journal of Obesity, 41*, 1710–1717. https://doi.org/10.1038/ijo.2017.158.

Russell-Mayhew, S., von Ranson, K. M., & Masson, P. C. (2010). How does Overeaters Anonymous help its members? A qualitative analysis. *European Eating Disorders Review, 18*, 33–42. https://doi.org/10.1002/erv.966.

Schulte, E. M., Avena, N., & Gearhardt, A. (2015). Which foods may be addictive? The roles of processing, fat content, and glycemic load. *PLoS One, 10*(2), e0117959. https://doi.org/10.1371/journal.pone.0117959.

Schulte, E. M., Potenza, M. N., & Gearhardt, A. N. (2017). A commentary on the "eating addiction" versus "food addiction" perspectives on addictive-like food consumption. *Appetite, 115*, 9–15. https://doi.org/10.1016/j.appet.2016.10.033.

Schulte, E. M., Potenza, M., & Gearhardt, A. (2018). How much does the Addiction-Like Eating Behavior Scale add to the debate regarding food versus eating addictions? *International Journal of Obesity, 42*, 946. https://doi.org/10.1038/ijo.2017.265.

Scott, D. W. (1983). Alcohol and food abuse: Some comparisons. *British Journal of Addiction, 78*, 339–349. https://doi.org/10.1111/j.1360-0443.1983.tb02521.x.

Slive, A., & Young, F. (1986). Bulimia as substance abuse: A metaphor for strategic treatment. *Journal of Strategic & Systemic Therapies, 5*, 71–84. https://doi.org/10.1521/jsst.1986.5.3.71.

Stoltz, S. G. (1984). Recovering from foodaholism. *Journal for Specialists in Group Work, 9*, 51–61. https://doi.org/10.1080/01933928408412515.

Stunkard, A. (1990). A description of eating disorders in 1932. *American Journal of Psychiatry, 147*, 263–268. https://doi.org/10.1176/ajp.147.3.263.

Swanson, D. W., & Dinello, F. A. (1970). Follow-up of patients starved for obesity. *Psychosomatic Medicine, 32*, 209–214. https://doi.org/10.1097/00006842-197003000-00007.

Szmukler, G. I., & Tantam, D. (1984). Anorexia nervosa: Starvation dependence. *British Journal of Medical Psychology, 57*, 303–310. https://doi.org/10.1111/j.2044-8341.1984.tb02595.x.

Thorner, H. A. (1970). On compulsive eating. *Journal of Psychsomatic Research, 14*, 321–325. https://doi.org/10.1016/0022-3999(70)90060-7.

Thornley, S., McRobbie, H., Eyles, H., Walker, N., & Simmons, G. (2008). The obesity epidemic: Is glycemic index the key to unlocking a hidden addiction? *Medical Hypotheses, 71*, 709–714. https://doi.org/10.1016/j.mehy.2008.07.006.

Tuomisto, T., Hetherington, M. M., Morris, M.-F., Tuomisto, M. T., Turjanmaa, V., & Lappalainen, R. (1999). Psychological and physiological characteristics of sweet food "addiction". *International Journal of Eating Disorders, 25*, 169–175. https://doi.org/10.1002/(SICI)1098-108X(199903)25:2<169.

Vainik, U., & Meule, A. (2018). Jangle fallacy epidemic in obesity research: A comment on Ruddock et al. (2017). *International Journal of Obesity, 42*, 585–586. https://doi.org/10.1038/ijo.2017.264.

Vandereycken, W. (1990). The addiction model in eating disorders: Some critical remarks and a selected bibliography. *International Journal of Eating Disorders, 9*, 95–101. https://doi.org/10.1002/1098-108X(199001)9:1<95.

Vandereycken, W. (1994). Emergence of bulimia nervosa as a separate diagnostic entity: Review of the literature from 1960 to 1979. *International Journal of Eating Disorders, 16*, 105–116. https://doi.org/10.1002/1098-108X(199409)16:2<105.

Volkow, N. D., Wang, G.-J., Fowler, J. S., & Telang, F. (2008). Overlapping neuronal circuits in addiction and obesity: Evidence of systems pathology. *Philosophical Transactions of the Royal Society B, 363*, 3191–3200. https://doi.org/10.1098/rstb.2008.0107.

Weiner, B., & White, W. (2007). The Journal of Inebriety (1876–1914): History, topical analysis, and photographic images. *Addiction, 102*, 15–23. https://doi.org/10.1111/j.1360-0443.2006.01680.x.

White, W. (2000). Addiction as a disease: Birth of a concept. *Counselor, 1*(1), 46–51.

Wilson, G. T. (1991). The addiction model of eating disorders: A critical analysis. *Advances in Behaviour Research and Therapy, 13*, 27–72. https://doi.org/10.1016/0146-6402(91)90013-Z.

Wilson, G. T. (2000). Eating disorders and addiction. *Drugs & Society, 15*, 87–101. https://doi.org/10.1300/J023v15n01_05.

Wilson, G. T. (2010). Eating disorders, obesity and addiction. *European Eating Disorders Review, 18*, 341–351. https://doi.org/10.1002/erv.1048.

Wise, J. (1981). Endorphins and metabolic control in the obese: A mechanism for food addiction. *Journal of Obesity & Weight Regulation, 1*, 165–181.

Wulff, M. (1932). Über einen interessanten oralen Symptomenkomplex und seine Beziehungen zur Sucht. *Internationale Zeitschrift für Psychoanalyse, 18*, 281–302.

Yau, Y. H. C., Gottlieb, C. D., Krasna, L. C., & Potenza, M. N. (2014). Food addiction: Evidence, evaluation, and treatment. In K. P. Rosenberg, & L. C. Feder (Eds.), *Behavioral Addictions* (pp. 143–184). London, UK: Academic Press.

Ziauddeen, H., & Fletcher, P. C. (2013). Is food addiction a valid and useful concept? *Obesity Reviews, 14*, 19–28. https://doi.org/10.1111/j.1467-789X.2012.01046.x.

Ziolko, H. U. (1966). Zur Psychodynamik der Ess- und Stehlsucht (Hyperorexie und Kleptomanie). *Psychotherapy and Psychosomatics, 14*, 226–236. https://doi.org/10.1159/000285827.

CHAPTER

Food addiction prevalence: development and validation of diagnostic tools

2

Ashley N. Gearhardt[a], Erica M. Schulte, Emma T. Schiestl

Department of Psychology, University of Michigan, Ann Arbor, MI, United States

There has been a steady increase in both public interest and scientific evaluation of the food addiction concept in the last 20 years (Davis, 2013; Meule, 2015). This has been driven by a number of factors, particularly the global pandemic of obesity that has accompanied the spread of highly rewarding, cheap, and accessible processed foods (e.g., fast food, pastries). Obesity is now a leading cause of preventable death, a major factor in reduced life expectancies, and a contributor to increasing burden on the medical system (Kelly, Yang, Chen, Reynolds, & He, 2008; Mokdad, Marks, Stroup, & Gerberding, 2004; Ng et al., 2014). As the public health consequences of excess body weight have become more apparent, basic science has identified striking parallels between the biopsychosocial mechanisms underpinning addictive disorders and excess food consumption (Ahmed, Guillem, & Vandaele, 2013; Avena, Rada, & Hoebel, 2008; Johnson & Kenny, 2010; Parylak, Koob, & Zorrilla, 2011), such as dysfunction in reward, motivation, stress, and inhibitory control systems. This has led to the hypothesis that an addictive process may be contributing to overeating, for at least a subset of individuals (Gold, Frost-Pineda, & Jacobs, 2003; Gold, Graham, Cocores, & Nixon, 2009). Given that food consumption is necessary for survival and eating past satiety is a relatively common occurrence, one of the challenges of this emerging research is how best to assess and determine the prevalence of clinically relevant food addiction. In the current chapter, we will (1) consider initial approaches to identifying addictive-like eating, (2) discuss the development of the most established measure of food addiction (i.e., Yale Food Addiction Scale (YFAS)) and the prevalence of its "diagnostic" threshold score, (3) consider an alternative framework for conceptualizing addictive-like eating, and (4) outline future research directions that may advance understanding of the clinical relevance of food addiction.

[a] Ashley N. Gearhardt is the Senior author.

Compulsive Eating Behavior and Food Addiction. https://doi.org/10.1016/B978-0-12-816207-1.00002-0
Copyright © 2019 Elsevier Inc. All rights reserved.

Early approaches to identifying food addiction

To survive in times of famine, humans have evolved to find certain tastes (e.g., sweetness associated with sugar) or mouthfeel (e.g., creaminess associated with fat content) that reflect high calorie content especially rewarding (Lieberman, 2006). For much of human history, these rewarding signals from food were confined to naturally occurring foods, such as fruits or nuts. Over time, human ingenuity allowed for the creation of new food products, such as chocolates and cakes, which contained artificially high levels of rewarding ingredients such as sugar and fat. The idea that these highly rewarding foods (e.g., chocolate) could be eaten in an addictive way has been present since the 1800s (Meule, 2015), but access to large quantities of highly rewarding foods was often restricted to affluent individuals, as ingredients such as sugar were expensive (Mintz, 1986). Over time, with changes in food science and economic policy, highly rewarding foods are no longer a rare treat but now dominate the modern food environment. Highly processed foods that have artificially high levels of refined carbohydrates (e.g., sugar, white flour) and fat (e.g., pizza, chocolate, chips) are easily accessible, inexpensive, and heavily marketed (Monteiro, Levy, Claro, de Castro, & Cannon, 2010). As these highly processed foods have become more integrated into the food environment around the globe, rates of obesity (particularly severe obesity), overeating, and diet-related disease have soon followed (Monteiro, Moubarac, Cannon, Ng, & Popkin, 2013).

Parallels between highly processed foods and drugs of abuse suggest that an addictive process may be contributing to the widespread overconsumption of these foods. The strongest evidence for this has been observed in basic science models, where exposure to highly processed foods has been related to both biological and behavioral indicators of addiction. Animals exposed to highly processed foods (e.g., cheesecake, Oreo cookies) have exhibited neurobiological (e.g., dysfunction in the mesolimbic dopamine system) and behavioral changes (e.g., binge behavior, heightened motivational drive) that are consistent with addiction (Avena et al., 2008; Johnson & Kenny, 2010; Oginsky, Goforth, Nobile, Lopez-Santiago, & Ferrario, 2016; Oswald, Murdaugh, King, & Boggiano, 2011; Parylak et al., 2011). Animal models have also demonstrated that the consumption of highly processed foods leads to changes indicative of the "dark" side of addiction by increasing the likelihood that these animals will experience negative affective states (e.g., anhedonia, anxiety) (Parylak et al., 2011). This enhances motivation to consume these foods in an effort to both reduce negative emotional states and avoid withdrawal-like states when access to these foods is limited (Parylak et al., 2011). Furthermore, a common ingredient in highly processed foods, sugar, appears to be such a powerful reinforcer that animals will overwhelmingly choose access to it over drugs of abuse (e.g., cocaine), even if they are exhibiting signs of dependence to the drug (Ahmed et al., 2013). Thus, in animals, these palatable, rewarding, and highly processed foods appear to have a notable addictive potential.

In humans, obesity has been associated with neural responses implicated in addiction. For example, obesity has been related to dysfunction in the mesolimbic dopamine systems, also observed in addiction (Volkow, Wang, Fowler, & Telang, 2008). Both obesity and addiction have also been associated with problems related to cue reactivity, habit learning, self-control, stress reactivity, and interoceptive awareness (Volkow, Wang, Tomasi, & Baler, 2013). However, there are concerns with using obesity as a proxy for addiction. Obesity is a medical condition that can result from a number of factors, including physical inactivity, medication side effects, and genetic conditions (Grundy, 1998). Although obesity can reflect elevated intake of highly caloric foods (Rosenheck, 2008), excess consumption is not necessarily indicative of addiction. For example, 40% of college students binge drink (O'Malley & Johnston, 2002), but only 6% meet criteria for alcohol dependence (Knight et al., 2002). Additionally, individuals who have an addictive-like relationship with food may be able to engage in behaviors (e.g., fasting, purging, excessive exercising) that may lead to a body mass index (BMI) in the normal range (Meule, 2012). Thus, the use of obesity as a proxy for food addiction may both over- and underidentify a phenotype consistent with an addictive response to highly processed foods. Furthermore, there have been other conceptualizations of addictive-like eating that have not relied on BMI but instead on self-identification as a "chocoholic" or "carb craver" (Spring et al., 2008; Tuomisto et al., 1999). However, these self-identified labels have not included an assessment of behavioral symptoms of addiction and thus may represent a strong desire for a certain food type rather than the construct of addiction as defined by the medical and scientific community.

Yale Food Addiction Scale

Given the lack of a standardized definition of food addiction, the original YFAS was developed in 2009 (Gearhardt, Corbin, & Brownell, 2009). The YFAS applied the Diagnostic and Statistical Manual of Mental Disorder (DSM) IV diagnostic criteria for substance dependence (e.g., loss of control, continued use despite negative consequences, withdrawal, tolerance) to the consumption of highly processed foods (e.g., chocolate, pizza) (see Table 2.1 for DSM-IV and DSM-5 diagnostic criteria) (American Psychiatric Association, 2000). The resulting 25-item measure has been found to be psychometrically sound demonstrating internal consistency, test–retest reliability, convergent, discriminant, incremental, and predictive utility (Meule & Gearhardt, 2014). The YFAS provides two scoring options: (1) a continuous symptom count that ranges from zero to seven symptoms met and (2) a "diagnostic" threshold of three of more symptoms plus clinically significant impairment or distress paralleling the DSM-IV threshold for substance dependence (Gearhardt et al., 2009). The YFAS has also been translated and validated in a number of other languages, including German, Chinese, and Spanish (Chen, Tang, Guo, Liu, & Xiao, 2015; Granero et al., 2014; Meule, Heckel, & Kübler, 2012).

Prevalence estimates of YFAS food addiction seem to vary by the characteristics of the sample. A metaanalysis places the weighted mean prevalence of YFAS food addiction at 19.9% (Pursey, Stanwell, Gearhardt, Collins, & Burrows, 2014),

Table 2.1 Alterations from the DSM-IV substance-dependence criteria to the DSM-5 substance-use disorder criteria.

DSM-IV substance-dependence criteria	DSM-5 substance-use disorder criteria
1. Substance taken in larger amount and for a longer period than intended **2.** Persistent desire or repeated attempt to quit **3.** Much time/activity to obtain, use, or recover **4.** Important social, occupational, or recreational activities given up or reduced **5.** Use continues despite knowledge of adverse consequences (e.g., failure to fulfill role obligation, use when physically hazardous) **6.** Tolerance (marked increase in amount; marked decrease in effect) **7.** Characteristic withdrawal symptoms; substance taken to relieve withdrawal	**1.** Substance taken in larger amounts or over a longer period than was intended **2.** Persistent desire of unsuccessful efforts to quit **3.** Much time/activity to obtain, use, or recover **4.** Craving or a strong desire/urge to use **5.** Recurrent use resulting in failure to fulfill major role obligations at work, school, or home. **6.** Continued used despite persistent social/interpersonal problems **7.** Important social, occupational, or recreational activities given up or reduced **8.** Recurrent use in situations where it is physical hazardous **9.** Substance use continued despite persistent/recurrent physical or psychological problems **10.** Tolerance (marked increase in amount; marked decrease in effect) **11.** Characteristic withdrawal symptoms; substance taken to relieve withdrawal

Note: *The YFAS and the YFAS-C are based on the DSM-IV criteria for substance dependence. The YFAS 2.0 and the YFAS-C 2.0 are based on the DSM-5 criteria for substance-use disorder.*

although this number may be a high estimate because of the number of clinical samples (e.g., persons with eating disorders) included in the study. Many studies have found that women relative to men endorse higher rates of YFAS food addiction, although this gender difference has not always been found in all studies (Meule & Gearhardt, 2014; Pursey et al., 2014). No consistent pattern of differences in food addiction rate by race/ethnicity was identified with the original YFAS (Pursey et al., 2014). However, food addiction has reliably differed by weight status and clinical severity. For example, relatively healthy college students with low rates of obesity exhibit food addiction rates around 5%–10% (Meule & Gearhardt, 2014), whereas individuals with obesity who are undergoing bariatric surgery exhibit significantly elevated rates around 14%–58% (Ivezaj, Wiedemann, & Grilo, 2017). Although rates of food addiction seem to be higher in individuals with obesity, it is important to note that not all individuals with obesity meet the food addiction criteria and not all individuals who meet the food addiction criteria are obese (Pursey et al., 2014). Thus, as discussed previously, solely relying on BMI as a proxy of addictive-like responses to food likely introduces measurement error.

One of the main questions about the concept of food addiction is whether it is sufficiently distinct from existent eating disorders (e.g., anorexia nervosa (AN), bulimia nervosa (BN), binge eating disorder (BED)). Some overlap between food addiction and binge-focused eating disorders (i.e., BN, BED) is to be expected, as similar criteria are common across both constructs, including loss of control over consumption and an inability to stop despite a desire to do so (Gearhardt, White, & Potenza, 2011). Additionally, addiction and eating disorder perspectives implicate dysfunction in many of the same mechanisms (e.g., impulsivity, reward dysfunction, emotion dysregulation) (Schulte, Grilo, & Gearhardt, 2016). The similar phenotypic presentations and mechanistic underpinnings of these disorders provide support for the contribution of addictive processes in at least some presentations of binge-focused eating disorders.

Food addiction rates have also differed by eating disorder status. Individuals with BED seem to exhibit elevated rates of food addiction, with about half of individuals meeting the original YFAS criteria (Gearhardt et al., 2012; Gearhardt, White, Masheb, & Grilo, 2013). The presentation of BED and food addiction is associated with a more severe clinical presentation than BED alone with more frequent episodes of binge eating, greater emotion dysregulation, and eating pathology (Gearhardt et al., 2012). Rates of food addiction are also elevated in BN. In patients with BN, prevalence rates ranging from 81.5% (Granero et al., 2014) to 100% (Meule, Rezori, & Blechert, 2014) have been noted. However, the rate of food addiction is lower (30%) in individuals with remitted BN (Meule et al., 2014). Many of the mechanisms implicated in addiction such as heightened impulsivity, emotion dysregulation, and risky substance use are more strongly linked to BN than other eating disorders (Dansky, Brewerton, & Kilpatrick, 2000; de Jonge, Van Furth, Lacey, & Waller, 2003; Fischer, Smith, & Anderson, 2003), which may contribute to the higher rates of YFAS food addiction. Longitudinal research has found that higher rates of YFAS food addiction at baseline predict poorer treatment response to a psychosocial intervention for individuals with BN (Hilker et al., 2016), which suggests that assessing YFAS food addiction in patients with BN may be helpful in identifying patients who may need additional clinical support.

One unexpected finding is that food addiction is also associated with AN, particularly the binge/purge subtype (Granero et al., 2014). Given that the core characteristic of AN is restricting food intake, it was not anticipated that food addiction would be associated with this disorder. One possibility is that for individuals with AN, the YFAS is assessing a different construct, such as subjective (but not objective) loss of control over food consumption. A similar issue arose in the context of binge eating, in which individuals with AN endorsed high levels of binge eating episodes. However, in contrast to the binge episodes diagnostic of BN and BED (i.e., loss of control and objectively large quantity of food consumed), many of the binge eating episodes in AN reflected a subjective sense of loss of control, but a small quantity of food consumed (Latner, Vallance, & Buckett, 2008). To distinguish between these subtypes, binge eating episodes are evaluated for whether they are subjective (i.e., feelings of loss of control, but small amount of food consumed) or objective (i.e., feelings of loss of control, but with objectively large amount of food consumed), and both have been found to be clinically meaningful (Latner et al., 2008). Thus, in AN samples, the food addiction construct may be similar to subjective binge episodes by assessing a subjective sense of loss of control without objective addictive-like food consumption. However, more research is needed in AN samples to further evaluation the presentation of addictive-like eating indicators in this group.

Based on the original YFAS, about half of individuals with food addiction meet for an existing eating disorder, whereas the other half does not (Gearhardt, Boswell, & White, 2014). Participants who meet for food addiction only exhibit similar levels of clinically significant distress and eating pathology as those who meet for an eating disorder only (Gearhardt et al., 2014). Thus, participants who meet for food addiction (but not other eating disorders) appear to be a clinically relevant sample who may not be receiving adequate clinical care. Importantly, there are distinct symptoms and mechanisms in addiction versus eating disorders that may be relevant for individuals with food addiction but without an eating disorder. One distinction between food addiction and binge-focused eating disorders is the pattern of consumption. In binge-focused eating disorders, individuals must consume an objectively large amount of food in a discrete period of time (e.g., 2 hours) and experience a subjective sense of loss of control (American Psychiatric Association, 2013a). However, in addiction, no specific pattern of intake is required and individuals can meet the diagnostic criteria if they exhibit discrete patterns of binge consumption (e.g., binge drinking) or if they steadily and repeatedly administer over long periods of time (e.g., chain smoking). They can also meet criteria despite a subjective sense that they are in control of their substance use (American Psychiatric Association, 2013b). These diagnostic differences may lead people who exhibit compulsive eating behavior, but do not have discrete binges, to be captured by food addiction where they may not receive a diagnosis of binge-focused eating disorders.

There are also mechanistic and diagnostic differences between an addiction and eating disorder perspectives that likely contribute to the nonshared variance between these constructs. From an addiction perspective, withdrawal and tolerance are both mechanistically important and are assessed as diagnostic criteria (American Psychiatric Association, 2013a). Withdrawal and tolerance are considered adaptations to heavy, repeated use that may reflect the "dark" side of addiction (Koob, 2009). As the body adapts to the use, more and more of the substance may be needed to achieve previous levels of reward (i.e., tolerance) and when substance use is reduced or discontinued, these adaptations can result in physically and/or psychologically aversive experience (i.e., withdrawal) that can increase the likelihood of relapse (Koob, 2015). The YFAS assesses both withdrawal and tolerance, whereas eating disorder measures do not. In contrast, eating disorder perspectives focus on shape and weight concerns as major driving factors in pathological eating (Fairburn, Cooper, Shafran, & Wilson, 2008), where there is less focus on these as driving factors in the context of addiction (Schulte, Grilo, et al., 2016).

However, the most significant point of contention between addiction and eating disorder perspectives is the role of the food. Addiction perspectives would consider that the rewarding attributes of the food would interact with individual risk factors in a manner that may drive forward compulsive patterns of intake (Schulte, Grilo, et al., 2016). Thus, certain highly rewarding foods, such as highly processed foods with artificially high levels of refined carbohydrates, fat, and salt, may be riskier to eat than more naturally occurring foods (e.g., fruits, vegetables). In contrast, eating disorder perspectives focus little on the role of the food itself, stating there is no good food or bad food, but instead focus on the attitudes toward the foods with a goal of eating all foods in moderation (see Schulte, Grilo, et al., 2016) for a review of how eating disorder and addiction perspectives overlap). Thus, addiction and eating disorder perspective not only share notable overlap but also assess distinct criteria and mechanisms that may contribute to the unique clinical utility of food addiction.

Modified original YFAS

After the development of the original YFAS, a 9-item abbreviated version, the modified original YFAS (mYFAS), was created as a briefer assessment of addictive-like eating (Flint et al., 2014). The mYFAS includes one question from each of the seven diagnostic criteria included in DSM-IV (see Table 2.1) and two questions that represent clinically significant impairment or distress. As with the original YFAS, the mYFAS can be scored as either a continuous symptom count (zero to seven) or as a categorical diagnosis of food addiction based on the same cutoffs as the original YFAS. The mYFAS has exhibited similar psychometric properties to the original YFAS (Flint et al., 2014) and may be particularly useful in samples where minimizing participant burden is of extreme importance (e.g., epidemiological studies with large test batteries) or as a brief screener to identify patients for further evaluation.

The mYFAS has been used in the Harvard Nurses' Health Study (NHS), a large, ongoing prospective cohort study of nurses (Flint et al., 2014). The mYFAS was completed by 134,175 women and 5.8% met the mYFAS "diagnostic" threshold score. Rates of food addiction were higher (8.4%) in younger women (45–64 years of age) than in older women (2.7%) who were 62–88 years of age, which is consistent with the lower prevalence of substance-use disorders (SUDs) in older adults (Flint et al., 2014). Food addiction was also associated with obesity and diet-related disease (e.g., hypercholesterolemia) (Flint et al., 2014). Similarly, in a sample of veterans, the mYFAS accounted for the largest amount of variance in BMI relative to other measures of eating pathology (Masheb, Ruser, Min, Bullock, & Dorflinger, 2018). mYFAS food addiction has also been associated with other forms of psychopathology, such as an increased prevalence of depression (Flint et al., 2014; Masheb et al., 2018) and with higher rates of PTSD and childhood abuse (Masheb et al., 2018; Mason et al., 2014; Mason, Flint, Field, Austin, & Rich-Edwards, 2013). Thus, screening of food addiction in clinical samples may be particularly important.

Interestingly, mYFAS food addiction was less prevalent in current smokers but higher in former smokers in the NHS (Flint et al., 2014). Similarly, in veterans, mYFAS food addiction was inversely correlated with alcohol-use disorders (Masheb et al., 2018). These findings are consistent with the idea of the drug–food competition theory (Cummings, Ray, & Tomiyama, 2015; Kleiner et al., 2004), which hypothesizes that drugs of abuse and palatable foods are in competition with one another because they activate similar neural systems implicated in reward and motivation. Thus, food addiction and SUDs may be less likely to co-occur. However, there may be an addiction transfer in which cutting down or stopping excessive consumption of one addictive substance (e.g., smoking, palatable foods) may increase the likelihood of intake of the other substance. Consistent with this idea, individuals who quit an addictive substance are often at elevated risk for weight gain, and desire for palatable, highly processed foods is enhanced (Kleiner et al., 2004). Similarly, higher food addiction scores have been related to greater substance-use problems following bariatric surgery (Clark & Saules, 2013). Thus, the associations between substance use and addictive-like eating are complex and likely vary over time.

Children's Yale Food Addiction Scale

As evidence for food addiction in adults has grown, interest in examining addictive-like eating in children and adolescents has also developed. In the context of traditional SUDs, examining substance use earlier in the life span has been particularly important, as earlier exposure to a substance may increase the risk of developing an SUD later in life (McGue, Iacono, Legrand, Malone, & Elkins, 2001) and secondary substance-related health consequences (e.g., liver disease, lung cancer) (Brook, Brook, Zhang, Cohen, & Whiteman, 2002). Additionally, owing to a relatively underdeveloped prefrontal cortex (associated with future-oriented planning), paired with a more active reward system (associated with pleasure seeking), adolescents may be especially at risk for abusing rewarding substances (Steinberg, 2010). Because children and adolescents are exposed to highly palatable foods at an even younger age than other substances of abuse, it is particularly important to examine food addiction in younger cohorts.

The Yale Food Addiction Scale for Children (YFAS-C) was developed in 2013 (Gearhardt, Roberto, Seamans, Corbin, & Brownell, 2013). Paralleling the adult version of the scale, the YFAS-C applies the DSM-IV criteria for substance dependence to the consumption of highly processed foods. To ensure that the scale was developmentally appropriate, items were altered to reflect age-appropriate activities (e.g., items referring to "work" were changed to "school"), and the reading level was lowered to be appropriate for children as young as second grade (Gearhardt, Roberto, et al., 2013). This 25-item scale has demonstrated good internal consistency, internal validity, and convergent validity in a variety of community samples ranging from normal weight to obese. Scoring for the YFAS-C is consistent with the adult version of the scale, yielding either a continuous symptom count ranging from zero to seven or a dichotomous "diagnosis" by endorsing at least three symptoms plus clinically significant impairment or distress.

As with adults, the prevalence of food addiction in children and adolescents has also varied depending on the characteristics of the sample. When the scale was originally validated on a community sample of children ages 4–16 years, approximately 7% of the sample met the "diagnostic" threshold for food addiction (Gearhardt, Roberto, et al., 2013). However, in a large-scale study of Dutch adolescents without a history of eating pathology, only 2.6% of participants met for food addiction (Mies et al., 2017), suggesting that food addiction rates in children and adolescents may vary by culture. Additionally, food addiction rates in younger cohorts appear to be associated with weight class, with children and adolescents with overweight or obesity demonstrating elevated food addiction symptomatology in comparison with normal weight children (Burrows, Kay-Lambkin, Pursey, Skinner, & Dayas, 2018; Gearhardt, Roberto, et al., 2013; Mies et al., 2017; Richmond, Roberto, & Gearhardt, 2017; Schulte, Jacques-Tiura, Gearhardt, & Naar, 2017a). For example, in a study of adolescents enrolled in a comprehensive weight loss program, 38% of participants met the "diagnostic" criteria for food addiction (Meule, Hermann, & Kübler, 2015). Furthermore, adolescents who met for food addiction demonstrated significantly greater eating pathology (e.g., more binge days, elevated cravings, increased weight and shape concerns, and increased motor and attentional impulsivity) in comparison with participants without food addiction (Meule et al., 2015). Finally, as with adult samples, girls may have elevated rates of food addiction in comparison with boys (Mies et al., 2017), although other studies have not demonstrated gender differences (Burrows, Kay-Lambkin, Pursey, Skinner, & Dayas, 2018). Thus, further research may be necessary to determine if gender-related differences in food addiction are present in younger cohorts or if these gender-related differences do not emerge until adulthood. No studies using the YFAS-C have demonstrated differences in the prevalence of food addiction in children and adolescents based on race or ethnicity.

In contrast to adult samples, relatively less research has examined the association of food addiction with eating disorders in children and adolescents. However, research has suggested that the YFAS-C may distinguish a unique pattern of maladaptive eating behavior apart from established eating disorders in children and adolescents. For example, in a study examining food addiction in a sample of adolescents in treatment for weight loss, 38% of the sample met "diagnostic" criteria for food addiction but only 12% reached criteria for BED (Meule et al., 2015), suggesting that the YFAS-C may capture unique patterns of maladaptive eating above and beyond binge eating behaviors. Additionally, in a study examining adolescents in a psychiatric inpatient unit, only 42.9% of individuals with eating disorders qualified for food addiction (Albayrak et al., 2017). Together, these studies suggest that the YFAS-C captures unique patterns of addictive-like eating apart from established forms of eating pathology. One reason research may be sparse in this area is because children and adolescents may be less likely to experience impairment or distress related to their eating behaviors because of the limited number of responsibilities earlier in the life span. Because children and adolescents may be less likely to experience impairment in day-to-day life, this prevents them from meeting the YFAS-C "diagnostic" score. In the context of food addiction, this has been demonstrated in one study where approximately 16% of adolescents reached threshold for three or more symptoms of food addiction but only 4% met criteria for clinical impairment or distress (Laurent & Sibold, 2016). As a result, only a very small portion of the adolescents experiencing symptoms of additive-like eating met the diagnostic threshold for food addiction. Thus, it may be challenging to find children and adolescents who demonstrate clinically significant food addiction, preventing researchers from comparing different forms of eating pathology in younger samples.

Yale Food Addiction Scale 2.0

In 2013, there was a significant update to the substance-related and addictive disorders section of the DSM (American Psychiatric Association, 2013b). In addition to the inclusion of the first behavioral addiction (i.e., gambling), a number of diagnostic criteria were also changed regarding the SUD section. First, substance abuse and substance dependence were combined into one single diagnosis of an SUD, which reflected research that did not support these as two distinct disorders (Hasin, Fenton, Beseler, Park, & Wall, 2012; Hasin et al., 2013). Second, the legal consequences of diagnostic criterion that had previously been a component of the abuse criteria were dropped because of concerns regarding bias and poor predictive utility (Hasin et al., 2013). Third, intense craving was added as a criterion, given the importance of this construct as a driving factor in addiction (Hasin et al., 2013). Fourth, reflecting the growing focus on more dimensional approaches to psychopathology, a continuum of severity was outlined for SUD that ranges from mild (two to three symptoms) to moderate (four to five symptoms) to severe (six or more symptoms; see Table 2.1 for the DSM-5 criteria) (American Psychiatric Association, 2013b).

The YFAS 2.0 was developed and validated in 2016 to reflect these diagnostic changes (Gearhardt, Corbin, & Brownell, 2016). Although a behavioral addiction was now included in the DSM-5, the YFAS 2.0 continues to reflect a substance-focused model. This is based on a number of pieces of evidence. First, animal models suggest that the behavioral act of eating (when the food is not highly processed) has been insufficient to trigger an addictive phenotype, even under behavioral conditions designed to evoke addictive behaviors (e.g., stress exposure, intermittency) (Schulte, Potenza, & Gearhardt, 2016). Exposure to highly processed foods is a necessary component of triggering an addictive-like eating phenotype, and, like with drugs of abuse, this can be enhanced under certain behavioral conditions (Schulte, Potenza, et al., 2016). Thus, the attributes of the food being consumed, in an analogous manner to drugs of abuse, seem to be essential. Second, in humans, foods differ in their likelihood of being associated with addictive-like eating behaviors. Foods that are highly processed (i.e., include added levels of refined carbohydrates and/or fat) are much more likely to be consumed in an addictive-like way than naturally occurring foods (Schulte, Avena, & Gearhardt, 2015; Schulte, Smeal, & Gearhardt, 2017d). Thus, the behavioral act of eating naturally occurring foods that do not have artificially high levels of reward appears to be less relevant to an addiction phenotype, and the rewarding nature of the food appears to be central to the development of the behavior (for a review, see Schulte, Potenza, et al., 2016). Given this evidence, a substance-focused (rather than behavioral) perspective was maintained for the YFAS 2.0, and the changes to this section were adapted (i.e., inclusion of abuse criteria except legal problems, addition of craving, dimensional scoring) (Gearhardt et al., 2016).

In a sample of 550 participants recruited from the community, 14.6% of participants met the threshold for food addiction based on the YFAS 2.0 (i.e., two or more symptoms plus clinically significant impairment or distress) (Gearhardt et al., 2016). On average, participants met for two symptoms with the most commonly endorsed symptoms being withdrawal (29.7%) and an inability to cut down (25.0%). In a separate sample where the original YFAS and the YFAS 2.0 were directly compared, more people met the prevalence for food addiction with the DSM-5 version (15.8%) relative to the original version (10.0%) (Gearhardt et al., 2016). This likely reflects the change in the SUD diagnostic criteria from DSM-IV to DSM-5, in which four symptoms were added, but the threshold was dropped to two (from three symptoms). This increase in prevalence has also been demonstrated with other SUDs (Peer et al., 2013). However, the YFAS 2.0 performed as well, or better, than the original YFAS on all psychometric examinations (Gearhardt et al., 2016). Thus, even with the increased prevalence of food addiction using the YFAS 2.0, the measure still appears to be sound. Interestingly, the majority of participants who met the threshold for food addiction were in the severe (11.5%, six or more symptoms plus impairment/distress) than the moderate (1.9%, four to five symptoms plus impairment/distress) or mild (2.4%, two to three symptoms plus impairment or distress) (Gearhardt et al., 2016). It is possible that higher numbers of food addiction symptoms must be present for individuals to begin to experience clinically significant impairment or distress given the ubiquity of potentially addictive foods and the lack of acute intoxication associated with intake.

YFAS 2.0 food addiction was associated with a number of clinically relevant outcomes. Individuals with YFAS 2.0 food addiction were over four times more likely to have obesity and demonstrated more pathological eating behavior (e.g., binge eating) (Gearhardt et al., 2016). Weight cycling (i.e., the number of times one has lost and regained 20 pounds excluding pregnancy) was also associated with YFAS 2.0 food addiction, perhaps demonstrating a tendency toward relapsing to prior patterns of food intake. As with the original YFAS, the overlap between eating disorders and YFAS 2.0 food addiction suggests that these are related but distinct constructs. Approximately half of participants with BN (42.6%) and BED (47.2%) also met for YFAS 2.0 food addiction and approximately half (43.6%) of individuals with YFAS 2.0 food addiction did not meet for an existing eating disorder (Gearhardt et al., 2016). The YFAS 2.0 has also now been validated in German

(Meule, Müller, Gearhardt, & Blechert, 2017), Italian (Aloi et al., 2017), and French (Brunault et al., 2017) and has been found to be associated with elevated eating pathology and food cravings. Thus, the YFAS 2.0 appears to be a psychometrically sound tool to operationalize food addiction that provides clinically meaningful information.

Modified YFAS 2.0

As with the original YFAS, an abbreviated, 13-item version of the YFAS 2.0, the modified YFAS 2.0 (mYFAS 2.0), was developed, consisting of one question to assess each of the 11 DSM-5 diagnostic criteria for SUDs and two items evaluating impairment and distress. Paralleling the YFAS 2.0, the mYFAS 2.0 can be scored continuously, ranging from 0 to 11 depending on the number of symptoms endorsed, or dichotomously, indicating a "diagnostic" threshold by reporting at least two symptoms and either impairment or distress. In the validation paper, the mYFAS 2.0 demonstrated good internal consistency (Kuder–Richardson alpha = 0.86) and performed similarly as the full YFAS 2.0 on indexes of convergent, discriminant, and incremental validity using other measures of eating behavior. The continuous, symptom, and dichotomous "diagnostic" threshold scores of the mYFAS 2.0 may be, on average, more modest estimates of food addiction than the full YFAS 2.0, as individuals have only one item for possible endorsement of each symptom. Thus, the mYFAS 2.0 may be ideal in studies prioritizing specificity over sensitivity. Furthermore, the mYFAS 2.0 may be useful in large epidemiological studies to reduce participant burden or approaches where a brief screening tool of food addiction is desired.

Recently, the scope of food addiction, as assessed by the mYFAS 2.0, was evaluated in a large online sample recruited to be nationally representative of the United States (Schulte & Gearhardt, 2018). In this sample, 15% of individuals met criteria for the "diagnostic" threshold score, which parallels prior prevalence rates estimated by the YFAS 2.0 or mYFAS 2.0 in convenience samples (Gearhardt et al., 2016; Schulte & Gearhardt, 2017). Only one other study, conducted in Germany, has assessed food addiction rates in a nationally representative sample and observed that 7.9% met criteria on the full YFAS 2.0. Interestingly, the findings in the United States sample were nearly double the prevalence in Germany. While multiple factors likely contribute to this difference, possible explanations include higher obesity rates (Mensink et al., 2013; Ogden, Carroll, Kit, & Flegal, 2014) and availability of highly processed foods (e.g., fast foods) (Jeffery, Baxter, McGuire, & Linde, 2006; Maddock, 2004) in the United States. Cross-cultural assessment of food addiction prevalence may be particularly helpful for elucidating what may exacerbate or protect a population from elevated rates of addictive-like eating. For instance, if the availability of highly processed foods is a significant contributor, it may be expected that countries that have greater access to these foods would have increased prevalence rates of food addiction. Thus, future research may replicate the nationally representative sample approach in various countries, as well as continue to track the scope of food addiction within the United States and Germany to explore possible associations with population trends (e.g., whether an increase in obesity rates also inflates food addiction).

In the sample recruited to be nationally representative of the United States (Schulte & Gearhardt, 2018), demographic associations with the mYFAS 2.0 were also investigated. Inconsistent with prior studies using various versions of the YFAS (Gearhardt et al., 2016; Hauck, Weiss, Schulte, Meule, & Ellrott, 2017; Pedram et al., 2013; Pursey et al., 2014), no significant association was observed with gender. This may suggest that addictive-like eating behavior warrants assessment in samples more representative of men in future studies, instead of previous approaches recruiting only women (Flint et al., 2014; Mason, Flint, Field, Austin, & Rich-Edwards, 2013). Consistent with the mYFAS 2.0 validation paper (Schulte & Gearhardt, 2017), food addiction prevalence was higher in individuals who were younger and/or identified as Hispanic. Furthermore, a positive association was observed between food addiction and income, suggesting higher prevalence rates in more affluent individuals. Thus, individuals who are younger, Hispanic, and/or affluent may be particularly prone to the development of addictive-like eating and may benefit from early screening efforts. Furthermore, future research is warranted in these groups to understand mechanisms that may contribute to these individual differences. For instance, younger individuals may have greater exposure to the modern food environment of highly processed foods, which may result in an elevated propensity to exhibit indicators of food addiction.

Schulte and colleagues (2018) also observed a nonlinear relationship between food addiction prevalence and weight class, with individuals who were underweight or obese exhibiting the highest rates. This was consistent with data from the sample recruited to be nationally representative of Germany (Hauck et al., 2017). On the one hand, increased food addiction in individuals who were obese suggests that addictive-like eating may represent a behavioral phenotype within obesity. This may inform more tailored treatment approaches, such that individuals with obesity and addictive-like eating may benefit from evidence-based interventions for SUDs (e.g., motivational enhancement) (Lundahl, Kunz, Brownell, Tollefson, & Burke, 2010). On the other hand, elevated reported food addiction indictors in persons with underweight may underscore a possible issue with construct validity of the YFAS in this sample. As described earlier, individuals with underweight may be reporting on subjective instances of addictive-like eating rather than objective addictive-like consumption of highly processed foods that may be endorsed by those with obesity. Thus, future work may consider utilizing qualitative methods to explore how persons with underweight are interpreting and rating the items on the mYFAS 2.0, as well as other versions of the YFAS.

YFAS 2.0 for children

Recently, the YFAS-C was updated to reflect changes to the DSM-5 SUD criteria. Like the original YFAS-C, the Yale Food Addiction Scale for Children 2.0 (YFAS-C 2.0) altered items from the adult version of the scale to appropriately reflect activities associated with childhood and adolescence (e.g., "work" was altered to "school"). Additionally, the YFAS-C 2.0 utilizes a lower reading level and is appropriate for children as young as third grade (Schiestl & Gearhardt, 2018). However, several concerns have recently been raised about the updated DSM-5 criteria for SUDs, particularly regarding assessment in younger cohorts. Specifically, critics have argued that the updated DSM-5 SUD criteria are too problem-focused, with diagnoses relying more heavily on the consequences (e.g., trouble at work, problems with friends/family) compared with the DSM-IV (Lane & Sher, 2014, 2015). The DSM-5 criteria for SUD contains five problem-focused criteria (i.e., use despite inability to fulfill major role obligations, use despite interpersonal problems, important activities given up, use in hazardous situations, and use despite physical or psychological problems), representing almost half of the total criteria (American Psychiatric Association, 2013b). In contrast, the DSM-IV substance-dependence criteria included only two problem-focused criteria out of seven total criteria (American Psychiatric Association, 2000). This may be problematic, as many of the consequences of substance use are dependent on contextual and demographic variables rather than the use of the substance itself (Lane & Sher, 2014). For example, while a college student may be able to arrange his schedule to accommodate excessive drinking behavior, an individual with a less flexible, full time job who consumes a similar amount of alcohol may be more likely to experience significant impairment at work. Furthermore, binge drinking may be more culturally normative in college settings making it less likely to cause interpersonal problems. Thus, similar patterns of consumption can result in vastly difference consequences based on a variety of contextual factors.

The significant reliance on problem-focused criteria in the DSM-5 SUD criteria may be especially problematic for assessing addictive behaviors in younger cohorts, particularly because children and adolescents may just be developing maladaptive patterns of consumption. Thus, their substance use may not yet interfere with school or social relationships in the same way as established patterns of substance use may impact adults. Additionally, adolescents may have fewer obligations than adults, giving them less opportunity to experience negative consequences associated with substance use. In support, prior research has observed that adolescents are unlikely to exhibit impairment in day-to-day functioning, suggesting possible issues with the problem-focused DSM-5 SUD criteria, whereas indicators focused more on psychological and behavioral mechanisms (e.g., tolerance, craving, loss of control) may be more commonly experienced (Hasin et al., 2013).

Challenges of the appropriateness of the problem-focused DSM-5 SUD criteria in adolescents have also been demonstrated in the context of food addiction. For example, one study found that in a community sample of adolescents, 16% reached the threshold for three or more food addiction symptoms. However, only 4% of the sample experienced significant impairment or distress related to their eating behaviors and as a result did not receive a diagnosis of food addiction (Laurent & Sibold, 2016). When the YFAS-C 2.0 was administered to a nonclinical community sample of adolescents (oversampled for overweight and obesity), participants did demonstrate some symptoms of food addiction (i.e., loss of control eating, the inability to cut down on certain foods, tolerance, withdrawal, craving, and spending excessive amounts of time eating or obtaining food). However, this sample had low endorsement rates of symptoms associated with the consequences of addictive-like eating (i.e., failure to fulfill role obligations, social or interpersonal problems, giving up important activities to eat, or eating in hazardous situations), and these problem-focused questions dropped from a final version of the scale (Schiestl & Gearhardt, 2018). Thus, the significant reliance on problem-focused criteria may hinder the ability to examine food addiction in younger cohorts. However, future research is needed to examine the validity and utility of applying the problem-focused DSM-5 criteria for SUDs to the examination of food addiction in clinical samples of children and adolescents seeking treatment for weight- and eating-related disorders.

One additional concern about using the DSM-5 SUD criteria to examine food addiction in children and adolescents pertains to the dichotomous nature of clinical assessments. Presently, individuals either meet the clinical threshold for a mental health disorder and receive a diagnosis or they do not (American Psychiatric Association, 2013b). However, this dichotomous approach to assessment does not reflect the vast variation in symptomatology and severity that people actually experience in daily life. Children and adolescents in particular may be just developing clinically meaningful patterns of substance use (Clark, Cuthbert, Lewis-Fernández, Narrow, & Reed, 2017). However, because these behaviors may be less severe at the onset of a disorder, children and adolescents may not yet reach threshold for a diagnosis and thus may not be identified for early intervention efforts (Isnard et al., 2003). As a result, dimensional approaches to scoring may better capture the variation in symptomatology and severity present during the onset of addictive-like eating behaviors.

Thus, in creating a developmentally appropriate DSM-5 version of the YFAS in a nonclinical adolescent sample, a dimensional approach to scoring was used and problem-focused questions were excluded. The resulting scale, the dimensional YFAS-C 2.0 (dYFAS-C 2.0), had good internal consistency reliability, strong convergent validity with emotional eating, external eating and BMI, and appropriate incremental validity in predicting unique variance in BMI above and beyond existing measures (Schiestl & Gearhardt, 2018). In contrast to the lack of association between restrained eating and the adult version of the YFAS 2.0, the dYFAS-C 2.0 was positively associated with restrained eating (Schiestl & Gearhardt, 2018). Thus, attempts to restrict eating may co-occur with food addiction in younger samples, but adults with food addiction may have had a number of failed attempts at controlling eating behavior and be less likely to engage in restrained eating. Longitudinal research may be able to determine if high scores on the dYFAS-C 2.0 in younger individuals predict "diagnostic" levels of food addiction in adulthood and how the association between food addiction and restrained eating develops over the life span.

Alternative model: eating addiction

The food addiction construct is based on an SUD perspective, positing that highly processed foods (e.g., pizza, chocolate, chips) may exhibit an addictive potential and interact with individual risk factors to trigger a food addiction phenotype. An alternative model was posed by Hebebrand et al. (2014), suggesting that addictive-like eating may be better conceptualized as a behavioral addiction to the act of eating. Broadly, the key difference between existing substance and behavioral addictions is that an addictive substance is ingested in SUDs. Thus, it would follow that a substance-based food addiction framework would be more appropriate if highly processed foods may directly contribute to the phenotype, akin to drugs of abuse, whereas a behavioral eating addiction may be a better model if all foods are relatively equally implicated in addictive-like consumption, suggesting that the behavior of compulsive eating may become addictive, akin to other addictive behaviors (e.g., gambling). While this remains an area of debate (Hebebrand et al., 2014; Ruddock, Christiansen, Halford, & Hardman, 2017; Schulte, Potenza, & Gearhardt, 2017b, 2017c), existing evidence suggests that highly processed foods may be uniquely associated with addictive-like intake, favoring the substance-based food addiction perspective.

As described earlier, animal studies have demonstrated that rats have exhibited biological (e.g., downregulation of dopamine) and behavioral (e.g., bingeing) features of addiction in response to intermittent access to highly processed foods (e.g., cheesecake) (Avena et al., 2008; Johnson & Kenny, 2010; Oginsky et al., 2016; Oswald et al., 2011; Parylak et al., 2011). Interestingly, despite implementing behavioral circumstances that may elevate the addictive potential of the food (e.g., intermittency, stress), rats do not experience these features in response to their nutritionally balanced chow. This may provide evidence that highly processed foods are directly influential in triggering the addictive-like responses, supporting the food addiction perspective.

The first systematic study to investigate which foods may be most associated with indicators of addiction, as assessed by the original YFAS, was conducted just 3 years ago (Schulte et al., 2015). The authors asked an undergraduate sample and individuals from an online community sample to complete the YFAS and then report how likely they were to experience problems, as described by the YFAS, in response to 35 nutritionally diverse foods (Schulte et al., 2015). In both samples, Schulte et al. (2015) observed that highly processed foods were closely implicated in addictive-like eating, whereas foods in a more natural state (e.g., fruits, vegetables, lean proteins) were minimally related. Furthermore, highly processed foods were reported to be more significantly problematic for individuals who met the "diagnostic" threshold score for YFAS food addiction, suggesting that highly processed foods may interact with personal risk factors to produce the addictive-like phenotype.

Several other studies have also examined the relationship between highly processed foods and the YFAS. Persons who meet for the "diagnostic" threshold of

YFAS food addiction have reported consuming greater quantities of highly processed foods (Pursey, Collins, Stanwell, & Burrows, 2015) and elevated amounts of added sugars and trans fat (Schulte et al., 2017a), which are ingredients commonly found in highly processed foods. Furthermore, a recent study in a large online community sample found that highly processed foods, relative to foods in a natural state, were more closely associated with subjective experience indicators of elevated addictive potential (e.g., greater craving, enjoyment), paralleling findings in drugs of abuse (Schulte et al., 2017d). Lastly, individuals endorsing multiple symptoms of food addiction on the original YFAS have exhibited similar patterns of neural reactivity to a highly processed food (chocolate milkshake) as individuals with an SUD in response to the drug, marked by greater reward-related responses during anticipation but blunted inhibitory control activity during consumption (Gearhardt et al., 2011). Collectively, these findings support that highly processed foods seem to be more implicated in addictive-like consumption than foods in a more natural state, which suggests the appropriateness of the substance-based food addiction framework. As the majority of evidence for the addictive potential of highly processed foods in humans has relied on self-report data, future research utilizing biological and behavioral methodology is warranted to build on existing findings.

While preliminary research in humans suggests that highly processed foods seem to directly contribute to the development of a food addiction phenotype, the behavioral contexts in which these foods are consumed may also elevate their addictive potential. Proponents of the behavioral eating addiction model (Hebebrand et al., 2014; Ruddock et al., 2017) have suggested that the importance of behaviors in the context of addictive-like eating, such as the YFAS operationalizing behavioral indicators of addiction (e.g., loss of control), provides support for a behavioral eating addiction perspective. However, this reflects a misconception about the shared importance of behaviors in both substance and behavioral addictions. Notably, all addictions are diagnosed using observable behavioral criteria (e.g., use despite negative consequences) (American Psychiatric Association, 2013c), although they may be adapted to fit the unique characteristics of the substance (e.g., lack of intoxication syndrome in tobacco-use disorder) or addictive behavior (e.g., unique criteria of chasing losses in gambling disorder). Furthermore, similar behavioral contexts can exacerbate the addictive nature of both substances and addictive behaviors, such as intermittent intake/engagement (e.g., alternating between periods of bingeing and restraint) and intake/engagement to cope with negative emotions (Hwa et al., 2011; Koob & Kreek, 2007; Sinha, 2001).

In summary, the importance of behaviors is not unique to behavioral addictions, but rather the ingestion of an addictive substance is the key differentiating factor between substance and behavioral addictions. Presently, existing research in animals and humans demonstrate that highly processed foods seem to be uniquely implicated in addictive-like consumption and thus supports the substance-based food addiction framework. However, additional research using biological and behavioral approaches in humans is needed to continue evaluating whether highly processed foods may be similarly addictive as drugs of abuse and to identify what the addictive agent may be, as has been established in existing SUDs (e.g., ethanol identified as the addictive substance in alcohol).

Next steps in the assessment of food addiction

The YFAS and its different iterations have provided the field with a psychometrically sound tool to operationalize food addiction based on an SUD framework. The YFAS has been useful in identifying individuals who may exhibit an addiction to highly processed foods (e.g., chocolate, French fries) to allow for a more rigorous investigation of the hypothesis that addictive processes can contribute to certain types of overeating. Based on the current research, the YFAS 2.0 is likely the best choice for assessing food addiction in adults. However, if concerns about participant burden are high, the mYFAS 2.0 is briefer and has similar psychometric properties. The best assessment approach in children is less clear, given concerns about the application of the DSM-5 criteria to younger samples. The YFAS-C is an appropriate choice, although it does not reflect the most up-to-date diagnostic criteria. A dimensional version of the YFAS 2.0-C has been validated in a nonclinical community sample, but the best choice for clinical samples is unknown.

In addition to scale development, food addiction is a burgeoning area of research that has a number of important future directions, including the identification of which foods or ingredients may be most addictive. This will be essential for understanding the mechanisms driving addictive-like eating, providing appropriate clinical care (e.g., considering harm reduction approaches that take into account the differing risk profiles of food) and implementing effective policy approaches (e.g., reformulation of addictive foods, restriction of certain types of advertising to minors). The identification of what foods may be addictive will also allow for a refinement of the label "food addiction," as all foods do not appear to be equally addictive (Schulte et al., 2015; Schulte et al., 2017d). The possibility of adapting a new term, such as "highly processed food addiction" or "highly palatable food addiction," will be important to consider as research progresses.

There are also important next steps to address in understanding the best way to diagnosis food addiction and estimate its prevalence. One important next step may be the development of a structured clinical interview. The role of a clinician in evaluating whether an individual exhibits an addiction to food may be particularly relevant given potential differences in interpretation related to subjective relative to objective levels of addictive-like overeating. Further research is also needed to investigate and refine some of the symptoms most commonly endorsed on the YFAS. For example, withdrawal when trying to reduce consumption of potentially addictive foods is one of the most commonly endorsed symptoms on the YFAS 2.0 (Gearhardt et al., 2016). However, there has been little experimental work to investigate what a withdrawal syndrome from certain types of food may look like and its potential role as an obstacle in meeting healthy goals. Although physical symptoms of withdrawal (e.g., gastrointestinal distress) do occur for some substances (most notably for severe alcohol and opioid-use disorders), they typically resolve relatively quickly and have less predictive utility for relapse than psychological symptoms (e.g., anhedonia, irritability), which are more common and last longer (Allsop, Saunders, & Phillips, 2000; Kenford et al., 2002; Poling, Kosten, & Sofuoglu, 2007). This is an important area of future research. Additional longitudinal research, especially in children and adolescents, is also essential to investigate the developmental course of food addiction. Finally, a movement toward more mechanistically informed and dimensional approaches to assessing psychopathology (e.g., RDOCS) is becoming an increasing focus of the field (Cuthbert & Insel, 2013). It will be important for the assessment of food addiction to reflect the overall changing notions of diagnostic assessment as they evolve.

References

Ahmed, S. H., Guillem, K., & Vandaele, Y. (2013). Sugar addiction: Pushing the drug-sugar analogy to the limit. *Current Opinion in Clinical Nutrition and Metabolic Care, 16*(4), 434–439.

Albayrak, Ö., Föcker, M., Kliewer, J., Esber, S., Peters, T., Zwaan, M., & Hebebrand, J. (2017). Eating-related psychopathology and food addiction in adolescent psychiatric inpatients. *European Eating Disorders Review, 25*(3), 214–220.

Allsop, S., Saunders, B., & Phillips, M. (2000). The process of relapse in severely dependent male problem drinkers. *Addiction, 95*(1), 95–106.

Aloi, M., Rania, M., Muñoz, R. C. R., Murcia, S. J., Fernández-Aranda, F., De Fazio, P., & Segura-Garcia, C. (2017). Validation of the Italian version of the Yale food addiction scale 2.0 (I-YFAS 2.0) in a sample of undergraduate students. *Eating and Weight Disorders-Studies on Anorexia, Bulimia and Obesity, 22*(3), 527–533.

American Psychiatric Association. (2000). *Diagnostic and statistical manual of mental disorders*. text revision (4th ed.) Washington DC.

American Psychiatric Association. (2013a). *Diagnostic and statistical manual of mental disorders* (5th ed.). Arlington, VA: American Psychiatric Association.

American Psychiatric Association. (2013b). *Diagnostic and statistical manual of mental disorders (DSM-5®)*. American Psychiatric Pub.

American Psychiatric Association. (2013c). *Diagnostic and statistical manual of mental disorders : DSM-5*. Retrieved from http://dsm.psychiatryonline.org/book.aspx?bookid=556.

Avena, N. M., Rada, P., & Hoebel, B. G. (2008). Evidence for sugar addiction: Behavioral and neurochemical effects of intermittent, excessive sugar intake. *Neuroscience & Biobehavioral Reviews, 32*(1), 20–39.

Brook, D. W., Brook, J. S., Zhang, C., Cohen, P., & Whiteman, M. (2002). Drug use and the risk of major depressive disorder, alcohol dependence, and substance use disorders. *Archives of General Psychiatry, 59*(11), 1039–1044.

Brunault, P., Courtois, R., Gearhardt, A. N., Gaillard, P., Journiac, K., Cathelain, S., & Ballon, N. (2017). Validation of the French version of the DSM-5 Yale food addiction scale in a nonclinical sample. *Canadian Journal of Psychiatry, 62*(3), 199–210. https://doi.org/10.1177/0706743716673320.

Burrows, T. L., Kay-Lambkin, F., Pursey, K. M., Skinner, J., & Dayas, C. (2018). Food addiction and associations with mental health symptoms: A systematic review with meta-analysis. *Journal of Human Nutrition and Dietetics*.

Chen, G., Tang, Z., Guo, G., Liu, X., & Xiao, S. (2015). The Chinese version of the Yale Food Addiction Scale: An examination of its validation in a sample of female adolescents. *Eating Behaviors, 18*, 97–102.

Clark, L. A., Cuthbert, B., Lewis-Fernández, R., Narrow, W. E., & Reed, G. M. (2017). Three approaches to understanding and classifying mental disorder: ICD-11, DSM-5, and the national institute of mental health's research domain criteria (RDoC). *Psychological Science in the Public Interest, 18*(2), 72–145.

Clark, S. M., & Saules, K. K. (2013). Validation of the Yale food addiction scale among a weight-loss surgery population. *Eating Behaviors, 14*(2), 216–219.

Cummings, J. R., Ray, L. A., & Tomiyama, A. J. (2015). Food–alcohol competition: As young females eat more food, do they drink less alcohol? *Journal of Health Psychology*, 1359105315611955.

Cuthbert, B. N., & Insel, T. R. (2013). Toward the future of psychiatric diagnosis: The seven pillars of RDoC. *BMC Medicine, 11*(1), 126.

Dansky, B. S., Brewerton, T. D., & Kilpatrick, D. G. (2000). Comorbidity of bulimia nervosa and alcohol use disorders: Results from the national women's study. *International Journal of Eating Disorders, 27*, 180–190, 10657891.

Davis, C. (2013). From passive overeating to "food addiction": A spectrum of compulsion and severity. *ISRN Obesity, 2013*.

Fairburn, C. G., Cooper, Z., Shafran, R., & Wilson, G. T. (2008). *Eating disorders: A transdiagnostic protocol*.

Fischer, S., Smith, G. T., & Anderson, K. G. (2003). Clarifying the role of impulsivity in bulimia nervosa. *International Journal of Eating Disorders, 33*(4), 406–411.

Flint, A. J., Gearhardt, A. N., Corbin, W. R., Brownell, K. D., Field, A. E., & Rimm, E. B. (2014). Food-addiction scale measurement in 2 cohorts of middle-aged and older women. *American Journal of Clinical Nutrition, 99*(3), 578–586. https://doi.org/10.3945/ajcn.113.068965.

Gearhardt, A. N., Boswell, R. G., & White, M. A. (2014). The association of "food addiction" with disordered eating and body mass index. *Eating Behaviors, 15*(3), 427–433.

Gearhardt, A. N., Corbin, W. R., & Brownell, K. D. (2009). Preliminary validation of the Yale food addiction scale. *Appetite, 52*(2), 430–436.

Gearhardt, A. N., Corbin, W. R., & Brownell, K. D. (2016). Development of the Yale food addiction scale version 2.0. *Psychology of Addictive Behaviors, 30*(1), 113.

Gearhardt, A. N., Roberto, C. A., Seamans, M. J., Corbin, W. R., & Brownell, K. D. (2013a). Preliminary validation of the Yale food addiction scale for children. *Eating Behaviors, 14*(4), 508–512.

Gearhardt, A. N., White, M. A., Masheb, R. M., & Grilo, C. M. (2013b). An examination of food addiction in a racially diverse sample of obese patients with binge eating disorder in primary care settings. *Comprehensive Psychiatry, 54*(5), 500–505.

Gearhardt, A. N., White, M. A., Masheb, R. M., Morgan, P. T., Crosby, R. D., & Grilo, C. M. (2012). An examination of the food addiction construct in obese patients with binge eating disorder. *International Journal of Eating Disorders, 45*(5), 657–663.

Gearhardt, A. N., White, M., & Potenza, M. (2011a). Binge eating disorder and food addiction. *Current Drug Abuse Reviews, 4*(3), 201–207.

Gearhardt, A. N., Yokum, S., Orr, P. T., Stice, E., Corbin, W. R., & Brownell, K. D. (2011b). Neural correlates of food addiction. *Archives of General Psychiatry, 68*(8), 808–816.

Gold, M. S., Frost-Pineda, K., & Jacobs, W. S. (2003). Overeating, binge eating, and eating disorders as addictions. *Psychiatric Annals, 33*, 117–122.

Gold, M. S., Graham, N. A., Cocores, J. A., & Nixon, S. J. (2009). Food addiction? *Journal of Addiction Medicine, 3*(1), 42–45.

Granero, R., Hilker, I., Agüera, Z., Jiménez-Murcia, S., Sauchelli, S., Islam, M. A., … Dieguez, C. (2014). Food addiction in a Spanish sample of eating disorders: DSM-5 diagnostic subtype differentiation and validation data. *European Eating Disorders Review, 22*(6), 389–396.

Grundy, S. M. (1998). Multifactorial causation of obesity: Implications for prevention. *American Journal of Clinical Nutrition, 67*(3), 563S–572S.

Hasin, D. S., Fenton, M. C., Beseler, C., Park, J. Y., & Wall, M. M. (2012). Analyses related to the development of DSM-5 criteria for substance use related disorders: 2. Proposed DSM-5 criteria for alcohol, cannabis, cocaine and heroin disorders in 663 substance abuse patients. *Drug and Alcohol Dependence, 122*(1), 28–37.

Hasin, D. S., O'Brien, C. P., Auriacombe, M., Borges, G., Bucholz, K., Budney, A., … Petry, N. M. (2013). DSM-5 criteria for substance use disorders: Recommendations and rationale. *American Journal of Psychiatry, 170*(8), 834–851.

Hauck, C., Weiss, A., Schulte, E. M., Meule, A., & Ellrott, T. (2017). Prevalence of 'food addiction' as measured with the Yale food addiction scale 2.0 in a representative German sample and its association with sex, age and weight categories. *Obes Facts, 10*(1), 12–24. https://doi.org/10.1159/000456013.

Hebebrand, J., Albayrak, O., Adan, R., Antel, J., Dieguez, C., de Jong, J., … Dickson, S. L. (2014). Eating addiction", rather than "food addiction", better captures addictive-like eating behavior. *Neuroscience & Biobehavioral Reviews, 47*, 295–306. https://doi.org/10.1016/j.neubiorev.2014.08.016.

Hilker, I., Sánchez, I., Steward, T., Jiménez-Murcia, S., Granero, R., Gearhardt, A. N., … Tolosa-Sola, I. (2016). Food addiction in bulimia nervosa: Clinical correlates and association with response to a brief psychoeducational intervention. *European Eating Disorders Review, 24*(6), 482–488.

Hwa, L. S., Chu, A., Levinson, S. A., Kayyali, T. M., DeBold, J. F., & Miczek, K. A. (2011). Persistent escalation of alcohol drinking in C57BL/6J mice with intermittent access to 20% ethanol. *Alcoholism: Clinical and Experimental Research, 35*(11), 1938–1947. https://doi.org/10.1111/j.1530-0277.2011.01545.x.

Isnard, P., Michel, G., Frelut, M. L., Vila, G., Falissard, B., Naja, W., & Mouren-Simeoni, M. C. (2003). Binge eating and psychopathology in severely obese adolescents. *International Journal of Eating Disorders, 34*(2), 235–243.

Ivezaj, V., Wiedemann, A. A., & Grilo, C. M. (2017). Food addiction and bariatric surgery: A systematic review of the literature. *Obesity Reviews, 18*(12), 1386–1397. https://doi.org/10.1111/obr.12600.

Jeffery, R. W., Baxter, J., McGuire, M., & Linde, J. (2006). Are fast food restaurants an environmental risk factor for obesity? *International Journal of Behavioral Nutrition and Physical Activity, 3*, 2. https://doi.org/10.1186/1479-5868-3-2.

Johnson, P. M., & Kenny, P. J. (2010). Dopamine D2 receptors in addiction-like reward dysfunction and compulsive eating in obese rats. *Nature Neuroscience, 13*, 635–641.

de Jonge, P. v. H., Van Furth, E., Lacey, J. H., & Waller, G. (2003). The prevalence of DSM-IV personality pathology among individuals with bulimia nervosa, binge eating disorder and obesity. *Psychological Medicine, 33*(7), 1311–1317.

Kelly, T., Yang, W., Chen, C.-S., Reynolds, K., & He, J. (2008). Global burden of obesity in 2005 and projections to 2030. *International Journal of Obesity, 32*(9), 1431.

Kenford, S. L., Smith, S. S., Wetter, D. W., Jorenby, D. E., Fiore, M. C., & Baker, T. B. (2002). Predicting relapse back to smoking: Contrasting affective and physical models of dependence. *Journal of Consulting and Clinical Psychology, 70*(1), 216.

Kleiner, K. D., Gold, M. S., Frostpineda, K., Lenzbrunsman, B., Perri, M. G., & Jacobs, W. S. (2004). Body mass index and alcohol use. *Journal of Addictive Diseases, 23*(3), 105–118.

Knight, J. R., Wechsler, H., Kuo, M., Seibring, M., Weitzman, E. R., & Schuckit, M. A. (2002). Alcohol abuse and dependence among US college students. *Journal of Studies on Alcohol and Drugs, 63*(3), 263.

Koob, G. F. (2009). Neurobiological substrates for the dark side of compulsivity in addiction. *Neuropharmacology, 56*, 18–31.

Koob, G. F. (2015). The dark side of emotion: The addiction perspective. *European Journal of Pharmacology, 753*, 73–87.

Koob, G. F., & Kreek, M. J. (2007). Stress, dysregulation of drug reward pathways, and the transition to drug dependence. *American Journal of Psychiatry, 164*(8), 1149–1159. https://doi.org/10.1176/appi.ajp.2007.05030503.

Lane, S. P., & Sher, K. J. (2014). Not all alcohol use disorder criteria are created equal: Implications for severity grading. *Alcoholism: Clinical and Experimental Research, 38*, 203A.

Lane, S. P., & Sher, K. J. (2015). Limits of current approaches to diagnosis severity based on criterion counts: An example with DSM-5 alcohol use disorder. *Clinical Psychological Science, 3*(6), 819–835.

Latner, J. D., Vallance, J. K., & Buckett, G. (2008). Health-related quality of life in women with eating disorders: Association with subjective and objective binge eating. *Journal of Clinical Psychology in Medical Settings, 15*(2), 148.

Laurent, J. S., & Sibold, J. (2016). Addictive-like eating, body mass index, and psychological correlates in a community sample of preadolescents. *Journal of Pediatric Health Care, 30*(3), 216–223.

Lieberman, L. S. (2006). Evolutionary and anthropological perspectives on optimal foraging in obesogenic environments. *Appetite, 47*(1), 3–9.

Lundahl, B. W., Kunz, C., Brownell, C., Tollefson, D., & Burke, B. L. (2010). A meta-analysis of motivational interviewing: Twenty-five years of empirical studies. *Research on Social Work Practice, 20*(2), 137–160.

Maddock, J. (2004). The relationship between obesity and the prevalence of fast food restaurants: State-level analysis. *American Journal of Health Promotion, 19*(2), 137–143.

Masheb, R. M., Ruser, C. B., Min, K. M., Bullock, A. J., & Dorflinger, L. M. (2018). Does food addiction contribute to excess weight among clinic patients seeking weight reduction? Examination of the modified Yale food addiction survey. *Comprehensive Psychiatry, 84*, 1–6.

Mason, S. M., Flint, A. J., Field, A. E., Austin, S. B., & Rich-Edwards, J. W. (2013). Abuse victimization in childhood or adolescence and risk of food addiction in adult women. *Obesity, 21*(12).

Mason, S. M., Flint, A. J., Roberts, A. L., Agnew-Blais, J., Koenen, K. C., & Rich-Edwards, J. W. (2014). Posttraumatic stress disorder symptoms and food addiction in women by timing and type of trauma exposure. *JAMA psychiatry, 71*(11), 1271–1278.

McGue, M., Iacono, W. G., Legrand, L. N., Malone, S., & Elkins, I. (2001). Origins and consequences of age at first drink. I. Associations with substance-use disorders, disinhibitory behavior and psychopathology, and P3 amplitude. *Alcoholism: Clinical and Experimental Research, 25*(8), 1156–1165.

Mensink, G. B., Schienkiewitz, A., Haftenberger, M., Lampert, T., Ziese, T., & Scheidt-Nave, C. (2013). Overweight and obesity in Germany: Results of the German health interview and examination survey for adults (DEGS1). *Bundesgesundheitsblatt - Gesundheitsforschung - Gesundheitsschutz, 56*(5–6), 786–794. https://doi.org/10.1007/s00103-012-1656-3.

Meule, A. (2012). Food addiction and body-mass-index: A non-linear relationship. *Medical Hypotheses, 79*, 508–511.

Meule, A. (2015). Focus: Addiction: Back by popular demand: A narrative review on the history of food addiction research. *Yale Journal of Biology & Medicine, 88*(3), 295.

Meule, A., & Gearhardt, A. N. (2014). Five years of the Yale food addiction scale: Taking stock and moving forward. *Current Addiction Reports*, 1–13.

Meule, A., Heckel, D., & Kübler, A. (2012). Factor structure and item analysis of the Yale Food Addiction Scale in obese candidates for bariatric surgery. *European Eating Disorders Review, 20*(5), 419–422.

Meule, A., Hermann, T., & Kübler, A. (2015). Food addiction in overweight and obese adolescents seeking weight-loss treatment. *European Eating Disorders Review, 23*(3), 193–198.

Meule, A., Müller, A., Gearhardt, A. N., & Blechert, J. (2017). German version of the Yale food addiction scale 2.0: Prevalence and correlates of 'food addiction'in students and obese individuals. *Appetite, 115*, 54–61.

Meule, A., Rezori, V., & Blechert, J. (2014). Food addiction and bulimia nervosa. *European Eating Disorders Review, 22*(5), 331–337.

Mies, G. W., Treur, J. L., Larsen, J. K., Halberstadt, J., Pasman, J. A., & Vink, J. M. (2017). The prevalence of food addiction in a large sample of adolescents and its association with addictive substances. *Appetite, 118*, 97–105.

Mintz, S. W. (1986). *Sweetness and power: The place of sugar in modern history.* Penguin.

Mokdad, A. H., Marks, J. S., Stroup, D. F., & Gerberding, J. L. (2004). Actual causes of death in the United States, 2000. *Journal of the American Medical Association: The Journal of the American Medical Association, 291*(10), 1238–1245.

Monteiro, C. A., Levy, R. B., Claro, R. M., de Castro, I. R. R., & Cannon, G. (2010). Increasing consumption of ultra-processed foods and likely impact on human health: Evidence from Brazil. *Public Health Nutrition, 14*(1), 5–13.

Monteiro, C. A., Moubarac, J. C., Cannon, G., Ng, S. W., & Popkin, B. (2013). Ultra-processed products are becoming dominant in the global food system. *Obesity Reviews, 14*(S2), 21–28.

Ng, M., Fleming, T., Robinson, M., Thomson, B., Graetz, N., Margono, C., ... Abera, S. F. (2014). Global, regional, and national prevalence of overweight and obesity in children and adults during 1980–2013: A systematic analysis for the global burden of disease study 2013. *The Lancet, 384*(9945), 766–781.

O'Malley, P. M., & Johnston, L. D. (2002). Epidemiology of alcohol and other drug use among American college students. *Journal of Studies on Alcohol and Drugs*, (14), 23.

Ogden, C. L., Carroll, M. D., Kit, B. K., & Flegal, K. M. (2014). Prevalence of childhood and adult obesity in the United States, 2011-2012. *Journal of the American Medical Association, 311*(8), 806–814. https://doi.org/10.1001/jama.2014.732.

Oginsky, M. F., Goforth, P. B., Nobile, C. W., Lopez-Santiago, L. F., & Ferrario, C. R. (2016). Eating 'junk-food'produces rapid and long-lasting increases in NAc CP-AMPA receptors: Implications for enhanced cue-induced motivation and food addiction. *Neuropsychopharmacology, 41*(13), 2977–2986.

Oswald, K. D., Murdaugh, D. L., King, V. L., & Boggiano, M. M. (2011). Motivation for palatable food despite consequences in an animal model of binge eating. *International Journal of Eating Disorders, 44*(3), 203–211.

Parylak, S. L., Koob, G. F., & Zorrilla, E. P. (2011). The dark side of food addiction. *Physiology & Behavior, 104*(1), 149–156.

Pedram, P., Wadden, D., Amini, P., Gulliver, W., Randell, E., Cahill, F., … Sun, G. (2013). Food addiction: Its prevalence and significant association with obesity in the general population. *PLoS One, 8*(9), e74832. https://doi.org/10.1371/journal.pone.0074832.

Peer, K., Rennert, L., Lynch, K. G., Farrer, L., Gelernter, J., & Kranzler, H. R. (2013). Prevalence of DSM-IV and DSM-5 alcohol, cocaine, opioid, and cannabis use disorders in a largely substance dependent sample. *Drug and Alcohol Dependence, 127*(1–3), 215–219. https://doi.org/10.1016/j.drugalcdep.2012.07.009.

Poling, J., Kosten, T. R., & Sofuoglu, M. (2007). Treatment outcome predictors for cocaine dependence. *The American Journal of Drug Alcohol Abuse, 33*(2), 191–206.

Pursey, K. M., Collins, C. E., Stanwell, P., & Burrows, T. L. (2015). Foods and dietary profiles associated with 'food addiction'in young adults. *Addictive Behaviors Reports, 2*, 41–48.

Pursey, K. M., Stanwell, P., Gearhardt, A. N., Collins, C. E., & Burrows, T. L. (2014). The prevalence of food addiction as assessed by the Yale food addiction scale: A systematic review. *Nutrients, 6*(10), 4552–4590.

Richmond, R. L., Roberto, C. A., & Gearhardt, A. N. (2017). The association of addictive-like eating with food intake in children. *Appetite, 117*, 82–90.

Rosenheck, R. (2008). Fast food consumption and increased caloric intake: A systematic review of a trajectory towards weight gain and obesity risk. *Obesity Reviews, 9*(6), 535–547.

Ruddock, H. K., Christiansen, P., Halford, J. C. G., & Hardman, C. A. (2017). The development and validation of the addiction-like eating behaviour scale. *International Journal of Obesity*. https://doi.org/10.1038/ijo.2017.158.

Schiestl, E. T., & Gearhardt, A. N. (2018). Preliminary validation of the Yale food addiction scale for children 2.0: A dimensional approach to scoring. *European Eating Disorders Review, 26*(6), 605–617.

Schulte, E. M., Avena, N. M., & Gearhardt, A. N. (2015). Which foods may Be addictive? The roles of processing, fat content, and glycemic load. *PLoS One, 10*(2), e0117959.

Schulte, E. M., & Gearhardt, A. N. (2017). Development of the modified Yale food addiction scale version 2.0. *European Eating Disorders Review*. https://doi.org/10.1002/erv.2515.

Schulte, E. M., & Gearhardt, A. N. (2018). Associations of food addiction in a sample recruited to Be nationally representative of the United States. *European Eating Disorders Review, 26*(2), 112–119. https://doi.org/10.1002/erv.2575.

Schulte, E. M., Grilo, C. M., & Gearhardt, A. N. (2016a). Shared and unique mechanisms underlying binge eating disorder and addictive disorders. *Clinical Psychology Review, 44*, 125–139.

Schulte, E. M., Potenza, M. N., & Gearhardt, A. N. (2016b). A commentary on the "eating addiction" versus "food addiction" perspectives on addictive-like food consumption. *Appetite*.

Schulte, E. M., Jacques-Tiura, A. J., Gearhardt, A. N., & Naar, S. (2017a). Food addiction prevalence and concurrent validity in african American adolescents with obesity. *Psychology of Addictive Behaviors: Journal of the Society of Psychologists in Addictive Behaviors*.

Schulte, E. M., Potenza, M. N., & Gearhardt, A. N. (2017b). A commentary on the "eating addiction" versus "food addiction" perspectives on addictive-like food consumption. *Appetite, 115*, 9–15. https://doi.org/10.1016/j.appet.2016.10.033.

Schulte, E. M., Potenza, M. N., & Gearhardt, A. N. (2017c). How much does the addiction-like eating behaviour scale add to the debate regarding food versus eating addictions? *International Journal of Obesity*.

Schulte, E. M., Smeal, J. K., & Gearhardt, A. N. (2017d). Foods are e effect report questions of abuse liability. *PLoS One, 12*(8), e0184220.

Sinha, R. (2001). How does stress increase risk of drug abuse and relapse? *Psychopharmacology, 158*(4), 343–359. https://doi.org/10.1007/s002130100917.

Spring, B., Schneider, K., Smith, M., Kendzor, D., Appelhans, B., Hedeker, D., ... Pagoto, S. (2008). Abuse potential of carbohydrates for overweight carbohydrate cravers. *Psychopharmacology, 197*(4), 637–647.

Steinberg, L. (2010). A dual systems model of adolescent risk-taking. *Developmental Psychobiology, 52*(3), 216–224.

Tuomisto, T., Hetherington, M. M., Morris, M. F., Tuomisto, M. T., Turjanmaa, V., & Lappalainen, R. (1999). Psychological and physiological characteristics of sweet food "addiction". *International Journal of Eating Disorders, 25*(2), 169–175.

Volkow, N. D., Wang, G. J., Fowler, J. S., & Telang, F. (2008). Overlapping neuronal circuits in addiction and obesity: Evidence of systems pathology. *Philosophical Transactions of the Royal Society B: Biological Sciences, 363*(1507), 3191–3200.

Volkow, N. D., Wang, G. J., Tomasi, D., & Baler, R. D. (2013). Obesity and addiction: Neurobiological overlaps. *Obesity Reviews, 14*, 2–18.

CHAPTER

3 Dissecting compulsive eating behavior into three elements

Catherine F. Moore[1,2], Valentina Sabino[1], George F. Koob[3], Pietro Cottone[1]

Laboratory of Addictive Disorders, Departments of Pharmacology and Psychiatry, Boston University School of Medicine, Boston, MA, United States[1]; Graduate Program for Neuroscience, Boston University School of Medicine, Boston, MA, United States[2]; National Institute on Alcohol Abuse and Alcoholism, National Institutes of Health, Bethesda, MD, United States[3]

Introduction

Excessive, uncontrollable eating behaviors, which are observed in forms of obesity, eating disorders, and the proposed construct of 'food addiction,' are increasingly being recognized as having characteristics of compulsivity, similar to what is described in substance-use disorders and behavioral addictions (American Psychiatric Association, 2013; Davis, 2013; Gearhardt, Corbin, & Brownell, 2009; Moore, Sabino, Koob, & Cottone, 2017b; Volkow, Wang, Tomasi, & Baler, 2013a, de Zwaan, 2001). By definition, compulsive behaviors are inappropriate, repetitive, and continue in the face of adverse consequences. These behaviors are 'ego-dystonic' or against one's will, but despite this, the individual feels compelled to perform them. Compulsive behaviors are not just persistent, repetitive, and perseverative, but engaged to *prevent* or *provide relief* from distress, anxiety, or stress (American Psychiatric Association, 2000; el-Guebaly, Mudry, Zohar, Tavares, & Potenza, 2012; Robbins, Curran, & de Wit, 2012).

Therefore, compulsivity can be approached as a transdiagnostic construct that embraces multiple disorders. In the substance- and alcohol-use disorder fields, addiction is described as being composed of three stages: *binge/intoxication* stage, *withdrawal/negative affect* stage, and *preoccupation—anticipation* stage, which reflect dysfunctions in incentive salience/pathological habits, reward/stress, and executive function, respectively (Koob & Le Moal, 1997; Koob & Volkow, 2016; Kwako, Bickel, & Goldman, 2018). Together, these stages drive compulsive drug-seeking behavior. These domains of dysfunction correspond to neuroadaptations that reflect allostatic changes in three key neurocircuits, respectively: basal ganglia, extended amygdala, and prefrontal cortex (PFC) (Koob & Volkow, 2016).

Within drug addiction, there has been a great deal of effort to systematically define compulsivity, a task that has proven elusive, stimulating much debate along the way (Belin-Rauscent, Fouyssac, Bonci, & Belin, 2015; Everitt, 2014; Hopf &

Compulsive Eating Behavior and Food Addiction. https://doi.org/10.1016/B978-0-12-816207-1.00003-2
Copyright © 2019 Elsevier Inc. All rights reserved.

Lesscher, 2014; Koob, 2013; Koob et al., 2014; Piazza & Deroche-Gamonet, 2013; Volkow & Fowler, 2000). As research into food addiction has ramped up in recent decades, it has become necessary to have similar debates as to what constitutes pathological, compulsive eating behavior, which would include discussions on how it should be measured and, ultimately, treated (Moore, Panciera, Sabino, & Cottone, 2018b; Moore, Sabino, Koob, & Cottone, 2017a, 2017b).

Throughout this chapter, we will describe compulsive eating behavior as seen across multiple eating disorders, including binge eating disorder (BED), forms of obesity, and food addiction. To do this, we modified commonly accepted definitions of compulsivity from the drug addiction literature to the extent that they are applicable to compulsive eating behavior (Moore et al., 2017b). In the drug addiction literature, compulsivity is often presented as encompassing one or more of three major elements: habitual drug taking (Everitt, 2014), drug use to relieve a negative affective state (Koob & Volkow, 2010), and drug use in spite of negative consequences (Deroche-Gamonet, Belin, & Piazza, 2004). Recent research indicates that these same processes underlie compulsive eating behavior. Thus, the elements of compulsive eating behavior have been defined as (1) *habitual overeating*; (2) *overeating to relieve a negative emotional state*; and (3) *overeating despite aversive consequences* (Moore et al., 2017b). Each element can be linked to discrete neurobiological processes and mechanisms, although these mechanisms also greatly intersect.

This chapter will serve as a guide to the elements of compulsive eating, including brief descriptions of the different eating disorders characterized by compulsive eating behavior, as well as rationalization and definitions of the three elements of compulsive eating. We will describe both preclinical and clinical studies pertaining to the behaviors, as well as their neurobiological substrates.

The prevalence and significance of compulsive eating

Compulsive eating behavior is still in the early stages of its conceptualization despite being a central feature of multiple eating disorders, including BED, certain forms of obesity, and the newly proposed construct of "food addiction" (Gearhardt, Corbin, & Brownell, 2016, 2009; Moore et al., 2017b; Volkow, Wang, Fowler, & Telang, 2008a). BED is a psychiatric condition defined by uncontrolled, intermittent overconsumption of food in brief periods of time (American Psychiatric Association, 2013; Kessler et al., 2013, 2016; de Zwaan, 2001). Obesity, defined by a body mass index (BMI) of 30 or above (WHO, 2000), is not an eating disorder *per se*, even though it may be a consequence of pathological or compulsive eating behavior (Davis, 2013). "Food addiction" is a recently proposed disorder characterized by compulsive eating which emerged from a multitude of research demonstrating the similarities between pathological forms of eating behaviors and addictions to drugs of abuse (Avena et al., 2004, 2008b; Gearhardt et al., 2009; Hernandez & Hoebel, 1988; Rada, Avena, & Hoebel, 2005). Using the Yale Food Addiction Scale

(YFAS), a recently developed tool (Gearhardt et al., 2009) subsequently updated to reflect changes to the substance-use disorder diagnosis in the Diagnostic and Statistical Manual fifth ed. (DSM-5), a diagnosis of food addiction is made when a patient displays clinically significant impairment or distress and meets criteria such as eating much more than intended and experiencing problems in ability to function because of food (Gearhardt et al., 2016). As most epidemiological data on food addiction are collected using the YFAS, in this chapter we will refer to food addiction in humans as the condition measured by this scale, even though it is important to emphasize that this emerging not fully established condition is different from well-recognized feeding-related pathologies and that further validation is therefore necessary (Long, Blundell, & Finlayson, 2015).

According to recent estimates, prevalence rates of obesity are 35%–40% among United States adults, a number that has been progressively increasing (Finkelstein, Trogdon, Cohen, & Dietz, 2009; Hales, Carroll, Fryar, & Ogden, 2017; Ogden, Carroll, & Flegal, 2014). BED affects an estimated 1%–3% of the general population (Cossrow et al., 2016; Spitzer et al., 1992; de Zwaan, 2001). Studies of food addiction in community samples have yielded prevalence rates of 5%–15% (Davis et al., 2011; Gearhardt et al., 2009, 2016). In overweight/obesity samples, the prevalence of BED and food addiction is significantly higher, and among individuals with BED and/or food addiction, 40%–70% are obese (Dingemans, Bruna, & van Furth, 2002; Dingemans & van Furth, 2012; Kessler et al., 2013; Pedram et al., 2013; Pursey, Stanwell, Gearhardt, Collins, & Burrows, 2014). In populations seeking treatment for obesity (i.e., bariatric surgery, weight loss management programs), dramatically increased rates of BED and food addiction are observed (e.g., BED prevalence of 15%–30% in hospital-based obesity treatment seeking individuals (Pursey et al., 2014; Spitzer et al., 1992)). Importantly, obese people who seek treatment have been described as showing distinct characteristics compared with nontreatment seeking obese individuals, including greater binge eating and associated emotional distress (Fitzgibbon, Stolley, & Kirschenbaum, 1993).

Comorbidity between obesity/BED/food addiction is incredibly high and likely because of strong links between compulsive eating behaviors and high BMI (Gearhardt, Boswell, & White, 2014). More than half (57%) of obese individuals with BED also have YFAS-diagnosed food addiction (Gearhardt et al., 2014; Pursey et al., 2014), and almost 75% of obese individuals with food addiction also have BED (Davis et al., 2011; Gearhardt et al., 2011, 2012).

Obesity, BED, and food addiction are all associated with significant distress and impairments. Individuals with BED and/or obesity have higher rates of comorbid psychiatric conditions (e.g., mood and anxiety disorders), lower health-related quality of life, and higher health-care costs (Agh et al., 2015; Dee et al., 2014; Halfon, Larson, & Slusser, 2013; Kolotkin & Andersen, 2017). A diagnosis of BED in obese patients is associated with lower weight-related quality of life and increased psychological distress compared with weight-matched non-BED controls (Kolotkin et al., 2004). In adults with obesity and/or BED, the addition of a food addiction diagnosis is also associated with worse quality of life, greater eating pathology, and a higher likelihood of comorbid psychiatric disorders compared with non-food-addicted peers (Davis et al., 2011; Gearhardt et al., 2012).

While some comorbidity among these eating disorders is to be expected, as some of the criteria are common across diagnoses (e.g., loss of control in BED and food addiction), substantial overlap is also likely indicative of transdiagnostic domains (i.e., elements of compulsive eating behaviors). Indeed, high comorbidity may reflect shared etiologies and common underlying mechanisms (e.g., habit, inhibitory control) of certain instances of food addiction, eating disorders, and forms of obesity. Furthermore, because comorbidity among disorders is associated with greater psychopathy and more distress, this represents a population with greater need for treatment.

Food addiction as a disorder of compulsive eating

The concept of food addiction has been met with criticism for several reasons, the most evident being perhaps the lack of direct and overlapping analogies between some of the diagnostic criteria of drug addiction. Specifically, whether the "existence" of food addiction would require the presence of diagnostic criteria such as tolerance and withdrawal (i.e., "dependence" in the traditional, pharmacological sense) is under debate. Tolerance is classically defined as reduced drug effect following repeated use, which drives the user to increase the dose to achieve the original effect; withdrawal is classically defined as the characteristic physical withdrawal syndrome that follows the removal of the substance (O'Brien, 2011b). Indeed, despite evidence that excessive consumption of food rich in sugar/fat has similar powerful effects of abused drugs on the brain reward and anti-reward systems (Cottone et al., 2009a; Johnson & Kenny, 2010; Koob, 2013; Volkow et al., 2013a), arguing that food can produce effects comparable with drugs of abuse remains controversial (Salamone & Correa, 2013; Ziauddeen, Farooqi, & Fletcher, 2012). In addition, behavioral addictions (i.e., non−substance-related addictions) have also been argued not to share the classically defined tolerance and withdrawal criteria with substance-use disorders (Holden, 2010; Ziauddeen et al., 2012). For example, pathological gambling, a disorder comprising feelings of loss of control over gambling with an emphasis on life disruptions and damage (Fauth-Buhler, Mann, & Potenza, 2016), is built on the concept that engagement in a behavior can become compulsive even in absence of substance-induced effects in the brain. However, these criticisms reflect more the (unfortunately rather common) misunderstanding of tolerance and withdrawal in addiction as purely "physical" phenomena (Benton, 2010; George, Koob, & Vendruscolo, 2014; Piazza & Deroche-Gamonet, 2013), different than simply behavioral manifestations of neuroadaptations. Note that neuroadaptations in modern terminology are "physiological" changes to the brain, which at some molecular/cellular/neurochemical level are also "physical." Indeed, tolerance (increased reward-seeking and taking to produce the same effect) (Koob, 1996) and withdrawal (a motivational syndrome characterized by dysphoria, anxiety, irritability when the desired reward is unavailable) (Koob & Le Moal, 1997; 2008b) are both present in all behavioral addictions (Tao et al., 2010; Wray & Dickerson, 1981). Similarly, tolerance and withdrawal are also observed in compulsive eating and food addiction, where individuals ingest more and more to produce the same hedonic effect, and dysphoria, depression, and irritability (so-called "motivational" withdrawal symptoms) occur following palatable food abstinence. Recognizing the evolving understanding of addictive disorders, the newest edition of the DSM (fifth ed. (APA, 2013)) has placed more emphasis on behavioral rather than physical aspects of addiction and shifted the terminology from "substance-dependence" to "substance-use disorders." This change also reflects the redirection of focus from dependence qualifiers to more behavioral dimensions that better reflect the underlying neurobiology, e.g., by including the addition of "craving" as a criterion (Badiani, 2014; Gearhardt et al., 2011; Hasin et al., 2013; O'Brien, 2011a). Therefore, we here argue that using old physical definitions of tolerance and withdrawal rather than physiological/motivational measures of tolerance and withdrawal is a red herring in the overall conceptual framework of addiction and that food addiction should be accepted as an addictive disorder, as it shares similar neurobiology, risk factors, etiology, and behavioral manifestations as other currently recognized behavioral and drug addictions (Carlier, Marshe, Cmorejova, Davis, & Muller, 2015; Schulte, Grilo, & Gearhardt, 2016). Additionally, understanding the unique characteristics of food addiction can inform current research tools, including the necessity to move away from diagnostic questions involving physical symptoms that allegedly only apply to some abused drugs (e.g., physical withdrawal and dependence). The utility of a food addiction diagnosis separate from, or in the absence of, these other conditions is not yet known; however, it holds potential for increasingly personalized treatment approaches for these disorders (Davis et al., 2011).

Dissecting compulsive eating behavior into three elements

It is important to systematically define and conceptualize what compulsive eating behavior is, as well as identify the multifaceted nature of how it can arise, manifest, and be therapeutically targeted. Compulsive eating behavior refers to the following elements: (1) *habitual overeating*, (2) *overeating to relieve a negative emotional state*, and (3) *overeating despite aversive consequences*. These elements were developed using observations of compulsive drug use from the drug addiction literature, where there is substantial theoretical and neurobiological evidence to support their existence (Moore et al., 2017b). The neurobiological processes underlying these processes encompass multiple brain regions within interconnected circuits; however, in this chapter, for simplicity we will focus on three key brain regions that are most critical for the elements of compulsive eating: *the basal ganglia, the extended amygdala*, and *the PFC*.

Habitual overeating

Maladaptive habit formation

Instrumental behavior is guided by goal-directed and habit systems. "Goal-directed" behavior refers to making behavioral choices that are advantageous, which includes the careful consideration of the effort of the action and value of the outcome. After repeated experience where an ordered, structured action sequence is established, habit systems are engaged to guide behavior and are extremely prone to evocation by an associated context or stimulus. This "stimulus-response" behavior is highly efficient, though at the expense of flexibility (Graybiel, 2008). In habit formation, reinforcers serve to strengthen the stimulus-response association but are not themselves encoded as the goal (Everitt & Robbins, 2005). Thus, the presence of the reward-paired stimulus alone can indeed elicit a response or enhance robustness of behavior (Everitt & Robbins, 2016), and alterations in the *actual* outcome value no longer effect behavioral adaptation (Corbit, 2016; Everitt & Robbins, 2005; Voon et al., 2015).

While habits are adaptive *per se*, they can become hijacked by high-magnitude rewards, such as addictive drugs and highly palatable food. In the context of palatable food, environmental food–associated stimuli or conditioned stimuli (e.g., fast food commercials, the smell of donuts from the break room, etc.) can robustly enhance craving and "food-seeking" behaviors regardless of the presence of food or hunger (Everitt & Robbins, 2005; Giuliano & Cottone, 2015; Robinson et al., 2015). Furthermore, in compulsive eating, reducing the *value* of the food (e.g., through satiation) no longer effectively reduces eating behavior (Corbit, 2016). Compulsive eating or drug-taking behavior can therefore be conceptualized as a maladaptive habit response, where behavior is not anymore directly under control of the goal, but it is rather governed by conditioned reinforcers (Everitt & Robbins, 2005).

Incentive salience

The incentive salience of palatable food-associated cues is exaggerated in compulsive eating individuals. Incentive salience refers to motivation for rewards that is driven by both physiological state and previously learned associations about a reward cue (Hyman, 2005). Similar to drug addiction, these cues can activate food-seeking and the development of compulsive habits (Koob & Volkow, 2016). Individuals with obesity and/or BED reliably show greater attentional biases and sensitivity to food cues (Voon, 2015), similarly seen in individuals with substance-use disorders when tested with drug-associated cues (Hester & Luijten, 2014). Attentional biases to food cues are observed both at a behavioral (i.e., eye fixation) and a neural level (i.e., increased activation of nucleus accumbens; NAc) (Hendrikse et al., 2015; Lawrence, Hinton, Parkinson, & Lawrence, 2012; Meule, Lutz, Vogele, & Kubler, 2012; Schmitz, Naumann, Trentowska, & Svaldi, 2014; Shank et al., 2015). Increased cue sensitivity has been positively associated with BMI (Shank et al., 2015), food-seeking and eating (Lawrence et al., 2012), and binge eating symptom severity (Schmitz et al., 2014). The enhanced attentional bias toward food cues may reflect an increase in the incentive salience of food stimuli and may also be exacerbated by failures in top-down inhibitory control processes, as discussed in further detail later in this chapter (Voon, 2015).

Clinical evidence of habitual eating

Compulsive eating behavior driven by habit learning mechanisms would suggest (1) a breakdown in goal-directed eating behavior; (2) enhanced stimulus-driven eating behavior; and (3) the persistence of eating behavior despite devaluation.

Several studies have investigated goal-directed learning deficits in obesity. In a study of lean versus obese participants' ability to adapt behavior, researchers found that obese participants showed deficits in learning from negative reward prediction errors (PEs) (Mathar, Neumann, Villringer, & Horstmann, 2017). In reward-related learning, dopamine encodes PEs, which refer to a difference in expected versus actual outcomes. Positive and negative PEs signal that a reward is larger or smaller than expected, respectively (Schultz, Dayan, & Montague, 1997). Obese subjects were found to have greater PEs than lean participants, although only positive PEs had an impact on behavior (Mathar et al., 2017). Negative PEs should cause a downward adjustment to the motivational value of the outcome and subsequent decreases in the action (McClure, Daw, & Montague, 2003). Therefore, obese subjects failing to use negative PE signals to adjust eating behavior efficiently likely contribute to habitual overeating. Furthermore, behavioral adaptation to positive PEs was in tact, which may then contribute to exaggerated incentive salience in addiction and compulsive eating (Hyman, 2005).

In an experiment designed to investigate whether people tend to make decisions based on likely outcomes (model-based; goal-directed) or based on previously reinforced behaviors (model-free; habitual), individuals with BED were more likely to shift to habitual behaviors compared with non-BED obese subjects (Voon et al., 2015). Furthermore, individuals with addiction or obsessive-compulsive disorder also displayed this bias toward engaging in habit learning devices (Voon et al., 2015). A favoring of habitual behaviors may represent a shared dysfunction in neural processing observed in multiple disorders characterized by compulsivity.

Similar to studies in patients with addiction (Ersche et al., 2016) or obsessive-compulsive disorder (Gillan et al., 2015), obese individuals also show greater resistance to devaluation (Horstmann et al., 2015; Janssen et al., 2017). Habitual overeating was tested in one human laboratory study through assessment of responding for a sweet or savory snack before and after devaluation (Horstmann et al., 2015). Obese participants learned to perform a task for food rewards and were then subsequently allowed to eat to satiation one of the two rewards (devalued condition). Subjects with higher BMI were less sensitive to devaluation, showing equal response for both devalued and nondevalued food rewards (Horstmann et al., 2015). A similar study by Janssen et al. (2017) also found reduced devaluation magnitude in obese individuals. These results suggest that difficulty in adapting behavior to fit goal-directed actions may contribute to food-seeking and overconsumption.

Preclinical evidence of habitual eating

Compulsive, habitual responding is measured in animal models as the inflexible responding that follows one of the several procedures designed to devalue the reward outcome. Devaluation methods include specific satiety, addition of the bitter tastant quinine, and postingestive malaise induced by treatment with lithium chloride. Similar to preclinical models of drug addiction, palatable food access accelerated the shift from flexible to habitual food-seeking, whereas rats with intermittent access to a palatable diet showed resistance to outcome devaluation by specific satiety (Furlong, Jayaweera, Balleine, & Corbit, 2014). Specifically, rats that had daily limited access to palatable food did not show differences in responding for either grain pellets or a sucrose solution (counterbalanced: nondevalued vs. devalued by free access before the test) (Furlong et al., 2014). In a similar study, rats that had 2 weeks of prior continuous access to a cafeteria diet (i.e., mixture of sweet and savory foods, high in calories) showed resistance to devaluation by free access before testing, an effect that was still present 1 week following removal of the home cage diet (Reichelt, Morris, & Westbrook, 2014). In animals with prolonged palatable diet exposure, resistance to food devaluation with the bitter tastant quinine was also associated with continued seeking in the presence of aversive consequences (de Jong, Meijboom, Vanderschuren, & Adan, 2013) and with inflexible, abnormal eating behaviors (Heyne et al., 2009). This effect mirrors what is seen in animal models of drug addiction, where drug-seeking behaviors become habitual (Everitt & Robbins, 2013), suggesting that both drugs of abuse and highly palatable food consumption more so predispose individuals to habitual responding (de Jong et al., 2013; Velazquez-Sanchez et al., 2015).

Behaviorally, habitual drug- or food-seeking can be operationalized as instrumental responding under a second-order schedule of reinforcement, an experimental procedure where seeking behavior is distinct and discernible from intake (Murray et al., 2015). In this task, responding on a seeking lever is motivated by the contingent presentation of a conditioned stimulus (Di Ciano & Everitt, 2005; Everitt & Robbins, 2000; Velazquez-Sanchez et al., 2015). The first interval of this procedure consists of animals responding for conditioned reinforcers (no reward), allowing for evaluation of conditioned stimulus effects on behavior without the presence of the primary reinforcer (Everitt & Robbins, 2000). In animals with a history of binge-like eating of palatable food, conditioned reinforcers maintained responding to a greater degree than animals that only self-administered control chow food (Velazquez-Sanchez et al., 2015). Seeking responses can also be devalued in this paradigm, via the introduction of unpredictable shock. In animal models of drug addiction, drug-seeking responses are resistant to devaluation with shock (Belin, Mar, Dalley, Robbins, & Everitt, 2008; Jonkman, Pelloux, & Everitt, 2012; Zapata, Minney, & Shippenberg, 2010), though devaluation in second-order schedules have not yet been tested on food-seeking behavior in compulsively eating animals.

Neurobiological basis of habitual overeating

Habitual behavior is primarily controlled by the basal ganglia, a group of subcortical nuclei that includes the ventral (i.e., NAc) and dorsal components (dorsal striatum; DS) of the striatum, key structures in reward and reinforcement. Specifically, the NAc plays a key role in reward and reinforcement, while the DS coordinate instrumental learning and habitual behavior (Everitt & Robbins, 2005).

Eating palatable food or using addictive drugs increases extracellular dopamine in the NAc (Day, Roitman, Wightman, & Carelli, 2007; Rada et al., 2005; Stuber et al., 2008). Following repeated pairings with a drug or food-associated cue, the cue alone (absent primary reinforcer) elicits NAc dopamine release (Schultz et al., 1997). This potentiated dopaminergic response to cues is hypothesized to increase incentive salience, likely contributing to accelerated habit formation and compulsive behavior (Everitt et al., 2008; Everitt & Robbins, 2016).

Structural differences in striatal habit learning systems may underlie dysfunctional, or intensified, habitual behavior. A study by Voon et al. (2015) observed that obese subjects with BED, who were found to be predisposed to engage in habit-based responding, had structural differences from weight-matched controls: reduced gray matter volume in ventral and dorsal striatum and associated prefrontal projection areas. Information encoding in prefrontal and striatal areas is critical for maintaining goal-directed responding (McNamee, Liljeholm, Zika, & O'Doherty, 2015). Therefore, accelerated habit formation may be related to impaired goal-directed learning because of lower gray matter volume in areas that encode action-outcome associations.

Habit formation is linked to a shift in ventral to dorsal basal ganglia control over behavior. From animal studies of addiction, it was established that the NAc is critical early in the goal-directed responding for a reward (i.e., early stage of drug use), while inactivation of the DS has no effect on acquisition of self-administration behavior. However, once a habit is formed, dopamine inactivation in the dorsolateral striatum (DLS) blocks compulsive responding and effectively restores sensitivity to devaluation (Belin & Everitt, 2008; Corbit, Nie, & Janak, 2012). This effect was replicated in rats exposed to long-term restricted access to palatable food, indicating a parallel loss of goal-directed food-seeking behavior owing to increased DLS activation (Furlong et al., 2014). Upstream brain regions, such as the amygdala and prefrontal areas, also influence habitual behavior via projections to the striatum. The central nucleus of the amygdala (CeA) projections to the DLS regulates habitual behavior, such that lesions of the anterior CeA prevent habit formation (Lingawi & Balleine, 2012). Prefrontal executive-inhibitory control functions coordinate flexible, goal-directed behavior. Various prefrontal (e.g., ventromedial PFC (vmPFC)) projections to the striatum underlie action-outcome, or goal-directed, learning (Balleine & O'Doherty, 2010). Thus, preexisting vulnerabilities and/or neuroadaptations in the amygdala and prefrontal areas caused by palatable food likely contribute to the development or the exacerbation of the predominance of habitual behavior (Everitt & Robbins, 2016). These influences of amygdala and prefronto-cortical systems (discussed in depth later on in this chapter) on striatal-mediated habit formation demonstrate an example of the overlapping neurobiological substrates of the different elements of compulsivity. Thus, an imperative area of research is to investigate the interactions of these brain systems and their effects on different aspects of compulsive behavior.

Dopaminergic signaling in the anterior DLS is critical for habit formation (Yin & Knowlton, 2006). The dopamine type-1 receptor (D1R) neurons of the direct, striatonigral pathway drive enhanced excitability (Surmeier, Ding, Day, Wang, & Shen, 2007). Accelerated habit formation by drugs of abuse and palatable food may be facilitated in part by an increased ratio of D1R to dopamine type-2 receptor (D2R) signaling (Furlong et al., 2014; Volkow, Wang, Tomasi, & Baler, 2013b). Animals that habitually respond for drugs (Park, Volkow, Pan, & Du, 2013) or palatable food (Furlong et al., 2014) show increased D1R neuronal activation. Furthermore, in rats fed a palatable diet, intra-DLS injections of the D1R antagonist SCH-23,390 blocked the acquired habitual eating (Furlong et al., 2014) and restored sensitivity to devaluation (i.e., increased goal-directed behavior). Glutamatergic signaling via amino-3-hydroxy-5-methyl-4-isoxazolepropionic acid (AMPA) receptors in the DLS may also contribute to habitual overeating. AMPA receptors in the DLS are also critical for expression of habitual overeating of palatable food; intra-DLS infusion of CNQX (6-cyano-7-nitroquinoxaline-2,3-dione; an AMPA/kainate receptor antagonist) restored sensitivity to devaluation of the palatable food, effectively blocking habitual intake (Furlong et al., 2014).

The μ-opioid system is also a regulator of incentive motivation for rewards and reward-associated cues (Giuliano & Cottone, 2015; Laurent, Morse, & Balleine, 2015; Wassum, Cely, Maidment, & Balleine, 2009), key factors in action-outcome versus stimulus-driven habitual overeating (Corbit, 2016). In rats responding for palatable food under a second-order schedule of reinforcement, GSK1521498, a selective μ-opioid receptor antagonist, and naltrexone, a mixed opioid receptor antagonist, reduced food-seeking responses as well as binge-like eating of palatable food (Giuliano, Robbins, Nathan, Bullmore, & Everitt, 2012). In subjects with BED, the selective μ-receptor antagonist GSK1521498 decreased attentional bias to food cues (Chamberlain et al., 2012). Similarly, naltrexone decreased neural responses to food cues in healthy subjects, indicated by a reduced activation of the anterior cingulate (ACC) and the DS (Murray et al., 2014). Through reducing the hedonic value attributed to palatable food-associated cues, μ-opioid antagonists may be working to restore an action-outcome (vs. stimulus-response or habitual) response toward palatable food in those that compulsively eat.

Overeating to relieve a negative emotional state

Emergence of a negative affect

Engaging in overeating behavior to alleviate a negative emotional state has been identified as a major domain of compulsive eating, also derived from the drug addiction literature (Koob, 2009; Koob & Le Moal, 2001; Parylak, Koob, & Zorrilla,

2011). This element is based on the conceptualization of compulsivity as being driven by negative reinforcement, which recognizes the stress/anxiety before the behavior and subsequent relief as a major driver of the repeated, uncontrollable behavior (Koob & Le Moal, 2008a). The construct of negative reinforcement, defined here as overeating that alleviates a negative emotional state created by palatable food abstinence, is particularly relevant as a driving force in the *withdrawal/negative* affect stage described in addiction. This framework has also been hypothesized to contribute to the compulsivity associated with obsessive-compulsive disorder (Berlin & Hollander, 2014). Therefore, in compulsive eating and drug addiction, the withdrawal-induced negative affect, which is characterized by dysphoria, irritability, and anxiety, emerges via two biopsychological processes: decreased reward function and increased stress. The emergence of decreased reward functioning manifests as "affective habituation" and as a decrease in the motivation for ordinary everyday rewards (Koob, 1996; Parylak et al., 2011). While initially sensitization of incentive salience may occur (Robinson & Berridge, 1993), subsequently neuroadaptations develop, caused by repeated, overstimulation of the reward system by palatable food (or drugs of abuse), and are thought to downregulate reward system functioning (Koob et al., 2014; Koob & Le Moal, 2001). Concurrently, negative affect arises via the recruitment of the brain stress systems. These systems, typically purposed for responding to dangerous conditions, are thought to be repeatedly engaged during withdrawal from drugs or palatable food (Cottone et al., 2009a; Koob, 1999), and this result in increased irritability and anxiety. While at first overeating behavior (as drug-taking behavior) is positively reinforced, the palatable food eventually acquires negative reinforcing properties (i.e., relief of the negative emotional symptoms of withdrawal) (Cottone et al., 2009a; Iemolo et al., 2012; Koob & Le Moal, 2001; Parylak et al., 2011; Teegarden & Bale, 2007). Therefore, both of these processes are hypothesized to reflect an allostatic change in mood, in which compulsive drug-taking/eating results in a "paradoxical" self-medication effect that improves the reward deficit and suppresses negative emotions (Cottone et al., 2009a; Koob & Le Moal, 2001; Parylak et al., 2011).

Clinical evidence of overeating to relieve a negative emotional state

Dietary restraint from palatable food may precipitate emotional withdrawal symptoms, driving a negatively reinforced mechanism. Physical signs of withdrawal, which are dependent on the specific pharmacological properties of the drug (and are therefore missing in feeding-related disturbances), may or may not be present following drug discontinuation as function of the type of drug of abuse, and they typically dissipate quickly (Koob & Le Moal, 2001). Negative emotional states such as dysphoria, irritability, and anxiety are instead long-lasting and common not only to all major drugs of abuse (Koob & Le Moal, 2001) but also to forms of obesity and eating disorders. There is evidence that obese individuals placed on a dieting regimen feel increased irritability, nervousness, and anxiety (Keys, Brozek, Henschel, Mickelson, & Taylor, 1950; Silverstone & Lascelles, 1966; Stunkard, 1957). Furthermore, in a study of the effects of diet on emotional states, switching from a high- to low-fat diet negatively impacted the subjects' mood (Wells, Read, Laugharne, & Ahluwalia, 1998). In people with BED, binge eating episodes are preceded by negative mood (Greeno, Wing, & Shiffman, 2000; Haedt-Matt & Keel, 2011), and levels of anxiety predict increased binge eating (Fitzsimmons-Craft, Bardone-Cone, Brownstone, & Harney, 2012). According to a theory called the restraint theory, originally developed by Herman and Mack (1975) and further elaborated by Herman and Polivy (1980), emotional "disinhibitors" (e.g., negative emotions) can trigger a binge episode. Dietary restraint is a predictor of stress and symptoms of depression (Eldredge, Wilson, & Whaley, 1990; Kagan & Squires, 1983; Rosen et al., 1987, 1990). In turn, negative emotional states are associated with overeating in response to stress (Greeno & Wing, 1994; Heatherton, Herman, & Polivy, 1991), likely reflecting an attempt to self-medicate with so-called "comfort" foods (Finch & Tomiyama, 2015; Macht, 2008; Tomiyama, Dallman, & Epel, 2011). And, eating of palatable food can effectually dampen the body's response to stress (Pecoraro, Reyes, Gomez, Bhargava, & Dallman, 2004). Furthermore, historical theories of overeating support a negatively reinforced mechanism, for instance, the "psychosomatic" concept developed by Kaplan and Kaplan (1957) which proposed that obese people overeat as a mechanism to cope with anxiety and that eating serves to reduce this anxiety. Similarly, Bruch (1973) theorized that overeating was caused by "emotional tension," "uncomfortable sensations and feelings," and the inability to differentiate hunger from emotional tension states.

Affective disorders, such as depression and anxiety, are characterized by negative emotional states and may predate, or be reciprocally caused by, forms of obesity and eating disorders, similar to addiction to drugs (Dallman et al., 2003; Fairburn et al., 1998; Rosenbaum & White, 2015). Rates of psychiatric diseases are more common in individuals with BED and/or obesity, especially mood and anxiety disorders (Barry, Pietrzak, & Petry, 2008; Galanti, Gluck, & Geliebter, 2007; Grilo, Ivezaj, & White, 2015; Peterson, Miller, Crow, Thuras, & Mitchell, 2005; Yanovski, Nelson, Dubbert, & Spitzer, 1993). Mood-related psychopathologies are associated with severity of binge eating (Wilfley et al., 2000) and worse treatment outcomes (Clark, Niaura, King, & Pera, 1996; Linde et al., 2004; McGuire, Wing, Klem, Lang, & Hill, 1999). It is possible that the increased rate of a mood disorder may reflect a preexisting state of allostasis or a deviation from a "normal" range of reward function (Koob & Le Moal, 2001). Therefore, abstinence from palatable food (i.e., dieting) precipitates a negative emotional state, or "affective withdrawal," which is in turn alleviated by overeating; thus, compulsive eating is maintained through negatively reinforced mechanisms.

However, it should be noted that the concept of "withdrawal" has a key difference in individuals affected by disordered eating compared with drug-addicted individuals. Indeed, while drug users may quit drugs, food abstinence cannot simply be achieved by quitting food (as food is needed for survival) but rather by dieting. Notably, dieting reflects a reduction in calories consumed typically coupled with a shift from energy-dense, palatable "forbidden" foods to energy-diffuse, less palatable "safe" foods (Gonzalez & Vitousek, 2004; Stirling & Yeomans, 2004). Caloric restriction on its own has well-known effects on binge eating and rebound weight gain, with evidence that fasting is predictive of later binge eating (Stice, Davis, Miller, & Marti, 2008) and that dieters often regain one- to two-thirds of the weight lost on the diet (Mann et al., 2007). It should be noted that food restriction is known to cause neuroadaptations that promote compulsive behavior (compulsive eating and compulsive drug use) (Carr, 2016; Sedki, Gardner Gregory, Luminare, D'Cunha, & Shalev, 2015; Shalev, 2012). While food restriction and consumption of highly palatable food often produce similar behavioral outcomes, the effects are mediated by differential neurobiological mechanisms (Cottone, Sabino, Steardo, & Zorrilla, 2009b, 2012; Smith et al., 2015).

Preclinical evidence of overeating to relieve a negative emotional state

There is an abundance of evidence from animal models showing that access to palatable food leads to reduced reward and emotional withdrawal symptoms (Avena, Bocarsly, Rada, Kim, & Hoebel, 2008a; Blasio, Rice, Sabino, & Cottone, 2014a; Colantuoni et al., 2002; Cottone, Sabino, Steardo, & Zorrilla, 2008b; 2009b; Iemolo et al., 2012; Johnson & Kenny, 2010; Sharma, Fernandes, & Fulton, 2013). Rats genetically predisposed to obesity (i.e., obesity prone) were found to have elevated brain stimulation reward thresholds (i.e., decreased reward) (Valenza, Steardo, Cottone, & Sabino, 2015), similar to rats made obese through continuous high-fat diet access (Johnson & Kenny, 2010). Thus, decreased reward system functioning is hypothesized to confer vulnerability to, as well as result from, overconsumption of palatable food. This may be in part because of neuroadaptations in the mesolimbic dopamine pathway. It is hypothesized that, similar to drug use, the rewarding properties of palatable food are lessened over time, likely because of a decrease in dopaminergic transmission in the ventral striatum (Bello, Lucas, & Hajnal, 2002; Bello, Sweigart, Lakoski, Norgren, and Hajnal, 2003; Hajnal & Norgren, 2002). Overeating behavior may therefore represent an attempt to overcome a hypofunctional reward response (Geiger et al., 2009; Wang et al., 2001).

Animals with a history of intermittent access to palatable food show multiple withdrawal-dependent anxiety-like behaviors, such as decreased open arm time in the elevated plus maze (Avena et al., 2008a; Colantuoni et al., 2002; Cottone et al., 2009a, 2009b), decreased time in the light compartment of a dark/light box (Iemolo et al., 2013), and increased defensive retreat (Cottone et al., 2009b). Cessation of withdrawal through renewed access to palatable food induces marked overeating of palatable food (Avena et al., 2008a; Colantuoni et al., 2002; Cottone et al., 2009b; Rossetti, Spena, Halfon, & Boutrel, 2014), and this renewed access alone is sufficient to relieve depressive- and anxiety-like behaviors induced by palatable food withdrawal (Iemolo et al., 2012). Thus, preclinical evidence strongly supports the emergence of a negative emotional state upon withdrawal from palatable food and that compulsive eating behavior can be driven via negative reinforcing mechanisms (i.e., relief of anxiety or stress).

Neurobiological basis of overeating to relieve a negative emotional state: decreased reward function

Dual neurobiological substrates underlie this element of compulsivity: within-system neuroadaptations (i.e., decreased function of reward pathways) (Koob & Bloom, 1988; Koob & Le Moal, 2001) and between-system neuroadaptations (i.e., recruitment of the brain "antireward" stress systems during food withdrawal) (Parylak et al., 2011).

Firstly, within-system neuroadaptations occur in compulsive eating via repeated stimulation by consumption of palatable food, eventually desensitizing the mesolimbic dopamine system, which may in turn lead to deficiencies in reward signaling. A study of individuals with BED found that subjects showed blunted activation of striatal and prefrontal brain regions during anticipation of a monetary reward, and this decrease was linked to increased binge eating (Balodis et al., 2014). Multiple neuroadaptations in the dopaminergic system have been observed in animals eating palatable diets, including downregulation of D2Rs in the striatum (Barry et al., 2018; Colantuoni et al., 2001; Johnson & Kenny, 2010), reduced basal levels of dopamine in the NAc (Rada, Bocarsly, Barson, Hoebel, & Leibowitz, 2010), and alterations in dopamine transport and turnover (Bello et al., 2003; Fordahl, Locke, & Jones, 2016; Hajnal & Norgren, 2002). Extended access to palatable food during adolescence results in decreased markers of dopaminergic signaling in the NAc in adulthood, demonstrating that certain palatable food-induced neuroadaptations are long-lasting (Teegarden, Scott, & Bale, 2009; Vendruscolo, Gueye, Darnaudery, Ahmed, & Cador, 2010).

Downregulation of striatal D2Rs has been directly linked to compulsive eating behavior in a rodent model of obesity. Following prolonged access to a high-fat diet, obese rats displayed reward deficits coupled to decreased striatal D2Rs (Johnson & Kenny, 2010). Viral knockdown of D2Rs within the striatum exacerbated reward deficits and compulsive eating behavior (Johnson & Kenny, 2010). These data suggest that compromised dopamine signaling may cause overeating to compensate for such reward deficits. Lisdexamfetamine (LDX), an FDA-approved medication for BED, has been shown to increase striatal dopamine in rats (Rowley et al., 2012). LDX also decreases compulsive eating in rats (Heal, Goddard, Brammer, Hutson, & Vickers, 2016) and humans, as measured by the Yale−Brown obsessive compulsive scale modified for binge eating (McElroy et al., 2016). Thus, LDX administration may exert effects on compulsive eating via restoration of dopamine signaling.

Neurobiological basis of overeating to relieve a negative emotional state: increased stress

Between-system neuroadaptations, which refer to concomitant recruitment of the stress systems in the extended amygdala, result in the emergence of a negative emotional state on withdrawal. The extended amygdala is a brain macrostructure consisting of the CeA, the bed nucleus of the stria terminalis (BNST), and the NAc shell (Alheid, 2003). Obese subjects display alterations in amygdalar food cue–induced activation (Stoeckel et al., 2008) and differential resting-state functional connectivity between amygdala and cortical areas (Lips et al., 2014) compared with lean subjects. However, the evidence of the amygdala's role in compulsive eating is largely derived from preclinical research. Amygdala circuits undergo many neuroplastic changes in addiction, most notably the engagement of corticotropin-releasing factor (CRF) and its type-1 receptor (CRF1). The CRF–CRF1 receptor system mediates responses to stressors, including relapse to addictive behaviors (Koob et al., 2014; Koob & Zorrilla, 2010; Shalev, Erb, & Shaham, 2010). Similar to drugs, extended access to palatable food recruits the CRF–CRF1 system (Cottone et al., 2009a; Iemolo et al., 2013; Koob & Zorrilla, 2010). Withdrawal from palatable food increases CRF expression and CRF1 receptor electrophysiological responsiveness in the CeA (Cottone et al., 2009a; Iemolo et al., 2013; Teegarden & Bale, 2007, 2008), which is accompanied by arousal and anxiety-like behavior. On resumption of the palatable diet, CRF expression decreases and withdrawal-dependent behaviors are reversed (Cottone et al., 2009a; Iemolo et al., 2013; Teegarden & Bale, 2007). Intra-CeA administration of CRF1 receptor antagonists blocks both compulsive eating and anxiety-like behavior seen during withdrawal (Cottone et al., 2009a; Iemolo et al., 2013; Koob et al., 2014).

The CRF–CRF1 receptor system in the BNST, known for its role in threat processing and responsiveness (Lebow & Chen, 2016), may also underlie stress-induced binge eating. The BNST was shown to be activated by intermittent access to palatable food in an animal model that uses cycles of stress to precipitate binge eating (Micioni Di Bonaventura et al., 2014). Indeed, intra-BNST infusion of a CRF1 receptor antagonist blocked stress-induced binge-like eating (Micioni Di Bonaventura et al., 2014). Another animal model of genetic susceptibility to stress-induced binge eating found increased CRF mRNA expression in the BNST of binge eating-prone but not binge eating-resistant rats following stress (Calvez, de Avila, Guevremont, & Timofeeva, 2016). This suggests that the hyperactivity of the CRF–CRF1 receptor system within multiple regions of the extended amygdala (CeA and BNST) contributes to compulsive eating driven by stressful conditions. Interestingly, excessive alcohol drinking engages CRF in the medial PFC (mPFC) and is accompanied by deficits in executive function which may facilitate the transition to compulsive-like responding and relapse (George et al., 2012); future studies will be important to determine whether prefronto-cortical CRF–CRF1 receptor systems also mediate deficits in executive function in compulsive eating behavior.

A recent clinical trial investigated CRF1 receptor antagonism with the drug pexacerfont on stress-induced eating in restrained eaters (i.e., people with a history of chronic, unsuccessful dieting). This study found promising reductions in food problems and preoccupations using the YFAS (Gearhardt et al., 2009), as well as reductions in food craving and eating (Epstein et al., 2016). The study was terminated early for reasons unrelated to pexacerfont safety, but even with a reduced sample size, this trial demonstrated the potential of CRF1 receptor antagonists on maladaptive eating behaviors (Epstein et al., 2016).

Along with the CRF—CRF1 receptor brain stress system, the endocannabinoid system in the extended amygdala is also recruited in withdrawal states, likely as a compensatory "buffer mechanism" to reinstate homeostasis within amygdala circuits (Hillard, Weinlander, & Stuhr, 2012; Koob, 2015; Koob et al., 2014; Patel, Cravatt, & Hillard, 2005; Sidhpura & Parsons, 2011). During withdrawal from palatable food, 2-arachidonoylglycerol (2-AG) and cannabinoid receptor 1 (CB1R) levels are increased in the CeA (Blasio et al., 2013). Intra-CeA infusion of rimonabant, a CB1R inverse agonist, precipitated anxiety-like behavior and hypophagia during withdrawal from palatable food (Blasio et al., 2013, 2014a). Thus, the endocannabinoid system of the amygdala is likely concurrently recruited during palatable food withdrawal to buffer the negative emotional state. Rimonabant may therefore precipitate a withdrawal-like syndrome in a subpopulation of obese individuals who abstain from palatable food as they attempt to lose weight (e.g., by dieting). This mechanism may therefore explain some of the incidents reported of the emergence of severe psychiatric side effects following rimonabant treatment in obese patients (Christensen, Kristensen, Bartels, Bliddal, & Astrup, 2007). Altogether, this evidence suggests that neuroadaptations in the brain stress systems are triggered by excessive palatable food intake, sensitized during food withdrawal and protracted abstinence, and contribute to addiction and compulsive eating behavior (Koob, 2013).

Overeating despite aversive consequences

Failure of inhibitory control

A major aspect of addictive behaviors is described as a subjective feeling of "loss of control" over eating and drug-taking, which results in continuation of these behaviors despite many incurring negative consequences (Deroche-Gamonet et al., 2004; Hopf & Lesscher, 2014; Pelloux, Everitt, & Dickinson, 2007; Smith et al., 2015; Vanderschuren & Everitt, 2004; Velazquez-Sanchez et al., 2014). In drug addiction, these negative consequences encompass incidence of legal issues, loss of personal relationships, health consequences, and so on; yet, those afflicted with addiction often continue to seek and take drugs (Sussman & Sussman, 2011). Similarly, overeating can result in a myriad of medical conditions associated with excessive weight gain (e.g., hypertension, type II diabetes, cancer), in addition to impaired social functioning, emotional disturbances, and psychiatric disorders (Klatzkin, Gaffney, Cyrus, Bigus, & Brownley, 2015; Warschburger, 2005; WHO, 2000). Despite negative consequences of overeating/drug-taking behavior, addicted individuals find it very difficult to stop. Deficits in inhibitory control mechanisms, which are responsible for suppressing inappropriate actions, cause "loss of control." Preexisting deficits may confer vulnerability to addictive behavior and/or deficits may emerge from continued drug use or palatable food overconsumption (Chen et al., 2013; Lubman, Yucel, & Pantelis, 2004; Volkow et al., 2013a). Compulsivity and impulsivity are two highly interrelated behaviors, both interpreted as resulting from failure of inhibitory "top-down" control mechanisms (Dalley, Everitt, & Robbins, 2011). Forms of impulsivity, broadly divided into impulsive action (impaired ability to withhold responding) and impulsive choice (impaired delayed gratification), have been found to be associated with or predictive of compulsive drug-taking and compulsive eating (Belin et al., 2008; Dalley et al., 2011; Pearson, Wonderlich, & Smith, 2015; Velazquez-Sanchez et al., 2014).

Clinical evidence of overeating despite negative consequences

Executive function and inhibitory control related to food are dysfunctional in individuals with disorders characterized by compulsive eating (Batterink, Yokum, & Stice, 2010; Hege et al., 2015; Svaldi, Naumann, Trentowska, & Schmitz, 2014; Wu, Hartmann, Skunde, Herzog, & Friederich, 2013). A meta analysis evaluated deficits in bulimic-type disorders in "inhibitory tasks" encompassing response inhibition (e.g., "go—no go") and cognitive inhibition (e.g., "Stroop" task). Bulimic-type eating disorder patients show consistent dysfunctions in both response and cognitive inhibitory control tasks to general stimuli, and these impairments worsen with food-specific stimuli (Wu et al., 2013). Furthermore, a meta analysis exploring cognitive flexibility deficits in eating disorders found that patients with anorexia nervosa and bulimia nervosa performed worse on all domains of the Wisconsin Card Sorting Task, including preservative errors, a measure of compulsivity (Tchanturia et al., 2012). Similar deficits in inhibitory control are seen in patients with BED and obesity (Batterink et al., 2010; Hege et al., 2015; Svaldi et al., 2014).

Typically, when the consequences of overeating (emotional and physical) outweigh the pleasurable effects, people attempt to diet (i.e., abstain from highly palatable food) (Curtis & Davis, 2014). However, an overwhelming majority will relapse into unhealthy eating habits when the diets inevitably fail (Halmi, 2013). Inhibitory control deficits predict further weight gain (Pauli-Pott, Albayrak, Hebebrand, & Pott, 2010; Seeyave et al., 2009) and worse response to treatment (Murdaugh, Cox, Cook, & Weller, 2012).

Preclinical evidence of overeating despite negative consequences

In preclinical research, compulsive behavior in addiction can be operationalized as continued responding for a reward in spite of adverse conditions or consequences (Barnea-Ygael, Yadid, Yaka, Ben-Shahar, & Zangen, 2012; Belin et al., 2008; Cottone et al., 2012; Deroche-Gamonet et al., 2004; Smith et al., 2015; Vanderschuren & Everitt, 2004). Numerous experimental paradigms have demonstrated compulsive eating behavior emerging following a history of palatable food consumption. Compulsive eating in animals can be defined as showing persistent consumption palatable food even in the presence of mild electric shock (i.e., resistance to punishment) (Rossetti et al., 2014) or persistent consumption in the presence of a conditioned stimulus signaling an electric shock (i.e., resistance to cue suppression of feeding) (Latagliata, Patrono, Puglisi-Allegra, & Ventura, 2010; Nieh et al., 2015; Velazquez-Sanchez et al., 2015). Compulsive animals will endure an aversive environment (i.e., cross a novel, bright, and potentially risky situation) to obtain palatable food (Calvez & Timofeeva, 2016; Cottone et al., 2012; Dore et al., 2014; Oswald, Murdaugh, King, & Boggiano, 2011; Smith et al., 2015; Velazquez-Sanchez et al., 2014). Through direct measurement of compulsive-like eating behavior in these animal models, researchers have begun to shed light on the underlying mechanisms of compulsive behavior as it specifically applies to overeating despite negative consequences.

Neurobiological basis of overeating despite negative consequences

The PFC is a key area implicated in inhibitory control over drug-seeking and taking (Riga et al., 2014). Prefronto-cortical regions include the mPFC (further subdivided into dorsomedial and vmPFC regions), dorsolateral PFC (dlPFC), ACC, and orbitofrontal (OFC) cortices. These prefronto-cortical areas control a range of cognitive functions, including decision-making and response inhibition via subcortical connections, including with the basal ganglia and the extended amygdala (George & Koob, 2013; Riga et al., 2014). It is hypothesized that within the PFC, two opposing systems drive craving and inhibit craving: a "GO" system, consisting of dlPFC, ACC, and OFC, and a "STOP" system consisting of the vmPFC (Koob & Volkow, 2016). In addiction, the two systems become unbalanced: the "GO" PFC areas become hyperresponsive to drug cues, while the "STOP" system is hypoactive, resulting in disinhibition of basal ganglia and amygdala regions (Koob & Volkow, 2016).

Prefrontal "GO" system regions are activated by drug-associated cues in addiction, which drives craving via striato-cortical circuits (Tomasi & Volkow, 2013; Volkow et al., 2013a). As in drug addiction (Tomasi & Volkow, 2013; Volkow et al., 1993; Volkow & Fowler, 2000), abnormal activation of prefrontal regions following cue exposure has been consistently seen in obese individuals and those with BED (Geliebter et al., 2006; Schienle, Schafer, Hermann, & Vaitl, 2009; Tomasi & Volkow, 2013; Volkow et al., 2008b; Wang et al., 2011). Transcranial direct current stimulation (tDCS) targeting the dlPFC reduced craving for palatable food and subsequent food intake in individuals with BED. While the mechanism of this effect is unknown, it is conceivable that tDCS may act by effectively attenuating the "cue-induced craving" circuit modulated by the dlPFC.

Medial prefrontal dysregulation, specifically hypofunctioning of "STOP" regions (vmPFC), is associated with deficits in inhibitory control (Balodis et al., 2013; Boeka & Lokken, 2011; Hege et al., 2015). In one study, BED patients had lower activation of vmPFC during an inhibitory control task, and this activation predicted the severity of impaired dietary restraint compared with weight-matched controls (Balodis et al., 2013). A study of obese patients found lower glucose metabolism in prefrontal areas, and this correlated with reductions in striatal D2R (Volkow et al., 2008b). Because reduced striatal D2R is associated with lower inhibitory control (i.e., higher impulsivity) (Klein et al., 2007), perhaps a hypofunctioning PFC causes disinhibition of ventral striatal impulsivity circuits (Volkow et al., 2008b).

A promising therapeutic target for compulsive eating is through modulation of mPFC glutamatergic projections to striatum, including either the "GO" or the "STOP" systems, or somehow both. Potential therapeutics have spanned a range of neurotransmitter systems that may work through cortico-striatal modulation, including μ-opioid and N-methyl-D-aspartate glutamate receptor (NMDAR)–targeted medications, as well as sigma-1 receptor (Sig1R) and trace amine-associated receptor-1 (TAAR1) agonists/antagonists that are still in the preclinical phase.

Preclinical evidence implicates the opioid system of the PFC in binge-like and compulsive eating (Blasio, Steardo, Sabino, & Cottone, 2014b; Giuliano & Cottone, 2015), and studies of opioid antagonists in humans have shown promise in treatment of eating disorders (Cambridge et al., 2013; McElroy et al., 2013). In an animal model of compulsive eating (Cottone et al., 2012; Smith et al., 2015; Velazquez-Sanchez et al., 2014), limited access to a palatable diet increased the opioid peptide proopiomelanocortin and prodynorphin mRNA expression levels in the mPFC (Blasio et al., 2014b). Site-specific, intra-mPFC administration of naltrexone, a mixed opioid receptor antagonist, reduced binge-like eating (Blasio et al., 2014b). The PFC opioid system is thought to modulate compulsive eating behaviors through inhibitory control of appetitively motivated behaviors, likely via glutamatergic projections to the NAc (Mena et al., 2011, 2013; Selleck et al., 2015). Dysregulation of glutamatergic plasticity of NAc neurons can be caused by prolonged access to either drugs or palatable food, and this dysregulation is linked to the development of addiction and to relapse (Brown et al., 2015; Kalivas, Lalumiere, Knackstedt, & Shen, 2009).

In an open-label, prospective trial of BED patients, memantine (an NMDAR uncompetitive antagonist) reduced binge eating as well as "disinhibition" of eating behaviors (Brennan et al., 2008). Moreover, in compulsive shoppers, a behavioral addiction with similarities to compulsive eating, memantine reduced impulsivity and enhanced cognitive control in the treatment group (Grant, Odlaug, Mooney, O'Brien, & Kim, 2012). Preclinical studies indicate a cortico-striatal mechanism for memantine's beneficial effects, as an intra-NAc shell microinfusion of memantine reduced binge-like eating in rats (Smith et al., 2015). Additionally, though not tested site-specifically, a systemic injection of memantine also reduced compulsive-like eating in a light-dark conflict test (Smith et al., 2015).

Sig1Rs are regarded as a promising target for multiple psychiatric disorders, including addiction, and are neuromodulators of the PFC glutamatergic system (Dong et al., 2007; Hayashi, Tsai, Mori, Fujimoto, & Su, 2011; van Waarde et al., 2011). Drugs targeting Sig1Rs, though not yet tested in humans, modulate alcohol and drug reinforcement (Blasio et al., 2015; Robson, Noorbakhsh, Seminerio, & Matsumoto, 2012; Sabino et al., 2009a, 2009b, 2011), as well as compulsive-like eating in animal models. In rats with a history of limited access to palatable food, antagonism of Sig1Rs with BD-1063 decreased compulsive-like eating (Cottone et al., 2012). In the same animal model, Sig1R expression levels were increased in prefronto-cortical regions (ACC) (Cottone et al., 2012). It is possible to speculate that increased prefronto-cortical Sig1R activity may drive glutamatergic signaling in cortico-striatal circuits, contributing to compulsive behavior (Cottone et al., 2012; Dong et al., 2007; Kalivas & Volkow, 2005).

The TAAR1 system has gained a lot of recent attention for its role in regulating not only the behavioral actions of psychostimulants (Grandy, Miller, & Li, 2016) but also impulsive-like behavior (Espinoza et al., 2015). TAAR1 is a G protein–coupled receptor that is activated by trace amines, dopamine, and serotonin, among other neurotransmitters (Borowsky et al., 2001), and it is thus well-positioned to modulate multiple neurotransmitter systems critically involved in compulsive eating behavior. A recent study found that systemic injections of the selective TAAR1 agonist RO5256390 blocked binge eating and other compulsive behaviors, including development of conditioned place preference for palatable food and compulsive-like eating in a light/dark conflict test (Ferragud et al., 2016). In binge eating animals, TAAR1 receptor protein expression was reduced in the mPFC (Ferragud et al., 2016). Microinfusions of RO5256390 into the infralimbic, but not prelimbic, cortex reduced bingeing of palatable food (Ferragud et al., 2016). From these results, TAAR1 has been hypothesized to have an inhibitory role over feeding behavior; therefore, a loss of its function may be responsible for compulsive, binge eating (Moore, Sabino, & Cottone, 2018a). As TAAR1s are also activated by amphetamine (Borowsky et al., 2001), the active metabolite in the BED therapeutic LDX (Goodman, 2010), LDX and TAAR1 agonism may both restore impaired prefrontal control over inhibitory behaviors through similar mechanisms.

The "STOP" and "GO" systems intersect in the PFC to maintain overeating despite negative consequences and likely interact with other elements. For example, prefrontal circuits can reengage striatal regions to activate craving and habitual overeating behaviors. Furthermore, prefrontal circuits also may cause disinhibition of both impulsive acts and stress reactions, via the basal ganglia and extended amygdala, respectively, contributing to overeating to relieve a negative emotional state. These projections to subcortical structures are modulated by multiple neurotransmitter and peptide systems within the prefrontal cortices. How these modulatory systems interact in disinhibition versus engagement of stress reactions/habits, as well as how to use this information to develop promising therapeutic targets for compulsive eating behavior, remains a challenge for future research.

Discussion

Compulsive eating behavior refers to a pathological form of eating that shares similar phenotype and neurobiology with the compulsive behavior associated with both drugs of abuse and behavioral addictions. Historically, preclinical feeding research has overwhelmingly consisted of studies on obesity, often relying on the predominant classical view of eating as simply energy-homeostatic behavior, ignoring overeating driven by psychobiological mechanisms. This limited view has thus far hindered understanding of a more complex behavioral expression of pathological eating and its underlying neurobiological bases.

Notably, eating behavior is under control of multiple mechanisms and, given its complexity, the scientific community is often compartmentalized along its two largest components: homeostatic and hedonic feeding. However, it is becoming increasingly clear that these two aspects are extremely interconnected, thus one cannot be studied without consideration of the other. For instance, appetite-regulating peptides (i.e., leptin, ghrelin, orexin, melanin-containing hormone) interact with ventral tegmental area dopamine signaling to affect food cue reactivity and cognitive control over food intake, among others (Reichelt, Westbrook, & Morris, 2015; Volkow, Wang, & Baler, 2011). Also receiving recent attention are the gut-brain pathways understood to act as modulators of reward and behavior (de Araujo et al., 2012). There has been a recent push toward investigation of how these two components interact in relation to the study of basic eating processes; however, more effort is needed to extend understanding to a context of compulsive eating (for more details, refer to chapter 11 "Homeostatic and Hedonic Regulation of Compulsive Eating Behavior").

Incentive for the dissection of compulsive eating into domains/elements derives in part from the National Institute of Mental Health's recent initiative encouraging researchers and clinicians to move toward a new approach for the classification of mental disorders using the Research Domain Criteria (RDoC) framework. The RDoC was implemented for the purpose of evaluating mental disorders through varying biologically validated high-level domains (The National Institute of Mental Health, 2013). The RDoC represents a push to identify fundamental components of disease that may span multiple disorders (i.e., inhibitory control dysfunction). Measurement of the core fundamental elements of the disorder in epidemiological, clinical, and even preclinical studies will eventually provide a clearer picture of the underlying mechanisms and a better temporal and cross-population picture of the disorder. The RDoC has been recently applied to the field of drug addiction, whereas functional domains are heuristically applied to different substance-use disorders (e.g., alcohol, opioid), with the long-term goal to identify meaningful subtypes of addictive disorders based on high-level domains (Kwako, Momenan, Litten, Koob, & Goldman, 2016, 2018). Similarly, the goal of this chapter was to evaluate compulsive eating as a transdiagnostic construct across eating disorders, as well as identify underlying psychobiological processes and associated neurobiology.

Summary

In drug abuse research, a focus on the behavioral expressions of compulsive drug use has helped achieve a better understanding of the neurobiology of the addiction process. Significant potential exists in the field of compulsive eating through focus on the interactions of maladaptive circuits. Through the integration of these elements (i.e., habitual/inflexible feeding responding, negatively reinforced feeding, and eating in spite of negative consequences), functionally anchored, complex animal models of compulsive eating can bring the feeding field to this next level. One clear example is to avoid paradigms that include conditions driven by hunger and homeostatic mechanisms, as these inevitably causes energy homeostasis experimental confounds that are less translatable to humans who most often have easy and reliable access to food (Corwin, Avena, & Boggiano, 2011; Cottone et al., 2009a, 2012). However, some experimental manipulations of diets are necessary for mimicking certain aspects of diet regulation seen in humans, such as the tendency for binge eaters to dichotomize food choices by bingeing with palatable food and selecting healthier alternatives for "nonbinge" meals (de Castro, 1995; Laessle, Tuschl, Kotthaus, & Pirke, 1989; Mela, 2001). This dichotomization is driven by highly cognitive processes (e.g., social norms, guilt, embarrassment) that cannot be modeled in rodents; thus, animals are given alternating, nonconcurrent access to differently preferred diets (Cottone, Sabino, Steardo, & Zorrilla, 2008a). Despite some inherent limitations, innovative animal models prove indispensable for understanding a disorder's neurobiology and discovery of novel pharmacological treatments (see chapter 10, "Modeling and Testing Compulsive Eating Behavior in Animals").

Investigations into how reward, stress, and cognitive function neurocircuits intersect and are disrupted to ultimately drive compulsive behavior have begun to lay the framework for novel diagnosis, treatment, and prevention of addiction disorders. The evidence provided in this chapter describes a conceptual framework of compulsive eating where, similar to drugs of abuse, palatable food intake initially activates the mesolimbic dopamine incentive salience pathways. However, repeated overstimulation by overconsumption of palatable food can lead to downstream neurobehavioral adaptations that comprise multiple neurocircuitries and neurotransmitter systems causing a predominance of dorsal striatal–driven habitual behavior, desensitization of the reward system, recruitment of the brain stress systems in the amygdala, and loss of inhibitory prefronto-cortical control over behavior. Dysregulation of these areas and neural circuits interact and potentially synergize one another, each contributing to the different elements of compulsive eating behavior. By focusing on the dynamic interactions between elements, we can start to identify how the elements influence each other, for example, how the relationship between negative emotional states on habit learning and decision-making processes may affect compulsivity. By reframing our conceptual and methodological approaches to study compulsive eating, better diagnosis, prevention, and treatment of disorders of pathological eating may be reached.

Funding and Disclosures

This work was supported by grants DA044664 (CM), DA030425 (PC), MH091945 (PC), MH093650 (VS), and AA024439 (VS) from NIDA, NIMH, and NIAAA, by the Peter Paul Career Development Professorship (PC), the McManus Charitable Trust (VS), and the Burroughs Wellcome Fund through the Transformative Training Program in Addiction Sciences (CM). Its contents are solely the responsibility of the authors and do not necessarily represent the official views of the National Institutes of Health. The authors declare no conflict of interest.

References

Agh, T., Kovacs, G., Pawaskar, M., Supina, D., Inotai, A., & Voko, Z. (2015). Epidemiology, health-related quality of life and economic burden of binge eating disorder: A systematic literature review. *Eating and Weight Disorders, 20*(1), 1–12. https://doi.org/10.1007/s40519-014-0173-9.

Alheid, G. F. (2003). Extended amygdala and basal forebrain. *Annals of the New York Academy of Sciences, 985*, 185–205. https://doi.org/10.1111/j.1749-6632.2003.tb07082.x.

American Psychiatric Association. (2000). *Diagnostic and statistical manual of mental disorders*. Arlington, Vol. A.

American Psychiatric Association. (2013). *Diagnostic and statistical manual of mental disorders*. Arlington, VA: American Psychiatric Publishing.

de Araujo, I. E., Ferreira, J. G., Tellez, L. A., Ren, X., & Yeckel, C. W. (2012). The gut-brain dopamine axis: A regulatory system for caloric intake. *Physiology & Behavior, 106*(3), 394–399. https://doi.org/10.1016/j.physbeh.2012.02.026.

Avena, N. M., Bocarsly, M. E., Rada, P., Kim, A., & Hoebel, B. G. (2008a). After daily bingeing on a sucrose solution, food deprivation induces anxiety and accumbens dopamine/acetylcholine imbalance. *Physiology & Behavior, 94*(3), 309–315. https://doi.org/10.1016/j.physbeh.2008.01.008.

Avena, N. M., Carrillo, C. A., Needham, L., Leibowitz, S. F., & Hoebel, B. G. (2004). Sugar-dependent rats show enhanced intake of unsweetened ethanol. *Alcohol, 34*(2–3), 203–209. https://doi.org/10.1016/j.alcohol.2004.09.006.

Avena, N. M., Rada, P., & Hoebel, B. G. (2008b). Evidence for sugar addiction: Behavioral and neurochemical effects of intermittent, excessive sugar intake. *Neuroscience & Biobehavioral Reviews, 32*(1), 20–39. https://doi.org/10.1016/j.neubiorev.2007.04.019.

Badiani, A. (2014). Is a 'general' theory of addiction possible? A commentary on: A multistep general theory of transition to addiction. *Psychopharmacology, 231*(19), 3923–3927. https://doi.org/10.1007/s00213-014-3627-x.

Balleine, B. W., & O'Doherty, J. P. (2010). Human and rodent homologies in action control: Corticostriatal determinants of goal-directed and habitual action. *Neuropsychopharmacology, 35*(1), 48–69. https://doi.org/10.1038/npp.2009.131.

Balodis, I. M., Grilo, C. M., Kober, H., Worhunsky, P. D., White, M. A., Stevens, M. C., et al. (2014). A pilot study linking reduced fronto-Striatal recruitment during reward processing to persistent bingeing following treatment for binge-eating disorder. *International Journal of Eating Disorders, 47*(4), 376–384. https://doi.org/10.1002/eat.22204.

Balodis, I. M., Molina, N. D., Kober, H., Worhunsky, P. D., White, M. A., Rajita, S., et al. (2013). Divergent neural substrates of inhibitory control in binge eating disorder relative to other manifestations of obesity. *Obesity, 21*(2), 367–377. https://doi.org/10.1002/oby.20068.

Barnea-Ygael, N., Yadid, G., Yaka, R., Ben-Shahar, O., & Zangen, A. (2012). Cue-induced reinstatement of cocaine seeking in the rat "conflict model": Effect of prolonged home-cage confinement. *Psychopharmacology, 219*(3), 875–883. https://doi.org/10.1007/s00213-011-2416-z.

Barry, R. L., Byun, N. E., Williams, J. M., Siuta, M. A., Tantawy, M. N., Speed, N. K., et al. (2018). Brief exposure to obesogenic diet disrupts brain dopamine networks. *PLoS One, 13*(4), e0191299. https://doi.org/10.1371/journal.pone.0191299.

Barry, D., Pietrzak, R. H., & Petry, N. M. (2008). Gender differences in associations between body mass index and DSM-IV mood and anxiety disorders: Results from the national epidemiologic survey on alcohol and related conditions. *Annals of Epidemiology, 18*(6), 458–466. https://doi.org/10.1016/j.annepidem.2007.12.009.

Batterink, L., Yokum, S., & Stice, E. (2010). Body mass correlates inversely with inhibitory control in response to food among adolescent girls: An fMRI study. *NeuroImage, 52*(4), 1696–1703. https://doi.org/10.1016/j.neuroimage.2010.05.059.

Belin-Rauscent, A., Fouyssac, M., Bonci, A., & Belin, D. (2015). How preclinical models evolved to resemble the diagnostic criteria of drug addiction. *Biological Psychiatry.* https://doi.org/10.1016/j.biopsych.2015.01.004.

Belin, D., & Everitt, B. J. (2008). Cocaine seeking habits depend upon dopamine-dependent serial connectivity linking the ventral with the dorsal striatum. *Neuron, 57*(3), 432–441. https://doi.org/10.1016/j.neuron.2007.12.019.

Belin, D., Mar, A. C., Dalley, J. W., Robbins, T. W., & Everitt, B. J. (2008). High impulsivity predicts the switch to compulsive cocaine-taking. *Science, 320*(5881), 1352–1355. https://doi.org/10.1126/science.1158136.

Bello, N. T., Lucas, L. R., & Hajnal, A. (2002). Repeated sucrose access influences dopamine D2 receptor density in the striatum. *NeuroReport, 13*(12), 1575–1578.

Bello, N. T., Sweigart, K. L., Lakoski, J. M., Norgren, R., & Hajnal, A. (2003). Restricted feeding with scheduled sucrose access results in an upregulation of the rat dopamine transporter. *American Journal of Physiology - Regulatory, Integrative and Comparative Physiology, 284*(5), R1260–R1268. https://doi.org/10.1152/ajpregu.00716.2002.

Benton, D. (2010). The plausibility of sugar addiction and its role in obesity and eating disorders. *Clinical Nutrition, 29*(3), 288–303. https://doi.org/10.1016/j.clnu.2009.12.001.

Berlin, G. S., & Hollander, E. (2014). Compulsivity, impulsivity, and the DSM-5 process. *CNS Spectrums, 19*(1), 62–68. https://doi.org/10.1017/S1092852913000722.

Blasio, A., Iemolo, A., Sabino, V., Petrosino, S., Steardo, L., Rice, K. C., et al. (2013). Rimonabant precipitates anxiety in rats withdrawn from palatable food: Role of the central amygdala. *Neuropsychopharmacology, 38*(12), 2498–2507. https://doi.org/10.1038/npp.2013.153.

Blasio, A., Rice, K. C., Sabino, V., & Cottone, P. (2014a). Characterization of a shortened model of diet alternation in female rats: Effects of the CB1 receptor antagonist rimonabant on food intake and anxiety-like behavior. *Behavioural Pharmacology, 25*(7), 609–617. https://doi.org/10.1097/FBP.0000000000000059.

Blasio, A., Steardo, L., Sabino, V., & Cottone, P. (2014b). Opioid system in the medial prefrontal cortex mediates binge-like eating. *Addiction Biology, 19*(4), 652–662. https://doi.org/10.1111/adb.12033.

Blasio, A., Valenza, M., Iyer, M. R., Rice, K. C., Steardo, L., Hayashi, T., et al. (2015). Sigma-1 receptor mediates acquisition of alcohol drinking and seeking behavior in alcohol-preferring rats. *Behavioural Brain Research, 287*, 315–322. https://doi.org/10.1016/j.bbr.2015.03.065.

Boeka, A. G., & Lokken, K. L. (2011). Prefrontal systems involvement in binge eating. *Eating and Weight Disorders, 16*(2), e121–126.

Borowsky, B., Adham, N., Jones, K. A., Raddatz, R., Artymyshyn, R., Ogozalek, K. L., et al. (2001). Trace amines: Identification of a family of mammalian G protein-coupled receptors. *Proceedings of the National Academy of Sciences of the United States of America, 98*(16), 8966–8971. https://doi.org/10.1073/pnas.151105198.

Brennan, B. P., Roberts, J. L., Fogarty, K. V., Reynolds, K. A., Jonas, J. M., & Hudson, J. I. (2008). Memantine in the treatment of binge eating disorder: An open-label, prospective trial. *International Journal of Eating Disorders, 41*(6), 520–526. https://doi.org/10.1002/eat.20541.

Brown, R. M., Kupchik, Y. M., Spencer, S., Garcia-Keller, C., Spanswick, D. C., Lawrence, A. J., et al. (2015). Addiction-like synaptic impairments in diet-induced obesity. *Biological Psychiatry.* https://doi.org/10.1016/j.biopsych.2015.11.019.

Bruch, H. (1973). *Eating disorders; obesity, anorexia nervosa, and the person within.* New York: Basic Books.

Calvez, J., de Avila, C., Guevremont, G., & Timofeeva, E. (2016). Stress differentially regulates brain expression of corticotropin-releasing factor in binge-like eating prone and resistant female rats. *Appetite, 107*, 585–595. https://doi.org/10.1016/j.appet.2016.09.010.

Calvez, J., & Timofeeva, E. (2016). Behavioral and hormonal responses to stress in binge-like eating prone female rats. *Physiology & Behavior, 157*, 28–38. https://doi.org/10.1016/j.physbeh.2016.01.029.

Cambridge, V. C., Ziauddeen, H., Nathan, P. J., Subramaniam, N., Dodds, C., Chamberlain, S. R., et al. (2013). Neural and behavioral effects of a novel mu opioid receptor antagonist in binge-eating obese people. *Biological Psychiatry, 73*(9), 887–894. https://doi.org/10.1016/j.biopsych.2012.10.022.

Carlier, N., Marshe, V. S., Cmorejova, J., Davis, C., & Muller, D. J. (2015). Genetic similarities between compulsive overeating and addiction phenotypes: A case for "food addiction"? *Current Psychiatry Reports, 17*(12), 96. https://doi.org/10.1007/s11920-015-0634-5.

Carr, K. D. (2016). Nucleus accumbens AMPA receptor trafficking upregulated by food restriction: An unintended target for drugs of abuse and forbidden foods. *Current Opinion in Behavioral Sciences, 9*, 32–39. https://doi.org/10.1016/j.cobeha.2015.11.019.

de Castro, J. M. (1995). The relationship of cognitive restraint to the spontaneous food and fluid intake of free-living humans. *Physiology & Behavior, 57*(2), 287–295.

Chamberlain, S. R., Mogg, K., Bradley, B. P., Koch, A., Dodds, C. M., Tao, W. X., et al. (2012). Effects of mu opioid receptor antagonism on cognition in obese binge-eating individuals. *Psychopharmacology, 224*(4), 501–509. https://doi.org/10.1007/s00213-012-2778-x.

Chen, B. T., Yau, H. J., Hatch, C., Kusumoto-Yoshida, I., Cho, S. L., Hopf, F. W., et al. (2013). Rescuing cocaine-induced prefrontal cortex hypoactivity prevents compulsive cocaine seeking. *Nature, 496*(7445), 359–362. https://doi.org/10.1038/nature12024.

Christensen, R., Kristensen, P. K., Bartels, E. M., Bliddal, H., & Astrup, A. V. (2007). A meta-analysis of the efficacy and safety of the anti-obesity agent Rimonabant. *Ugeskr Laeger, 169*(50), 4360–4363.

Clark, M. M., Niaura, R., King, T. K., & Pera, V. (1996). Depression, smoking, activity level, and health status: Pretreatment predictors of attrition in obesity treatment. *Addictive Behaviors, 21*(4), 509–513.

Colantuoni, C., Rada, P., McCarthy, J., Patten, C., Avena, N. M., Chadeayne, A., et al. (2002). Evidence that intermittent, excessive sugar intake causes endogenous opioid dependence. *Obesity Research, 10*(6), 478–488. https://doi.org/10.1038/oby.2002.66.

Colantuoni, C., Schwenker, J., McCarthy, J., Rada, P., Ladenheim, B., Cadet, J. L., et al. (2001). Excessive sugar intake alters binding to dopamine and mu-opioid receptors in the brain. *NeuroReport, 12*(16), 3549–3552.

Corbit, L. H. (2016). Effects of obesogenic diets on learning and habitual responding. *Current Opinion in Behavioral Sciences, 9*, 84–90.

Corbit, L. H., Nie, H., & Janak, P. H. (2012). Habitual alcohol seeking: Time course and the contribution of subregions of the dorsal striatum. *Biological Psychiatry, 72*(5), 389–395. https://doi.org/10.1016/j.biopsych.2012.02.024.

Corwin, R. L., Avena, N. M., & Boggiano, M. M. (2011). Feeding and reward: Perspectives from three rat models of binge eating. *Physiology & Behavior, 104*(1), 87–97. https://doi.org/10.1016/j.physbeh.2011.04.041.

Cossrow, N., Pawaskar, M., Witt, E. A., Ming, E. E., Victor, T. W., Herman, B. K., et al. (2016). Estimating the prevalence of binge eating disorder in a community sample from the United States: Comparing DSM-IV-TR and DSM-5 criteria. *Journal of Clinical Psychiatry, 77*(8), e968–974. https://doi.org/10.4088/JCP.15m10059.

Cottone, P., Sabino, V., Steardo, L., & Zorrilla, E. P. (2008a). Intermittent access to preferred food reduces the reinforcing efficacy of chow in rats. *American Journal of Physiology - Regulatory, Integrative and Comparative Physiology, 295*(4), R1066–R1076. https://doi.org/10.1152/ajpregu.90309.2008.

Cottone, P., Sabino, V., Steardo, L., & Zorrilla, E. P. (2008b). Opioid-dependent anticipatory negative contrast and binge-like eating in rats with limited access to highly preferred food. *Neuropsychopharmacology, 33*(3), 524–535. https://doi.org/10.1038/sj.npp.1301430.

Cottone, P., Sabino, V., Roberto, M., Bajo, M., Pockros, L., Frihauf, J. B., et al. (2009a). CRF system recruitment mediates dark side of compulsive eating. *Proceedings of the National Academy of Sciences of the United States of America, 106*(47), 20016–20020. https://doi.org/10.1073/pnas.0908789106.

Cottone, P., Sabino, V., Steardo, L., & Zorrilla, E. P. (2009b). Consummatory, anxiety-related and metabolic adaptations in female rats with alternating access to preferred food. *Psychoneuroendocrinology, 34*(1), 38–49. https://doi.org/10.1016/j.psyneuen.2008.08.010.

Cottone, P., Wang, X., Park, J. W., Valenza, M., Blasio, A., Kwak, J., et al. (2012). Antagonism of sigma-1 receptors blocks compulsive-like eating. *Neuropsychopharmacology, 37*(12), 2593–2604. https://doi.org/10.1038/npp.2012.89.

Curtis, C., & Davis, C. (2014). A qualitative study of binge eating and obesity from an addiction perspective. *Eating Disorders: The Journal of Treatment and Prevention, 22*(1), 19–32. https://doi.org/10.1080/10640266.2014.857515.

Dalley, J. W., Everitt, B. J., & Robbins, T. W. (2011). Impulsivity, compulsivity, and top-down cognitive control. *Neuron, 69*(4), 680–694. https://doi.org/10.1016/j.neuron.2011.01.020.

Dallman, M. F., Pecoraro, N., Akana, S. F., La Fleur, S. E., Gomez, F., Houshyar, H., et al. (2003). Chronic stress and obesity: A new view of "comfort food". *Proceedings of the National Academy of Sciences of the United States of America, 100*(20), 11696–11701. https://doi.org/10.1073/pnas.1934666100.

Davis, C. (2013). A narrative review of binge eating and addictive behaviors: Shared associations with seasonality and personality factors. *Frontiers in Psychiatry, 4*, 183. https://doi.org/10.3389/fpsyt.2013.00183.

Davis, C., Curtis, C., Levitan, R. D., Carter, J. C., Kaplan, A. S., & Kennedy, J. L. (2011). Evidence that 'food addiction' is a valid phenotype of obesity. *Appetite, 57*(3), 711–717. https://doi.org/10.1016/j.appet.2011.08.017.

Day, J. J., Roitman, M. F., Wightman, R. M., & Carelli, R. M. (2007). Associative learning mediates dynamic shifts in dopamine signaling in the nucleus accumbens. *Nature Neuroscience, 10*(8), 1020–1028. https://doi.org/10.1038/nn1923.

Dee, A., Kearns, K., O'Neill, C., Sharp, L., Staines, A., O'Dwyer, V., et al. (2014). The direct and indirect costs of both overweight and obesity: A systematic review. *BMC Research Notes, 7*, 242. https://doi.org/10.1186/1756-0500-7-242.

Deroche-Gamonet, V., Belin, D., & Piazza, P. V. (2004). Evidence for addiction-like behavior in the rat. *Science, 305*(5686), 1014−1017. https://doi.org/10.1126/science.1099020.

Di Ciano, P., & Everitt, B. J. (2005). Neuropsychopharmacology of drug seeking: Insights from studies with second-order schedules of drug reinforcement. *European Journal of Pharmacology, 526*(1−3), 186−198. https://doi.org/10.1016/j.ejphar.2005.09.024.

Dingemans, A. E., Bruna, M. J., & van Furth, E. F. (2002). Binge eating disorder: A review. *International Journal of Obesity and Related Metabolic Disorders, 26*(3), 299−307. https://doi.org/10.1038/sj.ijo.0801949.

Dingemans, A. E., & van Furth, E. F. (2012). Binge Eating Disorder psychopathology in normal weight and obese individuals. *International Journal of Eating Disorders, 45*(1), 135−138. https://doi.org/10.1002/eat.20905.

Dong, L. Y., Cheng, Z. X., Fu, Y. M., Wang, Z. M., Zhu, Y. H., Sun, J. L., et al. (2007). Neurosteroid dehydroepiandrosterone sulfate enhances spontaneous glutamate release in rat prelimbic cortex through activation of dopamine D1 and sigma-1 receptor. *Neuropharmacology, 52*(3), 966−974. https://doi.org/10.1016/j.neuropharm.2006.10.015.

Dore, R., Valenza, M., Wang, X., Rice, K. C., Sabino, V., & Cottone, P. (2014). The inverse agonist of CB1 receptor SR141716 blocks compulsive eating of palatable food. *Addiction Biology, 19*(5), 849−861. https://doi.org/10.1111/adb.12056.

Eldredge, K., Wilson, T., & Whaley, A. (1990). Failure, self-evaluation, and feeling fat in women. *International Journal of Eating Disorders, 9*(1), 37−50.

Epstein, D. H., Kennedy, A. P., Furnari, M., Heilig, M., Shaham, Y., Phillips, K. A., et al. (2016). Effect of the CRF1-receptor antagonist pexacerfont on stress-induced eating and food craving. *Psychopharmacology, 233*(23−24), 3921−3932. https://doi.org/10.1007/s00213-016-4424-5.

Ersche, K. D., Gillan, C. M., Jones, P. S., Williams, G. B., Ward, L. H., Luijten, M., et al. (2016). Carrots and sticks fail to change behavior in cocaine addiction. *Science, 352*(6292), 1468−1471. https://doi.org/10.1126/science.aaf3700.

Espinoza, S., Lignani, G., Caffino, L., Maggi, S., Sukhanov, I., Leo, D., et al. (2015). TAAR1 modulates cortical glutamate NMDA receptor function. *Neuropsychopharmacology, 40*(9), 2217−2227. https://doi.org/10.1038/npp.2015.65.

Everitt, B. J. (2014). Neural and psychological mechanisms underlying compulsive drug seeking habits and drug memories–indications for novel treatments of addiction. *European Journal of Neuroscience, 40*(1), 2163−2182. https://doi.org/10.1111/ejn.12644.

Everitt, B. J., Belin, D., Economidou, D., Pelloux, Y., Dalley, J. W., & Robbins, T. W. (2008). Review. Neural mechanisms underlying the vulnerability to develop compulsive drug-seeking habits and addiction. *Philosophical Transactions of the Royal Society of London B Biological Sciences, 363*(1507), 3125−3135. https://doi.org/10.1098/rstb.2008.0089.

Everitt, B. J., & Robbins, T. W. (2000). Second-order schedules of drug reinforcement in rats and monkeys: Measurement of reinforcing efficacy and drug-seeking behaviour. *Psychopharmacology, 153*(1), 17−30.

Everitt, B. J., & Robbins, T. W. (2005). Neural systems of reinforcement for drug addiction: From actions to habits to compulsion. *Nature Neuroscience, 8*(11), 1481−1489. https://doi.org/10.1038/nn1579.

Everitt, B. J., & Robbins, T. W. (2013). From the ventral to the dorsal striatum: Devolving views of their roles in drug addiction. *Neuroscience & Biobehavioral Reviews, 37*(9 Pt A), 1946–1954. https://doi.org/10.1016/j.neubiorev.2013.02.010.

Everitt, B. J., & Robbins, T. W. (2016). Drug addiction: Updating actions to habits to compulsions ten years on. *Annual Review of Psychology, 67*, 23–50. https://doi.org/10.1146/annurev-psych-122414-033457.

Fairburn, C. G., Doll, H. A., Welch, S. L., Hay, P. J., Davies, B. A., & O'Connor, M. E. (1998). Risk factors for binge eating disorder: A community-based, case-control study. *Archives of General Psychiatry, 55*(5), 425–432.

Fauth-Buhler, M., Mann, K., & Potenza, M. N. (2016). Pathological gambling: A review of the neurobiological evidence relevant for its classification as an addictive disorder. *Addiction Biology*. https://doi.org/10.1111/adb.12378.

Ferragud, A., Howell, A. D., Moore, C. F., Ta, T. L., Hoener, M. C., Sabino, V., et al. (2016). The trace amine-associated receptor 1 agonist RO5256390 blocks compulsive, binge-like eating in rats. *Neuropsychopharmacology*. https://doi.org/10.1038/npp.2016.233.

Finch, L. E., & Tomiyama, A. J. (2015). Comfort eating, psychological stress, and depressive symptoms in young adult women. *Appetite, 95*, 239–244. https://doi.org/10.1016/j.appet.2015.07.017.

Finkelstein, E. A., Trogdon, J. G., Cohen, J. W., & Dietz, W. (2009). Annual medical spending attributable to obesity: Payer-and service-specific estimates. *Health Affairs, 28*(5), w822–831. https://doi.org/10.1377/hlthaff.28.5.w822.

Fitzgibbon, M. L., Stolley, M. R., & Kirschenbaum, D. S. (1993). Obese people who seek treatment have different characteristics than those who do not seek treatment. *Health Psychology, 12*(5), 342–345.

Fitzsimmons-Craft, E. E., Bardone-Cone, A. M., Brownstone, L. M., & Harney, M. B. (2012). Evaluating the roles of anxiety and dimensions of perfectionism in dieting and binge eating using weekly diary methodology. *Eating Behaviors, 13*(4), 418–422. https://doi.org/10.1016/j.eatbeh.2012.06.006.

Fordahl, S. C., Locke, J. L., & Jones, S. R. (2016). High fat diet augments amphetamine sensitization in mice: Role of feeding pattern, obesity, and dopamine terminal changes. *Neuropharmacology, 109*, 170–182. https://doi.org/10.1016/j.neuropharm.2016.06.006.

Furlong, T. M., Jayaweera, H. K., Balleine, B. W., & Corbit, L. H. (2014). Binge-like consumption of a palatable food accelerates habitual control of behavior and is dependent on activation of the dorsolateral striatum. *Journal of Neuroscience, 34*(14), 5012–5022. https://doi.org/10.1523/JNEUROSCI.3707-13.2014.

Galanti, K., Gluck, M. E., & Geliebter, A. (2007). Test meal intake in obese binge eaters in relation to impulsivity and compulsivity. *International Journal of Eating Disorders, 40*(8), 727–732. https://doi.org/10.1002/eat.20441.

Gearhardt, A. N., Boswell, R. G., & White, M. A. (2014). The association of "food addiction" with disordered eating and body mass index. *Eating Behaviors, 15*(3), 427–433. https://doi.org/10.1016/j.eatbeh.2014.05.001.

Gearhardt, A. N., Corbin, W. R., & Brownell, K. D. (2009). Preliminary validation of the Yale food addiction scale. *Appetite, 52*(2), 430–436. https://doi.org/10.1016/j.appet.2008.12.003.

Gearhardt, A. N., Corbin, W. R., & Brownell, K. D. (2016). Development of the Yale food addiction scale version 2.0. *Psychology of Addictive Behaviors, 30*(1), 113–121. https://doi.org/10.1037/adb0000136.

Gearhardt, A. N., White, M. A., Masheb, R. M., Morgan, P. T., Crosby, R. D., & Grilo, C. M. (2012). An examination of the food addiction construct in obese patients with binge eating disorder. *International Journal of Eating Disorders, 45*(5), 657–663. https://doi.org/10.1002/eat.20957.

Gearhardt, A. N., White, M. A., & Potenza, M. N. (2011). Binge eating disorder and food addiction. *Current Drug Abuse Reviews, 4*(3), 201–207.

Geiger, B. M., Haburcak, M., Avena, N. M., Moyer, M. C., Hoebel, B. G., & Pothos, E. N. (2009). Deficits of mesolimbic dopamine neurotransmission in rat dietary obesity. *Neuroscience, 159*(4), 1193–1199. https://doi.org/10.1016/j.neuroscience.2009.02.007.

Geliebter, A., Ladell, T., Logan, M., Schneider, T., Sharafi, M., & Hirsch, J. (2006). Responsivity to food stimuli in obese and lean binge eaters using functional MRI. *Appetite, 46*(1), 31–35. https://doi.org/10.1016/j.appet.2005.09.002.

George, O., & Koob, G. F. (2013). Control of craving by the prefrontal cortex. *Proceedings of the National Academy of Sciences of the United States of America, 110*(11), 4165–4166. https://doi.org/10.1073/pnas.1301245110.

George, O., Koob, G. F., & Vendruscolo, L. F. (2014). Negative reinforcement via motivational withdrawal is the driving force behind the transition to addiction. *Psychopharmacology, 231*(19), 3911–3917. https://doi.org/10.1007/s00213-014-3623-1.

George, O., Sanders, C., Freiling, J., Grigoryan, E., Vu, S., Allen, C. D., et al. (2012). Recruitment of medial prefrontal cortex neurons during alcohol withdrawal predicts cognitive impairment and excessive alcohol drinking. *Proceedings of the National Academy of Sciences of the United States of America, 109*(44), 18156–18161. https://doi.org/10.1073/pnas.1116523109.

Gillan, C. M., Apergis-Schoute, A. M., Morein-Zamir, S., Urcelay, G. P., Sule, A., Fineberg, N. A., et al. (2015). Functional neuroimaging of avoidance habits in obsessive-compulsive disorder. *American Journal of Psychiatry, 172*(3), 284–293. https://doi.org/10.1176/appi.ajp.2014.14040525.

Giuliano, C., & Cottone, P. (2015). The role of the opioid system in binge eating disorder. *CNS Spectrums, 20*(6), 537–545. https://doi.org/10.1017/S1092852915000668.

Giuliano, C., Robbins, T. W., Nathan, P. J., Bullmore, E. T., & Everitt, B. J. (2012). Inhibition of opioid transmission at the mu-opioid receptor prevents both food seeking and binge-like eating. *Neuropsychopharmacology, 37*(12), 2643–2652. https://doi.org/10.1038/npp.2012.128.

Gonzalez, V. M., & Vitousek, K. M. (2004). Feared food in dieting and non-dieting young women: A preliminary validation of the food phobia survey. *Appetite, 43*(2), 155–173. https://doi.org/10.1016/j.appet.2004.03.006.

Goodman, D. W. (2010). Lisdexamfetamine dimesylate (vyvanse), a prodrug stimulant for attention-deficit/hyperactivity disorder. *PT, 35*(5), 273–287.

Grandy, D. K., Miller, G. M., & Li, J. X. (2016). "TAARgeting addiction"–the alamo bears witness to another revolution: An overview of the plenary symposium of the 2015 behavior, biology and chemistry conference. *Drug and Alcohol Dependence, 159*, 9–16. https://doi.org/10.1016/j.drugalcdep.2015.11.014.

Grant, J. E., Odlaug, B. L., Mooney, M., O'Brien, R., & Kim, S. W. (2012). Open-label pilot study of memantine in the treatment of compulsive buying. *Annals of Clinical Psychiatry, 24*(2), 119–126.

Graybiel, A. M. (2008). Habits, rituals, and the evaluative brain. *Annual Review of Neuroscience, 31*, 359–387. https://doi.org/10.1146/annurev.neuro.29.051605.112851.

Greeno, C. G., & Wing, R. R. (1994). Stress-induced eating. *Psychological Bulletin, 115*(3), 444–464.

Greeno, C. G., Wing, R. R., & Shiffman, S. (2000). Binge antecedents in obese women with and without binge eating disorder. *Journal of Consulting and Clinical Psychology, 68*(1), 95–102.

Grilo, C. M., Ivezaj, V., & White, M. A. (2015). Evaluation of the DSM-5 severity indicator for binge eating disorder in a community sample. *Behaviour Research and Therapy, 66*, 72–76. https://doi.org/10.1016/j.brat.2015.01.004.

el-Guebaly, N., Mudry, T., Zohar, J., Tavares, H., & Potenza, M. N. (2012). Compulsive features in behavioural addictions: The case of pathological gambling. *Addiction, 107*(10), 1726–1734.

Haedt-Matt, A. A., & Keel, P. K. (2011). Revisiting the affect regulation model of binge eating: A meta-analysis of studies using ecological momentary assessment. *Psychological Bulletin, 137*(4), 660–681. https://doi.org/10.1037/a0023660.

Hajnal, A., & Norgren, R. (2002). Repeated access to sucrose augments dopamine turnover in the nucleus accumbens. *NeuroReport, 13*(17), 2213–2216. https://doi.org/10.1097/01.wnr.0000044213.09266.38.

Hales, C. M., Carroll, M. D., Fryar, C. D., & Ogden, C. L. (2017). Prevalence of obesity among adults and youth: United States, 2015-2016. *NCHS Data Brief*, (288), 1–8.

Halfon, N., Larson, K., & Slusser, W. (2013). Associations between obesity and comorbid mental health, developmental, and physical health conditions in a nationally representative sample of US children aged 10 to 17. *Acad Pediatr, 13*(1), 6–13. https://doi.org/10.1016/j.acap.2012.10.007.

Halmi, K. A. (2013). Perplexities of treatment resistance in eating disorders. *BMC Psychiatry, 13*, 292. https://doi.org/10.1186/1471-244X-13-292.

Hasin, D. S., O'Brien, C. P., Auriacombe, M., Borges, G., Bucholz, K., Budney, A., et al. (2013). DSM-5 criteria for substance use disorders: Recommendations and rationale. *American Journal of Psychiatry, 170*(8), 834–851. https://doi.org/10.1176/appi.ajp.2013.12060782.

Hayashi, T., Tsai, S. Y., Mori, T., Fujimoto, M., & Su, T. P. (2011). Targeting ligand-operated chaperone sigma-1 receptors in the treatment of neuropsychiatric disorders. *Expert Opinion on Therapeutic Targets, 15*(5), 557–577. https://doi.org/10.1517/14728222.2011.560837.

Heal, D. J., Goddard, S., Brammer, R. J., Hutson, P. H., & Vickers, S. P. (2016). Lisdexamfetamine reduces the compulsive and perseverative behaviour of binge-eating rats in a novel food reward/punished responding conflict model. *Journal of Psychopharmacology, 30*(7), 662–675. https://doi.org/10.1177/0269881116647506.

Heatherton, T. F., Herman, C. P., & Polivy, J. (1991). Effects of physical threat and ego threat on eating behavior. *Journal of Personality and Social Psychology, 60*(1), 138–143.

Hege, M. A., Stingl, K. T., Kullmann, S., Schag, K., Giel, K. E., Zipfel, S., et al. (2015). Attentional impulsivity in binge eating disorder modulates response inhibition performance and frontal brain networks. *International Journal of Obesity, 39*(2), 353–360. https://doi.org/10.1038/ijo.2014.99.

Hendrikse, J. J., Cachia, R. L., Kothe, E. J., McPhie, S., Skouteris, H., & Hayden, M. J. (2015). Attentional biases for food cues in overweight and individuals with obesity: A systematic review of the literature. *Obesity Reviews, 16*(5), 424–432. https://doi.org/10.1111/obr.12265.

Herman, C. P., & Mack, D. (1975). Restrained and unrestrained eating. *Journal of Personality, 43*(4), 647–660.

Herman, C., & Polivy, J. (1980). Restrained eating. In A. J. Stunkard (Ed.), *Obesity (silver spring)* (pp. 208–225). Philadelphia: WB Saunders.

Hernandez, L., & Hoebel, B. G. (1988). Food reward and cocaine increase extracellular dopamine in the nucleus accumbens as measured by microdialysis. *Life Sciences, 42*(18), 1705–1712.

Hester, R., & Luijten, M. (2014). Neural correlates of attentional bias in addiction. *CNS Spectrums, 19*(3), 231–238. https://doi.org/10.1017/S1092852913000473.

Heyne, A., Kiesselbach, C., Sahun, I., McDonald, J., Gaiffi, M., Dierssen, M., et al. (2009). An animal model of compulsive food-taking behaviour. *Addiction Biology, 14*(4), 373–383. https://doi.org/10.1111/j.1369-1600.2009.00175.x.

Hillard, C. J., Weinlander, K. M., & Stuhr, K. L. (2012). Contributions of endocannabinoid signaling to psychiatric disorders in humans: Genetic and biochemical evidence. *Neuroscience, 204*, 207–229. https://doi.org/10.1016/j.neuroscience.2011.11.020.

Holden, C. (2010). Psychiatry. Behavioral addictions debut in proposed DSM-V. *Science, 327*(5968), 935. https://doi.org/10.1126/science.327.5968.935.

Hopf, F. W., & Lesscher, H. M. (2014). Rodent models for compulsive alcohol intake. *Alcohol, 48*(3), 253–264. https://doi.org/10.1016/j.alcohol.2014.03.001.

Horstmann, A., Dietrich, A., Mathar, D., Possel, M., Villringer, A., & Neumann, J. (2015). Slave to habit? Obesity is associated with decreased behavioural sensitivity to reward devaluation. *Appetite, 87*, 175–183. https://doi.org/10.1016/j.appet.2014.12.212.

Hyman, S. E. (2005). Addiction: A disease of learning and memory. *American Journal of Psychiatry, 162*(8), 1414–1422. https://doi.org/10.1176/appi.ajp.162.8.1414.

Iemolo, A., Blasio, A., St Cyr, S. A., Jiang, F., Rice, K. C., Sabino, V., et al. (2013). CRF-CRF1 receptor system in the central and basolateral nuclei of the amygdala differentially mediates excessive eating of palatable food. *Neuropsychopharmacology, 38*(12), 2456–2466. https://doi.org/10.1038/npp.2013.147.

Iemolo, A., Valenza, M., Tozier, L., Knapp, C. M., Kornetsky, C., Steardo, L., et al. (2012). Withdrawal from chronic, intermittent access to a highly palatable food induces depressive-like behavior in compulsive eating rats. *Behavioural Pharmacology, 23*(5–6), 593–602. https://doi.org/10.1097/FBP.0b013e328357697f.

Janssen, L. K., Duif, I., van Loon, I., Wegman, J., de Vries, J. H., Cools, R., et al. (2017). Loss of lateral prefrontal cortex control in food-directed attention and goal-directed food choice in obesity. *NeuroImage, 146*, 148–156. https://doi.org/10.1016/j.neuroimage.2016.11.015.

Johnson, P. M., & Kenny, P. J. (2010). Dopamine D2 receptors in addiction-like reward dysfunction and compulsive eating in obese rats. *Nature Neuroscience, 13*(5), 635–641. https://doi.org/10.1038/nn.2519.

de Jong, J. W., Meijboom, K. E., Vanderschuren, L. J., & Adan, R. A. (2013). Low control over palatable food intake in rats is associated with habitual behavior and relapse vulnerability: Individual differences. *PLoS One, 8*(9), e74645. https://doi.org/10.1371/journal.pone.0074645.

Jonkman, S., Pelloux, Y., & Everitt, B. J. (2012). Drug intake is sufficient, but conditioning is not necessary for the emergence of compulsive cocaine seeking after extended self-administration. *Neuropsychopharmacology, 37*(7), 1612–1619. https://doi.org/10.1038/npp.2012.6.

Kagan, D., & Squires, R. (1983). Dieting, compulsive eating, and feelings of failure among adolescents. *International Journal of Eating Disorders, 3*(1), 15–26.

Kalivas, P. W., Lalumiere, R. T., Knackstedt, L., & Shen, H. (2009). Glutamate transmission in addiction. *Neuropharmacology, 56*(Suppl. 1), 169–173. https://doi.org/10.1016/j.neuropharm.2008.07.011.

Kalivas, P. W., & Volkow, N. D. (2005). The neural basis of addiction: A pathology of motivation and choice. *American Journal of Psychiatry, 162*(8), 1403–1413. https://doi.org/10.1176/appi.ajp.162.8.1403.

Kaplan, H. I., & Kaplan, H. S. (1957). The psychosomatic concept of obesity. *The Journal of Nervous and Mental Disease, 125*(2), 181–201.

Kessler, R. C., Berglund, P. A., Chiu, W. T., Deitz, A. C., Hudson, J. I., Shahly, V., et al. (2013). The prevalence and correlates of binge eating disorder in the world health organization world mental health surveys. *Biological Psychiatry, 73*(9), 904–914. https://doi.org/10.1016/j.biopsych.2012.11.020.

Kessler, R. M., Hutson, P. H., Herman, B. K., & Potenza, M. N. (2016). The neurobiological basis of binge-eating disorder. *Neuroscience & Biobehavioral Reviews, 63*, 223–238. https://doi.org/10.1016/j.neubiorev.2016.01.013.

Keys, A., Brozek, J., Henschel, A., Mickelson, O., & Taylor, H. (1950). *The biology of human starvation, 2 vols*. Minneapolis, MN: University of Minnesota Press.

Klatzkin, R. R., Gaffney, S., Cyrus, K., Bigus, E., & Brownley, K. A. (2015). Binge eating disorder and obesity: Preliminary evidence for distinct cardiovascular and psychological phenotypes. *Physiology & Behavior, 142*, 20–27. https://doi.org/10.1016/j.physbeh.2015.01.018.

Klein, T. A., Neumann, J., Reuter, M., Hennig, J., von Cramon, D. Y., & Ullsperger, M. (2007). Genetically determined differences in learning from errors. *Science, 318*(5856), 1642–1645. https://doi.org/10.1126/science.1145044.

Kolotkin, R. L., & Andersen, J. R. (2017). A systematic review of reviews: Exploring the relationship between obesity, weight loss and health-related quality of life. *Clinical Obesity, 7*(5), 273–289. https://doi.org/10.1111/cob.12203.

Kolotkin, R. L., Westman, E. C., Ostbye, T., Crosby, R. D., Eisenson, H. J., & Binks, M. (2004). Does binge eating disorder impact weight-related quality of life? *Obesity Research, 12*(6), 999–1005. https://doi.org/10.1038/oby.2004.122.

Koob, G. F. (1996). Drug addiction: The yin and yang of hedonic homeostasis. *Neuron, 16*(5), 893–896.

Koob, G. F. (1999). The role of the striatopallidal and extended amygdala systems in drug addiction. *Annals of the New York Academy of Sciences, 877*, 445–460.

Koob, G. F. (2009). Neurobiological substrates for the dark side of compulsivity in addiction. *Neuropharmacology, 56*(Suppl. 1), 18–31. https://doi.org/10.1016/j.neuropharm.2008.07.043.

Koob, G. F. (2013). Addiction is a reward deficit and stress surfeit disorder. *Frontiers in Psychiatry, 4*, 72. https://doi.org/10.3389/fpsyt.2013.00072.

Koob, G. F. (2015). The dark side of emotion: The addiction perspective. *European Journal of Pharmacology, 753*, 73–87. https://doi.org/10.1016/j.ejphar.2014.11.044.

Koob, G. F., & Bloom, F. E. (1988). Cellular and molecular mechanisms of drug dependence. *Science, 242*(4879), 715–723.

Koob, G. F., Buck, C. L., Cohen, A., Edwards, S., Park, P. E., Schlosburg, J. E., et al. (2014). Addiction as a stress surfeit disorder. *Neuropharmacology, 76*(Pt B), 370–382. https://doi.org/10.1016/j.neuropharm.2013.05.024.

Koob, G. F., & Le Moal, M. (1997). Drug abuse: Hedonic homeostatic dysregulation. *Science, 278*(5335), 52–58.

Koob, G. F., & Le Moal, M. (2001). Drug addiction, dysregulation of reward, and allostasis. *Neuropsychopharmacology, 24*(2), 97–129. https://doi.org/10.1016/S0893-133X(00)00195-0.

Koob, G. F., & Le Moal, M. (2008a). Addiction and the brain antireward system. *Annual Review of Psychology, 59*, 29–53. https://doi.org/10.1146/annurev.psych.59.103006.093548.

Koob, G. F., & Le Moal, M. (2008b). Review. Neurobiological mechanisms for opponent motivational processes in addiction. *Philosophical Transactions of the Royal Society of London B Biological Sciences, 363*(1507), 3113–3123. https://doi.org/10.1098/rstb.2008.0094.

Koob, G. F., & Volkow, N. D. (2010). Neurocircuitry of addiction. *Neuropsychopharmacology, 35*(1), 217–238. https://doi.org/10.1038/npp.2009.110.

Koob, G. F., & Volkow, N. D. (2016). Neurobiology of addiction: A neurocircuitry analysis. *Lancet Psychiatry, 3*(8), 760–773.

Koob, G. F., & Zorrilla, E. P. (2010). Neurobiological mechanisms of addiction: Focus on corticotropin-releasing factor. *Current Opinion in Investigational Drugs, 11*(1), 63–71.

Kwako, L. E., Bickel, W. K., & Goldman, D. (2018). Addiction biomarkers: Dimensional approaches to understanding addiction. *Trends in Molecular Medicine, 24*(2), 121–128. https://doi.org/10.1016/j.molmed.2017.12.007.

Kwako, L. E., Momenan, R., Litten, R. Z., Koob, G. F., & Goldman, D. (2016). Addictions neuroclinical assessment: A neuroscience-based framework for addictive disorders. *Biological Psychiatry, 80*(3), 179–189. https://doi.org/10.1016/j.biopsych.2015.10.024.

Laessle, R. G., Tuschl, R. J., Kotthaus, B. C., & Pirke, K. M. (1989). Behavioral and biological correlates of dietary restraint in normal life. *Appetite, 12*(2), 83–94.

Latagliata, E. C., Patrono, E., Puglisi-Allegra, S., & Ventura, R. (2010). Food seeking in spite of harmful consequences is under prefrontal cortical noradrenergic control. *BMC Neuroscience, 11*, 15. https://doi.org/10.1186/1471-2202-11-15.

Laurent, V., Morse, A. K., & Balleine, B. W. (2015). The role of opioid processes in reward and decision-making. *British Journal of Pharmacology, 172*(2), 449–459. https://doi.org/10.1111/bph.12818.

Lawrence, N. S., Hinton, E. C., Parkinson, J. A., & Lawrence, A. D. (2012). Nucleus accumbens response to food cues predicts subsequent snack consumption in women and increased body mass index in those with reduced self-control. *NeuroImage, 63*(1), 415–422. https://doi.org/10.1016/j.neuroimage.2012.06.070.

Lebow, M. A., & Chen, A. (2016). Overshadowed by the amygdala: The bed nucleus of the stria terminalis emerges as key to psychiatric disorders. *Molecular Psychiatry, 21*(4), 450–463. https://doi.org/10.1038/mp.2016.1.

Linde, J. A., Jeffery, R. W., Levy, R. L., Sherwood, N. E., Utter, J., Pronk, N. P., et al. (2004). Binge eating disorder, weight control self-efficacy, and depression in overweight men and women. *International Journal of Obesity and Related Metabolic Disorders, 28*(3), 418–425. https://doi.org/10.1038/sj.ijo.0802570.

Lingawi, N. W., & Balleine, B. W. (2012). Amygdala central nucleus interacts with dorsolateral striatum to regulate the acquisition of habits. *Journal of Neuroscience, 32*(3), 1073–1081. https://doi.org/10.1523/JNEUROSCI.4806-11.2012.

Lips, M. A., Wijngaarden, M. A., van der Grond, J., van Buchem, M. A., de Groot, G. H., Rombouts, S. A., et al. (2014). Resting-state functional connectivity of brain regions involved in cognitive control, motivation, and reward is enhanced in obese females. *American Journal of Clinical Nutrition, 100*(2), 524–531. https://doi.org/10.3945/ajcn.113.080671.

Long, C. G., Blundell, J. E., & Finlayson, G. (2015). A systematic review of the application and correlates of YFAS-diagnosed 'food addiction' in humans: Are eating-related 'addictions' a cause for concern or empty concepts? *Obes Facts, 8*(6), 386–401. https://doi.org/10.1159/000442403.

Lubman, D. I., Yucel, M., & Pantelis, C. (2004). Addiction, a condition of compulsive behaviour? Neuroimaging and neuropsychological evidence of inhibitory dysregulation. *Addiction, 99*(12), 1491–1502. https://doi.org/10.1111/j.1360-0443.2004.00808.x.

Macht, M. (2008). How emotions affect eating: A five-way model. *Appetite, 50*(1), 1–11. https://doi.org/10.1016/j.appet.2007.07.002.

Mann, T., Tomiyama, A. J., Westling, E., Lew, A. M., Samuels, B., & Chatman, J. (2007). Medicare's search for effective obesity treatments: Diets are not the answer. *American Psychologist, 62*(3), 220–233. https://doi.org/10.1037/0003-066X.62.3.220.

Mathar, D., Neumann, J., Villringer, A., & Horstmann, A. (2017). Failing to learn from negative prediction errors: Obesity is associated with alterations in a fundamental neural learning mechanism. *Cortex, 95*, 222–237. https://doi.org/10.1016/j.cortex.2017.08.022.

McClure, S. M., Daw, N. D., & Montague, P. R. (2003). A computational substrate for incentive salience. *Trends in Neurosciences, 26*(8), 423–428.

McElroy, S. L., Guerdjikova, A. I., Blom, T. J., Crow, S. J., Memisoglu, A., Silverman, B. L., et al. (2013). A placebo-controlled pilot study of the novel opioid receptor antagonist ALKS-33 in binge eating disorder. *International Journal of Eating Disorders, 46*(3), 239–245. https://doi.org/10.1002/eat.22114.

McElroy, S. L., Mitchell, J. E., Wilfley, D., Gasior, M., Ferreira-Cornwell, M. C., McKay, M., et al. (2016). Lisdexamfetamine dimesylate effects on binge eating behaviour and obsessive-compulsive and impulsive features in adults with binge eating disorder. *European Eating Disorders Review, 24*(3), 223–231. https://doi.org/10.1002/erv.2418.

McGuire, M. T., Wing, R. R., Klem, M. L., Lang, W., & Hill, J. O. (1999). What predicts weight regain in a group of successful weight losers? *Journal of Consulting and Clinical Psychology, 67*(2), 177–185.

McNamee, D., Liljeholm, M., Zika, O., & O'Doherty, J. P. (2015). Characterizing the associative content of brain structures involved in habitual and goal-directed actions in humans: A multivariate FMRI study. *Journal of Neuroscience, 35*(9), 3764–3771. https://doi.org/10.1523/JNEUROSCI.4677-14.2015.

Mela, D. J. (2001). Determinants of food choice: Relationships with obesity and weight control. *Obesity Research, 9*(Suppl. 4), 249S–255S. https://doi.org/10.1038/oby.2001.127.

Mena, J. D., Sadeghian, K., & Baldo, B. A. (2011). Induction of hyperphagia and carbohydrate intake by mu-opioid receptor stimulation in circumscribed regions of frontal cortex. *Journal of Neuroscience, 31*(9), 3249–3260. https://doi.org/10.1523/JNEUROSCI.2050-10.2011.

Mena, J. D., Selleck, R. A., & Baldo, B. A. (2013). Mu-opioid stimulation in rat prefrontal cortex engages hypothalamic orexin/hypocretin-containing neurons, and reveals dissociable roles of nucleus accumbens and hypothalamus in cortically driven feeding. *Journal of Neuroscience, 33*(47), 18540–18552. https://doi.org/10.1523/JNEUROSCI.3323-12.2013.

Meule, A., Lutz, A., Vogele, C., & Kubler, A. (2012). Women with elevated food addiction symptoms show accelerated reactions, but no impaired inhibitory control, in response to pictures of high-calorie food-cues. *Eating Behaviors, 13*(4), 423–428. https://doi.org/10.1016/j.eatbeh.2012.08.001.

Micioni Di Bonaventura, M. V., Ciccocioppo, R., Romano, A., Bossert, J. M., Rice, K. C., Ubaldi, M., et al. (2014). Role of bed nucleus of the stria terminalis corticotrophin-releasing factor receptors in frustration stress-induced binge-like palatable food consumption in female rats with a history of food restriction. *Journal of Neuroscience, 34*(34), 11316–11324. https://doi.org/10.1523/JNEUROSCI.1854-14.2014.

Moore, C. F., Sabino, V., & Cottone, P. (2018a). Trace amine associated receptor 1 (TAAR1) modulation of food reward. *Frontiers in Pharmacology, 9*, 129. https://doi.org/10.3389/fphar.2018.00129.

Moore, C. F., Panciera, J. I., Sabino, V., & Cottone, P. (2018b). Neuropharmacology of compulsive eating. *Philosophical Transactions of the Royal Society B: Biological Sciences, 373*(1742), 20170024.

Moore, C. F., Sabino, V., Koob, G. F., & Cottone, P. (2017a). Neuroscience of compulsive eating behavior. *Frontiers in Neuroscience, 11*(469). https://doi.org/10.3389/fnins.2017.00469.

Moore, C. F., Sabino, V., Koob, G. F., & Cottone, P. (2017b). Pathological overeating: Emerging evidence for a compulsivity construct. *Neuropsychopharmacology, 42*(7), 1375–1389. https://doi.org/10.1038/npp.2016.269.

Murdaugh, D. L., Cox, J. E., Cook, E. W., 3rd, & Weller, R. E. (2012). fMRI reactivity to high-calorie food pictures predicts short- and long-term outcome in a weight-loss program. *NeuroImage, 59*(3), 2709–2721.

Murray, J. E., Belin-Rauscent, A., Simon, M., Giuliano, C., Benoit-Marand, M., Everitt, B. J., et al. (2015). Basolateral and central amygdala differentially recruit and maintain dorsolateral striatum-dependent cocaine-seeking habits. *Nature Communications, 6*, 10088. https://doi.org/10.1038/ncomms10088.

Murray, E., Brouwer, S., McCutcheon, R., Harmer, C. J., Cowen, P. J., & McCabe, C. (2014). Opposing neural effects of naltrexone on food reward and aversion: Implications for the treatment of obesity. *Psychopharmacology, 231*(22), 4323–4335. https://doi.org/10.1007/s00213-014-3573-7.

Nieh, E. H., Matthews, G. A., Allsop, S. A., Presbrey, K. N., Leppla, C. A., Wichmann, R., et al. (2015). Decoding neural circuits that control compulsive sucrose seeking. *Cell, 160*(3), 528–541. https://doi.org/10.1016/j.cell.2015.01.003.

O'Brien, C. (2011a). Addiction and dependence in DSM-V. *Addiction, 106*(5), 866–867. https://doi.org/10.1111/j.1360-0443.2010.03144.x.

O'Brien, C. P. (2011b). Chapter 24: Drug addiction. In Laurence L. Brunton, Randa Hilal-Dandan, & Björn C. Knollmann (Eds.), *Goodman & gilman's the pharmacological basis of therapeutics* (12 ed). New York, NY: McGraw-Hill.

Ogden, C. L., Carroll, M. D., & Flegal, K. M. (2014). Prevalence of obesity in the United States. *Journal of the American Medical Association, 312*(2), 189–190. https://doi.org/10.1001/jama.2014.6228.

Oswald, K. D., Murdaugh, D. L., King, V. L., & Boggiano, M. M. (2011). Motivation for palatable food despite consequences in an animal model of binge eating. *International Journal of Eating Disorders, 44*(3), 203–211. https://doi.org/10.1002/eat.20808.

Park, K., Volkow, N. D., Pan, Y., & Du, C. (2013). Chronic cocaine dampens dopamine signaling during cocaine intoxication and unbalances D1 over D2 receptor signaling. *Journal of Neuroscience, 33*(40), 15827–15836. https://doi.org/10.1523/JNEUROSCI.1935-13.2013.

Parylak, S. L., Koob, G. F., & Zorrilla, E. P. (2011). The dark side of food addiction. *Physiology & Behavior, 104*(1), 149–156. https://doi.org/10.1016/j.physbeh.2011.04.063.

Patel, S., Cravatt, B. F., & Hillard, C. J. (2005). Synergistic interactions between cannabinoids and environmental stress in the activation of the central amygdala. *Neuropsychopharmacology, 30*(3), 497–507. https://doi.org/10.1038/sj.npp.1300535.

Pauli-Pott, U., Albayrak, O., Hebebrand, J., & Pott, W. (2010). Association between inhibitory control capacity and body weight in overweight and obese children and adolescents: Dependence on age and inhibitory control component. *Child Neuropsychology, 16*(6), 592–603. https://doi.org/10.1080/09297049.2010.485980.

Pearson, C. M., Wonderlich, S. A., & Smith, G. T. (2015). A risk and maintenance model for bulimia nervosa: From impulsive action to compulsive behavior. *Psychological Review, 122*(3), 516–535. https://doi.org/10.1037/a0039268.

Pecoraro, N., Reyes, F., Gomez, F., Bhargava, A., & Dallman, M. F. (2004). Chronic stress promotes palatable feeding, which reduces signs of stress: Feedforward and feedback effects of chronic stress. *Endocrinology, 145*(8), 3754–3762. https://doi.org/10.1210/en.2004-0305.

Pedram, P., Wadden, D., Amini, P., Gulliver, W., Randell, E., Cahill, F., et al. (2013). Food addiction: Its prevalence and significant association with obesity in the general population. *PLoS One, 8*(9), e74832. https://doi.org/10.1371/journal.pone.0074832.

Pelloux, Y., Everitt, B. J., & Dickinson, A. (2007). Compulsive drug seeking by rats under punishment: Effects of drug taking history. *Psychopharmacology, 194*(1), 127–137. https://doi.org/10.1007/s00213-007-0805-0.

Peterson, C. B., Miller, K. B., Crow, S. J., Thuras, P., & Mitchell, J. E. (2005). Subtypes of binge eating disorder based on psychiatric history. *International Journal of Eating Disorders, 38*(3), 273–276. https://doi.org/10.1002/eat.20174.

Piazza, P. V., & Deroche-Gamonet, V. (2013). A multistep general theory of transition to addiction. *Psychopharmacology, 229*(3), 387–413. https://doi.org/10.1007/s00213-013-3224-4.

Pursey, K. M., Stanwell, P., Gearhardt, A. N., Collins, C. E., & Burrows, T. L. (2014). The prevalence of food addiction as assessed by the Yale food addiction scale: A systematic review. *Nutrients, 6*(10), 4552–4590. https://doi.org/10.3390/nu6104552.

Rada, P., Avena, N. M., & Hoebel, B. G. (2005). Daily bingeing on sugar repeatedly releases dopamine in the accumbens shell. *Neuroscience, 134*(3), 737–744. https://doi.org/10.1016/j.neuroscience.2005.04.043.

Rada, P., Bocarsly, M. E., Barson, J. R., Hoebel, B. G., & Leibowitz, S. F. (2010). Reduced accumbens dopamine in Sprague-Dawley rats prone to overeating a fat-rich diet. *Physiology & Behavior, 101*(3), 394–400. https://doi.org/10.1016/j.physbeh.2010.07.005.

Reichelt, A. C., Morris, M. J., & Westbrook, R. F. (2014). Cafeteria diet impairs expression of sensory-specific satiety and stimulus-outcome learning. *Frontiers in Psychology, 5*, 852. https://doi.org/10.3389/fpsyg.2014.00852.

Reichelt, A. C., Westbrook, R. F., & Morris, M. J. (2015). Integration of reward signalling and appetite regulating peptide systems in the control of food-cue responses. *British Journal of Pharmacology, 172*(22), 5225–5238. https://doi.org/10.1111/bph.13321.

Riga, D., Matos, M. R., Glas, A., Smit, A. B., Spijker, S., & Van den Oever, M. C. (2014). Optogenetic dissection of medial prefrontal cortex circuitry. *Frontiers in Systems Neuroscience, 8*, 230. https://doi.org/10.3389/fnsys.2014.00230.

Robbins, T. W., Curran, H. V., & de Wit, H. (2012). Special issue on impulsivity and compulsivity. *Psychopharmacology, 219*(2), 251–252.

Robinson, T. E., & Berridge, K. C. (1993). The neural basis of drug craving: An incentive-sensitization theory of addiction. *Brain Research Reviews, 18*(3), 247–291.
Robinson, M. J., Burghardt, P. R., Patterson, C. M., Nobile, C. W., Akil, H., Watson, S. J., et al. (2015). Individual differences in cue-induced motivation and striatal systems in rats susceptible to diet-induced obesity. *Neuropsychopharmacology, 40*(9), 2113–2123. https://doi.org/10.1038/npp.2015.71.
Robson, M. J., Noorbakhsh, B., Seminerio, M. J., & Matsumoto, R. R. (2012). Sigma-1 receptors: Potential targets for the treatment of substance abuse. *Current Pharmaceutical Design, 18*(7), 902–919.
Rosenbaum, D. L., & White, K. S. (2015). The relation of anxiety, depression, and stress to binge eating behavior. *Journal of Health Psychology, 20*(6), 887–898. https://doi.org/10.1177/1359105315580212.
Rosen, J. C., Gross, J., & Vara, L. (1987). Psychological adjustment of adolescents attempting to lose or gain weight. *Journal of Consulting and Clinical Psychology, 55*(5), 742–747.
Rosen, J., Tacy, B., & Howell, D. (1990). Life stress, psychological symptoms and weight reducing behavior in adolescent girls: A prospective analysis. *International Journal of Eating Disorders, 9*(1), 17–26.
Rossetti, C., Spena, G., Halfon, O., & Boutrel, B. (2014). Evidence for a compulsive-like behavior in rats exposed to alternate access to highly preferred palatable food. *Addiction Biology, 19*(6), 975–985. https://doi.org/10.1111/adb.12065.
Rowley, H. L., Kulkarni, R., Gosden, J., Brammer, R., Hackett, D., & Heal, D. J. (2012). Lisdexamfetamine and immediate release d-amfetamine - differences in pharmacokinetic/pharmacodynamic relationships revealed by striatal microdialysis in freely-moving rats with simultaneous determination of plasma drug concentrations and locomotor activity. *Neuropharmacology, 63*(6), 1064–1074. https://doi.org/10.1016/j.neuropharm.2012.07.008.
Sabino, V., Cottone, P., Blasio, A., Iyer, M. R., Steardo, L., Rice, K. C., et al. (2011). Activation of sigma-receptors induces binge-like drinking in Sardinian alcohol-preferring rats. *Neuropsychopharmacology, 36*(6), 1207–1218. https://doi.org/10.1038/npp.2011.5.
Sabino, V., Cottone, P., Zhao, Y., Iyer, M. R., Steardo, L., Jr., Steardo, L., et al. (2009a). The sigma-receptor antagonist BD-1063 decreases ethanol intake and reinforcement in animal models of excessive drinking. *Neuropsychopharmacology, 34*(6), 1482–1493. https://doi.org/10.1038/npp.2008.192.
Sabino, V., Cottone, P., Zhao, Y., Steardo, L., Koob, G. F., & Zorrilla, E. P. (2009b). Selective reduction of alcohol drinking in Sardinian alcohol-preferring rats by a sigma-1 receptor antagonist. *Psychopharmacology, 205*(2), 327–335. https://doi.org/10.1007/s00213-009-1548-x.
Salamone, J. D., & Correa, M. (2013). Dopamine and food addiction: Lexicon badly needed. *Biological Psychiatry, 73*(9), e15–24. https://doi.org/10.1016/j.biopsych.2012.09.027.
Schienle, A., Schafer, A., Hermann, A., & Vaitl, D. (2009). Binge-eating disorder: Reward sensitivity and brain activation to images of food. *Biological Psychiatry, 65*(8), 654–661. https://doi.org/10.1016/j.biopsych.2008.09.028.
Schmitz, F., Naumann, E., Trentowska, M., & Svaldi, J. (2014). Attentional bias for food cues in binge eating disorder. *Appetite, 80*, 70–80. https://doi.org/10.1016/j.appet.2014.04.023.
Schulte, E. M., Grilo, C. M., & Gearhardt, A. N. (2016). Shared and unique mechanisms underlying binge eating disorder and addictive disorders. *Clinical Psychology Review, 44*, 125–139. https://doi.org/10.1016/j.cpr.2016.02.001.

Schultz, W., Dayan, P., & Montague, P. R. (1997). A neural substrate of prediction and reward. *Science, 275*(5306), 1593–1599.

Sedki, F., Gardner Gregory, J., Luminare, A., D'Cunha, T. M., & Shalev, U. (2015). Food restriction-induced augmentation of heroin seeking in female rats: Manipulations of ovarian hormones. *Psychopharmacology, 232*(20), 3773–3782. https://doi.org/10.1007/s00213-015-4037-4.

Seeyave, D. M., Coleman, S., Appugliese, D., Corwyn, R. F., Bradley, R. H., Davidson, N. S., et al. (2009). Ability to delay gratification at age 4 years and risk of overweight at age 11 years. *Archives of Pediatrics and Adolescent Medicine, 163*(4), 303–308. https://doi.org/10.1001/archpediatrics.2009.12.

Selleck, R. A., Lake, C., Estrada, V., Riederer, J., Andrzejewski, M., Sadeghian, K., et al. (2015). Endogenous opioid signaling in the medial prefrontal cortex is required for the expression of hunger-induced impulsive action. *Neuropsychopharmacology.* https://doi.org/10.1038/npp.2015.97.

Shalev, U. (2012). Chronic food restriction augments the reinstatement of extinguished heroin-seeking behavior in rats. *Addiction Biology, 17*(4), 691–693. https://doi.org/10.1111/j.1369-1600.2010.00303.x.

Shalev, U., Erb, S., & Shaham, Y. (2010). Role of CRF and other neuropeptides in stress-induced reinstatement of drug seeking. *Brain Research, 1314*, 15–28. https://doi.org/10.1016/j.brainres.2009.07.028.

Shank, L. M., Tanofsky-Kraff, M., Nelson, E. E., Shomaker, L. B., Ranzenhofer, L. M., Hannallah, L. M., et al. (2015). Attentional bias to food cues in youth with loss of control eating. *Appetite, 87*, 68–75. https://doi.org/10.1016/j.appet.2014.11.027.

Sharma, S., Fernandes, M. F., & Fulton, S. (2013). Adaptations in brain reward circuitry underlie palatable food cravings and anxiety induced by high-fat diet withdrawal. *International Journal of Obesity, 37*(9), 1183–1191. https://doi.org/10.1038/ijo.2012.197.

Sidhpura, N., & Parsons, L. H. (2011). Endocannabinoid-mediated synaptic plasticity and addiction-related behavior. *Neuropharmacology, 61*(7), 1070–1087. https://doi.org/10.1016/j.neuropharm.2011.05.034.

Silverstone, J., & Lascelles, B. (1966). Dieting and depression. *The British Journal of Psychiatry, 112*(486), 513–519.

Smith, K. L., Rao, R. R., Velazquez-Sanchez, C., Valenza, M., Giuliano, C., Everitt, B. J., et al. (2015). The uncompetitive N-methyl-D-Aspartate antagonist memantine reduces binge-like eating, food-seeking behavior, and compulsive eating: Role of the nucleus accumbens shell. *Neuropsychopharmacology, 40*, 1163–1171. https://doi.org/10.1038/npp.2014.299.

Spitzer, R. L., Devlin, M., Walsh, B. T., Hasin, D., Wing, R., Marcus, M., et al. (1992). Binge eating disorder - a multisite field trial of the diagnostic-criteria. *International Journal of Eating Disorders, 11*(3), 191–203. https://doi.org/10.1002/1098-108x(199204)11:3<191::Aid-Eat2260110302>3.0.Co;2-S.

Stice, E., Davis, K., Miller, N. P., & Marti, C. N. (2008). Fasting increases risk for onset of binge eating and bulimic pathology: A 5-year prospective study. *Journal of Abnormal Psychology, 117*(4), 941–946. https://doi.org/10.1037/a0013644.

Stirling, L. J., & Yeomans, M. R. (2004). Effect of exposure to a forbidden food on eating in restrained and unrestrained women. *International Journal of Eating Disorders, 35*(1), 59–68. https://doi.org/10.1002/eat.10232.

Stoeckel, L. E., Weller, R. E., Cook, E. W., 3rd, Twieg, D. B., Knowlton, R. C., & Cox, J. E. (2008). Widespread reward-system activation in obese women in response to pictures of

high-calorie foods. *NeuroImage, 41*(2), 636–647. https://doi.org/10.1016/j.neuroimage.2008.02.031.
Stuber, G. D., Klanker, M., de Ridder, B., Bowers, M. S., Joosten, R. N., Feenstra, M. G., et al. (2008). Reward-predictive cues enhance excitatory synaptic strength onto midbrain dopamine neurons. *Science, 321*(5896), 1690–1692. https://doi.org/10.1126/science.1160873.
Stunkard, A. J. (1957). The dieting depression; incidence and clinical characteristics of untoward responses to weight reduction regimens. *The American Journal of Medicine, 23*(1), 77–86.
Surmeier, D. J., Ding, J., Day, M., Wang, Z., & Shen, W. (2007). D1 and D2 dopamine-receptor modulation of striatal glutamatergic signaling in striatal medium spiny neurons. *Trends in Neurosciences, 30*(5), 228–235. https://doi.org/10.1016/j.tins.2007.03.008.
Sussman, S., & Sussman, A. N. (2011). Considering the definition of addiction. *International Journal of Environmental Research and Public Health, 8*(10), 4025–4038. https://doi.org/10.3390/ijerph8104025.
Svaldi, J., Naumann, E., Trentowska, M., & Schmitz, F. (2014). General and food-specific inhibitory deficits in binge eating disorder. *International Journal of Eating Disorders, 47*(5), 534–542. https://doi.org/10.1002/eat.22260.
Tao, R., Huang, X., Wang, J., Zhang, H., Zhang, Y., & Li, M. (2010). Proposed diagnostic criteria for internet addiction. *Addiction, 105*(3), 556–564. https://doi.org/10.1111/j.1360-0443.2009.02828.x.
Tchanturia, K., Davies, H., Roberts, M., Harrison, A., Nakazato, M., Schmidt, U., et al. (2012). Poor cognitive flexibility in eating disorders: Examining the evidence using the Wisconsin card sorting task. *PLoS One, 7*(1), e28331. https://doi.org/10.1371/journal.pone.0028331.
Teegarden, S. L., & Bale, T. L. (2007). Decreases in dietary preference produce increased emotionality and risk for dietary relapse. *Biological Psychiatry, 61*(9), 1021–1029. https://doi.org/10.1016/j.biopsych.2006.09.032.
Teegarden, S. L., & Bale, T. L. (2008). Effects of stress on dietary preference and intake are dependent on access and stress sensitivity. *Physiology & Behavior, 93*(4–5), 713–723. https://doi.org/10.1016/j.physbeh.2007.11.030.
Teegarden, S. L., Scott, A. N., & Bale, T. L. (2009). Early life exposure to a high fat diet promotes long-term changes in dietary preferences and central reward signaling. *Neuroscience, 162*(4), 924–932. https://doi.org/10.1016/j.neuroscience.2009.05.029.
The National Institute of Mental Health. (2013). *Research domain criteria (RDoC)* [Online]. Available: https://www.nimh.nih.gov/research-priorities/rdoc/constructs/rdoc-matrix.shtml.
Tomasi, D., & Volkow, N. D. (2013). Striatocortical pathway dysfunction in addiction and obesity: Differences and similarities. *Critical Reviews in Biochemistry and Molecular Biology, 48*(1), 1–19. https://doi.org/10.3109/10409238.2012.735642.
Tomiyama, A. J., Dallman, M. F., & Epel, E. S. (2011). Comfort food is comforting to those most stressed: Evidence of the chronic stress response network in high stress women. *Psychoneuroendocrinology, 36*(10), 1513–1519. https://doi.org/10.1016/j.psyneuen.2011.04.005.
Valenza, M., Steardo, L., Cottone, P., & Sabino, V. (2015). Diet-induced obesity and diet-resistant rats: Differences in the rewarding and anorectic effects of D-amphetamine. *Psychopharmacology, 232*(17), 3215–3226. https://doi.org/10.1007/s00213-015-3981-3.
Vanderschuren, L. J., & Everitt, B. J. (2004). Drug seeking becomes compulsive after prolonged cocaine self-administration. *Science, 305*(5686), 1017–1019. https://doi.org/10.1126/science.1098975.

Velazquez-Sanchez, C., Ferragud, A., Moore, C. F., Everitt, B. J., Sabino, V., & Cottone, P. (2014). High trait impulsivity predicts food addiction-like behavior in the rat. *Neuropsychopharmacology, 39*(10), 2463–2472. https://doi.org/10.1038/npp.2014.98.

Velazquez-Sanchez, C., Santos, J. W., Smith, K. L., Ferragud, A., Sabino, V., & Cottone, P. (2015). Seeking behavior, place conditioning, and resistance to conditioned suppression of feeding in rats intermittently exposed to palatable food. *Behavioral Neuroscience, 129*(2), 219–224. https://doi.org/10.1037/bne0000042.

Vendruscolo, L. F., Gueye, A. B., Darnaudery, M., Ahmed, S. H., & Cador, M. (2010). Sugar overconsumption during adolescence selectively alters motivation and reward function in adult rats. *PLoS One, 5*(2), e9296. https://doi.org/10.1371/journal.pone.0009296.

Volkow, N. D., & Fowler, J. S. (2000). Addiction, a disease of compulsion and drive: Involvement of the orbitofrontal cortex. *Cerebral Cortex, 10*(3), 318–325.

Volkow, N. D., Fowler, J. S., Wang, G. J., Hitzemann, R., Logan, J., Schlyer, D. J., et al. (1993). Decreased dopamine D2 receptor availability is associated with reduced frontal metabolism in cocaine abusers. *Synapse, 14*(2), 169–177. https://doi.org/10.1002/syn.890140210.

Volkow, N. D., Wang, G. J., & Baler, R. D. (2011). Reward, dopamine and the control of food intake: Implications for obesity. *Trends in Cognitive Sciences, 15*(1), 37–46. https://doi.org/10.1016/j.tics.2010.11.001.

Volkow, N. D., Wang, G. J., Fowler, J. S., & Telang, F. (2008a). Overlapping neuronal circuits in addiction and obesity: Evidence of systems pathology. *Philosophical Transactions of the Royal Society of London B Biological Sciences, 363*(1507), 3191–3200. https://doi.org/10.1098/rstb.2008.0107.

Volkow, N. D., Wang, G. J., Telang, F., Fowler, J. S., Thanos, P. K., Logan, J., et al. (2008b). Low dopamine striatal D2 receptors are associated with prefrontal metabolism in obese subjects: Possible contributing factors. *NeuroImage, 42*(4), 1537–1543. https://doi.org/10.1016/j.neuroimage.2008.06.002.

Volkow, N. D., Wang, G. J., Tomasi, D., & Baler, R. D. (2013a). The addictive dimensionality of obesity. *Biological Psychiatry, 73*(9), 811–818. https://doi.org/10.1016/j.biopsych.2012.12.020.

Volkow, N. D., Wang, G. J., Tomasi, D., & Baler, R. D. (2013b). Unbalanced neuronal circuits in addiction. *Current Opinion in Neurobiology, 23*(4), 639–648. https://doi.org/10.1016/j.conb.2013.01.002.

Voon, V. (2015). Cognitive biases in binge eating disorder: The hijacking of decision making. *CNS Spectrums, 20*(6), 566–573. https://doi.org/10.1017/S1092852915000681.

Voon, V., Derbyshire, K., Ruck, C., Irvine, M. A., Worbe, Y., Enander, J., et al. (2015). Disorders of compulsivity: A common bias towards learning habits. *Molecular Psychiatry, 20*(3), 345–352. https://doi.org/10.1038/mp.2014.44.

van Waarde, A., Ramakrishnan, N. K., Rybczynska, A. A., Elsinga, P. H., Ishiwata, K., Nijholt, I. M., et al. (2011). The cholinergic system, sigma-1 receptors and cognition. *Behavioural Brain Research, 221*(2), 543–554. https://doi.org/10.1016/j.bbr.2009.12.043.

Wang, G. J., Geliebter, A., Volkow, N. D., Telang, F. W., Logan, J., Jayne, M. C., et al. (2011). Enhanced striatal dopamine release during food stimulation in binge eating disorder. *Obesity, 19*(8), 1601–1608. https://doi.org/10.1038/oby.2011.27.

Wang, G. J., Volkow, N. D., Logan, J., Pappas, N. R., Wong, C. T., Zhu, W., et al. (2001). Brain dopamine and obesity. *Lancet, 357*(9253), 354–357.

Warschburger, P. (2005). The unhappy obese child. *International Journal of Obesity, 29*(Suppl. 2), S127–S129.

Wassum, K. M., Cely, I. C., Maidment, N. T., & Balleine, B. W. (2009). Disruption of endogenous opioid activity during instrumental learning enhances habit acquisition. *Neuroscience, 163*(3), 770–780. https://doi.org/10.1016/j.neuroscience.2009.06.071.

Wells, A. S., Read, N. W., Laugharne, J. D., & Ahluwalia, N. S. (1998). Alterations in mood after changing to a low-fat diet. *British Journal of Nutrition, 79*(1), 23–30.

Wilfley, D. E., Friedman, M. A., Dounchis, J. Z., Stein, R. I., Welch, R. R., & Ball, S. A. (2000). Comorbid psychopathology in binge eating disorder: Relation to eating disorder severity at baseline and following treatment. *Journal of Consulting and Clinical Psychology, 68*(4), 641–649.

World Health Organization. (2000). *Obesity: Preventing and managing the global epidemic. Report of a WHO consultation* (pp. 1–253). World Health Organization Technical Report Series 894, i-xii.

Wray, I., & Dickerson, M. G. (1981). Cessation of high frequency gambling and "withdrawal' symptoms. *British Journal of Addiction, 76*(4), 401–405.

Wu, M., Hartmann, M., Skunde, M., Herzog, W., & Friederich, H. C. (2013). Inhibitory control in bulimic-type eating disorders: A systematic review and meta-analysis. *PLoS One, 8*(12), e83412. https://doi.org/10.1371/journal.pone.0083412.

Yanovski, S. Z., Nelson, J. E., Dubbert, B. K., & Spitzer, R. L. (1993). Association of binge eating disorder and psychiatric comorbidity in obese subjects. *American Journal of Psychiatry, 150*(10), 1472–1479. https://doi.org/10.1176/ajp.150.10.1472.

Yin, H. H., & Knowlton, B. J. (2006). The role of the basal ganglia in habit formation. *Nature Reviews Neuroscience, 7*(6), 464–476. https://doi.org/10.1038/nrn1919.

Zapata, A., Minney, V. L., & Shippenberg, T. S. (2010). Shift from goal-directed to habitual cocaine seeking after prolonged experience in rats. *Journal of Neuroscience, 30*(46), 15457–15463. https://doi.org/10.1523/JNEUROSCI.4072-10.2010.

Ziauddeen, H., Farooqi, I. S., & Fletcher, P. C. (2012). Obesity and the brain: How convincing is the addiction model? *Nature Reviews Neuroscience, 13*(4), 279–286. https://doi.org/10.1038/nrn3212.

de Zwaan, M. (2001). Binge eating disorder and obesity. *International Journal of Obesity and Related Metabolic Disorders, 25*(Suppl. 1), S51–S55. https://doi.org/10.1038/sj.ijo.0801699.

CHAPTER

Habitual overeating

4

Catherine F. Moore[1,2], Valentina Sabino[1], George F. Koob[3], Pietro Cottone[1]

Laboratory of Addictive Disorders, Departments of Pharmacology and Psychiatry, Boston University School of Medicine, Boston, MA, United States[1]; Graduate Program for Neuroscience, Boston University School of Medicine, Boston, MA, United States[2]; National Institute on Alcohol Abuse and Alcoholism, National Institutes of Health, Bethesda, MD, United States[3]

Introduction

Behavioral automaticity (i.e., habit formation) is a highly efficient process whereby actions that are repeatedly reinforced become rote, therefore requiring negligible mental energy. The process of habit formation refers to what develops from an initially conscious, effortful, goal-directed behavior and then later becomes automatic and stimulus-driven (Everitt & Robbins, 2005; Smith & Graybiel, 2016). High-value reinforcers, such as drugs of abuse and highly palatable food, are hypothesized to rapidly engage habit systems and accelerate habit formation (Everitt & Robbins, 2016). Therefore, addiction to drugs and compulsive eating (a behavioral feature of forms of obesity, binge eating disorder (BED), and food addiction) are characterized in part by habitual drug-taking or habitual overeating behavior, respectively (Everitt & Robbins, 2016; Moore, Sabino, Koob, & Cottone, 2017). In this chapter, we will discuss the behavioral and neurobiological evidence for habitual overeating as an element of compulsive eating behavior, often drawing similarities with habitual behaviors in drug addiction.

Overview of habit formation

Instrumental behaviors are determined by the balance between two distinct behavioral systems: goal-directed versus habitual. Initially, instrumental responses are goal-directed, meaning that they are performed based on the expectancy that the outcome has a value. Drugs, palatable food, and other rewards act as instrumental reinforcers, increasing the likelihood of the subject repeating the response. Simultaneously, when a positive or satisfying outcome is experienced, the association between any reward-associated stimulus (e.g., paired cue or environmental context) and the response (termed the S-R relationship) is strengthened (Everitt & Robbins, 2005). Over time, actions repeatedly performed in the same

Compulsive Eating Behavior and Food Addiction. https://doi.org/10.1016/B978-0-12-816207-1.00004-4
Copyright © 2019 Elsevier Inc. All rights reserved.

situation can become habitual. These habitual behaviors are based solely in the S-R relationship, without consideration of the outcome value (Everitt & Robbins, 2016; Graybiel, 2008). However, the goal-directed system remains available, and under normal circumstances can be intentionally reengaged and behavior adjusted accordingly (Vandaele & Janak, 2018). However, in compulsive habits there is a weakening of the evaluative processes that would normally allow for the switch from S-R-driven back to goal-directed actions when the value of the reward is reduced (Belin, Belin-Rauscent, Murray, & Everitt, 2013; Horstmann et al., 2015; Watson, Wiers, Hommel, & de Wit, 2014). Habits therefore reflect a breakdown in goal-directed control, as well as enhanced S-R responding (Corbit & Janak, 2016; Vandaele & Janak, 2018).

Compulsive eating driven by habit

Pathological habitual behavior has been observed in humans with eating disorders characterized by compulsive eating, as well as in animal models with a history of palatable food consumption. Current evidence that supports compulsive eating behavior is in part driven by habit mechanisms and includes persistence of eating behavior despite devaluation and a predominance of "model-based" versus "model-free" response strategies in disorders (and animal models) characterized by compulsive eating, which we will describe in detail below.

Reduced sensitivity to outcome devaluation in compulsive eating

A paradigm commonly used to evaluate habitual behavior is the outcome devaluation test, where the value of the outcome is reduced and the resulting frequency of behavior is measured. If a behavior is under goal-directed control, then its frequency should decrease when the value of the outcome is reduced. Habitual behavior, therefore, is characterized by the persistence of responding despite outcome devaluation. An inability to adapt eating behavior based on motivational value of the outcome may therefore reflect compulsive habitual eating (Ostlund & Balleine, 2008). These outcome devaluation tests were developed for laboratory assessments and notably have good translation from humans to animals (Balleine & O'Doherty, 2010).

Outcome-specific satiety is the most commonly used outcome devaluation method for studying habitual eating behavior (see Fig. 4.1). Very broadly, the outcome-specific satiety paradigm lowers the value of a *specific* food outcome through allowing the subject free access before testing, therefore inducing satiety (de Jong et al., 2013; Furlong, Jayaweera, Balleine, & Corbit, 2014). Multiple variations of this task exist; therefore, we will discuss specific methodologies as well as results obtained in preclinical models and clinical samples of patients with disorders characterized by compulsive eating.

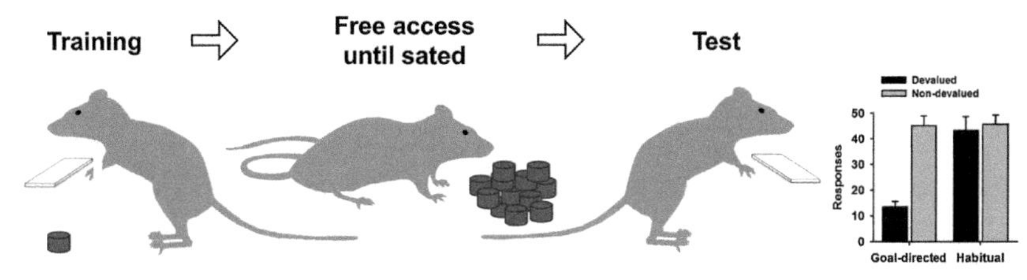

FIGURE 4.1

General schematic of the "outcome-specific satiety" devaluation procedure in animals. In this test, subjects are first trained to perform a response to receive a food outcome. Next, the outcome value of the food is manipulated (i.e., lessened) through allowing free consumption until satiety. Immediately following, subjects are tested under the same or similar conditions as training and allowed to respond for the food reward. Behavior that is goal-directed will reflect this change in outcome value; specifically, responding will be reduced for devalued outcomes. However, habitual behavior is insensitive to changes in outcome value, and responding will be maintained.

A human laboratory-based test of outcome-specific satiety was used to investigate habitual overeating in lean and obese participants (Horstmann et al., 2015). Initially subjects learn to associate cues with a food reward in an operant paradigm, where responding under a variable ratio schedule during the cue presentation resulted in presentation of the food reward (either M&M's or pretzels) (Horstmann et al., 2015). One of the two rewards was then devalued through free access for 30 min until satiety and an identical operant paradigm was employed postdevaluation, where the change in response rate for the outcomes was analyzed. Overall, higher responding for the devalued outcome was correlated with higher body mass index (BMI), indicating less behavioral adaptation following devaluation in obese subjects (Horstmann et al., 2015).

In rats, intermittent access to a diet of sweetened condensed milk (high in fat and sugar) accelerated habitual responding (Furlong et al., 2014). This experiment used an outcome-specific satiety procedure where animals were first trained in operant self-administration sessions to respond for one outcome (either grain pellets or a sucrose solution, counterbalanced between subjects) under a variable interval schedule of reinforcement. Animals were then allowed 1 h of access to their instrumental outcome before a 15-min session where they responded under extinction conditions (i.e., no outcome presentation). In this study, there were three groups: *ad libitum* chow (controls), continuous access, and intermittent (2 h daily) access to sweetened condensed milk. Intermittent palatable diet access, but not continuous access, resulted in accelerated habitual control over behavior when tested in the outcome devaluation procedure (Furlong et al., 2014).

A study by Tantot et al. (2017) found that rats continually fed a high-fat diet also showed resistance to devaluation in a procedure with methods similar to Furlong et al. (2014). This is in apparent contrast to Furlong et al. (2014), where habitual responding was not observed in rats with continuous, but only intermittent, palatable diet access. However, the discrepancy between these studies can most likely be attributed to slight variations in methodology, most notably that Tantot et al. (2017) used a greater number of training sessions (11–13 vs. 5–7 in Furlong et al. (2014)). Therefore, both continuous and intermittent access to a palatable diet may be presumed to result in habitual behavior, but intermittent access conditions likely allow for the emergence of habitual behavior at a faster rate.

Another preclinical study found that continuous cafeteria diet impaired goal-directed and/or enhanced habitual behavior in a nonoperant variation of outcome-specific satiety devaluation (Reichelt, Morris, & Westbrook, 2014). Rats were given access to both cherry sucrose and grape maltodextrin, which were calorie-matched and equally preferred food reinforcers. Animals had free access to each solution for 20 min over two daily familiarization sessions. Over two testing days, animals received access to one solution before a test session where both solutions were free to drink (counterbalanced across subjects). In control animals, preexposure to one solution greatly increased preference for the opposite solution. However, in rats with a history of access to a cafeteria diet, the preference remained equal for both solutions, regardless of preexposure (Reichelt et al., 2014). This test was repeated in a new cohort of animals to test devaluation after a 1-week withdrawal period, where cafeteria diet was replaced with *ad libitum* standard chow. The rats with a prior history of cafeteria diet also displayed habitual drinking of the devalued solution (Reichelt et al., 2014).

In a similar study, Reichelt et al. (2014) demonstrated that rats with a history of cafeteria diet access show habitual Pavlovian conditioned behavior for a devalued outcome. In this study, animals underwent Pavlovian conditioning sessions where the receipt of one of two distinct outcomes into a liquid magazine was presented with an associated cue (e.g., noise → grape maltodextrin, tone → cherry sucrose). Before the devaluation test session, rats were given 20 min of free access to one of the outcomes. During the test session, head entries into the liquid magazine (i.e., conditioned approach behavior) were measured under extinction conditions during alternating presentations of the previously associated cues. In these tests, control animals showed a *reduction* in their cue-invigorated responding to the devalued outcome, indicative of goal-directed action. On the contrary, animals with a history of cafeteria diet access showed persistent head entries during the cue presentation of both devalued and nondevalued outcomes (Reichelt et al., 2014).

Overall, both the clinical and preclinical studies presented above have found habitual responding in disorders of compulsive eating (obesity, BED) or animal models of compulsive eating.

Model-based and model-free learning in compulsive eating

Dual-system theories posit that all behavior is a result of goal-directed and habitual response systems (Dickinson & Balleine, 1993), and the balance between the two can become shifted toward habitual responding in disorders of compulsivity, including addiction and compulsive eating (Everitt & Robbins, 2016). A laboratory task designed to evaluate the balance between goal-directed and habitual response strategies was recently designed (Voon et al., 2015). This task is based on computational models of reinforcement learning, which have identified two types of decision models: "model-based" and "model-free" learning. "Model-based" learning refers to goal-directed decision-making, where choices are made depending on the outcomes predicted by the environment (i.e., prospective); "model-free" learning, representing habit-controlled behavior, refers instead to choices made based on prior reinforcement (i.e., retrospective); therefore, the decisions made are divorced from the outcome goals (Daw, Niv, & Dayan, 2005). Researchers developed a laboratory-based decision-making task that determines the contribution of "model-based" and "model-free" responding for each participant (Voon et al., 2015). A study investigating multiple disorders of compulsivity, including methamphetamine addiction, obsessive-compulsive disorder (OCD), and obese subjects with BED, determined that individuals with any one of these disorders showed a greater tendency to engage in "model-free" or habitual responding compared with healthy controls (Voon et al., 2015). Specifically, subjects with BED responded based on the computationally simpler mechanism, where decisions were made based on past reinforcement only, not related to future outcomes (i.e., habitual) (Voon et al., 2015). In contrast, healthy controls were more likely to utilize computationally "expensive" decision-making strategies that were based on continually adjusted predictions (i.e., goal-directed) (Voon et al., 2015). This study provided evidence that individuals with disorders of compulsivity, namely BED, methamphetamine addiction, or OCD, all display a neurocomputational bias toward "model-free" habitual responding (Voon et al., 2015).

Palatable food cues facilitate habitual behavior

Cues that are associated with palatable food are ubiquitous in the environment and can influence goal-directed behavior, potentiating habitual overeating. The influence of palatable food-associated cues on habitual behavior has been assessed using Pavlovian-instrumental transfer (PIT) paradigms paired with devaluation procedures. As discussed earlier, associations between palatable foods and their cues are strengthened over repeated experience, and these cues come to elicit food-seeking and eating behavior. One way of measuring this stimulus-driven behavior is through PIT methods (Watson & de Wit, 2018). This procedure involves two phases: Pavlovian training, with repeated associations between a stimulus and an outcome (e.g., cue light presentation with food reward), and instrumental training, where a response elicits the same reward (e.g., lever press results in food reward). During the transfer test, responding for the reward is measured with and without presentations of the associated stimulus. Importantly, when the PIT effect is observed even after these outcomes have been devalued, this suggests an S-R habit (Holland, 2004; Watson et al., 2014).

In a variation of a human laboratory test of PIT, experimenters observed that the presence of palatable food-associated cues resulted in insensitivity to devaluation (i.e., facilitated habitual responding) (van Steenbergen, Watson, Wiers, Hommel, & de Wit, 2017; Watson et al., 2014). Specifically in this test, subjects were first trained to respond for candy or popcorn on a variable ratio schedule. Next during Pavlovian training, visual cues were paired with each food (candy or popcorn). After a period of free access to one of the food outcomes (devalued condition), participants' response for food was tested either with or without previously associated cues present. When tested without cues, sensitivity to devaluation was intact (i.e., responding was low for the devalued outcome). However, the introduction of the cue previously paired with the devalued outcome enhanced the responding for the devalued food, suggesting that food cues interfered with goal-directed food-seeking behavior (van Steenbergen et al., 2017; Watson et al., 2014). Importantly, these experiments were performed in healthy control subjects, highlighting the power of food-associated stimuli even in the absence of compulsive eating behavior.

Similarly, palatable food-associated cues and contexts interfere with goal-directed behavior in rats (Kendig, Cheung, Raymond, & Corbit, 2016). Rats were trained to associate a junk food diet (high in fat and sugar) with a specific context and chow food with a separate context. Animals were then trained to instrumentally respond for two preferred food outcomes (grain pellets and sucrose solution) under a random ratio schedule. Devaluation testing was held in contexts previously paired with junk food or chow or in the same context as training (where both outcomes had been available). Before testing, one of the outcomes (grain pellets or sucrose solution, counterbalanced) was devalued through free access until subjects were sated. Animals were then moved to one of the three contexts and were allowed to respond for both outcomes. Animals that were tested in the chow-paired context displayed sensitivity to devaluation, demonstrating intact goal-directed responding. However, when those same rats were tested in the junk food context, responding for devalued outcomes remained high, indicating habitual behavior (Kendig et al., 2016). Therefore, environments previously paired with palatable food can impair goal-directed control over food-seeking behavior.

Our current environment is full of cues that have been associated with high-calorie, palatable food, such as television commercials and fast food logos. Evidence that palatable food-associated cues facilitate the development of habitual automatic actions demonstrates how our current food environment may be encouraging habitual overeating in vulnerable individuals, contributing to the increased prevalence in eating disorders and obesity.

Neurobiological habit systems

Habitual behavior is subserved by the basal ganglia, a group of subcortical nuclei that includes the nucleus accumbens (NAc) and the dorsal striatum (DS), key structures in reinforcement learning (Graybiel, 2008). The development of habits corresponds with a transition from an involvement of the ventral (i.e., NAc) to that of the dorsal components of the striatum in the control of behavior. Early studies investigating the contribution of these regions to goal-directed versus stimulus-driven responding found that the NAc, but not the DS, is critical for the acquisition of instrumental responding, while the DS, but not the NAc, underlies habitual responding (Belin & Everitt, 2008; Corbit, Nie, & Janak, 2012; Yin & Knowlton, 2006). In line with this notion, animals with a prolonged history of exposure to drugs of abuse or palatable food show increased habitual behavior accompanied by increased activation of the dorsolateral striatum (DLS) (Furlong et al., 2014). Furthermore, inactivation of the DLS restores goal-directed responding for drugs (Belin & Everitt, 2008) and food (Furlong et al., 2014).

Alterations to DS function have been observed in BED and/or obesity. In a study using positron emission tomography neuroimaging methods, BMI was found to be associated with lower dopamine 2 receptor (D_2R)—binding potential in ventral striatum and greater D_2R binding in DS. The D_2R-binding potential in the DS was also associated with opportunistic or "habitual" eating, as measured by the "Three-Factor Eating Questionnaire" (Guo, Simmons, Herscovitch, Martin, & Hall, 2014). Therefore, dopamine circuitry alterations in obese individuals may simultaneously increase susceptibility to opportunistic (i.e., habitual) overeating while the rewarding effects of palatable food intake are weakened.

Researchers have also investigated structural differences in brain regions responsible for habitual behavior in individuals with BED (visualized with magnetic resonance imaging techniques). Compared with non-BED controls, individuals with BED had lower gray matter volume in DS and associated prefrontal projection areas (Voon et al., 2015). In this study, the BED subjects also demonstrated increased model-free (i.e., habitual) responding (Voon et al., 2015), therefore linking accelerated habit formation to lower gray matter volumes in areas responsible for the balance between goal-directed and habitual responding.

Striatal projections to and from prefrontal and limbic brain regions also contribute to the development and maintenance of goal-directed versus habitual behavior. Specifically, outside of the striatum, goal-directed behaviors rely on the prelimbic cortex and the orbitofrontal cortex (OFC), while habitual behaviors depend on the infralimbic cortex and the amygdala (Balleine & O'Doherty, 2010; Lingawi & Balleine, 2012; Smith, Virkud, Deisseroth, & Graybiel, 2012; Smith & Graybiel, 2016) (Smith et al., 2012). In neuroimaging studies of obese subjects, deficits in goal-directed reinforcement learning were associated with lower functional coupling of striatal and prefrontal brain regions (Mathar, Neumann, Villringer, & Horstmann, 2017). A separate study observed higher functional connectivity between amygdala and DS in obese versus normal weight controls when viewing palatable food-associated stimuli (Nummenmaa et al., 2012), although this study did not explicitly measure habitual behavior.

Potential interactions of habit with other elements of compulsive eating

Aside from habitual overeating, compulsive eating behavior is also characterized by overeating to alleviate a negative emotional state, as well as overeating despite negative consequences (Moore et al., 2017). Underlying these two elements is the emergence of a negative emotional state as well as dysfunctions in inhibitory control processes, largely subserved by the amygdala and prefrontal cortex (PFC), respectively. There is evidence to suggest that these psychobiological processes and brain regions can, both directly and indirectly, facilitate the development of habitual behavior. Thus, the three elements of compulsive eating behavior have intersecting psychobiological processes and neurobiological substrates and therefore are not mutually exclusive.

In addiction and compulsive eating, the emergence of a negative emotional state occurs after repeated cycles of withdrawal and is driven by decreased reward and increased stress (Koob, 2013). Disorders of compulsive eating are incredibly comorbid with mood and anxiety disorders (Davis et al., 2011; Grilo, White, & Masheb, 2009; Halfon, Larson, & Slusser, 2013), and overeating of highly palatable diets can cause or exacerbate negative emotional states (i.e., increased stress and anxiety) on withdrawal of the diet (Cottone et al., 2009; Parylak, Koob, & Zorrilla, 2011). Eating to relieve a negative emotional state is driven by negative reinforcement, thereby increasing the likelihood of repeating the response (i.e., overeating), and may facilitate habit development. Similar concepts are well-studied in the OCD literature, where habits form through the negatively reinforcing relief of anxiety/stress on completion of a compulsive behavior (Figee et al., 2016; Kashyap et al., 2012). Additionally, as stress itself has been shown to promote habitual behavior in humans (Schwabe & Wolf, 2009), negative emotional states in compulsive eating may similarly bias behavior toward habit systems and away from goal-directed strategies. A major neurobiological substrate of the emergence of negative emotional states in palatable food withdrawal is the amygdala, a brain region important for emotional processing. The amygdala has been shown to influence habitual behaviors through indirect projections to the DLS (Lingawi & Balleine, 2012; Murray et al., 2015). Habitual behavior can be either induced or blocked with intra-amygdala injections of anxiogenic or anxiolytic drugs, respectively (Wingard & Packard, 2008).

Another key element of compulsive eating behavior, overeating despite negative consequences, derives from dysfunctions in prefronto-cortical inhibitory control processes (Moore et al., 2017). Prefrontal interactions with striatal regions are necessary for flexible, goal-directed actions (Killcross & Coutureau, 2003), and addiction has been characterized as a pathology of prefrontal regulation of habit circuitry (Kalivas, 2008). Within prefrontal circuitries, two opposing systems have been postulated—a "GO" and a "STOP" system (Koob & Volkow, 2016). The STOP system (ventromedial PFC) likely serves to inhibit craving in part through modulation of incentive value (Bechara, Damasio, Damasio, & Lee, 1999), while the GO system (dorsolateral PFC, OFC, and anterior cingulate cortex) may drive craving through reengagement of habit systems (Jasinska, Stein, Kaiser, Naumer, & Yalachkov, 2014). Inhibitory control networks in the PFC underlie impulsivity traits (i.e., the tendency to engage in unplanned actions), which have been found to be associated with greater habitual behavior (Dietrich, de Wit, & Horstmann, 2016). Furthermore, impulsivity has been identified as a risk factor for development of binge eating behaviors (Davis et al., 2011; Dawe & Loxton, 2004; Schag, Schonleber, Teufel, Zipfel, & Giel, 2013; Velazquez-Sanchez et al., 2014). Therefore, dysfunctions in prefrontal inhibitory control processes may simultaneously bias away from cognitive and behavioral control and toward stimulus-driven habitual responding.

One other example suggesting an intersection of all three compulsive eating elements is that during stress states, activity in PFC regions is dampened while limbic (i.e., amygdala) areas are hyperresponsive, thereby causing automatic, habitual behaviors to dominate (Li & Sinha, 2008). Therefore, preexisting vulnerabilities in and/or modulation of the amygdala and prefrontal brain regions by repeated palatable food exposure could theoretically facilitate the consolidation of S-R relationships in the striatum that occurs through repeated experience, contributing to the predominance of habitual behavior (Everitt & Robbins, 2016). Further research is needed to disentangle the relative contributions and interactions of overeating to relieve a negative emotional state and overeating despite negative consequences on habitual overeating behavior.

Summary and conclusions

Research into compulsive eating behavior as a construct similar to drug addiction is a new and emerging field. Habitual overeating, similar to habitual drug-taking in addiction, has been postulated as a key element of compulsive eating behavior (Moore et al., 2017). Overall, there is convergent evidence from clinical and preclinical research demonstrating that obese and binge eating subjects display habitual overeating behavior. Studies described here show that humans who are obese, or rats with a history of overeating a palatable diet, display reduced sensitivity to outcome devaluation measured through sensory-specific satiety methods. Additional clinical studies of habitual overeating behavior in subjects with BED and food addiction are warranted to fully elucidate the role of habit in disorders characterized by compulsive eating behavior.

The evidence of habitual overeating presented here highlights multiple behavioral and neurobiological similarities to habitual drug-taking. The demonstration that individuals with BED and methamphetamine addiction share a bias for "model-free" versus "model-based" response strategies supports habitual behavior as a common feature of both disorders (Voon et al., 2015). Similarly, habitual overeating rats and people show similar neuroadaptations to striatal systems associated with habitual behavior. Further inquiry into therapeutics targeted at habit system and strategies for improving control over goal-directed behavior therefore have implications for treating addictive behavior toward both drugs and food.

References

Balleine, B. W., & O'Doherty, J. P. (2010). Human and rodent homologies in action control: Corticostriatal determinants of goal-directed and habitual action. *Neuropsychopharmacology: Official Publication of the American College of Neuropsychopharmacology, 35*(1), 48–69. https://doi.org/10.1038/npp.2009.131.

Bechara, A., Damasio, H., Damasio, A. R., & Lee, G. P. (1999). Different contributions of the human amygdala and ventromedial prefrontal cortex to decision-making. *Journal of Neuroscience: The Official Journal of the Society for Neuroscience, 19*(13), 5473–5481.

Belin, D., Belin-Rauscent, A., Murray, J. E., & Everitt, B. J. (2013). Addiction: Failure of control over maladaptive incentive habits. *23*(4), 564–572. https://doi.org/10.1016/j.conb.2013.01.025.

Belin, D., & Everitt, B. J. (2008). Cocaine seeking habits depend upon dopamine-dependent serial connectivity linking the ventral with the dorsal striatum. *Neuron, 57*(3), 432–441. https://doi.org/10.1016/j.neuron.2007.12.019.

Corbit, L. H., & Janak, P. H. (2016). Habitual alcohol seeking: Neural bases and possible relations to alcohol use disorders. *Alcoholism: Clinical and Experimental Research, 40*(7), 1380–1389. https://doi.org/10.1111/acer.13094.

Corbit, L. H., Nie, H., & Janak, P. H. (2012). Habitual alcohol seeking: Time course and the contribution of subregions of the dorsal striatum. *Biological Psychiatry, 72*(5), 389–395. https://doi.org/10.1016/j.biopsych.2012.02.024.

Cottone, P., Sabino, V., Roberto, M., Bajo, M., Pockros, L., Frihauf, J. B., et al. (2009). CRF system recruitment mediates dark side of compulsive eating. *Proceedings of the National Academy of Sciences of the United States of America, 106*(47), 20016–20020. https://doi.org/10.1073/pnas.0908789106.

Davis, C., Curtis, C., Levitan, R. D., Carter, J. C., Kaplan, A. S., & Kennedy, J. L. (2011). Evidence that 'food addiction' is a valid phenotype of obesity. *Appetite, 57*(3), 711–717. https://doi.org/10.1016/j.appet.2011.08.017.

Dawe, S., & Loxton, N. J. (2004). The role of impulsivity in the development of substance use and eating disorders. *Neuroscience & Biobehavioral Reviews, 28*(3), 343–351. https://doi.org/10.1016/j.neubiorev.2004.03.007.

Daw, N. D., Niv, Y., & Dayan, P. (2005). Uncertainty-based competition between prefrontal and dorsolateral striatal systems for behavioral control. *Nature Neuroscience, 8*(12), 1704–1711. https://doi.org/10.1038/nn1560.

Dickinson, A., & Balleine, B. W. (1993). Actions and responses: The dual psychology of behaviour. In N. Eilan, R. McMarthy, & M. Brewer (Eds.), *Spatial representation: Problems in philosophy and psychology* (pp. 277–293). Oxford: Basil Blackwell Ltd.

Dietrich, A., de Wit, S., & Horstmann, A. (2016). General habit propensity relates to the sensation seeking subdomain of impulsivity but not obesity. *Frontiers in Behavioral Neuroscience, 10*, 213. https://doi.org/10.3389/fnbeh.2016.00213.

Everitt, B. J., & Robbins, T. W. (2005). Neural systems of reinforcement for drug addiction: From actions to habits to compulsion. *Nature Neuroscience, 8*(11), 1481–1489. https://doi.org/10.1038/nn1579.

Everitt, B. J., & Robbins, T. W. (2016). Drug addiction: Updating actions to habits to compulsions ten years on. *Annual Review of Psychology, 67*, 23–50. https://doi.org/10.1146/annurev-psych-122414-033457.

Figee, M., Pattij, T., Willuhn, I., Luigjes, J., van den Brink, W., Goudriaan, A., et al. (2016). Compulsivity in obsessive-compulsive disorder and addictions. *26*(5), 856–868. https://doi.org/10.1016/j.euroneuro.2015.12.003.

Furlong, T. M., Jayaweera, H. K., Balleine, B. W., & Corbit, L. H. (2014). Binge-like consumption of a palatable food accelerates habitual control of behavior and is dependent on activation of the dorsolateral striatum. *Journal of Neuroscience, 34*(14), 5012–5022. https://doi.org/10.1523/JNEUROSCI.3707-13.2014.

Graybiel, A. M. (2008). Habits, rituals, and the evaluative brain. *Annual Review of Neuroscience, 31*, 359–387. https://doi.org/10.1146/annurev.neuro.29.051605.112851.

Grilo, C. M., White, M. A., & Masheb, R. M. (2009). DSM-IV psychiatric disorder comorbidity and its correlates in binge eating disorder. *International Journal of Eating Disorders, 42*(3), 228–234. https://doi.org/10.1002/eat.20599.

Guo, J., Simmons, W. K., Herscovitch, P., Martin, A., & Hall, K. D. (2014). Striatal dopamine D2-like receptor correlation patterns with human obesity and opportunistic eating behavior. *Molecular Psychiatry, 19*(10), 1078–1084. https://doi.org/10.1038/mp.2014.102.

Halfon, N., Larson, K., & Slusser, W. (2013). Associations between obesity and comorbid mental health, developmental, and physical health conditions in a nationally representative sample of US children aged 10 to 17. *Academic Pediatrics, 13*(1), 6–13. https://doi.org/10.1016/j.acap.2012.10.007.

Holland, P. C. (2004). Relations between Pavlovian-instrumental transfer and reinforcer devaluation. *Journal of Experimental Psychology. Animal Behavior Processes, 30*(2), 104–117. https://doi.org/10.1037/0097-7403.30.2.104.

Horstmann, A., Dietrich, A., Mathar, D., Possel, M., Villringer, A., & Neumann, J. (2015). Slave to habit? Obesity is associated with decreased behavioural sensitivity to reward devaluation. *Appetite, 87*, 175–183. https://doi.org/10.1016/j.appet.2014.12.212.

Jasinska, A. J., Stein, E. A., Kaiser, J., Naumer, M. J., & Yalachkov, Y. (2014). Factors modulating neural reactivity to drug cues in addiction: A survey of human neuroimaging studies. *Neuroscience & Biobehavioral Reviews, 38*, 1–16. https://doi.org/10.1016/j.neubiorev.2013.10.013.

de Jong, J. W., Meijboom, K. E., Vanderschuren, L. J., & Adan, R. A. (2013). Low control over palatable food intake in rats is associated with habitual behavior and relapse vulnerability: Individual differences. *PLoS One, 8*(9), e74645. https://doi.org/10.1371/journal.pone.0074645.

Kalivas, P. W. (2008). Addiction as a pathology in prefrontal cortical regulation of corticostriatal habit circuitry. *Neurotoxicity Research, 14*(2–3), 185–189. https://doi.org/10.1007/BF03033809.

Kashyap, H., Fontenelle, L. F., Miguel, E. C., Ferrao, Y. A., Torres, A. R., Shavitt, R. G., et al. (2012). 'Impulsive compulsivity' in obsessive-compulsive disorder: A phenotypic marker of patients with poor clinical outcome. *46*(9), 1146–1152. https://doi.org/10.1016/j.jpsychires.2012.04.022.

Kendig, M. D., Cheung, A. M., Raymond, J. S., & Corbit, L. H. (2016). Contexts paired with junk food impair goal-directed behavior in rats: Implications for decision making in obesogenic environments. *Frontiers in Behavioral Neuroscience, 10*, 216. https://doi.org/10.3389/fnbeh.2016.00216.

Killcross, S., & Coutureau, E. (2003). Coordination of actions and habits in the medial prefrontal cortex of rats. *Cerebral Cortex, 13*(4), 400–408.

Koob, G. F. (2013). Addiction is a reward deficit and stress surfeit disorder. *Frontiers in Psychiatry, 4*, 72. https://doi.org/10.3389/fpsyt.2013.00072.

Koob, G. F., & Volkow, N. D. (2016). Neurobiology of addiction: A neurocircuitry analysis. *Lancet Psychiatry, 3*(8), 760–773.

Lingawi, N. W., & Balleine, B. W. (2012). Amygdala central nucleus interacts with dorsolateral striatum to regulate the acquisition of habits. *Journal of Neuroscience, 32*(3), 1073–1081. https://doi.org/10.1523/JNEUROSCI.4806-11.2012.

Li, C. S., & Sinha, R. (2008). Inhibitory control and emotional stress regulation: Neuroimaging evidence for frontal-limbic dysfunction in psycho-stimulant addiction. *Neuroscience & Biobehavioral Reviews, 32*(3), 581–597. https://doi.org/10.1016/j.neubiorev.2007.10.003.

Mathar, D., Neumann, J., Villringer, A., & Horstmann, A. (2017). Failing to learn from negative prediction errors: Obesity is associated with alterations in a fundamental neural learning mechanism. *Cortex; A Journal Devoted to The Study of the Nervous System and Behavior, 95*, 222–237. https://doi.org/10.1016/j.cortex.2017.08.022.

Moore, C. F., Sabino, V., Koob, G. F., & Cottone, P. (2017). Pathological overeating: Emerging evidence for a compulsivity construct. *Neuropsychopharmacology, 42*(7), 1375–1389. https://doi.org/10.1038/npp.2016.269.

Murray, J. E., Belin-Rauscent, A., Simon, M., Giuliano, C., Benoit-Marand, M., Everitt, B. J., et al. (2015). Basolateral and central amygdala differentially recruit and maintain dorsolateral striatum-dependent cocaine-seeking habits. *Nature Communications, 6*, 10088. https://doi.org/10.1038/ncomms10088.

Nummenmaa, L., Hirvonen, J., Hannukainen, J. C., Immonen, H., Lindroos, M. M., Salminen, P., et al. (2012). Dorsal striatum and its limbic connectivity mediate abnormal anticipatory reward processing in obesity. *PLoS One, 7*(2), e31089. https://doi.org/10.1371/journal.pone.0031089.

Ostlund, S. B., & Balleine, B. W. (2008). On habits and addiction: An associative analysis of compulsive drug seeking. *Drug Discovery Today: Disease Models, 5*(4), 235–245. https://doi.org/10.1016/j.ddmod.2009.07.004.

Parylak, S. L., Koob, G. F., & Zorrilla, E. P. (2011). The dark side of food addiction. *Physiology & Behavior, 104*(1), 149–156. https://doi.org/10.1016/j.physbeh.2011.04.063.

Reichelt, A. C., Morris, M. J., & Westbrook, R. F. (2014). Cafeteria diet impairs expression of sensory-specific satiety and stimulus-outcome learning. *Frontiers in Psychology, 5*, 852. https://doi.org/10.3389/fpsyg.2014.00852.

Schag, K., Schonleber, J., Teufel, M., Zipfel, S., & Giel, K. E. (2013). Food-related impulsivity in obesity and binge eating disorder–a systematic review. *Obesity Reviews, 14*(6), 477–495. https://doi.org/10.1111/obr.12017.

Schwabe, L., & Wolf, O. T. (2009). Stress prompts habit behavior in humans. *Journal of Neuroscience: The Official Journal of the Society for Neuroscience, 29*(22), 7191–7198. https://doi.org/10.1523/JNEUROSCI.0979-09.2009.

Smith, K. S., & Graybiel, A. M. (2016). Habit formation. *Dialogues in Clinical Neuroscience, 18*(1), 33–43.

Smith, K. S., Virkud, A., Deisseroth, K., & Graybiel, A. M. (2012). Reversible online control of habitual behavior by optogenetic perturbation of medial prefrontal cortex. *Proceedings of the National Academy of Sciences of the United States of America, 109*(46), 18932–18937. https://doi.org/10.1073/pnas.1216264109.

van Steenbergen, H., Watson, P., Wiers, R. W., Hommel, B., & de Wit, S. (2017). Dissociable corticostriatal circuits underlie goal-directed vs. cue-elicited habitual food seeking after satiation: Evidence from a multimodal MRI study. *46*(2), 1815–1827. https://doi.org/10.1111/ejn.13586.

Tantot, F., Parkes, S. L., Marchand, A. R., Boitard, C., Naneix, F., Laye, S., et al. (2017). The effect of high-fat diet consumption on appetitive instrumental behavior in rats. *Appetite, 108*, 203–211. https://doi.org/10.1016/j.appet.2016.10.001.

Vandaele, Y., & Janak, P. H. (2018). Defining the place of habit in substance use disorders. *Progress in Neuro-Psychopharmacology & Biological Psychiatry, 87*(Pt A), 22–32. https://doi.org/10.1016/j.pnpbp.2017.06.029.

Velazquez-Sanchez, C., Ferragud, A., Moore, C. F., Everitt, B. J., Sabino, V., & Cottone, P. (2014). High trait impulsivity predicts food addiction-like behavior in the rat. *Neuropsychopharmacology, 39*(10), 2463–2472. https://doi.org/10.1038/npp.2014.98.

Voon, V., Derbyshire, K., Ruck, C., Irvine, M. A., Worbe, Y., Enander, J., et al. (2015). Disorders of compulsivity: A common bias towards learning habits. *Molecular Psychiatry, 20*(3), 345–352. https://doi.org/10.1038/mp.2014.44.

Watson, P., & de Wit, S. (2018). Current limits of experimental research into habits and future directions. *20*, 33–39.

Watson, P., Wiers, R. W., Hommel, B., & de Wit, S. (2014). Working for food you don't desire. Cues interfere with goal-directed food-seeking. *Appetite, 79*, 139–148. https://doi.org/10.1016/j.appet.2014.04.005.

Wingard, J. C., & Packard, M. G. (2008). The amygdala and emotional modulation of competition between cognitive and habit memory. *Behavioural Brain Research, 193*(1), 126–131. https://doi.org/10.1016/j.bbr.2008.05.002.

Yin, H. H., & Knowlton, B. J. (2006). The role of the basal ganglia in habit formation. *Nature Reviews Neuroscience, 7*(6), 464–476. https://doi.org/10.1038/nrn1919.

CHAPTER

Reward deficits in compulsive eating

5

Paul J. Kenny

Nash Family Department of Neuroscience, Icahn School of Medicine at Mount Sinai, New York, NY, United States

The burden of disease and negative economic impact of obesity on society is considerable. According to the Center for Disease Control, obesity-related health care expenses in the United States were approximately $213 billion between 1998 and 2000 and $316 billion between 2005 and 2010 (Biener, Cawley, & Meyerhoefer, 2017). Over 300,000 deaths in the United States each year can be attributed to overweight and obesity-related diseases (Allison et al., 1999), with obesity the second leading cause of preventable death behind tobacco use. Thus, an understanding of the neurobiological mechanisms responsible for overeating and obesity will likely highlight novel strategies for the treatment of this devastating disorder. Reward pathways in the brain play a critical role in homeostatic regulation of feeding behavior. Hormonal signals of negative energy balance can increase the sensitivity of reward circuitries in the brain, which can increase the value of food. However, similar to drugs of abuse (Bardo & Bevins, 2000; Tzschentke, 1998), energy-dense palatable foods activate brain reward systems, even when energy requirements have been met, and this action contributes to their reinforcing effects (Grigson, 2002). Overconsumption of palatable food can trigger adaptive responses in brain reward systems in a manner similar to the actions of drugs of abuse. Such adaptive responses are hypothesized to provide a critical substrate for negative reinforcement processes that precipitate the emergence of compulsive overeating and obesity. In this chapter, I will provide an overview of the role for brain reward systems in regulating food intake. Then, I will describe procedures used to induce obesity in rodents and the use of such procedures to investigate obesity-related adaptive responses in brain reward systems. I will then compare the actions of drugs of abuse such as cocaine and morphine on brain reward systems with the actions of palatable high-energy food. As will be seen, overeating of palatable food can induce strikingly similar deficits in brain reward function to those seen in animals with a history of excessive consumption of drugs of abuse. It is hypothesized that reward dysfunction associated with weight gain contributes to the emergence of compulsive overeating that precipitates and sustains obesity.

Compulsive Eating Behavior and Food Addiction. https://doi.org/10.1016/B978-0-12-816207-1.00005-6
Copyright © 2019 Elsevier Inc. All rights reserved.

Food and brain reward systems

Palatable foods can enhance mood in humans (Davis, Strachan, & Berkson, 2004) and establish a conditioned place preference in rats (Delamater, Sclafani, & Bodnar, 2000; Reynolds & Berridge, 2002; Sclafani, Bodnar, & Delamater, 1998) and mice (Imaizumi et al., 2001; Takeda et al., 2001), which reflects the conditioned rewarding effects of palatable foods. Dopamine transmission in the mesoaccumbens pathway is thought to play an important role in reward processing, and the nucleus accumbens (NAcc) is considered a core component of brain reward systems (Wise, 2005, 2006). Accordingly, palatable sucrose solutions have been shown to increase dopamine-mediated transmission in the NAcc similar to drugs of abuse, albeit with a relatively lower magnitude of effect (Hajnal, Smith, & Norgren, 2004; Hernandez and Hoebel, 1988a, 1988b; Norgren, Hajnal, & Mungarndee, 2006; Rada, Avena, & Hoebel, 2005). Moreover, pharmacological antagonism of dopamine receptors or lesions of the mesoaccumbens dopamine system decreased free feeding and operant responses for food rewards by rats (Wise & Rompre, 1989). Tyrosine hydroxylase–deficient mice, which are unable to synthesize dopamine, gradually become aphagic and will die of starvation without daily treatment with L-3,4-dihydroxyphenylalanine (L-DOPA) to transiently restore brain dopamine levels (Szczypka et al., 1999a; Zhou & Palmiter, 1995). Spatially restricted restoration of dopamine in the NAcc selectively restored the ability of these dopamine-deficient mice to choose a palatable food over a nonpalatable food (Szczypka et al., 1999b). In addition to dopamine-mediated transmission, endogenous opiates that are known to enhance the activity of brain reward systems also play an important role in the reinforcing effects of food. Indeed, palatable food can stimulate endogenous opioid transmission at opioid receptors in the mesoaccumbens system (Tanda & Di Chiara, 1998). Administration of opiates directly into the ventral tegmental area (VTA) or NAcc stimulates food intake (Kelley et al., 2000), particularly energy-dense palatable food items that are preferred by the animals (Zhang, Gosnell, & Kelley, 1998). Conversely, the opioid receptor antagonist naloxone, which can induce profound deficits in the activity of brain reward systems (Hawkins & Stein, 1991), inhibits feeding behaviors in mammals (Brands et al., 1979; Brown & Holtzman, 1979; Reid, 1985; Trenchard & Silverstone, 1983; Wise & Raptis, 1986), particularly the consumption of palatable foods (Barbano & Cador, 2006; Levine et al., 1995). Human imaging studies have shown that appetizing foods activate brain regions known to play a role in reward processing, such as the orbitofrontal cortex (Simmons, Martin, & Barsalou, 2005) and mesocorticolimbic areas (Small et al, 2001, 2003), and have revealed considerable overlap in the areas activated by appetizing foods and drugs of abuse (Wang et al, 2004a, 2004b). It is also interesting to note that food restriction can increase the rewarding effects of drugs of abuse such as cocaine or opioids (Cabeza de Vaca & Carr, 1998; Carroll, France, & Meisch, 1979), supporting the notion that common brain circuitries regulate the reinforcing effects of palatable foods and drugs of abuse.

Emerging evidence suggests that compensatory adaptations may occur in endogenous opioid systems in the brain in response to their overstimulation by excessive consumption of palatable foods in a manner analogous to the development of opioid dependence (Colantuoni et al., 2002). Furthermore, such adaptive responses may play a role in the development of a drug-like dependence state on palatable foods. For example, Le Magnen examined the effects of naloxone in rats fed a highly palatable "cafeteria" diet, consisting of high-fat energy-dense food products commercially available at most cafeterias and vending machines for human consumption (Le Magnen et al., 1990). It was found that rats displayed opioid withdrawal-like physical symptoms, such as body shakes on administration of naloxone (Le Magnen et al., 1990). Hoebel et al. demonstrated that rats with extended daily access to a palatable sucrose solution displayed a spontaneous withdrawal syndrome after 24 h of enforced fasting consisting of physical symptoms very reminiscent of opiate withdrawal, including teeth chatter, tremor, and shakes (Colantuoni et al., 2002). Furthermore, administration of naloxone at doses that had no obvious behavioral effects in control rats could induce opioid withdrawal-like increases in anxiety-related behaviors in rats with extended daily access to sucrose solutions (Colantuoni et al., 2002). Furthermore, low naloxone doses could also induce marked decreases in accumbal dopamine overflow in rats with extended daily access to a sucrose solution (Colantuoni et al., 2002). Conversely, consumption of sucrose solutions has been shown to attenuate the expression of naloxone-precipitated morphine withdrawal (Jain, Mukherjee, & Singh, 2004). Together, these data suggest that neurotransmitter systems known to play a key role in controlling the activity of brain reward systems play a critical role in controlling food intake by regulating the motivational value of food items.

Overeating and diet-induced obesity in rodents: the "cafeteria" diet

Classically, overeating and obesity were often induced in rats by inducing lesions in the ventromedial hypothalamus (VMH) region of the brain, with VMH-lesioned rats being the most studied animal model of obesity (King, 2006; Kishi & Elmquist, 2005). Lesions to the VMH result initially in a very pronounced period of hyperphagia and rapid weight gain. This initial phase is followed by a more static phase in which food intake decreases to a level that maintains the newly established increases in body weight (Albert & Storlien, 1969; Cox, Kakolewski, & Valenstein, 1969). The relevance of centrally lesioned animals to the study of human obesity is questionable. Rodents with naturally occurring mutations have also been used extensively to study weight gain and obesity. In particular, mice with naturally occurring mutations in the gene coding for leptin (*ob/ob* mouse) or rodents with naturally occurring mutations in the machineries required to response to leptin (Zucker *fa/fa* rat, *cp/cp* rat, and *db/db* mouse) have been utilized extensively; see Zhang & Scarpace, 2006). Most recently, strains of mutant mice with experimenter-induced null mutations in components of feeding-related pathways have been developed (for reviews see Carroll, Voisey, & van Daal, 2004; Kishi & Elmquist, 2005). Rodents with spontaneous or experimenter-induced mutations in components of feeding and energy balance regulatory systems have shed important light on the mechanisms by which feeding behavior is regulated and may become dysregulated resulting in overeating and obesity (Carroll et al., 2004).

Perhaps more relevant to human obesity research, however, is the observation that laboratory rodents with access to palatable, calorically dense food products rapidly gain weight to the point of obesity (Sclafani, 1987, 1989; Sclafani & Springer, 1976). This diet-induced obesity results from increased caloric intake from energy-laden foods (positive energy balance), resulting in an increased proportion of energy stored as adipose tissues. Indeed, there is considerable evidence that hyperphagia associated with access to a palatable high-energy diet is directly responsible for the development and maintenance of an obese state in humans (Chang et al., 1990; Levin et al., 1987; but see Levin, Triscari, & Sullivan, 1986). Of the various procedures available to promote diet-induced obesity in rats, the cafeteria (or supermarket) diet (Sclafani & Springer, 1976) is a particularly attractive research tool (Rothwell & Stock, 1979; Sclafani & Springer, 1976). In the cafeteria diet model of diet-induced obesity, rats are permitted daily access to highly palatable food products commercially available at most cafeterias and vending machines for human consumption that are high in fat and/or carbohydrate content, thus recapitulating in laboratory animals many aspects of diet-induced obesity in humans. Repeated daily exposure to this diet results in the development of hyperphagia and obesity in rats (Rothwell & Stock, 1979; Sclafani & Springer, 1976). The cafeteria diet procedure for inducing overeating and obesity meets many criteria for an animal model with important utility for investigating the underlying neurobiology of diet-induced obesity. The cafeteria diet of overeating and obesity has excellent "*face validity*," defined as the phenomenological similarity between the behavior and pathophysiology exhibited by the animal model and the specific symptoms of the human condition (Geyer et al., 1995; Rothwell & Stock, 1982; Sclafani & Springer, 1976). In addition, the cafeteria diet demonstrates excellent "*construct validity*," defined as the accuracy with which the test or model measures that which it is intended to measure (i.e., overeating and obesity) (Balada et al., 1997; Geyer et al., 1995; Kretschmer et al., 2005). Finally, the cafeteria diet demonstrates established "*predictive validity*," defined as the extent to which the model allows one to make predictions and extrapolations about the human phenomenon based on the performance of the model (Fantino, Faion, & Rolland, 1986; Geyer et al., 1995; Miranda et al., 1988). Importantly, the food products that will be used as part of the cafeteria diet are the same palatable energy-dense products that are widely consumed by humans.

Assessing brain reward function: intracranial self-stimulation thresholds

Intracranial self-stimulation (ICSS) of certain brain areas, such as the lateral hypothalamus or VTA, is powerfully rewarding for humans (Bishop, Elder, & Heath, 1963) and laboratory animals such as dogs (Sadowski, 1972), cats (Wilkinson & Peele, 1963), rats (Olds & Milner, 1954), and mice (Cazala, Cazals, & Cardo, 1974). The powerfully rewarding properties of ICSS are reflected in the fact that subjects will readily learn to self-administer brief intracranial electrical pulses and endure painful stimuli such as electrical stock to the feet to obtain ICSS (Olds, 1958a). This has led to the hypothesis that ICSS may directly activate brain circuitries that regulate the rewarding effects of more conventional reinforcers such as food, water, and/or sex. The minimal stimulation intensity that maintains ICSS behavior is termed the reward threshold (Kornetsky & Esposito, 1979; Markou & Koob, 1992). Rats demonstrate little satiation to ICSS and no change in reward thresholds during extended or repeated testing sessions (Olds, 1958a, 1958b). Furthermore, repeated exposure to ICSS over days does not alter the baseline activity of brain reward systems (Annau, Heffner, & Koob, 1974). As such, the ICSS procedure has proved to be useful for monitoring the effects of drugs of abuse on brain reward systems across extended periods of time. Indeed, one of the most consistent findings with addictive drugs is their ability to lower ICSS thresholds. Experimenter-administered cocaine (Kenny et al., 2003), heroin (Kenny et al., 2006), nicotine (Kenny & Markou, 2006), amphetamine (Cazala, 1976; Olds, 1972; Phillips, Brooke, & Fibiger, 1975), and alcohol (De Witte & Bada, 1983) injections have all been shown to transiently lower ICSS reward thresholds in rats. This transient lowering of reward thresholds has been interpreted as a short-lasting increase in sensitivity to the rewarding effects of ICSS that arises through drug-induced amplification of reward signals in the brain. Conversely, withdrawal from drugs of abuse after prolonged exposure transiently elevates ICSS thresholds. For example, withdrawal from chronic experimenter-administered morphine (Schulteis et al., 1994), amphetamine (Lin, Koob, & Markou, 2000), and cocaine (Baldo, Koob, & Markou, 1999) elevates ICSS thresholds. This withdrawal-associated elevation of reward thresholds has been interpreted as a compensatory decrease in the basal activity of brain reward systems in response to their overstimulation by chronic drug exposure. Below, I will provide a summary of the actions of prolonged consumption of cocaine or heroin on ICSS thresholds in laboratory rats. These actions can then form a framework against which the actions of palatable food consumption and weight gain can be compared.

Effects of cocaine on brain reward function

To better identify adaptations in reward systems likely to occur in the brains of human cocaine addicts, Koob et al. completed a series of studies in which the effects of volitionally consumed cocaine injections on ICSS thresholds were examined. In these studies, rats were prepared ICSS electrodes positioned in the lateral hypothalamus and were permitted to intravenously self-administer a different number of cocaine infusions. It was found that self-administration of a small number of cocaine injections (10 or 20 cocaine injections; 0.25 mg per injection, equivalent to 4.94 ± 0.23 and 9.88 ± 0.46 mg/kg, self-administered over 40 ± 6.9 and 99 ± 11.9 min, respectively) lowered reward thresholds (i.e., increased reward) when assessed 15 min after cocaine was consumed (Kenny et al., 2003), consistent with cocaine-induced activation of brain reward systems. Importantly, no effects on ICSS thresholds were observed 2, 24, or 48 h after administration (Kenny et al., 2003), demonstrating that consumption of relatively low amounts of cocaine doses does not induce long-term alterations in the basal activity of brain reward systems. Self-administration of 40 cocaine injections (19.64 ± 0.94 mg/kg; self-administered over 185 ± 10.9 min) also transiently lowered reward thresholds 15 min after consumption (Kenny et al., 2003) (see preliminary study 1). However, significant elevations of reward thresholds were observed at 2 and 24 h after administration, indicating long-lasting withdrawal-like reward deficits. Finally, 80 cocaine self-injections (39.53 ± 1.84 mg/kg, self-administered over 376 ± 19.9 min) significantly elevated ICSS thresholds 2 and 48 h after self-administration and tended to elevate thresholds even 15 min after self-administration (Kenny et al., 2003). Thus, consumption of relatively large quantities of cocaine can induce long-lasting compensatory decreases in the activity of brain reward systems. These data suggest that two distinct motivational drivers contribute to cocaine self-administration behavior: First, consumption of low to moderate amounts of the drug induces robust activation of brain reward systems, likely an important source of positive reinforcement. In contrast, consumption of large quantities of cocaine induces profound decreases in the reward sensitivity, alleviation of which is likely to provide a potent source of negative reinforcement (Kenny et al., 2003). It is important to note that extended but not restricted daily access to cocaine precipitates the emergence of compulsive-like cocaine consumption, characterized by escalating amounts of cocaine use (Ahmed et al., 2002) and the emergence of cocaine use that becomes progressively more resistant to punishment-induced suppressing of intake (Vanderschuren & Everitt, 2004). Based on these findings, Koob et al. have hypothesized that cocaine-induced reward dysfunction provides a critical source of negative reinforcement that precipitates the emergence of compulsive use of the drug (Koob & Le Moal, 2005).

Effects of heroin on brain reward function

The effects of heroin self-administration on brain reward functioning have also been examined in laboratory rats. Specifically, the effects of restricted (1 h) or unlimited (23 h) daily access to heroin infusions on ICSS reward thresholds were characterized (Kenny et al., 2006). In the 1-h rats, baseline ICSS thresholds assessed before each daily heroin self-administration session remained stable and unaltered across days, whereas 1-h heroin access significantly lowered ICSS thresholds when assessed immediately after each self-administration session (Kenny et al., 2006). This lowering of reward thresholds observed in 1-h rats reflects transiently increased reward sensitivity through heroin-induced amplification of reward signals in the brain (Kenny et al., 2006). This limited daily access to heroin did not induce persistent alterations in the baseline sensitivity of brain reward systems, reflected in the fact that baseline reward thresholds assessed immediately before each 1-h self-administration session remained stable and unaltered over the duration of the experiment. In rats with extended (23 h) daily access to heroin, however, the large amounts of heroin consumed by these rats gradually escalated across days of access and associated with a progressive elevation of reward thresholds (Kenny et al., 2006). Thus, excessive heroin consumption can induce compensatory decreases in the basal sensitivity of brain reward systems, and this evolving reward deficit was similarly associated with compulsive-like escalation of daily heroin use (Kenny et al., 2006). These actions of heroin on brain reward function are similar to the effects of cocaine described above and suggest that drug-induced deficits in brain reward function can propel the emergence of compulsive drug use (Ahmed et al., 2002).

Effects of hunger on brain reward function

It is known that hunger is associated with hyperactivity in brain reward systems, reflected by lowering of ICSS thresholds, and this action is blocked by naloxone and other opioid receptor antagonists (Carr & Papadouka, 1994; Carr & Simon, 1984). This suggests that negative energy balance increases the function of endogenous opiate systems in food-related reward circuits, likely to increase the motivational value of food during periods of hunger, which in turn enhances the sensitivity of brain reward systems. Consistent with this hypothesis, food-restricted rats are more sensitive to the rewarding effects of exogenously administered opioids (Berman, Devi, & Carr, 1994; Carr, 1996; Carr, Kim, & Cabeza de Vaca, 2000; Carr & Wolinsky, 1993; Wolinsky et al., 1994) and to other classes of addictive drugs such as cocaine, amphetamine, and alcohol (Carr, 2002; Carr et al., 2000). Lesions of the VMH, which can induce overeating and weight gain, also lower ICSS thresholds in a manner similar to food restriction (Hernandez & Hoebel, 1989). VMH lesion-induced lowering of ICSS thresholds is not abolished until rats become obese and food intake is normalized (Hernandez & Hoebel, 1989). Thus, reward hypersensitivity in hungry rats is likely to increase the appetitive and incentive value of food, driven at least in part by increases in endogenous opioid transmission in food-related reward circuits. In addition to supporting ICSS behavior, electrical stimulation of the lateral hypothalamic during ICSS can elicit feeding behavior (Herberg & Blundell, 1967; Margules & Olds, 1962). Based on these observations it has been hypothesized that the rewarding properties of lateral hypothalamus self-stimulation are related to activation of reward pathways whose cognate function is regulating the appetitive and incentive properties of food (Margules & Olds, 1962). Indeed, it has been shown that lateral hypothalamus ICSS is inhibited by prior feeding of animals to satiety (Wilkinson & Peele, 1962), overfeeding of rats through intragastric feeding tube (Hoebel & Teitelbaum, 1962), and gastric distention or intravenous glucagon infusion that mimics postprandial satiety (Hoebel, 1969; Hoebel & Balagura, 1967; Mount & Hoebel, 1967). Furthermore, forced feeding or development of obesity can switch the motivational valence of lateral hypothalamus ICSS from positive to negative (Hoebel & Thompson, 1969). Specifically, rats that previously responded vigorously for rewarding lateral hypothalamus ICSS instead responded as if the stimulation were aversive after forced feeding or after development of obesity (Hoebel & Thompson, 1969). Shizgal et al. demonstrated that leptin, a hormone secreted by adipocytes that suppresses food intake and promotes weight loss, reversed the lowering of ICSS thresholds induced by chronic food restriction (Fulton, Woodside, & Shizgal, 2000). This suggests that hunger- and satiety-related hormones regulate food intake, at least in part, by controlling the sensitivity of circuitries in the lateral hypothalamus that regulate the rewarding properties of food. Together, these observations suggest that hunger and satiety profoundly modulate the sensitivity of reward circuitries, including those that originate in the lateral hypothalamus. Hence, it is possible that drugs of abuse such as cocaine and heroin modulate brain reward function by altering the activities of reward circuitries whose endogenous function is, to some degree at least, concerned with controlling the incentive and motivational properties of food.

Effects of weight gain on brain reward function

As noted above, cocaine, heroin, and other drugs of abuse can transiently lower ICSS thresholds in a manner similar to hunger states. Conversely, prolonged consumption of these drugs can markedly elevate ICSS thresholds, reflecting profound deficits in brain reward systems. Such reward deficits are hypothesized to serve as an important substrate for negative reinforcement processes that precipitate the emergence of compulsive drug use (Koob & Le Moal, 2005). Hence, an important question is whether overconsumption of palatable energy-dense food can similarly remodel brain reward systems such that weight gain and obesity are associated with deficits in brain reward function. Interesting in this regard is the observation that obese woman show higher levels of anhedonia, reflecting deficits in brain reward function, than those with lower body weights (Davis et al., 2004).

Stice et al. have shown that the degree to which the caudate region of the striatum was activated in response to palatable food was inversely related to body weight in obese individuals. Furthermore, women who gained weight over a 6-month period had a marked decrease in striatal activity in response to palatable food when compared with women who did not gain weight (Stice et al., 2010). These observations are consistent with the possibility that increases in body weight are associated with deficits in the activity and/or sensitivity of brain circuitries involved in processing the rewarding properties of food.

Brain circuitries that regulate hedonic eating

Our laboratory has utilized the ICSS procedure to assess brain reward activity in rats with limited (1 h) or extended (~18–23 h) daily access to a cafeteria-style diet that included cheesecake, bacon, sausage, and other appetizing food (Johnson & Kenny, 2010). We found that rats with extended but not with restricted access to the cafeteria diet gained significant amounts of body weight over a 40-day period and that weight gain in these animals was associated with a progressively worsening brain reward deficit, reflected by progressively elevating ICSS thresholds (Johnson & Kenny, 2010). This finding suggests that the development of diet-induced obesity is associated with brain reward deficits similar to those seen in rats extended daily access to cocaine or heroin infusions (Johnson & Kenny, 2010). These obesity-associated reward deficits likely reflect similar counteradaptive responses in food reward circuitries seen in animals with extended access to cocaine or heroin, which serve to oppose the overstimulation of these circuits by palatable food or drugs of abuse (Johnson & Kenny, 2010). Accumulating evidence suggests that drug-induced reward dysfunction contributes to the emergence of compulsive drug-taking by providing a new source of motivation (negative reinforcement) to consume the drug to alleviate this persistent state of diminished reward (Ahmed et al., 2002; Ahmed & Koob, 2005; Koob & Le Moal, 2008). Therefore, it is possible that reward deficits in obese animals (and humans) may precipitate the emergence of compulsive eating (Geiger et al., 2009; Johnson & Kenny, 2010; Stice et al., 2008; Wang et al., 2001).

Striatal D2 dopamine receptor signaling and brain reward deficits in obesity

The mechanism by which weight gain induces deficits in brain reward function is unclear, but it is likely to involve abnormalities in dopamine D2 receptor signaling in the striatum. Fasted individuals permitted to consume their favorite meal to satiety had lower levels of binding of the dopamine D2 receptor antagonist raclopride in the striatum (Small et al., 2003). Obese individuals show lower levels of striatal dopamine D2 receptor availability compared with lean controls (Barnard et al., 2009; Stice et al., 2008; Wang et al., 2001), and weight loss in obese patients increases dopamine D2 receptor density (Wang et al., 2008). These observations suggest that palatable food consumption may decrease dopamine D2 receptor signaling in the striatum and that weight gain may induce more persistent deficits in D2 signaling. Based on these observations, our laboratory investigated whether disruption of striatal D2 receptor transmission modulates the emergence of reward deficits in rats during the development of obesity. Specifically, we used viral-mediated RNA interference to knockdown striatal D2 receptor levels and assessed ICSS thresholds in rats with extended daily access to a cafeteria-style diet (Johnson & Kenny, 2010). We found that striatal D2 receptor knockdown markedly accelerated the emergence of reward deficits in rats with extended access to the cafeteria diet (Johnson & Kenny, 2010). Interestingly, knockdown of striatal D2 receptors did not alter ICSS thresholds in rats with access to chow only, suggesting that deficits in striatal D2 receptor signaling alone is unlikely to explain the emergence of elevated ICSS thresholds in rats as they gain weight (Johnson & Kenny, 2010). Instead, some other factor, likely a circulating hormone involved in transmitting adiposity state to brain reward circuits, interacts with D2 receptor signaling as animals, and presumably humans, gain weight to precipitate deficits in brain reward function (Johnson & Kenny, 2010).

Striatal D2 dopamine receptor signaling and compulsive eating despite negative consequences in obesity

Obesity often features overeating despite a stated desire to limit food intake, many failed efforts to do so, and awareness of the social, health, and other negative consequences for not doing so (Booth et al., 2008; Delin et al., 1997; Puhl et al., 2008). Obese patients will even undergo potentially dangerous bariatric (gastric bypass) surgery to control their weight (Yurcisin, Gaddor, & DeMaria, 2009), yet will often relapse to overeating (Kalarchian et al., 2002; Saunders, 2001). The conspicuous failure to moderate consumption despite an understanding of future deleterious consequences is one hallmark of substance-use disorders and other compulsive disorders. As noted above, deficits in brain reward function induced by overconsumption of drugs of abuse is hypothesized to drive the development of compulsive drug use by serving as a source of negative reinforcement that perpetuates continued drug use. Based on these observations, we hypothesized that brain reward deficits in obese animals may precipitate the emergence of compulsive eating, as measured by persistent consumption in laboratory animals despite negative consequences. Therefore, our laboratory investigated whether obese rats will consume palatable food in a compulsive-liked manner and if striatal dopamine D2 receptor signaling plays a role in this process (Johnson & Kenny, 2010).

We found that overweight rats with a history of extended access to a cafeteria diet persistent in their consumption of palatable food items even in the presence of a noxious stimulus (cue light) that predicted the imminent delivery of a noxious footshock (Johnson & Kenny, 2010). By contrast, palatable food intake was markedly decreased in lean animals with a limited history of consuming the palatable diet by the same noxious stimulus. Strikingly, striatal dopamine D2 receptor knockdown accelerated the emergence of compulsive-like consumption of palatable food that is resistant to suppression by a noxious stimulus.

Summary

Together, these findings suggest that overconsumption of palatable energy-dense food consumption can precipitate the emergence of marked deficits in brain reward function similar to those seen in animals with a history of extended access to cocaine or heroin. Furthermore, diet-induced deficits in brain reward function in obese rats were associated with compulsive-like consumption of palatable food, reflected by food intake that persisted despite the presence of cues in the environment associated with imminent negative consequences (aversive footshocks). Finally, deficient striatal dopamine D2 receptor signaling is detected in obese rats, and knockdown of striatal D2 receptors accelerates the emergence of diet-associated reward deficits and the emergence of compulsive-like eating. Overall, these findings suggest that obesity and drug addiction share common reward mechanisms that may contribute to compulsive consummatory behaviors.

References

Ahmed, S. H., et al. (2002). Neurobiological evidence for hedonic allostasis associated with escalating cocaine use. *Nature Neuroscience, 5*(7), 625–626.

Ahmed, S. H., & Koob, G. F. (2005). Transition to drug addiction: A negative reinforcement model based on an allostatic decrease in reward function. *Psychopharmacology, 180*(3), 473–490.

Albert, D. J., & Storlien, L. H. (1969). Hyperphagia in rats with cuts between the ventromedial and lateral hypothalamus. *Science, 165*(893), 599–600.

Allison, D. B., et al. (1999). Annual deaths attributable to obesity in the United States. *Jama, 282*(16), 1530–1538.

Annau, Z., Heffner, R., & Koob, G. F. (1974). Electrical self-stimulation of single and multiple loci: Long term observations. *Physiology & Behavior, 13*(2), 281–290.

Balada, F., et al. (1997). Effect of the slimming agent oleoyl-estrone in liposomes on the body weight of Zucker obese rats. *International Journal of Obesity and Related Metabolic Disorders, 21*(9), 789–795.

Baldo, B. A., Koob, G. F., & Markou, A. (1999). Role of adenosine A2 receptors in brain stimulation reward under baseline conditions and during cocaine withdrawal in rats. *Journal of Neuroscience, 19*(24), 11017–11026.

Barbano, M. F., & Cador, M. (2006). Differential regulation of the consummatory, motivational and anticipatory aspects of feeding behavior by dopaminergic and opioidergic drugs. *Neuropsychopharmacology, 31*(7), 1371–1381.

Bardo, M. T., & Bevins, R. A. (2000). Conditioned place preference: What does it add to our preclinical understanding of drug reward? *Psychopharmacology, 153*(1), 31–43.

Barnard, N. D., et al. (2009). D2 dopamine receptor Taq1A polymorphism, body weight, and dietary intake in type 2 diabetes. *Nutrition, 25*(1), 58–65.

Berman, Y., Devi, L., & Carr, K. D. (1994). Effects of chronic food restriction on prodynorphin-derived peptides in rat brain regions. *Brain Research, 664*(1–2), 49–53.

Biener, A., Cawley, J., & Meyerhoefer, C. (2017). The high and rising costs of obesity to the US health care system. *Journal of General Internal Medicine, 32*, 6–8.

Bishop, M. P., Elder, S. T., & Heath, R. G. (1963). Intracranial self-stimulation in man. *Science, 140*, 394–396.

Booth, M. L., et al. (2008). Perceptions of adolescents on overweight and obesity: The weight of opinion study. *Journal of Paediatrics and Child Health, 44*(5), 248–252.

Brands, B., et al. (1979). Suppression of food intake and body weight gain by naloxone in rats. *Life Sciences, 24*(19), 1773–1778.

Brown, D. R., & Holtzman, S. G. (1979). Suppression of deprivation-induced food and water intake in rats and mice by naloxone. *Pharmacology Biochemistry and Behavior, 11*(5), 567–573.

Cabeza de Vaca, S., & Carr, K. D. (1998). Food restriction enhances the central rewarding effect of abused drugs. *Journal of Neuroscience, 18*(18), 7502–7510.

Carr, K. D. (1996). Feeding, drug abuse, and the sensitization of reward by metabolic need. *Neurochemical Research, 21*(11), 1455–1467.

Carr, K. D. (2002). Augmentation of drug reward by chronic food restriction: Behavioral evidence and underlying mechanisms. *Physiology & Behavior, 76*(3), 353–364.

Carr, K. D., Kim, G. Y., & Cabeza de Vaca, S. (2000). Chronic food restriction in rats augments the central rewarding effect of cocaine and the delta1 opioid agonist, DPDPE, but not the delta2 agonist, deltorphin-II. *Psychopharmacology, 152*(2), 200–207.

Carroll, M. E., France, C. P., & Meisch, R. A. (1979). Food deprivation increases oral and intravenous drug intake in rats. *Science, 205*(4403), 319–321.

Carroll, L., Voisey, J., & van Daal, A. (2004). Mouse models of obesity. *Clinics in Dermatology, 22*(4), 345–349.

Carr, K. D., & Papadouka, V. (1994). The role of multiple opioid receptors in the potentiation of reward by food restriction. *Brain Research, 639*(2), 253–260.

Carr, K. D., & Simon, E. J. (1984). Potentiation of reward by hunger is opioid mediated. *Brain Research, 297*(2), 369–373.

Carr, K. D., & Wolinsky, T. D. (1993). Chronic food restriction and weight loss produce opioid facilitation of perifornical hypothalamic self-stimulation. *Brain Research, 607*(1–2), 141–148.

Cazala, P. (1976). Effects of d- and l-amphetamine on dorsal and ventral hypothalamic self-stimulation in three inbred strains of mice. *Pharmacology Biochemistry and Behavior, 5*(5), 505–510.

Cazala, P., Cazals, Y., & Cardo, B. (1974). Hypothalamic self-stimulation in three inbred strains of mice. *Brain Research, 81*, 159–167.

Chang, S., et al. (1990). Metabolic differences between obesity-prone and obesity-resistant rats. *American Journal of Physiology, 259*(6 Pt 2), R1103–R1110.

Colantuoni, C., et al. (2002). Evidence that intermittent, excessive sugar intake causes endogenous opioid dependence. *Obesity Research, 10*(6), 478–488.

Cox, V. C., Kakolewski, J. W., & Valenstein, E. S. (1969). Ventromedial hypothalamic lesions and changes in body weight and food consumption in male and female rats. *Journal of Comparative & Physiological Psychology, 67*(3), 320–326.

Davis, C., Strachan, S., & Berkson, M. (2004). Sensitivity to reward: Implications for overeating and overweight. *Appetite, 42*(2), 131–138.

De Witte, P., & Bada, M. F. (1983). Self-stimulation and alcohol administered orally or intraperitoneally. *Experimental Neurology, 82*(3), 675–682.

Delamater, A. R., Sclafani, A., & Bodnar, R. J. (2000). Pharmacology of sucrose-reinforced place-preference conditioning: Effects of naltrexone. *Pharmacology Biochemistry and Behavior, 65*(4), 697–704.

Delin, C. R., et al. (1997). Eating behavior and the experience of hunger following gastric bypass surgery for morbid obesity. *Obesity Surgery, 7*(5), 405–413.

Fantino, M., Faion, F., & Rolland, Y. (1986). Effect of dexfenfluramine on body weight set-point: Study in the rat with hoarding behaviour. *Appetite*, (7 Suppl. l), 115–126.

Fulton, S., Woodside, B., & Shizgal, P. (2000). Modulation of brain reward circuitry by leptin. *Science, 287*(5450), 125–128.

Geiger, B. M., et al. (2009). Deficits of mesolimbic dopamine neurotransmission in rat dietary obesity. *Neuroscience, 159*(4), 1193–1199.

Geyer, M. A., & Markou, A. (1995). Animal models of psychiatric disorders. In F. E. Bloom, & D. J. Kupfer (Eds.), *Psychopharmacology: The fourth generation of progress* (pp. 787–798). New York: Raven Press.

Grigson, P. S. (2002). Like drugs for chocolate: Separate rewards modulated by common mechanisms? *Physiology & Behavior, 76*(3), 389–395.

Hajnal, A., Smith, G. P., & Norgren, R. (2004). Oral sucrose stimulation increases accumbens dopamine in the rat. *American Journal of Physiology - Regulatory, Integrative and Comparative Physiology, 286*(1), R31–R37.

Hawkins, M., & Stein, E. A. (1991). Effects of continuous naloxone administration on ventral tegmental self-stimulation. *Brain Research, 560*(1–2), 315–320.

Herberg, L. J., & Blundell, J. E. (1967). Lateral hypothalamus: Hoarding behavior elicited by electrical stimulation. *Science, 155*(760), 349–350.

Hernandez, L., & Hoebel, B. G. (1988a). Food reward and cocaine increase extracellular dopamine in the nucleus accumbens as measured by microdialysis. *Life Sciences, 42*(18), 1705–1712.

Hernandez, L., & Hoebel, B. G. (1988b). Feeding and hypothalamic stimulation increase dopamine turnover in the accumbens. *Physiology & Behavior, 44*(4–5), 599–606.

Hernandez, L., & Hoebel, B. G. (1989). Food intake and lateral hypothalamic self-stimulation covary after medial hypothalamic lesions or ventral midbrain 6-hydroxydopamine injections that cause obesity. *Behavioral Neuroscience, 103*(2), 412–422.

Hoebel, B. G. (1969). Feeding and self-stimulation. *Annals of the New York Academy of Sciences, 157*(2), 758–778.

Hoebel, B. G., & Balagura, S. (1967). Self-stimulation of the lateral hypothalamus modified by insulin and glucagon. *Physiology & Behavior, 2*, 337–340.

Hoebel, B. G., & Teitelbaum, P. (1962). Hypothalamic control of feeding and self-stimulation. *Science, 135*, 375–377.

Hoebel, B. G., & Thompson, R. D. (1969). Aversion to lateral hypothalamic stimulation caused by intragastric feeding or obesity. *Journal of Comparative & Physiological Psychology, 68*(4), 536–543.

Imaizumi, M., et al. (2001). Preference for high-fat food in mice: Fried potatoes compared with boiled potatoes. *Appetite, 36*(3), 237–238.

Jain, R., Mukherjee, K., & Singh, R. (2004). Influence of sweet tasting solutions on opioid withdrawal. *Brain Research Bulletin, 64*(4), 319–322.

Johnson, P. M., & Kenny, P. J. (2010). Dopamine D2 receptors in addiction-like reward dysfunction and compulsive eating in obese rats. *Nature Neuroscience, 13*(5), 635–641.

Kalarchian, M. A., et al. (2002). Binge eating among gastric bypass patients at long-term follow-up. *Obesity Surgery, 12*(2), 270–275.

Kelley, A. E., et al. (2000). A pharmacological analysis of the substrates underlying conditioned feeding induced by repeated opioid stimulation of the nucleus accumbens. *Neuropsychopharmacology, 23*(4), 455–467.

Kenny, P. J., et al. (2003). Low dose cocaine self-administration transiently increases but high dose cocaine persistently decreases brain reward function in rats. *European Journal of Neuroscience, 17*(1), 191–195.

Kenny, P. J., et al. (2006). Conditioned withdrawal drives heroin consumption and decreases reward sensitivity. *Journal of Neuroscience, 26*(22), 5894–5900.

Kenny, P. J., & Markou, A. (2006). Nicotine self-administration acutely activates brain reward systems and induces a long-lasting increase in reward sensitivity. *Neuropsychopharmacology, 31*(6), 1203–1211.

King, B. M. (2006). The rise, fall, and resurrection of the ventromedial hypothalamus in the regulation of feeding behavior and body weight. *Physiology & Behavior, 87*(2), 221–244.

Kishi, T., & Elmquist, J. K. (2005). Body weight is regulated by the brain: A link between feeding and emotion. *Molecular Psychiatry, 10*(2), 132–146.

Koob, G. F., & Le Moal, M. (2005). Plasticity of reward neurocircuitry and the 'dark side' of drug addiction. *Nature Neuroscience, 8*(11), 1442–1444.

Koob, G. F., & Le Moal, M. (2008). Addiction and the brain antireward system. *Annual Review of Psychology, 59*, 29–53.

Kornetsky, C., & Esposito, R. U. (1979). Euphorigenic drugs: Effects on the reward pathways of the brain. *Federation Proceedings, 38*(11), 2473–2476.

Kretschmer, B. D., et al. (2005). Modulatory role of food, feeding regime and physical exercise on body weight and insulin resistance. *Life Sciences, 76*(14), 1553–1573.

Le Magnen, L. (1990). A role for opiates in food reward and food addiction. In E. D. Capaldi, & T. L. Powley (Eds.), *Taste, experience, and feeding* (pp. 241–254). Washington, D.C.: American Psychological Association.

Levine, A. S., et al. (1995). Naloxone blocks that portion of feeding driven by sweet taste in food-restricted rats. *American Journal of Physiology, 268*(1 Pt 2), R248–R252.

Levin, B. E., et al. (1987). Resistance to diet-induced obesity: Food intake, pancreatic sympathetic tone, and insulin. *American Journal of Physiology, 252*(3 Pt 2), R471–R478.

Levin, B. E., Triscari, J., & Sullivan, A. C. (1986). Metabolic features of diet-induced obesity without hyperphagia in young rats. *American Journal of Physiology, 251*(3 Pt 2), R433–R440.

Lin, D., Koob, G. F., & Markou, A. (2000). Time-dependent alterations in ICSS thresholds associated with repeated amphetamine administrations. *Pharmacology Biochemistry and Behavior, 65*(3), 407–417.

Margules, D. L., & Olds, J. (1962). Identical "feeding" and "rewarding" systems in the lateral hypothalamus of rats. *Science, 135*, 374–375.

Markou, A., & Koob, G. F. (1992). Construct validity of a self-stimulation threshold paradigm: Effects of reward and performance manipulations. *Physiology & Behavior, 51*(1), 111–119.

Miranda, G. F., et al. (1988). Reduction of normal food intake in rats and dogs and inhibition of experimentally induced hyperphagia in rats by CM 57373 and fenfluramine. *European Journal of Pharmacology, 150*(1–2), 155–161.

Mount, G., & Hoebel, B. G. (1967). Lateral hypothalamic self-stimulation: Self-determined threshold increased by food intake. *Psychonomic Science, 9*, 265–266.

Norgren, R., Hajnal, A., & Mungarndee, S. S. (2006). Gustatory reward and the nucleus accumbens. *Physiology & Behavior, 89*(4), 531–535.

Olds, J. (1958a). Self-stimulation of the brain; its use to study local effects of hunger, sex, and drugs. *Science, 127*(3294), 315–324.

Olds, J. (1958b). Satiation effects in self-stimulation of the brain. *Journal of Comparative & Physiological Psychology, 51*(6), 675–678.

Olds, M. E. (1972). Comparative effects of amphetamine, scopolamine and chlordiazepoxide on self-stimulation behavior. *Revue Canadienne de Biologie, 31*(Suppl. l), 25–47.

Olds, J., & Milner, P. M. (1954). Positive reinforcement produced by electrical stimulation of the septal area and other regions of rat brain. *Journal of Comparative & Physiological Psychology, 47*, 419–427.

Phillips, A. G., Brooke, S. M., & Fibiger, H. C. (1975). Effects of amphetamine isomers and neuroleptics on self-stimulation from the nucleus accumbens and dorsal noradrenergic bundle. *Brain Research, 85*(1), 13–22.

Puhl, R. M., et al. (2008). Weight stigmatization and bias reduction: Perspectives of overweight and obese adults. *Health Education Research, 23*(2), 347–358.

Rada, P., Avena, N. M., & Hoebel, B. G. (2005). Daily bingeing on sugar repeatedly releases dopamine in the accumbens shell. *Neuroscience, 134*(3), 737–744.

Reid, L. D. (1985). Endogenous opioid peptides and regulation of drinking and feeding. *American Journal of Clinical Nutrition, 42*(5 Suppl. l), 1099–1132.

Reynolds, S. M., & Berridge, K. C. (2002). Positive and negative motivation in nucleus accumbens shell: Bivalent rostrocaudal gradients for GABA-elicited eating, taste "liking"/"disliking" reactions, place preference/avoidance, and fear. *Journal of Neuroscience, 22*(16), 7308–7320.

Rothwell, N. J., & Stock, M. J. (1979). Effects of continuous and discontinuous periods of cafeteria feeding on body weight, resting oxygen consumption and noradrenaline sensitivity in the rat [proceedings]. *Journal of Physiology, 291*, 59P.

Rothwell, N. J., & Stock, M. J. (1982). Energy expenditure of 'cafeteria'-fed rats determined from measurements of energy balance and indirect calorimetry. *Journal of Physiology, 328*, 371–377.

Sadowski, B. (1972). Intracranial self-stimulation patterns in dogs. *Physiology & Behavior, 8*(2), 189–193.

Saunders, R. (2001). Compulsive eating and gastric bypass surgery: What does hunger have to do with it? *Obesity Surgery, 11*(6), 757–761.

Schulteis, G., et al. (1994). Relative sensitivity to naloxone of multiple indices of opiate withdrawal: A quantitative dose-response analysis. *Journal of Pharmacology and Experimental Therapeutics, 271*(3), 1391–1398.

Sclafani, A. (1987). Carbohydrate-induced hyperphagia and obesity in the rat: Effects of saccharide type, form, and taste. *Neuroscience & Biobehavioral Reviews, 11*(2), 155–162.

Sclafani, A. (1989). Dietary-induced overeating. *Annals of the New York Academy of Sciences, 575*, 281–289. discussion 290-1.

Sclafani, A., Bodnar, R. J., & Delamater, A. R. (1998). Pharmacology of food conditioned preferences. *Appetite, 31*(3), 406.

Sclafani, A., & Springer, D. (1976). Dietary obesity in adult rats: Similarities to hypothalamic and human obesity syndromes. *Physiology & Behavior, 17*(3), 461–471.

Simmons, W. K., Martin, A., & Barsalou, L. W. (2005). Pictures of appetizing foods activate gustatory cortices for taste and reward. *Cerebral Cortex, 15*(10), 1602–1608.
Small, D. M., et al. (2001). Changes in brain activity related to eating chocolate: From pleasure to aversion. *Brain, 124*(Pt 9), 1720–1733.
Small, D. M., Jones-Gotman, M., & Dagher, A. (2003). Feeding-induced dopamine release in dorsal striatum correlates with meal pleasantness ratings in healthy human volunteers. *NeuroImage, 19*(4), 1709–1715.
Stice, E., et al. (2008). Relation between obesity and blunted striatal response to food is moderated by TaqIA A1 allele. *Science, 322*(5900), 449–452.
Stice, E., et al. (2010). Weight gain is associated with reduced striatal response to palatable food. *Journal of Neuroscience, 30*(39), 13105–13109.
Szczypka, M. S., et al. (1999). Feeding behavior in dopamine-deficient mice. *Proceedings of the National Academy of Sciences of the United States of America, 96*(21), 12138–12143.
Szczypka, M. S., et al. (1999). Viral gene delivery selectively restores feeding and prevents lethality of dopamine-deficient mice. *Neuron, 22*(1), 167–178.
Takeda, M., et al. (2001). Preference for corn oil in olfactory-blocked mice in the conditioned place preference test and the two-bottle choice test. *Life Sciences, 69*(7), 847–854.
Tanda, G., & Di Chiara, G. (1998). A dopamine-mu1 opioid link in the rat ventral tegmentum shared by palatable food (Fonzies) and non-psychostimulant drugs of abuse. *European Journal of Neuroscience, 10*(3), 1179–1187.
Trenchard, E., & Silverstone, T. (1983). Naloxone reduces the food intake of normal human volunteers. *Appetite, 4*(1), 43–50.
Tzschentke, T. M. (1998). Measuring reward with the conditioned place preference paradigm: A comprehensive review of drug effects, recent progress and new issues. *Progress in Neurobiology, 56*(6), 613–672.
Vanderschuren, L. J., & Everitt, B. J. (2004). Drug seeking becomes compulsive after prolonged cocaine self-administration. *Science, 305*(5686), 1017–1019.
Wang, G. J., et al. (2001). Brain dopamine and obesity. *Lancet, 357*(9253), 354–357.
Wang, G. J., et al. (2004). Similarity between obesity and drug addiction as assessed by neurofunctional imaging: A concept review. *Journal of Addictive Diseases, 23*(3), 39–53.
Wang, G. J., et al. (2004). Exposure to appetitive food stimuli markedly activates the human brain. *NeuroImage, 21*(4), 1790–1797.
Wang, G. J., et al. (2008). Gastric distention activates satiety circuitry in the human brain. *NeuroImage, 39*(4), 1824–1831.
Wilkinson, H. A., & Peele, T. L. (1962). Modification of intracranial self-stimulation by hunger satiety. *American Journal of Physiology, 203*, 537–540.
Wilkinson, H. A., & Peele, T. L. (1963). Intracranial self-stimulation in cats. *The Journal of Comparative Neurology, 121*, 425–440.
Wise, R. A. (2005). Forebrain substrates of reward and motivation. *The Journal of Comparative Neurology, 493*(1), 115–121.
Wise, R. A. (2006). Role of brain dopamine in food reward and reinforcement. *Philosophical Transactions of the Royal Society of London B Biological Sciences, 361*(1471), 1149–1158.
Wise, R. A., & Raptis, L. (1986). Effects of naloxone and pimozide on initiation and maintenance measures of free feeding. *Brain Research, 368*(1), 62–68.
Wise, R. A., & Rompre, P. P. (1989). Brain dopamine and reward. *Annual Review of Psychology, 40*, 191–225.

Wolinsky, T. D., et al. (1994). Effects of chronic food restriction on mu and kappa opioid binding in rat forebrain: A quantitative autoradiographic study. *Brain Research, 656*(2), 274–280.

Yurcisin, B. M., Gaddor, M. M., & DeMaria, E. J. (2009). Obesity and bariatric surgery. *Clinics in Chest Medicine, 30*(3), 539–553. ix.

Zhang, M., Gosnell, B. A., & Kelley, A. E. (1998). Intake of high-fat food is selectively enhanced by mu opioid receptor stimulation within the nucleus accumbens. *Journal of Pharmacology and Experimental Therapeutics, 285*(2), 908–914.

Zhang, Y., & Scarpace, P. J. (2006). The role of leptin in leptin resistance and obesity. *Physiology & Behavior, 88*(3), 249–256.

Zhou, Q. Y., & Palmiter, R. D. (1995). Dopamine-deficient mice are severely hypoactive, adipsic, and aphagic. *Cell, 83*(7), 1197–1209.

CHAPTER

The dark side of compulsive eating and food addiction: affective dysregulation, negative reinforcement, and negative urgency

Eric P. Zorrilla[1], George F. Koob[2]

Department of Neuroscience, The Scripps Research Institute, La Jolla, CA, United States[1]; Neurobiology of Addiction Section, Intramural Research Program, National Institute on Drug Abuse, Baltimore, MD, United States[2]

Introduction

Behavior that is compulsive, from the Latin *compuls* ("driven or forced"), is defined as an action that results from or relates to an irresistible urge. A compulsive behavior often is experienced as outside of one's control and intrusive, unwanted, and egodystonic. The construct of compulsivity exploded into global use during the later 20th century (see Fig. 6.1) (Michel et al., 2011). As Table 6.1 shows, many different compulsive behaviors are commonly described, corresponding to the presence of compulsivity across many psychiatric disorders. Among specific behaviors, however, "compulsive eating" has been the most frequently used term in the modern English language; as much as 2% of the written use of "compulsive" involves "compulsive eating" (Table 6.1). This high use occurs despite the even more recent emergence of the related construct "food addiction" (Fig. 6.2). The study of compulsive eating and food addiction has lagged decades behind the cultural emergence of these constructs (Fig. 6.2B vs. Fig. 6.2A). Despite some controversy (Avena, Gearhardt, Gold, Wang, & Potenza, 2012; Ziauddeen & Fletcher, 2013), however, recent inroads have been made (Avena, Rada, & Hoebel, 2008; Brownell & Gold, 2012; Gold, Badgaiyan, & Blum, 2015; Kessler, 2009). To bridge this divide further, we here revisit (Cottone, Sabino, Roberto, et al., 2009; Moore, Sabino, Koob, & Cottone, 2017; Parylak, Koob, & Zorrilla, 2011) the role of negative emotional states in driving compulsive eating—the so-called "dark side" of compulsive eating.

Compulsive Eating Behavior and Food Addiction. https://doi.org/10.1016/B978-0-12-816207-1.00006-8
Copyright © 2019 Elsevier Inc. All rights reserved.

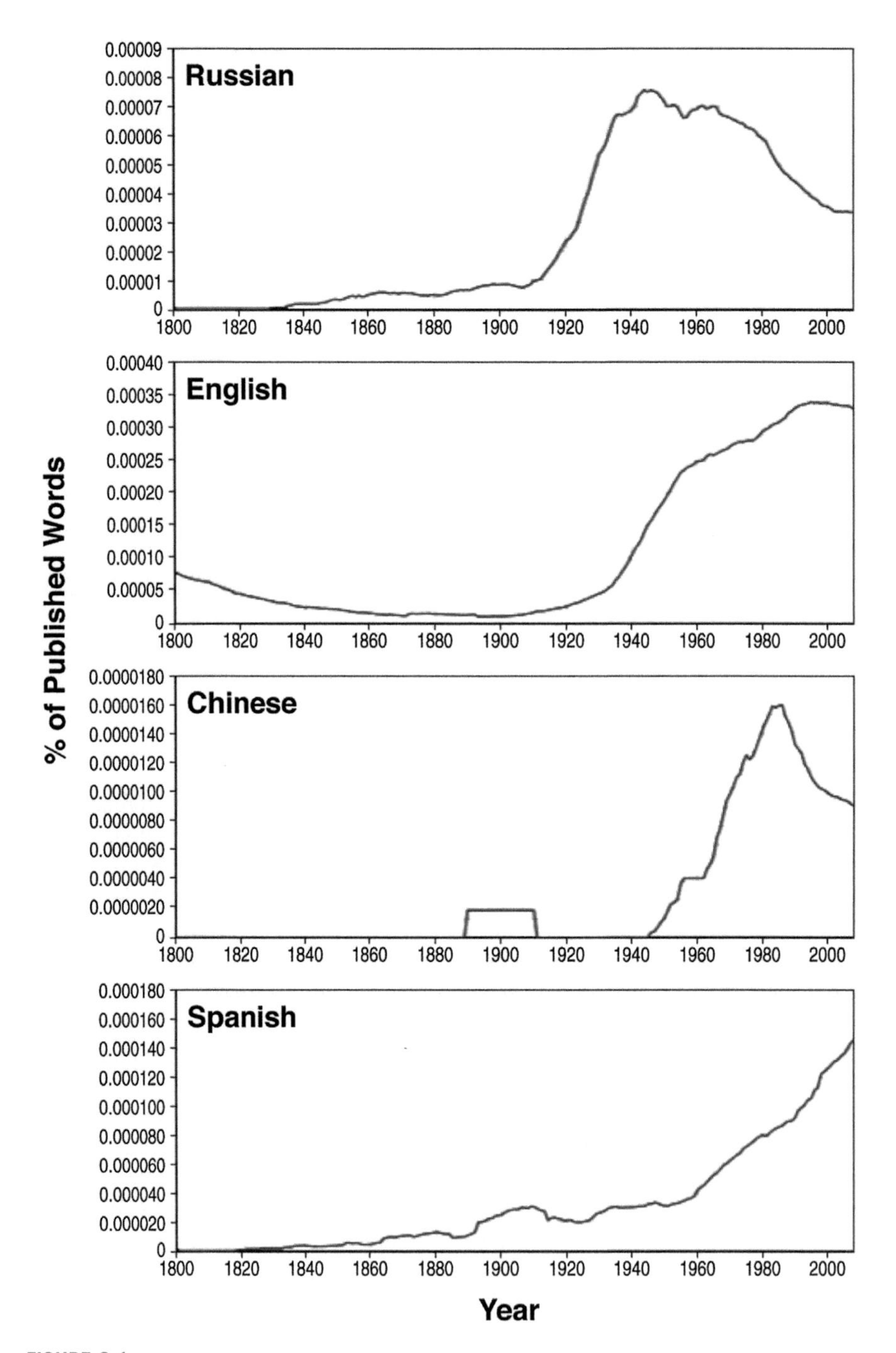

FIGURE 6.1

Standardized n-gram frequency relative to the corpus of published 1-grams in Google Books for that language (Michel et al., 2011). Across languages, n-grams that translate into "compulsive" sharply grew in relative use during the 20th century. Cultural adoption is ordered roughly from earliest (top) to latest (bottom). The n-gram datasets were generated in July 2012 by Google. Searches were performed during May 2018 with http://books.google.com/ngrams.

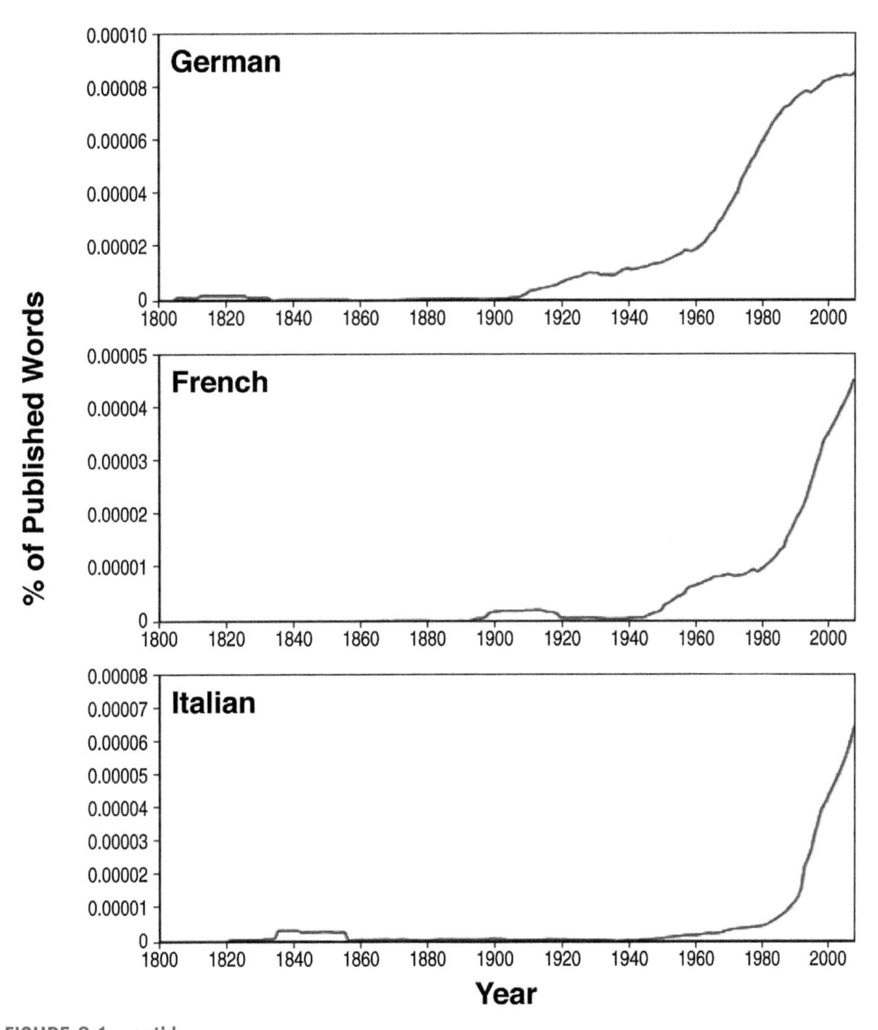

FIGURE 6.1 cont'd.

Yale Food Addiction Scale definition

One operationalization of human food addiction related to compulsive eating is the Yale Food Addiction Scale (YFAS). Developed per Diagnostic and Statistical Manual-IV (DSM-IV) criteria for substance-use disorders (Long, Blundell, & Finlayson, 2015; Pursey, Stanwell, Gearhardt, Collins, & Burrows, 2014) and revised per DSM-5 criteria (YFAS 2.0) (Carr, Catak, Pejsa-Reitz, Saules, & Gearhardt, 2017; Gearhardt, Corbin, & Brownell, 2016; Meule, Muller, Gearhardt, & Blechert, 2017), the YFAS scales are validated to measure one factor. Items assess (1) food reward tolerance, (2) escalation of intake, (3) increased effort/time to obtain food,

Table 6.1 Relative n-gram frequency (vs. "compulsive") in the English Corpus of Google Books.

Term	Peak use (1940–2008)	Year 2000
Compulsive eating	2.1% (1988)	1.5%
Compulsive gambling	1.8% (1992)	1.6%
Compulsive shopping[a]	1.4% (2008)	0.93%
Compulsive drinking[b]	0.64% (1958)	0.20%
Compulsive drug use[c]	0.70% (1984)	0.50%
Compulsive washing/cleaning[d]	0.26% (2008)	0.22%
Compulsive hoarding	0.25% (2008)	0.15%
Compulsive exercise	0.23% (2007)	0.20%
Compulsive checking	0.23% (2008)	0.17%
Compulsive sex	0.16% (1996)	0.16%
Compulsive grooming[e]	0.11% (2004)	0.10%
Compulsive internet use[f]	0.05% (2008)	0.03%

Note: Searches were performed during May 2018 with http://books.google.com/ngrams as ((term)/compulsive). The n-gram dataset was generated in July 2012 by Google (v20120701).
[a] Includes compulsive shopping, compulsive buying.
[b] Includes compulsive drug, compulsive substance, compulsive smoking, compulsive nicotine, compulsive tobacco, compulsive marijuana, compulsive caffeine, compulsive cocaine, compulsive opiate, compulsive heroin, compulsive morphine, compulsive psychostimulant, compulsive amphetamine, compulsive methamphetamine.
[c] Includes compulsive drinking, compulsive alcohol, compulsive ethanol.
[d] Includes compulsive washing, compulsive hand washing, compulsive cleaning.
[e] Includes compulsive grooming, compulsive hair, compulsive nail.
[f] Includes compulsive Internet, compulsive video game.

(4) loss of control over intake, (5) eating despite (risk of) adverse consequences, and (6) negative emotional symptoms on abstinence, with food used to relieve affective "withdrawal" symptoms (see also Avena, Bocarsly, Hoebel, & Gold, 2011; Corsica & Pelchat, 2010). Defined by YFAS criteria, compulsive eating is prevalent in binge-related eating disorders, including bulimia nervosa (BN) (~84% to 100%) and binge eating disorder (BED) (~47% to 57%) (Carr et al., 2017; Gearhardt et al., 2016; Long et al., 2015; Meule, Muller, et al., 2017; Pursey et al., 2014), which afflict ~8% to 10% of Western women (Micali et al., 2017; M. Perez, Ohrt, & Hoek, 2016; Smink, van Hoeken, & Hoek, 2013). YFAS-defined compulsive eating also is seen in about half of bariatric surgery patients (Carr et al., 2017; Gearhardt et al., 2016; Long et al., 2015; Meule, Muller, et al., 2017; Pursey et al., 2014) and a subset (~15% to 33%) of overweight/obese samples. Because three of four men and two of three women are overweight (Flegal, Kruszon-Moran, Carroll, Fryar, & Ogden, 2016), this implies that tens of millions of Americans show compulsive eating. Normal weight individuals without diagnosed eating disorders also meet YFAS criteria for food addiction.

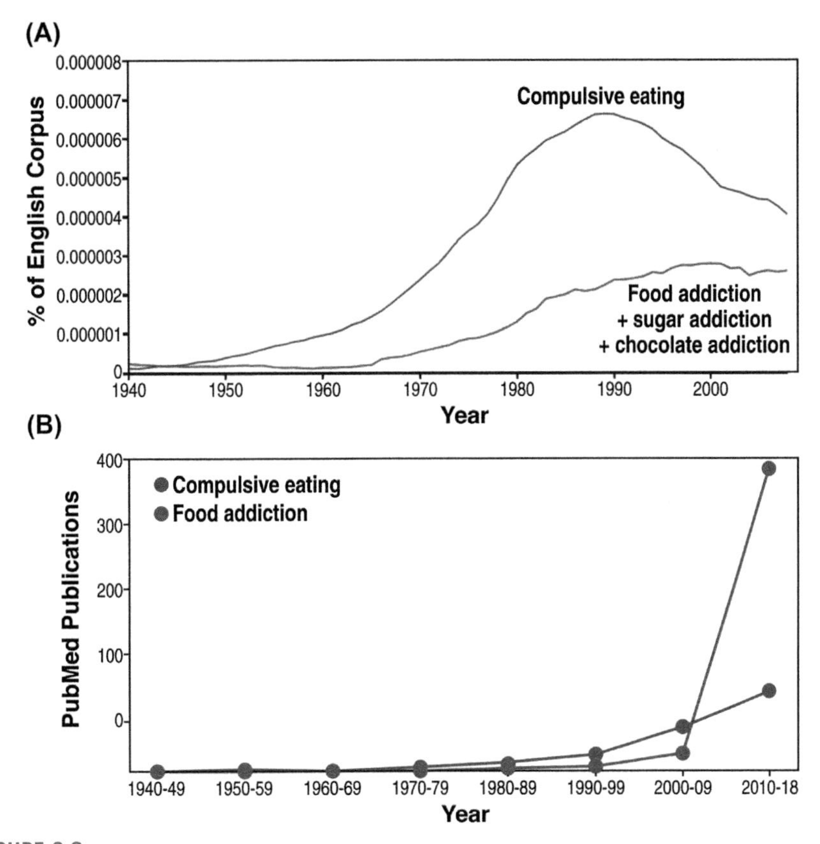

FIGURE 6.2

Modern cultural use of "compulsive eating" and "food addiction" significantly preceded their published study in biomedical science. (A) It shows standardized n-gram frequency for compulsive eating (blue) versus a compound search for food addiction + sugar addiction + chocolate addiction (red) relative to the corpus of published n-grams in English Google Books (Michel et al., 2011). (B) It shows incident number of scientific publications in PubMed for the same search terms during the indicated periods. The n-gram datasets were generated in July 2012 by Google. Searches were performed during May 2018 with http://books.google.com/ngrams and https://www.ncbi.nlm.nih.gov/pubmed/.

Compulsive eating is partly defined by "loss of control" in the YFAS and also is reflected in items that measure "escalated intake," "increased time/effort to obtain food," and "eating despite negative consequences" (Cottone, Sabino, Roberto, et al., 2009; Moore et al., 2017; Parylak et al., 2011; Spierling et al., 2018). But, this begs the question of what drives compulsive eating. Here, the "dark side" view emphasizes the neurobiology of "food reward tolerance" and "negative affective withdrawal and comfort eating" YFAS symptoms. Specifically, repeated exposure to highly rewarding stimuli, such as palatable food, is proposed to elicit opponent-process, counterregulatory adaptations (Solomon & Corbit, 1974) that diminish brain reward function and recruit brain stress circuitry. These adaptations are hypothesized to result in decreased food reward, requiring greater quantity and/or palatability of food than was previously needed to elicit a response, as well as a negative emotional state when the palatable food is not consumed. This deficit emotional state and its associated neurobiological substrates are what are conceptualized as underlying the "irresistible drive" to eat and as contributing to other YFAS symptoms of compulsive eating.

Opponent-process, negative reinforcement model of compulsive substance use

This opponent-process conceptualization of compulsive eating reflects a model that has been useful to understand the transition from casual to compulsive substance use. The "dark side" affective dysregulation hypothesis of addiction proposes that substances of abuse initially activate brain structures that elicit pleasurable emotional states ("a" process), such as euphoria, contentment, or well-being. To restore emotional homeostasis, however, counterregulatory, opponent-processes ("b" process) follow that decrement mood and increase vigilance/tension, as found experimentally (Ettenberg, Raven, Danluck, & Necessary, 1999; Jhou et al., 2013; Knackstedt, Samimi, & Ettenberg, 2002; Radke, Rothwell, & Gewirtz, 2011; Vargas-Perez, Ting, Heinmiller, Sturgess, & van der Kooy, 2007; Vargas-Perez, Ting, & van der Kooy, 2009; Wenzel et al., 2011). The opponent process is putatively subserved by within-system downregulation of brain reward circuitry and between-system recruitment of brain stress circuitry (George, Koob, & Vendruscolo, 2014; Koob & Bloom, 1988; Koob et al., 2014; Koob & Le Moal, 2005, 2008; Koob & Zorrilla, 2010; Zorrilla, Heilig, de Wit, & Shaham, 2013; Zorrilla, Logrip, & Koob, 2014).

With repeated cycles of intoxication/withdrawal, an allostatic hysteresis develops whereby the negative ("b") opponent process initiates earlier to a greater degree and more persistently than the rewarding ("a") process. The progressive outcome with repeated exposure to the substance of abuse is predominance of the opponent process. Consequently, a greater quantity and more frequent use of the previously rewarding substance are needed to maintain or approach euthymia. When the substance is not used, then negative emotional signs of withdrawal emerge, including irritability, anxiety, dysphoria, and subjective feelings of need. This deficit emotional state, or hyperkatifeia (Shurman, Koob, & Gutstein, 2010), is proposed to be dissociable from somatic signs of withdrawal and can sensitize and become persistent with repeated use. As a result, hypohedonia, negative emotional behavior, hyperarousability, and increased behavioral responses to stress may be seen despite months of sustained abstinence. Under this conceptualization, substance use is compulsively escalated or renewed (in relapse) via negative reinforcement mechanisms because it transiently prevents or relieves the negative emotional symptoms. Renewed use is ultimately self-defeating, however, because it perpetuates the vicious opponent-process cycle, even to the point where the substance begins to lose its negative reinforcing properties.

In a related negative reinforcement view of addiction, Baker and colleagues similarly proposed in 1986 that a "negative affect" network is one of two main determinants of drug "urges" (Baker, Morse, & Sherman, 1986). This model stipulated that the negative affect network is activated not only during withdrawal but also by conditioned predictors of withdrawal (e.g., drug cues), as well as by unappetitive consequences (e.g., punishment, frustrative nonreward) or their conditioned cues. Activation of this network by whatever cause was hypothesized to elicit withdrawal symptoms, negative affect, drug urges, and, thereby, drug seeking. Accordingly, in male drug-free individuals with addiction, heroin cues elicited craving, withdrawal symptoms, and a negative emotional state of interrelated anxiety, depression, fatigue, and anger that was dissociable from somatic "withdrawal sickness" (Sherman, Zinser, Sideroff, & Baker, 1989). Subsequently, in 2004, Baker and colleagues revised this hypothesis to propose that escape and avoidance of negative affect was the prepotent motive for compulsive drug use (Baker, Piper, McCarthy, Majeskie, & Fiore, 2004). In this reformulation, as had been proposed in the opponent-process "dark side" model (Koob, Caine, Parsons, Markou, & Weiss, 1997; Koob & Le Moal, 1997, 2001, 2005), negative affect, not physical withdrawal symptoms, is viewed as the motivational core of the abstinence syndrome. Baker et al., 2004 further proposed that through repeated use-abstinence cycles, addicted organisms learn to detect early interoceptive cues of negative affect preconsciously, such that the motivational basis to use drugs may be outside of awareness and not under cognitive control. In cases where stress or abstinence increases negative affect into consciousness, drug use-facilitating information-processing biases result. Both changes in processing would lead to compulsive drug use.

Recent clinical findings in compulsive alcohol use

Previously reviewed data support these affective dysregulation, negative reinforcement models (George et al., 2014; Koob et al., 2014; Koob & Zorrilla, 2010; Zorrilla et al., 2013, 2014), but two recent translational examples merit note here. First, negative affective state is a major category of alcohol relapse (Cannon, Leeka, Patterson, & Baker, 1990), wherein alcohol-related disturbances in sleep and negative affect can persist for months following cessation of alcohol use. Recently, gabapentin, a calcium-channel GABAergic modulator that mitigates these symptoms more than available treatments, showed efficacy to reduce alcohol craving and drinking in single-site studies. Thus, it may represent a novel therapeutic approach that targets "dark side" symptoms in alcohol-use disorders (AUDs) (Mason, Light, Williams, & Drobes, 2009; Mason et al., 2014; Mason, Quello, & Shadan, 2018).

Second, findings in a chronic intermittent ethanol (CIE) vapor exposure animal model have implicated increased activation of extrahypothalamic glucocorticoid receptor (GR) systems in compulsive-like ethanol self-administration. This model elicits escalated ethanol self-administration; increased anxiety-, depressive-, and irritability-like behavior; persistent ethanol self-administration despite adulteration with unappetitive quinine; and increased breakpoints to obtain ethanol in a progressive ratio (PR) schedule of reinforcement. Key signs of GR stress system activation in the CIE model include (1) increased peak daily corticosterone levels during CIE and early abstinence, (2) increased phosphorylation of GR in the central nucleus of the amygdala (CeA), (3) increased potency of systemic or intra-CeA GR antagonist administration to reduce the escalated ethanol self-administration of withdrawn dependent rats as compared with nondependent rats, (4) downregulation of GR in the medial prefrontal cortex (mPFC), nucleus accumbens (NAc), and bed nucleus of the stria terminalis (BNST) during acute withdrawal, followed by (5) upregulation of GR expression in the BNST, CeA, and NAc core during protracted abstinence, and (6) ability of chronic systemic mifepristone to reduce increased ethanol self-administration during protracted abstinence (Kimbrough et al., 2017; Sabino et al., 2006; Serrano et al., 2018; Somkuwar et al., 2017; Valdez et al., 2002; Vendruscolo et al., 2012; Vendruscolo et al., 2015; Zhao, Weiss, & Zorrilla, 2007). Accordingly, a double-blind, placebo-controlled study of 56 nontreatment-seeking, alcohol-dependent individuals found that daily treatment with the GR/progesterone receptor antagonist mifepristone reduced alcohol cue-induced craving in the laboratory during early abstinence as well as the number of drinks consumed through a 1-week follow-up period (Vendruscolo et al., 2015). Thus, GR antagonists are a proposed novel therapeutic approach to normalize the dark side–associated increase in extrahypothalamic GR signaling that may contribute to compulsive alcohol use.

Clinical findings in compulsive tobacco use

Clinical findings for compulsive tobacco use likewise support the affective dysregulation, negative reinforcement models. Smoking cessation elicits craving and a negative emotional state of irritability, anxiety, impatience, and depression (Jorenby et al., 1996) that uniquely predicts relapse vis-à-vis temporally and statistically dissociable craving symptoms (Baker et al., 2012; Piper, Schlam, et al., 2011; Vasilenko et al., 2014). Stress and tobacco withdrawal both elicit urges to smoke in direct relation to negative affect (Zinser, Baker, Sherman, & Cannon, 1992). Patients who experience significant negative affect or craving anhedonia on their quit day relapse more quickly than those who do not (Piper, Vasilenko, Cook, & Lanza, 2017). Anger also is a common context for early lapses (Deiches, Baker, Lanza, & Piper, 2013). Furthermore, smokers who later develop resurgent elevations in negative affect during abstinence are more likely to relapse than those who show gradual resolution of these symptoms (Piasecki et al., 2000; Piasecki, Fiore, & Baker, 1998; Piper, 2015). A history of depression, anxiety disorders, and personality disorders with negative emotionality, as well as baseline depressive symptoms, are all associated with increased smoking relapse (Bold et al., 2015; Japuntich et al., 2007; Leventhal et al., 2012; Piper, Cook, Schlam, Jorenby, & Baker, 2011; Piper et al., 2010). Anhedonia, both lifetime and withdrawal-associated, has particular prognostic validity (Cook et al., 2015; Cook, Lanza, Chu, Baker, & Piper, 2017; Leventhal, Piper, Japuntich, Baker, & Cook, 2014; Piper et al., 2017).

Consistent with a motivating primacy of negative affective symptoms in compulsive tobacco use, post quit negative affect better predicted treatment outcome in heavy smokers treated with a nicotine patch than did somatic measures of withdrawal/adaptation (Kenford et al., 2002). Accordingly, the antidepressant bupropion tripled 1-year nonsmoking rates in individuals with a history of depression (Hughes, Stead, Hartmann-Boyce, Cahill, & Lancaster, 2014; Piper et al., 2008; Smith et al., 2003). Mindfulness-based interventions also have resulted in less affective distress, biologically confirmed abstinence at 6 weeks in the majority of quit attempts, and superior 6-month abstinence versus usual care controls (Davis et al., 2014; Davis, Fleming, Bonus, & Baker, 2007).

Consistent with negative reinforcement, smoking lapses are followed by a reduction in negative affect (Piasecki, Jorenby, Smith, Fiore, & Baker, 2003), and smoking during withdrawal or after stress is associated with greater pleasure than smoking at other times (Zinser et al., 1992). Finally, consistent with opponent-process, affective dysregulation, individuals who were unable to quit and continued smoking developed higher rates of depression over a 3-year period; in contrast, depression rates ultimately diminished in successful quitters (Piper et al., 2013).

Conceptual extension to compulsive eating

We have proposed that cycles of overeating versus dietary restriction may be homologous to the chronic, relapsing cycles of drug use versus abstinence seen in addiction (Cottone, Sabino, Roberto, et al., 2009; Cottone, Sabino, Steardo, & Zorrilla, 2008a, 2008b, 2009; Kreisler, Garcia, Spierling, Hui, & Zorrilla, 2017; Parylak, Cottone, Sabino, Rice, & Zorrilla, 2012; Parylak et al., 2011; Spierling et al., 2018). Environments with palatable, energy-dense foods initially promote increased intake and weight gain through positive reinforcement mechanisms (Bray, Kim, Wilding, & World Obesity, 2017; Cobb et al., 2015; Pereira-Lancha, Coelho, de Campos-Ferraz, & Lancha, 2010). Accordingly, neuroimaging studies of healthy weight adolescents without disordered eating found that greater reward region activation in response to palatable foods predicts short-term weight gain (Shearrer, Stice, & Burger, 2018; Stice, Burger, & Yokum, 2015; Stice & Yokum, 2016a, 2016b; Winter,

Yokum, Stice, Osipowicz, & Lowe, 2017). But, weight-related health concerns, cultural norms for thinness or fitness, and unrealistic body goals (Bury, Tiggemann, & Slater, 2016; Robinson & Aveyard, 2017; Schaefer, Burke, & Thompson, 2018; Thompson, Schaefer, & Dedrick, 2018; Uhlmann, Donovan, Zimmer-Gembeck, Bell, & Ramme, 2018) then lead many to diet (Calder & Mussap, 2015; Vandervoort, Aime, & Green-Demers, 2015). A review of 72 studies with over 1 million respondents found that about half of people reported attempting to lose weight during the previous year. Even higher rates are seen in overweight individuals and women (Santos, Sniehotta, Marques, Carraca, & Teixeira, 2017) and young adults (60%−80% dieting), many despite having a "healthy" body mass index (BMI) (Ackard, Croll, & Kearney-Cooke, 2002; Fonseca, Matos, Guerra, & Pedro, 2009). Reflecting the influence of William Banting's dieting testimonial *Letter on Corpulence, Addressed to the Public,* many Western dieting philosophies advocate abstinence from certain "illicit" foods because of their "richness," macronutrient composition (e.g., sugary/starchy, fatty), or caloric content (Banting, 1993; Dansinger, Gleason, Griffith, Selker, & Schaefer, 2005; James et al., 2017; Peters, 1918; Santos et al., 2017). Instead, dieters often limit themselves to nominally "healthier," but typically less palatable, foods. As we and others have discussed (Corwin, 2011; Corwin & Grigson, 2009; Cottone, Sabino, Roberto, et al., 2009; Cottone, Sabino, Steardo, & Zorrilla, 2008b, 2008a; Cottone, Sabino, Steardo, et al., 2009; Kreisler et al., 2017; Parylak et al., 2012; Spierling et al., 2018), limiting intake in this way may increase the reinforcing value of "illicit" foods and devalue alternatives, thereby counterproductively promoting overconsumption of the palatable "illicit" food. Thus, Western diets often perpetuate intermittent overeating of palatable foods alternating with decreased, often restricted, intake of less palatable foods. Such intake "cycling" is a putative causal factor for BEDs (Goldschmidt, Wall, Loth, Le Grange, & Neumark-Sztainer, 2012; Mathes, Brownley, Mo, & Bulik, 2009; Polivy & Herman, 1985), weight gain (Cannon, 2005; Lowe, 2015; Lowe, Doshi, Katterman, & Feig, 2013), and adverse metabolic consequences (Montani, Schutz, & Dulloo, 2015).

Cycles of overeating (high reward) versus abstinence from palatable foods (low reward), akin to cycles of drug use/withdrawal, may promote the transition from casual to compulsive eating (Avena et al., 2008; Cottone, Sabino, Roberto, et al., 2009; Kreisler et al., 2017; Spierling et al., 2018). Opponent-process responses to highly palatable food combined with the "abstinence" of "low-reward" dieting are hypothesized to yield a hypohedonic negative emotional state with increased stress sensitivity (Parylak et al., 2011) that promotes "relapse" intake of palatable food via negative reinforcement mechanisms. As individuals progress toward compulsive intake, the prime motive for eating palatable food is proposed to shift from reward to relief—self-medicating the deficit (e.g., hypohedonia) and negative emotional states that emerge when preferred foods are not eaten (see Fig. 6.3).

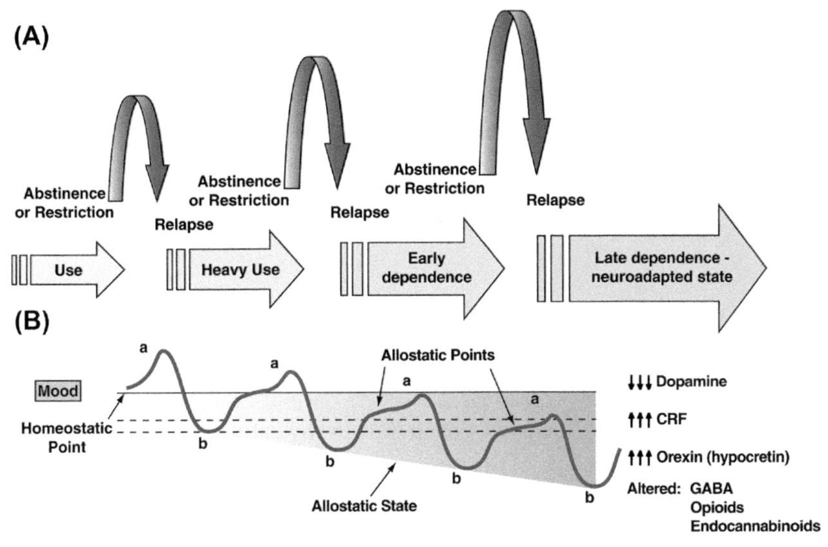

FIGURE 6.3

(A) The progression of compulsive eating over time. The schematic illustrates the hypothesized shift from positive reward motive, involving positively reinforcing, pleasurable effects of palatable food, to coping motive, involving negatively reinforcing relief from a negative emotional state, across cycles of intermittent use versus restriction/abstinence. (B) The *a*-process represents a positive hedonic or mood-enhancing effect, and the *b*-process represents a negative hedonic or mood-impairing effect. The affective stimulus (state) is proposed to be the net sum of the *a*-process and *b*-process. An individual who experiences a positive hedonic mood state with sufficient time between the next enhanced states is hypothesized to retain the *a*-process, with an appropriate counteradaptive opponent process (*b*-process) that balances the activational process (*a*-process) with no allostatic shift. In contrast, an individual who experiences repeated activational processes (*a*-processes) as a result of frequent intake of highly palatable food is hypothesized to undergo a change in opponent-processing, whereby the apparent *b*-process appears earlier and larger in magnitude relative to the diminishing *a*-process. In addition, a progressively greater allostatic state is seen where the steady state does not return to the original homeostatic level because the counteradaptive opponent-process (*b*-process) does not balance the activational process (*a*-process) but shows residual hysteresis. The changes shown are exaggerated and condensed over time but heuristically illustrate the hypothesis that the reward system bears allostatic changes in relation to frequent activation. The following definitions apply: allostasis, the process of achieving stability through change; allostatic state, a state of chronic deviation of the regulatory system from its normal (homeostatic) operating level; allostatic load, the cost to the brain and body of the deviation, accumulating over time and reflecting in pathological states.

(A) Taken with permission from Heilig, M. & Koob, G. F., (2007). A key role for corticotropin-releasing factor in alcohol dependence. Trends in neurosciences 30(8), 399–406. https://doi.org/10.1016/j.tins.2007.06.006. (B) Taken with permission from Koob, G. F., & Le Moal, M. (2001). Drug addiction, dysregulation of reward, and allostasis. Neuropsychopharmacology, 24(2), 97–129. https://doi.org/10.1016/S0893-133X(00)00195-0.

Evidence for the "dark side" from human studies

Yale Food Addiction Scale studies

The "dark side" is reflected in food addiction scales. Of the 35-items on the validated YFAS 2.0 scale (Gearhardt et al., 2016), 12 represent the "dark side" perspective. First, five items are consistent with the opponent-process idea of an increasingly aversive visceroemotional outcome after overeating palatable food. These include the following:

3. I ate to the point where I felt physically ill.
5. I spent a lot of time feeling sluggish or tired from overeating.
16. My eating behavior caused me a lot of distress.
18. I felt so bad about overeating....
22. I kept eating in the same way even though my eating caused emotional problems.

Second, two items resemble the opponent-process idea of antireward adaptations that manifest as food reward tolerance. These include the following:

24. Eating the same amount of food did not give me as much enjoyment as it used to.
26. I needed to eat more and more to get the feelings I wanted from eating....

Finally, six items represent the "dark side" view of a negative state during abstinence from palatable food that involves negative emotional symptoms and motivates intake via negative reinforcement.

11. When I cut down on or stopped eating certain foods, I felt irritable, nervous or sad.
12. If I had physical symptoms because I hadn't eaten certain foods, I would eat those foods to feel better.
13. If I had emotional problems because I hadn't eaten certain foods, I would eat those foods to feel better.
14. When I cut down on or stopped eating certain foods, I had physical symptoms. For example, I had headaches or fatigue.
15. When I cut down or stopped eating certain foods, I had strong cravings for them.
26. I needed to eat more and more to get the feelings I wanted....This included reducing negative emotions like sadness....

In the modified 13-item YFAS 2.0 (mYFAS 2.0), the above items even comprise the majority of symptoms (6 of 11, the remaining 2 items assess clinical levels of impairment). Factor analysis of the mYFAS 2.0 identified a 1-factor solution on which "dark side" items loaded from 0.73–0.80 (Schulte & Gearhardt, 2017). Thus, "dark side" symptoms are substantial components of the YFAS operationalization of food addiction.

The YFAS food addiction construct has predictive validity. YFAS food addiction is more prevalent in women and younger populations (Gearhardt et al., 2016; Magyar et al., 2018; Schulte & Gearhardt, 2017) and relates to greater binge eating, night eating, caloric and fat intake, and BMI in both obese and nonobese populations (Ayaz et al., 2018; Brunault et al., 2017; Burrows, Skinner, et al., 2017; Hauck, Weiss, Schulte, Meule, & Ellrott, 2017; Masheb, Ruser, Min, Bullock, & Dorflinger, 2018; Meule, Muller, et al., 2017; Nolan & Geliebter, 2017; Richmond, Roberto, & Gearhardt, 2017; Schulte, Jacques-Tiura, Gearhardt, & Naar, 2018). Relevant to the "dark side" model, both obese and nonobese adults with food addiction diagnoses or symptoms had greater comorbidity not only for BED and BN but also for depression, anxiety, and posttraumatic stress symptoms; emotional reactivity; and insomnia. They also more often used food to self-soothe (Berenson, Laz, Pohlmeier, Rahman, & Cunningham, 2015; Brewerton, 2017; Burmeister, Hinman, Koball, Hoffmann, & Carels, 2013; Burrows, Hides, Brown, Dayas, & Kay-Lambkin, 2017; Burrows, Kay-Lambkin, Pursey, Skinner, & Dayas, 2018; Ceccarini, Manzoni, Castelnuovo, & Molinari, 2015; Chao et al., 2017; Davis et al., 2011; de Vries & Meule, 2016; Gearhardt et al., 2012; Granero et al., 2014; Koball et al., 2016; Masheb et al., 2018; Meule, Heckel, Jurowich, Vogele, & Kubler, 2014; Meule, Hermann, & Kubler, 2015). Severe binge eating statistically mediated the relation between negative emotional symptoms and food addiction (Imperatori et al., 2014). Consistent with a shift toward negative reinforcement motive, individuals with more food addiction symptoms reported less anticipation of positive reinforcement from eating (Meule & Kubler, 2012).

Adolescents with food addiction symptoms were more likely also to have used alcohol, cannabis, and cigarettes (Mies et al., 2017), and conversely, food addiction was more prevalent in men with heroin-use disorder (Canan, Karaca, Sogucak, Gecici, & Kuloglu, 2017). These comorbidities may reflect transdiagnostic substrates for food and substance-use disorders (Gold et al., 2015; Parylak et al., 2011; Zhang, von Deneen, Tian, Gold, & Liu, 2011). Indeed, in animal models, sugar can cross-sensitize with psychostimulants (Avena et al., 2008) and increase subsequent alcohol intake (Avena, Carrillo, Needham, Leibowitz, & Hoebel, 2004); similarly, alcohol and dietary fat intake can reciprocally increase subsequent intake of one another and have shared peptidergic regulators (Barson et al., 2009; Karatayev, Baylan, & Leibowitz, 2009; Karatayev, Gaysinskaya, Chang, & Leibowitz, 2009).

With respect to health outcomes, food addiction was associated with increased visceral adiposity, a marker of cardiometabolic risk, even in women with normal BMI (Pursey, Gearhardt, & Burrows, 2016) and was more prevalent in type 2 diabetes (Raymond & Lovell, 2015; Yang et al., 2017), strongly predicting BMI and symptoms of depression, anxiety, and stress (Raymond, Kannis-Dymand, & Lovell, 2016, 2017; Raymond & Lovell, 2016). Food addiction was associated with poor weight intervention outcomes in some, but not all (Ivezaj, Wiedemann, & Grilo, 2017), studies, including increased attrition from weight management programs (Tompkins, Laurent, & Brock, 2017), less weight loss (Burmeister et al., 2013; Clark & Saules, 2013), and more "transfer" to other addictions postbariatric surgery (Clark & Saules, 2013). Of interest, an ongoing clinical trial is comparing the effectiveness of psychologically oriented interventions—adjunctive motivational counseling, combined treatment with sustained-release formulations of bupropion + naltrexone (Contrave) or Contrave + counseling—as compared to usual care alone (low-calorie diet plus physical activity) in obese patients with versus without diagnoses of food addiction (NCT03431831).

It remains uncertain how "dark side" factors are implicated in the above-reviewed relations because it is theoretically possible to meet criteria for "food addiction" or have elevated YFAS scores without endorsement of any "dark side" items. Thus, analyses should determine (1) the prevalence, temporal appearance, and natural history of "dark side" symptoms within the heterogeneous group of individuals who show overeating-related phenotypes (e.g., overweight, classes 1–2 vs. class 3 obesity, BED, BN), (2) the unique predictive validity of "dark side" (vs. non–dark side) food addiction symptoms for ponderal and health-related outcomes, psychiatric comorbidity, treatment response, or neurobiological or omics differences among these groups, (3) whether certain "dark side" symptoms have differential predictive validity versus others (e.g., food reward tolerance, posteating dysphoria, emotional vs. physical withdrawal symptoms, negative reinforcement use of food), and (4) whether "dark side" symptoms independently predict the progression of compulsive eating, measured by scales that do not include dark side pathology (Ruddock, Christiansen, Halford, & Hardman, 2017; Schroder, Sellman, & Adamson, 2017) or by instrumental measures of food reinforcement and seeking behavior (Abe, Fukuda, & Tokui, 2004; Epstein, Salvy, Carr, Dearing, & Bickel, 2010; L. H. Epstein, Stein, Paluch, MacKillop, & Bickel, 2018; Gearhardt et al., 2017; Kong et al., 2018).

Negative emotional states and psychiatric comorbidity in binge eating and obesity

Negative emotional traits of depression, low self-esteem, and neuroticism are associated with binge eating (Womble et al., 2001). Individuals with BED and BN also have increased prevalence of major depression, bipolar disorder, anxiety disorders, and alcohol or drug abuse (Hudson, Hiripi, Pope, & Kessler, 2007; J. E. Mitchell & Mussell, 1995; Swanson, Crow, Le Grange, Swendsen, & Merikangas, 2011). About 30%–80% of patients with BED show lifetime comorbid depression and anxiety disorders (Herzog, Keller, Sacks, Yeh, & Lavori, 1992), and both current (27.3% vs. 4.9%) and lifetime rates (52.3% vs. 23.0%) of mood disorders are greater in obese patients with BED versus obese patients without BED (Sheehan & Herman, 2015). The high depression symptomatology is not secondary to obesity because nonobese BED patients show depressive symptoms comparable with those of obese BED patients (Dingemans & van Furth, 2012). Over half of teenage bulimics and approximately one-third of those with BED report suicidal ideation; and one-third of teenage bulimics report attempting suicide (Brown, LaRose, & Mezuk, 2018; Carano et al., 2012; Swanson et al., 2011). Suicide attempts and completion are also high in BED (Runfola, Thornton, Pisetsky, Bulik, & Birgegard, 2014), especially with comorbid depression (Pisetsky, Thornton, Lichtenstein, Pedersen, & Bulik, 2013). Negative emotionality prospectively predicts the incidence of eating disordered symptoms (Leon, Fulkerson, Perry, Keel, & Klump, 1999; Pearson,

Zapolski, & Smith, 2015), and binge eating may reciprocally increase depression (Spoor et al., 2006; Stice, 1998, 2001). The psychiatric comorbidity is associated with poor treatment outcome (Fichter, Quadflieg, & Hedlund, 2008) and a greater frequency of binge eating (Hughes et al., 2013).

Depression and anxiety are well-reviewed comorbidities of human obesity (Singh, 2014), which was linked to increased suicidality in a large ($n = 14{,}497$), diverse representative sample from the Collaborative Psychiatric Epidemiologic Surveys both independent of and synergistically with binge eating (Brown et al., 2018).

Negative emotional states increase palatable food intake in vulnerable populations

Negative emotional states are putative triggers of both overeating (Carels, Douglass, Cacciapaglia, & O'Brien, 2004; Razzoli, Pearson, Crow, & Bartolomucci, 2017) and selecting palatable food (Wallis & Hetherington, 2009; Zellner et al., 2006) in vulnerable populations. During negative emotional states and under stress, normal and underweight individuals typically decrease their intake. In contrast, overweight individuals report eating more during negative states (Geliebter & Aversa, 2003). Self-identified individuals with food addiction report using food to self-medicate feeling tired, anxious, depressed, or irritable (Ifland et al., 2009). Ecological momentary assessment also indicates that increased global negative affect precedes binges (Berg et al., 2015, 2017; Dingemans, Danner, & Parks, 2017; Fischer, Wonderlich, Breithaupt, Byrne, & Engel, 2018; Haedt-Matt & Keel, 2011).

To identify nonhunger motives for eating palatable food, the Palatable Eating Motives Scale was adapted (PEMS) (Burgess, Turan, Lokken, Morse, & Boggiano, 2014) and revised (Boggiano, 2016) from the Drinking Motives Questionnaire-Revised, which assesses motives for drinking ethanol. The PEMS reflects real-life motives, as validated via momentary event sampling (Boggiano, Wenger, Turan, Tatum, Sylvester, et al., 2015). Eating palatable food to "cope" with problems, worries, and negative feelings is one of the four motives identified. Accordingly, one-third of people in the general population report eating and drinking to cope (Association, 2015), and women who report the highest levels of perceived stress report greater emotional eating (Tomiyama, Dallman, & Epel, 2011). Coping motives predicted binge eating in low-income adolescent African Americans (Boggiano, Wenger, Mrug, Burgess, & Morgan, 2015), undergraduates, and weight loss–seeking patients (Boggiano et al., 2014). Coping motives also show incremental validity over binge eating symptoms to predict BMI and are more common in college students with class 3 obesity (Boggiano, Wenger, Turan, Tatum, Sylvester, et al., 2015; Burgess et al., 2014). Increases in coping motives also predict subsequent increases in BMI (Boggiano, Wenger, Turan, Tatum, Morgan, et al., 2015). Individuals who ate palatable food to "cope" showed more emotion-triggered eating, perceived stress, and impaired emotion regulation (Boggiano et al., 2017; Orihuela, Mrug, & Boggiano, 2017).

Reward enhancement is a putatively different motive, defined by intake of palatable food for "pleasure, excitement, or increased fun" (Boggiano, 2016; Boggiano, Wenger, Turan, Tatum, Sylvester, et al., 2015), which also associates with binge eating and severe obesity (Boggiano, 2016). While this construct may seem to reflect a positive reinforcement motive, it could reflect seeking of pleasure, fun, and reward because of a deficit emotional state. Accordingly, reward enhancement motive correlated with anxiety- and depression eating in females and anger/frustration eating in males (Boggiano et al., 2017).

Consistent with the proposed etiologic importance of intermittent reward-abstinence cycles, dieting prospectively predicts increases in depressive symptoms (Stice & Bearman, 2001), binge eating (Goldschmidt et al., 2012; Mathes et al., 2009; Polivy & Herman, 1985), and weight gain (Cannon, 2005; Lowe, 2015; Lowe et al., 2013). Similarly, unlike nonrestrained eaters, restrained eaters, who have a past history of intermittent palatable feeding, overeat in response to stressful situations (Greeno & Wing, 1994), including public speaking (Heatherton, Polivy, & Herman, 1991) and subjectively stressful cognitive tasks (Rutledge & Linden, 1998).

Finally, laboratory studies support the view that heightened negative emotional states, which putatively result from opponent-processes, promote palatable food intake. Autobiographical recall of a sad memory increased the amount of snack food eaten (Fay & Finlayson, 2011). Obese binge eaters ate more chocolate after viewing a sad film than a neutral film (Chua, Touyz, & Hill, 2004); "depression" and "sadness" were reported triggers to eat. Overeating is greater when food is offered during, rather than after, the mood induction (Stice, 2002). In contrast, nonrestrained eaters *reduced* their snack food intake after viewing a sad film (Sheppard-Sawyer, McNally, & Fischer, 2000; Yeomans & Coughlan, 2009), again supporting the hypothesis that dietary restraint may alter responses to distressing stimuli.

Negative affect-driven food intake may also disrupt weight loss maintenance. Weight regain in the 6 months following successful weight loss is associated with eating in response to negative mood and the use of food to regulate mood (Elfhag & Rossner, 2005). Accordingly, cognitive therapy to reduce emotional eating in response to negative emotions can promote weight loss and reduce relapse to obesity (Jacob et al., 2018; Werrij et al., 2009). Other nonpharmacologic interventions that address negative emotions also show promise to promote healthy eating behaviors and improve weight management, including exercise, acupuncture, mindfulness, emotion reappraisal, and the relaxation response (Dunn et al., 2018; Katterman, Kleinman, Hood, Nackers, & Corsica, 2014; Laraia et al., 2018; Masih, Dimmock, Epel, & Guelfi, 2017; Peckmezian & Hay, 2017; Yeh et al., 2017).

Inhibitory influence of palatable food on negative mood

Foods that are consumed during binges or distress tend to be energy-dense and palatable (Singh, 2014). For example, binge eating women ate breads/pasta, sweets, high-fat meat items, and salty snacks (Allison & Timmerman, 2007). Similarly, the Nurses' Health Studies ($n = 123{,}688$ women) found that YFAS-defined food addiction was associated with an increased frequency of eating hamburgers and other red/processed meats, French fries, pizza, and other palatable, high-carbohydrate/high-fat foods, such as snacks and candy bars (Lemeshow et al., 2018). In normal weight individuals with non–binge eating patterns, palatable carbohydrate-rich food and drink reduce anger and tension (Benton & Owens, 1993; DeWall, Deckman, Gailliot, & Bushman, 2011) and increase calmness and emotional well-being within 1–2 h of consumption (Reid & Hammersley, 1999; Strahler & Nater, 2018); also see (Gibson, 2006). The ability of palatable food or sweet drinks to inhibit negative affect is not specific to irritability and aggression. Ecological momentary assessment of obese adults (Berg et al., 2015) and patients with anorexia nervosa (Engel et al., 2013) and BN (Smyth et al., 2007) found that global negative affect lessens after binges as do other emotional facets, including guilt (Berg et al., 2013; Smyth et al., 2007). Sweet drinks and sweets also reduce "feelings of being stressed out," the latter especially in emotional eaters (Strahler & Nater, 2018). Intake of a presumptively palatable, carbohydrate-rich, protein-poor food also prevented uncontrollable stress-induced increases in depressive symptoms in stress-prone subjects (Markus et al., 1998). Finally, after sad mood induction, intake of a high-carbohydrate beverage that was characteristically preferred and liked by subjects led to greater reductions in dysphoria than a less-preferred, isocaloric high-protein beverage (Corsica & Spring 2008; Spring et al., 2008). The results are consistent with a negative reinforcing action of palatable food in vulnerable populations (Agras & Telch, 1998).

A unique prediction of the opponent-process, affective dysregulation model, distinct from a simple affect regulation model (Abramson & Wunderlich, 1972; Kaplan & Kaplan, 1957), is that both the positive and negative reinforcing efficacy of a substance will decrease with repeated use. The reasons for this are the same—the opponent "b" process will progressively predominate in time and amplitude over the mood-enhancing/relieving "a" process. This prediction is reflected in the YFAS "tolerance" item, "I needed to eat more and more to get the feelings I wanted....This included reducing negative emotions like sadness" [emphasis added]. This food tolerance phenomenon, in addition to analytic issues reviewed elsewhere (Berg et al., 2017), may explain why some ecological momentary assessment studies still observe high levels of negative affect postbinge (Haedt-Matt & Keel, 2011). Similarly, the dysphoria-reducing efficacy of a preferred beverage diminished over repeated use (Spring et al., 2008). Analogously, many individuals with long-term cocaine addiction report not only that cocaine no longer gets them "high" but also that cocaine use increasingly does not "work" and ultimately is associated with increasing negative emotional symptoms (Small et al., 2009). Likewise, many individuals with long-term opioid addiction show increased hyperalgesia and negative mood disturbance even while under the influence of opioids, what has been interpreted as a prevailing drug-opponent process because symptoms ultimately remit with long-term abstinence (Athanasos, Ling, Bochner, White, & Somogyi, 2018; Mitchell, White, Somogyi, & Bochner, 2006; White, 2004). The results support an affective dysregulation model whereby substance use is initiated for its positive reinforcing properties, continue for its negative reinforcing properties as the allostatic, opponent-process begins, but ultimately begins to lose its comforting efficacy as well. "Perseverative" substance use may persist, despite decreasing duration and magnitude of hedonic relief, because of past learned contingencies (Bolles, 1972; Pearson, Wonderlich, & Smith, 2015; Walsh, 2013).

Consistent with this opponent-process, affective dysregulation prediction, high intake of processed palatable foods, such as "crisps or savory snacks; sweets ... or chocolate; biscuits; fried food, chips, samosas or bhajis ... and soft drinks," is associated with clinically significant anxiety and depression not only cross sectionally (Jacka, Kremer, et al., 2010; Jacka, Pasco, et al., 2010; Oddy et al., 2009) but also prospectively (Akbaraly et al., 2009; Baskin, Hill, Jacka, O'Neil, & Skouteris, 2017; Jacka, Cherbuin, Anstey, & Butterworth, 2014; Jacka et al., 2011; Sanchez-Villegas et al., 2009; Sanchez-Villegas et al., 2012; Sarris et al., 2015). While the role of third variables in these epidemiologic relations remain under debate (Lai et al., 2016; Winpenny, van Harmelen, White, van Sluijs, & Goodyer, 2018), recent causally oriented studies found that dietary interventions that promote long-term abstinence from such foods improve mental health (Adjibade et al., 2018), reducing major depression more effectively than social support controls (Jacka et al., 2017). We and others have proposed that certain foods may have more addictive-like kinetics of affective action (and thereby opponent-reaction) than others (Avena & Gold, 2011a; Freeman et al., 2018; Ifland et al., 2009; Parylak et al., 2012; Rahimlou, Morshedzadeh, Karimi, & Jafarirad, 2018; Schulte, Avena, & Gearhardt, 2015).

Negative affective symptoms during abstinence from palatable food

Some human data support the "dark side" prediction that abstinence from palatable food initially will increase motivating negative symptoms. Items 11–14 on the YFAS 2.0 measure this explicitly, but their prevalence, temporal appearance, and predictive validity remain unknown. Headache, irritability, and flu-like symptoms have been described in clinical accounts (Davis & Carter, 2009). As alluded to earlier, dieting, which involves decreased food reward, prospectively predicts increased self-reported "stress" (Rosen, Tacy, & Howell, 1990) and depressive symptoms (Stice & Bearman, 2001), linked to depressive symptoms in both overweight and nonoverweight individuals (Crow, Eisenberg, Story, & Neumark-Sztainer, 2006; Goldschmidt, Wall, Choo, Becker, & Neumark-Sztainer, 2016; Hinchliff, Kelly, Chan, Patton, & Williams, 2016; Stice, 2001). Increased depressive symptoms also are seen in weight management programs depending on procedural factors (Smoller, Wadden, & Stunkard, 1987). On a briefer, but recurrent, time scale, individuals who habitually skip breakfast have increased distress, depressive symptoms, and suicidal ideation (Kelly, Patalay, Montgomery, & Sacker, 2016; Khan, Ahmed, & Burton, 2017; Kwak & Kim, 2018; G. Lee, Han, & Kim, 2017; Lee, Park, et al., 2017; Lien, 2007; Tanihata et al., 2015) and then snack more on palatable foods high in saturated fat and sugar (Fayet-Moore, McConnell, Tuck, & Petocz, 2017; Mithra et al., 2018; Ramsay et al., 2018; Rodrigues et al., 2017).

Accordingly, they are at greater risk for increases in weight, BMI, and obesity (Kesztyus, Traub, Lauer, Kesztyus, & Steinacker, 2016; Niemeier, Raynor, Lloyd-Richardson, Rogers, & Wing, 2006; Okada, Tabuchi, & Iso, 2018; Sakurai et al., 2017; K. J. Smith et al., 2017; Tin, Ho, Mak, Wan, & Lam, 2011; Traub et al., 2018; van der Heijden, Hu, Rimm, & van Dam, 2007) and worse cardiovascular, metabolic, and all-cause mortality health outcomes (Azami et al., 2018; Bi et al., 2015; Cahill et al., 2013; Maugeri et al., 2018; Smith et al., 2010; Uemura et al., 2015; Uzhova et al., 2017; Yokoyama et al., 2016).

The portmanteau "hangry," added to the Oxford English Dictionary in 2018, recognizes increased irritability during abstinence from food. As an example, women with lower glucose levels more often stabbed pins into a voodoo doll that represented their spouse and delivered louder, longer aversive noise blasts to their spouse (Bushman, Dewall, Pond, & Hanus, 2014). Worsened mood may not require outright fasting; after eating a high-fat diet for 1 month, men and women who were switched to a presumptively less palatable, lower-fat diet reported greater anger, anxiety, and hostility during the subsequent month than did subjects who continued eating the high-fat diet (Wells, Read, Laugharne, & Ahluwalia, 1998). Many randomized controlled studies show that changes in diet lead to changes in mood (Bowen, Kestin, McTiernan, Carrell, & Green, 1995; Torres, Nowson, & Worsley, 2008; Weidner, Connor, Gerhard, Duell, & Connor, 2009). These effects often are attributed to physiochemical properties of the diet (e.g., fat, carbohydrate, specific fatty acids, cholesterol, electrolytes), while controlling calorie content. But, less attention has been given to the role of different individual preference, liking, or other subjectively evaluated properties of the diets, as proposed here (Wagner, Ahlstrom, Redden, Vickers, & Mann, 2014; Weltens, Zhao, & Van Oudenhove, 2014).

Cross-relapse data also suggest a shared "dark side" substrate during abstinence from food reward or substances of abuse. For example, abstinence from opiates, nicotine, alcohol, cocaine, and methamphetamine clinically often is followed by overeating of palatable, processed, salty foods, and weight gain in humans (Cocores & Gold, 2009; Edge & Gold, 2011; Emerson, Glovsky, Amaro, & Nieves, 2009; Gold, 2011; Jonas & Gold, 1986; Krahn et al., 2006; McGregor et al., 2005; Paczynski & Gold, 2011; Shriner, Graham, & Gold, 2010). Many treatment recovery programs address this liability explicitly via diet and exercise interventions, some of which show promise (Cowan & Devine, 2013; Kelly et al., 2015). Conversely, hunger increases vulnerability to smoking relapse (Leeman, O'Malley, White, & McKee, 2010) and is regarded as a relapse trigger in 12-step addiction recovery programs exemplified in the acronym H.A.L.T. (hunger, anger, loneliness, tired) (Nowinski, Baker, & Carroll, 1999). Similarly, sweets are often made available at Alcoholic Anonymous meetings (Edge & Gold, 2011) and are recommended in the Big Book to relieve relapse urges ("Many … have noticed a tendency to eat sweets and have found this practice beneficial; One of the many doctors … told us that the use of sweets was often helpful; occasionally … a vague craving arose which would be satisfied by candy") (Anonymous, 2001). Finally, a subgroup of bariatric surgery patients subsequently show increases in other addictive behaviors, including gambling, spending, exercise, sexuality, smoking, and alcohol, narcotic or psychostimulant use (Azam, Shahrestani, & Phan, 2018; Bak, Seibold-Simpson, & Darling, 2016; Conason et al., 2013; Dutta et al., 2006; Steffen, Engel, Wonderlich, Pollert, & Sondag, 2015; Wendling & Wudyka, 2011). A shared motivational substrate for negative reinforcement, a "dark side" conceptualization of the drug-food competition hypothesis (Gold, 2011), is consistent with a broadly inverse cross-sectional, but not lifetime, relationship between obesity (Warren & Gold, 2007) or BMI on the one hand (Gearhardt, Waller, Jester, Hyde, & Zucker, 2018) with substance-use disorders on the other.

Neuroadaptations in reward and antireward systems in human obesity and disordered eating

Consistent with the hypothesis of allostatic decrements in reward function, intracranial lateral hypothalamic self-stimulation thresholds increase in rats provided extended, but not limited, access to a palatable cafeteria diet (Johnson & Kenny, 2010; but see Iemolo et al., 2012). Elevated self-stimulation thresholds, an index of impaired brain reward function, develop with obesity and persist despite forced abstinence from the cafeteria diet for 2 weeks. These counteradaptive processes are hypothesized to reflect two mechanisms: within-system neuroadaptations and between-system neuroadaptations (Koob & Bloom, 1988). In a within-system neuroadaptation, "the primary cellular response element to the drug would itself adapt to neutralize the drug's effects; persistence of the opposing effects after the drug disappears would produce the withdrawal response" (Koob & Bloom, 1988). Thus, a within-system neuroadaptation is a molecular or cellular change within a given reward circuit to accommodate overactivity of hedonic processing, resulting in a decrease in reward function. In a between-systems neuroadaptation, "a different cellular system and separable molecular apparatus would be triggered by the changes in the primary drug response neurons and would produce the adaptation and tolerance" (Koob & Bloom, 1988). Thus, a between-system neuroadaptation is a circuitry change in which a different (antireward) circuit is activated by the reward circuit and has opposing actions, again limiting reward function.

Within-system neuroadaptations

The "dark side" model of compulsive eating proposes that overconsumption of palatable food downregulates mesolimbic dopaminergic reward circuitry, a within-system opponent-process adaptation. Mechanisms, similar to those seen in drug addiction, include reduced striatal dopamine D2 receptors and decreased basal or food-stimulated striatal dopamine release (Dunn et al., 2012; Haltia et al., 2007; Lindgren et al., 2018; Stice, Yokum, Zald, & Dagher, 2011; Volkow, Fowler, Wang, Baler, & Telang, 2009; Volkow, Fowler, Wang, Swanson, & Telang, 2007; Volkow, Wang, Fowler, & Telang, 2008; Volkow, Wang, Tomasi, & Baler, 2013; Wang, Volkow, Thanos, & Fowler, 2009). Food-deprived obese subjects showed smaller dorsal striatal extracellular dopamine responses to food stimulation (Wang et al., 2011) than normal weight subjects (Volkow et al., 2002). Accordingly, women whose BMI increased during a 6-month period subsequently showed reduced caudate activation to consumption of a chocolate milkshake than did women whose BMI remained stable (Stice, Yokum, Blum, & Bohon, 2010).

Reduced striatal dopamine D2 receptors and blunted dopamine release, whether as consequences or antecedents, are proposed markers of deficient mesolimbic reward function and hypohedonia that may promote compensatory overeating of palatable food (Bello & Hajnal, 2010; Blum, Thanos, & Gold, 2014; Blum et al., 2018; Gold et al., 2015; Gold et al., 2018; Volkow, Wang, Telang, et al., 2008). Consistent with this reward deficiency hypothesis, obese individuals show lower striatal dopamine D2 receptor levels than do nonobese controls in relation to their greater BMI (Volkow, Wang, Telang, et al., 2008; G. J. Wang et al., 2001). Caudate activation responses to a milkshake are also reduced in obese versus lean individuals (Stice, Spoor, Bohon, & Small, 2008), especially in individuals with the *Taq1* A1 polymorphism of the D2 receptor, which is linked to reduced D2 receptor expression (Stice et al., 2008, 2015). As reviewed by Gold and colleagues, this allele is increased in obesity with (vs. without) comorbid substance-use disorder (74% vs. 23%) as well as in overweight/obese subjects versus healthy controls (67% vs. 29%–33%) (Gold et al., 2015). Another polymorphism linked to reduced dopamine function, the 7R allele of the dopamine D4 receptor, was linked with higher lifetime maximum BMI in BN (Kaplan et al., 2008; Levitan, Kaplan, Davis, Lam, & Kennedy, 2010); binge eating, weight gain, and obesity in women with seasonal depression (Levitan et al., 2004, 2006); and risk for overweight/obesity in preschool children (Levitan et al., 2017). Relations between striatal DA function and binge eating frequency also have been seen in women with BN (Broft et al., 2012).

Similar to human findings, extended access to cafeteria diet reduced striatal D2 receptor levels in rats; lentivirus-mediated knockdown of D2 receptor expression accelerated diet-induced increases in reward thresholds, implicating a causal role in impaired brain reward function (Johnson & Kenny, 2010). Reductions in striatal D2 binding (Bello, Lucas, & Hajnal, 2002) and D2 receptor mRNA (Spangler et al., 2004) also were observed after daily, limited access to sucrose (Bello, Sweigart, Lakoski, Norgren, & Hajnal, 2003). Chronic palatable food exposure also blunted amphetamine-induced dopamine release in the NAc (Geiger et al., 2009; van de Giessen, Celik, Schweitzer, van den Brink, & Booij, 2014), decreased dopamine transporter expression and function (Hryhorczuk et al., 2016; but see; Bello et al., 2003), and decreased baseline extracellular dopamine in the NAc (Geiger et al., 2009; Zhang et al., 2015). Dampened mesolimbic dopaminergic transmission may promote weight gain because obesity-prone rats have lower basal and lipid-stimulated extracellular dopamine levels in the accumbens than do obesity-resistant rats even before their weights diverge (Geiger et al., 2008; Rada, Bocarsly, Barson, Hoebel, & Leibowitz, 2010).

In vivo microdialysis studies suggest dampened dopamine responses to less-preferred alternative rewards. Chronic cafeteria diet feeding decreased basal extracellular dopamine levels in the NAc, with lower stimulation-evoked dopamine release also in the NAc and dorsal striatum (Geiger et al., 2009). Chow increased dopamine efflux in chow-fed controls, but no longer in cafeteria diet-fed rats, a sign of food reward tolerance. Yet, the cafeteria diet still elicited dopamine efflux; thus, continued intake of the cafeteria diet had become required to prevent a chronic dopamine deficit. Intermittency of access to sucrose preserves its ability to sustain striatal dopamine release (Rada, Avena, & Hoebel, 2005). Similarly, rats with intermittent, extended access to a palatable diet show profoundly elevated daily intake and operant self-administration, whereas those with ad libitum access decrease their intake to that of chow controls (Kreisler et al., 2017; Spierling et al., 2018). Likewise, women who received a macaroni-and-cheese meal daily for 5 weeks decreased their intake more than those with weekly access (Avena & Gold, 2011b; Epstein, Carr, Cavanaugh, Paluch, & Bouton, 2011). Thus, repeated palatable food consumption may lead to food reward tolerance and persistent decrements in dopaminergic mesolimbic brain reward systems. A challenge for the field is to determine the stimulus properties of palatable food that drive these adaptations. While hedonic/affective kinetics are proposed, there may be direct roles for correlated physiochemical components of food (Akbaraly et al., 2009; Baskin et al., 2017; Jacka et al., 2011, 2014; Sanchez-Villegas et al., 2009, 2012; Sarris et al., 2015) or its metabolic/endocrine effects (Dunn et al., 2012; Michaelides et al., 2017).

Eating style also relates to distinct mesolimbic dopamine system profiles. Nonobese individuals who reported greater "emotional eating" during distress showed reduced baseline D2 receptor availability in the dorsal striatum versus nonemotional eaters; those high (vs. low) in dietary restraint had increased D2 binding in the dorsal striatum in response to food stimulation (Volkow & O'Brien, 2007; Volkow et al., 2003). The collective data support a "vicious circle" hypothesis that overeating drives down dopaminergic function and that overeating plays a compensatory role to remediate perceived reward deficit because of low striatal DAergic signaling.

The efficacy of some psychotropics to treat compulsive eating, in contrast to the marginal effectiveness of broad appetite suppressants (Avena, Murray, & Gold, 2013), may reflect remedial actions on within-system deficits in mesolimbic dopamine signaling. The psychostimulant lisdexamfetamine dimesylate, a slow-onset, long-acting prodrug of D-amphetamine, is an FDA-approved treatment for BED (Citrome, 2015; Gasior et al., 2017; Hudson, McElroy, Ferreira-Cornwell, Radewonuk, & Gasior, 2017; McElroy, Hudson, et al., 2016; McElroy et al., 2017; McElroy et al., 2015) in phase 2 trials for BN (NCT03397446) (Keshen & Helson, 2017). Consistent with D-amphetamine's actions on monoamine transporters, lisdexamfetamine elicits sustained increases in striatal dopamine efflux (Rowley et al., 2012) and positive affective responses in humans (Jasinski & Krishnan, 2009; Kaland & Klein-Schwartz, 2015). Likewise, animal models suggest that antiobesity actions of tesofensine (Astrup, Madsbad, et al., 2008; Astrup et al., 2008; Nielsen et al., 2009; Sjodin et al., 2010), a triple monoamine reuptake inhibitor, may reflect its ability to remedy palatable food/obesity-induced decrements in mesolimbic dopamine signaling (Appel, Bergstrom, Buus Lassen, & Langstrom, 2014; Axel, Mikkelsen,

& Hansen, 2010; Hansen, Jensen, Overgaard, Weikop, & Mikkelsen, 2013; van de Giessen, de Bruin, la Fleur, van den Brink, & Booij, 2012). Finally, Contrave, sustained-release bupropion (a dopamine/norepinephrine reuptake inhibitor) formulated with naltrexone, promotes weight loss in human obesity (Apovian, 2016) and reduced the severity of binge eating and depressive symptoms in open-label trials of obese patients with suspected BED (Guerdjikova et al., 2017; Halseth et al., 2018). Double-blind, placebo-controlled studies of Contrave in obese patients with versus without BED (NCT03045341, NCT03047005, NCT03063606) should soon clarify its efficacy to reduce binge eating. Finally, in an uncontrolled, chronic study, a complex nutraceutical that putatively enhances dopaminergic and opioidergic function by providing monoamine precursors and an enkephalinase inhibitor enhanced self-reported sleep quality, happiness, and energy while decreasing sugar craving, snacking, night bingeing, and body weight (Gold et al., 2015).

Between-system neuroadaptations

Much knowledge of the neurobiology of between-system neuroadaptations is derived from animal models, but some human neuroimaging findings are notable (see also Zhang et al., 2011). Increased food addiction symptoms and obesity are associated with greater amygdala reactivity to pictures of a milkshake (Gearhardt et al., 2011; Ng, Stice, Yokum, & Bohon, 2011). Similarly, obese women, as compared with healthy weight women, showed greater amygdala responses to pictures of palatable high-calorie (e.g., cheesecake), but not less-preferred, low-calorie foods (e.g., steamed vegetables) (Stoeckel et al., 2008). Drug and alcohol cues likewise elicit increased amygdala responses in people with substance-use disorders (Engelmann et al., 2012; Goudriaan, de Ruiter, van den Brink, Oosterlaan, & Veltman, 2010; Heinz, Beck, Grusser, Grace, & Wrase, 2009; Jasinska, Stein, Kaiser, Naumer, & Yalachkov, 2014; Mainz et al., 2012). There is increased functional connectivity of the amygdala to the insula and putamen during a milkshake cue in women with BN as compared with healthy volunteers (Bohon & Stice, 2012). Activation elicited by such “cue” trials is often interpreted as “anticipation” in a positive valence perspective. Within test sessions, however, most milkshake picture presentations are followed by a taste of the milkshake. In the context of ongoing (or prior real life) Pavlovian associations of seeing a milkshake and then enjoying milkshake reward, the picture-only cue trials may be processed as reward omission, which is aversive in animals (Calu, Roesch, Haney, Holland, & Schoenbaum, 2010; Do-Monte, Minier-Toribio, Quinones-Laracuente, Medina-Colon, & Quirk, 2017; Kawasaki, Annicchiarico, Glueck, Moron, & Papini, 2017). Moreover, drug cues in addiction can rapidly acquire use-motivating, aversive properties (Carelli & West, 2014; Colechio, Alexander, Imperio, Jackson, & Grigson, 2018; Colechio & Grigson, 2014; Colechio, Imperio, & Grigson, 2014; Nyland & Grigson, 2013). In addition to nonreward/negative contrasts, these may represent conditioned drug−opponent-processes (Childress, McLellan, Ehrman, & O’Brien, 1988; Childress, McLellan, & O’Brien, 1986; Colechio & Grigson, 2014; Colechio et al., 2014;

Grigson, 2008; McLellan, Childress, Ehrman, O'Brien, & Pashko, 1986; Topp, Lovibond, & Mattick, 1998) (see also Siegel & Ramos, 2002). The amygdala (Yacubian et al., 2006; Zorrilla & Koob, 2013), including the CeA and basolateral nucleus (Calu et al., 2010; Iordanova, Deroche, Esber, & Schoenbaum, 2016; Kawasaki et al., 2017; Kawasaki, Glueck, Annicchiarico, & Papini, 2015; Tye, Cone, Schairer, & Janak, 2010), mediates behavioral responses to reward omission. Likewise, the CeA is a key "between-system" substrate for opponent processes (see below and Koob, 2015; Koob & Le Moal, 2008; Roberto, Spierling, Kirson, & Zorrilla, 2017; Wenzel et al., 2011), and its activation amplifies and narrows motivation toward salient rewards at the expense of others (Robinson, Warlow, & Berridge, 2014; Warlow, Robinson, & Berridge, 2017). Thus, increased amygdala reactivity to palatable food cues might actually represent negative valence activation because of heightened sensitivity to omission of food reward or greater (conditioned) opponent-processing.

Consistent with this stress-like "dark side" interpretation of cue reactivity, rats with cyclic palatable food-restriction diet histories that were presented with the sight and smell of unobtainable palatable food showed HPA axis activation (Cifani, Zanoncelli, et al., 2009) and more cells expressing phosphorylated extracellular signal–regulated kinases in the CeA, paraventricular nucleus of hypothalamus, and dorsal and ventral BNST (Micioni Di Bonaventura, Lutz, et al., 2017). Once access was ultimately provided, "frustratively cued" rats doubled their binge-like intake versus days with no cues (Cifani, Polidori, Melotto, Ciccocioppo, & Massi, 2009; Micioni Di Bonaventura, Vitale, Massi, & Cifani, 2012). Antagonist studies showed that the binge-like intake was driven by activation of extrahypothalamic stress–related corticotropin-releasing factor-1 (CRF_1) systems, including in the CeA and BNST of the extended amygdala, in both of which CRF_1 mRNA was upregulated (Micioni Di Bonaventura et al., 2014; Micioni Di Bonaventura, Ubaldi, et al., 2017).

Also consistent with altered amygdala reactivity in compulsive eating, while healthy controls show reduced amygdala responses to sucrose when sated (vs. fasted), women recovered from BN do not. They thereby show greater amygdala responses to sucrose than controls when sated (Ely et al., 2017), as do obese children (Boutelle et al., 2015). Further work is needed to characterize these differences as antecedents versus neuroadaptations.

Negative withdrawal-like states after cessation of palatable food access

Behavioral signs of a negative emotional state have been seen in animals withdrawn from palatable food, including, early on, increased "nippiness" during withdrawal from sucrose solutions (Galic & Persinger, 2002). Hoebel and colleagues showed that rats with daily 12-hr access to high-sugar solutions plus chow alternated with 12-hr food deprivation led to somatic and anxiogenic-like signs of opiate withdrawal when challenged with the opioid antagonist naloxone (Colantuoni et al., 2002). Rats showed decreased preproenkephalin mRNA expression in the NAc (Spangler et al., 2004) as do rats with limited (but not continuous) daily access to a sweet-fat diet (Kelley, Will, Steininger, Zhang, & Haber, 2003). Without naloxone, somatic and anxiogenic-like signs of withdrawal also occurred following a 24–36 h fast after the last sugar–water access session (Avena et al., 2008). Perhaps as a human analog of precipitated opioid withdrawal, a placebo-controlled trial in obese women found that oral naltrexone elicited greater cortisol stress responses and aversive nausea in those who reported greater food addiction symptoms and reward-driven eating (Mason et al., 2015).

Hoebel and colleagues hypothesized that reduced reward function and increased anxiety-like behavior during withdrawal may originate from altered balance of striatal dopaminergic and acetylcholinergic (ACh) signaling. Naloxone elicited greater ACh release in the NAc of rats with a cyclic glucose + chow/deprivation history than in ad lib chow rats (Colantuoni et al., 2002) as well as reduced extracellular accumbens dopamine, similar to morphine withdrawal (Pothos, Rada, Mark, & Hoebel, 1991; Rada, Pothos, Mark, & Hoebel, 1991). After a 36-hr fast, cycled animals also had lower dopamine and higher ACh levels extracellularly in the NAc (Avena et al., 2008). Hoebel and colleagues proposed that the shift toward greater ACh release versus decreased dopamine corresponded to a shift away from approach behaviors and toward harm avoidance (Hoebel, Avena, & Rada, 2007).

We similarly found that rats that were diet-cycled with alternating 2-day access to a highly preferred high-sucrose, chocolate-flavored solid diet versus 5-day access to standard chow showed increased anxiety-like behavior in the elevated plus maze and defensive withdrawal tests when tested during the chow phase of their diet cycle (Cottone, Sabino, Roberto, et al., 2009; Cottone, Sabino, Steardo, et al., 2009; but see Rossetti, Spena, Halfon, & Boutrel, 2014). In this model, Cottone, Sabino, Iemolo, and colleagues also observed increased forced swim immobility, a depressive-related behavior in withdrawn diet-cycled rats (Iemolo et al., 2012). Boutrel and colleagues observed reduced locomotor activity in a novel open field, a measure of increased emotionality (Rossetti et al., 2014). When withdrawn from palatable food, rats in this model showed increased expression of the stress-related neuropeptide CRF in the CeA (Cottone, Sabino, Roberto, et al., 2009), which also was activated in mice 24 hr after withdrawal from high-fat diet (Teegarden & Bale, 2007). Similarly in animal models, CeA CRF systems are activated during withdrawal from alcohol (Funk, Zorrilla, Lee, Rice, & Koob, 2007; Roberto et al., 2010; Sommer et al., 2008; Zorrilla, Valdez, & Weiss, 2001), opiates (Heinrichs, Menzaghi, Schulteis, Koob, & Stinus, 1995; Maj, Turchan, Smialowska, & Przewlocka, 2003; McNally & Akil, 2002; Weiss et al., 2001), cocaine (Richter & Weiss, 1999), cannabinoids (Rodriguez de Fonseca, Carrera, Navarro, Koob, & Weiss, 1997), and nicotine (George et al., 2007; Marcinkiewcz et al., 2009). Systemic pretreatment with the selective CRF_1 antagonist R121919 blocked the food withdrawal–associated anxiety at doses that did not alter behavior of chow-fed controls (Cottone, Sabino, Roberto, et al., 2009). Analogously, CRF_1 antagonists ameliorate aversive- or anxiety-like states during withdrawal from alcohol (Knapp,

Overstreet, Moy, & Breese, 2004; Overstreet, Knapp, & Breese, 2004; Sommer et al., 2008), opiates (Skelton et al., 2007; Stinus, Cador, Zorrilla, & Koob, 2005), benzodiazepines (Skelton et al., 2007), cocaine (Basso, Spina, Rivier, Vale, & Koob, 1999; Sarnyai et al., 1995), and nicotine (George et al., 2007). CRF_1 antagonist pretreatment also blunted the degree to which diet-cycled animals overate the preferred diet on renewed access at doses that did not alter intake of chow controls or of rats fed the sucrose-rich diet, but without a history of diet cycling (Cottone, Sabino, Roberto, et al., 2009). Similarly, CRF_1 antagonists can reduce excessive intake of alcohol (Chu, Koob, Cole, Zorrilla, & Roberts, 2007; Funk et al., 2007; Gehlert et al., 2007; Gilpin, Richardson, & Koob, 2008; Richardson et al., 2008; Sabino et al., 2006; Valdez et al., 2002), cocaine (Specio et al., 2008), opiates (Greenwell et al., 2009), and nicotine (George et al., 2007) in animal models of dependence, while having less effect on self-administration of nondependent animals.

When diet-cycled animals were studied during access to the preferred, sucrose-rich diet, plus maze behavior, forced swim immobility, and CeA CRF levels normalized, supporting the hypothesis that increased activation of the amygdala CRF system and negative emotional behavior reflected an acute withdrawal-like state (Cottone, Sabino, Roberto, et al., 2009; Cottone, Sabino, Steardo, et al., 2009; Iemolo et al., 2012). Similar to findings during alcohol withdrawal (Roberto et al., 2010), diet-cycled rats also showed increased sensitivity of CeA GABAergic neurons to modulation by CRF_1 antagonism; R121919 reduced evoked inhibitory postsynaptic potentials to a greater degree in diet-cycled rats. Finally, Cottone, Sabino, Iemolo, and colleagues further showed that intra-CeA R121919 reduced the anxiogenic-like behavior and palatable diet intake of withdrawn diet-cycled rats (Iemolo et al., 2013), similar to the ameliorating effects of intra-CeA CRF antagonist administration on withdrawal-associated negative emotional symptoms and self-administration in models of substance dependence (Roberto et al., 2017). Thus, several results with diet-cycled rats resemble the between-system neuroadaptation of central extended amygdala CRF systems seen in animal models of dependence.

We also found that female rats with a history of highly limited (10 min/day) access to the same chocolate-flavored, sucrose-rich diet exhibited binge-like intake (eating >40% of their daily intake within 10 min) and an anxiogenic-like reduction in plus maze open arm time 24 h after their last access (Cottone et al., 2008b). Diet-cycled rats that spent the least time on the open arms were those that binged the most on the palatable diet, a correlation not evident in chow-fed controls. These results support the opponent-process, negative reinforcement prediction that greater dysphoria will associate reciprocally with greater palatable food intake.

Translationally consistent with the reviewed findings, a double-blind, placebo-controlled trial in individuals with restrained eating found promising results for the selective CRF_1 antagonist pexacerfont to reduce food craving and laboratory stress-induced eating (Epstein et al., 2016). Because the study was halted by the NIH IRB for reasons unrelated to adverse drug effects or efficacy (reinterpretation of the Common Rule for human subject protection under HHS, 45 CFR 46A), it only had ~30% power to detect a priori effect of interest. But, pexacerfont descriptively reduced laboratory stress-induced eating ($r = 0.30$) and craving for sweet foods ($r = 0.28-0.49$). In bogus taste tests, pexacerfont reduced palatable food intake across all imagery scripts ($r = 0.34$). Finally, nightly YFAS food addiction symptoms were lower in subjects that received pexacerfont by the evening after the first loading dose of pexacerfont ($r = 0.39$). Both Bayes factor and counter null analysis indicated a positive potential of pexacerfont (Epstein et al., 2016) and provide rationale for well-powered clinical trials of CRF_1 antagonists to reduce compulsive eating. A caveat is some have suggested that the reduction of YFAS scores within 24 h may be faster than pexacerfont's predicted time course of CNS exposure.

Negative emotional and sleep disturbance symptoms during alcohol withdrawal involve a hyperglutamatergic state. Accordingly, acamprosate, a neuromodulator that tends to normalize glutamatergic tone, reduces ethanol withdrawal-associated sleep disturbance and promotes abstinence (Higuchi & Japanese Acamprosate Study, 2015; Mason & Heyser, 2010; Mason & Lehert, 2012; Perney, Lehert, & Mason, 2012). Interestingly, a small placebo-controlled open-label study of outpatients with BED found that acamprosate yielded improvements in binge day frequency and measures of compulsiveness of binge eating and food craving in endpoint analysis, though these were not significant in longitudinal analysis (McElroy et al., 2011). Weight and BMI decreased in the acamprosate group, but increased in the placebo group. Memantine, a low-affinity, voltage-dependent NMDA receptor antagonist, also showed evidence of reducing binge-type eating in open-label trials (Brennan et al., 2008; Hermanussen & Tresguerres, 2005) and animal models (Popik, Kos, Zhang, & Bisaga, 2011; Smith et al., 2015). The relation of hyperglutamatergic tone to dark side symptoms in compulsive eating and the efficacy of antiglutamatergic treatments in double-blind studies remain to be determined.

Neuroadaptation in the brain endocannabinoid (eCB) system has also been seen. For example, we found that surinabant (SR147778) less potently reduced binge-like intake of rats with a history of highly limited access to sweet-fat diet than it did in *ad lib*–fed chow or palatable diet controls (Parylak et al., 2012). In the "dark side" domain, amygdala eCB-cannabinoid receptor 1 (CB1) signaling has been conceptualized as an antistress buffer that is recruited during stress in compensatory fashion (Lutz, Marsicano, Maldonado, & Hillard, 2015; Parsons & Hurd, 2015) and deficiency of which may increase vulnerability to negative emotional symptoms (Bluett et al., 2017). Consistent with a stress-like response, withdrawal from cyclic palatable food increased (compensatory) levels of the eCB 2-arachidonoylglycerol and its CB1R in the CeA (Blasio et al., 2013). Systemic or intra-CeA infusion of the CB1R inverse agonist rimonabant more potently precipitated anxiogenic-like behavior and anorexia in rats withdrawn from cyclic palatable food than in chow controls (Blasio et al., 2013; Blasio, Rice, Sabino, & Cottone, 2014). These findings may explain why rimonabant and taranabant, another CB1 inverse agonist, produced adverse psychiatric side effects in obese patients, including anxiety, depression, and irritability, which led to their withdrawal from the clinic and drug development

(D'Addario et al., 2014). In alcohol dependence, the stress-buffering amygdala eCB system becomes deficient (Serrano et al., 2018). Whether it similarly deteriorates in advanced forms of compulsive eating, leading to increased vulnerability to stress and dysphoria, remains to be determined.

Food reward tolerance

A hallmark of the "dark side" of addiction is tolerance, in which greater quantities of drug are required to produce the same effect. As alluded to above with mesolimbic dopamine responses, a similar decrement of hedonic response to food rewards may occur in animals with palatable food access. Hoebel and colleagues observed dramatic increases in glucose intake over successive days of 12-hr limited access; increasingly, rapid glucose consumption was seen during the first hour of access, consistent with the development of tolerance and a shift toward binge-like eating (Colantuoni et al., 2002). We and many other investigators have since replicated binge-like escalation of intake and self-administration that resemble tolerance using diverse palatable diets and schedules of intermittent access (Bello et al., 2009; Cooper, 2005; Corwin & Babbs, 2012; Cottone, Sabino, Roberto, et al., 2009; Cottone et al., 2008b; Cottone, Sabino, Steardo, et al., 2009; Kreisler et al., 2017; Parylak et al., 2012; Rossetti et al., 2014; Spierling et al., 2018). We also found that rats with intermittent extended access (24 h/day) developed increased binge-like intake and first 30-min operant self-administration comparable with that of highly restricted groups (30 min/day), affirming a key role for intermittent access, and not only restrictedness, in escalating consumption (Kreisler et al., 2017; Spierling et al., 2018).

Also, resembling tolerance, other previously acceptable rewards become less effective at supporting intake or operant responding. Rats that received intermittent access to a chocolate-flavored, sucrose-rich diet showed lower PR breakpoints than chow controls when responding for a less-preferred, but otherwise palatable, corn-syrup sweetened chow (Cottone, Sabino, Roberto, et al., 2009; Cottone, Sabino, Steardo, et al., 2009). Likewise, access to highly preferred diets led to underconsumption of otherwise acceptable chow even when it was the only food available, leading to voluntary weight loss (Cottone, Sabino, Roberto, et al., 2009; Cottone et al., 2008b, 2008a; Cottone, Sabino, Steardo, et al., 2009; Iemolo et al., 2013; Johnson & Kenny, 2010; Pickering, Alsio, Hulting, & Schioth, 2009; Rossetti et al., 2014). The chow hypophagia increases with longer durations of access to palatable food (24 h vs. 30 min/day) and is persistent (Kreisler et al., 2017); rats that received chronic ad lib access to a highly preferred diet continued to undereat chow for at least 2 weeks after it was the only diet available, despite having returned to normal body weight and adiposity (Kreisler et al., 2017). Both the chow hypophagia and the motivational deficits to obtain less-preferred food are mitigated by pretreatment with a CRF_1 antagonist (Cottone, Sabino, Roberto, et al., 2009), perhaps reflecting the ability of a CRF antagonist to reverse blunted reward function during nicotine or alcohol withdrawal (Bruijnzeel, Prado, & Isaac, 2009; Bruijnzeel, Small,

Pasek, & Yamada, 2010). Reduced intake of and breakpoints for moderately sweet solutions were also seen in rats with a history of access to sweet foods or solutions (Iemolo et al., 2012; Vendruscolo, Gueye, Darnaudery, Ahmed, & Cador, 2010). Food reward tolerance may also explain why healthy weight adolescents who were frequent ice cream eaters showed reduced fMRI BOLD putamen and caudate activation responses to receipt of an ice cream milkshake than less frequent eaters (Burger & Stice, 2012).

Individual differences in vulnerability

There are individual differences in the vulnerability to develop compulsive-like eating in animal models. We found stable individual differences in the degree to which rats with highly limited access (10–30 min/day) to nutritionally complete high-sucrose (Cottone et al., 2008b; Kreisler et al., 2017) or sweet-fat diets (Parylak et al., 2012) developed binge-like intake. Greater binge intake of high-sucrose diet correlated directly with greater anxiety-like behavior 24 h postaccess (Cottone et al., 2008b). Boggiano et al. (Boggiano et al., 2007) and Klump, Sisk and colleagues similarly found individual differences in vanilla frosting intake of rats with intermittent 4-hr access to the palatable food. From these, they classified "binge-prone" versus "binge-resistant" extremes; "binge-proneness" was inhibited by perinatal testosterone, developed with puberty and more prevalent in females (Culbert, Sinclair, Hildebrandt, Klump, & Sisk, 2018; Klump, Racine, Hildebrandt, & Sisk, 2013; Klump, Suisman, Culbert, Kashy, & Sisk, 2011). Binge-prone rats also tolerated greater footshock for palatable food (Oswald, Murdaugh, King, & Boggiano, 2011). Finally, we found strong, stable individual differences in the degree to which rats with intermittent, extended access to palatable diet developed not only binge-like and daily overeating but also the progressively greater "rejection" of regular chow on nonaccess days and weekends (Kreisler et al., 2017).

More recently, we have examined motivational measures, including highly elevated PR breakpoints, because reinforcement efficacy–based measures in humans may better predict ponderal outcomes than intake alone (Epstein, Leddy, Temple, & Faith, 2007; Epstein et al., 2010; Epstein et al., 2018; Epstein, Temple, et al., 2007; Gearhardt et al., 2017). PR performance has also been used to model compulsive-like responding for substances of abuse (Vendruscolo et al., 2012; Wade, Vendruscolo, Schlosburg, Hernandez, & Koob, 2015). We found strong individual differences in the development of compulsive-like PR breakpoints for palatable high-sucrose diet in both male ($ICC = 0.79$) and female rats ($ICC = 0.77$) under intermittent, extended access schedules (Spierling et al., 2018). Females were more likely than males to develop compulsive-like PR responding (>2 SD than ad lib controls). Validating the distinction, relative to lower PR intermittent rats, "compulsive-like" (high PR) intermittent rats showed greater (1) daily overeating of the preferred diet, (2) rejection of chow diet on nonaccess days, and (3) binge-like self-administration on renewed access. These differences emerged earlier in females than males. Compulsive-like female rats also showed higher respiratory exchange ratios in indirect calorimetry analysis (Spierling et al., 2018), a predictor of weight gain in humans (Piaggi, Thearle, Bogardus, & Krakoff, 2013) that indicates sparing of fat as a fuel substrate.

Finally, Pickering et al. found that only obesity-prone rats showed more anxiogenic-like behavior in an open field and increased PR breakpoints for palatable high-sucrose pellets 2 weeks after being switched to standard chow from chronic access to lard and 30% sucrose (Pickering et al., 2009). Obesity-prone animals more persistently underate the chow through 3 weeks of protracted withdrawal versus chow-only controls or obesity-resistant rats. Differences during withdrawal were seen despite no differences in preference, intake, or PR breakpoints when the diet was initially continuously available, suggesting a link to "dark side" withdrawal constructs from an obesity-vulnerability perspective. Consistent with within-system downregulation of reward systems, obesity-prone rats also differentially showed lower NAc D1, D2, and μ-opioid receptor expression during extended access to palatable diets (Alsio et al., 2010).

Stress-induced food-seeking and intake

Because palatable food can have negative reinforcing or "comforting" effects, environmental stress, and not only opponent-process affective dysregulation, may promote palatable food intake or relapse (Avena et al., 2008; Cottone, Sabino, Roberto, et al., 2009; Dallman et al., 2003; Ulrich-Lai et al., 2010). The initial ability of palatable food intake, under certain conditions, to attenuate exogenous activation of stress systems in animal models, seen in behavioral, autonomic, neuroendocrine, and neurochemical measures, strongly supports this possibility (Christiansen, Herman, & Ulrich-Lai, 2011; Dallman, Pecoraro, & la Fleur, 2005; Fachin et al., 2008; Kinzig, Hargrave, & Honors, 2008; Krolow et al., 2010; Maniam, Antoniadis, Le, & Morris, 2016; Maniam & Morris, 2010a, 2010b, 2010c; Nanni et al., 2003; Pecoraro, Reyes, Gomez, Bhargava, & Dallman, 2004; Teegarden & Bale, 2008; Ulrich-Lai et al., 2010; Ulrich-Lai, Ostrander, & Herman, 2011; Ulrich-Lai et al., 2007; Warne, 2009).

To model stress-induced relapse of compulsive eating, many studies have used the alpha-2 adrenergic antagonist yohimbine, which can produce high anxiety states (but see Chen et al., 2015) and reinstate responding for palatable food pellets and sucrose solutions (Ghitza, Gray, Epstein, Rice, & Shaham, 2006; Le et al., 2011; Richards et al., 2008). Yohimbine reinstates seeking of diverse palatable food pellets, but not of energy-devoid, less palatable, cellulose fiber pellets (Nair, Gray, & Ghitza, 2006). Studies have implicated CRF (Ghitza et al., 2006), orexin (hypocretin) (Richards et al., 2008), postsynaptic alpha-1 adrenoceptor (Le et al., 2011), and dorsal mPFC systems (Calu et al., 2013), including D1 receptors (Ball et al., 2016; Nair et al., 2011) in yohimbine-induced reinstatement.

Boggiano et al. followed by Cifani et al. have modeled the synergistic relationship between a history of cyclic, yo-yo-like intake, and stress triggers in promoting binge-like eating. In these models, weeks of intermittent restriction/refeeding combined with a stress trigger (footshock or frustrative nonrewarded presentation of an inaccessible palatable food, respectively) increase intake of palatable food (cookies or Nutella-mixed chow, respectively) (Boggiano & Chandler, 2006; Cifani, Polidori, et al., 2009; Hagan et al., 2002). Adaptation of endogenous opioid systems is implicated because naloxone challenge decreased and the μ/κ-agonist butorphanol increased palatable food intake in restricted + stressed rats specifically (Boggiano et al., 2005). Changes in the antistress or feeding-regulatory properties of endogenous nociception/orphanin FQ (N/OFQ) may be involved as well (Micioni Di Bonaventura et al., 2013; Pucci et al., 2016). In addition to the recruitment of CRF-CRF_1 systems in the central extended amygdala described earlier, recruitment of orexin-1 (OX_1) receptor systems is also implicated because the OX_1 receptor antagonists SB-649868 and GSK1059865 selectively reduced binge-like eating at doses that did not reduce chow intake or promote sleep (Piccoli et al., 2012).

Negative urgency and compulsive eating

The construct of negative urgency may play an integrating role in the "dark side" bridging impulsivity with compulsivity. Conceived as an emotion-based trait, negative urgency refers to acting rashly and impulsively when in extreme distress and involves impaired inhibitory control (Cyders & Smith, 2008). Not only negative affect but also negative urgency (Murphy, Stojek, & MacKillop, 2014; Pivarunas & Conner, 2015; Rose, Nadler, & Mackey, 2018; VanderBroek-Stice, Stojek, Beach, vanDellen, & MacKillop, 2017), as well as attentional ("decreased ability to focus") and motor ("acting without thinking") impulsivity, are comorbid in YFAS-defined food addiction and BN (Ceccarini et al., 2015; de Vries & Meule, 2016; Meule, de Zwaan, & Muller, 2017; Meule et al., 2015; Meule, Muller, et al., 2017). Impulsivity also associates with pediatric loss of control eating (Reinblatt et al., 2015). Conversely, many promising psychotropic (e.g., lisdexamfetamine dimesylate, tesofensine, dasotraline [(NCT03107026, NCT02684279)]) (Heal, Goddard, Brammer, Hutson, & Vickers, 2016; Hopkins et al., 2016; Koblan et al., 2015; McElroy, Mitchell, et al., 2016; Vickers, Goddard, Brammer, Hutson, & Heal, 2017) and nonpharmacological treatments (e.g., CBT, dialectical behavior therapy, integrated cognitive-affective therapy) not only target negative affect but also increased self-regulation and inhibitory control of impulses to binge during distress (Cancian, de Souza, Liboni, Machado, & Oliveira, 2017; Chen, Matthews, Allen, Kuo, & Linehan, 2008; Murray et al., 2015; Wallace, Masson, Safer, & von Ranson, 2014; Wonderlich et al., 2014) (see also Yarnell, Oscar-Berman, Avena, Blum, & Gold, 2013). Negative urgency relates more strongly to binge eating in patients with BN and to uncontrolled eating in adolescents than do other aspects of impulsivity, such as sensation seeking, lack of planning, or lack of persistence (Booth, Spronk, Grol, & Fox, 2018; Fischer, Smith, & Cyders, 2008; Pearson, Riley, Davis, & Smith, 2014). Furthermore, negative urgency prospectively predicts binge eating symptoms from elementary to middle school as well as in college students (Pearson, Zapolski, et al., 2015) and associates with more frequent snacking in adolescents (Coumans et al., 2018; G. T. Smith & Cyders, 2016). Negative urgency has transdiagnostic significance; it also is implicated in compulsive substance use, including for cigarettes (Billieux, Van der Linden, & Ceschi, 2007) and alcohol (Stautz & Cooper, 2013), as well as compulsive cell phone use, shopping, and gambling (Billieux, Rochat, Rebetz, & Van der Linden, 2008; Billieux, Van der Linden, D'Acremont, Ceschi, & Zermatten, 2007; Maclaren, Fugelsang, Harrigan, & Dixon, 2011). Accordingly, negative urgency predicts comorbid alcohol use in women with disordered eating (Fischer, Smith, Annus, & Hendricks, 2007).

Neurobiologically, negative urgency has been proposed to reflect impaired "top-down" cortical→amygdala/striatal processing, leading to reduced inhibitory control over potentially detrimental actions, as well as heightened "bottom-up" amygdala→cortical/striatal processing, leading to increased attention to, incentive salience of or cognitive resource interference by, emotion-evoking stimuli. These biases putatively reflect altered structure, function, and/or connectivity of reciprocal amygdala–orbitofrontal/ventromedial prefrontal cortical (OFC/vmPFC) projections (Cyders & Smith, 2008; Smith & Cyders, 2016) (also see Robbins, Gillan, Smith, de Wit, & Ersche, 2012). Supporting this hypothesis, trait urgency was related to the amplitude of resting-state low-frequency fluctuations in the lateral OFC and vmPFC in healthy volunteers (Zhao et al., 2017). Trait negative urgency also predicted increased vmPFC activation in response to an alcohol odor cue in social drinkers and mediated the relation between vmPFC activation and self-reported alcohol craving and problematic drinking (Cyders et al., 2014). Similarly, negative urgency predicted greater OFC and amygdala activation in response to negative visual stimuli and mediated the relation between activation and risky behavior (Cyders et al., 2015). In addition, negative urgency is also associated with resting-state and or inhibitory task–related activation (Go/No-go or gambling tasks) in other structures that subserve self-regulation and decision-making under risk. These included the dorsolateral and ventrolateral PFC, anterior insula, and anterior and posterior cingulate (Chester, Lynam, Milich, Powell, et al., 2016; L. Clark et al., 2008; Hoptman, Antonius, Mauro, Parker, & Javitt, 2014; Xue, Lu, Levin, & Bechara, 2010; Zhao et al., 2017). Greater insula activation prospectively predicted real-world substance use in subjects high in negative urgency (Chester, Lynam, Milich, Powell, et al., 2016). Negative urgency also predicted greater increases in mPFC activity during anticipation of a delayed incentive task (Weiland et al., 2014). Finally, in patients with schizophrenia, urgency was associated with reduced cortical thickness in such structures, including the vmPFC, orbitofrontal and inferior frontal gyri, and rostral anterior cingulate cortex (Hoptman et al., 2014).

Neurochemically, negative urgency may reflect deficient 5-HT and disinhibited DA activity within the OFC and vmPFC (Cyders & Smith, 2008; Floresco & Tse, 2007), leading to less inhibition of amygdala-subserved impulses. Accordingly, a composite polygenic 5-HT score predicted alcohol problems via increased negative urgency and not via other measures of impulsivity (Carver, Johnson, Joormann, Kim, & Nam, 2011; Wang & Chassin, 2018). Evidence of reduced 5-HT activity and responsiveness is likewise seen in BN in relation to the presence and severity of binge symptoms (Cyders &

Smith, 2008). Data also implicate genetic variation in the GABRA2 subunit, which relates to alcohol-use problems via urgency, in altered insula activation responses (Villafuerte et al., 2012) and dorsolateral mPFC GABA concentrations (Boy et al., 2011).

Traditionally, negative urgency has been conceptualized as a stable dispositional antecedent that potentiates responses to extreme situational distress (e.g., Engel et al., 2007; Fischer et al., 2018). For example, negative urgency predicted greater subsequent mood change, alcohol cue–induced craving, and intravenous alcohol self-administration after a negative mood induction (VanderVeen et al., 2016), as well as increased negative affect and relative reinforcing value after laboratory stressors (Owens, Amlung, Stojek, & MacKillop, 2018). Within a "dark side" view, trait negative urgency would be viewed as increasing the likelihood of using palatable food to compensate for opponent-process-associated hypohedonia and negative affect or to otherwise use food in comforting fashion to cope with life distress (Racine et al., 2013) (but see Davis-Becker, Peterson, & Fischer, 2014).

Negative urgency is not strictly trait-like, however. Situational factors can impact inhibitory signals from the OFC/vmPFC to the amygdala (Silbersweig et al., 2007). Relevant to compulsive eating, experimental skipping of breakfast, which increases the appeal of high (vs. low) calorie foods, increases fMRI BOLD activation in the amygdala, OFC, and anterior insula (Ely et al., 2017; Goldstone et al., 2009). Conversely, a high-protein breakfast intervention in "breakfast skippers," which reduces postmeal craving for sweet and savory foods (Hoertel, Will, & Leidy, 2014), reduces predinner amygdala and insula activation (Leidy, Ortinau, Douglas, & Hoertel, 2013). Moreover, experience can elicit enduring changes in urgency. For example, childhood abuse persistently increases amygdala activity and reduces PFC control over amygdalar responses (Teicher et al., 2003), and effective psychotherapy can elicit converse effects (D. L. Perez, Vago, et al., 2016) Accordingly, longitudinal increases in negative urgency scores also predict increases in bulimic symptoms (Anestis, Selby, & Joiner, 2007).

A novel hypothesis proposed here is that opponent-process adaptations to palatable food (or other substances of abuse) may increase both negative affect and negative urgency. Indeed, decreases in striatal D2 receptor availability in obese subjects correlate with reduced glucose metabolism in frontal cortical regions that subserve inhibitory control, including dorsolateral prefrontal, orbitofrontal, and anterior cingulate cortices (Michaelides, Thanos, Volkow, & Wang, 2012; Tomasi & Volkow, 2013; Volkow & Baler, 2015; Volkow, Wang, Telang, et al., 2008). Analogously, a direct correlation of striatal D2 availability with glucose metabolism in dorsolateral and anterior cingulate cortices also has been observed in individuals with an AUD (Volkow et al., 2007). Importantly, these relationships are not seen in non-AUD or nonobese controls, consistent with the hypothesis that they reflect circuitry changes. Perhaps accordingly, negative urgency also predicted greater caudate responses to alcohol images in alcohol-dependent individuals (Chester, Lynam, Milich, & DeWall, 2016). Similarly, in pathological gamblers, higher negative urgency correlates with reduced striatal D2 availability, as indexed by [11C]-raclopride binding potential (L. Clark et al., 2012).

More generally, the increased substance cue reactivity seen in the amygdala, OFC, cingulate cortex, vmPFC and dlPFC, and anterior insula of people with substance-use disorders and that predicts relapse (Engelmann et al., 2012; Goudriaan et al., 2010; Heinz et al., 2009; Jasinska et al., 2014; Mainz et al., 2012) may reflect adaptations within negative urgency circuits, and not (only) reward processing as is often interpreted (e.g., Schulte, Grilo, & Gearhardt, 2016; Stice et al., 2015; Winter et al., 2017). Similarly, as alluded to earlier, YFAS scores correlated directly with milkshake picture–induced activation of the amygdala, medial OFC, and anterior cingulate; individuals with categorically higher YFAS scores also showed greater dlPFC activation (Gearhardt et al., 2011). Likewise, obese, relative to lean, women showed increased activation of the amygdala, vmPFC, and inhibitory frontal operculum (Higo, Mars, Boorman, Buch, & Rushworth, 2011) in response to pictures or taste of a palatable milkshake (Ng et al., 2011) and of the amygdala, OFC, anterior cingulate, insula, and mPFC in response to pictures of palatable, high-calorie, but not less-preferred, low-calorie foods (Stoeckel et al., 2008). Resting-state functional connectivity of the amygdala is also increased in obese patients, including to the insula (Lips et al., 2014; Wijngaarden et al., 2015). Graph theory network analysis also showed decreased nodal degree/efficiency in the amygdala, medial OFC, rostral anterior cingulate, and insula of obese versus nonobese subjects. There was disruption of small-world (local) organization and a global reduction of higher-order network integration in obese volunteers. Within obese volunteers, greater BMI correlated with decreased global efficiency (Eglob) and especially decreased nodal degree/efficiency of the medial OFC (Meng et al., 2018). Finally, structural differences have been seen in this network. Obese patients showed decreased gray matter densities in the OFC, inferior and superior frontal gyri, rostral anterior cingulate, insula, and dmPFC, whereby reduced OFC gray matter/white matter ratios correlated with greater BMI and YFAS scores. Remarkably, many of these cortical structural differences normalized following bariatric surgery (Zhang et al., 2016).

As alluded to earlier, animal models based on intermittent access to palatable food also have shown adaptations in the amygdala and frontal cortex (Blasio, Steardo, Sabino, & Cottone, 2014; Cottone, Sabino, Roberto, et al., 2009; Iemolo et al., 2013). Thus, it is conceivable that increased negative urgency, because of plasticity in amygdala-vmPFC/OFC/insula circuits, may underlie the risky or otherwise less-inhibited eating-directed behaviors that develop in these models. Examples include food-seeking or self-administration despite threat of footshock; rapid emergence into exposed spaces where palatable food may be present; and vigorous food-seeking behavior despite decreasing reinforcement (Johnson & Kenny, 2010; Moore et al., 2017; Oswald et al., 2011; Parylak et al., 2011, 2012; Rossetti et al., 2014; Spierling et al., 2018; Teegarden & Bale, 2007). New experimental models that specifically address the construct of negative urgency, as opposed to other aspect of impulsivity (Cyders & Coskunpinar, 2011), would facilitate testing of this hypothesis.

Conclusion

Just as the transition from drug use to dependence is accompanied by a downregulation of brain reward circuitry, enhancement of "antireward"/stress circuitry, and dysfunction of negative urgency cortico–amygdala–cortical circuitry, both human data and animal models support the hypothesis that the transition to compulsive eating and food addiction involves an opponent-process "dark side" of affective dysregulation. However, significant challenges remain to address this culturally recognized global problem. Further work is needed to reach consensus on diagnostic criteria for food addiction in humans. Refinement of such criteria will further the development of suitable animal models to better study the most critical aspects of this disorder and identify the molecular loading on specific neurochemical circuits that convey vulnerability and resilience to compulsive eating.

Acknowledgments/Support

The authors thank Michael Arends for assistance with manuscript preparation. Financial support for this work was provided by the Pearson Center for Alcoholism and Addiction Research and National Institute on Alcohol Abuse and Alcoholism grant AA006420. The content is solely the responsibility of the authors and does not necessarily represent the official views of the National Institutes of Health.

Conflict of Interest.

EPZ and GFK are inventors on a patent filed for CRF_1 receptor antagonists (USPTO application no. 2010/0249138).

References

Abe, K., Fukuda, K., & Tokui, T. (2004). Marginal involvement of pyroglutamyl aminopeptidase I in metabolism of thyrotropin-releasing hormone in rat brain. *Biological and Pharmaceutical Bulletin, 27*(8), 1197–1201.

Abramson, E. E., & Wunderlich, R. A. (1972). Anxiety, fear and eating: A test of the psychosomatic concept of obesity. *Journal of Abnormal Psychology, 79*(3), 317–321.

Ackard, D. M., Croll, J. K., & Kearney-Cooke, A. (2002). Dieting frequency among college females: Association with disordered eating, body image, and related psychological problems. *Journal of Psychosomatic Research, 52*(3), 129–136.

Adjibade, M., Lemogne, C., Julia, C., Hercberg, S., Galan, P., Assmann, K. E., & Kesse-Guyot, E. (2018). Prospective association between adherence to dietary recommendations and incident depressive symptoms in the French NutriNet-Sante cohort. *British Journal of Nutrition*, 1–11. https://doi.org/10.1017/S0007114518000910.

Agras, W. S., & Telch, C. F. (1998). Effects of caloric deprivation and negative affect on binge eating in obese binge eating disordered women. *Behavior Therapy, 29*, 491–503.

Akbaraly, T. N., Brunner, E. J., Ferrie, J. E., Marmot, M. G., Kivimaki, M., & Singh-Manoux, A. (2009). Dietary pattern and depressive symptoms in middle age. *British Journal of Psychiatry, 195*(5), 408–413. https://doi.org/10.1192/bjp.bp.108.058925.

Allison, S., & Timmerman, G. M. (2007). Anatomy of a binge: Food environment and characteristics of nonpurge binge episodes. *Eating Behaviors, 8*(1), 31–38. https://doi.org/10.1016/j.eatbeh.2005.01.004.

Alsio, J., Olszewski, P. K., Norback, A. H., Gunnarsson, Z. E., Levine, A. S., Pickering, C., & Schioth, H. B. (2010). Dopamine D1 receptor gene expression decreases in the nucleus accumbens upon long-term exposure to palatable food and differs depending on diet-induced obesity phenotype in rats. *Neuroscience, 171*(3), 779–787. https://doi.org/10.1016/j.neuroscience.2010.09.046.

Anestis, M. D., Selby, E. A., & Joiner, T. E. (2007). The role of urgency in maladaptive behaviors. *Behaviour Research and Therapy, 45*(12), 3018–3029. https://doi.org/10.1016/j.brat.2007.08.012.

Anonymous, A. (2001). *Alcoholics anonymous: The story of how many thousands of men and women have recovered from alcoholism* (4th ed.). New York City: Alcoholics Anonymous World Services, Inc.

Apovian, C. M. (2016). Naltrexone/bupropion for the treatment of obesity and obesity with Type 2 diabetes. *Future Cardiology, 12*(2), 129–138. https://doi.org/10.2217/fca.15.79.

Appel, L., Bergstrom, M., Buus Lassen, J., & Langstrom, B. (2014). Tesofensine, a novel triple monoamine re-uptake inhibitor with anti-obesity effects: Dopamine transporter occupancy as measured by PET. *European Neuropsychopharmacology, 24*(2), 251–261. https://doi.org/10.1016/j.euroneuro.2013.10.007.

Association, A. P. (2015). *Stress in America™: Paying with our health*. Retrieved from https://www.apa.org/news/press/releases/stress/2014/stress-report.pdf.

Astrup, A., Madsbad, S., Breum, L., Jensen, T. J., Kroustrup, J. P., & Larsen, T. M. (2008). Effect of tesofensine on bodyweight loss, body composition, and quality of life in obese patients: A randomised, double-blind, placebo-controlled trial. *Lancet, 372*(9653), 1906–1913. https://doi.org/10.1016/S0140-6736(08)61525-1.

Astrup, A., Meier, D. H., Mikkelsen, B. O., Villumsen, J. S., & Larsen, T. M. (2008). Weight loss produced by tesofensine in patients with Parkinson's or Alzheimer's disease. *Obesity, 16*(6), 1363–1369. https://doi.org/10.1038/oby.2008.56.

Athanasos, P., Ling, W., Bochner, F., White, J. M., & Somogyi, A. A. (2018). Buprenorphine maintenance subjects are hyperalgesic and have No antinociceptive response to a very high morphine dose. *Pain Medicine*. https://doi.org/10.1093/pm/pny025.

Avena, N. M., Bocarsly, M. E., Hoebel, B. G., & Gold, M. S. (2011). Overlaps in the nosology of substance abuse and overeating: The translational implications of "food addiction". *Current Drug Abuse Reviews, 4*(3), 133–139.

Avena, N. M., Carrillo, C. A., Needham, L., Leibowitz, S. F., & Hoebel, B. G. (2004). Sugar-dependent rats show enhanced intake of unsweetened ethanol. *Alcohol, 34*(2–3), 203–209.

Avena, N. M., Gearhardt, A. N., Gold, M. S., Wang, G. J., & Potenza, M. N. (2012). Tossing the baby out with the bathwater after a brief rinse? The potential downside of dismissing food addiction based on limited data. *Nature Reviews Neuroscience, 13*(7), 514. https://doi.org/10.1038/nrn3212-c1. author reply 514.

Avena, N. M., & Gold, M. S. (2011a). Food and addiction - sugars, fats and hedonic overeating. *Addiction, 106*(7), 1214–1215. https://doi.org/10.1111/j.1360-0443.2011.03373.x. discussion 1219-1220.

Avena, N. M., & Gold, M. S. (2011b). Variety and hyperpalatability: Are they promoting addictive overeating? *American Journal of Clinical Nutrition, 94*(2), 367–368. https://doi.org/10.3945/ajcn.111.020164.

Avena, N. M., Murray, S., & Gold, M. S. (2013). The next generation of obesity treatments: Beyond suppressing appetite. *Frontiers in Psychology, 4*, 721. https://doi.org/10.3389/fpsyg.2013.00721.

Avena, N. M., Rada, P., & Hoebel, B. G. (2008). Evidence for sugar addiction: Behavioral and neurochemical effects of intermittent, excessive sugar intake. *Neuroscience & Biobehavioral Reviews, 32*(1), 20–39. https://doi.org/10.1016/j.neubiorev.2007.04.019.

Axel, A. M., Mikkelsen, J. D., & Hansen, H. H. (2010). Tesofensine, a novel triple monoamine reuptake inhibitor, induces appetite suppression by indirect stimulation of alpha1 adrenoceptor and dopamine D1 receptor pathways in the diet-induced obese rat. *Neuropsychopharmacology, 35*(7), 1464–1476. https://doi.org/10.1038/npp.2010.16.

Ayaz, A., Nergiz-Unal, R., Dedebayraktar, D., Akyol, A., Pekcan, A. G., Besler, H. T., & Buyuktuncer, Z. (2018). How does food addiction influence dietary intake profile? *PLoS One, 13*(4), e0195541. https://doi.org/10.1371/journal.pone.0195541.

Azami, Y., Funakoshi, M., Matsumoto, H., Ikota, A., Ito, K., Okimoto, H., … Miura, J. (2018). Long working hours and skipping breakfast concomitant with late evening meals are associated with suboptimal glycemic control among young male Japanese patients with type 2 diabetes. *J Diabetes Investig*. https://doi.org/10.1111/jdi.12852.

Azam, H., Shahrestani, S., & Phan, K. (2018). Alcohol use disorders before and after bariatric surgery: A systematic review and meta-analysis. *Annals of Translational Medicine, 6*(8), 148. https://doi.org/10.21037/atm.2018.03.16.

Baker, T. B., Morse, E., & Sherman, J. E. (1986). The motivation to use drugs: A psychobiological analysis of urges. *Current Theory and Research in Motivation, 34*, 257–323.

Baker, T. B., Piper, M. E., McCarthy, D. E., Majeskie, M. R., & Fiore, M. C. (2004). Addiction motivation reformulated: An affective processing model of negative reinforcement. *Psychological Review, 111*(1), 33–51. https://doi.org/10.1037/0033-295X.111.1.33.

Baker, T. B., Piper, M. E., Schlam, T. R., Cook, J. W., Smith, S. S., Loh, W. Y., & Bolt, D. (2012). Are tobacco dependence and withdrawal related amongst heavy smokers? Relevance to conceptualizations of dependence. *Journal of Abnormal Psychology, 121*(4), 909–921. https://doi.org/10.1037/a0027889.

Bak, M., Seibold-Simpson, S. M., & Darling, R. (2016). The potential for cross-addiction in post-bariatric surgery patients: Considerations for primary care nurse practitioners. *Journal of the American Association of Nurse Practitioners, 28*(12), 675–682. https://doi.org/10.1002/2327-6924.12390.

Ball, K. T., Miller, L., Sullivan, C., Wells, A., Best, O., Cavanaugh, B., & Vieweg, L. (2016). Effects of repeated yohimbine administration on reinstatement of palatable food seeking: Involvement of dopamine D1 -like receptors and food-associated cues. *Addiction Biology, 21*(6), 1140–1150. https://doi.org/10.1111/adb.12287.

Banting, W. (1993). Letter on corpulence, addressed to the public. 1869. *Obesity Research, 1*(2), 153–163.

Barson, J. R., Karatayev, O., Chang, G. Q., Johnson, D. F., Bocarsly, M. E., Hoebel, B. G., & Leibowitz, S. F. (2009). Positive relationship between dietary fat, ethanol intake, triglycerides, and hypothalamic peptides: Counteraction by lipid-lowering drugs. *Alcohol, 43*(6), 433–441. https://doi.org/10.1016/j.alcohol.2009.07.003.

Baskin, R., Hill, B., Jacka, F. N., O'Neil, A., & Skouteris, H. (2017). Antenatal dietary patterns and depressive symptoms during pregnancy and early post-partum. *Maternal and Child Nutrition, 13*(1). https://doi.org/10.1111/mcn.12218.

Basso, A. M., Spina, M., Rivier, J., Vale, W., & Koob, G. F. (1999). Corticotropin-releasing factor antagonist attenuates the "anxiogenic-like" effect in the defensive burying paradigm but not in the elevated plus-maze following chronic cocaine in rats. *Psychopharmacology, 145*(1), 21–30.

Bello, N. T., Guarda, A. S., Terrillion, C. E., Redgrave, G. W., Coughlin, J. W., & Moran, T. H. (2009). Repeated binge access to a palatable food alters feeding behavior, hormone profile, and hindbrain c-Fos responses to a test meal in adult male rats. *American Journal of Physiology – Regulatory, Integrative and Comparative Physiology, 297*(3), R622–R631. https://doi.org/10.1152/ajpregu.00087.2009.

Bello, N. T., & Hajnal, A. (2010). Dopamine and binge eating behaviors. *Pharmacology Biochemistry and Behavior, 97*(1), 25–33. https://doi.org/10.1016/j.pbb.2010.04.016.

Bello, N. T., Lucas, L. R., & Hajnal, A. (2002). Repeated sucrose access influences dopamine D2 receptor density in the striatum. *NeuroReport, 13*(12), 1575–1578.

Bello, N. T., Sweigart, K. L., Lakoski, J. M., Norgren, R., & Hajnal, A. (2003). Restricted feeding with scheduled sucrose access results in an upregulation of the rat dopamine transporter. *American Journal of Physiology – Regulatory, Integrative and Comparative Physiology, 284*(5), R1260–R1268. https://doi.org/10.1152/ajpregu.00716.2002.

Benton, D., & Owens, D. (1993). Is raised blood glucose associated with the relief of tension? *Journal of Psychosomatic Research, 37*(7), 723–735.

Berenson, A. B., Laz, T. H., Pohlmeier, A. M., Rahman, M., & Cunningham, K. A. (2015). Prevalence of food addiction among low-income reproductive-aged women. *Journal of Women's Health, 24*(9), 740–744. https://doi.org/10.1089/jwh.2014.5182.

Berg, K. C., Cao, L., Crosby, R. D., Engel, S. G., Peterson, C. B., Crow, S. J., … Wonderlich, S. A. (2017). Negative affect and binge eating: Reconciling differences between two analytic approaches in ecological momentary assessment research. *International Journal of Eating Disorders, 50*(10), 1222–1230. https://doi.org/10.1002/eat.22770.

Berg, K. C., Crosby, R. D., Cao, L., Crow, S. J., Engel, S. G., Wonderlich, S. A., & Peterson, C. B. (2015). Negative affect prior to and following overeating-only, loss of control eating-only, and binge eating episodes in obese adults. *International Journal of Eating Disorders, 48*(6), 641–653. https://doi.org/10.1002/eat.22401.

Berg, K. C., Crosby, R. D., Cao, L., Peterson, C. B., Engel, S. G., Mitchell, J. E., & Wonderlich, S. A. (2013). Facets of negative affect prior to and following binge-only, purge-only, and binge/purge events in women with bulimia nervosa. *Journal of Abnormal Psychology, 122*(1), 111–118. https://doi.org/10.1037/a0029703.

Bi, H., Gan, Y., Yang, C., Chen, Y., Tong, X., & Lu, Z. (2015). Breakfast skipping and the risk of type 2 diabetes: A meta-analysis of observational studies. *Public Health Nutrition, 18*(16), 3013–3019. https://doi.org/10.1017/S1368980015000257.

Billieux, J., Rochat, L., Rebetz, M. M. L., & Van der Linden, M. (2008). Are all facets of impulsivity related to self-reported compulsive buying behavior? *Personality and Individual Differences, 44*, 1432–1442.

Billieux, J., Van der Linden, M., & Ceschi, G. (2007a). Which dimensions of impulsivity are related to cigarette craving? *Addictive Behaviors, 32*(6), 1189–1199. https://doi.org/10.1016/j.addbeh.2006.08.007.

Billieux, J., Van der Linden, M., D'Acremont, M., Ceschi, G., & Zermatten, A. (2007b). Does impulsivity relate to perceived dependence on and actual use of the mobile phone? *Applied Cognitive Psychology, 21*, 527–537.

Blasio, A., Iemolo, A., Sabino, V., Petrosino, S., Steardo, L., Rice, K. C., … Cottone, P. (2013). Rimonabant precipitates anxiety in rats withdrawn from palatable food: Role of the central amygdala. *Neuropsychopharmacology, 38*(12), 2498–2507. https://doi.org/10.1038/npp.2013.153.

Blasio, A., Rice, K. C., Sabino, V., & Cottone, P. (2014a). Characterization of a shortened model of diet alternation in female rats: Effects of the CB1 receptor antagonist rimonabant on food intake and anxiety-like behavior. *Behavioural Pharmacology, 25*(7), 609–617. https://doi.org/10.1097/FBP.0000000000000059.

Blasio, A., Steardo, L., Sabino, V., & Cottone, P. (2014b). Opioid system in the medial prefrontal cortex mediates binge-like eating. *Addiction Biology, 19*(4), 652–662. https://doi.org/10.1111/adb.12033.

Bluett, R. J., Baldi, R., Haymer, A., Gaulden, A. D., Hartley, N. D., Parrish, W. P., ... Patel, S. (2017). Endocannabinoid signalling modulates susceptibility to traumatic stress exposure. *Nature Communications, 8*, 14782. https://doi.org/10.1038/ncomms14782.

Blum, K., Thanos, P. K., & Gold, M. S. (2014). Dopamine and glucose, obesity, and reward deficiency syndrome. *Frontiers in Psychology, 5*, 919. https://doi.org/10.3389/fpsyg.2014.00919.

Blum, K., Thanos, P. K., Wang, G. J., Febo, M., Demetrovics, Z., Modestino, E. J., ... Gold, M. S. (2018). The food and drug addiction epidemic: Targeting dopamine homeostasis. *Current Pharmaceutical Design, 23*(39), 6050–6061. https://doi.org/10.2174/1381612823666170823101713.

Boggiano, M. M. (2016). Palatable Eating Motives Scale in a college population: Distribution of scores and scores associated with greater BMI and binge-eating. *Eating Behaviors, 21*, 95–98. https://doi.org/10.1016/j.eatbeh.2016.01.001.

Boggiano, M. M., Artiga, A. I., Pritchett, C. E., Chandler-Laney, P. C., Smith, M. L., & Eldridge, A. J. (2007). High intake of palatable food predicts binge-eating independent of susceptibility to obesity: An animal model of lean vs obese binge-eating and obesity with and without binge-eating. *International Journal of Obesity, 31*(9), 1357–1367. https://doi.org/10.1038/sj.ijo.0803614.

Boggiano, M. M., Burgess, E. E., Turan, B., Soleymani, T., Daniel, S., Vinson, L. D., ... Morse, A. (2014). Motives for eating tasty foods associated with binge-eating. Results from a student and a weight-loss seeking population. *Appetite, 83*, 160–166. https://doi.org/10.1016/j.appet.2014.08.026.

Boggiano, M. M., & Chandler, P. C. (2006). Binge eating in rats produced by combining dieting with stress. *Current Protocols in Neuroscience*. https://doi.org/10.1002/0471142301.ns0923as36 (Chapter 9), Unit9 23A.

Boggiano, M. M., Chandler, P. C., Viana, J. B., Oswald, K. D., Maldonado, C. R., & Wauford, P. K. (2005). Combined dieting and stress evoke exaggerated responses to opioids in binge-eating rats. *Behavioral Neuroscience, 119*(5), 1207–1214. https://doi.org/10.1037/0735-7044.119.5.1207.

Boggiano, M. M., Wenger, L. E., Burgess, E. E., Tatum, M. M., Sylvester, M. D., Morgan, P. R., & Morse, K. E. (2017). Eating tasty foods to cope, enhance reward, socialize or conform: What other psychological characteristics describe each of these motives? *Journal of Health Psychology, 22*(3), 280–289. https://doi.org/10.1177/1359105315600240.

Boggiano, M. M., Wenger, L. E., Mrug, S., Burgess, E. E., & Morgan, P. R. (2015a). The kids-palatable eating motives scale: Relation to BMI and binge eating traits. *Eating Behaviors, 17*, 69–73. https://doi.org/10.1016/j.eatbeh.2014.12.014.

Boggiano, M. M., Wenger, L. E., Turan, B., Tatum, M. M., Morgan, P. R., & Sylvester, M. D. (2015b). Eating tasty food to cope. Longitudinal association with BMI. *Appetite, 87*, 365–370. https://doi.org/10.1016/j.appet.2015.01.008.

Boggiano, M. M., Wenger, L. E., Turan, B., Tatum, M. M., Sylvester, M. D., Morgan, P. R., … Burgess, E. E. (2015c). Real-time sampling of reasons for hedonic food consumption: Further validation of the palatable eating motives scale. *Frontiers in Psychology, 6*, 744. https://doi.org/10.3389/fpsyg.2015.00744.

Bohon, C., & Stice, E. (2012). Negative affect and neural response to palatable food intake in bulimia nervosa. *Appetite, 58*(3), 964–970. https://doi.org/10.1016/j.appet.2012.02.051.

Bold, K. W., Rasheed, A. S., McCarthy, D. E., Jackson, T. C., Fiore, M. C., & Baker, T. B. (2015). Rates and predictors of renewed quitting after relapse during a one-year follow-up among primary care patients. *Annals of Behavioral Medicine, 49*(1), 128–140. https://doi.org/10.1007/s12160-014-9627-6.

Bolles, R. C. (1972). Reinforcement, expectancy and learning. *Psychological Review, 79*, 394–409.

Booth, C., Spronk, D., Grol, M., & Fox, E. (2018). Uncontrolled eating in adolescents: The role of impulsivity and automatic approach bias for food. *Appetite, 120*, 636–643. https://doi.org/10.1016/j.appet.2017.10.024.

Boutelle, K. N., Wierenga, C. E., Bischoff-Grethe, A., Melrose, A. J., Grenesko-Stevens, E., Paulus, M. P., & Kaye, W. H. (2015). Increased brain response to appetitive tastes in the insula and amygdala in obese compared with healthy weight children when sated. *International Journal of Obesity, 39*(4), 620–628. https://doi.org/10.1038/ijo.2014.206.

Bowen, D. J., Kestin, M., McTiernan, A., Carrell, D., & Green, P. (1995). Effects of dietary fat intervention on mental health in women. *Cancer Epidemiology Biomarkers & Prevention, 4*(5), 555–559.

Boy, F., Evans, C. J., Edden, R. A., Lawrence, A. D., Singh, K. D., Husain, M., & Sumner, P. (2011). Dorsolateral prefrontal gamma-aminobutyric acid in men predicts individual differences in rash impulsivity. *Biological Psychiatry, 70*(9), 866–872. https://doi.org/10.1016/j.biopsych.2011.05.030.

Bray, G. A., Kim, K. K., Wilding, J. P. H., & World Obesity, F. (2017). Obesity: A chronic relapsing progressive disease process. A position statement of the world obesity federation. *Obesity Reviews, 18*(7), 715–723. https://doi.org/10.1111/obr.12551.

Brennan, B. P., Roberts, J. L., Fogarty, K. V., Reynolds, K. A., Jonas, J. M., & Hudson, J. I. (2008). Memantine in the treatment of binge eating disorder: An open-label, prospective trial. *International Journal of Eating Disorders, 41*(6), 520–526. https://doi.org/10.1002/eat.20541.

Brewerton, T. D. (2017). Food addiction as a proxy for eating disorder and obesity severity, trauma history, PTSD symptoms, and comorbidity. *Eating and Weight Disorders, 22*(2), 241–247. https://doi.org/10.1007/s40519-016-0355-8.

Broft, A., Shingleton, R., Kaufman, J., Liu, F., Kumar, D., Slifstein, M., … Walsh, B. T. (2012). Striatal dopamine in bulimia nervosa: A PET imaging study. *International Journal of Eating Disorders, 45*(5), 648–656. https://doi.org/10.1002/eat.20984.

Brownell, K. D., & Gold, M. S. (2012). *Food and addiction: A comprehensive handbook*. New York: Oxford University Press.

Brown, K. L., LaRose, J. G., & Mezuk, B. (2018). The relationship between body mass index, binge eating disorder and suicidality. *BMC Psychiatry, 18*(1), 196. https://doi.org/10.1186/s12888-018-1766-z.

Bruijnzeel, A. W., Prado, M., & Isaac, S. (2009). Corticotropin-releasing factor-1 receptor activation mediates nicotine withdrawal-induced deficit in brain reward function and stress-induced relapse. *Biological Psychiatry, 66*(2), 110–117. https://doi.org/10.1016/j.biopsych.2009.01.010. S0006-3223(09)00035-3 [pii].

Bruijnzeel, A. W., Small, E., Pasek, T. M., & Yamada, H. (2010). Corticotropin-releasing factor mediates the dysphoria-like state associated with alcohol withdrawal in rats. *Behavioural Brain Research, 210*(2), 288–291. https://doi.org/10.1016/j.bbr.2010.02.043.

Brunault, P., Courtois, R., Gearhardt, A. N., Gaillard, P., Journiac, K., Cathelain, S., … Ballon, N. (2017). Validation of the French version of the DSM-5 Yale food addiction scale in a nonclinical sample. *Canadian Journal of Psychiatry, 62*(3), 199–210. https://doi.org/10.1177/0706743716673320.

Burger, K. S., & Stice, E. (2012). Frequent ice cream consumption is associated with reduced striatal response to receipt of an ice cream-based milkshake. *American Journal of Clinical Nutrition, 95*(4), 810–817. https://doi.org/10.3945/ajcn.111.027003.

Burgess, E. E., Turan, B., Lokken, K. L., Morse, A., & Boggiano, M. M. (2014). Profiling motives behind hedonic eating. Preliminary validation of the palatable eating motives scale. *Appetite, 72*, 66–72. https://doi.org/10.1016/j.appet.2013.09.016.

Burmeister, J. M., Hinman, N., Koball, A., Hoffmann, D. A., & Carels, R. A. (2013). Food addiction in adults seeking weight loss treatment. Implications for psychosocial health and weight loss. *Appetite, 60*(1), 103–110. https://doi.org/10.1016/j.appet.2012.09.013.

Burrows, T., Hides, L., Brown, R., Dayas, C. V., & Kay-Lambkin, F. (2017a). Differences in dietary preferences, personality and mental health in Australian adults with and without food addiction. *Nutrients, 9*(3). https://doi.org/10.3390/nu9030285.

Burrows, T., Kay-Lambkin, F., Pursey, K., Skinner, J., & Dayas, C. (2018). Food addiction and associations with mental health symptoms: A systematic review with meta-analysis. *Journal of Human Nutrition and Dietetics*. https://doi.org/10.1111/jhn.12532.

Burrows, T., Skinner, J., Joyner, M. A., Palmieri, J., Vaughan, K., & Gearhardt, A. N. (2017b). Food addiction in children: Associations with obesity, parental food addiction and feeding practices. *Eating Behaviors, 26*, 114–120. https://doi.org/10.1016/j.eatbeh.2017.02.004.

Bury, B., Tiggemann, M., & Slater, A. (2016). Disclaimer labels on fashion magazine advertisements: Impact on visual attention and relationship with body dissatisfaction. *Body Image, 16*, 1–9. https://doi.org/10.1016/j.bodyim.2015.09.005.

Bushman, B. J., Dewall, C. N., Pond, R. S., Jr., & Hanus, M. D. (2014). Low glucose relates to greater aggression in married couples. *Proceedings of the National Academy of Sciences of the United States of America, 111*(17), 6254–6257. https://doi.org/10.1073/pnas.1400619111.

Cahill, L. E., Chiuve, S. E., Mekary, R. A., Jensen, M. K., Flint, A. J., Hu, F. B., & Rimm, E. B. (2013). Prospective study of breakfast eating and incident coronary heart disease in a cohort of male US health professionals. *Circulation, 128*(4), 337–343. https://doi.org/10.1161/CIRCULATIONAHA.113.001474.

Calder, R. K., & Mussap, A. J. (2015). Factors influencing women's choice of weight-loss diet. *Journal of Health Psychology, 20*(5), 612–624. https://doi.org/10.1177/1359105315573435.

Calu, D. J., Kawa, A. B., Marchant, N. J., Navarre, B. M., Henderson, M. J., Chen, B., … Shaham, Y. (2013). Optogenetic inhibition of dorsal medial prefrontal cortex attenuates stress-induced reinstatement of palatable food seeking in female rats. *Journal of Neuroscience, 33*(1), 214–226. https://doi.org/10.1523/JNEUROSCI.2016-12.2013.

Calu, D. J., Roesch, M. R., Haney, R. Z., Holland, P. C., & Schoenbaum, G. (2010). Neural correlates of variations in event processing during learning in central nucleus of amygdala. *Neuron, 68*(5), 991–1001. https://doi.org/10.1016/j.neuron.2010.11.019.

Canan, F., Karaca, S., Sogucak, S., Gecici, O., & Kuloglu, M. (2017). Eating disorders and food addiction in men with heroin use disorder: A controlled study. *Eating and Weight Disorders, 22*(2), 249–257. https://doi.org/10.1007/s40519-017-0378-9.

Cancian, A. C. M., de Souza, L. A. S., Liboni, R. P. A., Machado, W. L., & Oliveira, M. D. S. (2017). Effects of a dialectical behavior therapy-based skills group intervention for obese individuals: A Brazilian pilot study. *Eating and Weight Disorders*. https://doi.org/10.1007/s40519-017-0461-2.

Cannon, G. (2005). Dieting. Makes you fat? *British Journal of Nutrition, 93*(4), 569–570.

Cannon, D. S., Leeka, J. K., Patterson, E. T., & Baker, T. B. (1990). Principal components analysis of the inventory of drinking situations: Empirical categories of drinking by alcoholics. *Addictive Behaviors, 15*(3), 265–269.

Carano, A., De Berardis, D., Campanella, D., Serroni, N., Ferri, F., Di Iorio, G., … Di Giannantonio, M. (2012). Alexithymia and suicide ideation in a sample of patients with binge eating disorder. *Journal of Psychiatric Practice, 18*(1), 5–11. https://doi.org/10.1097/01.pra.0000410982.08229.99.

Carelli, R. M., & West, E. A. (2014). When a good taste turns bad: Neural mechanisms underlying the emergence of negative affect and associated natural reward devaluation by cocaine. *Neuropharmacology, 76 Pt B*, 360–369. https://doi.org/10.1016/j.neuropharm.2013.04.025.

Carels, R. A., Douglass, O. M., Cacciapaglia, H. M., & O'Brien, W. H. (2004). An ecological momentary assessment of relapse crises in dieting. *Journal of Consulting and Clinical Psychology, 72*(2), 341–348. https://doi.org/10.1037/0022-006X.72.2.341.

Carr, M. M., Catak, P. D., Pejsa-Reitz, M. C., Saules, K. K., & Gearhardt, A. N. (2017). Measurement invariance of the Yale food addiction scale 2.0 across gender and racial groups. *Psychological Assessment, 29*(8), 1044–1052. https://doi.org/10.1037/pas0000403.

Carver, C. S., Johnson, S. L., Joormann, J., Kim, Y., & Nam, J. Y. (2011). Serotonin transporter polymorphism interacts with childhood adversity to predict aspects of impulsivity. *Psychological Science, 22*(5), 589–595. https://doi.org/10.1177/0956797611404085.

Ceccarini, M., Manzoni, G. M., Castelnuovo, G., & Molinari, E. (2015). An evaluation of the Italian version of the Yale food addiction scale in obese adult inpatients engaged in a 1-month-weight-loss treatment. *Journal of Medicinal Food, 18*(11), 1281–1287. https://doi.org/10.1089/jmf.2014.0188.

Chao, A. M., Shaw, J. A., Pearl, R. L., Alamuddin, N., Hopkins, C. M., Bakizada, Z. M., … Wadden, T. A. (2017). Prevalence and psychosocial correlates of food addiction in persons with obesity seeking weight reduction. *Comprehensive Psychiatry, 73*, 97–104. https://doi.org/10.1016/j.comppsych.2016.11.009.

Chen, Y. W., Fiscella, K. A., Bacharach, S. Z., Tanda, G., Shaham, Y., & Calu, D. J. (2015). Effect of yohimbine on reinstatement of operant responding in rats is dependent on cue contingency but not food reward history. *Addiction Biology, 20*(4), 690–700. https://doi.org/10.1111/adb.12164.

Chen, E. Y., Matthews, L., Allen, C., Kuo, J. R., & Linehan, M. M. (2008). Dialectical behavior therapy for clients with binge-eating disorder or bulimia nervosa and borderline personality disorder. *International Journal of Eating Disorders, 41*(6), 505–512. https://doi.org/10.1002/eat.20522.

Chester, D. S., Lynam, D. R., Milich, R., & DeWall, C. N. (2016a). Craving versus control: Negative urgency and neural correlates of alcohol cue reactivity. *Drug and Alcohol Dependence, 163*(Suppl. 1), S25–S28. https://doi.org/10.1016/j.drugalcdep.2015.10.036.

Chester, D. S., Lynam, D. R., Milich, R., Powell, D. K., Andersen, A. H., & DeWall, C. N. (2016b). How do negative emotions impair self-control? A neural model of negative urgency. *NeuroImage, 132*, 43–50. https://doi.org/10.1016/j.neuroimage.2016.02.024.

Childress, A. R., McLellan, A. T., Ehrman, R., & O'Brien, C. P. (1988). Classically conditioned responses in opioid and cocaine dependence: A role in relapse? *NIDA Research Monograph, 84*, 25–43.

Childress, A. R., McLellan, A. T., & O'Brien, C. P. (1986). Conditioned responses in a methadone population. A comparison of laboratory, clinic, and natural settings. *Journal of Substance Abuse Treatment, 3*(3), 173–179.

Christiansen, A. M., Herman, J. P., & Ulrich-Lai, Y. M. (2011). Regulatory interactions of stress and reward on rat forebrain opioidergic and GABAergic circuitry. *Stress: The International Journal on the Biology of Stress, 14*(2), 205–215. https://doi.org/10.3109/10253890.2010.531331.

Chua, J. L., Touyz, S., & Hill, A. J. (2004). Negative mood-induced overeating in obese binge eaters: An experimental study. *International Journal of Obesity and Related Metabolic Disorders, 28*(4), 606–610. https://doi.org/10.1038/sj.ijo.0802595.

Chu, K., Koob, G. F., Cole, M., Zorrilla, E. P., & Roberts, A. J. (2007). Dependence-induced increases in ethanol self-administration in mice are blocked by the CRF1 receptor antagonist antalarmin and by CRF1 receptor knockout. *Pharmacology Biochemistry and Behavior, 86*(4), 813–821. https://doi.org/10.1016/j.pbb.2007.03.009.

Cifani, C., Polidori, C., Melotto, S., Ciccocioppo, R., & Massi, M. (2009a). A preclinical model of binge eating elicited by yo-yo dieting and stressful exposure to food: Effect of sibutramine, fluoxetine, topiramate, and midazolam. *Psychopharmacology, 204*(1), 113–125. https://doi.org/10.1007/s00213-008-1442-y.

Cifani, C., Zanoncelli, A., Tessari, M., Righetti, C., Di Francesco, C., Ciccocioppo, R., … Melotto, S. (2009b). Pre-exposure to environmental cues predictive of food availability elicits hypothalamic-pituitary-adrenal axis activation and increases operant responding for food in female rats. *Addiction Biology, 14*(4), 397–407. https://doi.org/10.1111/j.1369-1600.2009.00152.x.

Citrome, L. (2015). Lisdexamfetamine for binge eating disorder in adults: A systematic review of the efficacy and safety profile for this newly approved indication – what is the number needed to treat, number needed to harm and likelihood to be helped or harmed? *International Journal of Clinical Practice, 69*(4), 410–421. https://doi.org/10.1111/ijcp.12639.

Clark, L., Bechara, A., Damasio, H., Aitken, M. R., Sahakian, B. J., & Robbins, T. W. (2008). Differential effects of insular and ventromedial prefrontal cortex lesions on risky decision-making. *Brain, 131*(Pt 5), 1311–1322. https://doi.org/10.1093/brain/awn066.

Clark, S. M., & Saules, K. K. (2013). Validation of the Yale food addiction scale among a weight-loss surgery population. *Eating Behaviors, 14*(2), 216–219. https://doi.org/10.1016/j.eatbeh.2013.01.002.

Clark, L., Stokes, P. R., Wu, K., Michalczuk, R., Benecke, A., Watson, B. J., … Lingford-Hughes, A. R. (2012). Striatal dopamine D(2)/D(3) receptor binding in pathological gambling is correlated with mood-related impulsivity. *NeuroImage, 63*(1), 40–46. https://doi.org/10.1016/j.neuroimage.2012.06.067.

Cobb, L. K., Appel, L. J., Franco, M., Jones-Smith, J. C., Nur, A., & Anderson, C. A. (2015). The relationship of the local food environment with obesity: A systematic review of methods, study quality, and results. *Obesity, 23*(7), 1331–1344. https://doi.org/10.1002/oby.21118.

Cocores, J. A., & Gold, M. S. (2009). The Salted Food Addiction Hypothesis may explain overeating and the obesity epidemic. *Medical Hypotheses, 73*(6), 892–899. https://doi.org/10.1016/j.mehy.2009.06.049.

Colantuoni, C., Rada, P., McCarthy, J., Patten, C., Avena, N. M., Chadeayne, A., & Hoebel, B. G. (2002). Evidence that intermittent, excessive sugar intake causes endogenous opioid dependence. *Obesity Research, 10*(6), 478–488. https://doi.org/10.1038/oby.2002.66.

Colechio, E. M., Alexander, D. N., Imperio, C. G., Jackson, K., & Grigson, P. S. (2018). Once is too much: Early development of the opponent process in taste reactivity behavior is associated with later escalation of cocaine self-administration in rats. *Brain Research Bulletin, 138*, 88–95. https://doi.org/10.1016/j.brainresbull.2017.09.002.

Colechio, E. M., & Grigson, P. S. (2014). Conditioned aversion for a cocaine-predictive cue is associated with cocaine seeking and taking in rats. *International Journal of Comparative Psychology, 27*(3), 488–500.

Colechio, E. M., Imperio, C. G., & Grigson, P. S. (2014). Once is too much: Conditioned aversion develops immediately and predicts future cocaine self-administration behavior in rats. *Behavioral Neuroscience, 128*(2), 207–216. https://doi.org/10.1037/a0036264.

Conason, A., Teixeira, J., Hsu, C. H., Puma, L., Knafo, D., & Geliebter, A. (2013). Substance use following bariatric weight loss surgery. *JAMA Surgery, 148*(2), 145–150. https://doi.org/10.1001/2013.jamasurg.265.

Cook, J. W., Lanza, S. T., Chu, W., Baker, T. B., & Piper, M. E. (2017). Anhedonia: Its dynamic relations with craving, negative affect, and treatment during a quit smoking attempt. *Nicotine & Tobacco Research, 19*(6), 703–709. https://doi.org/10.1093/ntr/ntw247.

Cook, J. W., Piper, M. E., Leventhal, A. M., Schlam, T. R., Fiore, M. C., & Baker, T. B. (2015). Anhedonia as a component of the tobacco withdrawal syndrome. *Journal of Abnormal Psychology, 124*(1), 215–225. https://doi.org/10.1037/abn0000016.

Cooper, S. J. (2005). Palatability-dependent appetite and benzodiazepines: New directions from the pharmacology of GABA(A) receptor subtypes. *Appetite, 44*(2), 133–150. https://doi.org/10.1016/j.appet.2005.01.003.

Corsica, J. A., & Pelchat, M. L. (2010). Food addiction: True or false? *Current Opinion in Gastroenterology, 26*(2), 165–169. https://doi.org/10.1097/MOG.0b013e328336528d.

Corsica, J. A., & Spring, B. J. (2008). Carbohydrate craving: A double-blind, placebo-controlled test of the self-medication hypothesis. *Eating Behaviors, 9*(4), 447–454. https://doi.org/10.1016/j.eatbeh.2008.07.004.

Corwin, R. L. (2011). The face of uncertainty eats. *Current Drug Abuse Reviews, 4*(3), 174–181.

Corwin, R. L., & Babbs, R. K. (2012). Rodent models of binge eating: Are they models of addiction? *ILAR Journal, 53*(1), 23–34.

Corwin, R. L., & Grigson, P. S. (2009). Symposium overview–food addiction: Fact or fiction? *Journal of Nutrition, 139*(3), 617–619. https://doi.org/10.3945/jn.108.097691.

Cottone, P., Sabino, V., Steardo, L., & Zorrilla, E. P. (2008a). Intermittent access to preferred food reduces the reinforcing efficacy of chow in rats. *American Journal of Physiology – Regulatory, Integrative and Comparative Physiology, 295*(4), R1066–R1076. https://doi.org/10.1152/ajpregu.90309.2008.

Cottone, P., Sabino, V., Steardo, L., & Zorrilla, E. P. (2008b). Opioid-dependent anticipatory negative contrast and binge-like eating in rats with limited access to highly preferred food. *Neuropsychopharmacology, 33*(3), 524–535. https://doi.org/10.1038/sj.npp.1301430.

Cottone, P., Sabino, V., Roberto, M., Bajo, M., Pockros, L., Frihauf, J. B., ... Zorrilla, E. P. (2009a). CRF system recruitment mediates dark side of compulsive eating. *Proceedings of the National Academy of Sciences of the United States of America, 106*(47), 20016–20020. https://doi.org/10.1073/pnas.0908789106, 0908789106 [pii].

Cottone, P., Sabino, V., Steardo, L., & Zorrilla, E. P. (2009b). Consummatory, anxiety-related and metabolic adaptations in female rats with alternating access to preferred food. *Psychoneuroendocrinology, 34*(1), 38–49. https://doi.org/10.1016/j.psyneuen.2008.08.010.

Coumans, J. M. J., Danner, U. N., Intemann, T., De Decker, A., Hadjigeorgiou, C., Hunsberger, M., ... Consortium, I. F. (2018). Emotion-driven impulsiveness and snack food consumption of European adolescents: Results from the I.Family study. *Appetite, 123*, 152–159. https://doi.org/10.1016/j.appet.2017.12.018.

Cowan, J. A., & Devine, C. M. (2013). Diet and body composition outcomes of an environmental and educational intervention among men in treatment for substance addiction. *Journal of Nutrition Education and Behavior, 45*(2), 154–158. https://doi.org/10.1016/j.jneb.2011.10.011.

Crow, S., Eisenberg, M. E., Story, M., & Neumark-Sztainer, D. (2006). Psychosocial and behavioral correlates of dieting among overweight and non-overweight adolescents. *Journal of Adolescent Health, 38*(5), 569–574. https://doi.org/10.1016/j.jadohealth.2005.05.019.

Culbert, K. M., Sinclair, E. B., Hildebrandt, B. A., Klump, K. L., & Sisk, C. L. (2018). Perinatal testosterone contributes to mid-to-post pubertal sex differences in risk for binge eating in male and female rats. *Journal of Abnormal Psychology, 127*(2), 239–250. https://doi.org/10.1037/abn0000334.

Cyders, M. A., & Coskunpinar, A. (2011). Measurement of constructs using self-report and behavioral lab tasks: Is there overlap in nomothetic span and construct representation for impulsivity? *Clinical Psychology Review, 31*(6), 965–982. https://doi.org/10.1016/j.cpr.2011.06.001.

Cyders, M. A., Dzemidzic, M., Eiler, W. J., Coskunpinar, A., Karyadi, K., & Kareken, D. A. (2014). Negative urgency and ventromedial prefrontal cortex responses to alcohol cues: FMRI evidence of emotion-based impulsivity. *Alcoholism: Clinical and Experimental Research, 38*(2), 409–417. https://doi.org/10.1111/acer.12266.

Cyders, M. A., Dzemidzic, M., Eiler, W. J., Coskunpinar, A., Karyadi, K. A., & Kareken, D. A. (2015). Negative urgency mediates the relationship between amygdala and orbitofrontal cortex activation to negative emotional stimuli and general risk-taking. *Cerebral Cortex, 25*(11), 4094–4102. https://doi.org/10.1093/cercor/bhu123.

Cyders, M. A., & Smith, G. T. (2008). Emotion-based dispositions to rash action: Positive and negative urgency. *Psychological Bulletin, 134*(6), 807–828. https://doi.org/10.1037/a0013341.

D'Addario, C., Micioni Di Bonaventura, M. V., Pucci, M., Romano, A., Gaetani, S., Ciccocioppo, R., ... Maccarrone, M. (2014). Endocannabinoid signaling and food addiction. *Neuroscience & Biobehavioral Reviews, 47*, 203–224. https://doi.org/10.1016/j.neubiorev.2014.08.008.

Dallman, M. F., Pecoraro, N., Akana, S. F., La Fleur, S. E., Gomez, F., Houshyar, H., ... Manalo, S. (2003). Chronic stress and obesity: A new view of "comfort food". *Proceedings of the National Academy of Sciences of the United States of America, 100*(20), 11696–11701. https://doi.org/10.1073/pnas.1934666100.

Dallman, M. F., Pecoraro, N. C., & la Fleur, S. E. (2005). Chronic stress and comfort foods: Self-medication and abdominal obesity. *Brain, Behavior, and Immunity, 19*(4), 275–280. https://doi.org/10.1016/j.bbi.2004.11.004.

Dansinger, M. L., Gleason, J. A., Griffith, J. L., Selker, H. P., & Schaefer, E. J. (2005). Comparison of the atkins, ornish, weight watchers, and zone diets for weight loss and heart disease risk reduction: A randomized trial. *Journal of the American Medical Association, 293*(1), 43–53. https://doi.org/10.1001/jama.293.1.43.

Davis-Becker, K., Peterson, C. M., & Fischer, S. (2014). The relationship of trait negative urgency and negative affect to disordered eating in men and women. *Personality and Individual Differences, 56*, 9–14.

Davis, C., & Carter, J. C. (2009). Compulsive overeating as an addiction disorder. A review of theory and evidence. *Appetite, 53*(1), 1–8. https://doi.org/10.1016/j.appet.2009.05.018.

Davis, C., Curtis, C., Levitan, R. D., Carter, J. C., Kaplan, A. S., & Kennedy, J. L. (2011). Evidence that 'food addiction' is a valid phenotype of obesity. *Appetite, 57*(3), 711–717. https://doi.org/10.1016/j.appet.2011.08.017.

Davis, J. M., Fleming, M. F., Bonus, K. A., & Baker, T. B. (2007). A pilot study on mindfulness based stress reduction for smokers. *BMC Complementary and Alternative Medicine, 7*, 2. https://doi.org/10.1186/1472-6882-7-2.

Davis, J. M., Goldberg, S. B., Anderson, M. C., Manley, A. R., Smith, S. S., & Baker, T. B. (2014). Randomized trial on mindfulness training for smokers targeted to a disadvantaged population. *Substance Use & Misuse, 49*(5), 571–585. https://doi.org/10.3109/10826084.2013.770025.

Deiches, J. F., Baker, T. B., Lanza, S., & Piper, M. E. (2013). Early lapses in a cessation attempt: Lapse contexts, cessation success, and predictors of early lapse. *Nicotine & Tobacco Research, 15*(11), 1883–1891. https://doi.org/10.1093/ntr/ntt074.

DeWall, C. N., Deckman, T., Gailliot, M. T., & Bushman, B. J. (2011). Sweetened blood cools hot tempers: Physiological self-control and aggression. *Aggressive Behavior, 37*(1), 73–80. https://doi.org/10.1002/ab.20366.

Dingemans, A., Danner, U., & Parks, M. (2017). Emotion regulation in binge eating disorder: A review. *Nutrients, 9*(11). https://doi.org/10.3390/nu9111274.

Dingemans, A. E., & van Furth, E. F. (2012). Binge Eating Disorder psychopathology in normal weight and obese individuals. *International Journal of Eating Disorders, 45*(1), 135–138. https://doi.org/10.1002/eat.20905.

Do-Monte, F. H., Minier-Toribio, A., Quinones-Laracuente, K., Medina-Colon, E. M., & Quirk, G. J. (2017). Thalamic regulation of sucrose seeking during unexpected reward omission. *Neuron, 94*(2), 388–400. https://doi.org/10.1016/j.neuron.2017.03.036. e384.

Dunn, C., Haubenreiser, M., Johnson, M., Nordby, K., Aggarwal, S., Myer, S., & Thomas, C. (2018). Mindfulness approaches and weight loss, weight maintenance, and weight regain. *Current Obesity Reports, 7*(1), 37–49. https://doi.org/10.1007/s13679-018-0299-6.

Dunn, J. P., Kessler, R. M., Feurer, I. D., Volkow, N. D., Patterson, B. W., Ansari, M. S., … Abumrad, N. N. (2012). Relationship of dopamine type 2 receptor binding potential with fasting neuroendocrine hormones and insulin sensitivity in human obesity. *Diabetes Care, 35*(5), 1105–1111. https://doi.org/10.2337/dc11-2250.

Dutta, S., Morton, J., Shepard, E., Peebles, R., Farrales-Nguyen, S., Hammer, L. D., & Albanese, C. T. (2006). Methamphetamine use following bariatric surgery in an adolescent. *Obesity Surgery, 16*(6), 780–782. https://doi.org/10.1381/096089206777346646.

Edge, P. J., & Gold, M. S. (2011). Drug withdrawal and hyperphagia: Lessons from tobacco and other drugs. *Current Pharmaceutical Design, 17*(12), 1173–1179.

Elfhag, K., & Rossner, S. (2005). Who succeeds in maintaining weight loss? A conceptual review of factors associated with weight loss maintenance and weight regain. *Obesity Reviews, 6*(1), 67–85. https://doi.org/10.1111/j.1467-789X.2005.00170.x.

Ely, A. V., Wierenga, C. E., Bischoff-Grethe, A., Bailer, U. F., Berner, L. A., Fudge, J. L., ... Kaye, W. H. (2017). Response in taste circuitry is not modulated by hunger and satiety in women remitted from bulimia nervosa. *Journal of Abnormal Psychology, 126*(5), 519–530. https://doi.org/10.1037/abn0000218.

Emerson, M. H., Glovsky, E., Amaro, H., & Nieves, R. (2009). Unhealthy weight gain during treatment for alcohol and drug use in four residential programs for Latina and African American women. *Substance Use & Misuse, 44*(11), 1553–1565. https://doi.org/10.1080/10826080802494750.

Engel, S. G., Boseck, J. J., Crosby, R. D., Wonderlich, S. A., Mitchell, J. E., Smyth, J., ... Steiger, H. (2007). The relationship of momentary anger and impulsivity to bulimic behavior. *Behaviour Research and Therapy, 45*(3), 437–447. https://doi.org/10.1016/j.brat.2006.03.014.

Engelmann, J. M., Versace, F., Robinson, J. D., Minnix, J. A., Lam, C. Y., Cui, Y., ... Cinciripini, P. M. (2012). Neural substrates of smoking cue reactivity: A meta-analysis of fMRI studies. *NeuroImage, 60*(1), 252–262. https://doi.org/10.1016/j.neuroimage.2011.12.024.

Engel, S. G., Wonderlich, S. A., Crosby, R. D., Mitchell, J. E., Crow, S., Peterson, C. B., ... Gordon, K. H. (2013). The role of affect in the maintenance of anorexia nervosa: Evidence from a naturalistic assessment of momentary behaviors and emotion. *Journal of Abnormal Psychology, 122*(3), 709–719. https://doi.org/10.1037/a0034010.

Epstein, L. H., Carr, K. A., Cavanaugh, M. D., Paluch, R. A., & Bouton, M. E. (2011). Long-term habituation to food in obese and nonobese women. *American Journal of Clinical Nutrition, 94*(2), 371–376. https://doi.org/10.3945/ajcn.110.009035.

Epstein, D. H., Kennedy, A. P., Furnari, M., Heilig, M., Shaham, Y., Phillips, K. A., & Preston, K. L. (2016). Effect of the CRF1-receptor antagonist pexacerfont on stress-induced eating and food craving. *Psychopharmacology, 233*(23–24), 3921–3932. https://doi.org/10.1007/s00213-016-4424-5.

Epstein, L. H., Leddy, J. J., Temple, J. L., & Faith, M. S. (2007a). Food reinforcement and eating: A multilevel analysis. *Psychological Bulletin, 133*(5), 884–906. https://doi.org/10.1037/0033-2909.133.5.884.

Epstein, L. H., Salvy, S. J., Carr, K. A., Dearing, K. K., & Bickel, W. K. (2010). Food reinforcement, delay discounting and obesity. *Physiology & Behavior, 100*(5), 438–445. https://doi.org/10.1016/j.physbeh.2010.04.029.

Epstein, L. H., Stein, J. S., Paluch, R. A., MacKillop, J., & Bickel, W. K. (2018). Binary components of food reinforcement: Amplitude and persistence. *Appetite, 120*, 67–74. https://doi.org/10.1016/j.appet.2017.08.023.

Epstein, L. H., Temple, J. L., Neaderhiser, B. J., Salis, R. J., Erbe, R. W., & Leddy, J. J. (2007b). Food reinforcement, the dopamine D2 receptor genotype, and energy intake in obese and nonobese humans. *Behavioral Neuroscience, 121*(5), 877–886. https://doi.org/10.1037/0735-7044.121.5.877.

Ettenberg, A., Raven, M. A., Danluck, D. A., & Necessary, B. D. (1999). Evidence for opponent-process actions of intravenous cocaine. *Pharmacology Biochemistry and Behavior, 64*(3), 507–512.

Fachin, A., Silva, R. K., Noschang, C. G., Pettenuzzo, L., Bertinetti, L., Billodre, M. N., ... Dalmaz, C. (2008). Stress effects on rats chronically receiving a highly palatable diet are sex-specific. *Appetite, 51*(3), 592–598. https://doi.org/10.1016/j.appet.2008.04.016.

Fayet-Moore, F., McConnell, A., Tuck, K., & Petocz, P. (2017). Breakfast and breakfast cereal choice and its impact on nutrient and sugar intakes and anthropometric measures among a nationally representative sample of Australian children and adolescents. *Nutrients, 9*(10). https://doi.org/10.3390/nu9101045.

Fay, S. H., & Finlayson, G. (2011). Negative affect-induced food intake in non-dieting women is reward driven and associated with restrained-disinhibited eating subtype. *Appetite, 56*(3), 682–688. https://doi.org/10.1016/j.appet.2011.02.004.

Fichter, M. M., Quadflieg, N., & Hedlund, S. (2008). Long-term course of binge eating disorder and bulimia nervosa: Relevance for nosology and diagnostic criteria. *International Journal of Eating Disorders, 41*(7), 577–586. https://doi.org/10.1002/eat.20539.

Fischer, S., Smith, G. T., Annus, A. M., & Hendricks, M. (2007). The relationship of neuroticism and urgency to negative consequences of alcohol use in women with bulimic symptoms. *Personality and Individual Differences, 43*, 1199–1209.

Fischer, S., Smith, G. T., & Cyders, M. A. (2008). Another look at impulsivity: A meta-analytic review comparing specific dispositions to rash action in their relationship to bulimic symptoms. *Clinical Psychology Review, 28*(8), 1413–1425. https://doi.org/10.1016/j.cpr.2008.09.001.

Fischer, S., Wonderlich, J., Breithaupt, L., Byrne, C., & Engel, S. (2018). Negative urgency and expectancies increase vulnerability to binge eating in bulimia nervosa. *Eating Disorders, 26*(1), 39–51. https://doi.org/10.1080/10640266.2018.1418253.

Flegal, K. M., Kruszon-Moran, D., Carroll, M. D., Fryar, C. D., & Ogden, C. L. (2016). Trends in obesity among adults in the United States, 2005 to 2014. *Journal of the American Medical Association, 315*(21), 2284–2291. https://doi.org/10.1001/jama.2016.6458.

Floresco, S. B., & Tse, M. T. (2007). Dopaminergic regulation of inhibitory and excitatory transmission in the basolateral amygdala-prefrontal cortical pathway. *Journal of Neuroscience, 27*(8), 2045–2057. https://doi.org/10.1523/JNEUROSCI.5474-06.2007.

Fonseca, H., Matos, M. G., Guerra, A., & Pedro, J. G. (2009). Are overweight adolescents at higher risk of engaging in unhealthy weight-control behaviours? *Acta Paediatrica, 98*(5), 847–852.

Freeman, C. R., Zehra, A., Ramirez, V., Wiers, C. E., Volkow, N. D., & Wang, G. J. (2018). Impact of sugar on the body, brain, and behavior. *Frontiers in Bioscience, 23*, 2255–2266.

Funk, C. K., Zorrilla, E. P., Lee, M. J., Rice, K. C., & Koob, G. F. (2007). Corticotropin-releasing factor 1 antagonists selectively reduce ethanol self-administration in ethanol-dependent rats. *Biological Psychiatry, 61*(1), 78–86. https://doi.org/10.1016/j.biopsych.2006.03.063.

Galic, M. A., & Persinger, M. A. (2002). Voluminous sucrose consumption in female rats: Increased "nippiness" during periods of sucrose removal and possible oestrus periodicity. *Psychological Reports, 90*(1), 58–60. https://doi.org/10.2466/pr0.2002.90.1.58.

Gasior, M., Hudson, J., Quintero, J., Ferreira-Cornwell, M. C., Radewonuk, J., & McElroy, S. L. (2017). A phase 3, multicenter, open-label, 12-month extension safety and tolerability trial of lisdexamfetamine dimesylate in adults with binge eating disorder. *Journal of Clinical Psychopharmacology, 37*(3), 315–322. https://doi.org/10.1097/JCP.0000000000000702.

Gearhardt, A. N., Corbin, W. R., & Brownell, K. D. (2016). Development of the Yale food addiction scale version 2.0. *Psychology of Addictive Behaviors, 30*(1), 113–121. https://doi.org/10.1037/adb0000136.

Gearhardt, A. N., Miller, A. L., Sturza, J., Epstein, L. H., Kaciroti, N., & Lumeng, J. C. (2017). Behavioral associations with overweight in low-income children. *Obesity, 25*(12), 2123–2127. https://doi.org/10.1002/oby.22033.

Gearhardt, A. N., Waller, R., Jester, J. M., Hyde, L. W., & Zucker, R. A. (2018). Body mass index across adolescence and substance use problems in early adulthood. *Psychology of Addictive Behaviors, 32*(3), 309–319. https://doi.org/10.1037/adb0000365.

Gearhardt, A. N., White, M. A., Masheb, R. M., Morgan, P. T., Crosby, R. D., & Grilo, C. M. (2012). An examination of the food addiction construct in obese patients with binge eating disorder. *International Journal of Eating Disorders, 45*(5), 657–663. https://doi.org/10.1002/eat.20957.

Gearhardt, A. N., Yokum, S., Orr, P. T., Stice, E., Corbin, W. R., & Brownell, K. D. (2011). Neural correlates of food addiction. *Archives of General Psychiatry, 68*(8), 808–816. https://doi.org/10.1001/archgenpsychiatry.2011.32.

Gehlert, D. R., Cippitelli, A., Thorsell, A., Le, A. D., Hipskind, P. A., Hamdouchi, C., ... Heilig, M. (2007). 3-(4-Chloro-2-morpholin-4-yl-thiazol-5-yl)-8-(1-ethylpropyl)-2,6-dimethyl-imidazo [1,2-b]pyridazine: A novel brain-penetrant, orally available corticotropin-releasing factor receptor 1 antagonist with efficacy in animal models of alcoholism. *Journal of Neuroscience, 27*(10), 2718–2726. https://doi.org/10.1523/JNEUROSCI.4985-06.2007.

Geiger, B. M., Behr, G. G., Frank, L. E., Caldera-Siu, A. D., Beinfeld, M. C., Kokkotou, E. G., & Pothos, E. N. (2008). Evidence for defective mesolimbic dopamine exocytosis in obesity-prone rats. *The FASEB Journal, 22*(8), 2740–2746. https://doi.org/10.1096/fj.08-110759.

Geiger, B. M., Haburcak, M., Avena, N. M., Moyer, M. C., Hoebel, B. G., & Pothos, E. N. (2009). Deficits of mesolimbic dopamine neurotransmission in rat dietary obesity. *Neuroscience, 159*(4), 1193–1199. https://doi.org/10.1016/j.neuroscience.2009.02.007.

Geliebter, A., & Aversa, A. (2003). Emotional eating in overweight, normal weight, and underweight individuals. *Eating Behaviors, 3*(4), 341–347.

George, O., Ghozland, S., Azar, M. R., Cottone, P., Zorrilla, E. P., Parsons, L. H., ... Koob, G. F. (2007). CRF-CRF1 system activation mediates withdrawal-induced increases in nicotine self-administration in nicotine-dependent rats. *Proceedings of the National Academy of Sciences of the United States of America, 104*(43), 17198–17203. https://doi.org/10.1073/pnas.0707585104.

George, O., Koob, G. F., & Vendruscolo, L. F. (2014). Negative reinforcement via motivational withdrawal is the driving force behind the transition to addiction. *Psychopharmacology, 231*(19), 3911–3917. https://doi.org/10.1007/s00213-014-3623-1.

Ghitza, U. E., Gray, S. M., Epstein, D. H., Rice, K. C., & Shaham, Y. (2006). The anxiogenic drug yohimbine reinstates palatable food seeking in a rat relapse model: A role of CRF1 receptors. *Neuropsychopharmacology, 31*(10), 2188–2196. https://doi.org/10.1038/sj.npp.1300964.

Gibson, E. L. (2006). Emotional influences on food choice: Sensory, physiological and psychological pathways. *Physiology & Behavior, 89*(1), 53–61. https://doi.org/10.1016/j.physbeh.2006.01.024.

van de Giessen, E., Celik, F., Schweitzer, D. H., van den Brink, W., & Booij, J. (2014). Dopamine D2/3 receptor availability and amphetamine-induced dopamine release in obesity. *Journal of Psychopharmacology, 28*(9), 866–873. https://doi.org/10.1177/0269881114531664.

van de Giessen, E., de Bruin, K., la Fleur, S. E., van den Brink, W., & Booij, J. (2012). Triple monoamine inhibitor tesofensine decreases food intake, body weight, and striatal dopamine D2/D3 receptor availability in diet-induced obese rats. *European Neuropsychopharmacology, 22*(4), 290–299. https://doi.org/10.1016/j.euroneuro.2011.07.015.

Gilpin, N. W., Richardson, H. N., & Koob, G. F. (2008). Effects of CRF1-receptor and opioid-receptor antagonists on dependence-induced increases in alcohol drinking by alcohol-preferring (P) rats. *Alcoholism: Clinical and Experimental Research, 32*(9), 1535–1542. https://doi.org/10.1111/j.1530-0277.2008.00745.x.

Gold, M. S. (2011). From bedside to bench and back again: A 30-year saga. *Physiology & Behavior, 104*(1), 157–161. https://doi.org/10.1016/j.physbeh.2011.04.027.

Gold, M. S., Badgaiyan, R. D., & Blum, K. (2015). A shared molecular and genetic basis for food and drug addiction: Overcoming hypodopaminergic trait/state by incorporating dopamine agonistic therapy in psychiatry. *Psychiatric Clinics of North America, 38*(3), 419–462. https://doi.org/10.1016/j.psc.2015.05.011.

Gold, M. S., Blum, K., Febo, M., Baron, D., Modestino, E. J., Elman, I., & Badgaiyan, R. D. (2018). Molecular role of dopamine in anhedonia linked to reward deficiency syndrome (RDS) and anti- reward systems. *Frontiers in Bioscience, 10*, 309–325.

Goldschmidt, A. B., Wall, M., Choo, T. H., Becker, C., & Neumark-Sztainer, D. (2016). Shared risk factors for mood-, eating-, and weight-related health outcomes. *Health Psychology, 35*(3), 245–252. https://doi.org/10.1037/hea0000283.

Goldschmidt, A. B., Wall, M., Loth, K. A., Le Grange, D., & Neumark-Sztainer, D. (2012). Which dieters are at risk for the onset of binge eating? A prospective study of adolescents and young adults. *Journal of Adolescent Health, 51*(1), 86–92. https://doi.org/10.1016/j.jadohealth.2011.11.001.

Goldstone, A. P., Prechtl de Hernandez, C. G., Beaver, J. D., Muhammed, K., Croese, C., Bell, G., ... Bell, J. D. (2009). Fasting biases brain reward systems towards high-calorie foods. *European Journal of Neuroscience, 30*(8), 1625–1635. https://doi.org/10.1111/j.1460-9568.2009.06949.x.

Goudriaan, A. E., de Ruiter, M. B., van den Brink, W., Oosterlaan, J., & Veltman, D. J. (2010). Brain activation patterns associated with cue reactivity and craving in abstinent problem gamblers, heavy smokers and healthy controls: An fMRI study. *Addiction Biology, 15*(4), 491–503. https://doi.org/10.1111/j.1369-1600.2010.00242.x.

Granero, R., Hilker, I., Aguera, Z., Jimenez-Murcia, S., Sauchelli, S., Islam, M. A., ... Fernandez-Aranda, F. (2014). Food addiction in a Spanish sample of eating disorders: DSM-5 diagnostic subtype differentiation and validation data. *European Eating Disorders Review, 22*(6), 389–396. https://doi.org/10.1002/erv.2311.

Greeno, C. G., & Wing, R. R. (1994). Stress-induced eating. *Psychological Bulletin, 115*(3), 444–464.

Greenwell, T. N., Funk, C. K., Cottone, P., Richardson, H. N., Chen, S. A., Rice, K. C., ... Koob, G. F. (2009). Corticotropin-releasing factor-1 receptor antagonists decrease heroin self-administration in long- but not short-access rats. *Addiction Biology, 14*(2), 130–143. https://doi.org/10.1111/j.1369-1600.2008.00142.x.

Grigson, P. S. (2008). The state of the reward comparison hypothesis: Theoretical comment on huang and hsiao (2008). *Behavioral Neuroscience, 122*(6), 1383–1390. https://doi.org/10.1037/a0013968.

Guerdjikova, A. I., Walsh, B., Shan, K., Halseth, A. E., Dunayevich, E., & McElroy, S. L. (2017). Concurrent improvement in both binge eating and depressive symptoms with naltrexone/bupropion therapy in overweight or obese subjects with major depressive disorder in an open-label, uncontrolled study. *Advances in Therapy, 34*(10), 2307–2315. https://doi.org/10.1007/s12325-017-0613-9.

Haedt-Matt, A. A., & Keel, P. K. (2011). Revisiting the affect regulation model of binge eating: A meta-analysis of studies using ecological momentary assessment. *Psychological Bulletin, 137*(4), 660–681. https://doi.org/10.1037/a0023660.

Hagan, M. M., Wauford, P. K., Chandler, P. C., Jarrett, L. A., Rybak, R. J., & Blackburn, K. (2002). A new animal model of binge eating: Key synergistic role of past caloric restriction and stress. *Physiology & Behavior, 77*(1), 45–54.

Halseth, A., Shan, K., Gilder, K., Malone, M., Acevedo, L., & Fujioka, K. (2018). Quality of life, binge eating and sexual function in participants treated for obesity with sustained release naltrexone/bupropion. *Obesity Science & Practice, 4*(2), 141–152. https://doi.org/10.1002/osp4.156.

Haltia, L. T., Rinne, J. O., Merisaari, H., Maguire, R. P., Savontaus, E., Helin, S., … Kaasinen, V. (2007). Effects of intravenous glucose on dopaminergic function in the human brain in vivo. *Synapse, 61*(9), 748–756. https://doi.org/10.1002/syn.20418.

Hansen, H. H., Jensen, M. M., Overgaard, A., Weikop, P., & Mikkelsen, J. D. (2013). Tesofensine induces appetite suppression and weight loss with reversal of low forebrain dopamine levels in the diet-induced obese rat. *Pharmacology Biochemistry and Behavior, 110*, 265–271. https://doi.org/10.1016/j.pbb.2013.07.018.

Hauck, C., Weiss, A., Schulte, E. M., Meule, A., & Ellrott, T. (2017). Prevalence of 'food addiction' as measured with the Yale food addiction scale 2.0 in a representative German sample and its association with sex, age and weight categories. *Obesity Facts, 10*(1), 12–24. https://doi.org/10.1159/000456013.

Heal, D. J., Goddard, S., Brammer, R. J., Hutson, P. H., & Vickers, S. P. (2016). Lisdexamfetamine reduces the compulsive and perseverative behaviour of binge-eating rats in a novel food reward/punished responding conflict model. *Journal of Psychopharmacology, 30*(7), 662–675. https://doi.org/10.1177/0269881116647506.

Heatherton, T. F., Polivy, J., & Herman, C. P. (1991). Restraint, weight loss, and variability of body weight. *Journal of Abnormal Psychology, 100*(1), 78–83.

van der Heijden, A. A., Hu, F. B., Rimm, E. B., & van Dam, R. M. (2007). A prospective study of breakfast consumption and weight gain among U.S. men. *Obesity, 15*(10), 2463–2469. https://doi.org/10.1038/oby.2007.292.

Heinrichs, S. C., Menzaghi, F., Schulteis, G., Koob, G. F., & Stinus, L. (1995). Suppression of corticotropin-releasing factor in the amygdala attenuates aversive consequences of morphine withdrawal. *Behavioural Pharmacology, 6*(1), 74–80.

Heinz, A., Beck, A., Grusser, S. M., Grace, A. A., & Wrase, J. (2009). Identifying the neural circuitry of alcohol craving and relapse vulnerability. *Addiction Biology, 14*(1), 108–118. https://doi.org/10.1111/j.1369-1600.2008.00136.x.

Hermanussen, M., & Tresguerres, J. A. (2005). A new anti-obesity drug treatment: First clinical evidence that, antagonising glutamate-gated Ca^{2+} ion channels with memantine normalises binge-eating disorders. *Economics and Human Biology, 3*(2), 329–337. https://doi.org/10.1016/j.ehb.2005.04.001.

Herzog, D. B., Keller, M. B., Sacks, N. R., Yeh, C. J., & Lavori, P. W. (1992). Psychiatric comorbidity in treatment-seeking anorexics and bulimics. *Journal of the American Academy of Child & Adolescent Psychiatry, 31*(5), 810–818. https://doi.org/10.1097/00004583-199209000-00006.
Higo, T., Mars, R. B., Boorman, E. D., Buch, E. R., & Rushworth, M. F. (2011). Distributed and causal influence of frontal operculum in task control. *Proceedings of the National Academy of Sciences of the United States of America, 108*(10), 4230–4235. https://doi.org/10.1073/pnas.1013361108.
Higuchi, S., & Japanese Acamprosate Study, G. (2015). Efficacy of acamprosate for the treatment of alcohol dependence long after recovery from withdrawal syndrome: A randomized, double-blind, placebo-controlled study conducted in Japan (sunrise study). *Journal of Clinical Psychiatry, 76*(2), 181–188. https://doi.org/10.4088/JCP.13m08940.
Hinchliff, G. L. M., Kelly, A. B., Chan, G. C. K., Patton, G. C., & Williams, J. (2016). Risky dieting amongst adolescent girls: Associations with family relationship problems and depressed mood. *Eating Behaviors, 22*, 222–224. https://doi.org/10.1016/j.eatbeh.2016.06.001.
Hoebel, B. G., Avena, N. M., & Rada, P. (2007). Accumbens dopamine-acetylcholine balance in approach and avoidance. *Current Opinion in Pharmacology, 7*(6), 617–627. https://doi.org/10.1016/j.coph.2007.10.014.
Hoertel, H. A., Will, M. J., & Leidy, H. J. (2014). A randomized crossover, pilot study examining the effects of a normal protein vs. high protein breakfast on food cravings and reward signals in overweight/obese "breakfast skipping", late-adolescent girls. *Nutrition Journal, 13*, 80. https://doi.org/10.1186/1475-2891-13-80.
Hopkins, S. C., Sunkaraneni, S., Skende, E., Hing, J., Passarell, J. A., Loebel, A., & Koblan, K. S. (2016). Pharmacokinetics and exposure-response relationships of dasotraline in the treatment of attention-deficit/hyperactivity disorder in adults. *Clinical Drug Investigation, 36*(2), 137–146. https://doi.org/10.1007/s40261-015-0358-7.
Hoptman, M. J., Antonius, D., Mauro, C. J., Parker, E. M., & Javitt, D. C. (2014). Cortical thinning, functional connectivity, and mood-related impulsivity in schizophrenia: Relationship to aggressive attitudes and behavior. *American Journal of Psychiatry, 171*(9), 939–948. https://doi.org/10.1176/appi.ajp.2014.13111553.
Hryhorczuk, C., Florea, M., Rodaros, D., Poirier, I., Daneault, C., Des Rosiers, C., … Fulton, S. (2016). Dampened mesolimbic dopamine function and signaling by saturated but not monounsaturated dietary lipids. *Neuropsychopharmacology, 41*(3), 811–821. https://doi.org/10.1038/npp.2015.207.
Hudson, J. I., Hiripi, E., Pope, H. G., Jr., & Kessler, R. C. (2007). The prevalence and correlates of eating disorders in the National Comorbidity Survey Replication. *Biological Psychiatry, 61*(3), 348–358. https://doi.org/10.1016/j.biopsych.2006.03.040.
Hudson, J. I., McElroy, S. L., Ferreira-Cornwell, M. C., Radewonuk, J., & Gasior, M. (2017). Efficacy of lisdexamfetamine in adults with moderate to severe binge-eating disorder: A randomized clinical trial. *JAMA Psychiatry, 74*(9), 903–910. https://doi.org/10.1001/jamapsychiatry.2017.1889.
Hughes, E. K., Goldschmidt, A. B., Labuschagne, Z., Loeb, K. L., Sawyer, S. M., & Le Grange, D. (2013). Eating disorders with and without comorbid depression and anxiety: Similarities and differences in a clinical sample of children and adolescents. *European Eating Disorders Review, 21*(5), 386–394. https://doi.org/10.1002/erv.2234.

Hughes, J. R., Stead, L. F., Hartmann-Boyce, J., Cahill, K., & Lancaster, T. (2014). Antidepressants for smoking cessation. *Cochrane Database of Systematic Reviews, 1*, CD000031. https://doi.org/10.1002/14651858.CD000031.pub4.

Iemolo, A., Blasio, A., St Cyr, S. A., Jiang, F., Rice, K. C., Sabino, V., & Cottone, P. (2013). CRF-CRF1 receptor system in the central and basolateral nuclei of the amygdala differentially mediates excessive eating of palatable food. *Neuropsychopharmacology, 38*(12), 2456–2466. https://doi.org/10.1038/npp.2013.147.

Iemolo, A., Valenza, M., Tozier, L., Knapp, C. M., Kornetsky, C., Steardo, L., ... Cottone, P. (2012). Withdrawal from chronic, intermittent access to a highly palatable food induces depressive-like behavior in compulsive eating rats. *Behavioural Pharmacology, 23*(5–6), 593–602. https://doi.org/10.1097/FBP.0b013e328357697f.

Ifland, J. R., Preuss, H. G., Marcus, M. T., Rourke, K. M., Taylor, W. C., Burau, K., ... Manso, G. (2009). Refined food addiction: A classic substance use disorder. *Medical Hypotheses, 72*(5), 518–526. https://doi.org/10.1016/j.mehy.2008.11.035.

Imperatori, C., Innamorati, M., Contardi, A., Continisio, M., Tamburello, S., Lamis, D. A., ... Fabbricatore, M. (2014). The association among food addiction, binge eating severity and psychopathology in obese and overweight patients attending low-energy-diet therapy. *Comprehensive Psychiatry, 55*(6), 1358–1362. https://doi.org/10.1016/j.comppsych.2014.04.023.

Iordanova, M. D., Deroche, M. L., Esber, G. R., & Schoenbaum, G. (2016). Neural correlates of two different types of extinction learning in the amygdala central nucleus. *Nature Communications, 7*, 12330. https://doi.org/10.1038/ncomms12330.

Ivezaj, V., Wiedemann, A. A., & Grilo, C. M. (2017). Food addiction and bariatric surgery: A systematic review of the literature. *Obesity Reviews, 18*(12), 1386–1397. https://doi.org/10.1111/obr.12600.

Jacka, F. N., Cherbuin, N., Anstey, K. J., & Butterworth, P. (2014). Dietary patterns and depressive symptoms over time: Examining the relationships with socioeconomic position, health behaviours and cardiovascular risk. *PLoS One, 9*(1), e87657. https://doi.org/10.1371/journal.pone.0087657.

Jacka, F. N., Kremer, P. J., Berk, M., de Silva-Sanigorski, A. M., Moodie, M., Leslie, E. R., ... Swinburn, B. A. (2011). A prospective study of diet quality and mental health in adolescents. *PLoS One, 6*(9), e24805. https://doi.org/10.1371/journal.pone.0024805.

Jacka, F. N., Kremer, P. J., Leslie, E. R., Berk, M., Patton, G. C., Toumbourou, J. W., & Williams, J. W. (2010a). Associations between diet quality and depressed mood in adolescents: Results from the Australian healthy neighbourhoods study. *Australian and New Zealand Journal of Psychiatry, 44*(5), 435–442. https://doi.org/10.3109/00048670903571598.

Jacka, F. N., O'Neil, A., Opie, R., Itsiopoulos, C., Cotton, S., Mohebbi, M., ... Berk, M. (2017). A randomised controlled trial of dietary improvement for adults with major depression (the 'SMILES' trial). *BMC Medicine, 15*(1), 23. https://doi.org/10.1186/s12916-017-0791-y.

Jacka, F. N., Pasco, J. A., Mykletun, A., Williams, L. J., Hodge, A. M., O'Reilly, S. L., ... Berk, M. (2010b). Association of Western and traditional diets with depression and anxiety in women. *American Journal of Psychiatry, 167*(3), 305–311. https://doi.org/10.1176/appi.ajp.2009.09060881.

Jacob, A., Moullec, G., Lavoie, K. L., Laurin, C., Cowan, T., Tisshaw, C., ... Bacon, S. L. (2018). Impact of cognitive-behavioral interventions on weight loss and psychological outcomes: A meta-analysis. *Health Psychology, 37*(5), 417–432. https://doi.org/10.1037/hea0000576.

James, C., Harrison, A., Seixas, A., Powell, M., Pengpid, S., & Peltzer, K. (2017). "Safe foods" or "fear foods": The implications of food avoidance in college students from low- and middle-income countries. *Eating and Weight Disorders, 22*(3), 407–419. https://doi.org/10.1007/s40519-017-0407-8.

Japuntich, S. J., Smith, S. S., Jorenby, D. E., Piper, M. E., Fiore, M. C., & Baker, T. B. (2007). Depression predicts smoking early but not late in a quit attempt. *Nicotine & Tobacco Research, 9*(6), 677–686. https://doi.org/10.1080/14622200701365301.

Jasinska, A. J., Stein, E. A., Kaiser, J., Naumer, M. J., & Yalachkov, Y. (2014). Factors modulating neural reactivity to drug cues in addiction: A survey of human neuroimaging studies. *Neuroscience & Biobehavioral Reviews, 38*, 1–16. https://doi.org/10.1016/j.neubiorev.2013.10.013.

Jasinski, D. R., & Krishnan, S. (2009). Abuse liability and safety of oral lisdexamfetamine dimesylate in individuals with a history of stimulant abuse. *Journal of Psychopharmacology, 23*(4), 419–427. https://doi.org/10.1177/0269881109103113.

Jhou, T. C., Good, C. H., Rowley, C. S., Xu, S. P., Wang, H., Burnham, N. W., … Ikemoto, S. (2013). Cocaine drives aversive conditioning via delayed activation of dopamine-responsive habenular and midbrain pathways. *Journal of Neuroscience, 33*(17), 7501–7512. https://doi.org/10.1523/JNEUROSCI.3634-12.2013.

Johnson, P. M., & Kenny, P. J. (2010). Dopamine D2 receptors in addiction-like reward dysfunction and compulsive eating in obese rats. *Nature Neuroscience, 13*(5), 635–641. https://doi.org/10.1038/nn.2519.

Jonas, J. M., & Gold, M. S. (1986). Cocaine abuse and eating disorders. *Lancet, 1*(8477), 390–391.

Jorenby, D. E., Hatsukami, D. K., Smith, S. S., Fiore, M. C., Allen, S., Jensen, J., & Baker, T. B. (1996). Characterization of tobacco withdrawal symptoms: Transdermal nicotine reduces hunger and weight gain. *Psychopharmacology, 128*(2), 130–138.

Kaland, M. E., & Klein-Schwartz, W. (2015). Comparison of lisdexamfetamine and dextroamphetamine exposures reported to U.S. poison centers. *Clinical Toxicology, 53*(5), 477–485. https://doi.org/10.3109/15563650.2015.1027903.

Kaplan, H. I., & Kaplan, H. S. (1957). The psychosomatic concept of obesity. *The Journal of Nervous and Mental Disease, 125*(2), 181–201.

Kaplan, A. S., Levitan, R. D., Yilmaz, Z., Davis, C., Tharmalingam, S., & Kennedy, J. L. (2008). A DRD4/BDNF gene-gene interaction associated with maximum BMI in women with bulimia nervosa. *International Journal of Eating Disorders, 41*(1), 22–28. https://doi.org/10.1002/eat.20474.

Karatayev, O., Baylan, J., & Leibowitz, S. F. (2009a). Increased intake of ethanol and dietary fat in galanin overexpressing mice. *Alcohol, 43*(8), 571–580. https://doi.org/10.1016/j.alcohol.2009.09.025.

Karatayev, O., Gaysinskaya, V., Chang, G. Q., & Leibowitz, S. F. (2009b). Circulating triglycerides after a high-fat meal: Predictor of increased caloric intake, orexigenic peptide expression, and dietary obesity. *Brain Research, 1298*, 111–122. https://doi.org/10.1016/j.brainres.2009.08.001.

Katterman, S. N., Kleinman, B. M., Hood, M. M., Nackers, L. M., & Corsica, J. A. (2014). Mindfulness meditation as an intervention for binge eating, emotional eating, and weight loss: A systematic review. *Eating Behaviors, 15*(2), 197–204. https://doi.org/10.1016/j.eatbeh.2014.01.005.

Kawasaki, K., Annicchiarico, I., Glueck, A. C., Moron, I., & Papini, M. R. (2017). Reward loss and the basolateral amygdala: A function in reward comparisons. *Behavioural Brain Research, 331*, 205–213. https://doi.org/10.1016/j.bbr.2017.05.036.

Kawasaki, K., Glueck, A. C., Annicchiarico, I., & Papini, M. R. (2015). Function of the centromedial amygdala in reward devaluation and open-field activity. *Neuroscience, 303*, 73–81. https://doi.org/10.1016/j.neuroscience.2015.06.053.

Kelley, A. E., Will, M. J., Steininger, T. L., Zhang, M., & Haber, S. N. (2003). Restricted daily consumption of a highly palatable food (chocolate Ensure(R)) alters striatal enkephalin gene expression. *European Journal of Neuroscience, 18*(9), 2592–2598.

Kelly, P. J., Baker, A. L., Deane, F. P., Callister, R., Collins, C. E., Oldmeadow, C., … Keane, C. A. (2015). Study protocol: A stepped wedge cluster randomised controlled trial of a healthy lifestyle intervention for people attending residential substance abuse treatment. *BMC Public Health, 15*, 465. https://doi.org/10.1186/s12889-015-1729-y.

Kelly, Y., Patalay, P., Montgomery, S., & Sacker, A. (2016). BMI development and early adolescent psychosocial well-being: UK millennium cohort study. *Pediatrics, 138*(6). https://doi.org/10.1542/peds.2016-0967.

Kenford, S. L., Smith, S. S., Wetter, D. W., Jorenby, D. E., Fiore, M. C., & Baker, T. B. (2002). Predicting relapse back to smoking: Contrasting affective and physical models of dependence. *Journal of Consulting and Clinical Psychology, 70*(1), 216–227.

Keshen, A., & Helson, T. (2017). Preliminary evidence for the off-label treatment of bulimia nervosa with psychostimulants: Six case reports. *The Journal of Clinical Pharmacology, 57*(7), 818–822. https://doi.org/10.1002/jcph.868.

Kessler, D. A. (2009). *The end of overeating: Taking control of the insatiable American appetite*. New York: Rodale.

Kesztyus, D., Traub, M., Lauer, R., Kesztyus, T., & Steinacker, J. M. (2016). Correlates of longitudinal changes in the waist-to-height ratio of primary school children: Implications for prevention. *Preventive Medicine Reports, 3*, 1–6. https://doi.org/10.1016/j.pmedr.2015.11.005.

Khan, A., Ahmed, R., & Burton, N. W. (2017). Prevalence and correlates of depressive symptoms in secondary school children in Dhaka city, Bangladesh. *Ethnicity and Health*, 1–13. https://doi.org/10.1080/13557858.2017.1398313.

Kimbrough, A., de Guglielmo, G., Kononoff, J., Kallupi, M., Zorrilla, E. P., & George, O. (2017). CRF1 receptor-dependent increases in irritability-like behavior during abstinence from chronic intermittent ethanol vapor exposure. *Alcoholism: Clinical and Experimental Research, 41*(11), 1886–1895. https://doi.org/10.1111/acer.13484.

Kinzig, K. P., Hargrave, S. L., & Honors, M. A. (2008). Binge-type eating attenuates corticosterone and hypophagic responses to restraint stress. *Physiology & Behavior, 95*(1–2), 108–113. https://doi.org/10.1016/j.physbeh.2008.04.026.

Klump, K. L., Racine, S., Hildebrandt, B., & Sisk, C. L. (2013). Sex differences in binge eating patterns in male and female adult rats. *International Journal of Eating Disorders, 46*(7), 729–736. https://doi.org/10.1002/eat.22139.

Klump, K. L., Suisman, J. L., Culbert, K. M., Kashy, D. A., & Sisk, C. L. (2011). Binge eating proneness emerges during puberty in female rats: A longitudinal study. *Journal of Abnormal Psychology, 120*(4), 948–955. https://doi.org/10.1037/a0023600.

Knackstedt, L. A., Samimi, M. M., & Ettenberg, A. (2002). Evidence for opponent-process actions of intravenous cocaine and cocaethylene. *Pharmacology Biochemistry and Behavior, 72*(4), 931–936.

Knapp, D. J., Overstreet, D. H., Moy, S. S., & Breese, G. R. (2004). SB242084, flumazenil, and CRA1000 block ethanol withdrawal-induced anxiety in rats. *Alcohol, 32*(2), 101–111. https://doi.org/10.1016/j.alcohol.2003.08.007.

Koball, A. M., Clark, M. M., Collazo-Clavell, M., Kellogg, T., Ames, G., Ebbert, J., & Grothe, K. B. (2016). The relationship among food addiction, negative mood, and eating-disordered behaviors in patients seeking to have bariatric surgery. *Surgery for Obesity and Related Diseases, 12*(1), 165–170. https://doi.org/10.1016/j.soard.2015.04.009.

Koblan, K. S., Hopkins, S. C., Sarma, K., Jin, F., Goldman, R., Kollins, S. H., & Loebel, A. (2015). Dasotraline for the treatment of attention-deficit/hyperactivity disorder: A randomized, double-blind, placebo-controlled, proof-of-concept trial in adults. *Neuropsychopharmacology, 40*(12), 2745–2752. https://doi.org/10.1038/npp.2015.124.

Kong, K. L., Eiden, R. D., Anzman-Frasca, S., Stier, C. L., Paluch, R. A., Mendez, J., … Epstein, L. H. (2018). Repeatability of the infant food reinforcement paradigm: Implications of individual and developmental differences. *Appetite, 120*, 123–129. https://doi.org/10.1016/j.appet.2017.08.012.

Koob, G. F. (2015). The dark side of emotion: The addiction perspective. *European Journal of Pharmacology, 753*, 73–87. https://doi.org/10.1016/j.ejphar.2014.11.044.

Koob, G. F., & Bloom, F. E. (1988). Cellular and molecular mechanisms of drug dependence. *Science, 242*(4879), 715–723.

Koob, G. F., Buck, C. L., Cohen, A., Edwards, S., Park, P. E., Schlosburg, J. E., … George, O. (2014). Addiction as a stress surfeit disorder. *Neuropharmacology, 76 Pt B*, 370–382. https://doi.org/10.1016/j.neuropharm.2013.05.024.

Koob, G. F., Caine, S. B., Parsons, L., Markou, A., & Weiss, F. (1997). Opponent process model and psychostimulant addiction. *Pharmacology Biochemistry and Behavior, 57*(3), 513–521.

Koob, G. F., & Le Moal, M. (1997). Drug abuse: Hedonic homeostatic dysregulation. *Science, 278*(5335), 52–58.

Koob, G. F., & Le Moal, M. (2001). Drug addiction, dysregulation of reward, and allostasis. *Neuropsychopharmacology, 24*(2), 97–129. https://doi.org/10.1016/S0893-133X(00)00195-0. S0893-133X(00)00195-0 [pii].

Koob, G. F., & Le Moal, M. (2005). Plasticity of reward neurocircuitry and the 'dark side' of drug addiction. *Nature Neuroscience, 8*(11), 1442–1444.

Koob, G. F., & Le Moal, M. (2008). Review. Neurobiological mechanisms for opponent motivational processes in addiction. *Philosophical Transactions of the Royal Society of London B Biological Sciences, 363*(1507), 3113–3123. https://doi.org/10.1098/rstb.2008.0094. C54412M472226253 [pii].

Koob, G. F., & Zorrilla, E. P. (2010). Neurobiological mechanisms of addiction: Focus on corticotropin-releasing factor. *Current Opinion in Investigational Drugs, 11*(1), 63–71.

Krahn, D., Grossman, J., Henk, H., Mussey, M., Crosby, R., & Gosnell, B. (2006). Sweet intake, sweet-liking, urges to eat, and weight change: Relationship to alcohol dependence and abstinence. *Addictive Behaviors, 31*(4), 622–631. https://doi.org/10.1016/j.addbeh.2005.05.056.

Kreisler, A. D., Garcia, M. G., Spierling, S. R., Hui, B. E., & Zorrilla, E. P. (2017). Extended vs. brief intermittent access to palatable food differently promote binge-like intake, rejection of less preferred food, and weight cycling in female rats. *Physiology & Behavior, 177*, 305–316. https://doi.org/10.1016/j.physbeh.2017.03.039.

Krolow, R., Noschang, C. G., Arcego, D., Andreazza, A. C., Peres, W., Goncalves, C. A., & Dalmaz, C. (2010). Consumption of a palatable diet by chronically stressed rats prevents effects on anxiety-like behavior but increases oxidative stress in a sex-specific manner. *Appetite, 55*(1), 108–116. https://doi.org/10.1016/j.appet.2010.03.013.

Kwak, Y., & Kim, Y. (2018). Association between mental health and meal patterns among elderly Koreans. *Geriatrics and Gerontology International, 18*(1), 161–168. https://doi.org/10.1111/ggi.13106.

Lai, J. S., Oldmeadow, C., Hure, A. J., McEvoy, M., Byles, J., & Attia, J. (2016). Longitudinal diet quality is not associated with depressive symptoms in a cohort of middle-aged Australian women. *British Journal of Nutrition, 115*(5), 842–850. https://doi.org/10.1017/S000711451500519X.

Laraia, B. A., Adler, N. E., Coleman-Phox, K., Vieten, C., Mellin, L., Kristeller, J. L., … Epel, E. (2018). Novel interventions to reduce stress and overeating in overweight pregnant women: A feasibility study. *Maternal and Child Health Journal, 22*(5), 670–678. https://doi.org/10.1007/s10995-018-2435-z.

Lee, G., Han, K., & Kim, H. (2017). Risk of mental health problems in adolescents skipping meals: The Korean National Health and Nutrition Examination Survey 2010 to 2012. *Nursing Outlook, 65*(4), 411–419. https://doi.org/10.1016/j.outlook.2017.01.007.

Leeman, R. F., O'Malley, S. S., White, M. A., & McKee, S. A. (2010). Nicotine and food deprivation decrease the ability to resist smoking. *Psychopharmacology, 212*(1), 25–32. https://doi.org/10.1007/s00213-010-1902-z.

Lee, S. A., Park, E. C., Ju, Y. J., Lee, T. H., Han, E., & Kim, T. H. (2017). Breakfast consumption and depressive mood: A focus on socioeconomic status. *Appetite, 114*, 313–319. https://doi.org/10.1016/j.appet.2017.04.007.

Le, A. D., Funk, D., Juzytsch, W., Coen, K., Navarre, B. M., Cifani, C., & Shaham, Y. (2011). Effect of prazosin and guanfacine on stress-induced reinstatement of alcohol and food seeking in rats. *Psychopharmacology, 218*(1), 89–99. https://doi.org/10.1007/s00213-011-2178-7.

Leidy, H. J., Ortinau, L. C., Douglas, S. M., & Hoertel, H. A. (2013). Beneficial effects of a higher-protein breakfast on the appetitive, hormonal, and neural signals controlling energy intake regulation in overweight/obese, "breakfast-skipping," late-adolescent girls. *American Journal of Clinical Nutrition, 97*(4), 677–688. https://doi.org/10.3945/ajcn.112.053116.

Lemeshow, A. R., Rimm, E. B., Hasin, D. S., Gearhardt, A. N., Flint, A. J., Field, A. E., & Genkinger, J. M. (2018). Food and beverage consumption and food addiction among women in the Nurses' Health Studies. *Appetite, 121*, 186–197. https://doi.org/10.1016/j.appet.2017.10.038.

Leon, G. R., Fulkerson, J. A., Perry, C. L., Keel, P. K., & Klump, K. L. (1999). Three to four year prospective evaluation of personality and behavioral risk factors for later disordered eating in adolescent girls and boys. *Journal of Youth and Adolescence, 28*, 181.

Leventhal, A. M., Japuntich, S. J., Piper, M. E., Jorenby, D. E., Schlam, T. R., & Baker, T. B. (2012). Isolating the role of psychological dysfunction in smoking cessation: Relations of personality and psychopathology to attaining cessation milestones. *Psychology of Addictive Behaviors, 26*(4), 838–849. https://doi.org/10.1037/a0028449.

Leventhal, A. M., Piper, M. E., Japuntich, S. J., Baker, T. B., & Cook, J. W. (2014). Anhedonia, depressed mood, and smoking cessation outcome. *Journal of Consulting and Clinical Psychology, 82*(1), 122–129. https://doi.org/10.1037/a0035046.

Levitan, R. D., Jansen, P., Wendland, B., Tiemeier, H., Jaddoe, V. W., Silveira, P. P., … Meaney, M. (2017). A DRD4 gene by maternal sensitivity interaction predicts risk for overweight or obesity in two independent cohorts of preschool children. *Journal of Child Psychology and Psychiatry, 58*(2), 180–188. https://doi.org/10.1111/jcpp.12646.

Levitan, R. D., Kaplan, A. S., Davis, C., Lam, R. W., & Kennedy, J. L. (2010). A season-of-birth/DRD4 interaction predicts maximal body mass index in women with bulimia nervosa. *Neuropsychopharmacology, 35*(8), 1729–1733. https://doi.org/10.1038/npp.2010.38.

Levitan, R. D., Masellis, M., Basile, V. S., Lam, R. W., Kaplan, A. S., Davis, C., … Kennedy, J. L. (2004). The dopamine-4 receptor gene associated with binge eating and weight gain in women with seasonal affective disorder: An evolutionary perspective. *Biological Psychiatry, 56*(9), 665–669. https://doi.org/10.1016/j.biopsych.2004.08.013.

Levitan, R. D., Masellis, M., Lam, R. W., Kaplan, A. S., Davis, C., Tharmalingam, S., … Kennedy, J. L. (2006). A birth-season/DRD4 gene interaction predicts weight gain and obesity in women with seasonal affective disorder: A seasonal thrifty phenotype hypothesis. *Neuropsychopharmacology, 31*(11), 2498–2503. https://doi.org/10.1038/sj.npp.1301121.

Lien, L. (2007). Is breakfast consumption related to mental distress and academic performance in adolescents? *Public Health Nutrition, 10*(4), 422–428. https://doi.org/10.1017/S1368980007258550.

Lindgren, E., Gray, K., Miller, G., Tyler, R., Wiers, C. E., Volkow, N. D., & Wang, G. J. (2018). Food addiction: A common neurobiological mechanism with drug abuse. *Frontiers in Bioscience, 23*, 811–836.

Lips, M. A., Wijngaarden, M. A., van der Grond, J., van Buchem, M. A., de Groot, G. H., Rombouts, S. A., … Veer, I. M. (2014). Resting-state functional connectivity of brain regions involved in cognitive control, motivation, and reward is enhanced in obese females. *American Journal of Clinical Nutrition, 100*(2), 524–531. https://doi.org/10.3945/ajcn.113.080671.

Long, C. G., Blundell, J. E., & Finlayson, G. (2015). A systematic review of the application and correlates of YFAS-diagnosed 'food addiction' in humans: Are eating-related 'addictions' a cause for concern or empty concepts? *Obesity Facts, 8*(6), 386–401. https://doi.org/10.1159/000442403.

Lowe, M. R. (2015). Dieting: Proxy or cause of future weight gain? *Obesity Reviews, 16*(Suppl. 1), 19–24. https://doi.org/10.1111/obr.12252.

Lowe, M. R., Doshi, S. D., Katterman, S. N., & Feig, E. H. (2013). Dieting and restrained eating as prospective predictors of weight gain. *Frontiers in Psychology, 4*, 577. https://doi.org/10.3389/fpsyg.2013.00577.

Lutz, B., Marsicano, G., Maldonado, R., & Hillard, C. J. (2015). The endocannabinoid system in guarding against fear, anxiety and stress. *Nature Reviews Neuroscience, 16*(12), 705–718. https://doi.org/10.1038/nrn4036.

Maclaren, V. V., Fugelsang, J. A., Harrigan, K. A., & Dixon, M. J. (2011). The personality of pathological gamblers: A meta-analysis. *Clinical Psychology Review, 31*(6), 1057–1067. https://doi.org/10.1016/j.cpr.2011.02.002.

Magyar, E. E., Tenyi, D., Gearhardt, A., Jeges, S., Abaligeti, G., Toth, A. L., … Csabi, G. (2018). Adaptation and validation of the Hungarian version of the Yale food addiction scale for children. *Journal of Behavioral Addictions, 7*(1), 181–188. https://doi.org/10.1556/2006.7.2018.03.

Mainz, V., Druke, B., Boecker, M., Kessel, R., Gauggel, S., & Forkmann, T. (2012). Influence of cue exposure on inhibitory control and brain activation in patients with alcohol dependence. *Frontiers in Human Neuroscience, 6*, 92. https://doi.org/10.3389/fnhum.2012.00092.

Maj, M., Turchan, J., Smialowska, M., & Przewlocka, B. (2003). Morphine and cocaine influence on CRF biosynthesis in the rat central nucleus of amygdala. *Neuropeptides, 37*(2), 105–110.

Maniam, J., Antoniadis, C. P., Le, V., & Morris, M. J. (2016). A diet high in fat and sugar reverses anxiety-like behaviour induced by limited nesting in male rats: Impacts on hippocampal markers. *Psychoneuroendocrinology, 68*, 202–209. https://doi.org/10.1016/j.psyneuen.2016.03.007.

Maniam, J., & Morris, M. J. (2010a). Long-term postpartum anxiety and depression-like behavior in mother rats subjected to maternal separation are ameliorated by palatable high fat diet. *Behavioural Brain Research, 208*(1), 72–79. https://doi.org/10.1016/j.bbr.2009.11.005.

Maniam, J., & Morris, M. J. (2010b). Palatable cafeteria diet ameliorates anxiety and depression-like symptoms following an adverse early environment. *Psychoneuroendocrinology, 35*(5), 717–728. https://doi.org/10.1016/j.psyneuen.2009.10.013.

Maniam, J., & Morris, M. J. (2010c). Voluntary exercise and palatable high-fat diet both improve behavioural profile and stress responses in male rats exposed to early life stress: Role of hippocampus. *Psychoneuroendocrinology, 35*(10), 1553–1564. https://doi.org/10.1016/j.psyneuen.2010.05.012.

Marcinkiewcz, C. A., Prado, M. M., Isaac, S. K., Marshall, A., Rylkova, D., & Bruijnzeel, A. W. (2009). Corticotropin-releasing factor within the central nucleus of the amygdala and the nucleus accumbens shell mediates the negative affective state of nicotine withdrawal in rats. *Neuropsychopharmacology, 34*(7), 1743–1752. https://doi.org/10.1038/npp.2008.231.

Markus, C. R., Panhuysen, G., Tuiten, A., Koppeschaar, H., Fekkes, D., & Peters, M. L. (1998). Does carbohydrate-rich, protein-poor food prevent a deterioration of mood and cognitive performance of stress-prone subjects when subjected to a stressful task? *Appetite, 31*(1), 49–65. https://doi.org/10.1006/appe.1997.0155.

Masheb, R. M., Ruser, C. B., Min, K. M., Bullock, A. J., & Dorflinger, L. M. (2018). Does food addiction contribute to excess weight among clinic patients seeking weight reduction? Examination of the modified Yale food addiction survey. *Comprehensive Psychiatry, 84*, 1–6. https://doi.org/10.1016/j.comppsych.2018.03.006.

Masih, T., Dimmock, J. A., Epel, E. S., & Guelfi, K. J. (2017). Stress-induced eating and the relaxation response as a potential antidote: A review and hypothesis. *Appetite, 118*, 136–143. https://doi.org/10.1016/j.appet.2017.08.005.

Mason, B. J., & Heyser, C. J. (2010). Acamprosate: A prototypic neuromodulator in the treatment of alcohol dependence. *CNS & Neurological Disorders – Drug Targets, 9*(1), 23–32.

Mason, B. J., & Lehert, P. (2012). Acamprosate for alcohol dependence: A sex-specific meta-analysis based on individual patient data. *Alcoholism: Clinical and Experimental Research, 36*(3), 497–508. https://doi.org/10.1111/j.1530-0277.2011.01616.x.

Mason, B. J., Light, J. M., Williams, L. D., & Drobes, D. J. (2009). Proof-of-concept human laboratory study for protracted abstinence in alcohol dependence: Effects of gabapentin. *Addiction Biology, 14*(1), 73–83. https://doi.org/10.1111/j.1369-1600.2008.00133.x.

Mason, A. E., Lustig, R. H., Brown, R. R., Acree, M., Bacchetti, P., Moran, P. J., ... Epel, E. S. (2015). Acute responses to opioidergic blockade as a biomarker of hedonic eating among obese women enrolled in a mindfulness-based weight loss intervention trial. *Appetite, 91*, 311–320. https://doi.org/10.1016/j.appet.2015.04.062.

Mason, B. J., Quello, S., Goodell, V., Shadan, F., Kyle, M., & Begovic, A. (2014). Gabapentin treatment for alcohol dependence: A randomized clinical trial. *JAMA Intern Med, 174*(1), 70–77. https://doi.org/10.1001/jamainternmed.2013.11950.

Mason, B. J., Quello, S., & Shadan, F. (2018). Gabapentin for the treatment of alcohol use disorder. *Expert Opinion on Investigational Drugs, 27*(1), 113–124. https://doi.org/10.1080/13543784.2018.1417383.

Mathes, W. F., Brownley, K. A., Mo, X., & Bulik, C. M. (2009). The biology of binge eating. *Appetite, 52*(3), 545–553. https://doi.org/10.1016/j.appet.2009.03.005.

Maugeri, A., Kunzova, S., Medina-Inojosa, J. R., Agodi, A., Barchitta, M., Homolka, M., ... Vinciguerra, M. (2018). Association between eating time interval and frequency with ideal cardiovascular health: Results from a random sample Czech urban population. *Nutrition, Metabolism, and Cardiovascular Diseases*. https://doi.org/10.1016/j.numecd.2018.04.002.

McElroy, S. L., Guerdjikova, A. I., Winstanley, E. L., O'Melia, A. M., Mori, N., McCoy, J., ... Hudson, J. I. (2011). Acamprosate in the treatment of binge eating disorder: A placebo-controlled trial. *International Journal of Eating Disorders, 44*(1), 81–90. https://doi.org/10.1002/eat.20876.

McElroy, S. L., Hudson, J., Ferreira-Cornwell, M. C., Radewonuk, J., Whitaker, T., & Gasior, M. (2016a). Lisdexamfetamine dimesylate for adults with moderate to severe binge eating disorder: Results of two pivotal phase 3 randomized controlled trials. *Neuropsychopharmacology, 41*(5), 1251–1260. https://doi.org/10.1038/npp.2015.275.

McElroy, S. L., Hudson, J. I., Gasior, M., Herman, B. K., Radewonuk, J., Wilfley, D., & Busner, J. (2017). Time course of the effects of lisdexamfetamine dimesylate in two phase 3, randomized, double-blind, placebo-controlled trials in adults with binge-eating disorder. *International Journal of Eating Disorders, 50*(8), 884–892. https://doi.org/10.1002/eat.22722.

McElroy, S. L., Hudson, J. I., Mitchell, J. E., Wilfley, D., Ferreira-Cornwell, M. C., Gao, J., ... Gasior, M. (2015). Efficacy and safety of lisdexamfetamine for treatment of adults with moderate to severe binge-eating disorder: A randomized clinical trial. *JAMA Psychiatry, 72*(3), 235–246. https://doi.org/10.1001/jamapsychiatry.2014.2162.

McElroy, S. L., Mitchell, J. E., Wilfley, D., Gasior, M., Ferreira-Cornwell, M. C., McKay, M., ... Hudson, J. I. (2016b). Lisdexamfetamine dimesylate effects on binge eating behaviour and obsessive-compulsive and impulsive features in adults with binge eating disorder. *European Eating Disorders Review, 24*(3), 223–231. https://doi.org/10.1002/erv.2418.

McGregor, C., Srisurapanont, M., Jittiwutikarn, J., Laobhripatr, S., Wongtan, T., & White, J. M. (2005). The nature, time course and severity of methamphetamine withdrawal. *Addiction, 100*(9), 1320–1329. https://doi.org/10.1111/j.1360-0443.2005.01160.x.

McLellan, A. T., Childress, A. R., Ehrman, R., O'Brien, C. P., & Pashko, S. (1986). Extinguishing conditioned responses during opiate dependence treatment turning laboratory findings into clinical procedures. *Journal of Substance Abuse Treatment, 3*(1), 33–40.

McNally, G. P., & Akil, H. (2002). Role of corticotropin-releasing hormone in the amygdala and bed nucleus of the stria terminalis in the behavioral, pain modulatory, and endocrine consequences of opiate withdrawal. *Neuroscience, 112*(3), 605–617.

Meng, Q., Han, Y., Ji, G., Li, G., Hu, Y., Liu, L., … Wang, G. J. (2018). Disrupted topological organization of the frontal-mesolimbic network in obese patients. *Brain Imaging Behav.* https://doi.org/10.1007/s11682-017-9802-z.

Meule, A., de Zwaan, M., & Muller, A. (2017a). Attentional and motor impulsivity interactively predict 'food addiction' in obese individuals. *Comprehensive Psychiatry, 72*, 83–87. https://doi.org/10.1016/j.comppsych.2016.10.001.

Meule, A., Heckel, D., Jurowich, C. F., Vogele, C., & Kubler, A. (2014). Correlates of food addiction in obese individuals seeking bariatric surgery. *Clinical Obesity, 4*(4), 228–236. https://doi.org/10.1111/cob.12065.

Meule, A., Hermann, T., & Kubler, A. (2015). Food addiction in overweight and obese adolescents seeking weight-loss treatment. *European Eating Disorders Review, 23*(3), 193–198. https://doi.org/10.1002/erv.2355.

Meule, A., & Kubler, A. (2012). Food cravings in food addiction: The distinct role of positive reinforcement. *Eating Behaviors, 13*(3), 252–255. https://doi.org/10.1016/j.eatbeh.2012.02.001.

Meule, A., Muller, A., Gearhardt, A. N., & Blechert, J. (2017b). German version of the Yale Food Addiction Scale 2.0: Prevalence and correlates of 'food addiction' in students and obese individuals. *Appetite, 115*, 54–61. https://doi.org/10.1016/j.appet.2016.10.003.

Micali, N., Martini, M. G., Thomas, J. J., Eddy, K. T., Kothari, R., Russell, E., … Treasure, J. (2017). Lifetime and 12-month prevalence of eating disorders amongst women in mid-life: A population-based study of diagnoses and risk factors. *BMC Medicine, 15*(1), 12. https://doi.org/10.1186/s12916-016-0766-4.

Michaelides, M., Miller, M. L., DiNieri, J. A., Gomez, J. L., Schwartz, E., Egervari, G., … Hurd, Y. L. (2017). Dopamine D2 receptor signaling in the nucleus accumbens comprises a metabolic-cognitive brain interface regulating metabolic components of glucose reinforcement. *Neuropsychopharmacology, 42*(12), 2365–2376. https://doi.org/10.1038/npp.2017.112.

Michaelides, M., Thanos, P. K., Volkow, N. D., & Wang, G. J. (2012). Dopamine-related frontostriatal abnormalities in obesity and binge-eating disorder: Emerging evidence for developmental psychopathology. *International Review of Psychiatry, 24*(3), 211–218. https://doi.org/10.3109/09540261.2012.679918.

Michel, J. B., Shen, Y. K., Aiden, A. P., Veres, A., Gray, M. K., Google Books, T., … Aiden, E. L. (2011). Quantitative analysis of culture using millions of digitized books. *Science, 331*(6014), 176–182. https://doi.org/10.1126/science.1199644.

Micioni Di Bonaventura, M. V., Ciccocioppo, R., Romano, A., Bossert, J. M., Rice, K. C., Ubaldi, M., … Cifani, C. (2014). Role of bed nucleus of the stria terminalis corticotrophin-releasing factor receptors in frustration stress-induced binge-like palatable food consumption in female rats with a history of food restriction. *Journal of Neuroscience, 34*(34), 11316–11324. https://doi.org/10.1523/JNEUROSCI.1854-14.2014.

Micioni Di Bonaventura, M. V., Lutz, T. A., Romano, A., Pucci, M., Geary, N., Asarian, L., & Cifani, C. (2017). Estrogenic suppression of binge-like eating elicited by cyclic food restriction and frustrative-nonreward stress in female rats. *International Journal of Eating Disorders, 50*(6), 624–635. https://doi.org/10.1002/eat.22687.

Micioni Di Bonaventura, M. V., Ubaldi, M., Giusepponi, M. E., Rice, K. C., Massi, M., Ciccocioppo, R., & Cifani, C. (2017). Hypothalamic CRF1 receptor mechanisms are not sufficient to account for binge-like palatable food consumption in female rats. *International Journal of Eating Disorders, 50*(10), 1194–1204. https://doi.org/10.1002/eat.22767.

Micioni Di Bonaventura, M. V., Ubaldi, M., Liberati, S., Ciccocioppo, R., Massi, M., & Cifani, C. (2013). Caloric restriction increases the sensitivity to the hyperphagic effect of nociceptin/orphanin FQ limiting its ability to reduce binge eating in female rats. *Psychopharmacology, 228*(1), 53–63. https://doi.org/10.1007/s00213-013-3013-0.

Micioni Di Bonaventura, M. V., Vitale, G., Massi, M., & Cifani, C. (2012). Effect of *Hypericum perforatum* extract in an experimental model of binge eating in female rats. *Journal of Obesity, 2012*, 956137. https://doi.org/10.1155/2012/956137.

Mies, G. W., Treur, J. L., Larsen, J. K., Halberstadt, J., Pasman, J. A., & Vink, J. M. (2017). The prevalence of food addiction in a large sample of adolescents and its association with addictive substances. *Appetite, 118*, 97–105. https://doi.org/10.1016/j.appet.2017.08.002.

Mitchell, J. E., & Mussell, M. P. (1995). Comorbidity and binge eating disorder. *Addictive Behaviors, 20*(6), 725–732.

Mitchell, T. B., White, J. M., Somogyi, A. A., & Bochner, F. (2006). Switching between methadone and morphine for maintenance treatment of opioid dependence: Impact on pain sensitivity and mood status. *American Journal on Addictions, 15*(4), 311–315. https://doi.org/10.1080/10550490600754374.

Mithra, P., Unnikrishnan, B., Thapar, R., Kumar, N., Hegde, S., Mangaldas Kamat, A., … Kumar, A. (2018). Snacking behaviour and its determinants among college-going students in coastal south India. *J Nutr Metab, 2018*. https://doi.org/10.1155/2018/6785741, 6785741.

Montani, J. P., Schutz, Y., & Dulloo, A. G. (2015). Dieting and weight cycling as risk factors for cardiometabolic diseases: Who is really at risk? *Obesity Reviews, 16*(Suppl. 1), 7–18. https://doi.org/10.1111/obr.12251.

Moore, C. F., Sabino, V., Koob, G. F., & Cottone, P. (2017). Pathological overeating: Emerging evidence for a compulsivity construct. *Neuropsychopharmacology, 42*(7), 1375–1389. https://doi.org/10.1038/npp.2016.269.

Murphy, C. M., Stojek, M. K., & MacKillop, J. (2014). Interrelationships among impulsive personality traits, food addiction, and Body Mass Index. *Appetite, 73*, 45–50. https://doi.org/10.1016/j.appet.2013.10.008.

Murray, S. B., Anderson, L. K., Cusack, A., Nakamura, T., Rockwell, R., Griffiths, S., & Kaye, W. H. (2015). Integrating family-based treatment and dialectical behavior therapy for adolescent bulimia nervosa: Preliminary outcomes of an open pilot trial. *Eating Disorders, 23*(4), 336–344. https://doi.org/10.1080/10640266.2015.1044345.

Nair, S. G., Gray, S. M., & Ghitza, U. E. (2006). Role of food type in yohimbine- and pellet-priming-induced reinstatement of food seeking. *Physiology & Behavior, 88*(4–5), 559–566. https://doi.org/10.1016/j.physbeh.2006.05.014.

Nair, S. G., Navarre, B. M., Cifani, C., Pickens, C. L., Bossert, J. M., & Shaham, Y. (2011). Role of dorsal medial prefrontal cortex dopamine D1-family receptors in relapse to high-fat food seeking induced by the anxiogenic drug yohimbine. *Neuropsychopharmacology, 36*(2), 497–510. https://doi.org/10.1038/npp.2010.181.

Nanni, G., Scheggi, S., Leggio, B., Grappi, S., Masi, F., Rauggi, R., & De Montis, M. G. (2003). Acquisition of an appetitive behavior prevents development of stress-induced neurochemical modifications in rat nucleus accumbens. *Journal of Neuroscience Research, 73*(4), 573–580. https://doi.org/10.1002/jnr.10685.

Ng, J., Stice, E., Yokum, S., & Bohon, C. (2011). An fMRI study of obesity, food reward, and perceived caloric density. Does a low-fat label make food less appealing? *Appetite, 57*(1), 65–72. https://doi.org/10.1016/j.appet.2011.03.017.

Nielsen, A. L., Larsen, T. M., Madsbad, S., Breum, L., Jensen, T. J., Kroustrup, J. P., & Astrup, A. (2009). The effect of tesofensine on body weight and body composition in obese subjects–secondary publication. *Ugeskr Laeger, 171*(41), 2974–2977.

Niemeier, H. M., Raynor, H. A., Lloyd-Richardson, E. E., Rogers, M. L., & Wing, R. R. (2006). Fast food consumption and breakfast skipping: Predictors of weight gain from adolescence to adulthood in a nationally representative sample. *Journal of Adolescent Health, 39*(6), 842–849. https://doi.org/10.1016/j.jadohealth.2006.07.001.

Nolan, L. J., & Geliebter, A. (2017). Validation of the Night Eating Diagnostic Questionnaire (NEDQ) and its relationship with depression, sleep quality, "food addiction", and body mass index. *Appetite, 111*, 86–95. https://doi.org/10.1016/j.appet.2016.12.027.

Nowinski, J., Baker, S., & Carroll, K. (1999). *Twelve step facilitation therapy manual: A clinical Research guide for therapists treating individuals with alcohol abuse and dependence*. Retrieved from Rockville, MD.

Nyland, J. E., & Grigson, P. S. (2013). A drug-paired taste cue elicits withdrawal and predicts cocaine self-administration. *Behavioural Brain Research, 240*, 87–90. https://doi.org/10.1016/j.bbr.2012.10.057.

Oddy, W. H., Robinson, M., Ambrosini, G. L., O'Sullivan, T. A., de Klerk, N. H., Beilin, L. J., … Stanley, F. J. (2009). The association between dietary patterns and mental health in early adolescence. *Preventive Medicine, 49*(1), 39–44. https://doi.org/10.1016/j.ypmed.2009.05.009.

Okada, C., Tabuchi, T., & Iso, H. (2018). Association between skipping breakfast in parents and children and childhood overweight/obesity among children: A nationwide 10.5-year prospective study in Japan. *International Journal of Obesity.* https://doi.org/10.1038/s41366-018-0066-5.

Orihuela, C. A., Mrug, S., & Boggiano, M. M. (2017). Reciprocal relationships between emotion regulation and motives for eating palatable foods in African American adolescents. *Appetite, 117*, 303–309. https://doi.org/10.1016/j.appet.2017.07.008.

Oswald, K. D., Murdaugh, D. L., King, V. L., & Boggiano, M. M. (2011). Motivation for palatable food despite consequences in an animal model of binge eating. *International Journal of Eating Disorders, 44*(3), 203–211. https://doi.org/10.1002/eat.20808.

Overstreet, D. H., Knapp, D. J., & Breese, G. R. (2004). Modulation of multiple ethanol withdrawal-induced anxiety-like behavior by CRF and CRF1 receptors. *Pharmacology Biochemistry and Behavior, 77*(2), 405–413. S0091305703003587 [pii].

Owens, M. M., Amlung, M. T., Stojek, M., & MacKillop, J. (2018). Negative urgency moderates reactivity to laboratory stress inductions. *Journal of Abnormal Psychology, 127*(4), 385–393. https://doi.org/10.1037/abn0000350.

Paczynski, R. P., & Gold, M. S. (2011). Cocaine and crack. In P. Ruiz, & E. Strain (Eds.), *Lowinson & ruiz's substance abuse: A comprehensive textbook* (5th ed.). Baltimore, MD: Lippincott Williams & Wilkins.

Parsons, L. H., & Hurd, Y. L. (2015). Endocannabinoid signalling in reward and addiction. *Nature Reviews Neuroscience, 16*(10), 579–594. https://doi.org/10.1038/nrn4004.

Parylak, S. L., Cottone, P., Sabino, V., Rice, K. C., & Zorrilla, E. P. (2012). Effects of CB1 and CRF1 receptor antagonists on binge-like eating in rats with limited access to a sweet fat diet: Lack of withdrawal-like responses. *Physiology & Behavior, 107*(2), 231–242. https://doi.org/10.1016/j.physbeh.2012.06.017.

Parylak, S. L., Koob, G. F., & Zorrilla, E. P. (2011). The dark side of food addiction. *Physiology & Behavior, 104*(1), 149–156. https://doi.org/10.1016/j.physbeh.2011.04.063.

Pearson, C. M., Riley, E. N., Davis, H. A., & Smith, G. T. (2014). Two pathways toward impulsive action: An integrative risk model for bulimic behavior in youth. *Journal of Child Psychology and Psychiatry, 55*(8), 852–864. https://doi.org/10.1111/jcpp.12214.

Pearson, C. M., Wonderlich, S. A., & Smith, G. T. (2015a). A risk and maintenance model for bulimia nervosa: From impulsive action to compulsive behavior. *Psychological Review, 122*(3), 516–535. https://doi.org/10.1037/a0039268.

Pearson, C. M., Zapolski, T. C., & Smith, G. T. (2015b). A longitudinal test of impulsivity and depression pathways to early binge eating onset. *International Journal of Eating Disorders, 48*(2), 230–237. https://doi.org/10.1002/eat.22277.

Peckmezian, T., & Hay, P. (2017). A systematic review and narrative synthesis of interventions for uncomplicated obesity: Weight loss, well-being and impact on eating disorders. *J Eat Disord, 5*, 15. https://doi.org/10.1186/s40337-017-0143-5.

Pecoraro, N., Reyes, F., Gomez, F., Bhargava, A., & Dallman, M. F. (2004). Chronic stress promotes palatable feeding, which reduces signs of stress: Feedforward and feedback effects of chronic stress. *Endocrinology, 145*(8), 3754–3762. https://doi.org/10.1210/en.2004-0305.

Pereira-Lancha, L. O., Coelho, D. F., de Campos-Ferraz, P. L., & Lancha, A. H., Jr. (2010). Body fat regulation: Is it a result of a simple energy balance or a high fat intake? *Journal of the American College of Nutrition, 29*(4), 343–351.

Perez, M., Ohrt, T. K., & Hoek, H. W. (2016b). Prevalence and treatment of eating disorders among Hispanics/Latino Americans in the United States. *Current Opinion in Psychiatry, 29*(6), 378–382. https://doi.org/10.1097/yco.0000000000000277.

Perez, D. L., Vago, D. R., Pan, H., Root, J., Tuescher, O., Fuchs, B. H., … Stern, E. (2016a). Frontolimbic neural circuit changes in emotional processing and inhibitory control associated with clinical improvement following transference-focused psychotherapy in borderline personality disorder. *Psychiatry and Clinical Neurosciences, 70*(1), 51–61. https://doi.org/10.1111/pcn.12357.

Perney, P., Lehert, P., & Mason, B. J. (2012). Sleep disturbance in alcoholism: Proposal of a simple measurement, and results from a 24-week randomized controlled study of alcohol-dependent patients assessing acamprosate efficacy. *Alcohol and Alcoholism, 47*(2), 133–139. https://doi.org/10.1093/alcalc/agr160.

Peters, L. H. (1918). *Diet and health: With key to the calories*. Chicago: The Reilly and Lee Co.

Piaggi, P., Thearle, M. S., Bogardus, C., & Krakoff, J. (2013). Lower energy expenditure predicts long-term increases in weight and fat mass. *The Journal of Cinical Endocrinology and Metabolism, 98*(4), E703–E707. https://doi.org/10.1210/jc.2012-3529.

Piasecki, T. M., Fiore, M. C., & Baker, T. B. (1998). Profiles in discouragement: Two studies of variability in the time course of smoking withdrawal symptoms. *Journal of Abnormal Psychology, 107*(2), 238–251.

Piasecki, T. M., Jorenby, D. E., Smith, S. S., Fiore, M. C., & Baker, T. B. (2003). Smoking withdrawal dynamics: II. Improved tests of withdrawal-relapse relations. *Journal of Abnormal Psychology, 112*(1), 14–27.

Piasecki, T. M., Niaura, R., Shadel, W. G., Abrams, D., Goldstein, M., Fiore, M. C., & Baker, T. B. (2000). Smoking withdrawal dynamics in unaided quitters. *Journal of Abnormal Psychology, 109*(1), 74–86.

Piccoli, L., Micioni Di Bonaventura, M. V., Cifani, C., Costantini, V. J., Massagrande, M., Montanari, D., ... Corsi, M. (2012). Role of orexin-1 receptor mechanisms on compulsive food consumption in a model of binge eating in female rats. *Neuropsychopharmacology, 37*(9), 1999–2011. https://doi.org/10.1038/npp.2012.48.

Pickering, C., Alsio, J., Hulting, A. L., & Schioth, H. B. (2009). Withdrawal from free-choice high-fat high-sugar diet induces craving only in obesity-prone animals. *Psychopharmacology, 204*(3), 431–443. https://doi.org/10.1007/s00213-009-1474-y.

Piper, M. E. (2015). Withdrawal: Expanding a key addiction construct. *Nicotine & Tobacco Research, 17*(12), 1405–1415. https://doi.org/10.1093/ntr/ntv048.

Piper, M. E., Federmen, E. B., McCarthy, D. E., Bolt, D. M., Smith, S. S., Fiore, M. C., & Baker, T. B. (2008). Using mediational models to explore the nature of tobacco motivation and tobacco treatment effects. *Journal of Abnormal Psychology, 117*(1), 94–105. https://doi.org/10.1037/0021-843X.117.1.94.

Piper, M. E., Cook, J. W., Schlam, T. R., Jorenby, D. E., & Baker, T. B. (2011a). Anxiety diagnoses in smokers seeking cessation treatment: Relations with tobacco dependence, withdrawal, outcome and response to treatment. *Addiction, 106*(2), 418–427. https://doi.org/10.1111/j.1360-0443.2010.03173.x.

Piper, M. E., Schlam, T. R., Cook, J. W., Sheffer, M. A., Smith, S. S., Loh, W. Y., ... Baker, T. B. (2011b). Tobacco withdrawal components and their relations with cessation success. *Psychopharmacology, 216*(4), 569–578. https://doi.org/10.1007/s00213-011-2250-3.

Piper, M. E., Rodock, M., Cook, J. W., Schlam, T. R., Fiore, M. C., & Baker, T. B. (2013). Psychiatric diagnoses among quitters versus continuing smokers 3 years after their quit day. *Drug and Alcohol Dependence, 128*(1–2), 148–154. https://doi.org/10.1016/j.drugalcdep.2012.08.023.

Piper, M. E., Smith, S. S., Schlam, T. R., Fleming, M. F., Bittrich, A. A., Brown, J. L., ... Baker, T. B. (2010). Psychiatric disorders in smokers seeking treatment for tobacco dependence: Relations with tobacco dependence and cessation. *Journal of Consulting and Clinical Psychology, 78*(1), 13–23. https://doi.org/10.1037/a0018065.

Piper, M. E., Vasilenko, S. A., Cook, J. W., & Lanza, S. T. (2017). What a difference a day makes: Differences in initial abstinence response during a smoking cessation attempt. *Addiction, 112*(2), 330–339. https://doi.org/10.1111/add.13613.

Pisetsky, E. M., Thornton, L. M., Lichtenstein, P., Pedersen, N. L., & Bulik, C. M. (2013). Suicide attempts in women with eating disorders. *Journal of Abnormal Psychology, 122*(4), 1042–1056. https://doi.org/10.1037/a0034902.

Pivarunas, B., & Conner, B. T. (2015). Impulsivity and emotion dysregulation as predictors of food addiction. *Eating Behaviors, 19*, 9–14. https://doi.org/10.1016/j.eatbeh.2015.06.007.

Polivy, J., & Herman, C. P. (1985). Dieting and binging. A causal analysis. *American Psychologist, 40*(2), 193–201.

Popik, P., Kos, T., Zhang, Y., & Bisaga, A. (2011). Memantine reduces consumption of highly palatable food in a rat model of binge eating. *Amino Acids, 40*(2), 477–485. https://doi.org/10.1007/s00726-010-0659-3.

Pothos, E., Rada, P., Mark, G. P., & Hoebel, B. G. (1991). Dopamine microdialysis in the nucleus accumbens during acute and chronic morphine, naloxone-precipitated withdrawal and clonidine treatment. *Brain Research, 566*(1–2), 348–350.

Pucci, M., Micioni Di Bonaventura, M. V., Giusepponi, M. E., Romano, A., Filaferro, M., Maccarrone, M., ... D'Addario, C. (2016). Epigenetic regulation of nociceptin/orphanin FQ and corticotropin-releasing factor system genes in frustration stress-induced binge-like palatable food consumption. *Addiction Biology, 21*(6), 1168–1185. https://doi.org/10.1111/adb.12303.

Pursey, K. M., Gearhardt, A. N., & Burrows, T. L. (2016). The relationship between "food addiction" and visceral adiposity in young females. *Physiology & Behavior, 157*, 9–12. https://doi.org/10.1016/j.physbeh.2016.01.018.

Pursey, K. M., Stanwell, P., Gearhardt, A. N., Collins, C. E., & Burrows, T. L. (2014). The prevalence of food addiction as assessed by the Yale food addiction scale: A systematic review. *Nutrients, 6*(10), 4552–4590. https://doi.org/10.3390/nu6104552.

Racine, S. E., Keel, P. K., Burt, S. A., Sisk, C. L., Neale, M., Boker, S., & Klump, K. L. (2013). Exploring the relationship between negative urgency and dysregulated eating: Etiologic associations and the role of negative affect. *Journal of Abnormal Psychology, 122*(2), 433–444.

Rada, P., Avena, N. M., & Hoebel, B. G. (2005). Daily bingeing on sugar repeatedly releases dopamine in the accumbens shell. *Neuroscience, 134*(3), 737–744. https://doi.org/10.1016/j.neuroscience.2005.04.043.

Rada, P., Bocarsly, M. E., Barson, J. R., Hoebel, B. G., & Leibowitz, S. F. (2010). Reduced accumbens dopamine in Sprague-Dawley rats prone to overeating a fat-rich diet. *Physiology & Behavior, 101*(3), 394–400. https://doi.org/10.1016/j.physbeh.2010.07.005.

Rada, P., Pothos, E., Mark, G. P., & Hoebel, B. G. (1991). Microdialysis evidence that acetylcholine in the nucleus accumbens is involved in morphine withdrawal and its treatment with clonidine. *Brain Research, 561*(2), 354–356.

Radke, A. K., Rothwell, P. E., & Gewirtz, J. C. (2011). An anatomical basis for opponent process mechanisms of opiate withdrawal. *Journal of Neuroscience, 31*(20), 7533–7539. https://doi.org/10.1523/JNEUROSCI.0172-11.2011.

Rahimlou, M., Morshedzadeh, N., Karimi, S., & Jafarirad, S. (2018). Association between dietary glycemic index and glycemic load with depression: A systematic review. *European Journal of Nutrition*. https://doi.org/10.1007/s00394-018-1710-5.

Ramsay, S. A., Bloch, T. D., Marriage, B., Shriver, L. H., Spees, C. K., & Taylor, C. A. (2018). Skipping breakfast is associated with lower diet quality in young US children. *European Journal of Clinical Nutrition, 72*(4), 548–556. https://doi.org/10.1038/s41430-018-0084-3.

Raymond, K. L., Kannis-Dymand, L., & Lovell, G. P. (2016). A graduated food addiction classification approach significantly differentiates obesity among people with type 2 diabetes. *Journal of Health Psychology*. https://doi.org/10.1177/1359105316672096, 1359105316672096.

Raymond, K. L., Kannis-Dymand, L., & Lovell, G. P. (2017). A graduated food addiction classifications approach significantly differentiates depression, anxiety and stress among people with type 2 diabetes. *Diabetes Research and Clinical Practice, 132*, 95–101. https://doi.org/10.1016/j.diabres.2017.07.028.

Raymond, K. L., & Lovell, G. P. (2015). Food addiction symptomology, impulsivity, mood, and body mass index in people with type two diabetes. *Appetite, 95*, 383–389. https://doi.org/10.1016/j.appet.2015.07.030.

Raymond, K. L., & Lovell, G. P. (2016). Food addiction associations with psychological distress among people with type 2 diabetes. *Journal of Diabetic Complications, 30*(4), 651–656. https://doi.org/10.1016/j.jdiacomp.2016.01.020.

Razzoli, M., Pearson, C., Crow, S., & Bartolomucci, A. (2017). Stress, overeating, and obesity: Insights from human studies and preclinical models. *Neuroscience & Biobehavioral Reviews, 76*(Pt A), 154–162. https://doi.org/10.1016/j.neubiorev.2017.01.026.

Reid, M., & Hammersley, R. (1999). The effects of sucrose and maize oil on subsequent food intake and mood. *British Journal of Nutrition, 82*(6), 447–455.

Reinblatt, S. P., Mahone, E. M., Tanofsky-Kraff, M., Lee-Winn, A. E., Yenokyan, G., Leoutsakos, J. M., … Riddle, M. A. (2015). Pediatric loss of control eating syndrome: Association with attention-deficit/hyperactivity disorder and impulsivity. *International Journal of Eating Disorders, 48*(6), 580–588. https://doi.org/10.1002/eat.22404.

Richardson, H. N., Zhao, Y., Fekete, E. M., Funk, C. K., Wirsching, P., Janda, K. D., … Koob, G. F. (2008). MPZP: A novel small molecule corticotropin-releasing factor type 1 receptor (CRF1) antagonist. *Pharmacology Biochemistry and Behavior, 88*(4), 497–510. https://doi.org/10.1016/j.pbb.2007.10.008.

Richards, J. K., Simms, J. A., Steensland, P., Taha, S. A., Borgland, S. L., Bonci, A., & Bartlett, S. E. (2008). Inhibition of orexin-1/hypocretin-1 receptors inhibits yohimbine-induced reinstatement of ethanol and sucrose seeking in Long-Evans rats. *Psychopharmacology, 199*(1), 109–117. https://doi.org/10.1007/s00213-008-1136-5.

Richmond, R. L., Roberto, C. A., & Gearhardt, A. N. (2017). The association of addictive-like eating with food intake in children. *Appetite, 117*, 82–90. https://doi.org/10.1016/j.appet.2017.06.002.

Richter, R. M., & Weiss, F. (1999). In vivo CRF release in rat amygdala is increased during cocaine withdrawal in self-administering rats. *Synapse, 32*(4), 254–261. https://doi.org/10.1002/(SICI)1098-2396(19990615)32:4<254::AID-SYN2>3.0.CO;2-H.

Robbins, T. W., Gillan, C. M., Smith, D. G., de Wit, S., & Ersche, K. D. (2012). Neurocognitive endophenotypes of impulsivity and compulsivity: Towards dimensional psychiatry. *Trends in Cognitive Sciences, 16*(1), 81–91. https://doi.org/10.1016/j.tics.2011.11.009.

Roberto, M., Cruz, M. T., Gilpin, N. W., Sabino, V., Schweitzer, P., Bajo, M., … Parsons, L. H. (2010). Corticotropin releasing factor-induced amygdala gamma-aminobutyric Acid release plays a key role in alcohol dependence. *Biological Psychiatry, 67*(9), 831–839. https://doi.org/10.1016/j.biopsych.2009.11.007.

Roberto, M., Spierling, S. R., Kirson, D., & Zorrilla, E. P. (2017). Corticotropin-releasing factor (CRF) and addictive behaviors. *International Review of Neurobiology, 136*, 5–51. https://doi.org/10.1016/bs.irn.2017.06.004.

Robinson, E., & Aveyard, P. (2017). Emaciated mannequins: A study of mannequin body size in high street fashion stores. *Journal of Eating Disorders, 5*, 13. https://doi.org/10.1186/s40337-017-0142-6.

Robinson, M. J., Warlow, S. M., & Berridge, K. C. (2014). Optogenetic excitation of central amygdala amplifies and narrows incentive motivation to pursue one reward above another. *Journal of Neuroscience, 34*(50), 16567–16580. https://doi.org/10.1523/JNEUROSCI.2013-14.2014.

Rodrigues, P. R. M., Luiz, R. R., Monteiro, L. S., Ferreira, M. G., Goncalves-Silva, R. M. V., & Pereira, R. A. (2017). Adolescents' unhealthy eating habits are associated with meal skipping. *Nutrition, 42*, 114–120 e111. https://doi.org/10.1016/j.nut.2017.03.011.

Rodriguez de Fonseca, F., Carrera, M. R., Navarro, M., Koob, G. F., & Weiss, F. (1997). Activation of corticotropin-releasing factor in the limbic system during cannabinoid withdrawal. *Science, 276*(5321), 2050–2054.

Rose, M. H., Nadler, E. P., & Mackey, E. R. (2018). Impulse control in negative mood states, emotional eating, and food addiction are associated with lower quality of life in adolescents with severe obesity. *Journal of Pediatric Psychology, 43*(4), 443–451. https://doi.org/10.1093/jpepsy/jsx127.

Rosen, J. C., Tacy, B., & Howell, D. (1990). Life stress, psychological symptoms and weight reducing behavior in adolescent girls: A prospective analysis. *International Journal of Eating Disorders, 9*(1), 17–26.

Rossetti, C., Spena, G., Halfon, O., & Boutrel, B. (2014). Evidence for a compulsive-like behavior in rats exposed to alternate access to highly preferred palatable food. *Addiction Biology, 19*(6), 975–985. https://doi.org/10.1111/adb.12065.

Rowley, H. L., Kulkarni, R., Gosden, J., Brammer, R., Hackett, D., & Heal, D. J. (2012). Lisdexamfetamine and immediate release d-amfetamine - differences in pharmacokinetic/pharmacodynamic relationships revealed by striatal microdialysis in freely-moving rats with simultaneous determination of plasma drug concentrations and locomotor activity. *Neuropharmacology, 63*(6), 1064–1074. https://doi.org/10.1016/j.neuropharm.2012.07.008.

Ruddock, H. K., Christiansen, P., Halford, J. C. G., & Hardman, C. A. (2017). The development and validation of the addiction-like eating behaviour scale. *International Journal of Obesity, 41*(11), 1710–1717. https://doi.org/10.1038/ijo.2017.158.

Runfola, C. D., Thornton, L. M., Pisetsky, E. M., Bulik, C. M., & Birgegard, A. (2014). Self-image and suicide in a Swedish national eating disorders clinical register. *Comprehensive Psychiatry, 55*(3), 439–449. https://doi.org/10.1016/j.comppsych.2013.11.007.

Rutledge, T., & Linden, W. (1998). To eat or not to eat: Affective and physiological mechanisms in the stress-eating relationship. *Journal of Behavioral Medicine, 21*(3), 221–240.

Sabino, V., Cottone, P., Koob, G. F., Steardo, L., Lee, M. J., Rice, K. C., & Zorrilla, E. P. (2006). Dissociation between opioid and CRF1 antagonist sensitive drinking in Sardinian alcohol-preferring rats. *Psychopharmacology, 189*(2), 175–186. https://doi.org/10.1007/s00213-006-0546-5.

Sakurai, M., Yoshita, K., Nakamura, K., Miura, K., Takamura, T., Nagasawa, S. Y., … Nakagawa, H. (2017). Skipping breakfast and 5-year changes in body mass index and waist circumference in Japanese men and women. *Obesity Science & Practice, 3*(2), 162–170. https://doi.org/10.1002/osp4.106.

Sanchez-Villegas, A., Delgado-Rodriguez, M., Alonso, A., Schlatter, J., Lahortiga, F., Serra Majem, L., & Martinez-Gonzalez, M. A. (2009). Association of the mediterranean dietary pattern with the incidence of depression: The seguimiento universidad de Navarra/university of navarra follow-up (SUN) cohort. *Archives of General Psychiatry, 66*(10), 1090–1098. https://doi.org/10.1001/archgenpsychiatry.2009.129.

Sanchez-Villegas, A., Toledo, E., de Irala, J., Ruiz-Canela, M., Pla-Vidal, J., & Martinez-Gonzalez, M. A. (2012). Fast-food and commercial baked goods consumption and the risk of depression. *Public Health Nutrition, 15*(3), 424–432. https://doi.org/10.1017/S1368980011001856.

Santos, I., Sniehotta, F. F., Marques, M. M., Carraca, E. V., & Teixeira, P. J. (2017). Prevalence of personal weight control attempts in adults: A systematic review and meta-analysis. *Obesity Reviews, 18*(1), 32–50. https://doi.org/10.1111/obr.12466.

Sarnyai, Z., Biro, E., Gardi, J., Vecsernyes, M., Julesz, J., & Telegdy, G. (1995). Brain corticotropin-releasing factor mediates 'anxiety-like' behavior induced by cocaine withdrawal in rats. *Brain Research, 675*(1–2), 89–97.

Sarris, J., Logan, A. C., Akbaraly, T. N., Amminger, G. P., Balanza-Martinez, V., Freeman, M. P., ... International Society for Nutritional Psychiatry, R. (2015). Nutritional medicine as mainstream in psychiatry. *Lancet Psychiatry, 2*(3), 271–274. https://doi.org/10.1016/S2215-0366(14)00051-0.

Schaefer, L. M., Burke, N. L., & Thompson, J. K. (2018). Thin-ideal internalization: How much is too much? *Eating and Weight Disorders*. https://doi.org/10.1007/s40519-018-0498-x.

Schroder, R., Sellman, J. D., & Adamson, S. (2017). Development and validation of a brief measure of eating compulsivity (MEC). *Substance Use & Misuse, 52*(14), 1918–1924. https://doi.org/10.1080/10826084.2017.1343352.

Schulte, E. M., Avena, N. M., & Gearhardt, A. N. (2015). Which foods may be addictive? The roles of processing, fat content, and glycemic load. *PLoS One, 10*(2), e0117959. https://doi.org/10.1371/journal.pone.0117959.

Schulte, E. M., & Gearhardt, A. N. (2017). Development of the modified Yale food addiction scale version 2.0. *European Eating Disorders Review, 25*(4), 302–308. https://doi.org/10.1002/erv.2515.

Schulte, E. M., Grilo, C. M., & Gearhardt, A. N. (2016). Shared and unique mechanisms underlying binge eating disorder and addictive disorders. *Clinical Psychology Review, 44*, 125–139. https://doi.org/10.1016/j.cpr.2016.02.001.

Schulte, E. M., Jacques-Tiura, A. J., Gearhardt, A. N., & Naar, S. (2018). Food addiction prevalence and concurrent validity in African American adolescents with obesity. *Psychology of Addictive Behaviors, 32*(2), 187–196. https://doi.org/10.1037/adb0000325.

Serrano, A., Pavon, F. J., Buczynski, M. W., Schlosburg, J., Natividad, L. A., Polis, I. Y., ... Parsons, L. H. (2018). Deficient endocannabinoid signaling in the central amygdala contributes to alcohol dependence-related anxiety-like behavior and excessive alcohol intake. *Neuropsychopharmacology*. https://doi.org/10.1038/s41386-018-0055-3.

Shearrer, G. E., Stice, E., & Burger, K. S. (2018). Adolescents at high risk of obesity show greater striatal response to increased sugar content in milkshakes. *American Journal of Clinical Nutrition, 107*(6), 859–866. https://doi.org/10.1093/ajcn/nqy050.

Sheehan, D. V., & Herman, B. K. (2015). The psychological and medical factors associated with untreated binge eating disorder. *Prim Care Companion CNS Disord, 17*(2). https://doi.org/10.4088/PCC.14r01732.

Sheppard-Sawyer, C. L., McNally, R. J., & Fischer, J. H. (2000). Film-induced sadness as a trigger for disinhibited eating. *International Journal of Eating Disorders, 28*(2), 215–220.

Sherman, J. E., Zinser, M. C., Sideroff, S. I., & Baker, T. B. (1989). Subjective dimensions of heroin urges: Influence of heroin-related and affectively negative stimuli. *Addictive Behaviors, 14*(6), 611–623.

Shriner, R. L., Graham, N. A., & Gold, M. S. (2010). Smoking cessation and the risk for type 2 diabetes mellitus. *Annals of Internal Medicine, 152*(11), 755. https://doi.org/10.7326/0003-4819-152-11-201006010-00020. author reply 755-756.

Shurman, J., Koob, G. F., & Gutstein, H. B. (2010). Opioids, pain, the brain, and hyperkatifeia: A framework for the rational use of opioids for pain. *Pain Medicine, 11*(7), 1092–1098. https://doi.org/10.1111/j.1526-4637.2010.00881.x.

Siegel, S., & Ramos, B. M. (2002). Applying laboratory research: Drug anticipation and the treatment of drug addiction. *Experimental and Clinical Psychopharmacology, 10*(3), 162–183.

Silbersweig, D., Clarkin, J. F., Goldstein, M., Kernberg, O. F., Tuescher, O., Levy, K. N., ... Stern, E. (2007). Failure of frontolimbic inhibitory function in the context of negative emotion in borderline personality disorder. *American Journal of Psychiatry, 164*(12), 1832–1841. https://doi.org/10.1176/appi.ajp.2007.06010126.

Singh, M. (2014). Mood, food, and obesity. *Frontiers in Psychology, 5*, 925. https://doi.org/10.3389/fpsyg.2014.00925.

Sjodin, A., Gasteyger, C., Nielsen, A. L., Raben, A., Mikkelsen, J. D., Jensen, J. K., ... Astrup, A. (2010). The effect of the triple monoamine reuptake inhibitor tesofensine on energy metabolism and appetite in overweight and moderately obese men. *International Journal of Obesity, 34*(11), 1634–1643. https://doi.org/10.1038/ijo.2010.87.

Skelton, K. H., Oren, D., Gutman, D. A., Easterling, K., Holtzman, S. G., Nemeroff, C. B., & Owens, M. J. (2007). The CRF1 receptor antagonist, R121919, attenuates the severity of precipitated morphine withdrawal. *European Journal of Pharmacology, 571*(1), 17–24. https://doi.org/10.1016/j.ejphar.2007.05.041.

Small, A. C., Kampman, K. M., Plebani, J., De Jesus Quinn, M., Peoples, L., & Lynch, K. G. (2009). Tolerance and sensitization to the effects of cocaine use in humans: A retrospective study of long-term cocaine users in philadelphia. *Substance Use & Misuse, 44*(13), 1888–1898. https://doi.org/10.3109/10826080902961179.

Smink, F. R., van Hoeken, D., & Hoek, H. W. (2013). Epidemiology, course, and outcome of eating disorders. *Current Opinion in Psychiatry, 26*(6), 543–548. https://doi.org/10.1097/YCO.0b013e328365a24f.

Smith, G. T., & Cyders, M. A. (2016). Integrating affect and impulsivity: The role of positive and negative urgency in substance use risk. *Drug and Alcohol Dependence, 163*(Suppl. 1), S3–S12. https://doi.org/10.1016/j.drugalcdep.2015.08.038.

Smith, K. J., Gall, S. L., McNaughton, S. A., Blizzard, L., Dwyer, T., & Venn, A. J. (2010). Skipping breakfast: Longitudinal associations with cardiometabolic risk factors in the childhood determinants of adult health study. *American Journal of Clinical Nutrition, 92*(6), 1316–1325. https://doi.org/10.3945/ajcn.2010.30101.

Smith, K. J., Gall, S. L., McNaughton, S. A., Cleland, V. J., Otahal, P., Dwyer, T., & Venn, A. J. (2017). Lifestyle behaviours associated with 5-year weight gain in a prospective cohort of Australian adults aged 26-36 years at baseline. *BMC Public Health, 17*(1), 54. https://doi.org/10.1186/s12889-016-3931-y.

Smith, S. S., Jorenby, D. E., Leischow, S. J., Nides, M. A., Rennard, S. I., Johnston, J. A., ... Baker, T. B. (2003). Targeting smokers at increased risk for relapse: Treating women and those with a history of depression. *Nicotine & Tobacco Research, 5*(1), 99–109.

Smith, K. L., Rao, R. R., Velazquez-Sanchez, C., Valenza, M., Giuliano, C., Everitt, B. J., ... Cottone, P. (2015). The uncompetitive N-methyl-D-aspartate antagonist memantine reduces binge-like eating, food-seeking behavior, and compulsive eating: Role of the nucleus accumbens shell. *Neuropsychopharmacology, 40*(5), 1163–1171. https://doi.org/10.1038/npp.2014.299.

Smoller, J. W., Wadden, T. A., & Stunkard, A. J. (1987). Dieting and depression: A critical review. *Journal of Psychosomatic Research, 31*(4), 429–440.

Smyth, J. M., Wonderlich, S. A., Heron, K. E., Sliwinski, M. J., Crosby, R. D., Mitchell, J. E., & Engel, S. G. (2007). Daily and momentary mood and stress are associated with binge eating and vomiting in bulimia nervosa patients in the natural environment. *Journal of Consulting and Clinical Psychology, 75*(4), 629–638. https://doi.org/10.1037/0022-006X.75.4.629.

Solomon, R. L., & Corbit, J. D. (1974). An opponent-process theory of motivation. I. Temporal dynamics of affect. *Psychological Review, 81*(2), 119–145.

Somkuwar, S. S., Vendruscolo, L. F., Fannon, M. J., Schmeichel, B. E., Nguyen, T. B., Guevara, J., … Mandyam, C. D. (2017). Abstinence from prolonged ethanol exposure affects plasma corticosterone, glucocorticoid receptor signaling and stress-related behaviors. *Psychoneuroendocrinology, 84*, 17–31. https://doi.org/10.1016/j.psyneuen.2017.06.006.

Sommer, W. H., Rimondini, R., Hansson, A. C., Hipskind, P. A., Gehlert, D. R., Barr, C. S., & Heilig, M. A. (2008). Upregulation of voluntary alcohol intake, behavioral sensitivity to stress, and amygdala crhr1 expression following a history of dependence. *Biological Psychiatry, 63*(2), 139–145. https://doi.org/10.1016/j.biopsych.2007.01.010.

Spangler, R., Wittkowski, K. M., Goddard, N. L., Avena, N. M., Hoebel, B. G., & Leibowitz, S. F. (2004). Opiate-like effects of sugar on gene expression in reward areas of the rat brain. *Molecular Brain Research, 124*(2), 134–142. https://doi.org/10.1016/j.molbrainres.2004.02.013.

Specio, S. E., Wee, S., O'Dell, L. E., Boutrel, B., Zorrilla, E. P., & Koob, G. F. (2008). CRF(1) receptor antagonists attenuate escalated cocaine self-administration in rats. *Psychopharmacology, 196*(3), 473–482. https://doi.org/10.1007/s00213-007-0983-9.

Spierling, S. R., Kreisler, A. D., Williams, C. A., Fang, S. Y., Pucci, S. N., Kines, K. T., & Zorrilla, E. P. (2018). Intermittent, extended access to preferred food leads to escalated food reinforcement and cyclic whole-body metabolism in rats: Sex differences and individual vulnerability. *Physiology & Behavior*. https://doi.org/10.1016/j.physbeh.2018.04.001.

Spoor, S. T., Stice, E., Bekker, M. H., Van Strien, T., Croon, M. A., & Van Heck, G. L. (2006). Relations between dietary restraint, depressive symptoms, and binge eating: A longitudinal study. *International Journal of Eating Disorders, 39*(8), 700–707. https://doi.org/10.1002/eat.20283.

Spring, B., Schneider, K., Smith, M., Kendzor, D., Appelhans, B., Hedeker, D., & Pagoto, S. (2008). Abuse potential of carbohydrates for overweight carbohydrate cravers. *Psychopharmacology, 197*(4), 637–647. https://doi.org/10.1007/s00213-008-1085-z.

Stautz, K., & Cooper, A. (2013). Impulsivity-related personality traits and adolescent alcohol use: A meta-analytic review. *Clinical Psychology Review, 33*(4), 574–592. https://doi.org/10.1016/j.cpr.2013.03.003.

Steffen, K. J., Engel, S. G., Wonderlich, J. A., Pollert, G. A., & Sondag, C. (2015). Alcohol and other addictive disorders following bariatric surgery: Prevalence, risk factors and possible etiologies. *European Eating Disorders Review, 23*(6), 442–450. https://doi.org/10.1002/erv.2399.

Stice, E. (1998). Relations of restraint and negative affect to bulimic pathology: A longitudinal test of three competing models. *International Journal of Eating Disorders, 23*(3), 243–260.

Stice, E. (2001). A prospective test of the dual-pathway model of bulimic pathology: Mediating effects of dieting and negative affect. *Journal of Abnormal Psychology, 110*(1), 124–135.

Stice, E. (2002). Risk and maintenance factors for eating pathology: A meta-analytic review. *Psychological Bulletin, 128*(5), 825–848.

Stice, E., & Bearman, S. K. (2001). Body-image and eating disturbances prospectively predict increases in depressive symptoms in adolescent girls: A growth curve analysis. *Developmental Psychology, 37*(5), 597–607.

Stice, E., Burger, K. S., & Yokum, S. (2015). Reward region responsivity predicts future weight gain and moderating effects of the TaqIA allele. *Journal of Neuroscience, 35*(28), 10316–10324. https://doi.org/10.1523/JNEUROSCI.3607-14.2015.

Stice, E., Spoor, S., Bohon, C., & Small, D. M. (2008). Relation between obesity and blunted striatal response to food is moderated by TaqIA A1 allele. *Science, 322*(5900), 449–452. https://doi.org/10.1126/science.1161550.

Stice, E., & Yokum, S. (2016a). Gain in body fat is associated with increased striatal response to palatable food cues, whereas body fat stability is associated with decreased striatal response. *Journal of Neuroscience, 36*(26), 6949–6956. https://doi.org/10.1523/JNEUROSCI.4365-15.2016.

Stice, E., & Yokum, S. (2016b). Neural vulnerability factors that increase risk for future weight gain. *Psychological Bulletin, 142*(5), 447–471. https://doi.org/10.1037/bul0000044.

Stice, E., Yokum, S., Blum, K., & Bohon, C. (2010). Weight gain is associated with reduced striatal response to palatable food. *Journal of Neuroscience, 30*(39), 13105–13109. https://doi.org/10.1523/JNEUROSCI.2105-10.2010.

Stice, E., Yokum, S., Zald, D., & Dagher, A. (2011). Dopamine-based reward circuitry responsivity, genetics, and overeating. *Current Topics in Behavioral Neurosciences, 6*, 81–93. https://doi.org/10.1007/7854_2010_89.

Stinus, L., Cador, M., Zorrilla, E. P., & Koob, G. F. (2005). Buprenorphine and a CRF1 antagonist block the acquisition of opiate withdrawal-induced conditioned place aversion in rats. *Neuropsychopharmacology, 30*(1), 90–98. https://doi.org/10.1038/sj.npp.1300487.

Stoeckel, L. E., Weller, R. E., Cook, E. W., 3rd, Twieg, D. B., Knowlton, R. C., & Cox, J. E. (2008). Widespread reward-system activation in obese women in response to pictures of high-calorie foods. *NeuroImage, 41*(2), 636–647. https://doi.org/10.1016/j.neuroimage.2008.02.031.

Strahler, J., & Nater, U. M. (2018). Differential effects of eating and drinking on wellbeing-An ecological ambulatory assessment study. *Biological Psychology, 131*, 72–88. https://doi.org/10.1016/j.biopsycho.2017.01.008.

Swanson, S. A., Crow, S. J., Le Grange, D., Swendsen, J., & Merikangas, K. R. (2011). Prevalence and correlates of eating disorders in adolescents. Results from the national comorbidity survey replication adolescent supplement. *Archives of General Psychiatry, 68*(7), 714–723. https://doi.org/10.1001/archgenpsychiatry.2011.22.

Tanihata, T., Kanda, H., Osaki, Y., Ohida, T., Minowa, M., Wada, K., … Hayashi, K. (2015). Unhealthy lifestyle, poor mental health, and its correlation among adolescents: A nationwide cross-sectional survey. *Asia-Pacific Journal of Public Health, 27*(2), NP1557–1565. https://doi.org/10.1177/1010539512452753.

Teegarden, S. L., & Bale, T. L. (2007). Decreases in dietary preference produce increased emotionality and risk for dietary relapse. *Biological Psychiatry, 61*(9), 1021–1029. https://doi.org/10.1016/j.biopsych.2006.09.032.

Teegarden, S. L., & Bale, T. L. (2008). Effects of stress on dietary preference and intake are dependent on access and stress sensitivity. *Physiology & Behavior, 93*(4–5), 713–723. https://doi.org/10.1016/j.physbeh.2007.11.030.

Teicher, M. H., Andersen, S. L., Polcari, A., Anderson, C. M., Navalta, C. P., & Kim, D. M. (2003). The neurobiological consequences of early stress and childhood maltreatment. *Neuroscience & Biobehavioral Reviews, 27*(1–2), 33–44.

Thompson, J. K., Schaefer, L. M., & Dedrick, R. F. (2018). On the measurement of thin-ideal internalization: Implications for interpretation of risk factors and treatment outcome in eating disorders research. *International Journal of Eating Disorders, 51*(4), 363–367. https://doi.org/10.1002/eat.22839.

Tin, S. P., Ho, S. Y., Mak, K. H., Wan, K. L., & Lam, T. H. (2011). Breakfast skipping and change in body mass index in young children. *International Journal of Obesity, 35*(7), 899–906. https://doi.org/10.1038/ijo.2011.58.

Tomasi, D., & Volkow, N. D. (2013). Striatocortical pathway dysfunction in addiction and obesity: Differences and similarities. *Critical Reviews in Biochemistry and Molecular Biology, 48*(1), 1–19. https://doi.org/10.3109/10409238.2012.735642.

Tomiyama, A. J., Dallman, M. F., & Epel, E. S. (2011). Comfort food is comforting to those most stressed: Evidence of the chronic stress response network in high stress women. *Psychoneuroendocrinology, 36*(10), 1513–1519. https://doi.org/10.1016/j.psyneuen.2011.04.005.

Tompkins, C. L., Laurent, J., & Brock, D. W. (2017). Food addiction: A barrier for effective weight management for obese adolescents. *Childhood Obesity, 13*(6), 462–469. https://doi.org/10.1089/chi.2017.0003.

Topp, L., Lovibond, P. F., & Mattick, R. P. (1998). Cue reactivity in dependent amphetamine users: Can monistic conditioning theories advance our understanding of reactivity? *Drug and Alcohol Review, 17*(3), 277–288. https://doi.org/10.1080/09595239800187111.

Torres, S. J., Nowson, C. A., & Worsley, A. (2008). Dietary electrolytes are related to mood. *British Journal of Nutrition, 100*(5), 1038–1045. https://doi.org/10.1017/S0007114508959201.

Traub, M., Lauer, R., Kesztyus, T., Wartha, O., Steinacker, J. M., Kesztyus, D., & Research Group "Join the Healthy, B". (2018). Skipping breakfast, overconsumption of soft drinks and screen media: Longitudinal analysis of the combined influence on weight development in primary schoolchildren. *BMC Public Health, 18*(1), 363. https://doi.org/10.1186/s12889-018-5262-7.

Tye, K. M., Cone, J. J., Schairer, W. W., & Janak, P. H. (2010). Amygdala neural encoding of the absence of reward during extinction. *Journal of Neuroscience, 30*(1), 116–125. https://doi.org/10.1523/JNEUROSCI.4240-09.2010.

Uemura, M., Yatsuya, H., Hilawe, E. H., Li, Y., Wang, C., Chiang, C., … Aoyama, A. (2015). Breakfast skipping is positively associated with incidence of type 2 diabetes mellitus: Evidence from the aichi workers' cohort study. *Journal of Epidemiology, 25*(5), 351–358. https://doi.org/10.2188/jea.JE20140109.

Uhlmann, L. R., Donovan, C. L., Zimmer-Gembeck, M. J., Bell, H. S., & Ramme, R. A. (2018). The fit beauty ideal: A healthy alternative to thinness or a wolf in sheep's clothing? *Body Image, 25*, 23–30. https://doi.org/10.1016/j.bodyim.2018.01.005.

Ulrich-Lai, Y. M., Christiansen, A. M., Ostrander, M. M., Jones, A. A., Jones, K. R., Choi, D. C., … Herman, J. P. (2010). Pleasurable behaviors reduce stress via brain reward pathways. *Proceedings of the National Academy of Sciences of the United States of America, 107*(47), 20529–20534. https://doi.org/10.1073/pnas.1007740107.

Ulrich-Lai, Y. M., Ostrander, M. M., & Herman, J. P. (2011). HPA axis dampening by limited sucrose intake: Reward frequency vs. caloric consumption. *Physiology & Behavior, 103*(1), 104–110. https://doi.org/10.1016/j.physbeh.2010.12.011.

Ulrich-Lai, Y. M., Ostrander, M. M., Thomas, I. M., Packard, B. A., Furay, A. R., Dolgas, C. M., ... Herman, J. P. (2007). Daily limited access to sweetened drink attenuates hypothalamic-pituitary-adrenocortical axis stress responses. *Endocrinology, 148*(4), 1823–1834. https://doi.org/10.1210/en.2006-1241.

Uzhova, I., Fuster, V., Fernandez-Ortiz, A., Ordovas, J. M., Sanz, J., Fernandez-Friera, L., ... Penalvo, J. L. (2017). The importance of breakfast in atherosclerosis disease: Insights from the PESA study. *Journal of the American College of Cardiology, 70*(15), 1833–1842. https://doi.org/10.1016/j.jacc.2017.08.027.

Valdez, G. R., Roberts, A. J., Chan, K., Davis, H., Brennan, M., Zorrilla, E. P., & Koob, G. F. (2002). Increased ethanol self-administration and anxiety-like behavior during acute ethanol withdrawal and protracted abstinence: Regulation by corticotropin-releasing factor. *Alcoholism: Clinical and Experimental Research, 26*(10), 1494–1501. https://doi.org/10.1097/01.ALC.0000033120.51856.F0.

VanderBroek-Stice, L., Stojek, M. K., Beach, S. R., vanDellen, M. R., & MacKillop, J. (2017). Multidimensional assessment of impulsivity in relation to obesity and food addiction. *Appetite, 112*, 59–68. https://doi.org/10.1016/j.appet.2017.01.009.

VanderVeen, J. D., Plawecki, M. H., Millward, J. B., Hays, J., Kareken, D. A., O'Connor, S., & Cyders, M. A. (2016). Negative urgency, mood induction, and alcohol seeking behaviors. *Drug and Alcohol Dependence, 165*, 151–158. https://doi.org/10.1016/j.drugalcdep.2016.05.026.

Vandervoort, J., Aime, A., & Green-Demers, I. (2015). The monster in the mirror: Reasons for wanting to change appearance. *Eating and Weight Disorders, 20*(1), 99–107. https://doi.org/10.1007/s40519-014-0160-1.

Vargas-Perez, H., Ting, A. K. R. A., Heinmiller, A., Sturgess, J. E., & van der Kooy, D. (2007). A test of the opponent-process theory of motivation using lesions that selectively block morphine reward. *European Journal of Neuroscience, 25*(12), 3713–3718. https://doi.org/10.1111/j.1460-9568.2007.05599.x.

Vargas-Perez, H., Ting, A. K. R., & van der Kooy, D. (2009). Different neural systems mediate morphine reward and its spontaneous withdrawal aversion. *European Journal of Neuroscience, 29*(10), 2029–2034. https://doi.org/10.1111/j.1460-9568.2009.06749.x.

Vasilenko, S. A., Piper, M. E., Lanza, S. T., Liu, X., Yang, J., & Li, R. (2014). Time-varying processes involved in smoking lapse in a randomized trial of smoking cessation therapies. *Nicotine & Tobacco Research, 16*(Suppl. 2), S135–S143. https://doi.org/10.1093/ntr/ntt185.

Vendruscolo, L. F., Barbier, E., Schlosburg, J. E., Misra, K. K., Whitfield, T. W., Jr., Logrip, M. L., ... Koob, G. F. (2012). Corticosteroid-dependent plasticity mediates compulsive alcohol drinking in rats. *Journal of Neuroscience, 32*(22), 7563–7571. https://doi.org/10.1523/JNEUROSCI.0069-12.2012.

Vendruscolo, L. F., Estey, D., Goodell, V., Macshane, L. G., Logrip, M. L., Schlosburg, J. E., ... Mason, B. J. (2015). Glucocorticoid receptor antagonism decreases alcohol seeking in alcohol-dependent individuals. *Journal of Clinical Investigation, 125*(8), 3193–3197. https://doi.org/10.1172/jci79828.

Vendruscolo, L. F., Gueye, A. B., Darnaudery, M., Ahmed, S. H., & Cador, M. (2010). Sugar overconsumption during adolescence selectively alters motivation and reward function in adult rats. *PLoS One, 5*(2), e9296. https://doi.org/10.1371/journal.pone.0009296.

Vickers, S. P., Goddard, S., Brammer, R. J., Hutson, P. H., & Heal, D. J. (2017). Investigation of impulsivity in binge-eating rats in a delay-discounting task and its prevention by the d-amphetamine prodrug, lisdexamfetamine. *Journal of Psychopharmacology, 31*(6), 784–797. https://doi.org/10.1177/0269881117691672.

Villafuerte, S., Heitzeg, M. M., Foley, S., Yau, W. Y., Majczenko, K., Zubieta, J. K., ... Burmeister, M. (2012). Impulsiveness and insula activation during reward anticipation are associated with genetic variants in GABRA2 in a family sample enriched for alcoholism. *Molecular Psychiatry, 17*(5), 511–519. https://doi.org/10.1038/mp.2011.33.

Volkow, N. D., & Baler, R. D. (2015). NOW vs LATER brain circuits: Implications for obesity and addiction. *Trends in Neurosciences, 38*(6), 345–352. https://doi.org/10.1016/j.tins.2015.04.002.

Volkow, N. D., Fowler, J. S., Wang, G. J., Baler, R., & Telang, F. (2009). Imaging dopamine's role in drug abuse and addiction. *Neuropharmacology, 56*(Suppl. 1), 3–8. https://doi.org/10.1016/j.neuropharm.2008.05.022.

Volkow, N. D., Fowler, J. S., Wang, G. J., Swanson, J. M., & Telang, F. (2007). Dopamine in drug abuse and addiction: Results of imaging studies and treatment implications. *Archives of Neurology, 64*(11), 1575–1579. https://doi.org/10.1001/archneur.64.11.1575.

Volkow, N. D., & O'Brien, C. P. (2007). Issues for DSM-V: Should obesity be included as a brain disorder? *American Journal of Psychiatry, 164*(5), 708–710. https://doi.org/10.1176/ajp.2007.164.5.708.

Volkow, N. D., Wang, G. J., Fowler, J. S., Logan, J., Jayne, M., Franceschi, D., ... Pappas, N. (2002). "Nonhedonic" food motivation in humans involves dopamine in the dorsal striatum and methylphenidate amplifies this effect. *Synapse, 44*(3), 175–180. https://doi.org/10.1002/syn.10075.

Volkow, N. D., Wang, G. J., Fowler, J. S., & Telang, F. (2008a). Overlapping neuronal circuits in addiction and obesity: Evidence of systems pathology. *Philosophical Transactions of the Royal Society of London B Biological Sciences, 363*(1507), 3191–3200. https://doi.org/10.1098/rstb.2008.0107.

Volkow, N. D., Wang, G. J., Maynard, L., Jayne, M., Fowler, J. S., Zhu, W., ... Pappas, N. (2003). Brain dopamine is associated with eating behaviors in humans. *International Journal of Eating Disorders, 33*(2), 136–142. https://doi.org/10.1002/eat.10118.

Volkow, N. D., Wang, G. J., Telang, F., Fowler, J. S., Thanos, P. K., Logan, J., ... Pradhan, K. (2008b). Low dopamine striatal D2 receptors are associated with prefrontal metabolism in obese subjects: Possible contributing factors. *NeuroImage, 42*(4), 1537–1543. https://doi.org/10.1016/j.neuroimage.2008.06.002.

Volkow, N. D., Wang, G. J., Tomasi, D., & Baler, R. D. (2013). Obesity and addiction: Neurobiological overlaps. *Obesity Reviews, 14*(1), 2–18. https://doi.org/10.1111/j.1467-789X.2012.01031.x.

de Vries, S. K., & Meule, A. (2016). Food addiction and bulimia nervosa: New data based on the Yale food addiction scale 2.0. *European Eating Disorders Review, 24*(6), 518–522. https://doi.org/10.1002/erv.2470.

Wade, C. L., Vendruscolo, L. F., Schlosburg, J. E., Hernandez, D. O., & Koob, G. F. (2015). Compulsive-like responding for opioid analgesics in rats with extended access. *Neuropsychopharmacology, 40*(2), 421–428. https://doi.org/10.1038/npp.2014.188.

Wagner, H. S., Ahlstrom, B., Redden, J. P., Vickers, Z., & Mann, T. (2014). The myth of comfort food. *Health Psychology, 33*(12), 1552–1557. https://doi.org/10.1037/hea0000068.

Wallace, L. M., Masson, P. C., Safer, D. L., & von Ranson, K. M. (2014). Change in emotion regulation during the course of treatment predicts binge abstinence in guided self-help dialectical behavior therapy for binge eating disorder. *Journal of Eating Disorders, 2*(1), 35. https://doi.org/10.1186/s40337-014-0035-x.

Wallis, D. J., & Hetherington, M. M. (2009). Emotions and eating. Self-reported and experimentally induced changes in food intake under stress. *Appetite, 52*(2), 355–362. https://doi.org/10.1016/j.appet.2008.11.007.

Walsh, B. T. (2013). The enigmatic persistence of anorexia nervosa. *American Journal of Psychiatry, 170*(5), 477–484. https://doi.org/10.1176/appi.ajp.2012.12081074.

Wang, F. L., & Chassin, L. (2018). Negative urgency mediates the relation between genetically-influenced serotonin functioning and alcohol problems. *Clinical Psychological Science, 6*(1), 106–122. https://doi.org/10.1177/2167702617733817.

Wang, G. J., Geliebter, A., Volkow, N. D., Telang, F. W., Logan, J., Jayne, M. C., … Fowler, J. S. (2011). Enhanced striatal dopamine release during food stimulation in binge eating disorder. *Obesity, 19*(8), 1601–1608. https://doi.org/10.1038/oby.2011.27.

Wang, G. J., Volkow, N. D., Logan, J., Pappas, N. R., Wong, C. T., Zhu, W., … Fowler, J. S. (2001). Brain dopamine and obesity. *Lancet, 357*(9253), 354–357.

Wang, G. J., Volkow, N. D., Thanos, P. K., & Fowler, J. S. (2009). Imaging of brain dopamine pathways: Implications for understanding obesity. *Journal of Addiction Medicine, 3*(1), 8–18. https://doi.org/10.1097/ADM.0b013e31819a86f7.

Warlow, S. M., Robinson, M. J. F., & Berridge, K. C. (2017). Optogenetic central amygdala stimulation intensifies and narrows motivation for cocaine. *Journal of Neuroscience, 37*(35), 8330–8348. https://doi.org/10.1523/JNEUROSCI.3141-16.2017.

Warne, J. P. (2009). Shaping the stress response: Interplay of palatable food choices, glucocorticoids, insulin and abdominal obesity. *Molecular and Cellular Endocrinology, 300*(1–2), 137–146. https://doi.org/10.1016/j.mce.2008.09.036.

Warren, M. W., & Gold, M. S. (2007). The relationship between obesity and drug use. *American Journal of Psychiatry, 164*(8), 1268. https://doi.org/10.1176/appi.ajp.2007.07030388r. author reply 1268-1269.

Weidner, G., Connor, S. L., Gerhard, G. T., Duell, P. B., & Connor, W. E. (2009). The effects of dietary cholesterol-lowering on psychological symptoms: A randomised controlled study. *Psychology Health & Medicine, 14*(3), 255–261. https://doi.org/10.1080/13548500902730101.

Weiland, B. J., Heitzeg, M. M., Zald, D., Cummiford, C., Love, T., Zucker, R. A., & Zubieta, J. K. (2014). Relationship between impulsivity, prefrontal anticipatory activation, and striatal dopamine release during rewarded task performance. *Psychiatry Research, 223*(3), 244–252. https://doi.org/10.1016/j.pscychresns.2014.05.015.

Weiss, F., Ciccocioppo, R., Parsons, L. H., Katner, S., Liu, X., Zorrilla, E. P., … Richter, R. R. (2001). Compulsive drug-seeking behavior and relapse. Neuroadaptation, stress, and conditioning factors. *Annals of the New York Academy of Sciences, 937*, 1–26.

Wells, A. S., Read, N. W., Laugharne, J. D., & Ahluwalia, N. S. (1998). Alterations in mood after changing to a low-fat diet. *British Journal of Nutrition, 79*(1), 23–30.

Weltens, N., Zhao, D., & Van Oudenhove, L. (2014). Where is the comfort in comfort foods? Mechanisms linking fat signaling, reward, and emotion. *Neuro-Gastroenterology and Motility, 26*(3), 303–315. https://doi.org/10.1111/nmo.12309.

Wendling, A., & Wudyka, A. (2011). Narcotic addiction following gastric bypass surgery–a case study. *Obesity Surgery, 21*(5), 680–683. https://doi.org/10.1007/s11695-010-0177-0.

Wenzel, J. M., Waldroup, S. A., Haber, Z. M., Su, Z. I., Ben-Shahar, O., & Ettenberg, A. (2011). Effects of lidocaine-induced inactivation of the bed nucleus of the stria terminalis, the central or the basolateral nucleus of the amygdala on the opponent-process actions of self-administered cocaine in rats. *Psychopharmacology, 217*(2), 221–230. https://doi.org/10.1007/s00213-011-2267-7.

Werrij, M. Q., Jansen, A., Mulkens, S., Elgersma, H. J., Ament, A. J., & Hospers, H. J. (2009). Adding cognitive therapy to dietetic treatment is associated with less relapse in obesity. *Journal of Psychosomatic Research, 67*(4), 315–324. https://doi.org/10.1016/j.jpsychores.2008.12.011.

White, J. M. (2004). Pleasure into pain: The consequences of long-term opioid use. *Addictive Behaviors, 29*(7), 1311–1324. https://doi.org/10.1016/j.addbeh.2004.06.007.

Wijngaarden, M. A., Veer, I. M., Rombouts, S. A., van Buchem, M. A., Willems van Dijk, K., Pijl, H., & van der Grond, J. (2015). Obesity is marked by distinct functional connectivity in brain networks involved in food reward and salience. *Behavioural Brain Research, 287*, 127–134. https://doi.org/10.1016/j.bbr.2015.03.016.

Winpenny, E. M., van Harmelen, A. L., White, M., van Sluijs, E. M., & Goodyer, I. M. (2018). Diet quality and depressive symptoms in adolescence: No cross-sectional or prospective associations following adjustment for covariates. *Public Health Nutrition*, 1–9. https://doi.org/10.1017/S1368980018001179.

Winter, S. R., Yokum, S., Stice, E., Osipowicz, K., & Lowe, M. R. (2017). Elevated reward response to receipt of palatable food predicts future weight variability in healthy-weight adolescents. *American Journal of Clinical Nutrition, 105*(4), 781–789. https://doi.org/10.3945/ajcn.116.141143.

Womble, L. G., Williamson, D. A., Martin, C. K., Zucker, N. L., Thaw, J. M., Netemeyer, R., … Greenway, F. L. (2001). Psychosocial variables associated with binge eating in obese males and females. *International Journal of Eating Disorders, 30*(2), 217–221.

Wonderlich, S. A., Peterson, C. B., Crosby, R. D., Smith, T. L., Klein, M. H., Mitchell, J. E., & Crow, S. J. (2014). A randomized controlled comparison of integrative cognitive-affective therapy (ICAT) and enhanced cognitive-behavioral therapy (CBT-E) for bulimia nervosa. *Psychological Medicine, 44*(3), 543–553. https://doi.org/10.1017/S0033291713001098.

Xue, G., Lu, Z., Levin, I. P., & Bechara, A. (2010). The impact of prior risk experiences on subsequent risky decision-making: The role of the insula. *NeuroImage, 50*(2), 709–716. https://doi.org/10.1016/j.neuroimage.2009.12.097.

Yacubian, J., Glascher, J., Schroeder, K., Sommer, T., Braus, D. F., & Buchel, C. (2006). Dissociable systems for gain- and loss-related value predictions and errors of prediction in the human brain. *Journal of Neuroscience, 26*(37), 9530–9537. https://doi.org/10.1523/JNEUROSCI.2915-06.2006.

Yang, F., Liu, A., Li, Y., Lai, Y., Wang, G., Sun, C., … Teng, W. (2017). Food addiction in patients with newly diagnosed type 2 diabetes in northeast China. *Frontiers in Endocrinology, 8*, 218. https://doi.org/10.3389/fendo.2017.00218.

Yarnell, S., Oscar-Berman, M., Avena, N., Blum, K., & Gold, M. (2013). Pharmacotherapies for overeating and obesity. *Journal of Genetic Syndromes & Gene Therapy, 4*(3), 131. https://doi.org/10.4172/2157-7412.1000131.

Yeh, T. L., Chen, H. H., Pai, T. P., Liu, S. J., Wu, S. L., Sun, F. J., & Hwang, L. C. (2017). The effect of auricular acupoint stimulation in overweight and obese adults: A systematic review and meta-analysis of randomized controlled trials. *Evid Based Complement Alternat Med, 2017*, 3080547. https://doi.org/10.1155/2017/3080547.

Yeomans, M. R., & Coughlan, E. (2009). Mood-induced eating. Interactive effects of restraint and tendency to overeat. *Appetite, 52*(2), 290–298. https://doi.org/10.1016/j.appet.2008.10.006.

Yokoyama, Y., Onishi, K., Hosoda, T., Amano, H., Otani, S., Kurozawa, Y., & Tamakoshi, A. (2016). Skipping breakfast and risk of mortality from cancer, circulatory diseases and all causes: Findings from the Japan collaborative cohort study. *Yonago Acta Medica, 59*(1), 55–60.

Zellner, D. A., Loaiza, S., Gonzalez, Z., Pita, J., Morales, J., Pecora, D., & Wolf, A. (2006). Food selection changes under stress. *Physiology & Behavior, 87*(4), 789–793. https://doi.org/10.1016/j.physbeh.2006.01.014.

Zhang, Y., Ji, G., Xu, M., Cai, W., Zhu, Q., Qian, L., … Wang, G. J. (2016). Recovery of brain structural abnormalities in morbidly obese patients after bariatric surgery. *International Journal of Obesity, 40*(10), 1558–1565. https://doi.org/10.1038/ijo.2016.98.

Zhang, Y., von Deneen, K. M., Tian, J., Gold, M. S., & Liu, Y. (2011). Food addiction and neuroimaging. *Current Pharmaceutical Design, 17*(12), 1149–1157.

Zhang, C., Wei, N. L., Wang, Y., Wang, X., Zhang, J. G., & Zhang, K. (2015). Deep brain stimulation of the nucleus accumbens shell induces anti-obesity effects in obese rats with alteration of dopamine neurotransmission. *Neuroscience Letters, 589*, 1–6. https://doi.org/10.1016/j.neulet.2015.01.019.

Zhao, J., Tomasi, D., Wiers, C. E., Shokri-Kojori, E., Demiral, S. B., Zhang, Y., … Wang, G. J. (2017). Correlation between traits of emotion-based impulsivity and intrinsic default-mode network activity. *Neural Plasticity, 2017*, 9297621. https://doi.org/10.1155/2017/9297621.

Zhao, Y., Weiss, F., & Zorrilla, E. P. (2007). Remission and resurgence of anxiety-like behavior across protracted withdrawal stages in ethanol-dependent rats. *Alcoholism: Clinical and Experimental Research, 31*(9), 1505–1515.

Ziauddeen, H., & Fletcher, P. C. (2013). Is food addiction a valid and useful concept? *Obesity Reviews, 14*(1), 19–28. https://doi.org/10.1111/j.1467-789X.2012.01046.x.

Zinser, M. C., Baker, T. B., Sherman, J. E., & Cannon, D. S. (1992). Relation between self-reported affect and drug urges and cravings in continuing and withdrawing smokers. *Journal of Abnormal Psychology, 101*(4), 617–629.

Zorrilla, E. P., Heilig, M., de Wit, H., & Shaham, Y. (2013). Behavioral, biological, and chemical perspectives on targeting CRF(1) receptor antagonists to treat alcoholism. *Drug and Alcohol Dependence, 128*(3), 175–186. https://doi.org/10.1016/j.drugalcdep.2012.12.017.

Zorrilla, E. P., & Koob, G. F. (2013). Amygdalostriatal projections in the neurocircuitry for motivation: A neuroanatomical thread through the career of ann kelley. *Neuroscience & Biobehavioral Reviews, 37*(9 Pt A), 1932–1945. https://doi.org/10.1016/j.neubiorev.2012.11.019.

Zorrilla, E. P., Logrip, M. L., & Koob, G. F. (2014). Corticotropin releasing factor: A key role in the neurobiology of addiction. *Frontiers in Neuroendocrinology, 35*(2), 234–244. https://doi.org/10.1016/j.yfrne.2014.01.001.

Zorrilla, E. P., Valdez, G. R., & Weiss, F. (2001). Changes in levels of regional CRF-like-immunoreactivity and plasma corticosterone during protracted drug withdrawal in dependent rats. *Psychopharmacology, 158*(4), 374–381. https://doi.org/10.1007/s002130100773.

CHAPTER

Food addiction and self-regulation

7

Cara M. Murphy[1], James MacKillop[2]

Center for Alcohol and Addiction Studies, Brown University, Providence, RI, United States[1]; Peter Boris Centre for Addictions Research, McMaster University/St. Joseph's Healthcare Hamilton, Hamilton, ON, United States[2]

I count him braver who overcomes his desires than him who conquers his enemies, for the hardest victory is over self.

Aristotle

Brief introduction to food addiction

In the scientific study of addiction, motivation for addictive drugs and motivation for food have long been intertwined. At a conceptual level, drug self-administration and eating are both consumption behaviors, and, by extension, unhealthy levels of both behaviors have the common topography of being disorders of overconsumption. However, drug and food motivation are also conjoined at a deeper level, in the underlying basic neurobiology of reward. Early research revealed a common neural basis for drug addiction across a number of drug classes, the mesolimbic dopamine pathway running from the ventral tegmental area in the midbrain to the nucleus accumbens in the striatum and areas of the prefrontal cortex (Wise & Bozarth, 1987). The innate functionality of this neurocircuitry is the processing of natural rewards (Kelley & Berridge, 2002), and a foundational insight into the neurobiology of addiction is that addictive drugs hijack these evolutionarily ancient brain regions that subserve primary reinforcers to enhance classical fitness (i.e., food and sex) (Nesse & Berridge, 1997). By producing intense stimulation of these pathways, addictive drugs co-opt a critical pathway for motivating human behavior, although not toward its evolutionary end point. This is referred to as the mismatch hypothesis, meaning that addiction vulnerability can be understood as being, in part, due to the mismatch between biological systems that evolved over millennia to promote fitness in the historical ancestral niche and the immensely different environment in which we live today. The neurobiology of addiction has been vastly elaborated since these foundational insights more than 30 years ago, now implicating dysregulated neurotransmission of not only dopamine but also endogenous opioids, endocannabinoids, glutamate, γ-aminobutyric acid, acetylcholine, and corticotropin-releasing factor

Compulsive Eating Behavior and Food Addiction. https://doi.org/10.1016/B978-0-12-816207-1.00007-X
Copyright © 2019 Elsevier Inc. All rights reserved.

and dynorphin (Koob & Volkow, 2016). Nonetheless, the perspective that addictive drugs hijack the brain's ancient positive and negative reinforcement neurocircuitry remains a fundamentally germane.

The incongruity between neurobiological systems and the current environment may contribute not only to addiction to drugs but also to food. The neuronal circuits involved in addiction and obesity appear to be largely overlapping; ingestive behaviors may go awry because of adaptations in brain reward circuitry leading to increasingly compulsive consumption (Volkow, Wang, Tomasi, & Baler, 2013; Volkow & Wise, 2005). Unlike ancient foods, modern foods are designed to be "hyperpalatable" to speed the absorption of fat, sugar, and salt into the system (Moss, 2014) and to maximize hedonic impact and subsequent consumption, thereby increasing the potential to trigger compulsive motivational processes (Gearhardt, Davis, Kuschner, & Brownell, 2011). For example, Johnson and Kenny (2010) found that rodents that consumed a contemporary "junk food" diet exhibited compulsive eating behavior that was equivalent to rodent models of drug addiction and was similarly characterized by striatal downregulation of dopamine D_2 receptors, neuroadaptations that reflected a homeostatic response to dopaminergic overstimulation.

Importantly, common neurobiology extends to common behavioral processes also, particularly in the domain of learning and memory. For example, environmental cues are well-established antecedents of both drug-taking (Acker & MacKillop, 2013; Murphy & Mackillop, 2014) and eating behavior (Stojek, Fischer, & MacKillop, 2015). Similarly, the reinforcing value of psychoactive drugs and food appear to be critical determinants of consumption behavior (Epstein, Salvy, Carr, Dearing, & Bickel, 2010; MacKillop et al., 2008; Murphy & MacKillop, 2006; Stojek & MacKillop, 2017). Complementing these findings, there is also increasing evidence that similar higher-order neuropsychological processes are implicated in dysregulated motivation for psychoactive drugs and food, particularly implicating deficits in executive functioning (Davis, Patte, Curtis, & Reid, 2010).

Collectively, these accumulating findings suggest that overconsumption of psychoactive drugs and food is not just topographically or metaphorically similar, but substantively manifestations of common biobehavioral processes. Indeed, the science can be thought of as having come full circle insofar, as food motivation was initially implicated in the ancient elementary neural circuits on which addictive drugs commonly acted, but many modern foods are now recognized as having similarly hyperstimulating properties that may make them addictive in the same way. As a result, there is considerable contemporary interest in food addiction (FA). While the term "food addiction" was first introduced in the scientific literature in the 1950s, over the last decade, there has been increased interest and a proliferation of scientific articles published on the topic (Gearhardt et al., 2011; Meule, 2015). This is substantially because of the development and validation of the Yale Food Addiction Scale (YFAS), a tool that was designed to assess the construct of FA based on the diagnostic criteria for substance dependence in the Diagnostic and Statistical Manual of Mental Disorders [DSM-IV-TR] (American Psychiatric Association, 2000). Seven symptoms of FA were created to map onto the symptoms of substance dependence including a problematic pattern of eating leading to clinically significant impairment or distress occurring within the past 12 months: (1) eating larger amounts than intended (vs. using substances in greater quantities or over a longer period than intended); (2) worrying about cutting down or abstaining from certain types of food (vs. persistent desire or failed attempts to cut down or control substance use); (3) spending a good deal recovering from the effects of overeating (vs. spending a considerable amount of time in activities to obtain, use, or recover from the effects of a substance); (4) spending time dealing with negative feelings from overeating instead of working, spending time with family/friends, or engaging in recreational activities (vs. important social, occupational, or recreational activities given up or reduced because of substance use); (5) continuing to consume certain foods, even though it's causing problems in one's life or causing physical or psychological harm (vs. continued use of a substance despite knowledge of a physical or psychological problem caused or exacerbated by a substance); (6) needing to eat more to get the same effect over time (vs. tolerance: need for markedly increased amounts of a substance to achieve intoxication or desired effect or diminished effect with continued use of the same amount of a substance); and (7) experiencing withdrawal symptoms when one stopped eating certain foods (vs. a withdrawal syndrome is experienced shortly after cessation of or reduction in heavy and prolonged substance use). Diagnostic criteria were considered to be met if an individual endorsed three or more of the seven criteria and one of two additional items assessing clinically significant impairment or distress.

When the fifth edition of the Diagnostic and Statistical Manual of Mental Disorders [DSM-5] (American Psychiatric Association, 2013) was released, the YFAS 2.0 (Gearhardt, Corbin, & Brownell, 2016) was developed to maintain consistency with the revised diagnostic criteria. This included adding the following symptoms: (1) problems with or concerns from family/friends related to overeating and/or avoiding social situations because of others disapproving of eating (vs. continued substance use despite having persistent or recurrent social or interpersonal problems caused or exacerbated by the effects of a substance); (2) overeating resulting in not doing well at work or school or not taking care of one's family/household (vs. recurrent failure to fulfill major role obligations at work, school, or home because of substance use); (3) being so distracted by eating or thinking about food that one could have been hurt or eating certain foods despite dangers such as diabetes or heart disease (vs. recurrent substance use in situations in which it is physically hazardous); (4) strong or intense urges to eat certain foods (vs. craving or a strong desire or urge to use a substance). Diagnostic criteria on the YFAS 2.0 are considered to be met for FA if an individual endorses clinically significant impairment or distress and 2–3 (mild), 4–5 (moderate), or 6 or more (severe) of the 11 criteria. Although the YFAS and subsequent versions and modifications were created to mirror diagnostic criteria for other addictive disorders, it should be noted that the YFAS is a self-report measure. Thus, while cutoffs are defined based on definitive scoring criteria and determined based on optimal rates in the extant literature on substance-use disorders and eating disorders (Gearhardt, Corbin, & Brownell,

2009), a positive (or negative) diagnosis is based on self-report rather than a structured clinical interview. Moreover, the term "food addiction" represents a hypothesized condition and is not a diagnosis in the DSM-5 or International Statistical Classification of Diseases and Related Health Problems [ICD-10] (World Health Organization, 1992).

Growing concern regarding health risks associated with obesity and increases in the prevalence of obesity in the United States and globally (The G. B. D. Obesity Collaboration et al., 2014) have coincided with much of the renewed interest in FA, but it is important to note that obesity and FA represent distinct phenomena. Obesity refers to a medical condition listed in the ICD-10 and is based on body composition reflecting adiposity associated with health risk. A crude measure of obesity is determined using a standard formula based on the individual's weight (in kilograms) divided by the individual's squared height (in meters), with a body mass index (BMI) of $\geq$ 25 indicating overweight and $\geq$30 indicating obesity; BMI is thus an indicator of anthropometric body density.

Obesity is known to be highly multifactorial in nature (Qasim et al., 2018), and individuals who are obese do not necessarily exhibit FA. Several studies in North America have suggested that only 16%–25% of individuals with overweight or obesity meet the criteria for FA (Davis et al., 2011; Ivezaj, White, & Grilo, 2016; Oullette, Gerrard, Gibbons, & Reis-Bergan, 1999). Equally, individuals who report FA do not necessarily have to be obese (Murphy, Stojek, & MacKillop, 2014; VanderBroek-Stice, Stojek, Beach, & MacKillop, 2017). Thus, the two diagnoses are far from synonymous and should not be conflated. Indeed, it is possible FA maybe be a particular phenotype of obesity (Davis et al., 2011) or a precursor or risk factor for obesity. In addition, there may be a spectrum of severity of overeating behavior; just as all individuals who drink alcohol heavily would not be diagnosed with an alcohol-use disorder, all individuals who demonstrate notable overeating would not be diagnosed with FA (Davis, 2013b). Instead, in both instances, the severity and compulsivity of the behaviors must be taken into account. New to the DSM-5 was the diagnosis of binge eating disorder (BED), a disorder characterized by eating unusually large amounts of food in a discrete period of time with a sense of loss of control over one's eating (i.e., binge eating). As many of the symptoms of substance-related and addictive disorders correspond to a sense of lack of control during an episode (of substance use, gambling, etc.), it has been suggested that FA may actually reflect a more acute and severe form of BED, with psychopathological characteristics akin to other addictive disorders (Davis, 2013a; 2013b). Mapping the unique features, commonalities, and interrelationships among FA, obesity, and BED is a major priority in understanding the clinical manifestations of dysregulated eating behavior.

In the larger context of the broad line of scientific inquiry into the validity of FA as a psychiatric disorder is determining the ways in which FA may resemble substance-related and addictive disorders. Contemporary biopsychosocial models of addiction emphasize person-level biobehavioral determinants of drugs motivation and environment-level interactions with social factors and broader societal influences (MacKillop & Ray, 2018). For example, commonly identified biobehavioral determinants include dysregulated incentive salience, negative emotionality, and self-regulation (e.g., Kwako, Momenan, Litten, Koob, & Goldman, 2016), which, in turn, affect and are affected by social network characteristics and the larger sociocultural context. Within this framework, although the notion of an "addictive personality" in the sense of a commonly used personality framework has not been well supported (Nathan, 1988), discrete dispositional factors have, nonetheless, been robustly associated with drug addiction, particularly self-regulation and impulsivity (Amlung, Vedelago, Acker, Balodis, & MacKillop, 2017; Coskunpinar, Dir, & Cyders, 2013; MacKillop et al., 2011; Stanford et al., 2009), broadly referring to a person's ability to control arising impulses and desires. Thus, an important question is the extent to which FA is associated with critical constructs such as self-regulation and impulsivity. If FA exhibits parallel relationships to these psychological processes, it will provide further evidence that it is substantively similar to drug addiction. On the other hand, if it does not, it will suggest the construct is more metaphorically similar than an alternative manifestation of common processes. Addressing this question is the specific focus of this chapter, as part of the larger program of research investigating the validity of the construct of FA.

Definitions of impulsivity and self-regulation

In considering the relationships between FA and impulsivity/self-regulation, a critical issue is how to define these constructs. Historically, self-report psychological assessments used unidimensional scales that assessed impulsivity as a monolithic construct (e.g., Eysenck & Eysenck, 1978), but subsequent measures fractionated impulsivity into multiple subscales (Patton, Stanford, & Barratt, 1995; Whiteside & Lynam, 2003). These questionnaire-based assessments have subsequently been complemented with behavioral tasks that objectively operationalize performance and permit state-based assessment (Fillmore & Weafer, 2013; Madden, Bickel, & Jacobs, 1999). However, increasing evidence has suggested that impulsivity is a multidimensional construct that refers to a family of conceptually related processes (all reflecting facets of regulatory capacity) that are variably related to each other quantitatively (de Wit & Richards, 2004; Dick et al., 2010; Evenden, 1999). These associations range from high-magnitude levels of overlap to negligible links, suggesting distinct underlying biological substrates and psychological processes. Importantly, however, it is not the case that there are simply a proliferation of indicators, but that the many measures can be meaningfully organized into clusters. Several factor analyses have been done (Caswell, Bond, Duka, & Morgan, 2015; MacKillop et al., 2014; Sharma, Markon, & Clark, 2014), yielding somewhat variable solutions (in part based on sample composition and measurement selection), but one approach that has received recent support suggests a tripartite approach. Specifically, using a large number of self-report and behavioral measures of specific to impulsivity (not including measures of risk-taking or cognitive processes), the diverse indicators fell into three latent clusters: (1) *impulsive choice,* reflecting tasks measuring discounting of delayed rewards (i.e., orientation toward smaller immediate rewards compared with larger delayed rewards); (2) *impulsive action,* reflecting performance on tasks measuring capacity to inhibit a prepotent motor response; and (3) *impulsive personality traits,* reflecting self-reported self-regulatory capacity on questionnaires about dispositional tendencies (MacKillop et al., 2016). Importantly, this latent structure was proposed a priori and supported by confirmatory factor analysis, as opposed to simply being observed using exploratory factor analysis, and the sample comprised a large cohort of healthy non—substance-using volunteers (verified via urine drug screen) (MacKillop et al., 2016). This latter aspect is important because an ongoing question in the field is whether addiction itself may give rise to changes in self-regulatory capacity, which in turn may affect the interrelationships among facets of impulsivity in clinical populations. Although the latent structure of impulsivity remains an active research question, this chapter will use this framework and terminology to approach the links between FA and self-regulation.

Empirical research on self-regulation and food addiction

General approach to the literature review

As the chapter's goal is to be comprehensive, a PubMed database search was conducted to find all potentially relevant studies in peer-reviewed publications up to April 2018. The Boolean logic was "food addiction" AND "impulsivity" OR "self-regulation" OR "emotional regulation" OR "self-control" OR "discounting" OR "behavioral inhibition" OR "response inhibition" OR "impulsive" OR "urgency" OR "impulsiveness." Eligibility criteria for inclusion were (1) an empirical research study; (2) human participants; (3) published in English; and (4) reported relationships between FA and one or more measures of self-regulation in one of the three aforementioned categories. Of 42 articles identified, 18 were excluded for not meeting eligibility criteria. Information pertaining to the studies described below is reported in Table 7.1. In examining the relationships between FA and self-regulation, the goal was to characterize and distinguish between both uncorrected associations reported (typically zero-order correlations) and multivariate analyses that incorporated other factors that could be relevant (e.g., sex, age).

Impulsive choice

Impulsive choice tasks determine whether an individual will choose to wait for a larger reward available at some future date or to take a smaller reward available without any delay. Considered a measure of impulsivity, when the smaller reward is selected, it indicates that the larger, delayed reward is being discounted (i.e., loses value based on the delay), known as delay discounting. Only two studies examined delay discounting in relation to FA, one in the United States and one in Canada. Both

Table 7.1 Studies of food addiction and self-regulation.

Study	Sample (Country)	FA measure	SR measure	SR category
Brunault et al. (2018)	*N* = 188 OB bariatric surgery candidates (France)	YFAS	BIS-11	Trait
Burrows, Hides, Brown, Dayas, and Kay-Lambkin (2017)	*N* = 1344 adults, web-based (Australia)	YFAS 2.0	SURPS	Trait
Ceccarini, Manzoni, Castelnuovo, and Molinari (2015)	*N* = 88 OB inpatient adults in WLT (Italy)	YFAS-16	BIS-11; DERS	Trait; trait
Davis et al. (2011)	*N* = 72 OB adults (Canada)	YFAS	BIS; DDT; DG	Trait; choice; choice
Gearhardt et al. (2012)	*N* = *81* OB treatment-seeking patients with BED (USA)	YFAS	DERS	Trait
Hsu et al. (2017)	*N* = 40 females (OB/FA = 20; CTL = 20) (Taiwan)	YFAS	BIS-11; GNG	Trait; action
Ivezaj et al. (2016)	*N* = 502 overweight/obese adults (USA)	YFAS	BIS-11; BSCS	Trait; trait
Loxton and Tipman (2017)	*N* = 374 community women (Australia)	YFAS	BIS-11	Trait
Meadows, Nolan, and Higgs (2017)	*N* = 614 adults, web-based[a] (international)	YFAS	BIS-15	Trait
Meule, Lutz, Vögele, and Kübler (2012)	*N* = 50 female college students (Germany)	YFAS	BIS-15; GNG	Trait; action
Meule, Heckel, Jurowich, Vögele, and Kübler (2014)	*N* = 96 OB bariatric surgery candidates (Germany)	YFAS	BIS-15	Trait
Meule et al. (2017)	*N* = 455 university students[b] (international)	YFAS 2.0	BIS-15	Trait
Meule, Hermann, and Kübler (2015)	*N* = 50 inpatient adolescents in WLT (Germany)	YFAS	BIS-15	Trait
Meule, de Zwaan, & Müller (2017)	*N* = 138 OB bariatric surgery candidates (Germany)	YFAS 2.0	BIS-15	Trait
Murphy et al. (2014)	*N* = 233 undergraduate students (USA)	YFAS	UPPS-P	Trait
Nunes-Neto et al. (2018)	*N* = 7639 nonclinical adults, web-based (Brazil)	mYFAS 2.0	MIDI	Trait

Continued

Table 7.1 Studies of food addiction and self-regulation.—*cont'd*

Study	Sample (Country)	FA measure	SR measure	SR category
Ouellette et al. (2017)	*N* = 146 OB bariatric surgery candidates (Canada)	YFAS	BIS-brief	Trait
Pivarunas and Conner (2015)	*N* = 878 undergraduates (USA)	YFAS	UPPS-P; DERS	Trait; trait
Raymond and Lovell (2015)	*N* = 334 patients: Type 2 diabetes (Australia)	YFAS	BIS-11	Trait
Rose, Nadler, and Mackey (2018)	*N* = 69 adolescents bariatric surgery candidates (USA)	YFAS-C	UPPS-P	Trait
VanderBroek-Stice et al. (2017)	*N* = 208 adults (48% community; 52% undergraduate) (USA)	YFAS	UPPS-P; GNG; DDT	Trait; action; choice
Vries and Meule (2016)	*N* = 456 females (BN = 115; CTL = 341) (Austria)	YFAS 2.0	BIS-15	Trait
Wolz et al. (2016)	*N* = 278 patients with feeding or ED (Spain)	YFAS	UPPS-P	Trait
Wolz, Granero, and Fernández-Aranda (2017)	*N* = 315 patients: Binge eating spectrum ED (Spain)	YFAS-S	UPPS-P; DERS	Trait; trait

BED, *binge eating disorder;* BIS, *Barratt Impulsiveness Scale;* BN, *bulimia nervosa;* BSCS, *Brief Self-Control Scale;* CTL, *control condition;* DDT, *Delay Discounting Task;* DERS, *Difficulties in Emotion Regulation Scale;* DG, *Delay of Gratification;* ED, *eating disorder diagnoses;* FA, *food addiction;* GNG, *Go/No-go task;* MIDI, *Minnesota impulse disorders interview;* NU, *negative urgency;* OB, *obese;* SR, *self-regulation;* SURPS, *Substance Use Risk Profile Scale;* TTM, *trichotillomania;* UPPS, *UPPS Impulsive Behavior Scale;* WLT, *weight loss treatment;* YFAS, *Yale Food Addiction Scale.*
[a] *sample 2 only*
[b] *study 1 only*

reported significant uncorrected associations between discounting of monetary rewards and FA and one also reported a significant association adjusting for age and income (the other did not report corrected analyses). In both cases, the effects were of small magnitude effect size. One of the two studies reported that FA mediated the relationship between delay discounting and obesity. In addition to monetary delay discounting, one study examined behavioral capacity to delay gratification, which requires exhibiting self-regulated restraint and is conceptually similar to delay discounting, and found a positive uncorrected association (corrected analyses were not reported). Collectively, these findings suggest that there is preliminary evidence that FA is associated with impulsive discounting, which is compatible with the broader drug addiction literature (Bickel, Johnson, Koffarnus, MacKillop, & Murphy, 2014; MacKillop et al., 2011), but clearly this is a nascent area of the field.

Impulsive action

The ability to inhibit prepotent motor responses is thought to be an indicator of cognitive control, an important aspect of self-regulation (Fillmore & Weafer, 2013). Three studies assessed the relationship between FA and impulsive action using Go/No-go tasks (GNG; Kiehl, Liddle, & Hopfinger, 2000). In general, the GNG tasks required participants to press a button as quickly as possible in response to one type of stimulus/target (i.e., "Go") and to refrain from pressing the button in response to a different stimulus/target (i.e., "No-go"). However, there was notable heterogeneity in methods across studies. In one study, participants were instructed that the letter "X" was the "Go" signal and the letter "K" was the "No-go" signal. The percentage of "No-go" trials in which an individual failed to inhibit in response to the "No-go" signal (i.e., commission error rate), the percentage of "Go" trials in which the participant failed to respond (i.e., omission error rate), and the time it took for a participant to respond to the "Go" signal (i.e., reaction time) were calculated. Controlling for age and income, more excess weight was associated with higher omission error rates and slower reaction time on the "Go" trials, but not difficulty inhibiting on the "No-go" trials. There was not a significant relationship between FA symptoms and GNG responding for any performance indicators.

In another study, participants were shown either the letter "X" or the letter "Y" on each trial and had to press a button ("Go") every time that the letter differed from the previous one. When, on the other hand, the same letter appeared in two consecutive trials, participants had to inhibit the response ("No-go"). In between each letter, an image of either a high-calorie food (e.g., pizza) or a neutral object (e.g., umbrella) was presented, introducing a motivational component. A median split of YFAS symptom count was then used to divide participants into groups that were high and low in FA symptoms, but mixed (picture type × group) ANOVAs showed neither significant main effects nor significant interactions for commission and omission errors. There was, however, a main effect for reaction time with participants in the high FA group having faster reaction times in response to food relative to neutral cues. This suggests that the food stimuli had a motivating effect but that there were no inhibitory deficits.

The third study used a more complex task. Participants were instructed to press a button as quickly as possible when they saw the number 1, 3, 4, 5, 6, 7, 8, or 9 ("Go") but to refrain from pressing the button when they saw the number 2 ("No-go"). The total number of commissions errors during the task and perceived difficulty of the task (based on self-reported) were determined. Women who were obese and who met criteria for FA to sweet foods (based on the YFAS and confirmed by interview with a psychiatrist) did not significantly differ from control participants on either commission errors or perceived difficulty.

Taken together, preliminary evidence does not suggest that motor impulsivity deficits in self-regulatory capacity are related to FA. Despite considerable differences between GNG tasks used, there was no evidence that difficulty inhibiting inappropriate responses as measured by commission errors was associated with FA and minimal evidence to support differences in other aspects of motor response (e.g., processing speed, attention).

Impulsive personality traits

Barratt Impulsiveness Scale: The relationship between FA and the Barratt Impulsiveness Scale (BIS; Fossati, Di Ceglie, Acquarini, & Barratt, 2001; Meule, Vögele, & Kübler, 2011; Patton et al., 1995; Spinella, 2007; Stanford et al., 2009; Steinberg, Sharp, Stanford, & Tharp, 2013) was examined in 15 of the studies we identified, with participants from more than a dozen countries represented. Most studies involved clinical populations including individuals with obesity ($n = 3$), those referred for or pursuing bariatric surgery ($n = 4$) or other weight loss treatment ($n = 2$), or individuals with other medical or psychiatric conditions (i.e., type 2 diabetes ($n = 1$), bulimia nervosa ($n = 1$)), as well as a small number of studies community ($n = 2$) or university ($n = 2$) samples. The majority further characterized findings based on specific impulsivity domains including (1) nonplanning (e.g., rarely plan tasks, trips), (2) motor (e.g., acts on impulsive, without thinking, spur of the moment), and (3) attentional (e.g., distracted, racing thoughts, difficulty concentrating) impulsivity. Eight studies reported results of zero-order associations between impulsivity domains and FA. In all eight, there was a positive significant association between FA and attentional impulsivity, generally of medium effect size. Significant associations with motor and nonplanning impulsivity were less consistent, being reported in four and three studies, respectively, and varying in level of statistical significance and magnitude of the association. Finally, significant associations between FA and total impulsivity (i.e., not domain specific) on the BIS were reported in five of six studies, all with small magnitude effect sizes.

Eleven studies compared impulsivity levels among individuals categorically dichotomized as FA positive or negative based on the YFAS diagnostic criteria or divided categorically based on median YFAS score. Patterns observed were similar to zero-order findings, with the most consistent findings being differences in attentional impulsivity and total impulsivity rather than motor or nonplanning impulsivity. A final study assessed whether impulsivity levels predicted the likelihood of being diagnosed with FA, finding that attentional impulsivity positively predicted YFAS diagnosis, but only when levels of motor impulsivity were also high.

Collectively, these findings suggest that there is consistent evidence that FA is associated with impulsive personality traits as measured by the BIS and particularly in relation to difficulties with sustained attention and concentration.

UPPS Impulsive Behavior Scale: Associations between FA and trait level impulsivity as measured by the UPPS Impulsive Behavior Scale (UPPS; Cyders et al., 2007; Lynam, Smith, Whiteside, & Cyders, 2006; Whiteside & Lynam, 2001; Whiteside, Lynam, Miller, & Reynolds, 2005) were examined in six studies, four in the United States and two in Spain. Several studies reported findings for specific domains of impulsivity measured by the UPPS-P: (1) (lack of) perseverance (i.e., tendency to give up; lack of motivation to complete tasks), (2) (lack of) premeditation (i.e., tendency to act or speak without thinking), (3) sensation seeking (i.e., tendency to pursue novel or exciting experiences), (4) negative urgency, and (5) positive urgency (i.e., tendency to act rashly when experiencing negative and positive emotions, respectively).

Zero-order correlations were reported in five of these studies. In four of the five, significant positive associations were reported between negative urgency and FA, generally of medium effect size. A fifth study used a composite score that combined negative and positive urgency, also finding a significant association. Two studies reported associations between positive urgency and FA, with significant positive associations that were small in magnitude in both. Three studies reported on the relationship between UPPS domains other than urgency (i.e., sensation seeking, (lack of) premeditation, (lack of) perseverance). Lack of perseverance was positively associated with FA in all three studies. In contrast, only one study observed a significant positive relationship between (lack of) perseverance and food addition. YFAS score was also negatively associated with sensation seeking in one study (i.e., individuals with a greater tendency to seek out novel or exciting activities report fewer FA symptoms).

Two studies subsequently entered these impulsivity scales into a regression model, with negative urgency remaining a significant positive predictor in both. Lack of perseverance positively predicted FA in one of these models, whereas (lack of) premeditation negatively predicted FA symptoms in the other. Another one of the studies conducted regression, adjusting for age and income, and reported that urgency and (lack of) perseverance continued to significantly predict FA. An additional study, which adjusted for sex, age, and eating disorder subtype, indicated that the likelihood of having a positive YFAS diagnosis was predicted by high scores for negative urgency and low scores for (lack of) premeditation, with negative urgency the strongest predictor of FA. Two studies used structural equation modeling to test the relationship between impulsivity and FA. In one, eating disorder symptoms were found to mediate the relationship between negative urgency and FA. In the other, negative urgency was found to be associated with both emotional eating and FA, both of which were associated with poorer weight-related quality of life (i.e., negative impact of weight on quality of life).

Together, findings consistently suggest that negative urgency is associated with FA directly and when adjusting for other relevant factors. Past research has suggested that negative urgency is a strong predictor of many substance-use outcomes such as frequency of alcohol, marijuana, and tobacco use, problematic drinking, and experimentation with illicit substances, even when trait negative affect and distress tolerance were accounted for (Kaiser, Milich, Lynam, & Charnigo, 2012). Similarly, negative urgency has been shown to account for the relationship between depression and increased likelihood of lifetime use of alcohol, inhalants, and any substance use (Gullo, Loxton, & Dawe, 2014). More broadly, negative urgency is thought to be a common risk factor related to a variety of behaviors such as increased self-harm, problematic alcohol use, and disordered eating (Dir, Karyadi, & Cyders, 2013;

Fischer, Settles, Collins, Gunn, & Smith, 2012). Thus, the link between negative urgency and FA is compatible with an emerging perspective that it is a common maladaptive psychological mechanism for a broad spectrum of psychiatric conditions.

Difficulties in Emotion Regulation Scale: Five studies examined self-regulation difficulties pertaining to dysregulation of emotion assessed by the Difficulties in Emotion Regulation Scale (DERS; Gratz & Roemer, 2004), in the United States, Canada, Italy, and Spain. All reported significant positive relationships between difficulty controlling emotions and FA in uncorrected analyses, generally of medium effect size. Additionally, two of these studies analyzed differences in emotion regulation when participants were divided based on YFAS diagnostic status. In both studies, individuals who met YFAS criteria for FA endorsed more emotion dysregulation.

Several other studies using DERS total score bear mentioning individually. One study conducted multiple mediation analyses and determined that emotion dysregulation via the DERS mediated the relationship between harm avoidance (i.e., proneness to worry, fear, shyness, and pessimism) and FA. This extends previous findings linking harm avoidance and addictive disorders (Hosák, Preiss, Halíř, Čermáková, & Csémy, 2004; Milivojevic et al., 2012) and suggests that individuals prone to negative emotion may lead to having a greater number of FA symptoms by way of difficulty regulating emotions. In another, Poisson regression was conducted using the five UPPS-P factors and the DERS total score to determine significant predictors of number of FA symptoms endorsed. Only the DERS and the closely related construct, negative urgency, positively predicted FA symptoms. In a related study, a comprehensive model that included the DERS, negative urgency, and characteristics of disordered eating was tested using hypothesis-driven structural equation modeling. Results suggested that putative relationships between emotion dysregulation as measured by the DERS and FA were diminished when negative urgency and eating disorder psychopathology were taken into account. That finding, however, can also be thought of as providing evidence that the DERS and negative urgency are essentially tapping a similar construct.

Two studies examined DERS subscales thought to measure different aspects of emotion dysregulation: (1) nonacceptance of emotional responses, (2) goal-directed behavior difficulties, (3) impulse control difficulties, (4) (lack of) emotional awareness (awareness), (5) (limited access to) strategies for emotion regulation, and (6) (lack of) emotional clarity. In one study, of patients who were obese and met criteria for BED, five of the six DERS scales (all but [lack of] emotional awareness) were significantly positively associated with FA symptoms, generally of medium effect size. In another study, adults with obesity in inpatient weight loss treatment were compared on the DERS scales as a function of YFAS diagnosis status. Using a more stringent significance criterion based on the number of tests, statistically significant differences were reported only for the impulse control difficulties subscale. This scale reflects the perception of losing control over one's behavior when feeling upset, which is again highly consonant with the findings on the UPPS Impulsive Behavior Scale.

Collectively, these findings suggest that there is consistent evidence that FA is associated difficulties with emotion regulation, and this relationship appears largely driven by acting impulsively when experiencing negative emotions. As such, it is highly consistent with findings regarding negative urgency with both FA and other addictive disorders.

Other measures of impulsive personality traits: *The Substance Use Risk Profile Scale* (SURPS; Woicik, Stewart, Pihl, & Conrod, 2009) was used in one study to examining the relationship between personality traits and FA. The SURPS assesses four personality dimensions thought to be associated with substance use: hopelessness, anxiety sensitivity, impulsivity, and sensation seeking, with the latter two domains having commonalities with other personality trait–based measures of impulsivity. On the SURPS, the impulsivity scale is unidimensional but have considerable conceptual overlap with UPPS and BIS domains (e.g., SURPS: "*I often don't think things through before I speak*"; UPPS-P (lack of) premeditation: "*I am not one of those people who blurt out things without thinking*"; BIS-11 nonplanning: "*I say things without thinking*"). Likewise, the sensation seeking scale on the SURPS has high conceptual correspondence to the UPPS sensation seeking scale (e.g., SURPS: "*I enjoy new and exciting experiences even if they are unconventional*" and "*I like doing things that frighten me a little*"; UPPS-P: "*I welcome new and exciting experiences and sensations, even if they are a little frightening and unconventional*"). In a large web-based survey study, individuals who were categorized as having FA based on the YFAS had significantly higher impulsivity scores and significantly *lower* sensation seeking scores on the SURPS than individuals who did not. Consequently, SURPS findings provide additional evidence for a positive association between trait-based impulsivity and FA and additional support for a negative relationship between sensation seeking and FA (VanderBroek-Stice et al., 2017). In this latter case, it is an interesting hypothesis that FA may be a manifestation of externalizing overconsumption behavior in individuals for whom psychoactive drugs are either excessively stimulating or otherwise unappealing because of social opprobrium.

The Brief Self-Control Scale (BSCS; Tangney, Baumeister, & Boone, 2004) assesses self-control, conceptualized as "the ability to override or change one's inner responses, as well as to interrupt undesired behavioral tendencies (such as impulses) and refrain from acting on them" (p. 274). This measure of self-control is negatively associated with eating disorder symptoms and problematic drinking patterns and positively associated with emotional stability (Tangney et al., 2004). One study compared self-reported levels of self-control on the BSCS among adults who were grouped into four categories: (1) individuals who did not meet criteria for FA on the YFAS but did meet criteria for BED based on an eating disorder questionnaire, (2) individuals who did not meet BED criteria but met criteria for FA, (3) individuals who met criteria for both BED and FA, and (4) individuals who met criteria for neither (CTL). Relative to the control group, all three groups reported significantly poorer self-control. The effect size was large and there was minimal attenuation when controlling for BMI, but other potential confounders were not included.

Therefore, individuals with FA, both with and without binge eating, report having greater difficulty interrupting or inhibiting impulsive and other unwanted behaviors. These findings are most compatible with the findings linking the BIS total score to FA reported earlier.

The Minnesota Impulsive Disorders Interview (MIDI; Christenson et al., 1994; Odlaug & Grant, 2010) is self-report instrument that includes questions pertaining to impulse control psychiatric disorders (e.g., trichotillomania, kleptomania, pyromania, intermittent explosive disorder). A large web-based survey included the YFAS and the trichotillomania (hair-pulling disorder) module of the MIDI. Individuals who met the criteria for FA on the YFAS did not significantly differ from those who did not in prevalence of positive screenings for symptoms of trichotillomania. Obviously there has been little research using this instrument or contextualizing FA with impulse control disorders, but this study certainly does not suggest a connection is there. Interestingly, in DSM-5, gambling disorder moved from the impulse control disorders section to become the first behavioral addiction to be grouped with substance-use disorders. This was a reflection of the closer similarity to drug addiction and a distinction between disorders reflecting an impulsive element in the context of appetitive reinforcers (i.e., substance-use disorders and gambling disorder) as opposed to impulsive behaviors that are more distal to reward, if at all (impulse control disorders). Framed thus, FA not being related to the latter category can actually be thought of as a form of discriminant validity (i.e., meaningful absence of association). That is, FA is linked to forms of impulsivity associated with other addictive disorders, but not the separate category of impulse control disorders. Of course, this conjecture must be tempered by the limited investigation of this overlap.

Summary and conclusions

In less than a decade, a considerable amount of research has been conducted to better understand FA both generally and from an addiction-based framework. Of the three domains of self-regulatory capacity (i.e., impulsive choice, impulsive action, and impulsive personality traits), the current evidence suggests that *impulsive personality traits* are most closely related to symptoms of FA. In particular, difficulty controlling behavior when in an emotional state, particularly one with negative valence (i.e., feeling upset, rejected, or when in an argument), was most consistently and robustly associated with FA across multiple assessment measures. This was a converging finding, both from the literature on UPPS-P, the DERS, and studies using both measures simultaneously, making the level of evidence linking deficits in emotional regulation to FA notably higher than the other findings. These findings are also consistent with recent conceptualizations of negative emotionality being a core biobehavioral determinant of addiction (e.g., Kwako et al., 2016). However, from studies using the BIS, there was also consistent evidence to suggest that FA is associated with difficulties regulating attention (focusing attention, concentrating) and with task persistence (e.g., giving up, not seeing tasks through to the end). The evidence regarding a relationship between FA and lack of premeditation (i.e., failing to think and reflect on the consequences of behavior before acting) and nonplanning (i.e., lack of "futuring" or forethought) (Barratt, 1985) tendencies was observed less consistently. Collectively, these findings are substantively similar to those in the literature on impulsive personality traits and psychoactive drug addiction, suggesting that dispositional propensities to act out on arising impulses are common antecedents of both.

With regard to impulsive choice and impulsive action, there were many fewer studies. This made strong conclusions hard to draw, but the patterns were at least relatively consistent. For delay discounting, a tendency to overvalue immediate rewards exhibited a significant relationship with FA, albeit with small magnitude associations. Finally, for impulsive action (i.e., ability to inhibit a prepotent motor response), no significant relationships were observed. In the case of the former, the nascent findings are compatible to the literature linking impulsive delay discounting to drug addiction, although that is not the case in terms of behavioral inhibition. At this stage, it is too early to say whether this difference with studies on drug addiction is meaningful, but it is certainly of interest that there may not be perfect homology in terms of links.

While the scope of this chapter was limited to the relationship between self-regulation and FA, the intentionally narrow scope permits for greater acuity regarding the ways in which FA self-regulatory capacity difficulties may parallel relationships observed in other addictive disorders. Of the self-regulatory domains studied, difficulty controlling behavior in response to strong emotions emerged across multiple measures as associated with FA. Emotion dysregulation and dyscontrol are common to other addictive behaviors (Bekh Bradley et al., 2011; Cheetham, Allen, Yücel, & Lubman, 2010; Gonzalez, Zvolensky, Vujanovic, Leyro, & Marshall, 2008; Kober, 2014; Wills, Walker, Mendoza, & Ainette, 2006). When individuals are upset, attempts to reduce acute negative emotional experiences (i.e., regulate affect) may take precedence over exerting self-control, thereby favoring the short-term goals at the expense of long-term goals (Tice, Bratslavsky, & Baumeister, 2001). Deficits in emotion regulation skills have been shown to predict alcohol use during and after cognitive behavioral therapy among patients with alcohol-use disorders (Berking et al., 2011). Negative affect relief expectancies are a risk factor for the escalation of smoking behavior and for the development of nicotine dependence (Heinz, Kassel, Berbaum, & Mermelstein, 2010). Negative affect is associated with prescription opioid misuse mediated by opioid craving (Martel, Dolman, Edwards, Jamison, & Wasan, 2014). Coping-oriented marijuana use is associated with emotion dysregulation, suggesting use of marijuana as a short-term strategy to reduce negative affect or emotional distress (Bonn-Miller, Vujanovic, & Zvolensky, 2008). Taken together, acting rashly when experiencing negative affective states and/or in an effort to reduce acute negative affective states is related to addictive behavior and individuals may be particularly at risk when they believe that the substance will help them to cope, alleviate their distress, or have a strong craving for it. If FA parallels other addictive disorders, learned associations and expectancies of food's negative reinforcing capacity made lead to escalating tendency to use food in problematic ways that lead to symptoms of FA.

A number of interventions have been developed to improve attentional awareness and emotion regulation. Mindfulness, a characteristic that involves observing one's experiences in the present moment with an attitude of acceptance, including emotional and inner experiences (Baer, Smith, Hopkins, Krietemeyer, & Toney, 2006; Bishop et al., 2004), is associated with less emotional lability and emotion dysregulation (Hill & Updegraff, 2012). Mindfulness-based interventions have been shown to increase the ability to respond to negative affect and craving in appropriate ways and to reduce consumption of a variety of substances, including alcohol, cocaine, amphetamines, marijuana, cigarettes, and opiates (Bowen, Chawla, & Witkiewitz, 2014, pp. 141–157; Chiesa & Serretti, 2014). Similarly, mindfulness-based eating awareness training that emphasizes developing awareness of cues for hunger and satiety and regulating responses to emotional states has been shown to improve perceptions of self-control with regard to eating behavior (Kristeller & Wolever, 2010). In view of this, mindfulness-based interventions may be particularly effective at reducing FA symptoms among individuals with limited capacity for self-regulation in instances of difficulties regulating attention and regulating impulses in response to strong emotional experiences. Similarly, acceptance and commitment therapy (ACT)–based interventions, which encourage tolerating emotional distress without engaging in value inconsistent behavior and dialectical behavior therapy (DBT) and which teach us skills such as emotion regulation, distress tolerance, and mindfulness, have been shown to have favorable outcomes for reducing substance use (Dimeff & Linehan, 2008; Stotts & Northrup, 2015) and eating pathology (Juarascio, Forman, & Herbert, 2010; Telch, Agras, & Linehan, 2001). Accordingly, ACT or DBT skills may have potential for reducing symptoms of FA, particularly among individual with noted self-regulatory difficulties.

Finally, there is evidence that participating in behavioral weight loss treatment can decrease disinhibited eating behavior (Foster, Wadden, Kendall, Stunkard, & Vogt, 1996). Interventions based on self-regulatory theory have shown beneficial outcomes for preventing weight gain and maintaining weight loss (Wing, Tate, Espeland, & et al., 2016; Wing, Tate, Gorin, Raynor, & Fava, 2006). These interventions utilize self-awareness and changing behavioral repertoire to monitor behavior, detect problems, implement problem-solving skills, evaluate the success of implemented strategies, and rewarding oneself for success. It is possible that implementing strategies to improve self-regulation with regard to weight management could also contribute to reductions in symptoms of FA.

It is important to note a number of limitations to the conclusions are drawn in this chapter. First, although the literature is accumulating, it is worth noting that in absolute terms, the number of studies in this area remains relatively small. Second, within this modest pool, there was considerable heterogeneity in the samples recruited, the measures used, and the analyses reported. In addition, the findings were integrated qualitatively, not quantitatively. Although the existing research base is too small at present, a quantitative integration of these findings via a meta-analysis will certainly be warranted as more findings using consistent methods accumulate. Finally, it is important to note that all the studies reviewed were cross-sectional, fundamentally indicating correlational relationships, not causational relationships. Certainly, it is hypothesized that deficits in self-regulation give rise to subsequent addictive behavior, and this has reported for drug addiction (e.g., Audrain-McGovern et al., 2009; Fernie et al., 2013), but the same cannot be said for FA at this time.

In sum, a review of the current evidence suggests that self-regulatory difficulties may be a common risk factor for both FA and other addictive disorders. This parallel provides further support for the concept of FA as a valid manifestation of compulsive overconsumption that is substantively akin to drug addiction. Of course, these cross-sectional links alone are by no means definitive and a great deal of further investigation is needed, but the consistent general patterns suggest these links in the nomological network are relatively robust. As to our knowledge regarding the factors that contribute to the development and trajectory of FA increases, we are likely to have greater understanding of how to best help individuals who meet diagnostic criteria. Future work testing strategies designed to increase self-awareness, acceptance of emotions, and emotion regulation capacity may have promise for individuals with FA.

Acknowledgment

JM's contributions were partially supported by the Peter Boris Chair in Addictions Research.

Conflicts of Interest

CMM has no conflicts to declare; JM is a principal in BEAM Diagnostics, Inc.

References

Acker, J., & MacKillop, J. (2013). Behavioral economic analysis of cue-elicited craving for tobacco: A virtual reality study. *Nicotine & Tobacco Research*. https://doi.org/10.1093/ntr/nts341.

American Psychiatric Association. (2000). *Diagnostic criteria from DSM-IV-TR*. Washington, D.C.: American Psychiatric Association.

American Psychiatric Association. (2013). *Diagnostic and statistical manual of mental disorders (DSM-5®)*. American Psychiatric Pub.

Amlung, M., Vedelago, L., Acker, J., Balodis, I., & MacKillop, J. (2017). Steep delay discounting and addictive behavior: A meta-analysis of continuous associations. *Addiction, 112*. https://doi.org/10.1111/add.13535.

Audrain-McGovern, J., Rodriguez, D., Epstein, L. H., Cuevas, J., Rodgers, K., & Wileyto, E. P. (2009). Does delay discounting play an etiological role in smoking or is it a consequence of smoking? *Drug and Alcohol Dependence, 103*, 99–106. https://doi.org/10.1016/j.drugalcdep.2008.12.019.

Baer, R. A., Smith, G. T., Hopkins, J., Krietemeyer, J., & Toney, L. (2006). Using self-report assessment methods to explore facets of mindfulness. *Assessment, 13*(1), 27–45. https://doi.org/10.1177/1073191105283504.

Barratt, E. S. (1985). Impulsiveness subtraits: Arousal and information processing. *Motivation, emotion, and personality, 5*, 137–146.

Bekh Bradley, D., DeFife, J. A., Guarnaccia, C., Phifer, M. J., Fani, M. N., Ressler, K. J., & Westen, D. (2011). Emotion dysregulation and negative affect: Association with psychiatric symptoms. *Journal of Clinical Psychiatry, 72*(5), 685.

Berking, M., Margraf, M., Ebert, D., Wupperman, P., Hofmann, S. G., & Junghanns, K. (2011). Deficits in emotion-regulation skills predict alcohol use during and after cognitive–behavioral therapy for alcohol dependence. *Journal of Consulting and Clinical Psychology, 79*(3), 307.

Bickel, W. K., Johnson, M. W., Koffarnus, M. N., MacKillop, J., & Murphy, J. G. (2014). The behavioral economics of substance use disorders: Reinforcement pathologies and their repair. *Annual Review of Clinical Psychology, 10*, 641–677.

Bishop, S. R., Lau, M., Shapiro, S., Carlson, L., Anderson, N. D., Carmody, J., … Devins, G. (2004). Mindfulness: A proposed operational definition. *Clinical Psychology: Science and Practice, 11*(3), 230–241. https://doi.org/10.1093/clipsy/bph077.

Bonn-Miller, M. O., Vujanovic, A. A., & Zvolensky, M. J. (2008). Emotional dysregulation: Association with coping-oriented marijuana use motives among current marijuana users. *Substance Use & Misuse, 43*(11), 1653–1665.

Bowen, S., Chawla, N., & Witkiewitz, K. (2014). *Mindfulness-based relapse prevention for addictive behaviors Mindfulness-Based Treatment Approaches* (2nd ed.). Elsevier.

Brunault, P., Ducluzeau, P.-H., Courtois, R., Bourbao-Tournois, C., Delbachian, I., Réveillère, C., & Ballon, N. (2018). Food addiction is associated with higher neuroticism, lower conscientiousness, higher impulsivity, but lower extraversion in obese patient candidates for bariatric surgery. *Substance Use & Misuse*, 1–5.

Burrows, T., Hides, L., Brown, R., Dayas, C. V., & Kay-Lambkin, F. (2017). Differences in dietary preferences, personality and mental health in australian adults with and without food addiction. *Nutrients, 9*(3), 285.

Caswell, A. J., Bond, R., Duka, T., & Morgan, M. J. (2015). Further evidence of the heterogeneous nature of impulsivity. *Personality and Individual Differences, 76*, 68–74. https://doi.org/10.1016/j.paid.2014.11.059.

Ceccarini, M., Manzoni, G. M., Castelnuovo, G., & Molinari, E. (2015). An evaluation of the Italian version of the Yale Food Addiction Scale in obese adult inpatients engaged in a 1-month-weight-loss treatment. *Journal of Medicinal Food, 18*(11), 1281–1287.

Cheetham, A., Allen, N. B., Yücel, M., & Lubman, D. I. (2010). The role of affective dysregulation in drug addiction. *Clinical Psychology Review, 30*(6), 621–634.

Chiesa, A., & Serretti, A. (2014). Are mindfulness-based interventions effective for substance use disorders? A systematic review of the evidence. *Substance Use & Misuse, 49*(5), 492–512.

Christenson, G. A., Faber, R. J., De Zwaan, M., Raymond, N. C., Specker, S. M., Ekern, M. D., … Eckert, E. D. (1994). Compulsive buying: Descriptive characteristics and psychiatric comorbidity. *Journal of Clinical Psychiatry, 55*(1), 5–11.

Coskunpinar, A., Dir, A. L., & Cyders, M. A. (2013). Multidimensionality in impulsivity and alcohol use: A meta-analysis using the UPPS model of impulsivity. *Alcoholism: Clinical and Experimental Research, 37*, 1441–1450. https://doi.org/10.1111/acer.12131.

Cyders, M. A., Smith, G. T., Spillane, N. S., Fischer, S., Annus, A. M., & Peterson, C. (2007). Integration of impulsivity and positive mood to predict risky behavior: Development and validation of a measure of positive urgency. *Psychological Assessment, 19*(1), 107–118. https://doi.org/10.1037/1040-3590.19.1.107.

Davis, C. (2013a). Compulsive overeating as an addictive behavior: Overlap between food addiction and binge eating disorder. *Current Obesity Reports, 2*(2), 171–178.

Davis, C. (2013b). From passive overeating to "food addiction": A spectrum of compulsion and severity. *ISRN obesity, 2013*.

Davis, C., Curtis, C., Levitan, R. D., Carter, J. C., Kaplan, A. S., & Kennedy, J. L. (2011). Evidence that 'food addiction' is a valid phenotype of obesity. *Appetite, 57*(3), 711–717. https://doi.org/10.1016/j.appet.2011.08.017.

Davis, C., Patte, K., Curtis, C., & Reid, C. (2010). Immediate pleasures and future consequences. A neuropsychological study of binge eating and obesity. *Appetite, 54*, 208–213.

Dick, D. M., Smith, G., Olausson, P., Mitchell, S. H., Leeman, R. F., O'Malley, S. S., & Sher, K. (2010). Understanding the construct of impulsivity and its relationship to alcohol use disorders. *Addiction Biology, 15*, 217–226. https://doi.org/10.1111/j.1369-1600.2009.00190.x.

Dimeff, L. A., & Linehan, M. M. (2008). Dialectical behavior therapy for substance abusers. *Addiction Science & Clinical Practice, 4*(2), 39.

Dir, A. L., Karyadi, K., & Cyders, M. A. (2013). The uniqueness of negative urgency as a common risk factor for self-harm behaviors, alcohol consumption, and eating problems. *Addictive Behaviors, 38*(5), 2158–2162.

Epstein, L. H., Salvy, S. J., Carr, K. A., Dearing, K. K., & Bickel, W. K. (2010). Food reinforcement, delay discounting and obesity. *Physiology & Behavior, 100*, 438–445. https://doi.org/10.1016/j.physbeh.2010.04.029.

Evenden, J. L. (1999). Varieties of impulsivity. *Psychopharmacology, 146*, 348–361. https://doi.org/10.1007/PL00005481.

Eysenck, S. B., & Eysenck, H. J. (1978). Impulsiveness and venturesomeness: Their position in a dimensional system of personality description. *Psychological Reports, 43*, 1247–1255.

Fernie, G., Peeters, M., Gullo, M. J., Christiansen, P., Cole, J. C., Sumnall, H., & Field, M. (2013). Multiple behavioural impulsivity tasks predict prospective alcohol involvement in adolescents. *Addiction, 108*, 1916–1923. https://doi.org/10.1111/add.12283.

Fillmore, M. T., & Weafer, J. (2013). Behavioral inhibition and addiction. In J. MacKillop, & H. de Wit (Eds.), *The Wiley-Blackwell handbook of addiction psychopharmacology* (pp. 135–164).

Fischer, S., Settles, R., Collins, B., Gunn, R., & Smith, G. T. (2012). The role of negative urgency and expectancies in problem drinking and disordered eating: Testing a model of comorbidity in pathological and at-risk samples. *Psychology of Addictive Behaviors, 26*(1), 112.

Fossati, A., Di Ceglie, A., Acquarini, E., & Barratt, E. S. (2001). Psychometric properties of an Italian version of the Barratt Impulsiveness Scale-11 (BIS-11) in nonclinical subjects. *Journal of Clinical Psychology, 57*(6), 815–828.

Foster, G. D., Wadden, T. A., Kendall, P. C., Stunkard, A. J., & Vogt, R. A. (1996). Psychological effects of weight loss and regain: A prospective evaluation. *Journal of Consulting and Clinical Psychology, 64*(4), 752–757.

Gearhardt, A. N., Corbin, W. R., & Brownell, K. D. (2009). Preliminary validation of the Yale food addiction scale. *Appetite, 52*(2), 430–436. https://doi.org/10.1016/j.appet.2008.12.003.

Gearhardt, A. N., Corbin, W. R., & Brownell, K. D. (2016). Development of the Yale food addiction scale version 2.0. *Psychology of Addictive Behaviors, 30*(1), 113.

Gearhardt, A. N., Davis, C., Kuschner, R., & Brownell, K. (2011). The addiction potential of hyperpalatable foods. *Current Drug Abuse Reviews, 4*(3), 140–145.

Gearhardt, A. N., White, M. A., Masheb, R. M., Morgan, P. T., Crosby, R. D., & Grilo, C. M. (2012). An examination of the food addiction construct in obese patients with binge eating disorder. *International Journal of Eating Disorders, 45*(5), 657–663.

Gonzalez, A., Zvolensky, M. J., Vujanovic, A. A., Leyro, T. M., & Marshall, E. C. (2008). An evaluation of anxiety sensitivity, emotional dysregulation, and negative affectivity among daily cigarette smokers: Relation to smoking motives and barriers to quitting. *Journal of Psychiatric Research, 43*(2), 138–147.

Gratz, K. L., & Roemer, L. (2004). Multidimensional assessment of emotion regulation and dysregulation: Development, factor structure, and initial validation of the difficulties in emotion regulation scale. *Journal of Psychopathology and Behavioral Assessment, 26*(1), 41–54.

Gullo, M. J., Loxton, N. J., & Dawe, S. (2014). Impulsivity: Four ways five factors are not basic to addiction. *Addictive Behaviors, 39*(11), 1547–1556.

Heinz, A. J., Kassel, J. D., Berbaum, M., & Mermelstein, R. (2010). Adolescents' expectancies for smoking to regulate affect predict smoking behavior and nicotine dependence over time. *Drug and Alcohol Dependence, 111*(1), 128–135.

Hill, C. L., & Updegraff, J. A. (2012). Mindfulness and its relationship to emotional regulation. *Emotion, 12*(1), 81.

Hosák, L., Preiss, M., Halíř, M., Čermáková, E., & Csémy, L. (2004). Temperament and character inventory (TCI) personality profile in metamphetamine abusers: A controlled study. *European Psychiatry, 19*(4), 193–195.

Hsu, J.-S., Wang, P.-W., Ko, C.-H., Hsieh, T.-J., Chen, C.-Y., & Yen, J.-Y. (2017). Altered brain correlates of response inhibition and error processing in females with obesity and sweet food addiction: A functional magnetic imaging study. *Obesity Research & Clinical Practice, 11*(6), 677–686.

Ivezaj, V., White, M. A., & Grilo, C. M. (2016). Examining binge-eating disorder and food addiction in adults with overweight and obesity. *Obesity, 24*(10), 2064–2069.

Johnson, P. M., & Kenny, P. J. (2010). Dopamine D2 receptors in addiction-like reward dysfunction and compulsive eating in obese rats. *Nature Neuroscience, 13*, 635–641. https://doi.org/10.1038/nn.2519.

Juarascio, A. S., Forman, E. M., & Herbert, J. D. (2010). Acceptance and commitment therapy versus cognitive therapy for the treatment of comorbid eating pathology. *Behavior Modification, 34*(2), 175–190.

Kaiser, A. J., Milich, R., Lynam, D. R., & Charnigo, R. J. (2012). Negative urgency, distress tolerance, and substance abuse among college students. *Addictive Behaviors, 37*(10), 1075–1083.

Kelley, A. E., & Berridge, K. C. (2002). The neuroscience of natural rewards: Relevance to addictive drugs. *Journal of Neuroscience, 22*, 3306–3311, 20026361.

Kiehl, K. A., Liddle, P. F., & Hopfinger, J. B. (2000). Error processing and the rostral anterior cingulate: An event-related fMRI study. *Psychophysiology, 37*(2), 216–223.

Kober, H. (2014). Emotion regulation in substance use disorders. In J. J. Gross (Ed.), *Handbook of emotion regulation* (2 ed., pp. 428–446). New York, NY: The Guilford Press.

Koob, G. F., & Volkow, N. D. (2016). Neurobiology of addiction: A neurocircuitry analysis. *The Lancet Psychiatry, 3*, 760–773. https://doi.org/10.1016/S2215-0366(16)00104-8.

Kristeller, J. L., & Wolever, R. Q. (2010). Mindfulness-based eating awareness training for treating binge eating disorder: The conceptual foundation. *Eating Disorders, 19*(1), 49–61.

Kwako, L. E., Momenan, R., Litten, R. Z., Koob, G. F., & Goldman, D. (2016). Addictions neuroclinical assessment: A neuroscience-based framework for addictive disorders. *Biological Psychiatry, 80*, 179–189. https://doi.org/10.1016/j.biopsych.2015.10.024.

Loxton, N. J., & Tipman, R. J. (2017). Reward sensitivity and food addiction in women. *Appetite, 115*, 28–35.

Lynam, D. R., Smith, G. T., Whiteside, S. P., & Cyders, M. A. (2006). *The UPPS-P: Assessing five personality pathways to impulsive behavior.* West Lafayette: Purdue University.

MacKillop, J., Amlung, M. T., Few, L. R., Ray, L. A., Sweet, L. H., & Munafo, M. R. (2011). Delayed reward discounting and addictive behavior: A meta-analysis. *Psychopharmacology Series*. https://doi.org/10.1007/s00213-011-2229-0.

MacKillop, J., Miller, J. D., Fortune, E., Maples, J., Lance, C. E., … Goodie, A. S. (2014). Multidimensional examination of impulsivity in relation to disordered gambling. *Experimental and Clinical Psychopharmacology, 22*. https://doi.org/10.1037/a0035874.

MacKillop, J., Murphy, J. G., Ray, L. A., Eisenberg, D. T., Lisman, S. A., Lum, J. K., & Wilson, D. S. (2008). Further validation of a cigarette purchase task for assessing the relative reinforcing efficacy of nicotine in college smokers. *Experimental and Clinical Psychopharmacology, 16*, 57–65. https://doi.org/10.1037/1064-1297.16.1.57.

MacKillop, J., & Ray, L. A. (2018). The etiology of addiction: A contemporary biopsychosocial approach. In J. MacKillop, G. Kenna, L. Leggio, & L. Ray (Eds.), *Integrating psychological and pharmacological treatments for addictive disorders: An evidence-based guide* (pp. 32–53). New York, NY, US: Routledge/Taylor & Francis Group.

MacKillop, J., Weafer, J., Gray, J., Oshri, A., Palmer, A. A., & de Wit, H. (2016). The latent structure of impulsivity: Impulsive choice, impulsive action, and impulsive personality traits. *Psychopharmacology, 233*(18), 3361–3370.

Madden, G. J., Bickel, W. K., & Jacobs, E. A. (1999). Discounting of delayed rewards in opioid-dependent outpatients: Exponential or hyperbolic discounting functions? *Experimental and Clinical Psychopharmacology, 7*, 284–293.

Martel, M. O., Dolman, A. J., Edwards, R. R., Jamison, R. N., & Wasan, A. D. (2014). The association between negative affect and prescription opioid misuse in patients with chronic pain: The mediating role of opioid craving. *The Journal of Pain, 15*(1), 90–100.

Meadows, A., Nolan, L. J., & Higgs, S. (2017). Self-perceived food addiction: Prevalence, predictors, and prognosis. *Appetite, 114*, 282–298.

Meule, A. (2015). Back by popular demand: A narrative review on the history of food addiction research. *Yale Journal of Biology & Medicine, 88*(3), 295–302.

Meule, A., de Zwaan, M., & Müller, A. (2017a). Attentional and motor impulsivity interactively predict 'food addiction'in obese individuals. *Comprehensive Psychiatry, 72*, 83–87.

Meule, A., Heckel, D., Jurowich, C., Vögele, C., & Kübler, A. (2014). Correlates of food addiction in obese individuals seeking bariatric surgery. *Clinical obesity, 4*(4), 228–236.

Meule, A., Hermann, T., & Kübler, A. (2015). Food addiction in overweight and obese adolescents seeking weight-loss treatment. *European Eating Disorders Review, 23*(3), 193–198.

Meule, A., Lutz, A., Vögele, C., & Kübler, A. (2012). Women with elevated food addiction symptoms show accelerated reactions, but no impaired inhibitory control, in response to pictures of high-calorie food-cues. *Eating Behaviors, 13*(4), 423–428.

Meule, A., Müller, A., Gearhardt, A. N., & Blechert, J. (2017b). German version of the Yale food addiction scale 2.0: Prevalence and correlates of 'food addiction'in students and obese individuals. *Appetite, 115*, 54–61.

Meule, A., Vögele, C., & Kübler, A. (2011). Psychometrische evaluation der deutschen Barratt impulsiveness scale–Kurzversion (BIS-15). *Diagnostica, 57*(3), 126–133.

Milivojevic, D., Milovanovic, S. D., Jovanovic, M., Svrakic, D. M., Svrakic, N. M., Svrakic, S. M., & Cloninger, C. R. (2012). Temperament and character modify risk of drug addiction and influence choice of drugs. *American Journal on Addictions, 21*(5), 462–467.

Moss, M. (2014). *Salt sugar fat: How the food giants hooked us*. Signal.

Murphy, J. G., & MacKillop, J. (2006). Relative reinforcing efficacy of alcohol among college student drinkers. *Experimental and Clinical Psychopharmacology, 14*, 219–227.

Murphy, C. M., & Mackillop, J. (2014). Mindfulness as a strategy for coping with cue-elicited cravings for alcohol: An experimental examination. *Alcoholism: Clinical and Experimental Research, 38*. https://doi.org/10.1111/acer.12322.

Murphy, C. M., Stojek, M. K., & MacKillop, J. (2014). Interrelationships among impulsive personality traits, food addiction, and Body Mass Index. *Appetite, 73*, 45–50. https://doi.org/10.1016/j.appet.2013.10.008.

Nathan, P. E. (1988). The addictive personality is the behavior of the addict. *Journal of Consulting and Clinical Psychology, 56*, 183–188.

Nesse, R. M., & Berridge, K. C. (1997). Psychoactive drug use in evolutionary perspective. *Science, 278*, 63–66.

Nunes-Neto, P. R., Köhler, C. A., Schuch, F. B., Solmi, M., Quevedo, J., Maes, M., … McElroy, S. L. (2018). Food addiction: Prevalence, psychopathological correlates and associations with quality of life in a large sample. *Journal of Psychiatric Research, 96*, 145–152.

Odlaug, B. L., & Grant, J. E. (2010). Impulse-control disorders in a college sample: Results from the self-administered Minnesota impulse disorders interview (MIDI). *Primary Care Companion to the Journal of Clinical Psychiatry, 12*(2).

Ouellette, A.-S., Rodrigue, C., Lemieux, S., Tchernof, A., Biertho, L., & Bégin, C. (2017). An examination of the mechanisms and personality traits underlying food addiction among individuals with severe obesity awaiting bariatric surgery. *Eating and Weight Disorders-Studies on Anorexia, Bulimia and Obesity, 22*(4), 633–640.

Oullette, J. A., Gerrard, M., Gibbons, F., & Reis-Bergan, M. (1999). Parents, peers, and prototypes: Antecedents of adolescent alcohol expectancies, alcohol cosumption, and alcohol-related life problems in rural youth. *Psychology of Addictive Behaviors, 13*, 183–197.

Patton, J. H., Stanford, M. S., & Barratt, E. S. (1995). Factor structure of the Barratt impulsiveness scale. *Journal of Clinical Psychology, 51*(6), 768–774.

Pivarunas, B., & Conner, B. T. (2015). Impulsivity and emotion dysregulation as predictors of food addiction. *Eating Behaviors, 19*, 9–14.

Qasim, A., Turcotte, M., de Souza, R. J., Samaan, M. C., Champredon, D., Dushoff, J., … Meyre, D. (2018). On the origin of obesity: Identifying the biological, environmental and cultural drivers of genetic risk among human populations. *Obesity Reviews, 19*, 121–149. https://doi.org/10.1111/obr.12625.

Raymond, K.-L., & Lovell, G. P. (2015). Food addiction symptomology, impulsivity, mood, and body mass index in people with type two diabetes. *Appetite, 95*, 383–389.

Rose, M. H., Nadler, E. P., & Mackey, E. R. (2018). Impulse control in negative mood states, emotional eating, and food addiction are associated with lower quality of life in adolescents with severe obesity. *Journal of Pediatric Psychology, 43*(4), 443–451.

Sharma, L., Markon, K. E., & Clark, L. A. (2014). Toward a theory of distinct types of "impulsive" behaviors: A meta-analysis of self-report and behavioral measures. *Psychological Bulletin, 140*, 374–408. https://doi.org/10.1037/a0034418.

Spinella, M. (2007). Normative data and a short form of the Barratt impulsiveness scale. *International Journal of Neuroscience, 117*(3), 359–368.

Stanford, M. S., Mathias, C. W., Dougherty, D. M., Lake, S. L., Anderson, N. E., & Patton, J. H. (2009). Fifty years of the Barratt impulsiveness scale: An update and review. *Personality and Individual Differences, 47*(5), 385–395. https://doi.org/10.1016/j.paid.2009.04.008.

Steinberg, L., Sharp, C., Stanford, M. S., & Tharp, A. T. (2013). New tricks for an old measure: The development of the Barratt impulsiveness scale–brief (BIS-Brief). *Psychological Assessment, 25*(1), 216.

Stojek, M. K., Fischer, S., & MacKillop, J. (2015). Stress, cues, and eating behavior. Using drug addiction paradigms to understand motivation for food. *Appetite, 92*. https://doi.org/10.1016/j.appet.2015.05.027.

Stojek, M. M. K., & MacKillop, J. (2017). Relative reinforcing value of food and delayed reward discounting in obesity and disordered eating: A systematic review. *Clinical Psychology Review, 55*. https://doi.org/10.1016/j.cpr.2017.04.007.

Stotts, A. L., & Northrup, T. F. (2015). The promise of third-wave behavioral therapies in the treatment of substance use disorders. *Current Opinion in Psychology, 2*, 75–81.

Tangney, J. P., Baumeister, R. F., & Boone, A. L. (2004). High self-control predicts good adjustment, less pathology, better grades, and interpersonal success. *Journal of Personality, 72*(2), 272–322.

Telch, C. F., Agras, W. S., & Linehan, M. M. (2001). Dialectical behavior therapy for binge eating disorder. *Journal of Consulting and Clinical Psychology, 69*(6), 1061.

The G. B. D. Obesity Collaboration, Ng, M., Fleming, T., Robinson, M., Thomson, B., Graetz, N., … Gakidou, E. (2014). Global, regional and national prevalence of overweight and obesity in children and adults 1980-2013: A systematic analysis. *Lancet, 384*(9945), 766–781. https://doi.org/10.1016/S0140-6736(14)60460-8.

Tice, D. M., Bratslavsky, E., & Baumeister, R. F. (2001). Emotional distress regulation takes precedence over impulse control: If you feel bad, do it! *Journal of Personality and Social Psychology, 80*(1), 53.

VanderBroek-Stice, L., Stojek, M. K., Beach, S. R., & MacKillop, J. (2017). Multidimensional assessment of impulsivity in relation to obesity and food addiction. *Appetite, 112*, 59–68.

Volkow, N., Wang, G. J., Tomasi, D., & Baler, R. D. (2013). Obesity and addiction: Neurobiological overlaps. *Obesity Reviews, 14*(1), 2–18.

Volkow, N., & Wise, R. (2005). How can drug addiction help us understand obesity? *Nature Neuroscience, 8*(5), 555–560. https://doi.org/10.1038/nn1452.

Vries, S. K., & Meule, A. (2016). Food addiction and bulimia nervosa: New data based on the Yale food addiction scale 2.0. *European Eating Disorders Review, 24*(6), 518–522.

Whiteside, S. P., & Lynam, D. R. (2001). The Five Factor Model and impulsivity: Using a structural model of personality to understand impulsivity. *Personality and Individual Differences, 30*(4), 669–689.

Whiteside, S. P., & Lynam, D. R. (2003). Understanding the role of impulsivity and externalizing psychopathology in alcohol abuse: Application of the UPPS impulsive behavior scale. *Experimental and Clinical Psychopharmacology, 11*, 210–217.

Whiteside, S. P., Lynam, D. R., Miller, J. D., & Reynolds, S. K. (2005). Validation of the UPPS impulsive behavior scale: A four-factor model of impulsivity. *European Journal of Personality, 19*, 559–574.

Wills, T. A., Walker, C., Mendoza, D., & Ainette, M. G. (2006). Behavioral and emotional self-control: Relations to substance use in samples of middle and high school students. *Psychology of Addictive Behaviors, 20*(3), 265.

Wing, R. R., Tate, D. F., Espeland, M. A., et al. (2016). Innovative self-regulation strategies to reduce weight gain in young adults: The study of novel approaches to weight gain prevention (snap) randomized clinical trial. *JAMA Internal Medicine*. https://doi.org/10.1001/jamainternmed.2016.1236.

Wing, R. R., Tate, D. F., Gorin, A. A., Raynor, H. A., & Fava, J. L. (2006). A self-regulation program for maintenance of weight loss. *New England Journal of Medicine, 355*(15), 1563–1571. https://doi.org/10.1056/NEJMoa061883.

Wise, R. A., & Bozarth, M. A. (1987). A psychomotor stimulant theory of addiction. *Psychological Review, 94*, 469–492.

de Wit, H., & Richards, J. B. (2004). Dual determinants of drug use in humans: Reward and impulsivity. *Current Theory and Research in Motivation, 50*, 19–55.

Woicik, P. A., Stewart, S. H., Pihl, R. O., & Conrod, P. J. (2009). The substance use risk profile scale: A scale measuring traits linked to reinforcement-specific substance use profiles. *Addictive Behaviors, 34*(12), 1042–1055. https://doi.org/10.1016/j.addbeh.2009.07.001.

Wolz, I., Granero, R., & Fernández-Aranda, F. (2017). A comprehensive model of food addiction in patients with binge-eating symptomatology: The essential role of negative urgency. *Comprehensive Psychiatry, 74*, 118–124.

Wolz, I., Hilker, I., Granero, R., Jiménez-Murcia, S., Gearhardt, A. N., Dieguez, C., … Fernández-Aranda, F. (2016). "food addiction" in patients with eating disorders is associated with negative urgency and difficulties to focus on long-term goals. *Frontiers in Psychology, 7*, 61.

World Health Organization. (1992). *International statistical classification of diseases and related health problems* (10th revision. ed.) Geneva.

CHAPTER

Reward processing in food addiction and overeating

Katherine R. Naish, Iris M. Balodis

Peter Boris Centre for Addictions Research, Department of Psychiatry and Behavioural Neurosciences, McMaster University, St. Joseph's Healthcare Hamilton, Hamilton, ON, Canada

If feeding were controlled solely by homeostatic mechanisms, most of us would be at our ideal body weight, and people would consider feeding like breathing or elimination, a necessary but unexciting part of existence.

Saper, Chou, and Elmquist (2002).

Introduction

Eating behavior is determined by more than just energy needs (Saper et al., 2002). While food intake is critical for survival, it is also an enjoyable experience for most of us. Many people have a favorite food, and it is common to expend energy and money on a good meal. Food intake is influenced by a multitude of social and cognitive factors, from who we dine with (Cruwys, Bevelander, & Hermans, 2015) and perceived social norms around eating (Herman & Polivy, 2005), to our memory of a recent meal (Higgs, 2005). Eating is also affected by a person's emotional state (Macht, 2008). In addition, certain foods—particularly those high in fat and sugar (Drewnowski & Greenwood, 1983)—are enjoyed more than others, leading to overconsumption of these foods in some individuals (Drewnowski, 1997). While high motivation to seek out and eat high-energy foods is adaptive when food is scarce, it becomes problematic in the modern Western world where food is widely accessible and designed to be as attractive and palatable as possible.

Food intake is influenced by both homeostatic and hedonic factors, with both hunger and satiety signals along with hedonic properties of food guiding what we eat (Fig. 8.1). The homeostatic–hedonic interaction is biased toward a positive energy balance, such that nonhomeostatic factors often promote food intake beyond the point at which an individual's energy requirements have been met (Leigh & Morris, 2018; Saper et al., 2002). Eating food is intrinsically rewarding; individuals are motivated to expend energy and effort to receive food (Hernandez & Hoebel, 1988). Like other reinforcing stimuli, food is associated with enhanced activity in brain regions associated with reward (Volkow, Wang, & Baler, 2011). In a similar way to drugs of abuse, perceiving and consuming palatable food elicits activity in the brain's reward system (Volkow et al., 2008). The overlap between brain activation

Compulsive Eating Behavior and Food Addiction. https://doi.org/10.1016/B978-0-12-816207-1.00008-1
Copyright © 2019 Elsevier Inc. All rights reserved.

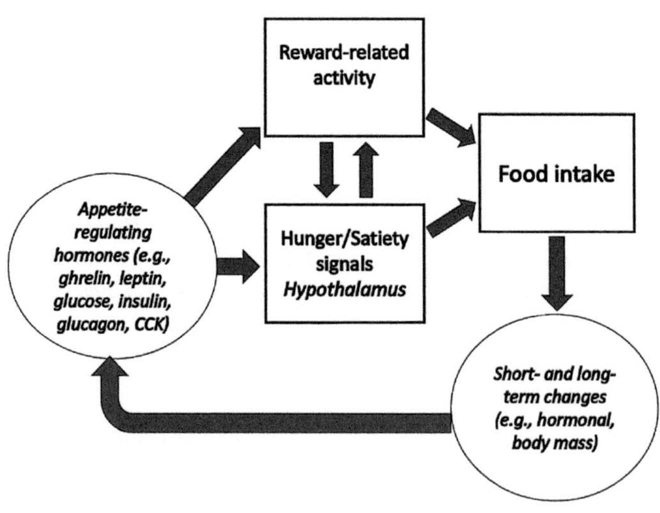

FIGURE 8.1

Schematic interacting influences of hedonic and homeostatic influences on food intake. Other factors known to influence food intake (e.g., social influences, mood, stress, inhibitory control) are not shown here.

associated with palatable food and addictive substances led to the notion that people might become addicted to specific foods in the same way as they can become addicted to drugs (e.g., Hernandez & Hoebel, 1988; Kenny, 2011; Volkow & Wise, 2005; Volkow et al., 2008; Wang et al., 2004a).

The tendency to eat beyond our metabolic needs is evident in the rising prevalence of obesity, which is largely driven by an imbalance between calories consumed and calories expended (World Health Organization, 2017). The wide availability and prominence of highly palatable foods in modern Western societies accounts partly for this, but what makes some people more susceptible to overeating than others? This chapter examines the role of reward processing in compulsive eating and "food addiction" by exploring what is known about reward processing in individuals with these and related conditions characterized by disordered eating. As a standardized measure of food addiction was developed only recently (Gearhardt, Corbin, & Brownell, 2009), few studies have assessed reward-related neural processing in relation to this construct. As such, our chapter draws from research on conditions associated with overeating or food addiction, including overweight and obesity, binge eating disorder (BED), and Prader–Willi syndrome (PWS).

In the "*Reward processing in eating behavior*" section below, we discuss the basic reward response to food stimuli and how this interacts with normal homeostatic regulation of eating. In the following section ("*Food addiction, obesity, and binge eating disorder*"), we define the term "food addiction" and discuss its relationship to obesity and BED. We then examine what is known about reward processing in each of these conditions and how this might contribute to eating psychopathology in each of them.

Reward processing in eating behavior

Food reward

The notion of "pleasure centers" in the brain was first proposed in the 1950s (Olds, 1956), when it was discovered that an animal will work to receive electrical stimulation of specific brain regions (Olds & Milner, 1954; for a review, see Berridge & Kringelbach, 2008). Animals are motivated to work for food reward in a similar way, and later work suggested the importance of dopamine in the reinforcing effects of food and addictive substances (Fibiger, 1978; German & Bowden, 1974; Wise, 1978; for a review, see; Chiara & Bassareo, 2007). Food and other reinforcing stimuli are associated with increased dopamine in the nucleus accumbens (NAc) in animals (Day, Roitman, Wightman, & Carelli, 2007; Hernandez & Hoebel, 1988; Mark, Smith, Rada, & Hoebel, 1994). Likewise, in humans, dopaminergic projections from the ventral tegmental area (VTA) to the striatum seem to play a particularly important role in food reward (for a review, see Berridge, 2009).

Visual and olfactory food cues elicit increased striatal dopamine transmission (Volkow et al., 2002), and dopamine release is positively associated with enjoyment of a meal (Small, Jones-Gotman, & Dagher, 2003). Interestingly, dopamine agonists *reduce* both food intake (Goldfield, Lorello, & Doucet, 2007; Leddy et al., 2004; Rush et al., 2005) and visual attention toward food cues (Nathan et al., 2012). Taken together, these findings could suggest that "reward-seeking" behavior (such as consuming palatable foods) serves the purpose of increasing striatal dopamine levels. This idea and the related "reward deficiency hypothesis" of obesity are discussed in more detail later in the chapter. Projections from the VTA to the dorsal striatum, lateral hypothalamus, orbitofrontal cortex (OFC), anterior cingulate cortex (ACC), and limbic regions (amygdala and hippocampus) are also implicated in reward processing (Volkow et al., 2011). For example, the OFC and amygdala are thought to represent the current reward value of a food based on homeostatic signals (Siep et al., 2009; Small, Zatorre, Dagher, Evans, & Jones-Gotman, 2001).

One of the most prominent biopsychological theories of addiction emphasizes neuroadaptations occurring with repeated drug use, in particular changes in dopamine transmission that concurrently imbue drug-related stimuli with salience (Berridge & Robinson, 2003; Robinson & Berridge, 1993). A main tenet of "incentive-sensitization" theory posits that different brain circuits may underlie specific reward components. For example, a hyperactive striatal responding to drug cues underlies the "wanting" or craving associated with drugs, while different neural substrates underlie the hedonic properties or "liking" of a drug (Berridge & Robinson, 2003).

Parsing reward into distinct components (i.e., anticipation vs. outcome) has shed light on the role of anticipatory processes influencing decision-making. In particular, alterations in anticipatory signaling could influence choice behavior by increasing sensitivity to some cues (e.g., drug cues), while blunting sensitivity to others (e.g., nondrug cues). Similar reward neurocircuitry is thought to underlie psychological aspects of food reward; for example, strong food cravings or "wanting" reported in individuals with disordered eating may also link with greater anticipatory signaling of food cues.

Indeed, distinct neural systems appear to underlie anticipation of a reward (e.g., food craving or anticipated receipt of food) and the actual receipt of the reward (e.g., consuming a palatable food). Animal and human studies show increased striatal activation during reward anticipation, but increased medial prefrontal cortex (PFC) activity during the outcome, or consummatory, phase of reward (Knutson, Fong, Adams, Varner, & Hommer, 2001; Rademacher et al., 2010). In food reward, the ventral striatum, amygdala, midbrain, and thalamus show greater involvement in anticipatory processing, while insula and OFC activity are implicated to a greater extent in reward receipt (Stice, Spoor, Ng, & Zald, 2009). It is important to distinguish anticipatory from consummatory reward processes in studies of food and other addictions because these processes likely play different roles in behavior. In particular, neural processes underlying anticipation are ideally positioned in time to influence decision-making around food intake. Anticipatory reward processing is associated with craving, impulsivity, and treatment outcomes (Balodis et al., 2014; Beck et al., 2009; Wrase et al., 2007), and it is believed to be a stronger predictor of actual food consumption than consummatory processing (Epstein et al., 2007). It has been suggested that an enhanced sensitivity to the rewarding properties of some food-associated cues could underlie overeating and obesity in humans (e.g., Rolls, 2011).

Interactions between homeostatic and reward-related mechanisms

Homeostatic processes guide food intake based on the body's energy and nutrient needs. Increases and decreases in appetite are driven by hormones released in the hypothalamus, which in turn are influenced by physiological changes that signal the body's energy supplies and needs (Marieb & Hoehn, 2010). Reward-related neural activity is closely interrelated with homeostatic processes (Fig. 8.1; Leigh & Morris, 2018; Lutter & Nestler, 2009; Volkow et al., 2011). For example, the peptide hormones ghrelin and leptin—perhaps best known for their role in appetite regulation—are also implicated in reward processing (Leigh & Morris, 2018). Ghrelin is produced in the stomach (and possibly the brain) and is involved in the *initiation* of eating (Schellekens, Finger, Dinan, & Cryan, 2012). Leptin, on the other hand, is an appetite-suppressing hormone. Leptin is involved primarily in the long-term regulation of eating, being released by adipose cells in response to increasing and decreasing fat stores. Receptors for both ghrelin and leptin are present in the VTA, and both affect dopamine transmission in the mesolimbic dopamine system. Ghrelin facilitates dopamine transmission in the VTA (Abizaid et al., 2006; Jerlhag et al., 2007), and direct ghrelin administration into the VTA increases striatal dopamine turnover (Abizaid et al., 2006). Ghrelin infusions are associated with enhanced activity in the amygdala, OFC, anterior insula, and striatum in healthy weight individuals (Lutter & Nestler, 2009), whereas ghrelin receptor antagonist administration suppresses reward-related activity associated with addictive substances (e.g., Jerlhag et al., 2009, 2010). In line with these findings, ghrelin is positively associated with reward-related neural activity elicited by viewing appetitive foods (Kroemer et al., 2013), as well as shifts in dietary preference toward foods high in fat and sugar (Disse et al., 2010; Shimbara et al., 2004). Thus, in addition to facilitating hunger via the hypothalamus, ghrelin might increase the perceived reward value of foods through its action in the dopaminergic reward system.

Leptin also modulates the reward value of food and other reinforcing stimuli (for reviews, see Figlewicz, 2003; Opland, Leinninger, & Myers, 2010). In contrast to ghrelin, leptin inhibits dopamine neuron activity in the VTA, leading to decreases in both basal dopamine levels and dopamine release associated with food intake (see Lutter & Nestler, 2009, for a review). When viewing food cues, plasma leptin concentrations are positively associated with striatal activation (Grosshans et al., 2012). In cases of congenital leptin deficiencies, leptin replacement therapy decreases reward-related neural activity in response to food cues, as well as hunger and liking ratings associated with viewing food images (Farooqi et al., 2007). Thus, in a similar way to ghrelin, leptin increases associated with increased fat stores might decrease the hedonic value of food through the reward system.

Homeostatic–reward system interactions are demonstrated in the effects of hunger and satiety on reward-related activity. In rats, food deprivation increases the rate of intracortical self-stimulation of the lateral hypothalamus, while satiety decreases motivation for stimulation of this area (Hoebel, 1969; Margules & Olds, 1962). The OFC receives dopaminergic VTA input and is implicated in the processing of food reward and other reinforcing stimuli (Kringelbach, 2005). The responsivity of some of its neurons to taste has led to the conclusion that the OFC houses the secondary taste cortex, which—unlike the primary taste cortex—is sensitive to hunger and satiety levels, rather than representing taste exclusively. Specifically, OFC responses to taste are diminished when the animal is satiated (Rolls, Sienkiewicz, & Yaxley, 1989). In addition to dopaminergic projections from the VTA, the OFC receives input from a range of brain regions including primary sensory regions, the hypothalamus, and the limbic system (Öngür & Price, 2000). As responses of the OFC to taste are modulated by hunger and satiety, this region might be responsible for integrating sensory and homeostatic information to produce a reward response based on an organism's energy needs (Small et al., 2007).

Human studies also show effects of hunger and satiety on processing of food cues. Activity in areas including the medial OFC and striatum is greatest for high-calorie food cues when an individual is hungry, but higher for low-calorie foods when the person is satiated (Siep et al., 2009). Likewise, OFC and insula activity associated with chocolate intake in humans are reduced as the level of satiety increases (Small et al., 2001), while greater striatal dopamine release during food tasting was associated with hunger (Volkow et al., 2002). Indeed, hunger enhances the pleasurable experience of eating food, while satiety decreases food's hedonic value (Rolls et al., 2011).

In summary, the reward system plays a central role in food intake, which could underlie or contribute to compulsive eating and food addiction. In the rest of this chapter, we discuss reward processing in conditions associated with overconsumption of food or dependence on certain food types.

Conditions associated with overeating or compulsive behavior toward food

The terms "obesity," "food addiction," and "BED" are often incorrectly used interchangeably, and most studies do not assess all three constructs in the same individuals. Notably, the conditions do not always co-occur (e.g., Fairburn et al., 1998; Gearhardt et al., 2012; Pedram et al., 2013), and the presence of food addiction or BED affects treatment outcomes in individuals attempting to lose weight or change their eating behavior. As such, it is important to draw clear distinctions between these conditions (Davis, 2017) both behaviorally and when examining patterns of reward processing.

Food addiction refers to a dependence on food or eating, analogous to the dependence on addictive drugs seen in substance-use disorders (SUDs). The Yale Food Addiction Scale (YFAS; Gearhardt et al., 2009) was developed to assess food addiction based on the DSM-IV criteria for substance dependence (American Psychiatric Association, 2000), with self-report items tailored to measure dependence on food rather than a drug. The seven characteristics of food addiction included on the original version of the YFAS are (1) food taken in larger amount for longer period than intended (e.g., "I find that when I start eating certain foods, I end up eating much more than planned"); (2) persistent desire or repeated unsuccessful attempt to quit ("Not eating certain types of food or cutting down on certain types of food is something I worry about"); (3) much time/activity to obtain, use, recover (e.g., "I spend a lot of time feeling sluggish or fatigued from overeating"); (4) important social, occupational, or recreational activities given up or reduced (e.g., "There have been times when I consumed certain foods so often or in such large quantities that I started to eat food instead of working, spending time with my family or friends, or engaging in other important activities or recreational activities I enjoy"); (5) use continues despite knowledge of adverse consequences (e.g., "I kept consuming the same types of food or the same amount of food even though I was having emotional and/or physical problems"; (6) withdrawal symptoms (e.g., "I have found that I have elevated desire for or urges when I cut down or stop eating them"); and (7) tolerance (e.g., "Over time, I have found that I need to eat more and more to get the feeling I want, such as reduced negative emotions or increased pleasure"). Food addiction threshold is met if an individual endorses at least three of these seven symptoms together with significant distress or impairment. Severity can also be measured using a continuous symptom count to reflect the number (0–7) of food addiction symptoms present in an individual.

Food addiction is distinct from generalized overeating and is not necessarily associated with weight gain or obesity (e.g., Pedram et al., 2013). The prevalence of food addiction in healthy weight individuals is around 10%, while in overweight or obese individuals it is just below 25% (Pursey, Stanwell, Gearhardt, Collins, & Burrows, 2014). Food addiction is present in nearly half (47.8%) of individuals seeking treatment for compulsive overeating (Bégin et al., 2012) and in 57.8% of bariatric surgery candidates (Sevinçer, Konuk, Bozkurt, & Coşkun, 2016). In treatment-seeking individuals, those who meet the criteria for food addiction show higher levels of depression, impulsivity, food cravings, and eating psychopathology (e.g., Bégin et al., 2012; Meule, Hermann, & Kübler, 2015), as well as lower levels of weight loss when completing a weight-loss treatment program (Burmeister, Hinman, Koball, Hoffmann, & Carels, 2013) or undergoing weight-loss surgery (Clark & Saules, 2013). Heightened cravings and impulsivity are also hallmarks of SUDs—consistent with the fact that food addiction is assessed based on the diagnostic criteria for substance dependence.

Food addiction is consistently associated with binge eating, both in individuals with BED and those with subclinical bingeing (Gearhardt, White, Masheb, & Grilo, 2013, 2014, 2009; Bégin et al., 2012; Davis et al., 2011; Meule et al., 2015). Food addiction is considerably more prevalent in individuals with BED and bulimia nervosa (BN) compared with those without an eating disorder (Pursey et al., 2014); approximately 40%–60% of individuals with BED meet the full criteria for food addiction (Gearhardt et al., 2012, 2013, 2014). BED is an eating disorder characterized by recurrent episodes of binge eating, in which the individual eats more than would normally be eaten within the same (discrete) period of time. Binge eating must be accompanied by a feeling of lack of control to meet the criteria for BED. Individuals with BED might also eat more quickly than usual during these periods, eat beyond the point of feeling uncomfortably full, eat alone because of embarrassment, and feel guilty or depressed following a binge episode (American Psychiatric Association, 2013). Individuals with BED combined with food addiction have higher levels of depression and emotion dysregulation compared with those with BED who do not meet criteria for food addiction (Gearhardt et al., 2013). Finally, in a sample of obese patients with BED, food addiction symptom score predicted binge eating frequency above and beyond all other measures (Gearhardt et al., 2013).

The focus of our chapter is on reward processing in food addiction, obesity, and BED; however, we also touch on what is known about reward processing in PWS. PWS is a genetic disorder that is characterized, in part, by compulsive eating. The syndrome follows an interesting developmental trajectory; children with PWS display feeding difficulties and failure to thrive in early life, but at age 2–5 years they develop a preoccupation with food accompanied by excessive overeating (McAllister, Whittington, & Holland, 2011). Obesity-related health problems are among the most common causes of death in individuals with PWS, and deaths resulting directly from overeating (e.g., choking during a binge episode, stomach rupture) also account for a considerable proportion of the mortality rate (Butler, Manzardo, Heinemann, Loker, & Loker, 2017; McAllister et al., 2011).

Reward processing in food addiction

Few studies to date have examined the neurobiology of reward processing related to food addiction as measured by the YFAS. The first study of neural processes in food addiction was reported by Gearhardt et al. (2011), who assessed brain activity during both anticipation and receipt of a chocolate milkshake. Food addiction severity was positively correlated with activity in the medial OFC, ACC, and amygdala during anticipatory processing. YFAS score did not correlate with activity in any brain region during consummatory processing; however, comparing individuals who were "low" and "high" in food addiction symptoms revealed different patterns of activity at both the anticipatory and consummatory phases. Compared with individuals with "low" food addiction ($\leq$1 symptom), "high" food addiction individuals ($\geq$3 YFAS symptoms) displayed greater dorsolateral prefrontal cortex (DLPFC) and caudate activity during anticipation of the milkshake, but lower lateral OFC activity during milkshake receipt (Gearhardt et al., 2011).

The association between food addiction severity and anticipatory processing in medial OFC, ACC, and amygdala (Gearhardt et al., 2011) might suggest heightened reward sensitivity and motivational salience of food in individuals with food addiction. Medial OFC and ACC activity correlate with trait reward sensitivity in individuals with BN and BED (Schienle, Schäfer, Hermann, & Vaitl, 2009), while amygdala activation is associated with hunger and desire for food during food cue exposure (Killgore et al., 2013). Heightened caudate activity in individuals with higher food addiction symptoms is consistent with findings on obesity; individuals who are obese show higher striatal activity during food cue exposure compared with their lower body mass index (BMI) counterparts (Rothermund et al., 2007; Stoeckel et al., 2008). As discussed in the following section, food receipt is associated with *lower* striatal activation in individuals with higher BMIs (Stice, Spoor, Bohon, & Small, 2008; Stice, Spoor, Bohon, Veldhuizen, & Small, 2008). If common mechanisms are involved in food addiction and obesity, this could explain why food addiction was associated with heightened striatal activity during anticipatory but not consummatory processing in Gearhardt and colleagues' study (Gearhardt et al., 2011).

The DLPFC responds to reward cues including high-calorie foods (Killgore et al., 2003; Wallis & Miller, 2003) and appears to play a role in reward expectancy (Wallis & Miller, 2003; Watanabe, Hikosaka, Sakagami, & Shirakawa, 2002). In monkeys, DLPFC activation during reward anticipation is predictive of subsequent behavioral responses to the reward (Wallis & Miller, 2003), suggesting a role for this region in reward-based responding. Interestingly, DLPFC responsivity to milkshake receipt is positively associated with dietary restraint (Burger & Stice, 2011). As the DLPFC is also implicated in inhibitory control, it is possible that food cue-related activity in this area reflects attempted control of responses to food cues. Indeed, upregulating activity in the DLPFC using neuromodulation reduces binge eating (Van den Eynde et al., 2010), food cravings (Goldman et al., 2011; Kekic et al., 2014), as well as both craving and consumption of addictive substances (for reviews, see Barr et al., 2011; Jansen et al., 2013).

Interestingly, food addiction was associated with heightened medial OFC activity during anticipation and lower lateral OFC activity during receipt of food (Gearhardt et al., 2011). Distinct roles of the medial and lateral OFC in reward processing have been described previously (O'Doherty et al., 2001; Small et al., 2001). In a study of chocolate consumption, Small et al. (2001) found that lateral OFC activity increased, but medial OFC activity decreased, as participants became more satiated and motivation for the chocolate decreased. In another study, rewarding outcomes on a task were associated with medial OFC activity, while punishing outcomes were associated with the lateral OFC (O'Doherty et al., 2001). In the SUD literature, drug cue exposure is associated with increases in medial OFC activity (e.g., Wang et al., 1999), but decreased activity in the lateral OFC (Goldstein et al., 2007). In this way, the lateral OFC may play an important role in suppressing previously rewarding responses and inhibitory control (Elliott, Dolan, & Frith, 2000).

There is some evidence of dopamine dysregulation in craving and overconsumption of foods associated with food addiction. Methylphenidate—a dopamine agonist—typically decreases food intake in both normal weight and obese individuals (Goldfield et al., 2007; Leddy et al., 2004; Rush et al., 2005). Individuals who meet food addiction criteria, however, fail to show the typical appetite-suppressant effects of this drug (Davis, Levitan, Kaplan, Kennedy, & Carter, 2014), perhaps because of underlying differences in the dopamine system in individuals with food addiction. In an earlier study, Davis et al. found that individuals with food addiction had higher scores on a composite genetic index associated with elevated striatal dopamine signaling. Scores on the genetic index were also positively correlated with emotional eating, binge eating, and food cravings (Davis et al., 2013). More prospective research is needed to clarify whether altered dopaminergic signaling represents a precursor to food addiction or whether specific food intake patterns influence these responses.

Reward processing in overweight and obesity

Striatal reward processing in overweight/obesity

The mesolimbic dopamine pathway from the VTA to the striatum is perhaps the most strongly linked to reward. Both the ventral and dorsal striatum receive dopaminergic input from the VTA and are implicated in both food and nonfood reward processing. Reward-related striatal responses are linked with BMI; compared with healthy weight controls, individuals with obesity show greater striatal activity when viewing images of high-calorie food (Rothemund et al., 2007; Stoeckel et al., 2008). Indeed, BMI has been identified as a significant predictor of reward-related activity during food cue exposure (Rothemund et al., 2007). It is possible that enhanced striatal activity to food cues and the anticipation of food reward contribute to overconsumption of high-calorie foods, leading to a higher BMI, in these individuals.

Altered striatal activity is also observed during the consummatory phase of reward. In contrast to the pattern observed for anticipatory processing, however, higher BMI is associated with *lower* striatal activity during receipt of reward (Stice, Spoor, Bohon, & Small, 2008; Stice, Spoor, Bohon, Veldhuizen et al., 2008). Compared with healthy weight controls, females with a BMI in the obese range showed less caudate activation in response to tasting a palatable chocolate milkshake. This attenuated striatal activation was accompanied by enhanced activation in the gustatory and somatosensory regions of the brain during food intake (Stice, Spoor, Bohon, Veldhuizen et al., 2008). Furthermore, BMI was positively correlated with activation in gustatory and somatosensory regions during food intake and anticipation of intake, but negatively correlated with activation in the caudate nucleus associated with tasting a palatable food (Stice, Spoor, Bohon, Veldhuizen et al., 2008; see also Stice, Spoor, Bohon, & Small, 2008). This combination of heightened anticipatory processing in gustatory and somatosensory areas combined with lower striatal reward activity during outcome processing could represent a risk factor for overeating and weight gain (Stice, Spoor, Bohon, Veldhuizen et al., 2008). There is evidence that the relationship between BMI and striatal responsivity is moderated by a specific allele associated with reduced striatal D2 receptors. A study by Stice, Yokum, Bohon, Marti, and Smolen (2010) found that striatal activity when participants imagine consuming a highly palatable food predicts increased BMI 1 year later, but only in those individuals with a genotype associated with hypofunction in the dopamine system. These findings suggest that altered dopamine and striatal signaling could contribute to overeating and weight gain and point to the existence of distinct subgroups at risk for obesity.

Both human and animal studies have shown altered dopamine transmission and dopamine receptor density associated with body weight (Michaelides, Thanos, Volkow, & Wang, 2012; Orosco, Rouch, & Nicolaidis, 1996; Thanos, Michaelides, Piyis, Wang, & Volkow, 2008; Volkow et al., 2008; Wang et al., 2001). More specifically, animal studies show that obesity and overconsumption of high-fat foods are associated with downregulation of striatal dopamine receptors (Johnson & Kenny, 2010) and decreased dopamine transmission in the VTA (Cordeira, Frank, Sena-Esteves, Pothos, & Rios, 2010). Human studies also demonstrate reduced striatal dopamine receptor levels associated with overweight and obesity (Volkow et al., 2008; Wang et al., 2001; de Weijer et al., 2011), and dopamine-blocking drugs are associated with weight gain (Baptista, 1999). Interestingly, increases in dopamine transmission to food cues have also been associated with *increased* food intake and responsivity to food stimuli. For example, striatal dopamine release in response to food stimuli is significantly higher in individuals with BED compared with those without BED and is positively correlated with binge eating severity (Wang et al., 2011). Furthermore, striatal dopamine receptor availability is positively associated with glucose metabolism in the DLPFC, medial OFC, anterior cingulate gyrus, and somatosensory cortex, suggesting that a reduction in dopamine transmission affects activity in prefrontal regions during food cue exposure (Volkow et al., 2008).

Dopamine plays a key role in reward sensitivity, with higher dopamine associated with a greater ability to experience pleasure and reward and lower dopamine availability linked to low motivation (Davis, Strachan, & Berkson, 2004). Reduced dopamine transmission and receptors in obesity have led to the "reward deficiency hypothesis," which posits that overeating (and other addictive behaviors) arises as a compensatory mechanism to counteract a sluggish reward system (Blum, Cull, Braverman, & Comings, 1996; Wang et al., 2001). Accordingly, individuals who experience pleasurable effects associated with a dopamine agonist (methylphenidate) have significantly lower levels of dopamine receptors compared with those who experience negative effects of the drug (Volkow et al., 1999). In this way, individuals with higher dopamine receptor availability might find too much stimulation (i.e., elicited by a highly rewarding stimulus) aversive, while individuals with lower receptor availability find the same stimulus pleasurable (Volkow et al., 1999).

In summary, in individuals classified as overweight or obese, there is evidence for diminished striatal dopamine receptor availability (e.g., Volkow et al., 2008; Wang et al., 2001; de Weijer et al., 2011) but *higher* reward-related activity in the dopaminergic system in response to food cues. At this point, it is unclear whether reduced dopamine availability is a precursor for obesity or a neuroadaptation from overeating and binge behavior. Furthermore, overweight and obesity are linked with heightened reward network activity during anticipatory processing to food cues but decreased response to actual food intake. Enhanced food cue responses may therefore increase motivation for food, but weaker responses associated with food intake may reflect a blunted consummatory reward response that could explain the overconsumption of palatable foods (Volkow et al., 2011).

Prefrontal and OFC reward processing in overweight/obesity

The OFC receives input from a number of brain regions, including primary sensory regions (e.g., primary taste cortex), the hypothalamus, and the limbic system (Öngür & Price, 2000). Activation in this region in response to food cues or food intake relies on both sensory information about the food and satiety signals reflecting the motivational state of the individual. Neuronal populations in the OFC respond to taste, texture, and temperature qualities of food, but—unlike primary sensory regions—respond differentially depending on whether an individual is hungry or satiated (Rolls et al., 1989). In monkeys, OFC lesions disrupt normal food preferences and reward-based learning (for a review, see Rolls, 2004). The OFC is also implicated in learned associations about stimuli, with damage to this region affecting an individual's ability to form associations between a stimulus and its outcome (Schoenbaum & Roesch, 2005).

In humans, higher BMI is associated with reduced OFC volume (Cohen, Yates, Duong, & Convit, 2011; Maayan, Hoogendoorn, Sweat, & Convit, 2011; Shott et al., 2015; Walther, Birdsill, Glisky, & Ryan, 2010). For example, one study found that overweight and obese individuals had lower OFC volume compared with lean individuals and that OFC volume was associated with the type of food that participants reported consuming (Cohen et al., 2011). More specifically, OFC volume positively correlated with the reported amounts of "high-quality" foods (defined as farm produce, fish, whole grains, and nuts) that participants reported eating. The association between OFC volume and dietary intake could suggest that food choice is a mediating factor between OFC structure and body weight, with dysfunction in the OFC influencing dietary choices that lead to weight gain. A study of adolescents reported by Maayan et al. (2011) also found lower OFC volume in obese compared with lean individuals, while a study of females aged 55 years and older found a negative correlation between BMI and gray matter volume in several regions including the left OFC (Walther et al., 2010). The study of adolescents also revealed an association between OFC volume and disinhibited eating, showing that individuals with a smaller OFC volume were more disinhibited in their eating behavior (Maayan et al., 2011). Notably, disinhibited eating is positively associated with food addiction (Gearhardt, Corbin, & Brownell, 2016).

While the aforementioned studies suggest that higher BMI is associated with smaller volume or weaker connectivity of the OFC, Killgore et al. (2013) found a *positive* association between BMI and OFC responses to food cue exposure. Participants showed larger regional responses in the OFC, amygdala, and insula when viewing high-calorie compared with low-calorie food images, and the activation in these regions significantly predicted body size. It should be noted that this association was seen only for females; male participants showed no relationship between BMI and neural responses to food cues (Killgore et al., 2013). Discrepancies in the findings of studies examining OFC structure and function in relation to BMI could reflect the diverse roles of different parts of the OFC and its connections. For example, one study found that viewing food images was associated with stronger OFC–striatum connectivity, but weaker OFC–amygdala connectivity, in individuals who were obese compared with those who were lean (Stoeckel et al., 2009). This activation pattern could reflect a stronger reward value associated with the food cues, accompanied by lower input from areas representing emotional and affective associations with the stimulus. As discussed in the previous section, it is well-documented that the medial and lateral OFC are differentially involved in reward processing (e.g., O'Doherty et al., 2001; Small et al., 2001).

The OFC is strongly implicated in reward-based learning, so it is possible that structural or functional differences seen in obesity are related to differences in how affective or physiological associations with food are formed. A study by Shott et al. (2015) found that OFC volume was related to taste response learning in healthy weight controls but not in people with obesity. In healthy weight controls, higher OFC volume was associated with greater activity in the frontal cortex, limbic regions, hypothalamus, ACC, and insula during a reward-learning task; however, this relationship was not present in individuals with obesity. This result might suggest weaker signaling between the OFC and other brain areas involved in taste reward learning in individuals with higher BMIs. Animal studies show that OFC lesions disrupt the ability to update conditioned responding when the value of a reward changes (e.g., Izquierdo, Suda, & Murray, 2004; Zeeb & Winstanley, 2013). While satiety signaling may be altered with OFC damage, there is some evidence suggesting more fundamental impairments in unlearning reward contingency outcomes (Izquierdo et al., 2004). It is possible that disturbances to OFC function or structure in humans promote overeating by disrupting the ability to "unlearn" associations between the perception (e.g., sight and smell) of a palatable food and its reward value as satiety is reached. That is, although an individual might feel satiated, a food that is strongly associated with reward at the beginning of an eating episode does not become devalued as satiety increases.

Reward processing in BED

Striatal reward processing in BED

Individuals with BED show distinct reward-related activity in response to both food and nonfood cues compared with individuals without BED. One study of nonfood reward used the monetary incentive delay task to examine neural activity during anticipation of monetary reward in individuals with BED and in lean and obese (BMI-matched) controls (Balodis et al., 2013a). During anticipation of a monetary reward, the BED group showed diminished ventral striatal responses relative to the non-BED obese group. These findings highlight distinct anticipatory processing between subgroups with obesity, suggesting that individuals with BED show reduced recruitment of networks involved in reward processing and self-regulation. These findings also related to treatment outcome in the treatment-seeking BED group: those patients who were still binge eating following treatment showed lower anticipatory activation of the ventral striatum and inferior frontal gyrus compared with those who had stopped bingeing at the end of treatment (Balodis et al., 2014). This finding implicates hypocorticostriatal functioning in bingeing outcomes in individuals with BED. The finding could indicate more severe pathology in individuals with lower anticipatory striatal processing of reward cues; however, it should be noted that the groups did not differ in binge eating frequency at the beginning of treatment. Notably, blunted anticipatory processing of monetary reward cues is noted in other disorders characterized by impulse control problems, including smoking (Peters et al., 2011), alcohol-use disorder (Beck et al., 2009), and gambling disorder (Balodis et al., 2012; Reuter et al., 2005). Reduced anticipatory ventral striatal activity is associated with higher impulsivity and craving in individuals with addictive disorders (Balodis et al., 2012; Beck et al., 2009; Wrase et al., 2007).

Persistent binge eating following treatment for BED is also associated with reduced medial PFC activity during reward outcome processing (Balodis et al.,

2014), which could reflect hypofunctioning in networks involved in inhibitory control in individuals with treatment-resistant binge eating. Indeed, higher BMI is associated with reduced prefrontal activation during inhibitory control tasks (Batterink et al., 2010), and successful dieters show enhanced activity in prefrontal regions during meal consumption compared with nondieters (DelParigi et al., 2007). In addition, dietary restraint is positively associated with increased activity in the dorsal PFC during consummatory food processing (Burger & Stice, 2011; DelParigi et al., 2007). Reduced inhibitory control activity in response to reward outcomes has also been noted in alcohol-use disorder (Bogg, Fukunaga, Finn, & Brown, 2012; Forbes, Rodriguez, Musselman, & Narendran, 2014). These findings implicate reduced inhibitory control in individuals who engage in overeating or addictive behaviors. Reduced striatal responding during anticipatory reward processing in individuals who continued bingeing following treatment, compared with those who stopped bingeing, highlights the existence of distinct subgroups of individuals with obesity and BED. These findings may clarify ambiguous striatal findings in research on obesity, particularly in studies where BED is not assessed. Importantly, the reward-related activity associated with BED treatment outcomes in Balodis et al. (2014) study was in response to monetary cues, suggesting that altered reward processing in BED is not restricted to food reward.

Other BED functional neuroimaging studies (which measure blood-oxygen-level-dependent or BOLD signal in the brain) also provide support for unique ventral striatal response patterns in BED populations. Weygandt, Schaefer, Schienle, and Haynes (2012) used multivariate pattern recognition to identify patterns of activation differentiating reward-related responses in individuals with BED from those with BN, as well as from healthy weight and overweight controls. The right ventral striatum differentiated individuals with BED from overweight controls; activation in the left ventral striatum best distinguished individuals with BED from those with BN. Specifically, in response to food cues, individuals with BED showed less ventral striatal activity compared with both the BN and obese control groups (Weygandt et al., 2012). BED is also associated with structural differences in the striatum; obese individuals with BED show a smaller brain volume in the left ventral striatum, bilateral caudate, and OFC compared with obese subjects without BED (Voon et al., 2015). The notion that reduced striatal signaling contributes to the symptoms of BED and is also supported by the observation that atrophy in the striatum, among other regions, is associated with compulsive overeating in individuals with frontotemporal dementia (Woolley et al., 2007).

However, there is also evidence of *heightened* striatal responses to food cues in individuals with BED compared with healthy controls. For example, Wang et al. (2011) found that individuals with BED had higher extracellular dopamine levels in the caudate nucleus during food stimuli perception compared with individuals without BED. Furthermore, dopamine release in the caudate nucleus is positively associated with binge eating symptom severity, but not BMI (Wang et al., 2011). This finding is important because it suggests a relationship between striatal dopamine and binge eating, but not weight per se, highlighting the importance of assessing BED and other eating psychopathology when studying samples of individuals with obesity. Similarly, Lee and colleagues found heightened activity in the ventral striatum associated with food cue exposure in individuals with BED compared with healthy control participants (Lee, Namkoong, & Jung, 2017). Finally, binge eating severity in individuals with moderate binge eating symptoms is positively related to functional connectivity within the reward system during high-calorie taste cue exposure (Filbey, Myers, & DeWitt, 2012), suggesting that binge eating is associated with heightened reward network communication when exposed to palatable food cues.

Although the findings on striatal functioning in BED are mixed, they provide evidence for altered striatal dopamine network processing in individuals with binge eating symptoms compared with those without. This is similar to findings on neural processing in SUDs, which strongly implicate frontostriatal reward pathways in addictive behaviors (e.g., Volkow, Fowler, Wang, & Swanson, 2004). It is possible that striatal alterations in both eating and SUDs occur with symptom severity and illness duration. Blunted reward responses in addictive disorders might result from repeated activation of the system, leading to neuroadaptations to rewarding stimuli (Davis et al., 2004). Behaviorally, this manifests as tolerance, whereby individuals with chronic addictive behaviors require increasingly great stimulation (e.g., more of a palatable food or addictive substance) to elicit the same rewarding effect. Particular alterations may occur at specific phases of the disorder; severe or chronic food addiction or compulsive eating might be associated with a dampened reward response, while individuals with milder symptoms might show enhanced responses.

In line with work on overweight and obesity in individuals without eating disorders, the findings on BED suggest that binge eating might be driven partly by altered striatal signaling associated with food cues. Differences between individuals with BED and those with high BMIs but no eating disorder symptoms demonstrate that reward function alterations are not a product of body mass alone. Nevertheless, comparisons of healthy individuals in different BMI ranges indicate that both body weight and binge eating pathology are associated with changes in the striatal reward system.

Prefrontal and OFC reward processing in BED

A number of studies implicate OFC alterations in BED (Balodis et al., 2013a,b; Schäfer, Vaitl, & Schienle, 2010; Schienle et al., 2009; Voon et al., 2015). Schienle et al. (2009) explored the relationship between reward sensitivity and neural responses to food stimuli in individuals with BED, BN, and both obese and normal weight controls. In line with previous imaging work (e.g., Beaver et al., 2006; Wang et al., 2004b), food cues activate the OFC, ACC, and insula across all participants. Compared with healthy weight and obese controls, food cues produce greater medial OFC activity in BED participants. Furthermore, medial OFC and ACC activity positively correlate with reward sensitivity—as indexed by scores on the Behavioral Activation Scale—in both the BN and BED patients, suggesting that these regions underlie heightened sensitivity to reward in these conditions.

A potential role for the medial OFC in BED was also highlighted by Schafer et al. (2010), who found that both individuals with BED or BN had greater gray matter volume in the medial OFC compared with healthy, normal weight controls. In contrast, another study of BED found smaller volume of the OFC (in addition to the left ventral striatum and bilateral caudate) in individuals with obesity and BED compared with individuals with obesity only (Voon et al., 2015). Voon et al. findings are consistent with the results of studies in overweight and obese populations (Cohen et al., 2011; Maayan et al., 2011; Shott et al., 2015; Walther et al., 2010). Reduced OFC volume is also in line with findings by Woolley et al. (2007) demonstrating an association between OFC atrophy and compulsive overeating in patients with frontotemporal dementia, suggesting a relationship between compulsive overconsumption of food and decreased OFC volume.

In the earliest reported study examining neural responses to food in individuals with BED (Karhunen et al., 2000), exposure to a freshly cooked meal was associated with increased blood flow in left prefrontal and frontal brain regions in obese females who endorsed binge eating. In contrast, non-BED obese participants showed the greatest increase in activity in the right hemisphere, while non-BED healthy weight participants showed no change in cerebral blood flow associated with food exposure. In addition, increased activity in the left frontal and prefrontal regions was positively associated with self-reported hunger and desire to eat, but only in participants with BED. It is worth noting that self-reported hunger is also associated with increases in striatal dopamine transmission in individuals without an eating disorder (Volkow et al., 2002), and hunger is known to enhance taste responses in the OFC (e.g., Rolls et al., 1989). The increased activation in frontal regions in BED could reflect greater activation of the OFC, which responds to sensory properties of food (e.g., taste) as well as information about hunger and satiety levels (Rolls et al., 2004). The correlation between hunger and frontal activation in response to food cue exposure is therefore not surprising, but it is interesting that this pattern was seen only in individuals with BED. Karhunen et al. (2000) finding could suggest a number of things; individuals with BED may be more susceptible to the influence of hunger signals when exposed to food cues, leading to greater activation in the OFC and perhaps a greater tendency to engage in overconsumption of food in the presence of hunger signals. Future research could examine how the OFC may be differentially activated by hunger and satiety signals in BED; one possibility is that hypothalamic signals of hunger and satiety are more influential but less distinguished from each other in individuals with the disorder.

A second possible explanation of Karhunen et al. (2000) findings is that the stronger signal in left frontal and prefrontal regions in BED reflects greater activation of areas involved in cognitive control. Given that the central symptoms of BED are overconsumption of foods associated with a feeling of loss of control, individuals with BED might exert heightened cognitive control (compared with individuals without BED) when they are in the presence of food and do not wish to overeat. Indeed, if binge eating is driven by a heightened reward response to food cue exposure, it is possible that additional inhibitory control is warranted to control the desire to engage in eating. The association between frontal/prefrontal activation and self-reported hunger and desire to eat might reflect a stronger activation of cognitive control networks to counteract the stronger hunger and desire to eat.

The results of an electroencephalography (EEG) study by Tammela et al. (2010) also implicate frontal brain regions in BED. This study examined brain activity during a resting state, exposure to a picture of a landscape, and exposure to food stimuli in obese females with and without BED. Across all participants, frontal beta activity was stronger during the food compared to the landscape exposure condition. While no group differences specific to food exposure were found, frontal beta activity was greater in individuals with BED compared with those without in all three conditions and was positively correlated with both binge eating symptoms and disinhibited eating. As this group difference was present across all stimulus conditions, this finding could reflect nonspecific enhancement of arousal in individuals with BED. Enhanced activity in all participants for the food compared with other stimuli could reflect the greater motivational salience of food stimuli compared with the landscape (and resting state condition). While it is not possible to infer from these experimental conditions alone, the fact that participants with BED showed greater frontal activity during the experiment could have relevance to their eating pathology. For example, a greater arousal response to experimental conditions could reflect heightened stress levels or generalized differences in responses to external stimuli, which could influence reward responsivity to palatable food cues. Indeed, another EEG study (Imperatori et al., 2015) found enhanced resting state connectivity in the beta frequency band in individuals with BED compared with those without, which was significantly correlated with severity of binge eating symptoms. These findings suggest differences in neural processing in individuals with BED, which are not specific to the processing of food or other reward-related stimuli.

In a neuroimaging study, Geliebter et al. (2006) presented visual and auditory food cues and nonfood cues to obese and lean participants with the presence or absence of binge eating. Obese participants with subclinical binge eating showed heightened activation in the right premotor cortex when perceiving binge-type food. This premotor activity could reflect motor preparation (e.g., of mouth movements) in the obese, binge eating group (Geliebter et al., 2006). Importantly, this increased activity was not evident in a group of participants who were lean and reported binge eating or in participants who were obese but did not endorse binge eating. While this heightened premotor activity might not reflect the reward system directly, it might indicate a lack of inhibitory control over the motor system in individuals with binge eating tendencies.

In summary, BED is associated with altered activity in the striatum and medial OFC during anticipatory processing of both food and nonfood reward stimuli. Across the literature, there is evidence of both increased and decreased activity and volume associated with binge eating. Individuals with BED also show heightened frontal and prefrontal activity during both stimulus processing and resting state and enhanced premotor activity during food cue exposure.

Reward processing in Prader–Willi syndrome

Neuroimaging studies show differences in reward-related brain regions in individuals with PWS compared with healthy controls. In particular, individuals with PWS show greater activation of the OFC and hypothalamus in response to high-calorie food cues compared with healthy controls (Dimitropoulos & Schultz, 2008). Holsen et al. (2012) compared neural responses to food stimuli before and after eating in individuals with PWS and individuals of similar (obese) BMI without PWS. Before eating, participants with PWS showed greater activity in the striatum and amygdala and lower activity in the hypothalamus and hippocampus, compared with the obese control group. After eating, the PWS group showed greater activity in the hypothalamus, hippocampus, and amygdala but lower activity in the OFC and DLPFC. Heightened activation of the NAc specifically before eating is consistent with this region's role in anticipatory reward processing and could reflect enhanced anticipatory processing in PWS. The finding also suggests a parallel with processing in BED, which is also characterized by alterations in ventral striatum processing during anticipatory reward processing (e.g., Balodis et al., 2013a; Weygandt et al., 2012). The decreased activity in DLPFC and OFC could reflect less reward-related activity associated with the receipt of food in PWS, which could underlie a motivation to continue eating.

The researchers suggest that heightened activity in the amygdala both before and after a meal could contribute to the impaired satiety that characterizes PWS (Holsen et al., 2012). The fact that individuals with PWS show less hypothalamus activity compared with controls *before* eating, but more activity after eating (Holsen et al., 2012), could reflect disruption of hunger and satiety signals influencing responses to food cues in PWS. Individuals with PWS also show a delayed response of the hypothalamus to glucose intake (Shapira et al., 2005), which could account for the differences in activation after the meal reported by Holsen et al. (2012). Indeed, Shapira found modulation of activity in the hypothalamus occurred 25 min after glucose ingestion in individuals with PWS, compared with at 10 or 15 min in normal weight and obese (respectively) controls. Because the postmeal scan occurred within 15 min of eating in Holsen and colleagues' study, it is possible that these differences reflect differences in the onset, rather than the magnitude, of hypothalamic activity following eating. Nonetheless, these findings suggest differences in how the hypothalamus responds to food cues, which could indicate deficits in the integration of homeostatic and hedonic processes that govern food intake. Indeed, a number of features of PWS are attributed to the hypothalamus (Swaab, 1997). In addition to the excessive appetite and compulsive eating, PWS is also associated with short stature, decreased levels of sex hormones, and problems with temperature regulation—all of which could be related to hypothalamic processes (Swaab, 1997).

Finally, it is well known that PWS is associated with elevated ghrelin levels; according to one study, ghrelin levels are four and a half times higher in individuals with PWS compared with weight-matched controls (Cummings et al., 2002). As discussed previously, ghrelin stimulates appetite through its action on orexin-releasing neurons in the lateral hypothalamus, but may also act on dopaminergic transmission in the reward system. Receptors for ghrelin are found in the VTA, and ghrelin is associated with increased dopamine transmission in this brain region. Thus, it is possible that elevated ghrelin in PWS enhances the motivation for eating by increasing reward value of food.

Comorbidities and addiction transfer

As discussed earlier in the chapter, animal research suggests interaction or overlap between the processes underlying food reward and those associated with direct self-stimulation and intake of addictive drugs. Both drug and palatable food intake are associated with striatal dopamine release (e.g., Hernandez & Hoebel, 1988), and food restriction can also modulate drug self-administration (for a review, see Carroll & Meisch, 1984). Furthermore, the effect of food deprivation on drug self-administration seems to be mediated by a change in reward responsivity to the drug, suggesting that food deprivation enhances sensitivity of the reward system for drugs and food (de Vaca & Carr, 1988). If eating behavior and other addictive behaviors are influenced by common reward pathways, we might predict that food addiction and compulsive eating would be associated with other addictive behaviors. Indeed, comorbidities between addictive disorders and the phenomenon of substitution behaviors are consistent with the notion of common reward pathways. While we discuss data that speak to reward system involvement in comorbidities and addiction transfer, it is important to note that mechanisms unrelated to reward processing might also contribute to these phenomena, but are not discussed in detail here.

At least a quarter of individuals with BED have a current or past SUD (Grilo, White, & Masheb, 2009; Hudson, Hiripi, Pope, & Kessler, 2007; Wilfley et al., 2000), and SUDs are also higher in bulimia and anorexia nervosa populations compared with the general population (Holderness, Brooks-Gunn, & Warren, 1994; Krahn, 1991). Binge behavior in particular is linked to comorbidity: the prevalence of SUD is greater in individuals with BN and the binge/purge subtype of anorexia nervosa compared with the restrictive subtype of anorexia (Holderness et al., 1994). A family history of alcohol-use disorder is also a risk factor for obesity (Grucza et al., 2010), and individuals who quit smoking commonly show increased food intake and weight gain (Filozof, Pinilla, & Fernández-Cruz, 2004). Finally, liver transplant patients with a history of alcohol-use disorder are at greater risk of metabolic syndrome (associated with obesity) compared with patients with no history of alcohol-use disorder (Anastácio et al., 2011; Laryea et al., 2007). The phenomenon of addiction transfer is also seen in individuals who undergo bariatric surgery for weight loss. 2%–6% of patients admitted for treatment for an SUD have previously undergone bariatric surgery (Saules et al., 2010; Wiedemann, Saules, & Ivezaj, 2013). Assessing the prevalence of SUDs in bariatric patient populations, two studies observed that SUDs were significantly more common 2 years after bariatric surgery compared to before surgery (Conason et al., 2013; King et al., 2012).

Importantly, it is estimated that 43%–61% of individuals receiving treatment for an SUD following bariatric surgery have no previous history of substance-use problems (Saules et al., 2010; Wiedemann et al., 2013), suggesting SUD development after their surgery and dietary changes.

It should be noted that bariatric surgery affects how substances are absorbed and metabolized in the body, which could contribute to the increased risk of SUDs following treatment. However, research using an animal model of the Roux-en-Y bariatric procedure shows increased alcohol intake following surgery even when alcohol is administered intravenously (Polston et al., 2013), suggesting that altered metabolism does not fully account for postsurgery changes in alcohol use. Furthermore, bariatric patients sometimes develop non–substance-related addictive behaviors such as compulsive gambling or shopping (Blum et al., 2011). Given the overlap in reward processes associated with food intake and other addictive behaviors, one explanation for addiction transfer is that the new addictive behavior is driven by the same reward-related processing alterations that drove overeating.

Blum et al. (1996) found that individuals who were obese and had a comorbid SUD had a heightened prevalence of an allele associated with low levels of the dopamine D2 receptor compared with obese individuals without comorbid SUD. As lower levels of dopamine receptor availability are documented in both SUDs (for a review, see Volkow et al., 2004) and obesity (Volkow et al., 2008; Wang et al., 2001; de Weijer et al., 2011), this finding is consistent with the hypothesis that comorbid SUD in obesity relates to decreased reward functioning. Fowler, Ivezaj, and Saules (2014) examined food addiction traits in individuals who had developed an SUD following bariatric surgery compared with those who did not develop an SUD. Specifically, these researchers used the YFAS to determine the food items that were most problematic (i.e., experienced as most "addictive") for patients before they had undergone surgery. Individuals who had developed an SUD since their weight-loss surgery were more likely to endorse high-sugar/low-fat foods and foods with a high glycemic index as problematic. Both sugar and fat in foods promote overeating and increases in body weight and elicit reward-related activity in the brain, but it has been suggested that foods high in sugar but low in fat may show the greatest addiction potential (Avena, Rada, & Hoebel, 2009). In support of this, a neuroimaging study in humans showed that intake of a milkshake with a high-sugar/low-fat composition elicited greater activation of reward areas compared with milkshakes with high-sugar/high-fat, low-sugar/low-fat, or low-sugar/high-fat compositions (Stice, Burger, & Yokum, 2013). Likewise, foods with a high glycemic index activate the reward system to a greater extent than do foods with a low glycemic index (Lennerz et al., 2013). It is possible that individuals who show addictive tendencies toward high-sugar/low-fat foods are most susceptible to developing an SUD post–weight-loss surgery when their intake of these foods—and associated reward-related activation—is reduced.

Interestingly, while the prevalence of lifetime SUD is higher in individuals with BED, the prevalence of current SUD in individuals with current BED is considerably smaller (Grilo et al., 2009), and studies have found a lower risk of SUDs in individuals who are currently overweight or obese (e.g., Simon et al., 2006). This might suggest that the same reward mechanisms underlie the *development* of SUDs and overeating (leading to overweight or obesity), but overeating high-energy foods precludes the drive to engage in alternative addictive behaviors such as substance use. This is consistent with animal work showing reduced drug self-administration and self-stimulation of reward areas when animals are in a satiated state (Hoebel, 1969; Margules & Olds, 1962).

General discussion

This chapter explored the role of the reward system in food addiction and overeating. We examined what is known about reward processing in conditions characterized by these behaviors (namely, food addiction, obesity, BED, and PWS), as well as relationships between overeating/food addiction and other addictive disorders.

The OFC and striatum are consistently implicated in food addiction, obesity, and BED, suggesting an important role of these regions in disordered eating behaviors. Studies demonstrate altered sensitivity to food related to specific reward phases (e.g., anticipation vs. outcome). Food addiction (as assessed using the YFAS) is associated with enhanced medial OFC, ACC, and amygdala activity during anticipatory food processing, but decreased lateral OFC activity during food intake (Gearhardt et al., 2011). Similar heightened medial OFC anticipatory signaling to food cues is seen in BED, which is further positively associated with reward sensitivity (Schienle et al., 2009). Studies find decreased OFC volume in obesity (Cohen et al., 2011; Maayan et al., 2011; Shott et al., 2015; Walther et al., 2010), while both increased and decreased OFC volume are observed in BED (Schäfer et al., 2010; Voon et al., 2015). Compared with healthy controls, individuals with PWS show decreased OFC activity after eating a meal, possibly reflecting dysfunctional consummatory reward processing in this condition.

Both the ventral and dorsal striatum are implicated in overeating and food addiction. Individuals with food addiction show higher caudate activity during food reward anticipation (Gearhardt et al., 2011), while individuals with BED show smaller caudate volume compared with controls (Shott et al., 2015). Interestingly, dopamine release in the caudate nucleus is significantly associated with binge eating severity but not BMI (Wang et al., 2011), suggesting a role for this region in eating behavior rather than body weight per se. Individuals with BED have a smaller ventral striatum volume (Voon et al., 2015) and decreased striatal responsivity to food cues compared with controls (Weygandt et al., 2012). BED is associated with altered ventral striatal function for both food and nonfood rewards, with individuals with BED showing reduced striatal activity during anticipation of monetary rewards (Balodis et al., 2013a). Obesity is associated with heightened striatal activity during anticipation of food receipt (Rothemund et al., 2007; Stoeckel et al., 2008), with BMI being a significant predictor of reward-related responses to food cues (Rothemund et al., 2007). Conversely, striatal activation associated with consummatory processing is lower in obesity and negatively correlated with BMI (Stice, Spoor, Bohon, & Small, 2008; Stice, Spoor, Bohon, Veldhuizen et al., 2008; Volkow et al., 2008; de Weijer et al., 2011).

Divergent findings across BED and obesity studies highlight important biobehavioral differences in these conditions. To date, few studies apply measures of food addiction in these populations; therefore, varying levels of food addiction within these samples may obscure findings. Food addiction is present in 40%−60% of individuals with BED (Gearhardt et al., 2012, 2013, 2014) and around a quarter of individuals who are overweight or obese (Pursey et al., 2014). As the YFAS was developed relatively recently (Gearhardt et al., 2009), many studies have not assessed food addiction in samples of individuals with BED or obesity. Individuals with food addiction show altered reward processing compared with BMI-matched individuals without food addiction (Davis et al., 2014, 2013; Geardhardt et al., 2011). Thus, it is possible that varying levels of food addiction within high-BMI samples could explain some of the heterogeneity across studies. In the same way, many studies in individuals who are overweight or obese do not screen for BED, so it is possible that different findings are driven by interindividual and intersample differences in binge eating pathology. Future studies should implement assessment of BED and food addiction to accurately characterize the sample and identify meaningful subgroups.

While differences in reward processing are often posited as an explanation for food addiction and overeating, it is important to note that altered reward processing could also be a result of disordered eating in some individuals. Indeed, animal studies have demonstrated changes in brain responsivity and hormone levels related to long-term dietary changes (e.g., Avena et al., 2008). Furthermore, hormones such as leptin and ghrelin are affected by dietary intake and adiposity and appear to play a role in reward; thus, it is possible that dietary patterns associated with food addiction, obesity, and BED precede and cause altered reward processing. Longitudinal studies are needed to adequately capture potential bidirectional relationships between reward processing and disordered eating. The effects of dietary intake and patterns of eating behavior on reward processing could also be assessed by examining reward processing in individuals of a wide range of BMIs and individuals at varying stages and severities of disordered eating.

References

Abizaid, A., Liu, Z. W., Andrews, Z. B., Shanabrough, M., Borok, E., Elsworth, J. D., ... Horvath, T. L. (2006). Ghrelin modulates the activity and synaptic input organization of midbrain dopamine neurons while promoting appetite. *Journal of Clinical Investigation, 116*(12), 3229−3239.

American Psychiatric Association. (2000). *Diagnostic and statistical manual of mental disorders* (revised 4th ed.). Washington, DC: Author.

American Psychiatric Association. (2013). *Diagnostic and statistical manual of mental disorders* (5th ed.). Arlington, VA: American Psychiatric Publishing.

Anastácio, L. R., Ferreira, L. G., de Sena Ribeiro, H., Liboredo, J. C., Lima, A. S., & Correia, M. I. (2011). Metabolic syndrome after liver transplantation: Prevalence and predictive factors. *Nutrition, 27*(9), 931–937.

Avena, N. M., Rada, P., & Hoebel, B. G. (2008). Evidence for sugar addiction: behavioral and neurochemical effects of intermittent, excessive sugar intake. *Neuroscience Biobehavioral Reviews, 32*(1), 20–39.

Avena, N. M., Rada, P., & Hoebel, B. G. (2009). Sugar and fat bingeing have notable differences in addictive-like behavior. *Journal of Nutrition, 139*(3), 623–628.

Balodis, I. M., Kober, H., Worhunsky, P. D., Stevens, M. C., Pearlson, G. D., & Potenza, M. N. (2012). Diminished frontostriatal activity during processing of monetary rewards and losses in pathological gambling. *Biological Psychiatry, 71*(8), 749–757.

Balodis, I. M., Kober, H., Worhunsky, P. D., White, M. A., Stevens, M. C., Pearlson, G. D., ... Potenza, M. N. (2013a). Monetary reward processing in obese individuals with and without binge eating disorder. *Biological Psychiatry, 73*(9), 877–886.

Balodis, I. M., Molina, N. D., Kober, H., Worhunsky, P. D., White, M. A., Sinha, R., ... Potenza, M. N. (2013b). Divergent neural substrates of inhibitory control in binge eating disorder relative to other manifestations of obesity. *Obesity, 21*(2), 367–377.

Balodis, I. M., Grilo, C. M., Kober, H., Worhunsky, P. D., White, M. A., Stevens, M. C., ... Potenza, M. N. (2014). A pilot study linking reduced fronto–Striatal recruitment during reward processing to persistent bingeing following treatment for binge-eating disorder. *International Journal of Eating Disorders, 47*(4), 376–384.

Baptista, T. (1999). Body weight gain induced by antipsychotic drugs: Mechanisms and management. *Acta Psychiatrica Scandinavica, 100*(1), 3–16.

Barr, M. S., Farzan, F., Wing, V. C., George, T. P., Fitzgerald, P. B., & Daskalakis, Z. J. (2011). Repetitive transcranial magnetic stimulation and drug addiction. *International Review of Psychiatry, 23*(5), 454–466.

Batterink, L., Yokum, S., & Stice, E. (2010). Body mass correlates inversely with inhibitory control in response to food among adolescent girls: an fMRI study. *Neuroimage, 52*(4), 1696–1703.

Beaver, J. D., Lawrence, A. D., van Ditzhuijzen, J., Davis, M. H., Woods, A., & Calder, A. J. (2006). Individual differences in reward drive predict neural responses to images of food. *Journal of Neuroscience, 26*(19), 5160–5166.

Beck, A., Schlagenhauf, F., Wüstenberg, T., Hein, J., Kienast, T., Kahnt, T., ... Wrase, J. (2009). Ventral striatal activation during reward anticipation correlates with impulsivity in alcoholics. *Biological Psychiatry, 66*(8), 734–742.

Bégin, C., St-Louis, M. E., Turmel, S., Tousignant, B., Marion, L. P., Ferland, F., ... Gagnon-Girouard, M. P. (2012). Does food addiction distinguish a specific subgroup of overweight/obese overeating women. *Health, 4*(12A), 1492–1499.

Berridge, K. C., & Kringelbach, M. L. (2008). Affective neuroscience of pleasure: Reward in humans and animals. *Psychopharmacology, 199*(3), 457–480.

Berridge, K. C., & Robinson, T. E. (2003). Parsing reward. *Trends in Neurosciences, 26*(9), 507–513.

Berridge, K. C. (2009). 'Liking'and 'wanting'food rewards: Brain substrates and roles in eating disorders. *Physiology & Behavior, 97*(5), 537–550.

Blum, K., Cull, J. G., Braverman, E. R., & Comings, D. E. (1996). Reward deficiency syndrome. *American Scientist, 84*(2), 132–145.

Blum, K., Bailey, J., Gonzalez, A. M., Oscar-Berman, M., Liu, Y., Giordano, J., ... Gold, M. (2011). Neuro-genetics of reward deficiency syndrome (RDS) as the root cause of "addiction transfer": A new phenomenon common after bariatric surgery. *Journal of Genetic Syndromes & Gene Therapy, 2012*(1).

Bogg, T., Fukunaga, R., Finn, P. R., & Brown, J. W. (2012). Cognitive control links alcohol use, trait disinhibition, and reduced cognitive capacity: Evidence for medial prefrontal cortex dysregulation during reward-seeking behavior. *Drug and Alcohol Dependence, 122*(1–2), 112–118.

Burger, K. S., & Stice, E. (2011). Relation of dietary restraint scores to activation of reward-related brain regions in response to food intake, anticipated intake, and food pictures. *NeuroImage, 55*(1), 233–239.

Burmeister, J. M., Hinman, N., Koball, A., Hoffmann, D. A., & Carels, R. A. (2013). Food addiction in adults seeking weight loss treatment. Implications for psychosocial health and weight loss. *Appetite, 60*, 103–110.

Butler, M. G., Manzardo, A. M., Heinemann, J., Loker, C., & Loker, J. (2017). Causes of death in Prader-Willi syndrome: Prader-Willi syndrome association (USA) 40-year mortality survey. *Genetics in Medicine, 19*(6), 635.

Carroll, M. E., & Meisch, R. A. (1984). Increased drug-reinforced behavior due to food deprivation. In *Advances in Behavioral Pharmacology* (Vol. 4, pp. 47–88). Elsevier.

Chiara, G. D., & Bassareo, V. (2007). Reward system and addiction: What dopamine does and doesn't do. *Current Opinion in Pharmacology, 7*, 69–76.

Clark, S. M., & Saules, K. K. (2013). Validation of the yale food addiction scale among a weight-loss surgery population. *Eating Behaviors, 14*(2), 216–219.

Cohen, J. I., Yates, K. F., Duong, M., & Convit, A. (2011). Obesity, orbitofrontal structure and function are associated with food choice: A cross-sectional study. *BMJ open, 1*(2), e000175.

Conason, A., Teixeira, J., Hsu, C. H., Puma, L., Knafo, D., & Geliebter, A. (2013). Substance use following bariatric weight loss surgery. *JAMA Surgery, 148*(2), 145–150.

Cordeira, J. W., Frank, L., Sena-Esteves, M., Pothos, E. N., & Rios, M. (2010). Brain-derived neurotrophic factor regulates hedonic feeding by acting on the mesolimbic dopamine system. *Journal of Neuroscience, 30*(7), 2533–2541.

Cruwys, T., Bevelander, K. E., & Hermans, R. C. (2015). Social modeling of eating: A review of when and why social influence affects food intake and choice. *Appetite, 86*, 3–18.

Cummings, D. E., Clement, K., Purnell, J. Q., Vaisse, C., Foster, K. E., Frayo, R. S., ... Weigle, D. S. (2002). Elevated plasma ghrelin levels in Prader–Willi syndrome. *Nature Medicine, 8*(7), 643.

Davis, C., Strachan, S., & Berkson, M. (2004). Sensitivity to reward: Implications for overeating and overweight. *Appetite, 42*(2), 131–138.

Davis, C., Curtis, C., Levitan, R. D., Carter, J. C., Kaplan, A. S., & Kennedy, J. L. (2011). Evidence that "food addiction" is a valid phenotype of obesity. *Appetite, 57*(3), 711–717.

Davis, C., Levitan, R. D., Kaplan, A. S., Kennedy, J. L., & Carter, J. C. (2014). Food cravings, appetite, and snack-food consumption in response to a psychomotor stimulant drug: The moderating effect of "food-addiction". *Frontiers in Psychology, 5*, 403.

Davis, C. (2013). Compulsive overeating as an addictive behavior: Overlap between food addiction and binge eating disorder. *Current Obesity Reports, 2*(2), 171–178.

Davis, C. (2017). A commentary on the associations among 'food addiction', binge eating disorder, and obesity: Overlapping conditions with idiosyncratic clinical features. *Appetite, 115*, 3–8.

Day, J. J., Roitman, M. F., Wightman, R., & Carelli, R. M. (2007). Associative learning mediates dynamic shifts in dopamine signaling in the nucleus accumbens. *Nature Neuroscience, 10*(8), 1020–1028.

De Vaca, S. C., & Carr, K. D. (1998). Food restriction enhances the central rewarding effect of abused drugs. *Journal of Neuroscience, 18*(18), 7502–7510.

de Weijer, B. A., van de Giessen, E., van Amelsvoort, T. A., Boot, E., Braak, B., Janssen, I. M., ... Booij, J. (2011). Lower striatal dopamine D 2/3 receptor availability in obese compared with non-obese subjects. *EJNMMI Research, 1*(1), 37.

DelParigi, A., Chen, K., Salbe, A. D., Hill, J. O., Wing, R. R., Reiman, E. M., & Tataranni, P. A. (2007). Successful dieters have increased neural activity in cortical areas involved in the control of behavior. *International Journal of Obesity, 31*(3), 440.

Dimitropoulos, A., & Schultz, R. T. (2008). Food-related neural circuitry in Prader-Willi syndrome: Response to high-versus low-calorie foods. *Journal of Autism and Developmental Disorders, 38*(9), 1642–1653.

Disse, E., Bussier, A. L., Veyrat-Durebex, C., Deblon, N., Pfluger, P. T., Tschöp, M. H., ... Rohner-Jeanrenaud, F. (2010). Peripheral ghrelin enhances sweet taste food consumption and preference, regardless of its caloric content. *Physiology & Behavior, 101*(2), 277–281.

Drewnowski, A., & Greenwood, M. R. (1983). Cream and sugar: Human preferences for high-fat foods. *Physiology & Behavior, 30*(4), 629–633.

Drewnowski, A. (1997). Taste preferences and food intake. *Annual Review of Nutrition, 17*(1), 237–253.

Elliott, R., Dolan, R. J., & Frith, C. D. (2000). Dissociable functions in the medial and lateral orbitofrontal cortex: Evidence from human neuroimaging studies. *Cerebral Cortex, 10*(3), 308–317.

Epstein, L. H., Temple, J. L., Neaderhiser, B. J., Salis, R. J., Erbe, R. W., & Leddy, J. J. (2007). Food reinforcement, the dopamine D_2 receptor genotype, and energy intake in obese and nonobese humans. *Behavioral Neuroscience, 121*(5), 877.

Fairburn, C. G., Doll, H. A., Welch, S. L., Hay, P. J., Davies, B. A., & O'connor, M. E. (1998). Risk factors for binge eating disorder: A community-based, case-control study. *Archives of General Psychiatry, 55*(5), 425–432.

Farooqi, I. S., Bullmore, E., Keogh, J., Gillard, J., O'rahilly, S., & Fletcher, P. C. (2007). Leptin regulates striatal regions and human eating behavior. *Science, 317*(5843), 1355.

Fibiger, H. C. (1978). Drugs and reinforcement mechanisms: A critical review of the catecholamine theory. *Annual Review of Pharmacology and Toxicology, 18*(1), 37–56.

Figlewicz, D. P. (2003). Adiposity signals and food reward: Expanding the CNS roles of insulin and leptin. *American Journal of Physiology - Regulatory, Integrative and Comparative Physiology, 284*(4), R882–R892.

Filbey, F. M., Myers, U. S., & DeWitt, S. (2012). Reward circuit function in high BMI individuals with compulsive overeating: Similarities with addiction. *NeuroImage, 63*(4), 1800–1806.

Filozof, C., Pinilla, F., & Fernández-Cruz, A. (2004). Smoking cessation and weight gain. *Obesity Reviews, 5*(2), 95–103.

Forbes, E. E., Rodriguez, E. E., Musselman, S., & Narendran, R. (2014). Prefrontal response and frontostriatal functional connectivity to monetary reward in abstinent alcohol-dependent young adults. *PLoS One, 9*(5), e94640.

Fowler, L., Ivezaj, V., & Saules, K. K. (2014). Problematic intake of high-sugar/low-fat and high glycemic index foods by bariatric patients is associated with development of post-surgical new onset substance use disorders. *Eating Behaviors, 15*(3), 505–508.

Gearhardt, A. N., Corbin, W. R., & Brownell, K. D. (2009). Preliminary validation of the Yale food addiction scale. *Appetite, 52*(2), 430–436.

Gearhardt, A. N., Yokum, S., Orr, P. T., Stice, E., Corbin, W. R., & Brownell, K. D. (2011). Neural correlates of food addiction. *Archives of General Psychiatry, 68*(8), 808–816.

Gearhardt, A. N., White, M. A., Masheb, R. M., Morgan, P. T., Crosby, R. D., & Grilo, C. M. (2012). An examination of the food addiction construct in obese patients with binge eating disorder. *International Journal of Eating Disorders, 45*(5), 657–663.

Gearhardt, A. N., White, M. A., Masheb, R. M., & Grilo, C. M. (2013). An examination of food addiction in a racially diverse sample of obese patients with binge eating disorder in primary care settings. *Comprehensive Psychiatry, 54*(5), 500–505.

Gearhardt, A. N., Boswell, R. G., & White, M. A. (2014). The association of "food addiction" with disordered eating and body mass index. *Eating Behaviors, 15*(3), 427–433.

Gearhardt, A. N., Corbin, W. R., & Brownell, K. D. (2016). Development of the yale food addiction scale version 2.0. *Psychology of Addictive Behaviors, 30*(1), 113.

Geliebter, A., Ladell, T., Logan, M., Schweider, T., Sharafi, M., & Hirsch, J. (2006). Responsivity to food stimuli in obese and lean binge eaters using functional MRI. *Appetite, 46*(1), 31–35.

German, D. C., & Bowden, D. M. (1974). Catecholamine systems as the neural substrate for intracranial self-stimulation: A hypothesis. *Brain Research, 73*(3), 381–419.

Goldfield, G. S., Lorello, C., & Doucet, E. (2007). Methylphenidate reduces energy intake and dietary fat intake in adults: A mechanism of reduced reinforcing value of food? *American Journal of Clinical Nutrition, 86*(2), 308–315.

Goldman, R. L., Borckardt, J. J., Frohman, H. A., O'Neil, P. M., Madan, A., Campbell, L. K., … George, M. S. (2011). Prefrontal cortex transcranial direct current stimulation (tDCS) temporarily reduces food cravings and increases the self-reported ability to resist food in adults with frequent food craving. *Appetite, 56*(3), 741–746.

Goldstein, R. Z., Tomasi, D., Rajaram, S., Cottone, L. A., Zhang, L., Maloney, T. E. E. A., … Volkow, N. D. (2007). Role of the anterior cingulate and medial orbitofrontal cortex in processing drug cues in cocaine addiction. *Neuroscience, 144*(4), 1153–1159.

Grilo, C. M., White, M. A., & Masheb, R. M. (2009). DSM-IV psychiatric disorder comorbidity and its correlates in binge eating disorder. *International Journal of Eating Disorders, 42*(3), 228–234.

Grosshans, M., Vollmert, C., Vollstädt-Klein, S., Tost, H., Leber, S., Bach, P., … Kiefer, F. (2012). Association of leptin with food cue–induced activation in human reward pathways. *Archives of General Psychiatry, 69*(5), 529–537.

Grucza, R. A., Krueger, R. F., Racette, S. B., Norberg, K. E., Hipp, P. R., & Bierut, L. J. (2010). The emerging link between alcoholism risk and obesity in the United States. *Archives of General Psychiatry, 67*(12), 1301–1308.

Herman, C. P., & Polivy, J. (2005). Normative influences on food intake. *Physiology & Behavior, 86*(5), 762–772.

Hernandez, L., & Hoebel, B. G. (1988). Food reward and cocaine increase extracellular dopamine in the nucleus accumbens as measured by microdialysis. *Life Sciences, 42*(18), 1705–1712.

Higgs, S. (2005). Memory and its role in appetite regulation. *Physiology & Behavior, 85*(1), 67–72.

Hoebel, B. G. (1969). Feeding and self-stimulation. *Annals of the New York Academy of Sciences, 157*(1), 758–778.

Holderness, C. C., Brooks-Gunn, J., & Warren, M. P. (1994). Co-morbidity of eating disorders and substance abuse review of the literature. *International Journal of Eating Disorders, 16*(1), 1–34.

Holsen, L. M., Savage, C. R., Martin, L. E., Bruce, A. S., Lepping, R. J., Ko, E., … Goldstein, J. M. (2012). Importance of reward and prefrontal circuitry in hunger and satiety: Prader–Willi syndrome vs simple obesity. *International Journal of Obesity, 36*(5), 638.

Hudson, J. I., Hiripi, E., Pope, H. G., Jr., & Kessler, R. C. (2007). The prevalence and correlates of eating disorders in the National Comorbidity Survey Replication. *Biological Psychiatry, 61*(3), 348–358.

Imperatori, C., Fabbricatore, M., Farina, B., Innamorati, M., Quintiliani, M. I., Lamis, D. A., … Speranza, A. M. (2015). Alterations of EEG functional connectivity in resting state obese and overweight patients with binge eating disorder: A preliminary report. *Neuroscience Letters, 607*, 120–124.

Izquierdo, A., Suda, R. K., & Murray, E. A. (2004). Bilateral orbital prefrontal cortex lesions in rhesus monkeys disrupt choices guided by both reward value and reward contingency. *Journal of Neuroscience, 24*(34), 7540–7548. https://doi.org/10.1523/JNEUROSCI.1921-04.2004.

Jansen, J. M., Daams, J. G., Koeter, M. W., Veltman, D. J., van den Brink, W., & Goudriaan, A. E. (2013). Effects of non-invasive neurostimulation on craving: A meta-analysis. *Neuroscience & Biobehavioral Reviews, 37*(10), 2472–2480.

Jerlhag, E., Egecioglu, E., Dickson, S. L., Douhan, A., Svensson, L., & Engel, J. A. (2007). Preclinical study: Ghrelin administration into tegmental areas stimulates locomotor activity and increases extracellular concentration of dopamine in the nucleus accumbens. *Addiction Biology, 12*(1), 6–16.

Jerlhag, E., Egecioglu, E., Landgren, S., Salomé, N., Heilig, M., Moechars, D., … Engel, J. A. (2009). Requirement of central ghrelin signaling for alcohol reward. *Proceedings of the National Academy of Sciences, 106*(27), 11318–11323.

Jerlhag, E., Egecioglu, E., Dickson, S. L., & Engel, J. A. (2010). Ghrelin receptor antagonism attenuates cocaine-and amphetamine-induced locomotor stimulation, accumbal dopamine release, and conditioned place preference. *Psychopharmacology, 211*(4), 415–422.

Johnson, P. M., & Kenny, P. J. (2010). Dopamine D2 receptors in addiction-like reward dysfunction and compulsive eating in obese rats. *Nature Neuroscience, 13*(5), 635.

Karhunen, L. J., Vanninen, E. J., Kuikka, J. T., Lappalainen, R. I., Tiihonen, J., & Uusitupa, M. I. (2000). Regional cerebral blood flow during exposure to food in obese binge eating women. *Psychiatry Research: Neuroimaging, 99*(1), 29–42.

Kekic, M., McClelland, J., Campbell, I., Nestler, S., Rubia, K., David, A. S., & Schmidt, U. (2014). The effects of prefrontal cortex transcranial direct current stimulation (tDCS) on food craving and temporal discounting in women with frequent food cravings. *Appetite, 78*, 55–62.

Kenny, P. J. (2011). Reward mechanisms in obesity: New insights and future directions. *Neuron, 69*(4), 664–679.

Killgore, W. D., Young, A. D., Femia, L. A., Bogorodzki, P., Rogowska, J., & Yurgelun-Todd, D. A. (2003). Cortical and limbic activation during viewing of high-versus low-calorie foods. *NeuroImage, 19*(4), 1381–1394.

Killgore, W. D., Weber, M., Schwab, Z. J., Kipman, M., DelDonno, S. R., Webb, C. A., & Rauch, S. L. (2013). Cortico-limbic responsiveness to high-calorie food images predicts weight status among women. *International Journal of Obesity, 37*(11), 1435.

King, W. C., Chen, J. Y., Mitchell, J. E., Kalarchian, M. A., Steffen, K. J., Engel, S. G., ... Yanovski, S. Z. (2012). Prevalence of alcohol use disorders before and after bariatric surgery. *JAMA, 307*(23), 2516–2525.

Knutson, B., Fong, G. W., Adams, C. M., Varner, J. L., & Hommer, D. (2001). Dissociation of reward anticipation and outcome with event-related fMRI. *NeuroReport, 12*(17), 3683–3687.

Krahn, D. D. (1991). The relationship of eating disorders and substance abuse. *Journal of Substance Abuse, 3*(2), 239–253.

Kringelbach, M. L. (2005). The human orbitofrontal cortex: Linking reward to hedonic experience. *Nature Reviews Neuroscience, 6*(9), 691.

Kroemer, N. B., Krebs, L., Kobiella, A., Grimm, O., Pilhatsch, M., Bidlingmaier, M., ... Smolka, M. N. (2013). Fasting levels of ghrelin covary with the brain response to food pictures. *Addiction Biology, 18*(5), 855–862.

Laryea, M., Watt, K. D., Molinari, M., Walsh, M. J., McAlister, V. C., Marotta, P. J., ... Peltekian, K. M. (2007). Metabolic syndrome in liver transplant recipients: Prevalence and association with major vascular events. *Liver Transplantation, 13*(8), 1109–1114.

Leddy, J. J., Epstein, L. H., Jaroni, J. L., Roemmich, J. N., Paluch, R. A., Goldfield, G. S., & Lerman, C. (2004). Influence of methylphenidate on eating in obese men. *Obesity Research, 12*(2), 224–232.

Lee, J. E., Namkoong, K., & Jung, Y. C. (2017). Impaired prefrontal cognitive control over interference by food images in binge-eating disorder and bulimia nervosa. *Neuroscience Letters, 651*, 95–101.

Leigh, S. J., & Morris, M. J. (2018). The role of reward circuitry and food addiction in the obesity epidemic: An update. *Biological Psychology, 131*, 31–42.

Lennerz, B. S., Alsop, D. C., Holsen, L. M., Stern, E., Rojas, R., Ebbeling, C. B., ... Ludwig, D. S. (2013). Effects of dietary glycemic index on brain regions related to reward and craving in men. *The American Journal of Clinical Nutrition, 98*(3), 641–647.

Lutter, M., & Nestler, E. J. (2009). Homeostatic and hedonic signals interact in the regulation of food intake. *Journal of Nutrition, 139*(3), 629–632.

Maayan, L., Hoogendoorn, C., Sweat, V., & Convit, A. (2011). Disinhibited eating in obese adolescents is associated with orbitofrontal volume reductions and executive dysfunction. *Obesity, 19*(7), 1382–1387.

Macht, M. (2008). How emotions affect eating: A five-way model. *Appetite, 50*, 1–11.

Margules, D. L., & Olds, J. (1962). Identical" feeding" and" rewarding" systems in the lateral hypothalamus of rats. *Science, 135*(3501), 374–375.

Marieb, E. N., & Hoehn, K. (2010). *Human anatomy and physiology* (8th ed.). San Francisco: Benjamin Cummings.

Mark, G. P., Smith, S. E., Rada, P. V., & Hoebel, B. G. (1994). An appetitively conditioned taste elicits a preferential increase in mesolimbic dopamine release. *Pharmacology Biochemistry and Behavior, 48*(3), 651–660.

McAllister, C. J., Whittington, J. E., & Holland, A. J. (2011). Development of the eating behaviour in Prader–Willi syndrome: Advances in our understanding. *International Journal of Obesity, 35*(2), 188.

Meule, A., Hermann, T., & Kübler, A. (2015). Food addiction in overweight and obese adolescents seeking weight-loss treatment. *European Eating Disorders Review, 23*(3), 193–198.

Michaelides, M., Thanos, P. K., Volkow, N. D., & Wang, G. J. (2012). Dopamine-related frontostriatal abnormalities in obesity and binge-eating disorder: Emerging evidence for developmental psychopathology. *International Review of Psychiatry, 24*(3), 211–218.

Nathan, P. J., O'Neill, B. V., Mogg, K., Bradley, B. P., Beaver, J., Bani, M., ... Dodds, C. M. (2012). The effects of the dopamine D3 receptor antagonist GSK598809 on attentional bias to palatable food cues in overweight and obese subjects. *International Journal of Neuropsychopharmacology, 15*(2), 149–161.

O'Doherty, J., Kringelbach, M. L., Rolls, E. T., Hornak, J., & Andrews, C. (2001). Abstract reward and punishment representations in the human orbitofrontal cortex. *Nature Neuroscience, 4*(1), 95.

Olds, J., & Milner, P. (1954). Positive reinforcement produced by electrical stimulation of septal area and other regions of rat brain. *Journal of Comparative & Physiological Psychology, 47*(6), 419.

Olds, J. (1956). Pleasure centers in the brain. *Scientific American, 195*(4), 105–117.

Öngür, D., & Price, J. L. (2000). The organization of networks within the orbital and medial prefrontal cortex of rats, monkeys and humans. *Cerebral Cortex, 10*(3), 206–219.

Opland, D. M., Leinninger, G. M., & Myers, M. G., Jr. (2010). Modulation of the mesolimbic dopamine system by leptin. *Brain Research, 1350*, 65–70.

Orosco, M., Rouch, C., & Nicolaidis, S. (1996). Rostromedial hypothalamic monoamine changes in response to intravenous infusions of insulin and glucose in freely feeding obese zucker rats: A microdialysis study. *Appetite, 26*(1), 1–20.

Pedram, P., Wadden, D., Amini, P., Gulliver, W., Randell, E., Cahill, F., ... Sun, G. (2013). Food addiction: Its prevalence and significant association with obesity in the general population. *PLoS One, 8*(9), e74832.

Peters, J., Bromberg, U., Schneider, S., Brassen, S., Menz, M., Banaschewski, T., ... Heinz, A. (2011). Lower ventral striatal activation during reward anticipation in adolescent smokers. *American Journal of Psychiatry, 168*(5), 540–549.

Polston, J. E., Pritchett, C. E., Tomasko, J. M., Rogers, A. M., Leggio, L., Thanos, P. K., ... Hajnal, A. (2013). Roux-en-Y gastric bypass increases intravenous ethanol self-administration in dietary obese rats. *PLoS One, 8*(12), e83741.

Pursey, K. M., Stanwell, P., Gearhardt, A. N., Collins, C. E., & Burrows, T. L. (2014). The prevalence of food addiction as assessed by the yale food addiction scale: A systematic review. *Nutrients, 6*(10), 4552–4590.

Rademacher, L., Krach, S., Kohls, G., Irmak, A., Gründer, G., & Spreckelmeyer, K. N. (2010). Dissociation of neural networks for anticipation and consumption of monetary and social rewards. *NeuroImage, 49*(4), 3276–3285.

Reuter, J., Raedler, T., Rose, M., Hand, I., Gläscher, J., & Büchel, C. (2005). Pathological gambling is linked to reduced activation of the mesolimbic reward system. *Nature Neuroscience, 8*(2), 147.

Robinson, T. E., & Berridge, K. C. (1993). The neural basis of drug craving: An incentive-sensitization theory of addiction. *Brain Research Reviews, 18*(3), 247–291.

Rolls, E. T., Sienkiewicz, Z. J., & Yaxley, S. (1989). Hunger modulates the responses to gustatory stimuli of single neurons in the caudolateral orbitofrontal cortex of the macaque monkey. *European Journal of Neuroscience, 1*(1), 53–60.

Rolls, E. T. (2004). The functions of the orbitofrontal cortex. *Brain and Cognition, 55*(1), 11–29.

Rolls, E. T. (2011). Taste, olfactory and food texture reward processing in the brain and obesity. *International Journal of Obesity, 35*(4), 550.

Rothemund, Y., Preuschhof, C., Bohner, G., Bauknecht, H. C., Klingebiel, R., Flor, H., & Klapp, B. F. (2007). Differential activation of the dorsal striatum by high-calorie visual food stimuli in obese individuals. *NeuroImage, 37*(2), 410–421.

Rush, C. R., Higgins, S. T., Vansickel, A. R., Stoops, W. W., Lile, J. A., & Glaser, P. E. (2005). Methylphenidate increases cigarette smoking. *Psychopharmacology, 181*(4), 781–789.

Saper, C. B., Chou, T. C., & Elmquist, J. K. (2002). The need to feed: Homeostatic and hedonic control of eating. *Neuron, 36*(2), 199–211.

Saules, K. K., Wiedemann, A., Ivezaj, V., Hopper, J. A., Foster-Hartsfield, J., & Schwarz, D. (2010). Bariatric surgery history among substance abuse treatment patients: Prevalence and associated features. *Surgery for Obesity and Related Diseases, 6*(6), 615–621.

Schäfer, A., Vaitl, D., & Schienle, A. (2010). Regional grey matter volume abnormalities in bulimia nervosa and binge-eating disorder. *NeuroImage, 50*(2), 639–643.

Schellekens, H., Finger, B. C., Dinan, T. G., & Cryan, J. F. (2012). Ghrelin signalling and obesity: At the interface of stress, mood and food reward. *Pharmacology & Therapeutics, 135*(3), 316–326.

Schienle, A., Schäfer, A., Hermann, A., & Vaitl, D. (2009). Binge-eating disorder: Reward sensitivity and brain activation to images of food. *Biological Psychiatry, 65*(8), 654–661.

Schoenbaum, G., & Roesch, M. (2005). Orbitofrontal cortex, associative learning, and expectancies. *Neuron, 47*(5), 633–636.

Sevinçer, G. M., Konuk, N., Bozkurt, S., & Coşkun, H. (2016). Food addiction and the outcome of bariatric surgery at 1-year: Prospective observational study. *Psychiatry Research, 244*, 159–164.

Shapira, N. A., Lessig, M. C., He, A. G., James, G. A., Driscoll, D. J., & Liu, Y. (2005). Satiety dysfunction in Prader-Willi syndrome demonstrated by fMRI. *Journal of Neurology, Neurosurgery & Psychiatry, 76*(2), 260–262.

Shimbara, T., Mondal, M. S., Kawagoe, T., Toshinai, K., Koda, S., Yamaguchi, H., … Nakazato, M. (2004). Central administration of ghrelin preferentially enhances fat ingestion. *Neuroscience Letters, 369*(1), 75–79.

Shott, M. E., Cornier, M. A., Mittal, V. A., Pryor, T. L., Orr, J. M., Brown, M. S., & Frank, G. K. (2015). Orbitofrontal cortex volume and brain reward response in obesity. *International Journal of Obesity, 39*(2), 214.

Siep, N., Roefs, A., Roebroeck, A., Havermans, R., Bonte, M. L., & Jansen, A. (2009). Hunger is the best spice: An fMRI study of the effects of attention, hunger and calorie content on food reward processing in the amygdala and orbitofrontal cortex. *Behavioural Brain Research, 198*(1), 149–158.

Simon, G. E., Von Korff, M., Saunders, K., Miglioretti, D. L., Crane, P. K., Van Belle, G., & Kessler, R. C. (2006). Association between obesity and psychiatric disorders in the US adult population. *Archives of General Psychiatry, 63*(7), 824–830.

Small, D. M., Zatorre, R. J., Dagher, A., Evans, A. C., & Jones-Gotman, M. (2001). Changes in brain activity related to eating chocolate: From pleasure to aversion. *Brain, 124*(9), 1720–1733.

Small, D. M., Jones-Gotman, M., & Dagher, A. (2003). Feeding-induced dopamine release in dorsal striatum correlates with meal pleasantness ratings in healthy human volunteers. *NeuroImage, 19*(4), 1709–1715.

Small, D. M., Bender, G., Veldhuizen, M. G., Rudenga, K., Nachtigal, D., & Felsted, J. (2007). The role of the human orbitofrontal cortex in taste and flavor processing. *Annals of the New York Academy of Sciences, 1121*(1), 136–151.

Stice, E., Spoor, S., Bohon, C., & Small, D. M. (2008). Relation between obesity and blunted striatal response to food is moderated by TaqIA A1 allele. *Science, 322*(5900), 449–452.

Stice, E., Spoor, S., Bohon, C., Veldhuizen, M. G., & Small, D. M. (2008). Relation of reward from food intake and anticipated food intake to obesity: A functional magnetic resonance imaging study. *Journal of Abnormal Psychology, 117*(4), 924.

Stice, E., Spoor, S., Ng, J., & Zald, D. H. (2009). Relation of obesity to consummatory and anticipatory food reward. *Physiology & Behavior, 97*(5), 551–560.

Stice, E., Yokum, S., Bohon, C., Marti, N., & Smolen, A. (2010). Reward circuitry responsivity to food predicts future increases in body mass: Moderating effects of DRD2 and DRD4. *NeuroImage, 50*(4), 1618–1625.

Stice, E., Burger, K. S., & Yokum, S. (2013). Relative ability of fat and sugar tastes to activate reward, gustatory, and somatosensory regions–. *American Journal of Clinical Nutrition, 98*(6), 1377–1384.

Stoeckel, L. E., Weller, R. E., Cook, E. W., III, Twieg, D. B., Knowlton, R. C., & Cox, J. E. (2008). Widespread reward-system activation in obese women in response to pictures of high-calorie foods. *NeuroImage, 41*(2), 636–647.

Stoeckel, L. E., Kim, J., Weller, R. E., Cox, J. E., Cook, E. W., III, & Horwitz, B. (2009). Effective connectivity of a reward network in obese women. *Brain Research Bulletin, 79*(6), 388–395.

Swaab, D. F. (1997). Prader-Willi syndrome and the hypothalamus. *Acta Paediatrica, 86*(S423), 50–54.

Tammela, L. I., Pääkkönen, A., Karhunen, L. J., Karhu, J., Uusitupa, M. I., & Kuikka, J. T. (2010). Brain electrical activity during food presentation in obese binge-eating women. *Clinical Physiology and Functional Imaging, 30*(2), 135–140.

Thanos, P. K., Michaelides, M., Piyis, Y. K., Wang, G. J., & Volkow, N. D. (2008). Food restriction markedly increases dopamine D2 receptor (D2R) in a rat model of obesity as assessed with in-vivo μPET imaging ([11C] raclopride) and in-vitro ([3H] spiperone) autoradiography. *Synapse, 62*(1), 50–61.

Van den Eynde, F., Claudino, A. M., Mogg, A., Horrell, L., Stahl, D., Ribeiro, W., ... Schmidt, U. (2010). Repetitive transcranial magnetic stimulation reduces cue-induced food craving in bulimic disorders. *Biological Psychiatry, 67*(8), 793–795.

Volkow, N. D., Wang, G. J., Fowler, J. S., Logan, J., Gatley, S. J., Gifford, A., ... Pappas, N. (1999). Prediction of reinforcing responses to psychostimulants in humans by brain dopamine D2 receptor levels. *American Journal of Psychiatry, 156*(9), 1440–1443.

Volkow, N. D., Wang, G. J., Fowler, J. S., Logan, J., Jayne, M., Franceschi, D., ... Pappas, N. (2002). "Nonhedonic" food motivation in humans involves dopamine in the dorsal striatum and methylphenidate amplifies this effect. *Synapse, 44*(3), 175–180.

Volkow, N. D., Fowler, J. S., Wang, G. J., & Swanson, J. M. (2004). Dopamine in drug abuse and addiction: Results from imaging studies and treatment implications. *Molecular Psychiatry, 9*(6), 557.

Volkow, N. D., & Wise, R. A. (2005). How can drug addiction help us understand obesity? *Nature Neuroscience, 8*(5), 555–560.

Volkow, N. D., Wang, G. J., Telang, F., Fowler, J. S., Thanos, P. K., Logan, J., ... Pradhan, K. (2008). Low dopamine striatal D2 receptors are associated with prefrontal metabolism in obese subjects: Possible contributing factors. *NeuroImage, 42*(4), 1537–1543.

Volkow, N. D., Wang, G. J., & Baler, R. D. (2011). Reward, dopamine and the control of food intake: Implications for obesity. *Trends in Cognitive Sciences, 15*(1), 37–46.

Voon, V., Derbyshire, K., Rück, C., Irvine, M. A., Worbe, Y., Enander, J., ... Bullmore, E. T. (2015). Disorders of compulsivity: A common bias towards learning habits. *Molecular Psychiatry, 20*(3), 345.

Wallis, J. D., & Miller, E. K. (2003). Neuronal activity in primate dorsolateral and orbital prefrontal cortex during performance of a reward preference task. *European Journal of Neuroscience, 18*(7), 2069–2081.

Walther, K., Birdsill, A. C., Glisky, E. L., & Ryan, L. (2010). Structural brain differences and cognitive functioning related to body mass index in older females. *Human Brain Mapping, 31*(7), 1052–1064.

Wang, G. J., Volkow, N. D., Fowler, J. S., Cervany, P., Hitzemann, R. J., Pappas, N. R., ... Felder, C. (1999). Regional brain metabolic activation during craving elicited by recall of previous drug experiences. *Life Sciences, 64*(9), 775–784.

Wang, G. J., Volkow, N. D., Logan, J., Pappas, N. R., Wong, C. T., Zhu, W., ... Fowler, J. S. (2001). Brain dopamine and obesity. *The Lancet, 357*(9253), 354–357.

Wang, G.-J., Volkow, N. D., Thanos, P. K., & Fowler, J. S. (2004a). Similarity Between Obesity and Drug Addiction as Assessed by Neurofunctional Imaging. *Journal of Addictive Diseases, 23*(3), 39–53.

Wang, G. J., Volkow, N. D., Telang, F., Jayne, M., Ma, J., Rao, M., ... Fowler, J. S. (2004b). Exposure to appetitive food stimuli markedly activates the human brain. *NeuroImage, 21*(4), 1790–1797.

Wang, G. J., Geliebter, A., Volkow, N. D., Telang, F. W., Logan, J., Jayne, M. C., ... Fowler, J. S. (2011). Enhanced striatal dopamine release during food stimulation in binge eating disorder. *Obesity, 19*(8), 1601–1608.

Watanabe, M., Hikosaka, K., Sakagami, M., & Shirakawa, S. I. (2002). Coding and monitoring of motivational context in the primate prefrontal cortex. *Journal of Neuroscience, 22*(6), 2391–2400.

Weygandt, M., Schaefer, A., Schienle, A., & Haynes, J. D. (2012). Diagnosing different binge-eating disorders based on reward-related brain activation patterns. *Human Brain Mapping, 33*(9), 2135–2146.

Wiedemann, A. A., Saules, K. K., & Ivezaj, V. (2013). Emergence of New Onset substance use disorders among post-weight loss surgery patients. *Clinical Obesity, 3*(6), 194–201.

Wilfley, D. E., Friedman, M. A., Dounchis, J. Z., Stein, R. I., Welch, R. R., & Ball, S. A. (2000). Comorbid psychopathology in binge eating disorder: Relation to eating disorder severity at baseline and following treatment. *Journal of Consulting and Clinical Psychology, 68*(4), 641.

Wise, R. A. (1978). Catecholamine theories of reward: A critical review. *Brain Research, 152*(2), 215–247.

Woolley, J. D., Gorno-Tempini, M. L., Seeley, W. W., Rankin, K., Lee, S. S., Matthews, B. R., & Miller, B. L. (2007). Binge eating is associated with right orbitofrontal-insular-striatal atrophy in frontotemporal dementia. *Neurology, 69*(14), 1424–1433.

World Health Organization. (2017). *Obesity and overweight.* Retrieved from: http://www.who.int/news-room/fact-sheets/detail/obesity-and-overweight.

Wrase, J., Schlagenhauf, F., Kienast, T., Wüstenberg, T., Bermpohl, F., Kahnt, T., ... Heinz, A. (2007). Dysfunction of reward processing correlates with alcohol craving in detoxified alcoholics. *NeuroImage, 35*(2), 787–794.

Zeeb, F. D., & Winstanley, C. A. (2013). Functional disconnection of the orbitofrontal cortex and basolateral amygdala impairs acquisition of a rat gambling task and disrupts animals' ability to alter decision-making behavior after reinforcer devaluation. *Journal of Neuroscience, 33*(15), 6434–6443.

Further reading

Davis, C., & Carter, J. C. (2009). Compulsive overeating as an addiction disorder. A review of theory and evidence. *Appetite, 53*(1), 1–8.

Kessler, R. M., Hutson, P. H., Herman, B. K., & Potenza, M. N. (2016). The neurobiological basis of binge-eating disorder. *Neuroscience & Biobehavioral Reviews, 63*, 223–238.

Miller, J. L., James, G. A., Goldstone, A. P., Couch, J. A., He, G., Driscoll, D. J., & Liu, Y. (2007). Enhanced activation of reward mediating prefrontal regions in response to food stimuli in Prader–Willi syndrome. *Journal of Neurology, Neurosurgery & Psychiatry, 78*(6), 615–619.

Tschöp, M., Wawarta, R., Riepl, R. L., Friedrich, S., Bidlingmaier, M., Landgraf, R., & Folwaczny, C. (2001). Post-prandial decrease of circulating human ghrelin levels. *Journal of Endocrinological Investigation, 24*(6), RC19–RC21.

Volkow, N. D., Wang, G. J., Fowler, J. S., & Telang, F. (2008). Overlapping neuronal circuits in addiction and obesity: Evidence of systems pathology. *Philosophical Transactions of the Royal Society B: Biological Sciences, 363*(1507), 3191–3200.

Volkow, N., Wang, G. J., Fowler, J. S., Tomasi, D., & Baler, R. (2011). Food and drug reward: Overlapping circuits in human obesity and addiction. In *Brain imaging in behavioral neuroscience* (pp. 1–24). Berlin, Heidelberg: Springer.

Volkow, N. D., Wang, G. J., Tomasi, D., & Baler, R. D. (2013). Obesity and addiction: Neurobiological overlaps. *Obesity Reviews, 14*(1), 2–18.

CHAPTER

Interactions of hedonic and homeostatic systems in compulsive overeating

9

Clara Rossetti[1,2], Benjamin Boutrel[1,2,*]

Center for Psychiatric Neuroscience, Department of Psychiatry, Lausanne University Hospital, Switzerland[1]; Division of Adolescent and Child Psychiatry, Department of Psychiatry, Lausanne University Hospital, University of Lausanne, Switzerland[2]

Introduction

The decision to eat is not only influenced by caloric needs but also by nonhomeostatic factors, including food palatability and environmental cues known to trigger conditioned responses (Lutter & Nestler, 2009; Williams & Elmquist, 2012). Hence, while the internal state powerfully drives feeding behaviors through the integration of peripheral signals (peptides, hormones, nutrients) within autonomic, hypothalamic, and limbic brain regions, other intermingled layers of reward processing concur to invigorate feeding behaviors, notably through dopamine signaling and its ability to pair food consumption to the context predicting its availability (Volkow, Wang, Tomasi, & Baler, 2013a). In an evolutionary perspective, the reinforcing properties of palatable foods used to be critical for adapting to poor environments where food sources were scarce. The competence to better anticipate sources of wealthy (palatable) food and the capacity to eat larger amounts when possible (stored in the body as fat for future use) or the profound discomfort associated with caloric restriction contributed to optimize survival. However, food habits have profoundly changed over the past decades. Energy-dense foods, especially high-fat/high-sugar diets, have both a reduced satiety capacity and a higher hedonic value (compared with that of meals richer in fibers, proteins, and/or complex carbohydrates), which may explain their excessive consumption and their role in promoting overweight and obesity (Blundell & Macdiarmid, 1997; Lawton, Burley, Wales, & Blundell, 1993). Moreover, in modern societies, food has become plentiful, ubiquitous, and quite aggressively marketed. As a consequence, the evolutionary adaptation favoring energy storing has become a threatening weakness promoting disinhibited and uncontrolled food-seeking habits. Energy-dense foods are potentially harmful for human health not only for their unbalanced contents but also for their capacity to promote overeating behaviors (Berthoud et al., 2011; Egecioglu et al., 2011).

* Senior author.

Compulsive Eating Behavior and Food Addiction. https://doi.org/10.1016/B978-0-12-816207-1.00009-3
Copyright © 2019 Elsevier Inc. All rights reserved.

The complexity of eating behavior depends on the intrinsic properties of food, which affect simultaneously metabolic and hedonic processes of feeding (Berthoud, 2002; Morton, Meek, & Schwartz, 2014). Consequently, the decision to eat is the result of a tight cooperation between the homeostatic and the hedonic regulation of food intake. Energy homeostasis mostly depends on the hypothalamus, which integrates hunger and satiety stimuli generated by peptide hormones released by peripheral organs. Hypothalamic homeostatic circuits are extensively interconnected with hedonic circuits, which respond to the rewarding properties of food and provide the necessary motivation required for scavenging behaviors, and with executive functions that coordinate decision-making processes (Berthoud, 2011).

Excessive and uncontrolled overeating is a pathological eating behavior observed in eating disorders and some forms of obesity. Binge eating disorder (BED), now classified in the fifth version of the Diagnostic and Statistical Manual of mental disorders (DSM-5) as a separated eating disorder, and bulimia nervosa (BN) are both characterized by recurrent episodes of compulsive overeating (aka binge eating episodes) (APA, 2013). However, unlike BN, BED is not accompanied by compensatory behaviors (e.g., vomiting, laxative abuse, dieting, or excessive physical activity) and therefore often comorbid with obesity. Binge eating refers to rapid consumption of large amount of highly palatable food without hunger. This eating pattern persists despite physical discomfort and is associated with marked distress and dysphoric mood. Additionally, binge eaters experience loss of control over eating and dedicate a great deal of time in procuring food. After binge episodes, these subjects manifest frequently a feeling of guilt and disgust, and all their efforts to give up binging are often unsuccessful (Citrome, 2015).

Recently, increased attention has been brought to certain forms of obesity that are characterized by the overconsumption of palatable foods presenting very high reinforcing properties. Although it is not recognized as an official disorder in the DSM-5, "food addiction" can be diagnosed through the Yale Food Addiction Scale (YFAS), recently introduced and validated by Gearhardt et al. (Gearhardt, Corbin, & Brownell, 2009). This scale has been modified from that used for substance-use disorders in the DSM-5 because binge eating and food addiction may present some commonalities. Both are associated with the incapacity to reduce the quantity of palatable food eaten and the frequency of overeating episodes, despite negative consequences on physical and mental health. However, loss of control eating occurs when large amounts of food are consumed during a discrete period of time in BED and BN, followed by breaks or even purge, while food addiction is defined by frequent episodes of (and almost constant) overwhelming urge for food consumption.

BED is the most frequent eating disorder among adult populations with a lifetime prevalence between 0.2% and 4.7% followed by BN with a lifetime prevalence between 1.0% and 2.0% (Kessler et al., 2013). A recent metaanalysis, mainly based on American studies, has also revealed that 25% of obese people screened with the YFAS fulfilled food addiction criteria. Moreover, this incidence increased to 57.6% among patients with a previous diagnosis of BED or BN (Pursey, Stanwell, Gearhardt, Collins, & Burrows, 2014). The former two observations must be interpreted in a context of public health concern: if post-2000 trends continue, global obesity prevalence will reach 18% in men and surpass 21% in women by 2025, while severe obesity will surpass 6% and 9%, respectively (Collaboration NCDRF, 2016).

The current treatments for obesity, including the promotion of body weight loss, are largely ineffective for alleviating signs of compulsive food intake. The development of tailored therapeutic options specifically dedicated to reducing compulsive overeating are necessary. This requires an improved and deeper understanding of the homeostatic–hedonic system interaction.

Ingestive behavior can be seen as the succession of different phases that start by switching the attention toward food and finish with its procurement and consumption. Both the decision to eat and the termination of a meal result from a balanced integration of homeostatic and hedonic signals (Berthoud, 2002). This integration occurs mainly in the hypothalamus, which is directly or indirectly connected to different brain regions belonging to the hedonic system and, at the same time, receives neural and chemical signals coming from peripheral organs.

Despite undeniable progresses in the understanding of the biological mechanisms underlying compulsive overeating, the intricate regulation of the homeostatic and hedonic systems in driving eating behaviors remains partly elusive. It has long been acknowledged that overeating likely depends on an imbalance between opposite circuits, those motivating food consumption and those limiting it. Converging evidence now suggests that recurrent consumption of palatable food disrupts circuits controlling (1) palatability/hedonic response to food; (2) motivation; (3) conditioning/habits formation; and (4) inhibitory control/emotional regulation. Impairments in these circuits would have two main consequences: the rise of the reinforcing value of palatable food and the weakening of control processes.

This review will discuss findings supporting the assumption according to which concomitant impairments of the homeostatic and hedonic regulation may contribute to converting an initial excessive but controlled overeating into a chronic, compulsive, and long-lasting maladaptive eating behavior.

Homeostatic regulation of food intake

The hypothalamus is considered the most critical hub in the brain, where neural and hormonal inputs converge, which ultimately contributes to regulating energy balance. Autonomic signals (sympathetic and parasympathetic) reach the hypothalamus through brainstem nuclei and convey both information about physical and chemical composition of food and mechanical stimuli reflecting the distension of the gastrointestinal tract. Hormonal signals are produced and released in the bloodstream by peripheral tissues before reaching the brain through the blood–brain barrier (Cooke & Bloom, 2006). Depending on their effect on feeding and energy balance, these peripheral factors can be divided into short-term signals, controlling meal size and satiety threshold, and long-term signals that fluctuate in proportion to changes in the basal energy stores. Among the several distinct nuclei and areas that constitute the mammalian hypothalamus, the arcuate nucleus (ARC) is located between the third ventricle and the median eminence, a region where the blood–brain barrier is relatively permeable. Therefore, the ARC is well situated to receive a large array of hormone signals that promote or inhibit food intake. Several studies, including immunohistochemical, functional gene expression, and retrograde tracing techniques, revealed two relevant and distinct neural populations inside the ARC. The first population is orexigenic and releases neuropeptide Y (NPY), agouti-related peptide (AgRP), and gamma-aminobutyric acid (GABA). The second class of neurons is anorexigenic and expresses proopiomelanocortin (POMC) and cocaine- and amphetamine-regulated transcript (CART). These neurons reduce food intake by releasing the melanocortin α-MSH (α-melanocyte–stimulating hormone), whereas NPY neurons can induce food intake by blocking the firing of POMC/CART neurons through direct inhibitory GABAergic projections. NPY exerts its orexigenic activity by binding Y1 and Y5 receptors, whereas α-MSH and AgRP behave, respectively, as agonist and antagonist of melanocortin receptors (MCR3 and MCR4), whose activation on second-order hypothalamic neurons leads to food intake suppression and energy expenditure elevation.

The activity of these two classes of neurons is strongly modulated by adiposity signal peptides (leptin and insulin), satiety peptides (peptide YY [PYY] and glucagon-like peptide 1 [GLP-1]), and the hunger peptide ghrelin, all released from peripheral organs.

Arcuate neurons exert their effect on food intake by modulating the activity of second-order neurons located in the other hypothalamic nuclei. ARC projections to the paraventricular nucleus of the hypothalamus (PVH) allow the communication of orexigenic (NPY/AgRP) neurons with the hypothalamic–pituitary–adrenal (HPA) axis and the hypothalamic–pituitary–thyroid axis, whereas ARC projections to the lateral hypothalamus (LHA) connect anorexigenic (POMC/CART) neurons to orexin/hypocretin (ORX/HCRT) and melanin-concentrating hormone (MCH) neurons. ORX/HCRT and MCH neurons, in turn, project very broadly to cortical (orbitofrontal, piriform sensorimotor, and motor cortex) and mesolimbic (nucleus accumbens [NAc], striatum, and hippocampus) structures. In summary, the hypothalamus represents a neural interface able to collect, sort, and integrate different signals and stimulates hypothalamic–brainstem circuitry regulating behavioral, visceral, and endocrine output components. Although far from being exhaustive, this brief description is intended to give a preliminary introduction about the functioning of the homeostatic system, whereas its interaction between different components of the hedonic system will be discussed in the following paragraphs.

Food palatability and motivation

The pleasurable effect induced by food (or drug) during its consumption (in particular during the first experiences) has been referred to as a "liking" response by Berridge (Berridge, 1996). Many brain regions are involved in this "liking" response, including subcortical forebrain structures (ventral pallidum, NAc, amygdala) and neocortical structures (orbitofrontal cortex [OFC], insular cortex, and anterior cingulate cortex [ACC]) (Cardinal, Parkinson, Hall, & Everitt, 2002; Kringelbach, 2005). However, it is worth noting that not all brain areas activated by pleasure generate pleasure. In particular, two brain structures have this peculiarity: the NAc and the ventral pallidum. Inside these structures, small regions called hedonic hotspots (about 1 mm^3 in the rat and 1 cm^3 in human brain) are responsible for generating unconscious feelings of pleasure (Berridge & Kringelbach, 2008; Pecina, Smith, & Berridge, 2006). The activation of these hotspots is subjected to a complex regulation exerted by different neurotransmitters, such as opioids, endocannabinoids, GABA, and orexin, which can amplify the palatability (or pleasurable effect) of food (Smith & Berridge, 2005, 2007).

An increased sensitivity for palatable food may therefore facilitate overeating. In line with this prediction, individuals from general population showing strong explicit liking response to sweet taste exhibited higher "trait binge eating" score when measured by the binge eating scale (Dalton & Finlayson, 2014; Finlayson, Arlotti, Dalton, King, & Blundell, 2011). Despite the relevance of sweet preference for developing binge eating, only two studies have so far assessed this feature in BED patients. Results were contradictory because one work reported greater sweet preference in BED subjects with high binge eating frequency versus non-BED subjects (Goodman et al., 2018), whereas the second observed decreased sucrose sensitivity in overweight BED subjects relative to overweight controls (Arlt, Smutzer, & Chen, 2017). Differences in the procedure used to assess sweet preference may have contributed to these inconsistent outcomes.

The endogenous opioid system, in particular through the stimulation of μ-receptor (Castro & Berridge, 2014), is considered the most important modulator of food palatability (or "liking" response) independently from hunger and caloric need. Indeed, the pharmacological activation of NAc and ventral pallidum hotspots by DAMGO (D-Ala2, NMe-Phe4, Glyol5-enkephalin), a μ-opioid receptor agonist, has been reported to double the orofacial liking reactions in rats to sweetness solution (Baldo & Kelley, 2007; Smith, Berridge, & Aldridge, 2011) and to enhance consumption of high-fat food (Caref & Nicola, 2018; Zhang, Gosnell, & Kelley, 1998). Moreover, rodents treated with μ-opioid agonists display increased intake of palatable food (Blasio, Steardo, Sabino, & Cottone, 2014; Woolley, Lee, & Fields, 2006). Conversely, μ-opioid antagonists do reduce binge eating behavior, confirming the crucial role played by the opioid system in the "liking" response to palatable food (Cambridge et al., 2013; Giuliano, Robbins, Nathan, Bullmore, & Everitt, 2012). Human findings support preclinical observations: two studies using the selective μ-receptor antagonist GSK1521498 reported reduced sensory hedonic ratings of high-sugar and high-fat dairy products and caloric intake of high-fat/high-sucrose snack foods in obese BED subjects (Nathan et al., 2012; Ziauddeen et al., 2013).

Positive reinforcers (such as palatable food and drugs of abuse) have the intrinsic property to enhance the probability of repeating the specific behaviors required to consume them. The extensive work of Berridge and collaborators has clearly permitted to dissociate the motivational drive (also named as "wanting" response) from the "liking" response. The "wanting" response represents the incentive salience that the brain attributes to rewards and reward-predicting cues. The neurotransmitter that has first been involved in this brain response is dopamine. Dopaminergic projections from the ventral tegmental area (VTA) to the NAc represent the most important component of the brain reward function (Volkow, Wise, & Baler, 2017).

The striatum is mainly composed of medium spiny neurons (MSNs), divided into those expressing dopamine type-1 receptors (D1), forming the direct pathway, and those expressing dopamine type-2 receptors (D2; indirect pathway). While D1-MSNs are considered to mediate reinforcement and reward, D2-MSNs have rather been associated with aversion and avoidance. Consequently, it has been proposed that D1-MSNs may facilitate the selection of rewarding actions encoded in the cortex, while D2-MSNs would rather help to suppress cortical patterns that encode maladaptive or nonrewarding actions. In other words, positive reinforcement learning would be modulated through the D1 direct signaling pathway, while negative reinforcement learning would be modulated by the D2 indirect pathway (Cox et al., 2015; Soares-Cunha, Coimbra, Sousa, & Rodrigues, 2016; Volkow, Wang, Tomasi, & Baler, 2013b). Therefore, phasic release of dopamine would preferentially activate low-affinity D1 dopamine receptors that are involved in the attribution of incentive salience to the reward and to reward-predicting cues (Dreyer, Herrik, Berg, & Hounsgaard, 2010; Richfield, Penney, & Young, 1989), while D2 dopamine receptors would be activated by both phasic and tonic dopamine release; they would not be deemed necessary for reward intake but would rather limit seeking/taking behaviors (Caine et al., 2002).

The balance between D2 and D1 signaling in the ventral and dorsal striatal regions is of high importance for the correct processing of rewards in general and food in particular (Park, Volkow, Pan, & Du, 2013). Although converging observations have shown that the transition from occasional to compulsive drug use is associated with an imbalanced dopamine signaling, with a reduced expression of striatal D2 receptors (Volkow et al., 1996, 2001; Volkow, Fowler, Wang, Swanson, & Telang, 2007) supposedly reflecting neuroadaptation induced by sustained drug-induced dopamine release (Blum et al., 2000) in both drug-addicted subjects (Volkow, Wang, Fowler, & Telang, 2008) and laboratory animals (Markou & Koob, 1991; Paterson, Myers, & Markou, 2000; Schulteis, Markou, Cole, & Koob, 1995), recent reports claimed that obese rats exhibiting binge-like behavior displayed similar neuroadaptations concomitant with altered brain reward function. Moreover, they even demonstrate a causal mechanism between an artificial reduction of D2R expression and compulsive eating (Colantuoni et al., 2001; Johnson & Kenny, 2010).

Confirming the former observations, obese subjects with lower D2R striatal availability displayed a reduced activation of reward circuits when exposed to palatable food (Stice, Spoor, Bohon, Veldhuizen, & Small, 2008; Volkow, Wang, Telang, et al., 2008; Wang et al., 2001; de Weijer et al., 2011). Despite a trend for reduced levels of dopamine in their brain (Bello & Hajnal, 2010), imaging studies reflecting the availability of striatal D2R in obese patients led to mixed results (Karlsson et al., 2015; Steele et al., 2010; Wang et al., 2011), most likely because of the heterogeneity in the recruitment of patients and their individual capacity to manage stress and negative emotions during the execution of the tests.

It is important to highlight that projections from feeding centers to the mesolimbic dopaminergic system and endocrine signals have direct effects on dopamine release. Indeed, orexin/hypocretin and MCH neurons of the LHA project directly to both the VTA and NAc and are able to modulate dopamine release. Similar modulation results by the binding of feeding-related peptides (ghrelin, insulin, leptin) on their receptors located on VTA dopaminergic neurons (Gutierrez, Lobo, Zhang, & de Lecea, 2011; Labouebe et al., 2013; Liu & Borgland, 2015).

Besides dopamine, recent evidence points out the participation of the opioid system in the incentive motivational processes underlying palatable food seeking (Castro & Berridge, 2014; Giuliano & Cottone, 2015; Pecina & Berridge, 2013). Evidence supporting the involvement of opioid system in the *wanting* response to food comes from laboratory animal (Giuliano et al., 2012; Levine & Billington, 2004) and human studies in which opioid receptors have been pharmacologically stimulated inside and outside the NAc (Pecina & Berridge, 2005; Yeomans & Gray, 2002; Ziauddeen et al., 2013).

Binge eaters, as well as animals submitted to specific feeding protocols inducing binge eating patterns, frequently exhibit loss of motivation for standard and less preferred food regularly consumed before the onset of the uncontrolled overeating. This phenomenon is called "anticipatory negative contrast," and it is believed to be dependent on opioid neurotransmission according to a recent report in which nalmefene, a nonselective opioid receptor antagonist, abolished the anticipatory negative contrast and the hyperphagia for highly palatable food. In the same study, an even more specific effect was found after the administration of the selective μ-opioid receptor antagonist GSK1521498 (Cottone et al., 2012). Similar outcomes were obtained in a different anticipatory contrast paradigm in which rats were treated with naltrexone and β-funaltrexamine (Katsuura & Taha, 2014).

According to animal studies, human positron-emission tomography (PET) imaging found reduced [^{11}C]-carfentanil binding of μ-opioid receptor in obese subjects (Karlsson et al., 2015). These results fit with the downregulation of μ-opioid receptor showed in patients addicted to opiates (Koch & Hollt, 2008; Whistler, 2012) but are in contrast with the increase of μ-opioid receptors displayed by cocaine and alcohol abusers (Gorelick et al., 2005; Heinz et al., 2005; Weerts et al., 2011).

An interesting association between binge eating, dopamine, and opioid neurotransmission emerges from a genetic study comparing BED subjects with obese controls. In this work, two different polymorphisms have been investigated: the polymorphism *Taq1A* of the D2 dopamine receptor gene, which leads to a reduction of the D2 receptor density in the striatum (A1$^-$), and the polymorphism A118G of the μ-opioid receptor gene, which determines "a gain of function" of the μ-receptor (G$^+$). The analysis revealed that the 80% of subjects carrying the double polymorphisms ($A1^-/G^+$) were in the BED group (Davis et al., 2009). Although further clinical researches are needed, the evidence of the involvement of the opioid system in both the rewarding and motivational properties of palatable food has raised great interest for the potential use of μ-receptor antagonists as a treatment for uncontrolled overeating.

Conditioned learning and habit formation

It is believed that during the escalation of drug intake (binge/intoxication stage), physiological learning processes can become maladaptive and sustain the progression of addiction. The exposure to a reinforcer stimulates associative learning processes between the reinforcer and the environmental cues that signal its availability. Physiologically, these learning processes affect motivational responses and are required for reward prediction, cue recognition, and goal-directed actions.

The first exposure to a reinforcer drives the firing of dopaminergic neurons in the VTA, which, in turn, release dopamine in the NAc. However, with subsequent and repeated exposure to the same reward, dopaminergic neurons stop firing for reward availability and fire instead for stimuli that predict reward delivery (Norgren, Hajnal, & Mungarndee, 2006; Schultz, 1998) increasing their incentive salience. Therefore, when a reward (unconditioned stimulus) is repeatedly paired with a neutral environmental cue (conditioned stimulus), this latter is learned and can itself trigger reward seeking and taking (Weingarten, 1983) through an associative learning process (Pavlovian conditioning).

In the case of repeated exposure to the same reward, conditioned learning may guide behavior and induce habits formation (Robbins & Costa, 2017).

This hypothesis has been extensively studied in animals demonstrating that the reinforcement of stimulus-reward associations can lead to a transition from a flexible (sensitive to devaluation) voluntary behavior (goal-directed behavior) to an automatic behavior that is habitual and unresponsive to devaluation (dissociated from the outcome) (Belin & Everitt, 2008; Robbins & Costa, 2017).

Referring to food, it is evident that not only external cues related to food, such as the smell or the sight of food, but also environmental cues or mental food representations can trigger compulsive eating even in a satiated state (Higgs, 2016; Petrovich, 2011; Reppucci & Petrovich, 2012).

The ability of high-calorie food to promote overeating depends, therefore, also on its capacity to strengthen the learned association with external incentive stimuli (Ferriday & Brunstrom, 2008). This mechanism, which was advantageous during human evolution in helping to locate the scarce food in the environment, has nowadays the potential to facilitate uncontrolled overeating. Recently, it has been hypothesized that compulsive overeating, similarly to drug addiction, may be, at least in part, sustained by a sort of "overlearning" (Carr, 2011).

A relevant feature of habitual behaviors is that they persist despite devaluation of the outcome. Devaluation procedures have been applied in animal research using bitter-tasting solution to alter the palatability of food. In these conditions, it has been observed that rats exposed to prolonged alternate access to palatable food did not respond to devaluation and lost goal-directed control of responding (Furlong, Jayaweera, Balleine, & Corbit, 2014). Similarly, in humans, recent evidence suggests that obese people are less sensitive to the devaluation procedure and exhibit more habitual responses (Horstmann et al., 2015) and show stronger food cue reactivity and attentional biases when exposed to environmental cues representing high-energy-dense foods (Carnell, Benson, Pantazatos, Hirsch, & Geliebter, 2014; Schmitz, Naumann, Trentowska, & Svaldi, 2014).

In addition, a recent study found that BED obese participants were more prone to habitual learning relative to non-BED obese subjects, and this behavioral predisposition was associated with decreased striatal and OFC gray matter volume (Voon et al., 2015).

Converging evidence suggests that the amygdala plays a key role in processing associative memories, whereby neutral stimuli acquire reinforcing properties through repeated pairings with primary reinforcers. In particular, the basolateral amygdala (BLA) is considered critical in potentiating feeding behaviors despite satiety through an amygdalo-hypothalamic circuit (Petrovich, Setlow, Holland, & Gallagher, 2002). Projections of the BLA to the LHA are both direct and indirect, and disruption of this connection prevents potentiating eating (Holland, Petrovich, & Gallagher, 2002). Indirect pathways that link BLA to LHA through the ventral medial prefrontal cortex are essential in cue-induced feeding as well because neurotoxic lesions of the medial prefrontal cortex are sufficient to abolish appetitive conditioned eating (Petrovich, Ross, Holland, & Gallagher, 2007).

In sharp contrast to the BLA, the central nucleus of amygdala (CeA) plays an important role in the formation of fear conditioning that inhibits feeding even during fasting (Petrovich, Ross, Mody, Holland, & Gallagher, 2009). The central region of the amygdala provides a dense innervation to the LHA. Anterograde tracing and immunohistochemical techniques revealed that inputs from CeA to LHA are mainly inhibitory, suggesting that CeA would inhibit feeding by directly blocking the activity of hypothalamic orexigenic neurons (Nakamura, Tsumori, Yokota, Oka, & Yasui, 2009).

Other brain regions, those involved in the brain reward function particularly, are implicated in the "overlearning" process that drives automatic nonhomeostatic food intake. In particular, the reward stimulation of dopaminergic VTA neurons induces the release of dopamine not only in the NAc but also in other limbic and cortical structures such as the dorsal striatum, the amygdala, the hippocampus, LHA, and the prefrontal cortex (PFC) (Goldstein & Volkow, 2011). In particular, it has been hypothesized that the phasic release of dopamine from VTA to BLA and ventral hippocampus would determine salience attribution to food-related cues and would promote associative learning enhancing hippocampus-dependent long-term memory and amygdala-dependent emotional memory (Everitt & Robbins, 2005). Once established, these memories would affect reward seeking, modulating the activation of dopamine motive pathways.

In support of this assertion, there is evidence that both hippocampus and amygdala send back extensive glutamatergic projections to dopaminergic and GABAergic neurons of VTA and NAc, respectively (Geisler & Wise, 2008). On a molecular basis, the reinforcement of conditioned responses to predictive cues would result from glutamate-dependent plasticity changes, such as alterations in the subunit composition of AMPA and NMDA receptors, in these neurons (Kauer & Malenka, 2007). In summary, these synaptic plasticity processes would render dopamine neurons more sensitive to the excitatory inputs of amygdala and hippocampus and would affect striatal dopamine signaling.

These impairments in striatal dopamine neurotransmission seem also to determine a transition from the ventral to the dorsal striatum control of behavior and ultimately the formation of maladaptive habits (Everitt & Robbins, 2013). The dominance of the dorsal striatum over the ventral striatum in habitual behavior is also supported by the observation that the pharmacological blocking of dopamine release in the dorsal part suppresses compulsive-like responding and restores devaluation in rodents (Belin & Everitt, 2008). Moreover, in a recent PET study, Wang et al., found that BED obese subjects showed higher dopamine release in dorsal striatum (caudate) than obese controls after exposure to food stimuli. Dopamine release was positively correlated with binge behavior, but not with body weight, suggesting that impairment in dorsal striatum functions may be relevant for conditioned cue responses (Wang et al., 2011).

Hence, the strengthening of associative learning processes, occurring during repeated consumption of palatable food, may enhance the incentive value of food-related cues and stimulate hedonic eating over homeostatic needs.

Stress and negative emotional state

In drug addiction, the initial hedonic ("liking") response to drug rapidly loses its intensity and shows tolerance. Instead, the negative emotional state, triggered by opponent motivational processes associated with drug withdrawal, increases slowly and becomes stronger and stronger with repeated drug exposures. This affective state is characterized by anxiety, irritability, agitation, emotional discomfort, and stress responses before committing compulsive actions, as well as a loss of motivation for ordinary rewards. This negative emotional state has been suggested by Koob and Le Moal to be a negative reinforcer able to promote the transition from occasional to compulsive drug use (Koob & Le Moal, 2005). Their hypothesis suggests that this withdrawal-induced negative affect results from concomitant decrease of reward sensitivity and progressive recruitment of stress systems.

The stress response is a complex coordination of multiple brain structures and peripheral organs (Tsigos & Chrousos, 2002) that invigorates the organism to face stressors of any kind. The amygdala and locus coeruleus are key components of the stress system, but we will focus here on the HPA axis. PVH neurons release corticotropin-releasing factor (CRF) in the anterior pituitary gland, triggering the release of the adrenocorticotropic hormone (ACTH) into the bloodstream (Charmandari, Tsigos, & Chrousos, 2005). This hormone stimulates the release of glucocorticoid (GC) hormones from the cortex of the adrenal gland (cortisol in humans and corticosterone in rodents), which, in turn, bind onto glucocorticoid receptors (GRs). Those receptors, widely distributed in the brain and the body, are implicated in complex intracellular signaling cascades leading to the activation/repression of a plethora of genes that control the biological mechanisms able to restore the depleted energy (food intake, gluconeogenesis, fat deposition) (Dostert & Heinzel, 2004). Stress-induced temporary inhibition of food intake is mediated by the inactivation of NPY/AgRP neurons by CRF release in the ARC and by the secretion of ACTH that, such as α-MSH, can stimulate melanocortin receptors (Heinrichs & Richard, 1999; Schulz, Paulus, Lobmann, Dallman, & Lehnert, 2010).

When stress exposure is time-limited, the stress response has short duration because of the strong feedback inhibition exerted by GR binding on the HPA axis activity. GRs in the hippocampus also are critical in the regulation of inhibitory GABAergic projections to PVH neurons which further block the CRF secretion (Herman, Tasker, Ziegler, & Cullinan, 2002). However, chronic stress disrupts the feedback control exerted on the HPA axis and unleashes its activity. As a consequence, the initial stress-induced temporary inhibition of food intake gets reversed by the prolonged action of GCs in favor of facilitating energy storage to face the sustained exposure to stress. In this adapted situation, GCs stimulate food intake by increasing the release of NPY/AgRP and suppressing POMC gene expression in the ARC (Savontaus, Conwell, & Wardlaw, 2002; Shimizu et al., 2008). GCs also alter the activity of appetite-related peptides (e.g., leptin, insulin, and ghrelin) and facilitate fat deposition in white adipose tissue (Rebuffe-Scrive et al., 1990). Not surprisingly, obese people exhibit increased cortisol secretion (Bjorntorp & Rosmond, 2000).

Meanwhile, repeated consumption of high-fat diets alters CRF and GCs release and increases sensitization of reward pathways by potentiating dopaminergic transmission. More precisely, it has been proposed that repeated dopamine release in the NAc stimulates a cascade of events that subsequently activate CRF and dynorphin systems in the extended amygdala increasing brain stress responses. Dynorphins are opioids peptides that bind preferentially the κ-opioid receptors producing analgesic effects similar to those induced by μ-receptors but opposite motivational response. Indeed, dynorphins produce aversive dysphoric-like effects in animal and humans, possibly by reducing dopamine and glutamate release in the NAc and by enhancing the activity of the CRF system (Knoll & Carlezon, 2010; Wee & Koob, 2010).

Therefore, during periods of abstinence (withdrawal), the impairment of the mesocorticolimbic dopamine system (because of the D2 dopamine receptors downregulation) associated with a stress-induced negative emotional state would perpetuate drug consumption in the effort to suppress anxiety and depressive symptoms (Koob & Volkow, 2016).

It is important to mention that the concept of abstinence (withdrawal) in eating disorders is different from that of drug addiction. Indeed, while drug-addicted individuals can quit drugs, compulsive overeaters cannot completely suppress food intake but rather tend to avoid palatable foods or try dieting. Whatever the behavior adopted, abstention from palatable food is a source of motivational withdrawal syndrome, characterized by anxiety, dysphoria, and apathy, which has been frequently observed in humans and laboratory animals (Parylak, Koob, & Zorrilla, 2011; Stice et al., 2008; Teegarden & Bale, 2007). In this frame, palatable food overeating would be a self-therapy producing a "comfort" effect that would help in coping with stress and negative emotionality.

Many reports describe increased palatable food consumption in humans under stressful conditions (Macht, 2008; Tomiyama, Dallman, & Epel, 2011), and stress or intermittent access to palatable food (supposed to induce stress indirectly) has been demonstrated to stimulate palatable food overeating and binge eating in different animal models (Cottone, Sabino, Steardo, & Zorrilla, 2009; Sharma, Fernandes, & Fulton, 2013). The activation of the CRF system in the extended amygdala, similar to what was observed in drug addiction, seems to be a crucial event in stress-induced palatable food overeating. A few human studies only reported impairments within the amygdala in obese patients during palatable food withdrawal (Oltmanns et al., 2012; Stoeckel et al., 2008), but more evidence comes from preclinical research. In particular, increased recruitment of the CRF system, mainly through the activation of the receptor type-1 (CRFR1), has been described in the central amygdala (CeA) in response to palatable food withdrawal, and the injection of selective CRFR1 antagonists directly in the CeA blocked stress-induced binge eating (Cottone, Sabino, Roberto, et al., 2009; Iemolo et al., 2013; Micioni Di Bonaventura et al., 2014).

Consistent with a possible role of negative emotionality in palatable food overeating, binge eaters show high rates of psychiatric comorbidity. Major depression, bipolar disorders, anxiety disorders, and drug abuse are the most prevalent. Although frequently reported in obese patients, major depression is much more frequent in BED patients relative to weight-matched obese subjects. Moreover, mental comorbidities are recognized as the cause of extremely high rates of suicidal ideation in BED patients, and the prescription of antidepressants can reduce the frequency and the severity of binge eating episodes alleviating dysphoric symptoms.

Binge eating and impaired executive functions

A typical feature of drug addiction and compulsive overeating is the incapacity to refrain despite aversive consequences. This loss of control over drug or food consumption is related to deficits in executive functions, notably those underlying inhibitory control. Preliminary evidence suggests that executive functions, broadly defined as self-regulatory skills enabling goal-directed behaviors, are compromised in drug-addicted (Koob & Volkow, 2010) as well as in obese and BED individuals (Fitzpatrick, Gilbert, & Serpell, 2013; Voon et al., 2015). Whether reduced inhibitory control is a vulnerability factor or rather a consequence of prolonged drug use or palatable food overconsumption is still debated (Chen et al., 2013; Lubman, Yucel, & Pantelis, 2004; Volkow, Wang, Fowler, et al., 2008).

In animal models, compulsive-like behavior is usually modeled as perseverative reward-seeking/taking despite negative consequences or under adverse conditions. Using different paradigms, compulsive overeating of palatable food has been observed in presence of a mild electric shock (Rossetti, Spena, Halfon, & Boutrel, 2014), in presence of a stimulus signaling an electric shock (Latagliata, Patrono, Puglisi-Allegra, & Ventura, 2010), or when food taking requires entry into a potentially dangerous environment (Cottone et al., 2012).

Inhibitory control over behavior is largely regulated by the PFC, and dysfunctions in different neural pathways connecting PFC and striatum are thought to underlie the loss of control over reward seeking. Accordingly, neurobiological investigations have shown that two functional opposite systems operate in the PFC and regulate motivated behaviors suppressing prepotent, impulsive responses (Koob & Volkow, 2016).

The first system, implicating the ventromedial prefrontal cortex (vmPFC), would represent a "stop" signal able to reassign the appropriate salience value to drug/food-related cues and suppress the emotional-driven consumption of drug and food. Therefore, impairments in this part of the prefrontal cortex would increase cue salience and motivation for rewards. In line with this hypothesis, low activation of the mPFC increases impulsivity and craving for drugs and food (Tomasi & Volkow, 2013), while elevated metabolic activity of this region is associated with the loss of interest for rewards in depressive patients (Ferenczi et al., 2016).

The second system, involving the dorsolateral prefrontal cortex (dlPFC), the ACC, and the OFC, would represent a "go" signal that controls stimulus-reinforcement associations and drives motivation (Volkow & Fowler, 2000). Among the regions constituting this "go" signal, particular interest has been addressed to the OFC. This prefrontal structure plays a crucial role in analyzing the salience of reward-stimulus associations and in correcting behavioral responses when their salience changes to block impulsive choices (Schoenbaum, Roesch, Stalnaker, & Takahashi, 2009). Consequently, damage of the OFC can lead to perseverative behaviors (Rolls, 2004) or to inappropriate salience attribution to reward-related cues and increased motivation for the reward (Holland & Petrovich, 2005).

Reciprocal connections link the OFC to the NAc. The NAc projects to the OFC via the striato-thalamo-orbitofrontal circuit and receives, in turn, dense projections from the OFC. Moreover, the OFC is functionally connected to other limbic brain regions, such as ACC, amygdala, and hippocampus (Ray & Price, 1993).

Extensive research in the field of drug addiction and obesity has established that the activities of the OFC, ACC, and dlPFC are notably under dopamine D2 receptor–mediated regulation. More precisely, it has been observed, first in drug addiction and then in compulsive overeating, that the downregulation of dopamine D2 receptors in the striatum correlated with decreased metabolic activity in these prefrontal structures (Volkow, Wang, Fowler, et al., 2008). Therefore, changes in D2 receptor striatal expression would affect the activity of prefrontal cortices facilitating impulsive and compulsive behavior. In sum, these findings corroborate the hypothesis that the craving for drugs and food, induced by the reduction of striatal dopamine D2 receptors, and the concomitant weakening of inhibitory control exerted by prefrontal structures would be responsible for the loss of control over drug and food consumption. Although scarce, some research works on binge eaters support this assumption. Cognitive studies on binge eaters have shown impaired executive functions in obese BED individual relative to BMI-matched controls (Boeka & Lokken, 2011; Duchesne et al., 2010), and two fMRI neuroimaging studies found reduced mPFC activation in BED subjects compared with obese subject without BED during the execution of a monetary incentive delay task (Balodis, Kober, et al., 2013) or a Stroop color-word interference task (Balodis, Molina, et al., 2013). Increased activation of the OFC has also been observed in BED patients in association with the presentation of pictorial food stimuli (Geliebter et al., 2006; Schienle, Schafer, Hermann, & Vaitl, 2009). Finally, an imaging morphometric study found increased gray matter volume in OFC of BED patients relative to healthy controls (Schafer, Vaitl, & Schienle, 2010).

Impulsivity is a personality trait that has received particular attention in addictive behaviors (de Wit, 2009) and recently in food addiction. Impulsivity is a multidimensional construct composed of an attentional (inability to focus attention or concentrate) and a motor component (acting without thinking). Interestingly, both attentional and motor impulsivity predict addiction-like eating in severe obese patients fulfilling YFAS criteria (Meule, de Zwaan, & Muller, 2017).

Altered serotonin (5-HT) neurotransmission has been often reported in eating disorders. Meanwhile, serotonin signaling has long been known to regulate impulsivity. Therefore, the lower cerebrospinal fluid concentration of 5-HT reported in bulimic patients might not be a coincidence (Jimerson, Lesem, Kaye, & Brewerton, 1992). Similarly, the decreased binding of 5-HT transporter found in binge eating women, possibly reflecting a quantitative reduction of 5-HT uptake (Kuikka et al., 2001), suggests an association between serotonin transmission, impulsive trait, and the vulnerability to develop eating disorders. In line with this assumption, several human polymorphisms of serotonin receptors (5-HT1A, 5-HT2A, 5-HT3) and of the serotonin transporter have demonstrated a tight association with the development of compulsive eating behavior (Bailer & Kaye, 2011).

In 2007, heavy smokers with insula damages have been reported to easily quit tobacco smoking without experiencing any form of craving, withdrawal, or relapse (Naqvi, et al., 2007). Later on, an increasing interest has been brought to the insula in the context of heavy drug and food consumption. The insular cortex can be anatomically divided into three portions: the anterior insula (agranular), the middle insula (dysgranular), and the posterior insula (granular). While this latter is referred as the primary interoceptive cortex, the anterior part would assign awareness to interoceptive signals and would integrate these signals with emotional and motivational inputs. This integrative function of the anterior insula is assured by its reciprocal connections with the BLA, NAc, hippocampal formation, ACC, and OFC (Berthoud, 2002). The anterior insula has been shown to regulate impulse control and risky decision-making (under uncertainty). According to the model proposed by Menon and Uddin (Menon & Uddin, 2010), the insula would play a role in executive processes because of its connection with ACC and OFC; its impairment would facilitate the shift to automatic habit formation.

Taste is one of the interoceptive signals processed by the anterior insula, and, in humans, tasting palatable food activates the insula and midbrain areas. Consistently, during PET imaging studies, obese subjects exhibited greater insular activation than control subjects when tasting a liquid meal composed of sugar and fat (DelParigi, Chen, Salbe, Reiman, & Tataranni, 2005). Similarly, obese patients showed enhanced activation of the insula in response to palatable food (Stice et al., 2008; Stice, Yokum, Burger, Epstein, & Small, 2011), whereas bulimic patients exhibited increased insula activation to high-caloric food pictures in comparison to overweight and normal weight control subjects (Schienle et al., 2009). Given the capacity of the insula to incorporate interoceptive signals in the planning of behavioral responses to food, the relevance of this structure in maladaptive behaviors leading to compulsive overeating deserves deeper investigation.

How homeostatic and hedonic regulations of feeding may concur to drive maladaptive pattern of food intake

Food intake is mostly regulated by hunger and satiety signals that converge into central homeostatic pathways. The hypothalamus plays a critical role in modulating food intake and energy expenditure to stabilize the caloric equation. Several hormones and peptides, released by peripheral organs following meal consumption, are in constant crosstalk with the different nuclei of the hypothalamus (Badman & Flier, 2005).

Unlike obesity and BN, so far very few studies have investigated signaling factors underlying appetite control and energy homeostasis in BED patients. The paucity of data in this field is mostly because of the recent separation of BED from other eating disorders in the DSM-5. The difficulty of recruiting appropriate control subjects is another source of complication.

Satiety depends on stomach distention and the release of satiety peptides, such as cholecystokinin (CCK), peptide YY3-36, and GLP-1. Obese subjects and binge eaters have a larger stomach capacity than healthy controls (Geliebter & Hashim, 2001), and surgical reduction in gastric capacity has been shown to abolish binge eating in BED patients, indicating that large stomach contribute to maintain overeating (Adami, Meneghelli, & Scopinaro, 1999).

Cholecystokinin

CCK was the first member of the gut–brain family of peptide hormones to be associated with meal-induced satiety (Smith & Gibbs, 1975). CCK is mainly released in the duodenum, but cortical neurons expressing CCK as neurotransmitter have been shown to send projection to the mPFC and amygdala. Rats maintained on a high-fat diet exhibit blunted circulating levels of CCK and reduced effect of this peptide on gastric emptying, compared with rats fed with a low-fat diet (Covasa, Grahn, & Ritter, 2000). In addition, an animal model lacking the CCK1 receptor and exhibiting early-onset obesity and binge eating behaviors (the Otsuka Long-Evans Tokushima Fatty—OLETF rats) exhibited increased preference for sucrose and reduction of D2 dopamine expression in the NAc (De Jonghe, Hajnal, & Covasa, 2005). Further confirming a role for CCK in the regulation of food reward, it has been reported that both the "liking" and "wanting" responses for food were compromised in OLEFT rats (Marco, Schroeder, & Weller, 2012). In humans, blunted postprandial release of CCK has been detected in obese and bulimic patients (Keel, Wolfe, Liddle, De Young, & Jimerson, 2007), whereas, unexpectedly, CCK blood levels measured during fasting or following postprandial conditions did not reveal any significant difference in BED patients relative to healthy controls (Geliebter, Gluck, & Hashim, 2005).

Glucagon-like peptide 1

GLP-1 is a gut hormone produced from proglucagon by intestinal cells and by neurons of the nucleus tractus solitarius (NTS). Besides its incretin activity (stimulation of insulin release), GLP-1 is involved in nutrients assimilation and energy homeostasis, by promoting satiety. The capacity of GLP-1 to inhibit food intake is probably because of a combined effect of gastric dilatation and increased blood release of serotonin, which is associated with satiety (Owji et al., 2002). Obese patients show attenuated postprandial increase of GLP-1, and a genetic linkage with the proglucagon gene has been found in families with morbid obesity (Clement et al., 1999; Ranganath, Norris, Morgan, Wright, & Marks, 1999). While BN is generally associated with blunted postprandial GLP-1 release, mixed data have been found in BED with either similar levels (Geliebter, Ochner, & Aviram-Friedman, 2008) or reduced GLP-1 release in BED individuals versus non-BED obese individuals (Dossat, Bodell, Williams, Eckel, & Keel, 2015).

In the brain, GPL-1 receptors are expressed in multiple regions, such as NTS, LHA, VTA, NAc, and ventral hippocampus. Furthermore, systemic administration of a GLP-1 analog, exendin-4, induced Fos activation in NAc (Gu et al., 2013), whereas its direct administration into the VTA reduced both the "liking" and "wanting" responses for sucrose and palatable food (Alhadeff, Rupprecht, & Hayes, 2012; Dickson et al., 2012). The activation of GLP-1 receptors in VTA and NAc, via another GLP-1 agonist (liraglutide), has shown to selectively reduce intake of highly palatable energy-dense food without producing a significant suppression of intake of a standard diet in rats presented with both diets simultaneously (Dossat, Lilly, Kay, & Williams, 2011).

In agreement with these effects on food reward, the activation of GLP-1 receptors by systemic injection of exendin-4 reduced the size of the binge eating episodes in a mouse model of binge eating behavior (Lutter et al., 2017). A pilot study in human binge eaters also reported improvement of binge eating behavior after 3 months of liraglutide treatment (Robert et al., 2015). This encouraging result suggests that modulation of GLP-1 activity may represent a promising approach to treat maladaptive eating behaviors.

Peptide YY

The PYY is a short-time appetite regulator belonging to the NPY family released in proportion to the energy content of a meal. PYY reduces food intake in rodents and in humans through the inhibitions of NPY neurons of the ARC (Batterham et al., 2002). Endogenous postprandial levels of PYY were significantly lower in obese subject compared with lean controls (Batterham et al., 2003), and in bulimic patients lower PYY release was associated with greater binge eating frequency (Rigamonti et al., 2014). Unexpectedly, PYY blood levels measured during fasting or following postprandial conditions did not reveal any significant difference between BED patients and obese subjects without BED (Geliebter et al., 2005).

Leptin

Leptin, a hunger-suppressant signal, is an adipocyte hormone released in proportion to the volume of the adipose tissue. This adipokine exerts its homeostatic function mainly in the ARC of the hypothalamus where it stimulates anorexigenic neurons expressing POMC/CART and inhibits orexigenic neurons expressing NPY and AgRP. People with congenital deficiency of leptin are morbidly obese, and body weight gain in obese patients is associated with high leptin serum levels and leptin resistance.

Although leptin disturbances in BED have been frequently linked to the obese status rather than to the symptomatology (Calandra, Musso, & Musso, 2003), higher serum leptin concentration was found in obese BED patients compared with non-binging counterparts (Adami, Campostano, Cella, & Scopinaro, 2002), suggesting that variations of leptin plasmatic levels do not depend solely on the stored fat. Unlikely, in BN patients, reduced basal levels of leptin have been observed compared with healthy controls (Monteleone, Martiadis, Colurcio, & Maj, 2002). The reduction of the hunger-suppressant signal of leptin associated with a blunted release of CCK and GLP-1 may contribute to the binge eating behavior of BN patients.

The simplistic view that peptide hormone alterations affect only energy metabolism is changing. An increasing number of investigations are now showing that these homeostatic factors can influence the activity of the reward system and can also impact cognitive functions (Volkow et al., 2013a). Recent evidence contributed to a better comprehension of the cellular and molecular mechanisms allowing the reciprocal interaction between the homeostatic and reward system. "Liking" and "wanting" responses to food are dynamic processes that are permanently influenced by the activity of peripheral peptide hormones. Generally, hunger increases the pleasantness and the motivation associated with food, while satiety reduces them. This hedonic shift, which can occur both in "linking" and "wanting" responses, has been called "alliesthesia" (Cabanac, 1971). Leptin participates in this shift decreasing "liking" and "wanting" responses induced by food. Accordingly, leptin-deficient patients rated food as more desirable and ob/ob mice (that do not produce leptin) showed enhanced preference for sucrose ("liking" response) (Domingos et al., 2011; Farooqi et al., 2007).

Leptin's reward modulation occurs mainly in the VTA through the inhibition of dopamine neurons expressing leptin receptors (LepRb) (Fulton et al., 2006; Hommel et al., 2006). In fact, in animal studies, leptin injection directly in the brain lowered dopamine release and suppressed food intake (Figlewicz & Benoit, 2009), whereas adenoviral knockdown of the leptin receptor in VTA increased preference for palatable food and enhanced the rewarding properties of food (Davis et al., 2011). Besides its direct effect on VTA neurons, leptin reduces the rewarding properties of food also by acting on ARC, NTS, and LHA. In the LHA, leptin receptors are expressed on a subpopulation of neural cells that send inhibitory projections to orexin and MCH-expressing neurons (Leinninger et al., 2011) with a subsequent reduction of food intake. Inside the ARC, the inhibition of NPY/AgRP neurons by leptin would prevent the firing of orexin neurons of the LHA, the resulting stimulation of opioid neurons in the NAc, and thus would reduce the palatability of food (Krugel, Schraft, Kittner, Kiess, & Illes, 2003). Finally, leptin can also modulate the motivation to eat by acting on the NTS. Indeed, leptin injected into the NTS has been shown to reduce progressive ratio breakpoint for sucrose in both food-restricted and nonrestricted rats (Kanoski, Alhadeff, Fortin, Gilbert, & Grill, 2014).

Leptin receptors are also highly expressed in the hippocampus, suggesting that leptin signaling is involved in learning and memory processes. Indeed, leptin administered in vitro in hippocampal neurons facilitates both long-term potentiation and long-term depression (Durakoglugil, Irving, & Harvey, 2005; Oomura et al., 2006). Unlike dorsal hippocampus, which is mainly involved in spatial memories, the ventral hippocampus contributes to neural processes related to learning and remembering the context where food was consumed (Fanselow & Dong, 2010). Recently, Kanoski and collaborators revealed the capacity of leptin to inhibit the recall of food-related memories by contextual cues when administrated directly in the ventral hippocampus of the rat (Kanoski et al., 2011).

Both animals and humans exhibit increased consumption of palatable food following stress or negative emotionality, even when not hungry (Pecoraro, Reyes, Gomez, Bhargava, & Dallman, 2004). A large body of experimental evidence points out to a role of leptin in the regulation of HPA axis functioning (Roubos, Dahmen, Kozicz, & Xu, 2012). Leptin-deficient (ob/ob) mice were found to be hypersecreting corticosterone (De Vos, Saladin, Auwerx, & Staels, 1995), possibly as a tentative to increase leptin release, as corticosterone can bind the promoter of leptin gene and increase its transcription (Gong, Bi, Pratley, & Weintraub, 1996). Likewise, HPA axis hyperactivation has been also observed in mice lacking the leptin receptor (db/db), suggesting that the normal functioning of leptin receptors is required for the HPA axis regulation (Chen et al., 1996). More recently, it has been proposed that the relationship between leptin and the HPA axis is bidirectional. Thus, under stress condition, the increased release of corticosterone promotes the secretion of leptin that negatively modulates the response of HPA axis to counteract the effect of the stressor (Roubos et al., 2012). Another mechanism by which leptin may suppress emotional, stress-induced feeding is the stimulation of VTA dopaminergic neurons that project to amygdala (Leshan et al., 2010).

In sum, changes in leptin levels might be a relevant explanation for stress-induced eating. Stress-induced cortisol (or corticosterone) elevation may increase leptin levels over the physiological range and promote brain leptin resistance. Accordingly, in humans, psychosocial stress has been associated with higher blood leptin levels (Michels, Sioen, Ruige, & De Henauw, 2017; Otsuka et al., 2006).

Alteration of leptin levels may also account for mood worsening, which is frequently associated with obesity and eating disorders. Mutant mice lacking leptin signaling develop depressive symptoms (Yamada et al., 2011), whereas systemic and central administration of leptin in wild-type mice reduces anxiety and depressive behavior (Guo, Huang, Garza, Chua, & Lu, 2013). In humans, a negative correlation between plasma leptin levels and major depression was reported (Lawson et al., 2012) and in healthy women, psychosocial stress–induced moderate rise of plasma leptin was shown to reduce palatable food consumption (Tomiyama et al., 2012). Collectively, these findings strongly support the assumption according to which impaired leptin signaling may exacerbate the negative emotional effects of palatable food withdrawal and therefore sustain maladaptive patterns of food intake.

The relation between leptin levels and cognitive abilities is less clear, but a recent work showed that larger leptin concentrations were associated, in the general population, with lower total score for the Montreal Cognitive Assessment scale (Warren, Hynan, & Weiner, 2012). Moreover, another study found that adolescents with loss of control over eating exhibited higher plasmatic levels of leptin relative to body weight–matched adolescents without compulsive eating (Miller et al., 2014). Thus, leptin resistance induced by compulsive overeating may, at least in part, be associated with the impairment of some executive functions observed in obesity and with the observation that bariatric surgery can restore them (Alosco et al., 2015).

Ghrelin

Ghrelin, the only orexigenic hormone to be identified, is mainly produced by the enteroendocrine cells of the oxyntic mucosa of the fundus of the stomach. Smaller amounts are also released by the gut and other peripheral organs. Centrally, ghrelin-immunoreactive neurons are found in the medial hypothalamus, close to the third ventricle. Ghrelin released by these cells, and produced peripherally, stimulates the firing of NPY/AgRP neurons of the ARC, increasing food intake through its receptors (Nakazato et al., 2001). Plasma ghrelin levels, in humans and rodents, fluctuate according with food intake. Indeed, they rise during fasting and immediately before meals, whereas they fall 1 h after food consumption, suggesting a role of ghrelin in meal anticipation (Cummings et al., 2001). Besides appetite stimulation, ghrelin also mediates additional biological activities that promote positive energy balance, including the release of growth hormone and the metabolism of glucose and lipids (Schellekens, Dinan, & Cryan, 2010).

As ghrelin increases appetite and food intake, it has been generally expected to find elevated levels of ghrelin in binge eaters and in obese people. However, contrary to this expectation, obese patients showed lower ghrelin levels under fasting conditions compared with healthy controls (Cummings et al., 2002; Tschop et al., 2001) but higher postprandial levels (le Roux et al., 2005). In BN patients, no substantial difference was found at baseline and after a meal, whereas BED subjects exhibited lower ghrelin plasmatic concentration at both time points (Geliebter et al., 2005; Gluck, Yahav, Hashim, & Geliebter, 2014). Interestingly, when ghrelin was measured after a sham-feeding paradigm (e.g., chewing without swallowing, procedure that allows to focus on ghrelin's stimulation of appetite), BN patients had increased ghrelin secretion (Monteleone, Serritella, Scognamiglio, & Maj, 2010), suggesting stronger hunger and propensity to binge.

The abundant central expression of ghrelin receptors outside the hypothalamus reveals the involvement of this hormone in other physiological functions. Accordingly, ghrelin participates in the modulation of learning and memory processes and food intake motivation (Schellekens, Finger, Dinan, & Cryan, 2012).

Ghrelin, unlike leptin, increases hedonic feeding. In rodents, central and peripheral administration of ghrelin stimulated dopamine release in the NAc and food intake. According to the presence of ghrelin receptors on VTA neurons, intra-VTA administration of ghrelin increased high-fat diet consumption in both satiated and fasted rats (Schele, Bake, Rabasa, & Dickson, 2016). Likewise, in rodents, intra-VTA injection of ghrelin raised dopaminergic cell activity, whereas the administration of a ghrelin antagonist blocked ghrelin-induced food intake (Egecioglu et al., 2010). Evidence that ghrelin can alter the rewarding value of food comes from rodent experiments of conditioned place preference (CPP). In this procedure, the animal learns to associate one compartment of the test arena with palatable food and another compartment with regular chow. The animals will consequently spend more time in the compartment associated with palatable food, even in the absence of food. In this paradigm, ghrelin has been shown to increase the amount of time spent in the palatable food-paired chamber, while the pharmacological or genetic blockage of ghrelin signaling attenuated CPP in both rats and mice (Egecioglu et al., 2010; Perello et al., 2010). In addition, ghrelin was found to raise motivation for palatable food by increasing the amount of effort required to obtain reward in self-administration procedures (Skibicka, Hansson, Egecioglu, & Dickson, 2012). In parallel, imaging human studies revealed that intravenously ghrelin-activated brain areas, such as striatum, amygdala, insula, and OFC, are involved in the attribution of rewarding value to food and food cues (Malik, McGlone, Bedrossian, & Dagher, 2008). This array of behavioral evidence strongly supports a key role for ghrelin in food reward–associated behaviors.

Given the role of ghrelin in food anticipation (release of ghrelin before food ingestion), some investigations have been addressed to understand the ability of this hormone to evoke feeding behavior in presence of food-associated cues. They revealed that ghrelin drastically increases cue-related feeding in satiated rats (Perello et al., 2010) and that in the opposite way a ghrelin receptor antagonist administration disrupts cue-potentiated feeding in mice (Walker, Ibia, & Zigman, 2012). Moreover, ventral hippocampus neurons, which are thought to link food-cue memories to the energy balance status, express ghrelin receptors, and around 85% of these neurons project to orexin cells of the LHA (Hsu et al., 2015; Kanoski, Fortin, Ricks, & Grill, 2013).

Accumulating data also suggest the involvement of ghrelin system in stress-induced food intake (Chuang et al., 2011; Diz-Chaves, 2011), and elevated levels of plasmatic ghrelin have been reported in humans and in laboratory animals subjected to stress procedures (Chuang & Zigman, 2010; Kristenssson et al., 2006; Lutter et al., 2008).

Interestingly, ghrelin levels have been proposed to correlate with the excessive consumption of "comfort food," suggesting a mechanism allowing to cope with stressful situations. Therefore, ghrelin signaling may facilitate hedonic feeding, in particular because "nonemotional eaters," defined as individuals whose food intake was suppressed or unchanged by stress, exhibited reduced levels of ghrelin after food consumption, in sharp contrast to "emotional eaters," defined as individuals prone to consume palatable food during stress (Raspopow, Abizaid, Matheson, & Anisman, 2010).

Compulsive overeating and substance-use disorders

The concept of "food addiction" proposed by Randolph in 1956 has been ignored until the last decade when it has received special attention because of the worldwide increase of obesity rate.

The YFAS, validated by Gearhardt and colleagues (Gearhardt, Appetite, 2009), is the first questionnaire aiming at assessing compulsive overeating on the base of diagnostic criteria used for substance dependence. Although the name of this questionnaire refers to food addiction, it is clearly more focused on assessing maladaptive eating behaviors rather than the addictive property of food.

At the present, obesity is not considered as a mental disorder and therefore not included in the DSM-5. Meanwhile, BED and bulimia are acknowledged in the "*Feeding and Eating Disorder*" section of the diagnostic and statistical manual of mental disorders. When applied to obese and BED patients, the YFAS recognizes as "food addicted" only a subset of obese and BED patients (Gearhardt et al., 2012; Pedram et al., 2013). This raises a problem with classifying those patients who do not correspond to the YFAS, such as nonobese BED patients and BN patients, but who exhibit recurrent episodes of compulsive overeating.

Some authors have highlighted that compulsive overeaters fulfilling the YFAS criteria should be included in the *Substance-Use Disorder* section of DSM-5 because of the frequent overlap of their behavioral symptoms with drug addiction. This assertion is supported by the fact that compulsive overeaters, such as drug abusers, present (1) consumption of larger amounts of food than intended, great deal of time spent in getting and eating food, persistent desire for food and unsuccessful attempts to reduce the amount of food eaten (criteria for *impaired control* in substance addiction); (2) isolation from social activities because of obesity (criteria for *social impairment* in substance addiction); (3) maintenance of overeating despite knowledge of adverse physical and psychological consequences caused by excessive food consumption (criteria for *risky use* in substance addiction); and (4) distress and dysphoria after episodes of binge eating (withdrawal criterion for *pharmacological effects* in substance addiction) (Gearhardt, White, & Potenza, 2011; Meule & Gearhardt, 2014; Ziauddeen, Farooqi, & Fletcher, 2012).

Furthermore, the link with *Substance-Use Disorders* has been justified by the fact that compulsive overeaters not only share behavioral similarities but also anatomical and neurochemical similarities with patients suffering from drug addiction. However, the inclusion of compulsive overeaters in the *Substance-Use Disorder* section would imply that certain types of food have addictive properties. Although many animal studies report the addictive potential of high-sugar and high-fat food, there is not convincing evidence that ingredients or macronutrients could be labeled as addictive in humans, apart a single case study of sugar addiction (Thornley & McRobbie, 2009). Rather, diets of people with compulsive overeating often contain a broad range of different palatable energy-dense foods.

Nevertheless, excessive consumption of palatable foods severely impacts the functioning of homeostatic and hedonic pathways, possibly triggering counteradaptations and leading to uncontrolled seeking and taking behaviors, quite similarly to what is reported in drug addiction.

Opening a relevant debate, a few authors have very recently proposed to consider compulsive overeating as a behavioral addiction, and they suggest to favor the concept of "eating addiction" rather than that of "food addiction" (DiLeone, Taylor, & Picciotto, 2012; Hebebrand et al., 2014).

Improving the current knowledge on the biological underpinnings of food intake may provide useful information about drug addiction. For instance, an intriguing literature reports that the caloric equation can affect drug consumption. In particular, food deprivation strengthens the effect of many drug of abuse, possibly because of the anorexigenic properties of drugs of abuse, but how explaining that, in return, food intake reduces the acquisition of cocaine and alcohol self-administration (Carroll & Lac, 1998). In the last years, there are also an increasing number of studies reporting the capacity of peripheral hormones to modulate drug-induced effects.

One of these hormones, ghrelin, was found to interact with alcohol, stimulants, nicotine, and cannabinoids (for review, see Panagopoulos & Ralevski, 2014). Compelling evidence has been observed for alcohol in both animal and human studies notably. Ghrelin was found to increase alcohol intake in a free-choice paradigm after intracerebroventricular administration in the rat, whereas a ghrelin receptor antagonist induced the opposite effect (Jerlhag et al., 2009). Likewise, in humans, ghrelin administration has been shown to exacerbate cue-induced craving in heavy drinkers (Haass-Koffler et al., 2015).

Emerging data point out to a role for GLP-1 in modulating motivation for psychostimulant drugs (for review, see Reddy, Stanwood, & Galli, 2014). Interestingly, it has been reported that pretreatment with the GLP-1 agonist exendin-4 attenuated both cocaine and amphetamine-induced locomotion in rodents (Egecioglu, Engel, & Jerlhag, 2013). A similar inhibitory effect was found in a CPP test, where exendin-4 reduced the rewarding effects of both amphetamine and cocaine (Graham, Erreger, Galli, & Stanwood, 2013). The ability of GLP-1 to modulate the rewarding properties of drug seems to be relevant also for alcohol. Indeed, exendin-4 diminished alcohol preference (in a CPP paradigm), alcohol motivation (in a progressive ratio operant conditioning paradigm), and withdrawal symptoms after alcohol abstinence (in an incubation craving paradigm) (Egecioglu, Steensland, et al., 2013).

Although very preliminary, some studies also involve leptin in the response to addictive drugs. In humans, there is increasing evidence that plasmatic levels of leptin are associated with alcohol (Bach et al., 2018; Hillemacher et al., 2007), nicotine (von der Goltz et al., 2010), and cocaine (Martinotti et al., 2017) craving. In another work, the inhibition of leptin signaling in VTA after administration of a leptin analogous antagonist or by viral-mediated suppression of leptin receptors induced increased preference for cocaine in a rodent CPP procedure (Shen, Jiang, Liu, Wang, & Ma, 2016).

Conclusion

The increasing prevalence of obesity and BED has prompted the scientific community to deeply investigate the various and intricate biological mechanisms that regulate eating behavior. The extensive number of researches conducted in this field has highlighted that compulsive overeating shares many aspects with compulsive drug use. The clinical and preclinical observations reported in this chapter support the assumption according to which recurrent palatable food consumption may induce brain adaptations similar to those observed following repeated drug administration. More precisely, palatable food overconsumption would strongly stimulate the brain reward function (including the mesocorticolimbic dopaminergic system), which, in turn, would facilitate the learning of food-cue associations and the shift from a goal-directed behavior to habitual behavior. These adaptations would progressively be followed by a desensitization of the reward system, a stress-induced negative emotionality due to the recruitment of the opponent motivational processes and finally, a loss of control over eating because of functional impairments occurring in the prefrontal cortices.

However, the emerging evidence of a regulatory role played by peripheral peptide hormones on the hedonic system highlights the need for a broader view on the neurobiological underpinnings of compulsive eating. Real progresses in this field may come from studies addressed to understand the biological mechanisms linking the homeostatic and hedonic systems. In this perspective, further efforts are necessary to uncover the interaction between the two systems at molecular, cellular, and circuit levels, which also may contribute to better understand whether compulsive overeating may predispose to compulsive drug use and vice versa, given the frequent comorbidity existing between compulsive overeating and drug abuse. Of critical importance would also be the focus on how developmental processes, including gene–environment interaction studies, may corrupt the hedonic and homeostatic regulations of food intake. Indeed, genetic and environmental factors affecting the neurodevelopment of brain structures involved in the cognitive control of behavior may confer higher risk for both drug and food abuse during adolescence, considered a critical period for the emergence of these disorders. In this frame, studies on the effect of maternal eating (such as excessive food intake, excessive consumption of palatable food, or, in contrast, food deprivation) on brain development and epigenetic regulation are strongly encouraged. Furthermore, synergistic efforts between preclinical and clinical research are required for better translating basic research knowledge into clinical investigations and possibly may pave the way toward the development of novel therapeutic approaches aimed at improving diagnosis, prevention, and treatment of compulsive overeating.

References

Adami, G. F., Campostano, A., Cella, F., & Scopinaro, N. (2002). Serum leptin concentration in obese patients with binge eating disorder. *International Journal of Obesity and Related Metabolic Disorders, 26*(8), 1125–1128. https://doi.org/10.1038/sj.ijo.0802010.

Adami, G. F., Meneghelli, A., & Scopinaro, N. (1999). Night eating and binge eating disorder in obese patients. *International Journal of Eating Disorders, 25*(3), 335–338.

Alhadeff, A. L., Rupprecht, L. E., & Hayes, M. R. (2012). GLP-1 neurons in the nucleus of the solitary tract project directly to the ventral tegmental area and nucleus accumbens to control for food intake. *Endocrinology, 153*(2), 647–658. https://doi.org/10.1210/en.2011-1443.

Alosco, M. L., Spitznagel, M. B., Strain, G., Devlin, M., Cohen, R., Crosby, R. D., … Gunstad, J. (2015). Improved serum leptin and ghrelin following bariatric surgery predict better postoperative cognitive function. *Journal of Clinical Neurology, 11*(1), 48–56. https://doi.org/10.3988/jcn.2015.11.1.48.

American Psychiatric Association (APA). (2013). *Diagnostic and statistical manual of mental disorders* (5th ed.). American Psychiatric Publishing.

Arlt, J. M., Smutzer, G. S., & Chen, E. Y. (2017). Taste assessment in normal weight and overweight individuals with co-occurring Binge Eating Disorder. *Appetite, 113*, 239–245. https://doi.org/10.1016/j.appet.2017.02.034.

Bach, P., Bumb, J. M., Schuster, R., Vollstadt-Klein, S., Reinhard, I., Rietschel, M., … Koopmann, A. (2018). Effects of leptin and ghrelin on neural cue-reactivity in alcohol addiction: Two streams merge to one river? *Psychoneuroendocrinology, 100*, 1–9. https://doi.org/10.1016/j.psyneuen.2018.09.026.

Badman, M. K., & Flier, J. S. (2005). The gut and energy balance: Visceral allies in the obesity wars. *Science, 307*(5717), 1909–1914. https://doi.org/10.1126/science.1109951.

Bailer, U. F., & Kaye, W. H. (2011). Serotonin: Imaging findings in eating disorders. *Current Topics in Behavioral Neurosciences, 6*, 59–79. https://doi.org/10.1007/7854_2010_78.

Baldo, B. A., & Kelley, A. E. (2007). Discrete neurochemical coding of distinguishable motivational processes: Insights from nucleus accumbens control of feeding. *Psychopharmacology, 191*(3), 439–459. https://doi.org/10.1007/s00213-007-0741-z.

Balodis, I. M., Kober, H., Worhunsky, P. D., White, M. A., Stevens, M. C., Pearlson, G. D., … Potenza, M. N. (2013). Monetary reward processing in obese individuals with and without binge eating disorder. *Biological Psychiatry, 73*(9), 877–886. https://doi.org/10.1016/j.biopsych.2013.01.014.

Balodis, I. M., Molina, N. D., Kober, H., Worhunsky, P. D., White, M. A., Rajita, S., … Potenza, M. N. (2013). Divergent neural substrates of inhibitory control in binge eating disorder relative to other manifestations of obesity. *Obesity, 21*(2), 367–377. https://doi.org/10.1002/oby.20068.

Batterham, R. L., Cohen, M. A., Ellis, S. M., Le Roux, C. W., Withers, D. J., Frost, G. S., … Bloom, S. R. (2003). Inhibition of food intake in obese subjects by peptide YY3-36. *New England Journal of Medicine, 349*(10), 941–948. https://doi.org/10.1056/NEJMoa030204.

Batterham, R. L., Cowley, M. A., Small, C. J., Herzog, H., Cohen, M. A., Dakin, C. L., … Bloom, S. R. (2002). Gut hormone PYY(3-36) physiologically inhibits food intake. *Nature, 418*(6898), 650–654. https://doi.org/10.1038/nature02666.

Belin, D., & Everitt, B. J. (2008). Cocaine seeking habits depend upon dopamine-dependent serial connectivity linking the ventral with the dorsal striatum. *Neuron, 57*(3), 432–441. https://doi.org/10.1016/j.neuron.2007.12.019.

Bello, N. T., & Hajnal, A. (2010). Dopamine and binge eating behaviors. *Pharmacology Biochemistry and Behavior, 97*(1), 25–33. https://doi.org/10.1016/j.pbb.2010.04.016.

Berridge, K. C. (1996). Food reward: Brain substrates of wanting and liking. *Neuroscience & Biobehavioral Reviews, 20*(1), 1–25.

Berridge, K. C., & Kringelbach, M. L. (2008). Affective neuroscience of pleasure: Reward in humans and animals. *Psychopharmacology, 199*(3), 457–480. https://doi.org/10.1007/s00213-008-1099-6.

Berthoud, H. R. (2002). Multiple neural systems controlling food intake and body weight. *Neuroscience & Biobehavioral Reviews, 26*(4), 393–428.

Berthoud, H. R. (2011). Metabolic and hedonic drives in the neural control of appetite: Who is the boss? *Current Opinion in Neurobiology, 21*(6), 888–896. https://doi.org/10.1016/j.conb.2011.09.004.

Berthoud, H. R., Lenard, N. R., & Shin, A. C. (2011). Food reward, hyperphagia, and obesity. *American Journal of Physiology. Regulatory, Integrative and Comparative Physiology, 300*(6), R1266–R1277.

Bjorntorp, P., & Rosmond, R. (2000). Obesity and cortisol. *Nutrition, 16*(10), 924–936.

Blasio, A., Steardo, L., Sabino, V., & Cottone, P. (2014). Opioid system in the medial prefrontal cortex mediates binge-like eating. *Addiction Biology, 19*(4), 652–662. https://doi.org/10.1111/adb.12033.

Blum, K., Braverman, E. R., Holder, J. M., Lubar, J. F., Monastra, V. J., Miller, D., … Comings, D. E. (2000). Reward deficiency syndrome: A biogenetic model for the diagnosis and treatment of impulsive, addictive, and compulsive behaviors. *Journal of Psychoactive Drugs, 32*(Suppl. i–iv), 1–112.

Blundell, J. E., & Macdiarmid, J. I. (1997). Passive overconsumption. Fat intake and short-term energy balance. *Annals of the New York Academy of Sciences, 827*, 392–407.

Boeka, A. G., & Lokken, K. L. (2011). Prefrontal systems involvement in binge eating. *Eating and Weight Disorders, 16*(2), e121–126.

Cabanac, M. (1971). Physiological role of pleasure. *Science, 173*(4002), 1103–1107.

Caine, S. B., Negus, S. S., Mello, N. K., Patel, S., Bristow, L., Kulagowski, J., … Borrelli, E. (2002). Role of dopamine D2-like receptors in cocaine self-administration: Studies with D2 receptor mutant mice and novel D2 receptor antagonists. *Journal of Neuroscience, 22*(7), 2977–2988.

Calandra, C., Musso, F., & Musso, R. (2003). The role of leptin in the etiopathogenesis of anorexia nervosa and bulimia. *Eating and Weight Disorders, 8*(2), 130–137.

Cambridge, V. C., Ziauddeen, H., Nathan, P. J., Subramaniam, N., Dodds, C., Chamberlain, S. R., … Fletcher, P. C. (2013). Neural and behavioral effects of a novel mu opioid receptor antagonist in binge-eating obese people. *Biological Psychiatry, 73*(9), 887–894. https://doi.org/10.1016/j.biopsych.2012.10.022.

Cardinal, R. N., Parkinson, J. A., Hall, J., & Everitt, B. J. (2002). Emotion and motivation: The role of the amygdala, ventral striatum, and prefrontal cortex. *Neuroscience & Biobehavioral Reviews, 26*(3), 321–352.

Caref, K., & Nicola, S. M. (2018). Endogenous opioids in the nucleus accumbens promote approach to high-fat food in the absence of caloric need. *Elife, 7*. https://doi.org/10.7554/eLife.34955.

Carnell, S., Benson, L., Pantazatos, S. P., Hirsch, J., & Geliebter, A. (2014). Amodal brain activation and functional connectivity in response to high-energy-density food cues in obesity. *Obesity, 22*(11), 2370–2378. https://doi.org/10.1002/oby.20859.

Carr, K. D. (2011). Food scarcity, neuroadaptations, and the pathogenic potential of dieting in an unnatural ecology: Binge eating and drug abuse. *Physiology & Behavior, 104*(1), 162–167. https://doi.org/10.1016/j.physbeh.2011.04.023.

Carroll, M. E., & Lac, S. T. (1998). Dietary additives and the acquisition of cocaine self-administration in rats. *Psychopharmacology, 137*(1), 81–89.

Castro, D. C., & Berridge, K. C. (2014). Opioid hedonic hotspot in nucleus accumbens shell: Mu, delta, and kappa maps for enhancement of sweetness "liking" and "wanting". *Journal of Neuroscience, 34*(12), 4239–4250. https://doi.org/10.1523/jneurosci.4458-13.2014.

Charmandari, E., Tsigos, C., & Chrousos, G. (2005). Endocrinology of the stress response. *Annual Review of Physiology, 67*, 259–284. https://doi.org/10.1146/annurev.physiol.67.040403.120816.

Chen, H., Charlat, O., Tartaglia, L. A., Woolf, E. A., Weng, X., Ellis, S. J., … Morgenstern, J. P. (1996). Evidence that the diabetes gene encodes the leptin receptor: Identification of a mutation in the leptin receptor gene in db/db mice. *Cell, 84*(3), 491–495.

Chen, B. T., Yau, H. J., Hatch, C., Kusumoto-Yoshida, I., Cho, S. L., Hopf, F. W., & Bonci, A. (2013). Rescuing cocaine-induced prefrontal cortex hypoactivity prevents compulsive cocaine seeking. *Nature, 496*(7445), 359–362. https://doi.org/10.1038/nature12024.

Chuang, J. C., Perello, M., Sakata, I., Osborne-Lawrence, S., Savitt, J. M., Lutter, M., & Zigman, J. M. (2011). Ghrelin mediates stress-induced food-reward behavior in mice. *Journal of Clinical Investigation, 121*(7), 2684–2692. https://doi.org/10.1172/jci57660.

Chuang, J. C., & Zigman, J. M. (2010). Ghrelin's roles in stress, mood, and anxiety regulation. *International Journal of Peptide International Journal of Peptides, 2010*. https://doi.org/10.1155/2010/460549.

Citrome, L. (2015). A primer on binge eating disorder diagnosis and management. *CNS Spectrums, 20*(Suppl. 1), 44–50. https://doi.org/10.1017/s1092852915000772. quiz 51.

Clement, K., Dina, C., Basdevant, A., Chastang, N., Pelloux, V., Lahlou, N., ... Froguel, P. (1999). A sib-pair analysis study of 15 candidate genes in French families with morbid obesity: Indication for linkage with islet 1 locus on chromosome 5q. *Diabetes, 48*(2), 398–402.

Colantuoni, C., Schwenker, J., McCarthy, J., Rada, P., Ladenheim, B., Cadet, J. L., ... Hoebel, B. G. (2001). Excessive sugar intake alters binding to dopamine and mu-opioid receptors in the brain. *NeuroReport, 12*(16), 3549–3552.

Collaboration NCDRF. (2016). Trends in adult body-mass index in 200 countries from 1975 to 2014: A pooled analysis of 1698 population-based measurement studies with 19.2 million participants. *Lancet, 387*, 1377–1396.

Cooke, D., & Bloom, S. (2006). The obesity pipeline: Current strategies in the development of anti-obesity drugs. *Nature Reviews Drug Discovery, 5*(11), 919–931. https://doi.org/10.1038/nrd2136.

Cottone, P., Sabino, V., Roberto, M., Bajo, M., Pockros, L., Frihauf, J. B., ... Zorrilla, E. P. (2009). CRF system recruitment mediates dark side of compulsive eating. *Proceedings of the National Academy of Sciences of the United States of America, 106*(47), 20016–20020. https://doi.org/10.1073/pnas.0908789106.

Cottone, P., Sabino, V., Steardo, L., & Zorrilla, E. P. (2009). Consummatory, anxiety-related and metabolic adaptations in female rats with alternating access to preferred food. *Psychoneuroendocrinology, 34*(1), 38–49. https://doi.org/10.1016/j.psyneuen.2008.08.010.

Cottone, P., Wang, X., Park, J. W., Valenza, M., Blasio, A., Kwak, J., ... Sabino, V. (2012). Antagonism of sigma-1 receptors blocks compulsive-like eating. *Neuropsychopharmacology, 37*(12), 2593–2604. https://doi.org/10.1038/npp.2012.89.

Covasa, M., Grahn, J., & Ritter, R. C. (2000). High fat maintenance diet attenuates hindbrain neuronal response to CCK. *Regulatory Peptides, 86*(1–3), 83–88.

Cox, S. M., Frank, M. J., Larcher, K., Fellows, L. K., Clark, C. A., Leyton, M., & Dagher, A. (2015). Striatal D1 and D2 signaling differentially predict learning from positive and negative outcomes. *NeuroImage, 109*, 95–101. https://doi.org/10.1016/j.neuroimage.2014.12.070.

Cummings, D. E., Purnell, J. Q., Frayo, R. S., Schmidova, K., Wisse, B. E., & Weigle, D. S. (2001). A preprandial rise in plasma ghrelin levels suggests a role in meal initiation in humans. *Diabetes, 50*(8), 1714–1719.

Cummings, D. E., Weigle, D. S., Frayo, R. S., Breen, P. A., Ma, M. K., Dellinger, E. P., & Purnell, J. Q. (2002). Plasma ghrelin levels after diet-induced weight loss or gastric bypass surgery. *New England Journal of Medicine, 346*(21), 1623–1630. https://doi.org/10.1056/NEJMoa012908.

Dalton, M., & Finlayson, G. (2014). Psychobiological examination of liking and wanting for fat and sweet taste in trait binge eating females. *Physiology & Behavior, 136*, 128–134. https://doi.org/10.1016/j.physbeh.2014.03.019.

Davis, J. F., Choi, D. L., Schurdak, J. D., Fitzgerald, M. F., Clegg, D. J., Lipton, J. W., ... Benoit, S. C. (2011). Leptin regulates energy balance and motivation through action at distinct neural circuits. *Biological Psychiatry, 69*(7), 668–674. https://doi.org/10.1016/j.biopsych.2010.08.028.

Davis, C. A., Levitan, R. D., Reid, C., Carter, J. C., Kaplan, A. S., Patte, K. A., ... Kennedy, J. L. (2009). Dopamine for "wanting" and opioids for "liking": A comparison of obese adults with and without binge eating. *Obesity, 17*(6), 1220–1225. https://doi.org/10.1038/oby.2009.52.

De Jonghe, B. C., Hajnal, A., & Covasa, M. (2005). Increased oral and decreased intestinal sensitivity to sucrose in obese, prediabetic CCK-A receptor-deficient OLETF rats. *American Journal of Physiology - Regulatory, Integrative and Comparative Physiology, 288*(1), R292–R300. https://doi.org/10.1152/ajpregu.00481.2004.

De Vos, P., Saladin, R., Auwerx, J., & Staels, B. (1995). Induction of ob gene expression by corticosteroids is accompanied by body weight loss and reduced food intake. *Journal of Biological Chemistry, 270*(27), 15958–15961.

DelParigi, A., Chen, K., Salbe, A. D., Reiman, E. M., & Tataranni, P. A. (2005). Sensory experience of food and obesity: A positron emission tomography study of the brain regions affected by tasting a liquid meal after a prolonged fast. *NeuroImage, 24*(2), 436–443. https://doi.org/10.1016/j.neuroimage.2004.08.035.

Dickson, S. L., Shirazi, R. H., Hansson, C., Bergquist, F., Nissbrandt, H., & Skibicka, K. P. (2012). The glucagon-like peptide 1 (GLP-1) analogue, exendin-4, decreases the rewarding value of food: A new role for mesolimbic GLP-1 receptors. *Journal of Neuroscience, 32*(14), 4812–4820. https://doi.org/10.1523/jneurosci.6326-11.2012.

DiLeone, R. J., Taylor, J. R., & Picciotto, M. R. (2012). The drive to eat: Comparisons and distinctions between mechanisms of food reward and drug addiction. *Nature Neuroscience, 15*(10), 1330–1335. https://doi.org/10.1038/nn.3202.

Diz-Chaves, Y. (2011). Ghrelin, appetite regulation, and food reward: Interaction with chronic stress. *International Journal of Peptide International Journal of Peptides, 2011*, 898450. https://doi.org/10.1155/2011/898450.

Domingos, A. I., Vaynshteyn, J., Voss, H. U., Ren, X., Gradinaru, V., Zang, F., … Friedman, J. (2011). Leptin regulates the reward value of nutrient. *Nature Neuroscience, 14*(12), 1562–1568. https://doi.org/10.1038/nn.2977.

Dossat, A. M., Bodell, L. P., Williams, D. L., Eckel, L. A., & Keel, P. K. (2015). Preliminary examination of glucagon-like peptide-1 levels in women with purging disorder and bulimia nervosa. *International Journal of Eating Disorders, 48*(2), 199–205. https://doi.org/10.1002/eat.22264.

Dossat, A. M., Lilly, N., Kay, K., & Williams, D. L. (2011). Glucagon-like peptide 1 receptors in nucleus accumbens affect food intake. *Journal of Neuroscience, 31*(41), 14453–14457. https://doi.org/10.1523/jneurosci.3262-11.2011.

Dostert, A., & Heinzel, T. (2004). Negative glucocorticoid receptor response elements and their role in glucocorticoid action. *Current Pharmaceutical Design, 10*(23), 2807–2816.

Dreyer, J. K., Herrik, K. F., Berg, R. W., & Hounsgaard, J. D. (2010). Influence of phasic and tonic dopamine release on receptor activation. *Journal of Neuroscience, 30*(42), 14273–14283. https://doi.org/10.1523/jneurosci.1894-10.2010.

Duchesne, M., Mattos, P., Appolinario, J. C., de Freitas, S. R., Coutinho, G., Santos, C., & Coutinho, W. (2010). Assessment of executive functions in obese individuals with binge eating disorder. *Revista Brasileira de Psiquiatria, 32*(4), 381–388.

Durakoglugil, M., Irving, A. J., & Harvey, J. (2005). Leptin induces a novel form of NMDA receptor-dependent long-term depression. *Journal of Neurochemistry, 95*(2), 396–405. https://doi.org/10.1111/j.1471-4159.2005.03375.x.

Egecioglu, E., Engel, J. A., & Jerlhag, E. (2013). The glucagon-like peptide 1 analogue, exendin-4, attenuates the rewarding properties of psychostimulant drugs in mice. *PLoS One, 8*(7), e69010. https://doi.org/10.1371/journal.pone.0069010.

Egecioglu, E., Jerlhag, E., Salome, N., Skibicka, K. P., Haage, D., Bohlooly, Y. M., … Dickson, S. L. (2010). Ghrelin increases intake of rewarding food in rodents. *Addiction Biology, 15*(3), 304–311. https://doi.org/10.1111/j.1369-1600.2010.00216.x.

Egecioglu, E., Steensland, P., Fredriksson, I., Feltmann, K., Engel, J. A., & Jerlhag, E. (2013). The glucagon-like peptide 1 analogue Exendin-4 attenuates alcohol mediated behaviors in rodents. *Psychoneuroendocrinology, 38*(8), 1259–1270. https://doi.org/10.1016/j.psyneuen.2012.11.009.

Egecioglu, E., Skibicka, K. P., Hansson, C., Alvarez-Crespo, M., Friberg, P. A., Jerlhag, E., Engel, J. A., & Dickson, S. L. (2011). Hedonic and incentive signals for body weight control. *Reviews in Endocrine and Metabolic Disorders, 12*(3), 141–151.

Everitt, B. J., & Robbins, T. W. (2005). Neural systems of reinforcement for drug addiction: From actions to habits to compulsion. *Nature Neuroscience, 8*(11), 1481–1489. https://doi.org/10.1038/nn1579.

Everitt, B. J., & Robbins, T. W. (2013). From the ventral to the dorsal striatum: Devolving views of their roles in drug addiction. *Neuroscience & Biobehavioral Reviews, 37*(9 Pt A), 1946–1954. https://doi.org/10.1016/j.neubiorev.2013.02.010.

Fanselow, M. S., & Dong, H. W. (2010). Are the dorsal and ventral hippocampus functionally distinct structures? *Neuron, 65*(1), 7–19. https://doi.org/10.1016/j.neuron.2009.11.031.

Farooqi, I. S., Bullmore, E., Keogh, J., Gillard, J., O'Rahilly, S., & Fletcher, P. C. (2007). Leptin regulates striatal regions and human eating behavior. *Science, 317*(5843), 1355. https://doi.org/10.1126/science.1144599.

Ferenczi, E. A., Zalocusky, K. A., Liston, C., Grosenick, L., Warden, M. R., Amatya, D., … Deisseroth, K. (2016). Prefrontal cortical regulation of brainwide circuit dynamics and reward-related behavior. *Science, 351*(6268), aac9698. https://doi.org/10.1126/science.aac9698.

Ferriday, D., & Brunstrom, J. M. (2008). How does food-cue exposure lead to larger meal sizes? *British Journal of Nutrition, 100*(6), 1325–1332. https://doi.org/10.1017/S0007114508978296.

Figlewicz, D. P., & Benoit, S. C. (2009). Insulin, leptin, and food reward: Update 2008. *American Journal of Physiology - Regulatory, Integrative and Comparative Physiology, 296*(1), R9–r19. https://doi.org/10.1152/ajpregu.90725.2008.

Finlayson, G., Arlotti, A., Dalton, M., King, N., & Blundell, J. E. (2011). Implicit wanting and explicit liking are markers for trait binge eating. A susceptible phenotype for overeating. *Appetite, 57*(3), 722–728. https://doi.org/10.1016/j.appet.2011.08.012.

Fitzpatrick, S., Gilbert, S., & Serpell, L. (2013). Systematic review: Are overweight and obese individuals impaired on behavioural tasks of executive functioning? *Neuropsychology Review, 23*(2), 138–156. https://doi.org/10.1007/s11065-013-9224-7.

Fulton, S., Pissios, P., Manchon, R. P., Stiles, L., Frank, L., Pothos, E. N., … Flier, J. S. (2006). Leptin regulation of the mesoaccumbens dopamine pathway. *Neuron, 51*(6), 811–822. https://doi.org/10.1016/j.neuron.2006.09.006.

Furlong, T. M., Jayaweera, H. K., Balleine, B. W., & Corbit, L. H. (2014). Binge-like consumption of a palatable food accelerates habitual control of behavior and is dependent on activation of the dorsolateral striatum. *Journal of Neuroscience, 34*(14), 5012–5022. https://doi.org/10.1523/jneurosci.3707-13.2014.

Gearhardt, A. N., Corbin, W. R., & Brownell, K. D. (2009). Preliminary validation of the Yale food addiction scale. *Appetite, 52*(2), 430–436. https://doi.org/10.1016/j.appet.2008.12.003.

Gearhardt, A. N., White, M. A., Masheb, R. M., Morgan, P. T., Crosby, R. D., & Grilo, C. M. (2012). An examination of the food addiction construct in obese patients with binge eating disorder. *International Journal of Eating Disorders, 45*(5), 657–663. https://doi.org/10.1002/eat.20957.

Gearhardt, A. N., White, M. A., & Potenza, M. N. (2011). Binge eating disorder and food addiction. *Current Drug Abuse Reviews, 4*(3), 201–207.

Geisler, S., & Wise, R. A. (2008). Functional implications of glutamatergic projections to the ventral tegmental area. *Reviews in the Neurosciences, 19*(4–5), 227–244.

Geliebter, A., Gluck, M. E., & Hashim, S. A. (2005). Plasma ghrelin concentrations are lower in binge-eating disorder. *Journal of Nutrition, 135*(5), 1326–1330. https://doi.org/10.1093/jn/135.5.1326.

Geliebter, A., & Hashim, S. A. (2001). Gastric capacity in normal, obese, and bulimic women. *Physiology & Behavior, 74*(4–5), 743–746.

Geliebter, A., Ladell, T., Logan, M., Schneider, T., Sharafi, M., & Hirsch, J. (2006). Responsivity to food stimuli in obese and lean binge eaters using functional MRI. *Appetite, 46*(1), 31–35. https://doi.org/10.1016/j.appet.2005.09.002.

Geliebter, A., Ochner, C. N., & Aviram-Friedman, R. (2008). Appetite-related gut peptides in obesity and binge eating disorder. *American Journal of Lifestyle Medicine, 2*(4), 305–314. https://doi.org/10.1177/1559827608317358.

Giuliano, C., & Cottone, P. (2015). The role of the opioid system in binge eating disorder. *CNS Spectrums, 20*(6), 537–545. https://doi.org/10.1017/s1092852915000668.

Giuliano, C., Robbins, T. W., Nathan, P. J., Bullmore, E. T., & Everitt, B. J. (2012). Inhibition of opioid transmission at the mu-opioid receptor prevents both food seeking and binge-like eating. *Neuropsychopharmacology, 37*(12), 2643–2652. https://doi.org/10.1038/npp.2012.128.

Gluck, M. E., Yahav, E., Hashim, S. A., & Geliebter, A. (2014). Ghrelin levels after a cold pressor stress test in obese women with binge eating disorder. *Psychosomatic Medicine, 76*(1), 74–79. https://doi.org/10.1097/psy.0000000000000018.

Goldstein, R. Z., & Volkow, N. D. (2011). Dysfunction of the prefrontal cortex in addiction: Neuroimaging findings and clinical implications. *Nature Reviews Neuroscience, 12*(11), 652–669. https://doi.org/10.1038/nrn3119.

von der Goltz, C., Koopmann, A., Dinter, C., Richter, A., Rockenbach, C., Grosshans, M., … Kiefer, F. (2010). Orexin and leptin are associated with nicotine craving: A link between smoking, appetite and reward. *Psychoneuroendocrinology, 35*(4), 570–577. https://doi.org/10.1016/j.psyneuen.2009.09.005.

Gong, D. W., Bi, S., Pratley, R. E., & Weintraub, B. D. (1996). Genomic structure and promoter analysis of the human obese gene. *Journal of Biological Chemistry, 271*(8), 3971–3974.

Goodman, E. L., Breithaupt, L., Watson, H. J., Peat, C. M., Baker, J. H., Bulik, C. M., & Brownley, K. A. (2018). Sweet taste preference in binge-eating disorder: A preliminary investigation. *Eating Behaviors, 28*, 8–15. https://doi.org/10.1016/j.eatbeh.2017.11.005.

Gorelick, D. A., Kim, Y. K., Bencherif, B., Boyd, S. J., Nelson, R., Copersino, M., … Frost, J. J. (2005). Imaging brain mu-opioid receptors in abstinent cocaine users: Time course and relation to cocaine craving. *Biological Psychiatry, 57*(12), 1573–1582. https://doi.org/10.1016/j.biopsych.2005.02.026.

Graham, D. L., Erreger, K., Galli, A., & Stanwood, G. D. (2013). GLP-1 analog attenuates cocaine reward. *Molecular Psychiatry, 18*(9), 961–962. https://doi.org/10.1038/mp.2012.141.

Guo, M., Huang, T. Y., Garza, J. C., Chua, S. C., & Lu, X. Y. (2013). Selective deletion of leptin receptors in adult hippocampus induces depression-related behaviours. *The International Journal of Neuropsychopharmacology, 16*(4), 857–867. https://doi.org/10.1017/s1461145712000703.

Gu, G., Roland, B., Tomaselli, K., Dolman, C. S., Lowe, C., & Heilig, J. S. (2013). Glucagon-like peptide-1 in the rat brain: Distribution of expression and functional implication. *The Journal of Comparative Neurology, 521*(10), 2235−2261. https://doi.org/10.1002/cne.23282.

Gutierrez, R., Lobo, M. K., Zhang, F., & de Lecea, L. (2011). Neural integration of reward, arousal, and feeding: Recruitment of VTA, lateral hypothalamus, and ventral striatal neurons. *IUBMB Life, 63*(10), 824−830. https://doi.org/10.1002/iub.539.

Haass-Koffler, C. L., Aoun, E. G., Swift, R. M., de la Monte, S. M., Kenna, G. A., & Leggio, L. (2015). Leptin levels are reduced by intravenous ghrelin administration and correlated with cue-induced alcohol craving. *Translational Psychiatry, 5*, e646. https://doi.org/10.1038/tp.2015.140.

Hebebrand, J., Albayrak, O., Adan, R., Antel, J., Dieguez, C., de Jong, J., … Dickson, S. L. (2014). "Eating addiction", rather than "food addiction", better captures addictive-like eating behavior. *Neuroscience & Biobehavioral Reviews, 47*, 295−306. https://doi.org/10.1016/j.neubiorev.2014.08.016.

Heinrichs, S. C., & Richard, D. (1999). The role of corticotropin-releasing factor and urocortin in the modulation of ingestive behavior. *Neuropeptides, 33*(5), 350−359. https://doi.org/10.1054/npep.1999.0047.

Heinz, A., Reimold, M., Wrase, J., Hermann, D., Croissant, B., Mundle, G., … Mann, K. (2005). Correlation of stable elevations in striatal mu-opioid receptor availability in detoxified alcoholic patients with alcohol craving: A positron emission tomography study using carbon 11-labeled carfentanil. *Archives of General Psychiatry, 62*(1), 57−64. https://doi.org/10.1001/archpsyc.62.1.57.

Herman, J. P., Tasker, J. G., Ziegler, D. R., & Cullinan, W. E. (2002). Local circuit regulation of paraventricular nucleus stress integration: Glutamate-gaba connections. *Pharmacology Biochemistry and Behavior, 71*(3), 457−468.

Higgs, S. (2016). Cognitive processing of food rewards. *Appetite, 104*, 10−17. https://doi.org/10.1016/j.appet.2015.10.003.

Hillemacher, T., Bleich, S., Frieling, H., Schanze, A., Wilhelm, J., Sperling, W., … Kraus, T. (2007). Evidence of an association of leptin serum levels and craving in alcohol dependence. *Psychoneuroendocrinology, 32*(1), 87−90. https://doi.org/10.1016/j.psyneuen.2006.09.013.

Holland, P. C., & Petrovich, G. D. (2005). A neural systems analysis of the potentiation of feeding by conditioned stimuli. *Physiology & Behavior, 86*(5), 747−761. https://doi.org/10.1016/j.physbeh.2005.08.062.

Holland, P. C., Petrovich, G. D., & Gallagher, M. (2002). The effects of amygdala lesions on conditioned stimulus-potentiated eating in rats. *Physiology & Behavior, 76*(1), 117−129.

Hommel, J. D., Trinko, R., Sears, R. M., Georgescu, D., Liu, Z. W., Gao, X. B., … DiLeone, R. J. (2006). Leptin receptor signaling in midbrain dopamine neurons regulates feeding. *Neuron, 51*(6), 801−810. https://doi.org/10.1016/j.neuron.2006.08.023.

Horstmann, A., Dietrich, A., Mathar, D., Possel, M., Villringer, A., & Neumann, J. (2015). Slave to habit? Obesity is associated with decreased behavioural sensitivity to reward devaluation. *Appetite, 87*, 175−183. https://doi.org/10.1016/j.appet.2014.12.212.

Hsu, T. M., Hahn, J. D., Konanur, V. R., Noble, E. E., Suarez, A. N., Thai, J., … Kanoski, S. E. (2015). Hippocampus ghrelin signaling mediates appetite through lateral hypothalamic orexin pathways. *Elife, 4*. https://doi.org/10.7554/eLife.11190.

Iemolo, A., Blasio, A., St Cyr, S. A., Jiang, F., Rice, K. C., Sabino, V., & Cottone, P. (2013). CRF-CRF1 receptor system in the central and basolateral nuclei of the amygdala differentially mediates excessive eating of palatable food. *Neuropsychopharmacology, 38*(12), 2456–2466. https://doi.org/10.1038/npp.2013.147.

Jerlhag, E., Egecioglu, E., Landgren, S., Salome, N., Heilig, M., Moechars, D., … Engel, J. A. (2009). Requirement of central ghrelin signaling for alcohol reward. *Proceedings of the National Academy of Sciences of the United States of America, 106*(27), 11318–11323. https://doi.org/10.1073/pnas.0812809106.

Jimerson, D. C., Lesem, M. D., Kaye, W. H., & Brewerton, T. D. (1992). Low serotonin and dopamine metabolite concentrations in cerebrospinal fluid from bulimic patients with frequent binge episodes. *Archives of General Psychiatry, 49*(2), 132–138.

Johnson, P. M., & Kenny, P. J. (2010). Dopamine D2 receptors in addiction-like reward dysfunction and compulsive eating in obese rats. *Nature Neuroscience, 13*(5), 635–641. https://doi.org/10.1038/nn.2519.

Kanoski, S. E., Alhadeff, A. L., Fortin, S. M., Gilbert, J. R., & Grill, H. J. (2014). Leptin signaling in the medial nucleus tractus solitarius reduces food seeking and willingness to work for food. *Neuropsychopharmacology, 39*(3), 605–613. https://doi.org/10.1038/npp.2013.235.

Kanoski, S. E., Fortin, S. M., Ricks, K. M., & Grill, H. J. (2013). Ghrelin signaling in the ventral hippocampus stimulates learned and motivational aspects of feeding via PI3K-Akt signaling. *Biological Psychiatry, 73*(9), 915–923. https://doi.org/10.1016/j.biopsych.2012.07.002.

Kanoski, S. E., Hayes, M. R., Greenwald, H. S., Fortin, S. M., Gianessi, C. A., Gilbert, J. R., & Grill, H. J. (2011). Hippocampal leptin signaling reduces food intake and modulates food-related memory processing. *Neuropsychopharmacology, 36*(9), 1859–1870. https://doi.org/10.1038/npp.2011.70.

Karlsson, H. K., Tuominen, L., Tuulari, J. J., Hirvonen, J., Parkkola, R., Helin, S., … Nummenmaa, L. (2015). Obesity is associated with decreased mu-opioid but unaltered dopamine D2 receptor availability in the brain. *Journal of Neuroscience, 35*(9), 3959–3965. https://doi.org/10.1523/jneurosci.4744-14.2015.

Katsuura, Y., & Taha, S. A. (2014). Mu opioid receptor antagonism in the nucleus accumbens shell blocks consumption of a preferred sucrose solution in an anticipatory contrast paradigm. *Neuroscience, 261*, 144–152. https://doi.org/10.1016/j.neuroscience.2013.12.004.

Kauer, J. A., & Malenka, R. C. (2007). Synaptic plasticity and addiction. *Nature Reviews Neuroscience, 8*(11), 844–858. https://doi.org/10.1038/nrn2234.

Keel, P. K., Wolfe, B. E., Liddle, R. A., De Young, K. P., & Jimerson, D. C. (2007). Clinical features and physiological response to a test meal in purging disorder and bulimia nervosa. *Archives of General Psychiatry, 64*(9), 1058–1066. https://doi.org/10.1001/archpsyc.64.9.1058.

Kessler, R. C., Berglund, P. A., Chiu, W. T., Deitz, A. C., Hudson, J. I., Shahly, V., … Xavier, M. (2013). The prevalence and correlates of binge eating disorder in the world health organization world mental health surveys. *Biological Psychiatry, 73*(9), 904–914. https://doi.org/10.1016/j.biopsych.2012.11.020.

Knoll, A. T., & Carlezon, W. A., Jr. (2010). Dynorphin, stress, and depression. *Brain Research, 1314*, 56–73. https://doi.org/10.1016/j.brainres.2009.09.074.

Koch, T., & Hollt, V. (2008). Role of receptor internalization in opioid tolerance and dependence. *Pharmacology & Therapeutics, 117*(2), 199–206. https://doi.org/10.1016/j.pharmthera.2007.10.003.

Koob, G. F., & Le Moal, M. (2005). Plasticity of reward neurocircuitry and the 'dark side' of drug addiction. *Nature Neuroscience, 8*(11), 1442–1444. https://doi.org/10.1038/nn1105-1442.

Koob, G. F., & Volkow, N. D. (2010). Neurocircuitry of addiction. *Neuropsychopharmacology, 35*(1), 217–238. https://doi.org/10.1038/npp.2009.110.

Koob, G. F., & Volkow, N. D. (2016). Neurobiology of addiction: A neurocircuitry analysis. *Lancet Psychiatry, 3*(8), 760–773. https://doi.org/10.1016/s2215-0366(16)00104-8.

Kringelbach, M. L. (2005). The human orbitofrontal cortex: Linking reward to hedonic experience. *Nature Reviews Neuroscience, 6*(9), 691–702. https://doi.org/10.1038/nrn1747.

Kristenssson, E., Sundqvist, M., Astin, M., Kjerling, M., Mattsson, H., Dornonville de la Cour, C., ... Lindstrom, E. (2006). Acute psychological stress raises plasma ghrelin in the rat. *Regulatory Peptides, 134*(2–3), 114–117. https://doi.org/10.1016/j.regpep.2006.02.003.

Krugel, U., Schraft, T., Kittner, H., Kiess, W., & Illes, P. (2003). Basal and feeding-evoked dopamine release in the rat nucleus accumbens is depressed by leptin. *European Journal of Pharmacology, 482*(1–3), 185–187.

Kuikka, J. T., Tammela, L., Karhunen, L., Rissanen, A., Bergstrom, K. A., Naukkarinen, H., ... Uusitupa, M. (2001). Reduced serotonin transporter binding in binge eating women. *Psychopharmacology, 155*(3), 310–314.

Labouebe, G., Liu, S., Dias, C., Zou, H., Wong, J. C., Karunakaran, S., ... Borgland, S. L. (2013). Insulin induces long-term depression of ventral tegmental area dopamine neurons via endocannabinoids. *Nature Neuroscience, 16*(3), 300–308. https://doi.org/10.1038/nn.3321.

Latagliata, E. C., Patrono, E., Puglisi-Allegra, S., & Ventura, R. (2010). Food seeking in spite of harmful consequences is under prefrontal cortical noradrenergic control. *BMC Neuroscience, 11*, 15. https://doi.org/10.1186/1471-2202-11-15.

Lawson, E. A., Miller, K. K., Blum, J. I., Meenaghan, E., Misra, M., Eddy, K. T., ... Klibanski, A. (2012). Leptin levels are associated with decreased depressive symptoms in women across the weight spectrum, independent of body fat. *Clinical Endocrinology, 76*(4), 520–525. https://doi.org/10.1111/j.1365-2265.2011.04182.x.

Lawton, C. L., Burley, V. J., Wales, J. K., & Blundell, J. E. (1993). Dietary fat and appetite control in obese subjects: Weak effects on satiation and satiety. *International Journal of Obesity and Related Metabolic Disorders, 17*(7), 409–416.

Leinninger, G. M., Opland, D. M., Jo, Y. H., Faouzi, M., Christensen, L., Cappellucci, L. A., ... Myers, M. G., Jr. (2011). Leptin action via neurotensin neurons controls orexin, the mesolimbic dopamine system and energy balance. *Cell Metabolism, 14*(3), 313–323. https://doi.org/10.1016/j.cmet.2011.06.016.

Leshan, R. L., Opland, D. M., Louis, G. W., Leinninger, G. M., Patterson, C. M., Rhodes, C. J., ... Myers, M. G., Jr. (2010). Ventral tegmental area leptin receptor neurons specifically project to and regulate cocaine- and amphetamine-regulated transcript neurons of the extended central amygdala. *Journal of Neuroscience, 30*(16), 5713–5723. https://doi.org/10.1523/jneurosci.1001-10.2010.

Levine, A. S., & Billington, C. J. (2004). Opioids as agents of reward-related feeding: A consideration of the evidence. *Physiology & Behavior, 82*(1), 57–61. https://doi.org/10.1016/j.physbeh.2004.04.032.

Liu, S., & Borgland, S. L. (2015). Regulation of the mesolimbic dopamine circuit by feeding peptides. *Neuroscience, 289*, 19–42. https://doi.org/10.1016/j.neuroscience.2014.12.046.

Lubman, D. I., Yucel, M., & Pantelis, C. (2004). Addiction, a condition of compulsive behaviour? Neuroimaging and neuropsychological evidence of inhibitory dysregulation. *Addiction, 99*(12), 1491–1502. https://doi.org/10.1111/j.1360-0443.2004.00808.x.

Lutter, M., Bahl, E., Hannah, C., Hofammann, D., Acevedo, S., Cui, H., … Michaelson, J. J. (2017). Novel and ultra-rare damaging variants in neuropeptide signaling are associated with disordered eating behaviors. *PLoS One, 12*(8), e0181556. https://doi.org/10.1371/journal.pone.0181556.

Lutter, M., & Nestler, E. J. (2009). Homeostatic and hedonic signals interact in the regulation of food intake. *Journal of Nutrition, 139*(3), 629–632. https://doi.org/10.3945/jn.108.097618.

Lutter, M., Sakata, I., Osborne-Lawrence, S., Rovinsky, S. A., Anderson, J. G., Jung, S., … Zigman, J. M. (2008). The orexigenic hormone ghrelin defends against depressive symptoms of chronic stress. *Nature Neuroscience, 11*(7), 752–753. https://doi.org/10.1038/nn.2139.

Macht, M. (2008). How emotions affect eating: A five-way model. *Appetite, 50*(1), 1–11. https://doi.org/10.1016/j.appet.2007.07.002.

Malik, S., McGlone, F., Bedrossian, D., & Dagher, A. (2008). Ghrelin modulates brain activity in areas that control appetitive behavior. *Cell Metabolism, 7*(5), 400–409. https://doi.org/10.1016/j.cmet.2008.03.007.

Marco, A., Schroeder, M., & Weller, A. (2012). Feeding and reward: Ontogenetic changes in an animal model of obesity. *Neuropharmacology, 62*(8), 2447–2454. https://doi.org/10.1016/j.neuropharm.2012.02.019.

Markou, A., & Koob, G. F. (1991). Postcocaine anhedonia. An animal model of cocaine withdrawal. *Neuropsychopharmacology, 4*(1), 17–26.

Martinotti, G., Montemitro, C., Baroni, G., Andreoli, S., Alimonti, F., Di Nicola, M., … Janiri, L. (2017). Relationship between craving and plasma leptin concentrations in patients with cocaine addiction. *Psychoneuroendocrinology, 85*, 35–41. https://doi.org/10.1016/j.psyneuen.2017.08.004.

Menon, V., & Uddin, L. Q. (2010). Saliency, switching, attention and control: A network model of insula function. *Brain Structure and Function, 214*(5–6), 655–667. https://doi.org/10.1007/s00429-010-0262-0.

Meule, A., de Zwaan, M., & Muller, A. (2017). Attentional and motor impulsivity interactively predict 'food addiction' in obese individuals. *Comprehensive Psychiatry, 72*, 83–87. https://doi.org/10.1016/j.comppsych.2016.10.001.

Meule, A., & Gearhardt, A. N. (2014). Food addiction in the light of DSM-5. *Nutrients, 6*(9), 3653–3671. https://doi.org/10.3390/nu6093653.

Michels, N., Sioen, I., Ruige, J., & De Henauw, S. (2017). Children's psychosocial stress and emotional eating: A role for leptin? *International Journal of Eating Disorders, 50*(5), 471–480. https://doi.org/10.1002/eat.22593.

Micioni Di Bonaventura, M. V., Ciccocioppo, R., Romano, A., Bossert, J. M., Rice, K. C., Ubaldi, M., … Cifani, C. (2014). Role of bed nucleus of the stria terminalis corticotrophin-releasing factor receptors in frustration stress-induced binge-like palatable food consumption in female rats with a history of food restriction. *Journal of Neuroscience, 34*(34), 11316–11324. https://doi.org/10.1523/jneurosci.1854-14.2014.

Miller, R., Tanofsky-Kraff, M., Shomaker, L. B., Field, S. E., Hannallah, L., Reina, S. A., … Yanovski, J. A. (2014). Serum leptin and loss of control eating in children and adolescents. *International Journal of Obesity, 38*(3), 397–403. https://doi.org/10.1038/ijo.2013.126.

Monteleone, P., Martiadis, V., Colurcio, B., & Maj, M. (2002). Leptin secretion is related to chronicity and severity of the illness in bulimia nervosa. *Psychosomatic Medicine, 64*(6), 874–879.

Monteleone, P., Serritella, C., Scognamiglio, P., & Maj, M. (2010). Enhanced ghrelin secretion in the cephalic phase of food ingestion in women with bulimia nervosa. *Psychoneuroendocrinology, 35*(2), 284–288. https://doi.org/10.1016/j.psyneuen.2009.07.001.

Morton, G. J., Meek, T. H., & Schwartz, M. W. (2014). Neurobiology of food intake in health and disease. *Nature Reviews Neuroscience, 15*(6), 367–378. https://doi.org/10.1038/nrn3745.

Nakamura, S., Tsumori, T., Yokota, S., Oka, T., & Yasui, Y. (2009). Amygdaloid axons innervate melanin-concentrating hormone- and orexin-containing neurons in the mouse lateral hypothalamus. *Brain Research, 1278*, 66–74. https://doi.org/10.1016/j.brainres.2009.04.049.

Nakazato, M., Murakami, N., Date, Y., Kojima, M., Matsuo, H., Kangawa, K., & Matsukura, S. (2001). A role for ghrelin in the central regulation of feeding. *Nature, 409*(6817), 194–198. https://doi.org/10.1038/35051587.

Naqvi, N. H., Rudrauf, D., Damasio, H., & Bechara, A. (2007). Damage to the insula disrupts addiction to cigarette smoking. *Science, 315*(5811), 531–534. https://doi.org/10.1126/science.1135926.

Nathan, P. J., O'Neill, B. V., Bush, M. A., Koch, A., Tao, W. X., Maltby, K., … Bullmore, E. T. (2012). Opioid receptor modulation of hedonic taste preference and food intake: A single-dose safety, pharmacokinetic, and pharmacodynamic investigation with GSK1521498, a novel mu-opioid receptor inverse agonist. *The Journal of Clinical Pharmacology, 52*(4), 464–474. https://doi.org/10.1177/0091270011399577.

Norgren, R., Hajnal, A., & Mungarndee, S. S. (2006). Gustatory reward and the nucleus accumbens. *Physiology & Behavior, 89*(4), 531–535. https://doi.org/10.1016/j.physbeh.2006.05.024.

Oltmanns, K. M., Heldmann, M., Daul, S., Klose, S., Rotte, M., Schafer, M., … Lehnert, H. (2012). Sibutramine promotes amygdala activity under fasting conditions in obese women. *Psychopharmacology, 221*(4), 693–700. https://doi.org/10.1007/s00213-011-2615-7.

Oomura, Y., Hori, N., Shiraishi, T., Fukunaga, K., Takeda, H., Tsuji, M., … Sasaki, K. (2006). Leptin facilitates learning and memory performance and enhances hippocampal CA1 long-term potentiation and CaMK II phosphorylation in rats. *Peptides, 27*(11), 2738–2749. https://doi.org/10.1016/j.peptides.2006.07.001.

Otsuka, R., Yatsuya, H., Tamakoshi, K., Matsushita, K., Wada, K., & Toyoshima, H. (2006). Perceived psychological stress and serum leptin concentrations in Japanese men. *Obesity, 14*(10), 1832–1838. https://doi.org/10.1038/oby.2006.211.

Owji, A. A., Khoshdel, Z., Sanea, F., Panjehshahin, M. R., Shojaee Fard, M., Smith, D. M., … Bloom, S. R. (2002). Effects of intracerebroventricular injection of glucagon like peptide-1 and its related peptides on serotonin metabolism and on levels of amino acids in the rat hypothalamus. *Brain Research, 929*(1), 70–75.

Panagopoulos, V. N., & Ralevski, E. (2014). The role of ghrelin in addiction: A review. *Psychopharmacology, 231*(14), 2725–2740. https://doi.org/10.1007/s00213-014-3640-0.

Park, K., Volkow, N. D., Pan, Y., & Du, C. (2013). Chronic cocaine dampens dopamine signaling during cocaine intoxication and unbalances D1 over D2 receptor signaling.

Journal of Neuroscience, 33(40), 15827–15836. https://doi.org/10.1523/jneurosci.1935-13.2013.

Parylak, S. L., Koob, G. F., & Zorrilla, E. P. (2011). The dark side of food addiction. *Physiology & Behavior, 104*(1), 149–156. https://doi.org/10.1016/j.physbeh.2011.04.063.

Paterson, N. E., Myers, C., & Markou, A. (2000). Effects of repeated withdrawal from continuous amphetamine administration on brain reward function in rats. *Psychopharmacology, 152*(4), 440–446.

Pecina, S., & Berridge, K. C. (2005). Hedonic hot spot in nucleus accumbens shell: Where do mu-opioids cause increased hedonic impact of sweetness? *Journal of Neuroscience, 25*(50), 11777–11786. https://doi.org/10.1523/jneurosci.2329-05.2005.

Pecina, S., & Berridge, K. C. (2013). Dopamine or opioid stimulation of nucleus accumbens similarly amplify cue-triggered 'wanting' for reward: Entire core and medial shell mapped as substrates for PIT enhancement. *European Journal of Neuroscience, 37*(9), 1529–1540. https://doi.org/10.1111/ejn.12174.

Pecina, S., Smith, K. S., & Berridge, K. C. (2006). Hedonic hot spots in the brain. *The Neuroscientist, 12*(6), 500–511. https://doi.org/10.1177/1073858406293154.

Pecoraro, N., Reyes, F., Gomez, F., Bhargava, A., & Dallman, M. F. (2004). Chronic stress promotes palatable feeding, which reduces signs of stress: Feedforward and feedback effects of chronic stress. *Endocrinology, 145*(8), 3754–3762. https://doi.org/10.1210/en.2004-0305.

Pedram, P., Wadden, D., Amini, P., Gulliver, W., Randell, E., Cahill, F., … Sun, G. (2013). Food addiction: Its prevalence and significant association with obesity in the general population. *PLoS One, 8*(9), e74832. https://doi.org/10.1371/journal.pone.0074832.

Perello, M., Sakata, I., Birnbaum, S., Chuang, J. C., Osborne-Lawrence, S., Rovinsky, S. A., … Zigman, J. M. (2010). Ghrelin increases the rewarding value of high-fat diet in an orexin-dependent manner. *Biological Psychiatry, 67*(9), 880–886. https://doi.org/10.1016/j.biopsych.2009.10.030.

Petrovich, G. D. (2011). Learning and the motivation to eat: Forebrain circuitry. *Physiology & Behavior, 104*(4), 582–589. https://doi.org/10.1016/j.physbeh.2011.04.059.

Petrovich, G. D., Ross, C. A., Holland, P. C., & Gallagher, M. (2007). Medial prefrontal cortex is necessary for an appetitive contextual conditioned stimulus to promote eating in sated rats. *Journal of Neuroscience, 27*(24), 6436–6441. https://doi.org/10.1523/jneurosci.5001-06.2007.

Petrovich, G. D., Ross, C. A., Mody, P., Holland, P. C., & Gallagher, M. (2009). Central, but not basolateral, amygdala is critical for control of feeding by aversive learned cues. *Journal of Neuroscience, 29*(48), 15205–15212. https://doi.org/10.1523/jneurosci.3656-09.2009.

Petrovich, G. D., Setlow, B., Holland, P. C., & Gallagher, M. (2002). Amygdalo-hypothalamic circuit allows learned cues to override satiety and promote eating. *Journal of Neuroscience, 22*(19), 8748–8753.

Pursey, K. M., Stanwell, P., Gearhardt, A. N., Collins, C. E., & Burrows, T. L. (2014). The prevalence of food addiction as assessed by the Yale food addiction scale: A systematic review. *Nutrients, 6*(10), 4552–4590. https://doi.org/10.3390/nu6104552.

Randolph, T. G. (1956). The descriptive features of food addiction; addictive eating and drinking. *Quarterly Journal of Studies on Alcohol - Part A, 17*(2), 198–224.

Ranganath, L., Norris, F., Morgan, L., Wright, J., & Marks, V. (1999). Inhibition of carbohydrate-mediated glucagon-like peptide-1 (7-36)amide secretion by circulating non-esterified fatty acids. *Clinical Science, 96*(4), 335–342.

Raspopow, K., Abizaid, A., Matheson, K., & Anisman, H. (2010). Psychosocial stressor effects on cortisol and ghrelin in emotional and non-emotional eaters: Influence of anger and shame. *Hormones and Behavior, 58*(4), 677–684. https://doi.org/10.1016/j.yhbeh.2010.06.003.

Ray, J. P., & Price, J. L. (1993). The organization of projections from the mediodorsal nucleus of the thalamus to orbital and medial prefrontal cortex in macaque monkeys. *The Journal of Comparative Neurology, 337*(1), 1–31. https://doi.org/10.1002/cne.903370102.

Rebuffe-Scrive, M., Bronnegard, M., Nilsson, A., Eldh, J., Gustafsson, J. A., & Bjorntorp, P. (1990). Steroid hormone receptors in human adipose tissues. *The Journal of Cinical Endocrinology and Metabolism, 71*(5), 1215–1219. https://doi.org/10.1210/jcem-71-5-1215.

Reddy, I. A., Stanwood, G. D., & Galli, A. (2014). Moving beyond energy homeostasis: New roles for glucagon-like peptide-1 in food and drug reward. *Neurochemistry International, 73*, 49–55. https://doi.org/10.1016/j.neuint.2013.10.003.

Reppucci, C. J., & Petrovich, G. D. (2012). Learned food-cue stimulates persistent feeding in sated rats. *Appetite, 59*(2), 437–447. https://doi.org/10.1016/j.appet.2012.06.007.

Richfield, E. K., Penney, J. B., & Young, A. B. (1989). Anatomical and affinity state comparisons between dopamine D1 and D2 receptors in the rat central nervous system. *Neuroscience, 30*(3), 767–777.

Rigamonti, A. E., Sartorio, A., Scognamiglio, P., Bini, S., Monteleone, A. M., Mastromo, D., … Monteleone, P. (2014). Different effects of cholestyramine on postprandial secretions of cholecystokinin and peptide YY in women with bulimia nervosa. *Neuropsychobiology, 70*(4), 228–234. https://doi.org/10.1159/000368160.

Robbins, T. W., & Costa, R. M. (2017). Habits. *Current Biology, 27*(22), R1200–r1206. https://doi.org/10.1016/j.cub.2017.09.060.

Robert, S. A., Rohana, A. G., Shah, S. A., Chinna, K., Wan Mohamud, W. N., & Kamaruddin, N. A. (2015). Improvement in binge eating in non-diabetic obese individuals after 3 months of treatment with liraglutide - a pilot study. *Obesity Research & Clinical Practice, 9*(3), 301–304. https://doi.org/10.1016/j.orcp.2015.03.005.

Rolls, E. T. (2004). The functions of the orbitofrontal cortex. *Brain and Cognition, 55*(1), 11–29. https://doi.org/10.1016/s0278-2626(03)00277-x.

Rossetti, C., Spena, G., Halfon, O., & Boutrel, B. (2014). Evidence for a compulsive-like behavior in rats exposed to alternate access to highly preferred palatable food. *Addiction Biology, 19*(6), 975–985. https://doi.org/10.1111/adb.12065.

Roubos, E. W., Dahmen, M., Kozicz, T., & Xu, L. (2012). Leptin and the hypothalamo-pituitary adrenal stress axis. *General and Comparative Endocrinology, 177*(1), 28–36. https://doi.org/10.1016/j.ygcen.2012.01.009.

le Roux, C. W., Patterson, M., Vincent, R. P., Hunt, C., Ghatei, M. A., & Bloom, S. R. (2005). Postprandial plasma ghrelin is suppressed proportional to meal calorie content in normal-weight but not obese subjects. *The Journal of Cinical Endocrinology and Metabolism, 90*(2), 1068–1071. https://doi.org/10.1210/jc.2004-1216.

Savontaus, E., Conwell, I. M., & Wardlaw, S. L. (2002). Effects of adrenalectomy on AGRP, POMC, NPY and CART gene expression in the basal hypothalamus of fed and fasted rats. *Brain Research, 958*(1), 130–138.

Schafer, A., Vaitl, D., & Schienle, A. (2010). Regional grey matter volume abnormalities in bulimia nervosa and binge-eating disorder. *NeuroImage, 50*(2), 639–643. https://doi.org/10.1016/j.neuroimage.2009.12.063.

Schele, E., Bake, T., Rabasa, C., & Dickson, S. L. (2016). Centrally administered ghrelin acutely influences food choice in rodents. *PLoS One, 11*(2), e0149456. https://doi.org/10.1371/journal.pone.0149456.

Schellekens, H., Dinan, T. G., & Cryan, J. F. (2010). Lean mean fat reducing "ghrelin" machine: Hypothalamic ghrelin and ghrelin receptors as therapeutic targets in obesity. *Neuropharmacology, 58*(1), 2−16. https://doi.org/10.1016/j.neuropharm.2009.06.024.

Schellekens, H., Finger, B. C., Dinan, T. G., & Cryan, J. F. (2012). Ghrelin signalling and obesity: At the interface of stress, mood and food reward. *Pharmacology & Therapeutics, 135*(3), 316−326. https://doi.org/10.1016/j.pharmthera.2012.06.004.

Schienle, A., Schafer, A., Hermann, A., & Vaitl, D. (2009). Binge-eating disorder: Reward sensitivity and brain activation to images of food. *Biological Psychiatry, 65*(8), 654−661. https://doi.org/10.1016/j.biopsych.2008.09.028.

Schmitz, F., Naumann, E., Trentowska, M., & Svaldi, J. (2014). Attentional bias for food cues in binge eating disorder. *Appetite, 80*, 70−80. https://doi.org/10.1016/j.appet.2014.04.023.

Schoenbaum, G., Roesch, M. R., Stalnaker, T. A., & Takahashi, Y. K. (2009). A new perspective on the role of the orbitofrontal cortex in adaptive behaviour. *Nature Reviews Neuroscience, 10*(12), 885−892. https://doi.org/10.1038/nrn2753.

Schulteis, G., Markou, A., Cole, M., & Koob, G. F. (1995). Decreased brain reward produced by ethanol withdrawal. *Proceedings of the National Academy of Sciences of the United States of America, 92*(13), 5880−5884.

Schultz, W. (1998). Predictive reward signal of dopamine neurons. *Journal of Neurophysiology, 80*(1), 1−27. https://doi.org/10.1152/jn.1998.80.1.1.

Schulz, C., Paulus, K., Lobmann, R., Dallman, M., & Lehnert, H. (2010). Endogenous ACTH, not only alpha-melanocyte-stimulating hormone, reduces food intake mediated by hypothalamic mechanisms. *American Journal of Physiology. Endocrinology and Metabolism, 298*(2), E237−E244. https://doi.org/10.1152/ajpendo.00408.2009.

Sharma, S., Fernandes, M. F., & Fulton, S. (2013). Adaptations in brain reward circuitry underlie palatable food cravings and anxiety induced by high-fat diet withdrawal. *International Journal of Obesity, 37*(9), 1183−1191. https://doi.org/10.1038/ijo.2012.197.

Shen, M., Jiang, C., Liu, P., Wang, F., & Ma, L. (2016). Mesolimbic leptin signaling negatively regulates cocaine-conditioned reward. *Translational Psychiatry, 6*(12), e972. https://doi.org/10.1038/tp.2016.223.

Shimizu, H., Arima, H., Watanabe, M., Goto, M., Banno, R., Sato, I., ... Oiso, Y. (2008). Glucocorticoids increase neuropeptide Y and agouti-related peptide gene expression via adenosine monophosphate-activated protein kinase signaling in the arcuate nucleus of rats. *Endocrinology, 149*(9), 4544−4553. https://doi.org/10.1210/en.2008-0229.

Skibicka, K. P., Hansson, C., Egecioglu, E., & Dickson, S. L. (2012). Role of ghrelin in food reward: Impact of ghrelin on sucrose self-administration and mesolimbic dopamine and acetylcholine receptor gene expression. *Addiction Biology, 17*(1), 95−107. https://doi.org/10.1111/j.1369-1600.2010.00294.x.

Smith, K. S., & Berridge, K. C. (2005). The ventral pallidum and hedonic reward: Neurochemical maps of sucrose "liking" and food intake. *Journal of Neuroscience, 25*(38), 8637−8649. https://doi.org/10.1523/jneurosci.1902-05.2005.

Smith, K. S., & Berridge, K. C. (2007). Opioid limbic circuit for reward: Interaction between hedonic hotspots of nucleus accumbens and ventral pallidum. *Journal of Neuroscience, 27*(7), 1594−1605. https://doi.org/10.1523/jneurosci.4205-06.2007.

Smith, K. S., Berridge, K. C., & Aldridge, J. W. (2011). Disentangling pleasure from incentive salience and learning signals in brain reward circuitry. *Proceedings of the National Academy of Sciences of the United States of America, 108*(27), E255–E264. https://doi.org/10.1073/pnas.1101920108.

Smith, G. P., & Gibbs, J. (1975). Cholecystokinin: A putative satiety signal. *Pharmacology Biochemistry and Behavior, 3*(1 Suppl. l), 135–138.

Soares-Cunha, C., Coimbra, B., Sousa, N., & Rodrigues, A. J. (2016). Reappraising striatal D1- and D2-neurons in reward and aversion. *Neuroscience & Biobehavioral Reviews, 68*, 370–386. https://doi.org/10.1016/j.neubiorev.2016.05.021.

Steele, K. E., Prokopowicz, G. P., Schweitzer, M. A., Magunsuon, T. H., Lidor, A. O., Kuwabawa, H., ... Wong, D. F. (2010). Alterations of central dopamine receptors before and after gastric bypass surgery. *Obesity Surgery, 20*(3), 369–374. https://doi.org/10.1007/s11695-009-0015-4.

Stice, E., Spoor, S., Bohon, C., Veldhuizen, M. G., & Small, D. M. (2008). Relation of reward from food intake and anticipated food intake to obesity: A functional magnetic resonance imaging study. *Journal of Abnormal Psychology, 117*(4), 924–935. https://doi.org/10.1037/a0013600.

Stice, E., Yokum, S., Burger, K. S., Epstein, L. H., & Small, D. M. (2011). Youth at risk for obesity show greater activation of striatal and somatosensory regions to food. *Journal of Neuroscience, 31*(12), 4360–4366. https://doi.org/10.1523/jneurosci.6604-10.2011.

Stoeckel, L. E., Weller, R. E., Cook, E. W., 3rd, Twieg, D. B., Knowlton, R. C., & Cox, J. E. (2008). Widespread reward-system activation in obese women in response to pictures of high-calorie foods. *NeuroImage, 41*(2), 636–647. https://doi.org/10.1016/j.neuroimage.2008.02.031.

Teegarden, S. L., & Bale, T. L. (2007). Decreases in dietary preference produce increased emotionality and risk for dietary relapse. *Biological Psychiatry, 61*(9), 1021–1029. https://doi.org/10.1016/j.biopsych.2006.09.032.

Thornley, S., & McRobbie, H. (2009). Carbohydrate withdrawal: Is recognition the first step to recovery? *N Z Med J, 122*(1290), 133–134.

Tomasi, D., & Volkow, N. D. (2013). Striatocortical pathway dysfunction in addiction and obesity: Differences and similarities. *Critical Reviews in Biochemistry and Molecular Biology, 48*(1), 1–19. https://doi.org/10.3109/10409238.2012.735642.

Tomiyama, A. J., Dallman, M. F., & Epel, E. S. (2011). Comfort food is comforting to those most stressed: Evidence of the chronic stress response network in high stress women. *Psychoneuroendocrinology, 36*(10), 1513–1519. https://doi.org/10.1016/j.psyncuen.2011.04.005.

Tomiyama, A. J., Schamarek, I., Lustig, R. H., Kirschbaum, C., Puterman, E., Havel, P. J., & Epel, E. S. (2012). Leptin concentrations in response to acute stress predict subsequent intake of comfort foods. *Physiology & Behavior, 107*(1), 34–39. https://doi.org/10.1016/j.physbeh.2012.04.021.

Tschop, M., Weyer, C., Tataranni, P. A., Devanarayan, V., Ravussin, E., & Heiman, M. L. (2001). Circulating ghrelin levels are decreased in human obesity. *Diabetes, 50*(4), 707–709.

Tsigos, C., & Chrousos, G. P. (2002). Hypothalamic-pituitary-adrenal axis, neuroendocrine factors and stress. *Journal of Psychosomatic Research, 53*(4), 865–871.

Volkow, N. D., Chang, L., Wang, G. J., Fowler, J. S., Ding, Y. S., Sedler, M., ... Pappas, N. (2001). Low level of brain dopamine D2 receptors in methamphetamine abusers:

Association with metabolism in the orbitofrontal cortex. *American Journal of Psychiatry, 158*(12), 2015–2021. https://doi.org/10.1176/appi.ajp.158.12.2015.

Volkow, N. D., & Fowler, J. S. (2000). Addiction, a disease of compulsion and drive: Involvement of the orbitofrontal cortex. *Cerebral Cortex, 10*(3), 318–325.

Volkow, N. D., Fowler, J. S., Wang, G. J., Swanson, J. M., & Telang, F. (2007). Dopamine in drug abuse and addiction: Results of imaging studies and treatment implications. *Archives of Neurology, 64*(11), 1575–1579. https://doi.org/10.1001/archneur.64.11.1575.

Volkow, N. D., Wang, G. J., Fowler, J. S., Logan, J., Hitzemann, R., Ding, Y. S., … Piscani, K. (1996). Decreases in dopamine receptors but not in dopamine transporters in alcoholics. *Alcoholism: Clinical and Experimental Research, 20*(9), 1594–1598.

Volkow, N. D., Wang, G. J., Fowler, J. S., & Telang, F. (2008). Overlapping neuronal circuits in addiction and obesity: Evidence of systems pathology. *Philosophical Transactions of the Royal Society of London B Biological Sciences, 363*(1507), 3191–3200. https://doi.org/10.1098/rstb.2008.0107.

Volkow, N. D., Wang, G. J., Telang, F., Fowler, J. S., Thanos, P. K., Logan, J., … Pradhan, K. (2008). Low dopamine striatal D2 receptors are associated with prefrontal metabolism in obese subjects: Possible contributing factors. *NeuroImage, 42*(4), 1537–1543. https://doi.org/10.1016/j.neuroimage.2008.06.002.

Volkow, N. D., Wang, G. J., Tomasi, D., & Baler, R. D. (2013a). The addictive dimensionality of obesity. *Biological Psychiatry, 73*(9), 811–818. https://doi.org/10.1016/j.biopsych.2012.12.020.

Volkow, N. D., Wang, G. J., Tomasi, D., & Baler, R. D. (2013b). Unbalanced neuronal circuits in addiction. *Current Opinion in Neurobiology, 23*(4), 639–648. https://doi.org/10.1016/j.conb.2013.01.002.

Volkow, N. D., Wise, R. A., & Baler, R. (2017). The dopamine motive system: Implications for drug and food addiction. *Nature Reviews Neuroscience, 18*(12), 741–752. https://doi.org/10.1038/nrn.2017.130.

Voon, V., Derbyshire, K., Ruck, C., Irvine, M. A., Worbe, Y., Enander, J., … Bullmore, E. T. (2015). Disorders of compulsivity: A common bias towards learning habits. *Molecular Psychiatry, 20*(3), 345–352. https://doi.org/10.1038/mp.2014.44.

Walker, A. K., Ibia, I. E., & Zigman, J. M. (2012). Disruption of cue-potentiated feeding in mice with blocked ghrelin signaling. *Physiology & Behavior, 108*, 34–43. https://doi.org/10.1016/j.physbeh.2012.10.003.

Wang, G. J., Geliebter, A., Volkow, N. D., Telang, F. W., Logan, J., Jayne, M. C., … Fowler, J. S. (2011). Enhanced striatal dopamine release during food stimulation in binge eating disorder. *Obesity, 19*(8), 1601–1608. https://doi.org/10.1038/oby.2011.27.

Wang, G. J., Volkow, N. D., Logan, J., Pappas, N. R., Wong, C. T., Zhu, W., … Fowler, J. S. (2001). Brain dopamine and obesity. *Lancet, 357*(9253), 354–357.

Warren, M. W., Hynan, L. S., & Weiner, M. F. (2012). Leptin and cognition. *Dementia and Geriatric Cognitive Disorders, 33*(6), 410–415. https://doi.org/10.1159/000339956.

Wee, S., & Koob, G. F. (2010). The role of the dynorphin-kappa opioid system in the reinforcing effects of drugs of abuse. *Psychopharmacology, 210*(2), 121–135. https://doi.org/10.1007/s00213-010-1825-8.

Weerts, E. M., Wand, G. S., Kuwabara, H., Munro, C. A., Dannals, R. F., Hilton, J., … McCaul, M. E. (2011). Positron emission tomography imaging of mu- and delta-opioid receptor binding in alcohol-dependent and healthy control subjects. *Alcoholism: Clinical and Experimental Research, 35*(12), 2162–2173. https://doi.org/10.1111/j.1530-0277.2011.01565.x.

de Weijer, B. A., van de Giessen, E., van Amelsvoort, T. A., Boot, E., Braak, B., Janssen, I. M., ... Booij, J. (2011). Lower striatal dopamine D2/3 receptor availability in obese compared with non-obese subjects. *EJNMMI Research, 1*(1), 37. https://doi.org/10.1186/2191-219x-1-37.

Weingarten, H. P. (1983). Conditioned cues elicit feeding in sated rats: A role for learning in meal initiation. *Science, 220*(4595), 431–433.

Whistler, J. L. (2012). Examining the role of mu opioid receptor endocytosis in the beneficial and side-effects of prolonged opioid use: From a symposium on new concepts in mu-opioid pharmacology. *Drug and Alcohol Dependence, 121*(3), 189–204. https://doi.org/10.1016/j.drugalcdep.2011.10.031.

Williams, K. W., & Elmquist, J. K. (2012). From neuroanatomy to behavior: Central integration of peripheral signals regulating feeding behavior. *Nature Neuroscience, 15*(10), 1350–1355. https://doi.org/10.1038/nn.3217.

de Wit, H. (2009). Impulsivity as a determinant and consequence of drug use: A review of underlying processes. *Addiction Biology, 14*(1), 22–31. https://doi.org/10.1111/j.1369-1600.2008.00129.x.

Woolley, J. D., Lee, B. S., & Fields, H. L. (2006). Nucleus accumbens opioids regulate flavor-based preferences in food consumption. *Neuroscience, 143*(1), 309–317. https://doi.org/10.1016/j.neuroscience.2006.06.067.

Yamada, N., Katsuura, G., Ochi, Y., Ebihara, K., Kusakabe, T., Hosoda, K., & Nakao, K. (2011). Impaired CNS leptin action is implicated in depression associated with obesity. *Endocrinology, 152*(7), 2634–2643. https://doi.org/10.1210/en.2011-0004.

Yeomans, M. R., & Gray, R. W. (2002). Opioid peptides and the control of human ingestive behaviour. *Neuroscience & Biobehavioral Reviews, 26*(6), 713–728.

Zhang, M., Gosnell, B. A., & Kelley, A. E. (1998). Intake of high-fat food is selectively enhanced by mu opioid receptor stimulation within the nucleus accumbens. *Journal of Pharmacology and Experimental Therapeutics, 285*(2), 908–914.

Ziauddeen, H., Chamberlain, S. R., Nathan, P. J., Koch, A., Maltby, K., Bush, M., ... Bullmore, E. T. (2013). Effects of the mu-opioid receptor antagonist GSK1521498 on hedonic and consummatory eating behaviour: A proof of mechanism study in binge-eating obese subjects. *Molecular Psychiatry, 18*(12), 1287–1293. https://doi.org/10.1038/mp.2012.154.

Ziauddeen, H., Farooqi, I. S., & Fletcher, P. C. (2012). Obesity and the brain: How convincing is the addiction model? *Nature Reviews Neuroscience, 13*(4), 279–286. https://doi.org/10.1038/nrn3212.

CHAPTER

Genetics and epigenetics of food addiction

10

Caroline Davis, Revi Bonder
York University, Toronto, ON, Canada

Introduction

Compulsive intake is a hallmark characteristic of any substance-use disorder. Intrinsically, therefore, *food addiction* is a syndrome characterized by excessive and loss-of-control (LOC) overeating and is predicated on the notion that foods high in fat, sugar, and salt have the potential to foster neuroadaptations and behavioral changes similar to those that occur from overuse of potent (direct or indirect) dopaminergic agonists such as cocaine, alcohol, and nicotine (Gearhardt, Davis, Kushner, & Brownell, 2011). To date, food addiction research has focused largely on binge eating as a key symptom of this condition, although other patterns of overeating, such as compulsive grazing, are also emerging as important components of the food addiction construct (Bonder, Davis, Kuk, & Loxton, 2018).There is also good evidence—derived mainly from animal research—of a positive association between high-sugar consumption and the abuse of conventional addictive substances, which has prompted the conclusion that the same neural circuitry is involved in the development of both conditions (Ahmed, Guillem, & Vandaele, 2013). A recent human twin study has also confirmed an association between elevated sugar intake and high substance use and attests to the conjoint heritability of these behaviors (Treur, Boomsma, Ligthart, Willemsen, & Vink, 2016). In other words, it appears that a propensity to engage excessively in any type of addictive behavior is rooted—at least to some degree—in an inherent biological susceptibility.

The purpose of the present chapter is to provide an overview of the genetics and epigenetics of food addiction—and related conditions—to shed light on the mechanisms and processes that contribute to the development of addictive tendencies toward highly pleasurable energy resources. However, as there is no formal and accredited diagnosis of food addiction and because the field has only had a working research tool to fill this gap since the development of the *Yale Food Addiction Scale* in 2009 (see Chapter 3 for full details), there is a pronounced dearth of *direct* genetic-based research. What we must rely on mostly for evidence-based conclusions are genetic studies of intermediate phenotypes of food addiction and/or related conditions such as binge eating and other patterns of LOC consumption.

At the biological heart of the addictive process is the brain dopamine circuitry (viz. the mesocorticolimbic neural network), which involves both subcortical and

Compulsive Eating Behavior and Food Addiction. https://doi.org/10.1016/B978-0-12-816207-1.00010-X
Copyright © 2019 Elsevier Inc. All rights reserved.

cortical neuroanatomical connections that influence reward responsiveness (e.g., Volkow, Wang, Fowler, Tomasi, & Telang, 2011), low hedonic tone (i.e., negative emotionality or anhedonia: Volkow et al., 2014), and executive dysfunction (Miranda, Rodrigue, & Kennedy, 2019). Collectively, these factors comprise the central intermediate phenotypes of all addiction disorders (Kwako, Bickel, & Goldman, 2018).

Interestingly, it has been noted that humans are more likely to become dependent on addictive substances, and to remain addicted, than any other creature on earth (Siegel, 2005). It is well-established that our capacity for pleasure and reward and our motivation to engage in reinforcing stimuli is regulated by the dopamine signal strength in this key brain area—that is, dopamine is a "go-and-get it neurotransmitter" that initiates myriad behaviors essential for human survival (Beeler, Cools, Luciana, Ostlund, & Petzinger, 2014). Evolutionary biology has also identified another, more complex, role for the brain reward pathways. For instance, while healthy dopamine levels are able to adjust reward-related behaviors in the pursuit of adaptive flexibility to environment changes, compulsive behavioral pathology tends to reflect the diminution of dopamine's proficiency to adapt our appetitive behaviors to significant changes in our environment, both in utero and following birth (Calvey, 2017).

At the onset of this chapter on genetics and epigenetics, it is important to emphasize that a multitude of biological systems have been implicated in the regulation of food intake, both centrally and peripherally. It is therefore well beyond the scope of this review to discuss all the genetic markers that may contribute, in some way, to disordered overeating in general and food addiction specifically. We have chosen, therefore, to confine our appraisal of the research to the best understood, and most frequently studied, genetic contributors to human diversity in the appetitive motivational system, and in so doing, to risk factors that may contribute to excessive consumption and addictive tendencies to rewarding behaviors.

While dopamine is regarded as the principal regulating neurotransmitter in this system—and has generated the most prolific genetic research on addictive and overeating behaviors—many other chemical messengers are also implicated. Considerable interest has been directed to the endogenous opiate system in the context of food reward. Soon after the discovery of brain opioid peptides in 1975, a flurry of studies established their important role in overeating induced by highly palatable foods and by stress (see Le Magnen, 1990 for a review of this early research). Opioid neural circuitry in the nucleus accumbens and the ventral palladium is known to regulate taste-reactivity responses to palatable stimuli and thereby to promote food intake (Bodnar, 2015), primarily by strengthening the hedonic properties of many pleasing substances (Pecina, Smith, & Berridge, 2006)—or, as Berridge (2004) has written so poignantly, by "painting a pleasure gloss" on rewarding experiences.

It would seem that the first hypothesized link between endogenous opiates and a possible "addiction" to food came in a letter published in the *Lancet*, describing the enkephalin reward system (McCloy & McCloy, 1979a). Later the same year, the authors published a second letter in response to a request for clarification of their theory (see below).

> *We propose that if there should be an overproduction of enkephalin, or a failure to inactivate it with sufficient rapidity in the neurons in the gut, or at any point in the enkephalingergic chain in the central nervous system, a condition of tolerance would develop. In consequence, to produce the enkephalin reward needed to reduce the hunger drive, greater and greater quantities of food would be eaten, leading inexorably to obesity. There would be dependence, and withdrawal symptoms on abstinence. We regard this as auto-addiction (McCloy & McCloy, 1979a, pg 156).*

> *The strength of the reward depends upon the number of cells recruited … by the influx of food in the duodenum. Regular* excess *of enkephalin, over and above the amount required to reduce the activity of each neuron would have the same addictive effect (autoaddiction) as the equivalent amount of circulating morphine (McCloy & McCloy, 1979b, pg 753).*

The μ-opioid receptors in the mesolimbic regions have prompted special research interest because their activation tends to promote hyperphagia and the preference for a high-fat diet (Olszewski & Levine, 2007; Pecina & Berridge, 2005). Poor impulsive control and the tendency to act on cravings, rather than delaying gratification, have also been associated with higher μ-receptor density and greater opioid system activation (Love, Stohler, & Zubieta, 2009). As a consequence, it has been suggested that higher μ-receptor levels in the brain may increase the risk for obesity (Nogueiras et al., 2012).

It is interesting to note, in regard to these findings, that the three most-studied vulnerability factors in addiction research (both drug-seeking and drug-taking behaviors) are impulsive responding, food preferences, and sex (see Davis & Moghimi, 2017). Specifically, highly impulsive, sweet-preferring, females define the greatest risk phenotype (Carroll & Lynch, 2016). For instance, when given daily access to a sucrose solution, the female rats had a higher breaking point on a progressive ratio lever press task for a cherry-flavored sucrose reward than their male counterparts (Reichelt, Abbott, Westbrook, & Morris, 2016). Moreover, the sweet-food "craving" seen in high-sugar preferring female rats reflects a behavioral dysregulation that parallels to what is observed in cases of drug addiction (Carroll & Lynch, 2016). Such sex differences—albeit based mostly on animal models—are congruent with evidence of higher rates of obesity, binge eating disorder (BED), and food addiction in women than in men (Davis, 2015; Pedram et al., 2013). Women are also more ardent consumers of the most common addictive drugs such as alcohol, cocaine, cannabis, nicotine, and opioids (Carroll & Smethells, 2016).

Oxytocin is also a highly conserved, and evolutionarily significant, neuropeptide that is pivotally involved in the regulation of many reward-related behaviors, which are important for survival of the species (Carter, 2014). In recent years, there has been emerging interest in the role of oxytocin for food choice and appetite regulation. In general, oxytocin neurons are activated by the consumption of sugar and other sweet carbohydrates; however, increases in oxytocin also preferentially inhibit their intake (Leng & Sabatier, 2017). It is believed that peripheral oxytocin sites (e.g., the taste buds) help to regulate the preference for sweet taste, while central/brain oxytocin processes food reward (Hochheimer et al., 2014).

In addition, the serotonin (5-HT) system is known to be involved in regulation of the negative-valence brain system, which includes several constructs such as fear and anxiety—psychological characteristics, which are commonly dysfunctional in most types of disordered eating, as well as in excessive use of addictive behaviors (Lutter, Croghan, & Cui, 2016). This should not be surprising because activation of the 5-HT system tends to suppress feeding behaviors (Voigt & Fink, 2015). In a recent study, the serotonin receptor (5-HT_{2c}) was activated in male rats by a selective serotonin receptor agonist. Results showed a significant diminution in episodes of high-fat binge intake but no effect on the meal size of standard rat chow (Price, Anastasio, Stutz, Hommel, & Cunningham, 2018). It was concluded that a suppression effect occurred via an attenuation of the hedonic and motivational properties of the high-fat food.

Elevated levels of this neurotransmitter are also associated with increased vulnerability for anxiety. As extracellular 5-HT is primarily regulated by the serotonin transporter, which recycles 5-HT back into the presynaptic neuron for future use, the SERT gene has been much studied in this context and associated with many neuropsychiatric disorders (Gingrich & Hen, 2001). For example, a study of 5-HT transporter knockout rats found that the homozygous group ($SERT^{-/-}$) showed increased signs of anxiety on various behavioral measures compared with the $SERT^{+/+}$ group (Olivier et al., 2008). Unexpectedly, however, there were no sex differences despite the fact that women are more at risk for developing anxiety and depression than men are. The authors concluded that SERT gene deletion has such a profound impact on behavior that it effectively obscured any sex differences.

Genetic influences and methodological considerations

There are myriad and differing viewpoints on best approaches to assess whether, and how, genetic factors play a role in the etiology of, and risk for, compulsive overeating and addiction disorders in general. These issues are complicated by the fact that the genetic contributions to complex behaviors and conditions, such as addictions, are polygenic and heterogeneous, and the effect size of each individual genetic marker is typically very small (Plomin, Haworth, & Davis, 2009). Compounding the difficulties is that the definition and diagnosis of psychiatric disorders are variable and typically based on the severity or duration of symptoms, which are themselves normally subjective in nature and self-reported or observational (Breen et al., 2016).

Early efforts to understand the biological mechanisms underlying these complex conditions relied mainly on family/adoption/twin studies, which concluded that the heritability of addiction disorders varied greatly across studies, with estimates between 0.3 and 0.7 (Agrawal & Lynskey, 2008). In the 1970s, with important advances in molecular biology and the ability to sequence DNA, the field of medical genetics began to burgeon. Although the genetic linkage and candidate gene studies that ensued have the great advantage of being theory-driven and can be evaluated on how they converge with other biological-based methodologies that characterize the mechanisms underlying a particular disease state, they also have some shortcomings. For example, many of the initial genetic studies were based on small, underpowered samples, and hence, there was often a failure to replicate findings (Munn-Chernoff & Baker, 2016). It has been noted, however, that because the genotypic contribution of each individual locus is generally small and heterogeneous, it is misguided to expect identical results across similar studies. Instead, replication should still occur across numerous studies at higher-than-chance rates—and this has indeed occurred (Muskiewicz, Uhl, & Hall, 2018). Moreover, metaanalyses of underpowered studies also have the potential to uncover significant associations.

With the development of more sophisticated genotyping procedures and advances in the field of statistical genetics, genome-wide association studies (GWAS) have become *au courant* and can provide some advantages over previous methods. They do not, for instance, require a priori causal hypotheses or specific knowledge about the function of a gene, or more precisely the function of polymorphic markers—such as single-nucleotide polymorphisms (SNPs)—on a gene. A clear downside to these approaches, however, is the enormity of the sample size that is required to achieve sufficient power for detection because "GWAS exact a punishing correction for multiple testing ($p < 5 \times 10^{-8}$)" (Munn-Chernoff & Baker, 2016, p. 94). Moreover, the sheer size of an ideal GWAS sample makes replication difficult, if not impossible, especially in the case of conditions with a relatively low population base rate. In addition, while GWAS are able to identify genetic variants ("loci"), they neither necessarily identify the genes themselves nor is the functional nature of their alleles typically obvious (Breen et al., 2016). And finally, while GWAS of complex neuropsychiatric disorders are able to detect significant individual variants, even collectively these variants do not generally account for a sizable proportion of the heritability estimated from family studies (Mufford et al., 2017).

A relatively novel, and rapidly developing, methodology is improving our ability to understand the neurobiological mechanisms and the genetic contribution underlying a range of neuropsychiatric disorders including those related to compulsive and addictive tendencies toward food. *Neuroimaging genomics* assimilates individual-level genetic markers with brain phenotypes derived from sophisticated brain imaging procedures, such as positron emission tomography and magnetic resonance imaging, to corroborate risk factor investigations and to identify biomarkers that can improve diagnosis, prognosis, and therapeutic outcomes (Mufford et al., 2017). Neuroimaging procedures can also highlight brain structures and functions that mediate the links from genetic influence to disease/disorder. In other words, these brain "endophenotypes" can be viewed as the functional links between a particular genetic factor(s) and the clinical phenotype (Klein at al., 2017). In recent years, imaging genomics has also incorporated a GWAS approach to the original candidate gene association studies because of the emerging availability of large, multicenter collaborative studies, which have the power for detecting variants with small effect sizes. It is important to acknowledge, however, that neuroimaging measurements are not a perfect reflection of the core and essential biology of the brain. They are analogous to photographic images—simply a "snapshot" that captures a moment in time.

In 1942, when the field of genetics was still in its prime, the term epigenetics—literally translated to mean *above* (epi) genetics—was first coined by Conrad Waddington, an embryologist who used an often-quoted metaphor of marbles rolling down a hill with diverging canals to illustrate the pathways involved in cell differentiation (Deans & Maggert, 2015; Deichmann, 2016). Through a series of genetic assimilation experiments, Waddington showed that certain phenotypes—mainly developmental trait–related—were heritable in variable environments and genotypes (Deans & Maggert, 2015).

As the field developed, a divide between the ideologies of notable epigenetic pioneers led to various working definitions, both disconnecting and extending from what was visualized originally (see Greally, 2018; Haig, 2004). Currently, the term epigenetics—which still lacks some clarity—is commonly used to reflect changes in gene expression that occur without DNA sequence modification (Greally, 2018); this includes chromatin modifications, DNA methylation, and noncoding RNA silencing (Allis & Jenuwein, 2016 for a review). Epigenetic modifications regulate the phenotypic expression of each cell by turning "on" or "off" certain genes. In other words, certain cells are selected either to express or inhibit (Allis & Jenuwein, 2016). Importantly, some epigenetic changes are also reversible.

While it was originally assumed that heritability was only possible via DNA transmission, evidence is now indicating that epigenetic changes can also be inherited via cellular memory—a process which "remembers" and incorporates environmental and developmental signals (Allis & Jenuwein, 2016). However, it is important to note that human epigenetic inheritance remains controversial, particularly because such an inheritance would mean that changes in somatic cells need to then also be expressed in the germline (see Horsthemke, 2018).

The study of epigenetics has recently gained considerable popularity—particularly in the 21st century as indicated by a search of the term on PubMed, where it is now being referred to as a "fashionable topic" (Deichmann, 2016). As such, epigenetic research is a useful endeavor for scientists to better establish the pathogenesis of certain diseases. For example, it has now been recognized that metabolic disorders are largely driven by epigenetic differentiations involving the chromatin (Allis & Jenuwein, 2016). However, although the studies of epigenetic mechanisms are a popular aspect of research in developmental biology, its role in psychiatric disorder has lagged and is still a relatively new research focus (Vaillancourt, Ernst, Mash, & Turecki, 2017).

Family and twin studies

To date, there are no twin or family studies that have directly examined the genetics of food addiction. However, the use of this methodology to investigate other overeating conditions such as binge eating and obesity—as well as substance-use disorders in general—is able to shed some indirect light on this research field.

Family studies have consistently demonstrated a link between biological parent and offspring overweight and obesity, with higher correlations established between mothers and their children (Whitaker, Jarvis, Beeken, Boniface, & Wardle, 2010). When both parents have obesity, however, the offspring is more likely to develop the condition compared with those who have only one parent with obesity (Whitaker et al., 2010; Zalbahar, Najman, McIntyre, & Mamun, 2017). Although family studies are effective in establishing relationships between parental and offspring body mass index (BMI), they cannot identify whether these correlations have a genetic or environmental basis (Albuquerque, Nóbrega, Manco, & Padez, 2017). The heritability of obesity has been further clarified by adoption and twin studies, which have shown that the population variance in adult BMI is largely explained by genetic factors (Albuquerque et al., 2017). There is good evidence that the BMIs of adopted children are more closely related to their biological than adoptive parents (Pigeyre et al., 2016). The same patterns are seen in adoption studies of identical twins separated at birth, who had similar weights throughout their life course despite being raised in different households (Llewellyn & Wardle, 2015). A systematic review of twin and adoption studies also concluded that genetic influences substantially outweigh environmental factors in relation to adult BMI (Silventoinen, Rokholm, Kaprio, & Sørensen, 2010).

Twin studies—which involve comparisons of genetically identical monozygotic twins to nonidentical dizygotic twins—have been popular in establishing the heritability of various physical and psychological traits. Feinleib et al. (1977) were the first to compare monozygotic with dizygotic twins in an attempt to establish the genetic basis of obesity. Their findings were later replicated in a 25-year longitudinal study, which found greater genetic similarities in fat mass in monozygotic twins compared with dizygotic twins (Stunkard et al., 1986). Specifically, while nonidentical dizygotic twins were 35%–45% congruent in their fat-mass profiles, this statistic increased to 70%–90% in monozygotic twins (Stunkard et al., 1986). More recent evidence has confirmed the earlier estimates of BMI heritability (Elks et al., 2012 for a review). For example, using a cohort of over 11,000 pairs of monozygotic and dizygotic twins, it has been estimated that genetic polymorphisms may account for as much as 37% of the variance in BMI (Llewellyn, Trzaskowski, Plomin, & Wardle, 2013).

Perhaps the most noteworthy finding from twin studies is the low influence of the shared environment—the environment common to both twins, such as food type and availability, and parenting style—on weight (Llewellyn & Wardle, 2015; Silventoinen et al., 2010). A systematic review of twin and adoption studies found that effects of the shared environment disappear in late adolescence and that the shared environment has a small effect on population variance in risk for obesity (Silventoinen et al., 2010). Similarly, others have described a very modest (10%) shared environment influence on obesity heritability (Wardle, Carnell, Haworth, & Plomin, 2008). In contrast, the nonshared environment effects—the environment unique to the individual not shared within the family—seem to have a greater influence on BMI, in the range of 14% (Wardle et al., 2008). Therefore, upbringing—such as family diet and observing parents' eating habits—appears to not play a key role in determining the development of obesity. Instead, it is more likely that the *external* obesogenic environment accentuates an individual's inherent susceptibility to obesity.

A twin study whose purpose was to determine the degree of overlap between environmental factors and genetic factors that contribute to the lifetime risk for obesity and binge eating (with or without compensatory behaviors) used a large longitudinal population-based sample of female twins. The data were obtained from the first and third interview waves collected 6 years apart (Bulik, Sullivan, & Kendler, 2003). Their results indicated a substantial contribution of additive genetic effects for obesity, but only a moderate genetic contribute for binge eating. In addition, there was little influence of common environmental factors for either condition. And finally, there appeared to be only a small subset of genetic factors that were relevant to the etiology of both conditions. Another family history study of BED found significantly higher rates of many mood and anxiety disorders—as well as BED itself—in the female relatives of probands with BED compared with control probands without BED (Lilenfeld, Ringham, Kalarchian, & Marcus, 2008). However, with the important exception of substance-use disorders among the female relatives of women with BED, almost all of the other comorbid disorders followed a pattern of independent transmission from BED.

A recent study of 18,000 adult twins from the Swedish Twin registry estimated that the heritability for lifetime binge eating behavior was a high 0.65 (Capusan et al., 2017). Moreover, the findings from this study are reasonably congruent with an earlier Norwegian twin study with a heritability estimate of 41% for binge eating without compensatory behaviors (Reichborn-Kjennerud, Bulik, Tambs, & Harris, 2004) and with a case–control family study of BED with an estimate of 57% (Javaras et al., 2008). Because Capusan et al. (2017) also found significant common genetic risk factors between binge eating behavior and attention-deficit hyperactivity disorder, they speculated that the genetic links may reflect shared neurocognitive deficits related to suboptimal executive and cognitive functioning and emotional regulation, which manifest as poor inhibitory control, poor decision-making, and a diminished ability to delay gratification. According to a large longitudinal twin study, poor executive function (across a range of different executive functions) is a common genetic risk factor for increased polysubstance use in late adolescence, but that other factors play a larger role in the progression from use to abuse (Gustavson et al., 2017).

Results of twin and family studies have also converged in suggesting that genetic influences are shared across a variety of substance-use disorders, although there has been limited success in establishing individual genetic markers that are common across all addictive substances (Palmer et al., 2015). As such, behavioral geneticists have developed multivariate genetic models that establish the heritability of a single latent trait referred to as "substance dependence vulnerability"—a factor that was found to be 64% heritable across males and females (Palmer et al., 2013). This value is commensurate with heritability estimates from twin, family, and adoption studies carried out independently for nicotine dependence and alcohol dependence (see Ma, Yuan, Jiang, Cui, & Li, 2015). It was also found that at least 20% of the variance in this vulnerability factor can be attributed to common SNPs (Palmer et al., 2015).

Candidate gene studies

In the early days of candidate gene studies, when genotyping was costly and tedious, only a few obesity-related genetic markers were examined simultaneously—due largely to the a priori nature of these studies and the few available markers of known functionality at that time (Loos, 2009). Although hundreds of loci were investigated eventually, many results were inconsistent across studies, and/or not replicated, frequently because of small sample sizes with insufficient statistical power (Loos, 2009). Some exceptions comprise findings on the melanocortin 4 receptor (*MC4R*) and brain-derived neurotrophic factor (*BDNF*) genes, whose importance has been confirmed in various studies (Goodarzi, 2018). Both of these genes are expressed in neurons of the paraventricular nucleus (PVN)—an important hypothalamic regulator of food intake—which suggests that the genetic predisposition to obesity may be mediated by neural mechanisms connected to appetite and satiety (Pigeyre et al., 2016). Congruently, research has shown a relationship between genetic obesity susceptibility and LOC and emotional overeating (Konttinen et al., 2015).

A heterozygous *MC4R* mutation (*MC4R* +/−) was first discovered in humans in 1998 and is still considered the most common cause of monogenic obesity—that is, obesity in which a single genetic mutation is sufficient to cause disease onset (Yazdie, Clee, & Meyre, 2015). *MC4R* mutations account for up to 6% of early-onset monogenic obesity, although not everyone with this mutation develops the condition (Yazdie et al., 2015). The *MC4R* gene has also been strongly implicated in the hypothalamic regulation of energy homeostasis and food intake (Loos, 2018). An SNP (rs17782313) near the *MC4R* gene has recently been shown to influence long-term postbariatric surgery outcomes (Resende et al., 2018). In particular, the number of women who acquired surgical success (i.e., a weight loss >50%) was significantly lower in the CC + CT group compared with the homozygous T group. The former also had higher presurgical BMI. Other research has demonstrated that the C allele of this SNP is positively associated with BMI via its influence on food cravings and emotional overeating (Yilmaz et al., 2015). It has also been implicated with elevated food enjoyment scores and sweet snack consumption (Ho-Urriola et al., 2014).

A candidate gene study in humans identified that the Val66Met polymorphism in the *BDNF* gene was also associated with obesity, where adults with the Val−Met and Val−Val genotypes had increased BMI (Shugart et al., 2009). These variants have also been associated with the development of BED and an increased susceptibility to anxiety and depression (see Notaras, Hill, & van den Buuse, 2015). Such findings mesh with evidence of a relationship between the minor Met allele and lower BMI in children and adolescents (Kalenda et al., 2018). On the other hand, there is evidence that in women with bulimia nervosa, those with a combination of the Met66 allele and the hypofunctional 7-repeat allele of the *DRD4* dopamine receptor gene reported significantly higher maximal lifetime BMI than those patients in the other gene–gene interaction groups (Kaplan et al., 2008). A recent systematic review of markers in the *BDNF* gene, and their association with BMI, also concluded that other polymorphisms including rs925946, rs10501087, and rs988712 can be considered as genetic determinants of obesity (Akbarian et al., 2018)

Dopamine genes and functional polymorphisms

The D2 dopamine receptor (*DRD2*) has been most notably associated with affinity for reward and addictive behaviors, including overeating (Davis et al., 2013). Of all *DRD2* polymorphisms, the Taq1A polymorphism (rs1800497) has been most frequently studied and is associated with allelic differences in dopamine D2 receptor densities, where those with the A1 allele are known to have about 30%–40% fewer receptors (Benton & Young, 2016). Parallels between individuals with obesity and illicit substance-use disorders have also been made. For example, using neuroimaging techniques, Wang et al. (2001) found that those with lower D2 receptor densities in the striatum were more likely to have obesity. Similarly, lower D2 density has been found in those with illicit substance disorders such as alcohol, nicotine, cocaine, and opiates (Benton & Young, 2016). However, in these studies, using neuroimaging "snap shots," it is difficult to separate cause from consequence, as the lower densities may reflect an innate biological vulnerability or they may reflect the process of downregulation caused by excessive stimulation of the striatal area from excessive food and drug consumption.

Morton et al. (2006) reported a significant association between the A1 allele and elevated BMI, while others (Carpenter, Wong, Li, Noble, & Heber, 2013) found the same relationship in a group of morbidly obese adults. Blum et al. (1996) also reported a positive relationship between the A1 allele and obesity, in those with obesity and comorbid illicit substance-use disorders. Recently, however, Benton and Young (2016) carried out a systematic review and concluded that most of the 33 studies they evaluated did not show evidence for increased obesity in those with the A1 allele. In particular, none of the studies conducted on children and adolescents found increased BMI in A1 carriers.

To date, there are virtually no studies that have directly examined dopamine genetic markers of food addiction, due in large part to the relative recency of human research in this field. One exception is a study that used a novel genetic methodology (see Nikolova, Ferrell, Manuck, & Harari, 2011) to create a quantitative index of multiple relevant genetic markers (Davis et al., 2013). Based on the well-established premise that complex human behaviors are regulated by many genetic indicators, it has been argued that relevant genetic markers should be aggregated to reflect a polygenetic risk or association, instead of considering each locus individually (Plomin et al., 2009). To this end, a multilocus genetic profile (MLGP) score was calculated for each participant. Davis et al. (2013) targeted six genetic markers of the dopamine system, each with known functionality, which had been associated, in previous research, with dopamine signaling strength in the striatal region of the brain. For each marker, a score of 1 was given to the genotype associated with the greatest signal strength, while a score of 0 was given to the genotype with the lowest association. When a genotype was associated with a statistically significant intermediate signal strength, a score of 0.5 was allocated. Each participant's MLGP score was thereby the sum of their score at each locus. Results of this study indicated that those with YFAS-diagnosed food addiction had higher MLGP scores than their age- and weight-matched counterparts, implying that risk for food addiction is association with a *higher* sensitivity to reward. The MLGP score was also positively correlated with food cravings, emotional overeating, and binge eating, suggesting that these factors mediated the relationship between genetic risk and the outcome variable.

Using an exome sequencing method in combination with a candidate gene association approach, Pedram et al. (2017) also found that the major A allele of rs2511521 located in *DRD2* was significantly associated with increased risk for food addiction. These authors pointed out that previous research had identified another *DRD2* SNP (rs1076560), which is significantly associated with increased risk for drug dependence. They note with interest that although rs2511521 is currently of unknown functionality, it is on the same linkage disequilibrium block as rs1076560—findings which suggest an overlap in the neurological circuitry regulating food addiction and other addictive substances.

A considerably larger body of genetic research has investigated dopamine links with BED as a common overlapping phenotype of those with food addiction (Davis, 2015; Kessler, Hutson, Herman, & Potenza, 2016). Evidence of elevated brain dopamine signaling has been found in obese adults with BED compared with their obese counterparts without BED. Specifically, this has been indicated by a greater frequency of the A2 allele of the *ANKK1* Taq1A SNP (rs1800497) and of the homozygous T group of *DRD2* rs6277 SNP in the BED group (Davis et al., 2012). Based on the known functionality of these allelic variants and their respective association with greater density of D2 receptors and greater binding potential of the D2 receptors, these authors concluded that BED was distinguished by a higher sensitivity to reward and greater hedonic capacity. It should be noted, however, that chronic and excessive consumption of high-calorie food may eventually cause deficits in impulsive control and in the strength of striatal dopamine signaling, as is seen in response to chronic drug and alcohol abuse (see Davis, 2017 for evidence). In other words, elevated reward sensitivity may play an important role in fostering engagement in preference for, and elevated intake of, sweet and fatty foods, while chronic consumption of these substances may contribute to the excessive and compulsive nature of the behaviors when the system becomes downregulated. Addiction disorders are dynamic conditions where the very behaviors that define the disorder contribute neurophysiologically to the severity and compulsiveness of the condition.

It should be noted, however, that while the Taq1A SNP has received the most attention in studies linking the dopamine D2 receptor to various psychiatric disorders, its association with addiction disorders has been largely in the field of alcoholism. In that context, a recent metaanalysis indicated that the A1 allele was the third most significant SNP associated with alcohol-related phenotypes (Buhler et al., 2015). The Taq1A SNP has had weaker, or nonexistent, associations with other addictions such as nicotine dependence (Gorwood et al., 2012) and cannabis or cocaine addiction (Buhler et al., 2015).

In a related study, it was also found that a combination of the "gain-of-function" A2 allele of Taq1A and the "gain-of-function" G allele of the μ-opioid receptor A118G SNP was significantly more prevalent in individuals with BED compared with their weight-matched controls (Davis et al., 2009). In summary, and while these studies need replication, their findings converge with other neurobiological and behavioral evidence that BED is a reward-responsive phenotype of obesity and compulsive overeating (Kessler et al., 2016).

Oxytocin, opiates, and overeating

In a current body of research using the same quantitative genetic methodology described above, an MLGP related to the *oxytocin* system was positively associated with addictive tendencies toward overeating (Davis, 2018). Based on another study, which had investigated 12 SNPs on the oxytocin receptor gene (*OXTR*) and 2 SNPs on the cluster-of-differentiation 38 gene (*CD38*), we found that 6 of these were positively related to reward-based personality traits. Findings confirmed links between the oxytocin system and reward responsiveness—an important relationship in the preservation of evolutionarily significant survival behaviors such as food intake, especially of high-calorie carbohydrates (Davis, Zai, Adams, Bonder, & Kennedy, 2019). Of particular interest to these findings is the evidence of long-term changes ensuing from the abuse of addictive drugs. For instance, following chronic exposure, the compensatory neuroadaptive changes that occur in the central oxytocin system include reductions in endogenous levels of oxytocin and its receptor densities (Lee & Weerts, 2016). In light of the evidence that highly palatable, sugar-dense foods have addictive properties similar to drugs of abuse, it follows that overconsumption of sugar may also downregulate the oxytocin system. A recent study found that a 4-marker *OXTR* haplotype (G-T-A-G) of rs237885, rs2268493, rs2268494, and rs54298 was significantly related to sugary and fatty food preferences (Davis, Patte, Zai, & Kennedy, 2017). In light of the evidence that low oxytocin is related to increased food consumption, especially of sweet carbohydrates, as well as to prosocial deficits, and poor stress regulation, we have proposed that central "oxytocin deficiency" may be a risk factor for compulsive and addictive overeating (Davis & Moghimi, 2017).

A few studies have also examined associations between brain opioid signaling strength and aspects of overeating. Most relevant to the food addiction construct is a study which predicted that a stronger activation potential of the opioid circuitry—as indicated by a functional marker (A118G) of the μ-opioid receptor gene (*OPRM1*)—would provide elevated risk for food addiction via its influence on one's hedonic responsiveness to a highly palatable diet (Davis & Loxton, 2014). The A118G (rsrs1799971) SNP is the most highly studied *OPRM1* marker, and in vivo evidence indicates that the G allele serves as a "gain of function" for those possessing this minor allele because it appears to cause a threefold binding affinity for endogenous beta endorphins (Barr et al., 2008). The A118G marker was significantly related to a composite index of hedonic responsiveness to food, which in turn was a strong predictor of YFAS symptom scores. These results support the view that high responsiveness of the opioid pathway may foster a proneness to overeating and ultimately to compulsive and addictive tendencies toward highly palatable foods (Davis & Loxton, 2014). An earlier study also found that the homozygous G allele group reported significantly higher preference for sweet and fatty foods (Davis et al., 2011).

Relatedly, a case–control study of men with alcohol dependence demonstrated that a preference for spicy food was significantly higher in the alcohol-dependent patients than their control counterparts, but only among those who carried the G allele in A118G (Park et al., 2017). It is also relevant that a substantial body of research has demonstrated that the opioid system—the μ-opioid receptor in particular—is central to the addictive process and to the risk for addiction disorders. For example, knockout mice lacking the OPRM1 gene lost the analgesia effects of morphine as well as its rewarding and addiction-inducing impact (Matthes et al., 1996). Human studies have also linked the 118G allele on *OPRM1* with increased risk of opioid and alcohol addiction in addition to an elevated striatal dopamine response to alcohol and increased pleasurable nicotine effects (see Darcq & Kieffer, 2018).

Serotonin (5-HT) and appetite

Genes of the 5-HT system have also been of interest in disordered eating research because of their well-established role in both appetite and mood regulation (Lucki, 1998). One candidate gene of particular historic interest has been the serotonin transporter gene (*SLC6A4*) because it contributes importantly to the regulation of extracellular 5-HT in the brain. In particular, the serotonin transporter protein is centrally implicated in the mechanism of action for serotonin reuptake inhibitor drugs (SSRIs)—used popularly in the treatment for depression, and with some new evidence from an animal model of depression showing that females demonstrate a stronger response to SSRI treatment than male animals (Sanchez, Khoury, Hassan, Wegener, & Mathe, 2018). SSRI treatment has also produced some modest effects in the treatment of BED (Ghaderi et al., 2018). Early genetic evidence indicated, however, that polymorphisms in, and near, the *SLC6A4* gene, including *5HTTLPR*, did not emerge as risk factors for key symptoms of disordered eating, including weight and shape concerns and binge eating (Munn-Chernoff et al., 2012). However, a more recent study investigated whether the SLC6A4-promoter variant (*5HTTLPR*) was implicated in the reduced ability to control food intake (viz. disinhibition) and a reduction in the capacity to lose weight. Results indicated that S (low-expressing) carriers showed greater disinhibition and demonstrated 1.6 times more failures to control the amount of food they ate compared with LL carriers (Bonnet et al., 2017). Other factors, such as overeating at night and cravings for specific pleasurable foods, were also significantly associated with the S genotype.

Given that the drive to eat beyond caloric need (aka *food reinforcement*) has been associated with overeating and future weight gain and that serotonin's role in this reward-driven process is well-established, it is of interest to consider genetic associations of 5-HT with variables that have relevance to the food addiction construct. To date, however, relatively little genetic research has been carried out with this neurotransmitter system. This may reflect an absence of compelling findings that are specific to compulsive overeating or to the addiction process in general. However, one study examined the rs6311 marker in the regulatory regions of the serotonin 2A receptor gene (*5-HTR2A*) and found that AA participants had a lower BMI and waist perimeter than G participants, although there were significant differences in rates of obesity and other anthropometric parameters (Sorli et al., 2008). It was revealed in a later study as well that another SNP (rs6314) on the same gene served as a susceptibility marker for obesity, whereby genotype group moderated the reinforcing value of food and its effect on BMI (Carr et al., 2013). Those with at least one copy of the minor A allele (AA + AG) and *low* food reinforcement had significantly lower BMI than the other three groups, while those in the AA + AG group and *high* food reinforcement had the statistically highest levels of BMI. Level of food reinforcement was not moderated by genotype in the GG group. It is important to note, however, that the sample size was small and that the number of markers which were examined was large (n = 44). Therefore, these results can only be seen as suggestive.

Neuroimaging genetics

The simplest explanation for weight gain is the consumption of excess calories—that is, more calories than are needed for daily expenditure requirements. More complicated are the many reasons—environmental and biological—contributing to a chronic, positive energy balance. As such, overeating and obesity clearly have a complex polygenetic structure, whereby myriad risk alleles confer susceptibility (as reviewed earlier in this chapter). In recent years, there has been a substantial focus on the dopaminergic neurogenetic contributions, related to the brain's common reward circuitry (Stanfill et al., 2015), in the risk factor profile for obesity.

A recent genetic neuroimaging study examined BMI and reward-related function using blood oxygen level dependency (BOLD) during the processing of a rewarding monetary stimulus (Lancaster, Ihssen, Brindley, & Linden, 2018). In addition, BMI genetic risk was estimated from saliva DNA and targeted SNPs based on findings from an international GWAS study. Results indicated that both BMI and ventral striatal BOLD were significantly heritable, positively related, and had a shared genetic etiology. There was, however, no evidence of a shared environmental etiology. The authors emphasize that an important strength of their study was that it comprised a sample of young adults, thereby reducing the likelihood that the positive associations were the consequence of neuroadaptations that occurred as the result of long-lasting obesity.

In light of the previously described associations between the functional A118G SNP on the *OPRM1* gene and preference for sugary and fatty foods, recent neuroimaging genetics research also lends support to the biological mechanisms underlying these links. Studies have shown hyperactivity in mesocorticolimbic brain structures in G carriers with excessive alcohol use compared with their AA counterparts (Filbey et al., 2008). However, there was uncertainty about whether such findings reflected a genetic vulnerability or the consequence of drug abuse. A more recent study of G-carrying adolescents with only moderate alcohol use demonstrated reduced prefrontal activation to alcohol taste cues, but higher connectivity from the ventral striatum to the frontal regions in this group compared with AA carriers, suggesting that G carriers may be at greater risk because of a diminished control capacity to regulate cravings (Korucuoglu at al., 2017).

Genome-wide association studies

According to a recent review of the genetics of eating disorders, GWAS studies of BED and related concepts have lagged woefully behind those of anorexia nervosa (Bulik, Kleiman, & Yilmaz, 2016). Indeed, at the time of writing this chapter, there has only been one food addiction study using this methodology. The purpose of that study was to assess whether genetic determinants of food addiction (assessed by the short version of the YFAS) overlapped with those of drug addiction (Cornelis et al., 2016). Two loci (rs75038630 and 74902201) met GW-significance, although their locations included genes with no known role in eating behaviors. This study also provided very limited support for shared pathways between food addiction and drug addiction. As the authors noted, however, the study likely had modest power to detect novel genetic markers and overlap with other addiction disorders because of the relatively small sample (<10,000 participants), the measurement error in food addiction assessment, and the low prevalence of food addiction symptoms in the sample. To date, GWAS methodology has not been carried out with other forms of compulsive overeating such as BED. Hence, in this area of genetics, we must rely on work carried out in the field of obesity and food preferences.

In the decade or so since GWAS emerged as a novel methodology for uncovering genetic associations with disease and disorders, more than 500 obesity-associated loci have been identified (Loos, 2018). Readers are directed to Dong et al. (2018) for a current and updated list of all obesity-related SNPs derived from GWAS analyses. A study estimating BMI heritability using simulations of whole-genome sequencing data concluded that genetic markers likely represent 30%–40% of variance in BMI (Yang et al., 2015). On the other hand, a genome-wide complex trait analysis study—which is used to detect polymorphisms with small casual effects that GWAS are not able to detect because of their large *P*-value thresholds (Albuquerque et al., 2017; Munn-Chernoff & Baker, 2016, p. 94)—predicted that only 17% of BMI variation is likely due to genetic variants (Yang et al., 2011).

Frayling et al. (2007) were the first to determine an association between obesity and the *FTO* gene in humans via GWAS when trying to locate the genetic variants associated with type II diabetes (T2D). They discovered that the rs9939609 SNP on the first intron of the *FTO* gene was positively associated with BMI and a risk for T2D (Frayling et al., 2007). Specifically, they demonstrated that individuals with the "at-risk" homozygous genotype of the minor allele (AA) weighed on average 3 kg more and had a 1.67-fold greater chance of developing obesity than individuals carrying the homozygous protective allele (TT). Today, FTO remains the strongest predictor of obesity in individuals of varying ancestries (Goodarzi, 2018) and has been associated with various obesity-related traits (e.g., waist-to-hip circumference ratio, abdominal obesity, and body fat percentage; see Albuquerque et al., 2017). However, it still only explains about 0.34% of population variation in BMI (Loos & Yeo, 2014).

FTO-related GWAS have been mostly conducted on Caucasian samples—with more than 80% of the loci initially identified in Caucasians (Loos, 2018)—despite there being substantial differences in allele frequencies across ethnic groups (see Albuquerque, Stice, Rodríguez-López, Manco, & Nóbrega, 2015). For example, while the frequency of BMI-associated FTO SNPs is around 42% in Caucasian Europeans, it is about 12%–20% and 12% in those of Asian and African origin, respectively (Loos & Yeo, 2014 for a review). Similarly, another study found that while the rs56137030 polymorphism on the *FTO* gene was most relevant to BMI variation, this polymorphism was among a cluster of 103 polymorphisms identified in Caucasian Europeans, while only 29 were found in those of African descent (Peters et al., 2013). Interestingly, a meta-analysis examined obesity susceptibility in populations of varying ethnicities and found that 35 polymorphisms at the FTO locus were associated with Caucasian, but not Chinese, African American, or Hispanic obesity (Tan et al., 2014).

The biological mechanisms underpinning *FTO* function in obesity are still unknown (Loos & Yeo, 2014; Yazdi, Clee, & Meyre, 2015). However, evidence is emerging that the *FTO* gene may play a role in food-intake regulation because of its overexpression in the hypothalamic region of the brain (Loos & Yeo, 2014; Yazdi et al., 2015). Interestingly, when studying *FTO* polymorphic variation in energy intake and expenditure, Speakman, Rance, and Johnston (2008) found no significant difference in energy expenditure, but significantly greater average daily intake in those with at least one copy of the at-risk allele (i.e., AA + AT) compared with the TT protective genotype at FTO rs9939609. That is, while the basal metabolic rate and maximal oxygen consumption were the same in both groups, those in the AA + AT genotype groups had an average daily intake of 9500 kilojoules (kJ) per day, while those with the protective TT form had an average daily intake of 9000 kJ (Speakman, Rance, & Johnstone, 2008). Similarly, global deletion of the

FTO in the hypothalamus in rodents showed small reductions in food intake and weight, while having no effect on energy expenditure (McMurray et al., 2013). A recent study examined the relationship between the FTO rs9939609 polymorphism and food cravings and found that individuals carrying the at-risk A allele had higher levels of food cravings and greater BMI, compared with those with the protective homozygous form (Dang et al., 2018). In another study, it was reported that when compared with individuals with the AT or TT FTO genotype, individuals with the AA genotype showed decreased subjective satiety, increased affinity for palatable foods, and consumed about 350 more kilocalories on average (Melhorn et al., 2018). Interestingly, when looking at the associations between LOC eating and the rs9939609 FTO gene polymorphism in children and adolescents, Tanofsky-Kraff et al. (2009) found that those with the AA and AT variants displayed significantly greater LOC eating and high-fat food selection at a lunch buffet than those with the TT form. These findings are especially relevant to the etiology of food addiction.

A relationship between OPRM1 and obesity has also been established via GWAS. As various polymorphisms on the OPRM1 gene have been associated with greater adiposity, it has been concluded that variation in opiate signaling strength may be linked to proneness to weight gain, likely because of its role in reward sensitivity (Haghighi et al., 2014). For instance, this study also demonstrated an association between the rs2281617 SNP and fat intake, which highlights the role of *OPRM1* in hedonic responsiveness to highly palatable foods (Haghighi et al., 2014). Specifically, the minor allele carriers (i.e., TT and TC) appear to have a protective effect because individuals in these genotypic groups had significantly lower fat intake, body fat mass, and BMI. These results are especially interesting because an earlier study investigating OPRM1 SNPs and response to psychomotor stimulants found corroborating evidence—that is, the CC genotype of rs2281617 reported significantly greater feeling of "euphoria" and "energy and stimulation" in response to a dose of amphetamine than their counterpart genotypes (Dlugos et al., 2011).

Gene–gene and gene–environment interactions

Although it is clearly established that BMI is substantially heritable, it has now widely agreed that the modern-day obesogenic environment has also contributed to the rampant rise in population weight gain—particularly because the global prevalence of obesity has increased much more quickly than the rate of evolution (Goodarzi, 2018). Therefore, it is likely that certain environmental factors can magnify obesity development in those with an increased genetic susceptibility (Goodarzi, 2018).

A Swedish population study found that from the 1950s to the 1980s, the heritability of BMI increased from 75% to 78.8%, suggesting that as the environment becomes increasingly obesogenic, those with an elevated genetic predisposition to obesity become more susceptible to its development (Rokholm et al., 2011). Another large-scale longitudinal study following 8788 individuals from 1900 to 1958 found that a 29-SNP BMI genetic risk score—which uses the weighted sum of at-risk alleles at SNPs associated with BMI—was greater in individuals born in more recent birth cohorts (Walter, Mejía-Guevara, Estrada, Liu, & Glymour, 2016). Similarly, a study conducted from 1901 to 1986 tracking 907 individuals, and using a 32-SNP genetic risk score, found a positive relationship between birth year and BMI, skinfold thickness (linked with abdominal adiposity), waist circumference, and waist–hip ratio (Demerath et al., 2013). Interestingly, MC4R deficiency—a determinant of monogenic obesity—was shown to be more prevalent in younger age cohorts, suggesting a generational, environmentally-induced effect. Specifically, while MC4R deficiency was present in 40% of adults with obesity over 52 years of age, it was present in 60% of adults with obesity between 18 and 52 years of age and in 79% of children with obesity (Stutzmann et al., 2008). If it is true that the modern-day environment contributes to weight gain, then it is arguable that urban rather than rural environments also exacerbate obesity susceptibility. Accordingly, studies conducted on South Indian populations found increased FTO- and MC4R-driven obesity-related traits such as BMI, waist circumference, hip circumference, and waist–hip ratio in urban versus rural settings (Taylor et al., 2011; Vasan et al., 2012). Similarly, a South Korean population study found increased skinfold thickness in those living in urban settings (Kim, Lee, Lee, & Kim, 2014).

Research has shown that lifestyle factors have the ability to increase or diminish the genetic susceptibility to obesity. For example, it has been demonstrated that physical activity tends to attenuate obesity heritability (Guo, Liu, Wang, Shen, & Hu, 2015) and that a physically active lifestyle is associated with a 40% reduction in the genetic predisposition to common obesity (Li et al., 2010). Similarly, a 6-month exercise intervention was shown to increase DNA methylation and reduce transcriptional activity of certain genes associated with adipocyte metabolism (Rönn et al., 2013). Other studies have also shown that physical activity can reduce the genetic susceptibility to obesity in the FTO gene (Kilpeläinen et al., 2011).

Alternatively, exposure to psychosocial stress increases the development of visceral obesity (Speaker & Fleshner, 2012). For example, a gene-by-stress interaction has been identified between psychosocial stress and five SNPs on the early B-cell factor 1 (EBF1) gene, which plays a critical role in adipose tissue regulation and affects waist circumference, BMI, and T2D status (Singh et al., 2015). It has also been reported that deviations from mean sleep duration increases the association between the FTO gene and BMI (Young, Wauthier, & Donnelly, 2016). Tyrell et al. (2017) identified a negative relationship between socioeconomic status and a 69-variant genetic risk score for obesity, even after adjusting for physical activity, a calorically dense "Westernized" diet, and television watching.

In Westernized nations, there is now an overabundant availability of hyperpalatable foods containing high proportions of added sugars (Popkin & Hawkes, 2016). A prospective cohort study of over 120,000 female and 50,000 male adults found that a genetic predisposition to obesity—via a 32-SNP BMI genetic risk score—became more pronounced with higher intake of sugar-sweetened beverages (Qi et al.,

2012). Specifically, those who consumed the most sugar-sweetened beverages had almost a twofold greater genetic association with BMI compared with their low consumption control counterparts (Qi et al., 2012). In a subsequent study, the same sample was also used to show that the effect of a genetic predisposition to obesity becomes stronger with greater fried food intake (Qi et al., 2014). Other associations between a high-fat diet and FTO amplification on obesity development have also been noted (Choquet & Meyre, 2011). However, a sample of 177,330 adults showed no significant relationship between dietary intake of protein, carbohydrate, fat and total energy, and the FTO-rs9939609 polymorphism, which as previously mentioned regulates food intake and is correlated with BMI (Qi et al., 2014).

Epigenetics

Currently, there are no studies directly examining the epigenetics of food addiction. However, given the many similarities between conventional substance-use disorders and compulsive overeating—especially in the neurobiology and regulatory role of the mesolimbic pathways—we will begin with a short summary of some of the recent research in the epigenetics of the broader addiction field. It has been argued that in the early evolution of *Homo sapiens*, key phenotypes were positively selected for, which now separate, and elevate, us from all other species—viz. flexibility, adaptability, and innovation. These characteristics not only reflect the hallmark neurobiological mechanisms that promote survival behaviors but are also central to the development of addictions (Calvey, 2017). Epigenetic change is a key component of our ability to adapt rapidly to our changing environment. However, potent addictive behaviors can usurp the neurochemical mechanisms that regulate flexibility and innovation and therefore become "the price we pay for adaptability" (Calvey, 2017, p. 1).

As the reinforcing effect of addictive drugs is mediated by dopamine signaling strength in the nucleus accumbens, considerable research has focused on genetic variation in the dopamine receptor genes (e.g., *DRD2*) and the dopamine transporter gene (*DAT1/SLC6A3*). One study compared alcohol-dependent adults with healthy controls and found hypermethylation of the dopamine transporter promoter in the former group, which was also negatively associated with alcohol craving (Hillemacher et al., 2009). It was concluded that because methylation inhibits the expression of the dopamine transporter, this process elevates dopamine levels in the synapse, which may then reduce cravings. A more recent, and corroborating, study examined five functional regions of the SLC6A3 gene and found that decreased methylation of the promoter region predicted nucleus accumbens activation in anticipation of monetary loss—an effect that was not present in alcohol-dependent participants (Muench, Wiens, Cortes, Momenan & Lohoff, 2018).

Research has also begun to study DNA methylation dynamics in the context of psychostimulant overuse, such as cocaine-use disorders. Similar to other drug dependences, overuse of stimulants fosters an "addiction cycle" characterized by excessive use and intoxication, withdrawal and negative affect, and strong cravings which encourage reinstatement (Vaillancourt et al., 2017). While there is good evidence that DNA methylation is dysregulated in brain reward pathways following chronic exposure to cocaine, it is now understood to be a dynamic process that can either silence or promote gene expression (Sadri-Vakili, 2015). Evidence is also accumulating that exposure to drugs such as cocaine during gestation can induce a permanent epigenetic change in the germline of the offspring that increases risk for addiction in adulthood (Sadri-Vakili, 2015).

Epigenetic processes also appear to play a role in the risk for behavioral addictions such as pathological gambling. A study of alterations of DNA methylation in the dopamine D2 receptor gene (*DRD2*) showed significantly lower levels of methylation in those with a lifetime history of gambling disorder who had been abstinent for at least 12 months compared with their control counterparts who were still actively gambling. In addition, those who had never utilized treatment options for their gambling also had higher methylation of the DRD2 gene compared with those who had sought treatment (Hillemacher et al., 2015). Similar findings have been observed in a preclinical investigation using a laboratory animal model (Zoratto et al., 2017). Following a 3-week evaluation for the propensity to gamble—operationalized by an operant task in male rats—an increase in DNA methylation at one site on the serotonin transporter gene was found.

Epigenetics and obesity

Emerging and converging evidence is showing that epigenetic changes—which rely on environmental cues—are also contributing to the pathogenesis of obesity (Albuquerque et al., 2017). This notion was first popularized by the Dutch Famine Cohort Studies, which found associations between famine during early gestation and obesity and its comorbidities (e.g., diabetes and heart disease; Lumey et al., 2007). While the biological mechanisms underlying these findings were, at the time, not yet conceptualized, they created a foundation for the view that the environment plays an important role in fetal programming and the development of obesity later in life (Pigeyre et al., 2016). To date, numerous epigenetic DNA methylation candidate gene and epigenome-wide association studies on obesity have been conducted. Based on this research, it has been estimated that epigenetic differentiations explain about 26%–29% of the variation in BMI and waist circumference (Sayols-Baixeras et al., 2017).

DNA methylation is the most widely studied epigenetic mechanism and involves the addition of a methyl group—retrieved from the DNA methyltransferases—to a cytosine base at the carbon-5 position of the 5′-C-phosphate-G-3′ (CpG) sites (Moore, Le, & Fan, 2013). Methylation in the promoter region of genes within CpG sites results in gene silencing (Moore et al., 2013). One of the first examples of an epigenetic methylation link between maternal and offspring obesity is seen in a seminal animal study, where a change to the diet of Agouti mice mothers determines methylation of the murine agouti gene, which in turn determines coat color pigment and adult-onset obesity (Wolff, Kodell, Moore, & Cooney, 1998). Similarly, later human studies have also indicated that maternal weight gain during gestation is positively associated with offspring BMI in adulthood (Albuquerque et al., 2017; Lopomo, Burgio, & Migliore, 2016). Interestingly, children born to mothers who lost weight before their pregnancy seem to be at a lower risk for obesity later in life than children born to mothers who lost weight postgestation (Kral et al., 2006).

A DNA methylation candidate gene study, which used umbilical cord tissue samples to examine 68 CpGs in five genes, found that hypermethylation within the promoter region of the retinoid X receptor and endothelial nitric oxide synthase genes were associated with increased levels of adiposity in childhood (Godfrey et al., 2011). Soubrey et al. (2013) also found that methylation levels on human imprinted genes were different in children born to parents with obesity compared with those whom were not. For example, hypomethylation at the insulin-like growth factor 2 gene (*IGF2*) was associated with greater paternal obesity (Soubry et al., 2013). As well, Sharp et al. (2015) found that compared with babies of normal weight mothers, babies of mothers with obesity had hypermethylation in several CpG sites leading to greater BMI. Indeed, it has been claimed that maternal obesity is the strongest predictor of offspring obesity (Schwartz et al., 2017). Interestingly, maternal undernutrition can also affect the offspring's risk for obesity—and related conditions—and the responses to in utero fetal programming appears to be sexually dimorphic. Females appear to be particularly susceptible to developing increased adipose tissue as a result of gestational exposure to maternal undernutrition compared with male offspring. They are also at greater risk than males for developing disrupted glucose homeostasis following in utero exposure to high levels of maternal sugar intake (Dearden, Bouret, & Ozanne, 2018).

Hormonal mechanisms too are involved in appetite, food intake, and body weight. Leptin and insulin, in particular, have been implicated in the hypothalamic regulation of energy homeostasis (Song, Johnson, & Tamashiro, 2017), and evidence shows the involvement of epigenetic modifications in genes involved with the signaling of these hormones (Song et al., 2017). For example, maternal obesity and gestational diabetes have been associated with increased leptin gene (*LEP*) methylation in the placenta—potentially resulting in lower leptin production and contributing to postnatal offspring obesity (Lesseur et al., 2014). Interestingly, longer breastfeeding duration has been associated with decreases in *LEP* methylation in infants (Obermann-Borst et al., 2013). The insulin receptor promoter (*IRP*) is also involved with hypothalamic regulation of energy homeostasis via DNA methylation, as seen by the increase in adult-onset diabetes and obesity because of hypermethylation of the IRP and subsequent hypothalamic insulin resistance following prenatal and neonatal overfeeding in rats (Plagemann et al., 2010).

The hypothalamic "feeding center" of the brain—which regulates energy homeostasis and food intake—includes the arcuate nucleus (ARC), which sends neuronal projections to other regions in the hypothalamus such as the ventromedial (VMN), dorsomedial, and PVN (Shewchuk, 2017). In rodent models, epigenetic differentiations in the hypothalamic feeding center contribute to obesity and T2D as indicated by the finding that offspring of dams with obesity and gestational diabetes had reduced neuronal density in the ARC and VMN (Breton, 2013). Similarly, leptin-deficient mice have disrupted neural projection pathways between the ARC and PVN (Bouret, Draper, & Simerly, 2004); and the offspring of dams on a high-fat, high-sucrose diet had increased body weight, insulin sensitivity, and glucose intolerance and showed increased proopiomelanocortin (*POMC*) expression—an anorexigenic hypothalamic neurohormone, which has been linked to obesity and diabetes in humans (Stevens, Begum, & White, 2011; Zheng et al., 2015). Likewise, neonatal overfeeding has been associated with *POMC* hypermethylation near the Sp1 transcription factor binding site (Plagemann et al., 2010). A study conducted on children has also shown that when compared with their normal weight counterparts, those with obesity have hypermethylation at CpG sites in the *POMC* (Kuehnen et al., 2012).

As previously discussed, food intake and appetite can be influenced by the brain's reward circuitry, where a hypersensitivity toward reinforcing stimuli—mainly via the dopaminergic pathway—has the potential to contribute to weight gain. Dams on a high-fat diet three months before gestation and during lactation produced offspring that had an increased preference for palatable foods (i.e., sucrose and fat) and showed differential dopamine- and opioid-related gene expressions and promoter-region hypomethylation in reward-related regions of the brain (Vucetic, Kimmel, Totoki, Hollenbeck, & Reyes, 2010). In a subsequent study, Vucetic, Kimmel, and Reyes (2011) found that a high-fat diet from weaning to about 18 weeks of age was associated with epigenetic methylation and subsequent transcriptional repression of the μ-opioid receptor gene.

It is well-established that a preference for, and consumption of, a high-fat diet is a potent risk factor for obesity (Hill & Peters, 1998). It is also known that preference for fatty foods is a complex trait regulated both by environmental factors and genetic factors, including markers of *OPRM1*—in particular, the T allele of rs2281617 appears to have a protective effect, as described earlier in this chapter. Environmental factors such as prenatal exposure to maternal cigarette smoking (PEMCS) have also been associated with fatty food preference (Haghighi et al., 2014). One study investigated whether these two factors were interactive, and if so, whether an epigenetic modification of the gene may mediate such an interaction (Lee et al., 2015). Results confirmed an interaction of the 2 main effects by demonstrating that the T allele was only protective in non-PEMCS offspring. Furthermore, PEMCS was associated with lower DNA methylation CpG cytosines across *OPRM1*. The authors concluded that quieting the protective T allele in PEMCS offspring might reflect an epigenetic modification of the *OPRM1* gene.

It is important to note that owing to the early stages of epigenetics, the studies discussed above are not without methodological shortcomings, and therefore, although there are compelling findings, these should still be interpreted with caution (Albuquerque et al., 2017; Pigeyre et al., 2016). For instance, while most human epigenetic studies use blood samples, DNA methylation is tissue specific, meaning that the epigenetic changes in other tissues, which are heavily implicated with obesity (e.g., adipose tissue), are often overlooked (Pigeyre et al., 2016). Additionally, because of pronounced differences between the hypothalamic development in rodents and those in humans—particularly because of developmental variation between the two species—it is still unclear whether epigenetic findings from rodent studies can be adequately generalized to the human condition (Breton, 2013). On the whole, however, the available studies still provide good evidence for an epigenetic basis of obesity.

Summary and conclusions

Following our review of the genetic and epigenetic research in the field of food addiction and related conditions, we provide below a broad summary of our main findings. First of all, we were obliged to conclude that insufficient direct research exists for us to draw more than preliminary conclusions about relevant genetic risk indicators for food addiction (as defined by the YFAS). Even when we consider compulsive overeating in general, there remains a paucity of reliable genetic information. And, it is difficult to imagine this situation changing in the near future, largely because the concept of food addiction has no formal status in the lexicon of psychiatry and because of the very large samples required to carry out sophisticated genetic analyses. In addition—and although GWAS was once heralded as the biological key to causal disease susceptibility loci—this methodology has rather become the "holy grail" of genetics in the decade since its widespread appearance in biomedical research, owing to emerging limitations of this approach. One issue is that common variants that have appeared in many disease-related GWAS typically account for only a small fraction of the heritability reported from twin and family studies, provoking questions about the missing heritability (Dube & Hegele, 2013). In addition, GWAS normally tests for associations of variants with a minor allele frequency greater than 5%, while overlooking rarer variants, which may have important impact on certain disease states (Pociot, 2017). Importantly, it is also possible that SNPs that have a low/moderate risk themselves may interact with other SNPs and thereby together may confer a significant composite effect—a possibility not assessable by the current GWAS methodology (Cocchi, Drago, Fabbri, & Serretti, 2015). Finally, many GWAS have been carried out on samples of individuals of European Caucasian descent without exploring whether the findings from these studies are transferable to other non-European samples (Pociot, 2017).

Our review has also highlighted the evidence that substance-use disorders in their various forms, as well as behavioral addictions, share a common biologically based liability. By inference, therefore, compulsive overeating and addictive tendencies toward highly palatable food are likely to reflect many of the same genetics risk markers found in conventional addiction disorders. As a consequence, we can say with some confidence that an enhanced sensitivity to rewarding stimuli is likely to be a biologically based characteristic of those who are predisposed to LOC overeating. Coupled with this aspect of the risk profile—and also highly relevant to addictive tendencies toward food—is the preference for sweet tastes, which tends to compromise our ability to inhibit consumption. Indeed, this risk factor is more likely to be expressed in an environment where tasty energy resources are readily available, relatively inexpensive, and flavor enhanced by added sugar, fat, and salt to increase palatability. It has recently been estimated that almost 60% of current energy intake in the United States comes from ultraprocessed foods and that the consumption of such a diet is significantly associated with excess weight gain, especially among adult women (Juul, Martinez-Steele, Parekh, Monteiro, & Chang, 2018). In this regard, it is hardly surprising that the bulk of genetic research in the field of compulsive overeating has focused on neurotransmitters/hormones such as dopamine and the opiates, which are known to influence the activation potential of the brain dopamine systems and to regulate the strength of its "pleasure" signal.

References

Agrawal, A., & Lynskey, M. T. (2008). Are there genetic influences on addiction: Evidence from family, adoption and twin studies? *Addiction, 103*, 1069–1081.

Ahmed, S. H., Guillem, K., & Vandaele, Y. (2013). Sugar addiction: Pushing the drug sugar analogy to the limit. *Current Opinion in Clinical Nutrition and Metabolic Care, 16*, 434–439.

Akbarian, S. A., Salehi-Abargouei, A., Pourmasoumi, M., Kelishadi, R., Nikpour, P., & Heidari-Beni, M. (2018). Association of brain-derived neurotrophic factor gene polymorphisms with body mass index: A systematic review and meta-analysis. *Advances in Medical Sciences, 63*(1), 43–56.

Albuquerque, D., Stice, E., Rodríguez-López, R., Manco, L., & Nóbrega, C. (2015). Current review of genetics of human obesity: From molecular mechanisms to an evolutionary perspective. *Molecular Genetics and Genomics, 290*(4), 1191–1221.

Albuquerque, D., Nóbrega, C., Manco, L., & Padez, C. (2017). The contribution of genetics and environment to obesity. *British Medical Bulletin, 123*(1), 159–173.

Allis, C. D., & Jenuwein, T. (2016). The molecular hallmarks of epigenetic control. *Nature Reviews Genetics, 17*, 487–500.

Barr, C. S., Schwandt, M. L., Lindell, S. G., Higley, J. D., Maestropien, D., Goldman, D., et al. (2008). Variation at the mu-opioid receptor gene (OPRM1) influences attachment behavior in infant primates. *Proceedings of the National Academy of Science USA, 105*, 5277–5281.

Beeler, J. A., Cools, R., Luciana, M., Ostlund, S. B., & Petzinger, G. (2014). A kinder, gentler dopamine...highlighting dopamine's role in behavioral flexibility. *Frontiers in Neuroscience, 8*. Article 4.

Benton, D., & Young, H. A. (2016). March). A meta-analysis of the relationship between brain dopamine receptors and obesity: A matter of changes in behavior rather than food addiction. *International Journal of Obesity, 40*, S12–S21.

Berridge, K. C. (2004). Pleasure, unconscious affect, and irrational desire. In A. S. R. Manstead, N. H. Frijda, & A. H. Fischer (Eds.), *Feelings and emotions: The Amsterdam symposium* (pp. 43–62). Cambridge (UK): Cambridge University Press.

Blum, K., Sheridan, P. J., Wood, R. C., Braverman, E. R., Chen, T. J. H., Cull, J. G., & Comings, D. E. (1996). The D2 dopamine receptor gene as a determinant of reward deficiency syndrome. *Journal of the Royal Society of Medicine, 89*, 396–400.

Bodnar, R. J. (2015). Endogenous opioids and feeding behavior: A decade of further progress (2004–2014). A festschrift to Dr. Abba Kastin. *Peptides, 72*, 20–33.

Bonder, R., Davis, C., Kuk, J. L., & Loxton, N. J. (2018). Compulsive "grazing" and addictive tendencies towards food. *European Eating Disorders Review, 26*(6), 569–573.

Bonnet, G., Gomez-Abellan, P., Vera, B., Sanchez-Romera, J. F., Hernandez-Martinez, A. M., Sookoian, S., et al. (2017). Serotonin-transporter promotor polymorphism modulates the ability to control food intake: Effects on total weight loss. *Molecular Nutrition & Food Research, 61*. Article Number 1700494.

Bouret, S. G., Draper, S. J., & Simerly, R. B. (2004). Trophic action of leptin on hypothalamic neurons that regulate feeding. *Science, 304*(5667), 108–110.

Breen, G., Li, Q., Roth, B. L., O'Donnell, P., Didriksen, M., Dolmetsch, R., et al. (2016). Translating genome-wide association findings into new therapeutics for psychiatry. *Nature Neuroscience, 19*, 1392–1396.

Breton, C. (2013). The hypothalamus–adipose axis is a key target of developmental programming by maternal nutritional manipulation. *Journal of Endocrinology, 216*(2), R19–R31.

Buhler, K.-M., Gine, E., Echeverry-Alzate, V., Calleja-Conde, J., Rodriquez de Fonseca, F., & Lopez-Moreno, J. A. (2015). Common single nucleotide variants underlying drug addiction: More than a decade of research. *Addiction Biology, 20*, 845–871.

Bulik, C. M., Sullivan, P. F., & Kendler, K. S. (2003). Genetic and environmental contributions to obesity and binge eating. *International Journal of Eating Disorders, 33*, 293–298.

Bulik, C. M., Kleiman, S. C., & Yilmaz, Z. (2016). Genetic epidemiology of eating disorders. *Current Opinion in Psychiatry, 29*, 383–388.

Calvey, T. (2017). The extended evolutionary synthesis and addiction: The price we pay for adaptability. In T. Calvey, & W. M. U. Daniels (Eds.), *Brain research in addiction. Progress in brain research* (Vol. 235, pp. 1–18).

Capusan, A. J., Yao, S., Kuja-Halkola, R., Bulik, C. M., Thornton, L. M., Bendtsen, P., … Larsson, H. (2017). Genetic and environmental aspects in the association between attention-deficit hyperactivity disorder symptoms and binge-eating behavior in adults: A twin study. *Psychological Medicine, 47*(16), 2866–2878.

Carpenter, C. L., Wong, A. M., Li, Z., Noble, E. P., & Heber, D. (2013). Association of dopamine D 2 receptor and leptin receptor genes with clinically severe obesity. *Obesity, 21*, 467–473.

Carr, K. A., Lin, H., Fletcher, K. D., Sucheston, L., Singh, P. K., Salis, R. J., et al. (2013). Two functional serotonin polymorphisms moderate the effect of food reinforcement on BMI. *Behavioral Neuroscience, 127*, 387–399.

Carroll, M. E., & Lynch, W. J. (2016). How to study sex differences in addiction using animal models. *Addiction Biology, 21*, 1007–1029.

Carroll, M. E., & Smethells, J. R. (2016). Sex differences in behavioral dyscontrol: Role in drug addiction and novel treatments. *Frontiers in Psychiatry, 6*. article 175.

Carter, C. S. (2014). Oxytocin pathways and the evolution of human behavior. *Annual Review of Psychology, 65*, 17–39.

Choquet, H., & Meyre, D. (2011). Genetics of obesity: What have we learned? *Current Genomics, 12*(3), 169–179.

Cocchi, E., Drago, A., Fabbri, C., & Serretti, A. (2015). A model to investigate SNPs' interaction in GWAS studies. *Journal of Neural Transmission, 122*, 145–153.

Cornelis, M. C., Flint, A., Field, A. E., Kraft, P., Han, J., Rimm, E. B., et al. (2016). A genome-wide investigation of food addiction. *Obesity, 24*, 1336–1341.

Dang, L. C., Samanez-Larkin, G. R., Smith, C. T., Castrellon, J. J., Perkins, S. F., Cowan, R. L., … Zald, D. H. (2018). FTO affects food cravings and interacts with age to influence age-related decline in food cravings. *Physiology & Behavior, 192*, 188–193.

Darcq, E., & Kieffer, B. L. (2018). Opioid receptors: Drivers to addiction? *Nature Reviews Neuroscience, 19*, 499–514.

Davis, C., Levitan, R. D., Reid, C., Carter, J. C., Kaplan, A. S., Patte, K. A., & Kennedy, J. L. (2009). Dopamine for "wanting" and opioids for "liking": A comparison of obese adults with and without binge eating. *Obesity, 17*(6), 1220–1225.

Davis, C., & Loxton, N. J. (2014). A psycho-genetic study of hedonic responsiveness in relation to "food addiction". *Nutrients, 6*, 4338–4353.

Davis, C., & Moghimi, E. (2017). *'Oxytocin deficiency'*: Implications for the food-addiction construct. *Current Addiction Reports, 4*, 158–164.

Davis, C., Zai, C., Levitan, R. D., Kaplan, A. S., Carter, J. C., Reid-Westoby, C., … Kennedy, J. L. (2011). Opiates, overeating and obesity: A psychogenetic analysis. *International Journal of Obesity, 35*, 1347–1354.

Davis, C., Levitan, R. D., Yilmaz, Z., Kaplan, A. S., Carter, J. C., & Kennedy, J. L. (2012). Binge eating disorder and the dopamine D2 receptor: Genotypes and sub-phenotypes. *Progress in Neuropsychopharmacology & Biological Psychiatry, 38*(2), 328–335.

Davis, C., Loxton, N. J., Levitan, R. D., Kaplan, A. S., Carter, J. C., & Kennedy, J. L. (2013). 'Food addiction' and its association with a dopaminergic multilocus genetic profile. *Physiology & Behavior, 118*, 63–69.

Davis, C., Patte, K., Zai, C., & Kennedy, J. L. (2017). Polymorphisms of the oxytocin receptor gene and overeating: The intermediary role of endophenotypic risk factors. *Nutrition & Diabetes, 7*(5), e279.

Davis, C., Zai, C. C., Adams, N., Bonder, R., & Kennedy, J. L. (2019). Oxytocin and its association with reward-based personality traits: A multilocus genetic profile (MLGP) approach. *Personality and Individual Differences, 138*, 231–236.

Davis, C. (2015). The epidemiology and genetics of binge eating disorder. *CNS Spectrums, 20*, 522–529.

Davis, C. (2017). A commentary on the associations among 'food addiction', binge eating disorder, and obesity: Overlapping conditions with idiosyncratic clinical features. *Appetite, 115*, 3–8.

Davis, C. (2018). Links between oxytocin and binge eating are mediated by addictive personality traits and preference for sugar and fat: A multilocus genetic profile (MLGP) paradigm. In *Presentation at the annual meeting of the society for the study of ingestive behaviors (SSIB), Bonita springs, Florida, July.*

Deans, C., & Maggert, K. A. (2015). What do you mean, "epigenetic"? *Genetics, 199*, 887–896.

Dearden, L., Bouret, S. G., & Ozanne, S. E. (2018). Sex and gender differences in developmental programming of metabolism. *Molecular Metabolism, 15*, 8e–19.

Deichmann, U. (2016). Epigenetics: The origins and evolution of a fashionable topic. *Developmental Biology, 416*, 249–254.

Demerath, E. W., Choh, A. C., Johnson, W., Curran, J. E., Lee, M., Bellis, C., … Towne, B. (2013). The positive association of obesity variants with adulthood adiposity strengthens over an 80-year period: A gene-by-birth year interaction. *Human Heredity, 75*(2–4), 175–185.

Dlugos, A. M., Hamidovic, A., Hodgkinson, C., Shen, P. H., Goldman, D., Palmer, A. A., et al. (2011). *OPRM1* gene variants modulate amphetamine-induced euphoria in humans. *Genes, Brain and Behavior, 10*, 199–209.

Dong, S. S., Zhang, Y. J., Chen, Y. X., Yao, S., Hao, R. H., Rong, Y., … Yang, T. L. (2018). Comprehensive review and annotation of susceptibility SNPs associated with obesity-related traits. *Obesity Reviews, 19*, 917–930.

Dube, J. B., & Hegele, R. A. (2013). Genetics 100 for cardiologists: Basics of genome-wide association studies. *Canadian Journal of Cardiology, 29*, 10–17.

Elks, C. E., den Hoed, M., Zhao, J. H., Sharp, S. J., Wareham, N. J., Loos, R. J. F., & Ong, K. K. (2012). Variability in the heritability of body mass index: A systematic review and meta-regression. *Frontiers in Endocrinology, 3*, 29.

Feinleib, M., Garrison, R., Fabsitz, R., Christian, J., Hrubec, Z., Borhani, N., … Wagner, J. (1977). The NHLBI twin study of cardiovascular disease risk factors: Methodology and summary of results. *American Journal of Epidemiology, 106*(4), 284–295.

Filbey, F. M., Ray, L., Smolen, A., Claus, E. D., Audette, A., & Hutchison, K. E. (2008). Differential neural response to alcohol priming and alcohol taste cues is associated with DRD4 VNTR and OPRM1 genotypes. *Alcoholism: Clinical and Experimental Research, 32*, 1–11.

Frayling, T. M., Timpson, N. J., Weedon, M. N., Freathy, R. M., Lindgren, C. M., Perry, J. R. B., … Mccarthy, M. I. (2007). A common variant in the FTO gene is associated with body mass index and predisposes to childhood and adult obesity. *Science, 316*(5826), 889–894.

Gearhardt, A., Davis, C., Kushner, R., & Brownell, K. (2011). The addiction potential of hyperpalatable foods. *Current Drug Abuse Reviews, 4*, 140–145.

Ghaderi, A., Odeberg, J., Gustafsson, S., Rastam, M., Brolund, A., Pettersson, A., et al. (2018). Psychological, pharmacological, and combined treatments for binge eating disorder: A systematic review and meta-analysis. *PeerJ*, 6ae5113.

Gingrich, J. A., & Hen, R. (2001). Dissecting the role of the serotonin system in neuropsychiatric disorders using knockout mice. *Psychopharmacology, 155*, 1–10.

Godfrey, K. M., Sheppard, A., Gluckman, P. D., Lillycrop, K. A., Burdge, G. C., McLean, C., … Hanson, M. A. (2011). Epigenetic gene promoter methylation at birth is associated with child's later adiposity. *Diabetes, 60*(5), 1528–1534.

Goodarzi, M. O. (2018). Genetics of obesity: What genetic association studies have taught us about the biology of obesity and its complications. *The Lancet Diabetes and Endocrinology, 6*(3), 223–236.

Gorwood, P., Le Strat, Y., Ramoz, N., Dubertret, C., Moalic, J. M., & Simonneau, M. (2012). Genetics of dopamine receptors and drug addiction. *Human Genetics, 131*, 803–822.

Greally, J. M. (2018). A user's guide to the ambiguous word "epigenetics." *Nature Reviews Molecular Cell Biology, 19*, 207–208.

Guo, G., Liu, H., Wang, L., Shen, H., & Hu, W. (2015). The genome-wide influence on human BMI depends on physical activity, life course, and historical period. *Demography, 52*(5), 1651–1670.

Gustavson, D. E., Stallings, M. C., Corley, R. P., Miyake, A., Hewitt, J. K., & Friedman, N. P. (2017). Executive functions and substance use: Relations in late adolescence and early adulthood. *Journal of Abnormal Psychology, 126*(2), 257–270.

Haghighi, A., Melka, M. G., Abrahamowicz, M., Leonard, G. T., Richer, L., Perron, M., et al. (2014). Opioid receptor mu 1 gene, fat intake and obesity in adolescence. *Molecular Psychiatry, 19*, 63–68.

Haig, D. (2004). The (dual) origin of epigenetics. *Cold Spring Harbor Symposia on Quantitative Biology, 69*, 67–70.

Hill, J. O., & Peters, J. C. (1998). Environmental contributions to the obesity epidemic. *Science, 280*, 1371–1374.

Hillemacher, T., Frieling, H., Hartl, T., Wilhelm, J., Kornhuber, J., & Bleich, S. (2009). Promoter specific methylation of the dopamine transporter gene is altered in alcohol dependence and associated with craving. *Journal of Psychiatric Research, 43*, 388–392.

Hillemacher, T., Frieling, H., Bucholz, V., Hussein, R., Bleich, S., Meyer, C., et al. (2015). Alternations in DNA-methylation of the dopamine-receptor 2 gene are associated with abstinence and health care utilization in individuals with a lifetime history of pathological gambling. *Progress in Neuro-Psychopharmacology and Biological Psychiatry, 63*, 30–34.

Hochheimer, A., Krohn, M., Rudert, K., Riedel, K., Becker, S., Thirion, C., et al. (2014). Endogenous gustatory responses and gene expression profile of stably proliferating human taste cells isolated from fungiform papillae. *Chemical Senses, 39*, 359–377.

Horsthemke, B. (2018). A critical view on transgenerational epigenetic inheritance in humans. *Nature Communications, 9*, 2973.

Ho-Urriola, J., Guzmán-Guzmán, I. P., Smalley, S. V., González, A., Weisstaub, G., Domínguez-Vásquez, P., ... Santos, J. L. (2014). Melanocortin-4 receptor polymorphism rs17782313: Association with obesity and eating in the absence of hunger in Chilean children. *Nutrition, 30*(2), 145–149.

Javaras, K., Laird, N., Reichborn-Kjennerud, T., Bulik, C., Pope, H., & Hudson, J. (2008). Familiality and heritability of binge eating disorder: Results of a case-control family study and a twin study. *International Journal of Eating Disorders, 41*(2), 174–179.

Juul, F., Martinez-Steele, E., Parekh, N., Monteiro, C. A., & Chang, V. W. (2018). Ultra-processed food consumption and excess weight among US adults. *British Journal of Nutrition, 120*, 90–100.

Kalenda, A., Landgraf, K., Löffler, D., Kovacs, P., Kiess, W., & Körner, A. (2018). The BDNF Val66Met polymorphism is associated with lower BMI, lower postprandial glucose levels and elevated carbohydrate intake in children and adolescents. *Pediatric Obesity, 13*(3), 159–167.

Kaplan, A. S., Levitan, R. D., Yilmaz, Z., Davis, C., Tharmalingam, S., & Kennedy, J. L. (2008). A DRD4/BDNF gene-gene interaction associated with maximum BMI in women with bulimia nervosa. *International Journal of Eating Disorders, 41*, 22–28.

Kessler, R. M., Hutson, P. H., Herman, B. K., & Potenza, M. N. (2016). The neurobiological basis of binge-eating disorder. *Neuroscience & Biobehavioral Reviews, 63*, 223–238.

Kilpeläinen, T. O., Qi, L., Brage, S., Sharp, S. J., Sonestedt, E., Demerath, E., ... Loos, R. J. F. (2011). Physical activity attenuates the influence of FTO variants on obesity risk: A meta-analysis of 218,166 adults and 19,268 children. *PLoS Medicine, 8*(11), e1001116.

Kim, J., Lee, T., Lee, H.-J., & Kim, H. (2014). Genotype-environment interactions for quantitative traits in Korea Associated Resource (KARE) cohorts. *BMC Genetics, 15*(1), 18.

Klein, M., Onnick, M., van Donkelaar, M., Wolfers, T., Harich, B., Shi, Y., et al. (2017). Brain imaging genetics in ADHD and beyond – mapping pathways from gene to disorder at different levels of complexity. *Neuroscience & Biobehavioral Reviews, 80*, 115–155.

Konttinen, H., Llewellyn, C., Wardle, J., Silventoinen, K., Joensuu, A., Männistö, S., ... Haukkala, A. (2015). Appetitive traits as behavioural pathways in genetic susceptibility to obesity: A population-based cross-sectional study. *Scientific Reports, 5*, 14726.

Korucuoglu, O., Gladwin, T. E., Baas, F., Mocking, R. J. T., Ruhe, H. G., Groot, P. F. C., et al. (2017). Neural response to alcohol taste cues in youth: Effects of the *OPRM1* gene. *Addiction Biology, 22*, 1562–1575.

Kral, J. G., Biron, S., Simard, S., Hould, F.-S., Lebel, S., Marceau, S., & Marceau, P. (2006). Large maternal weight loss from obesity surgery prevents transmission of obesity to children who were followed for 2 to 18 years. *Pediatrics, 118*(6), e1644–e1649.

Kuehnen, P., Mischke, M., Wiegand, S., Sers, C., Horsthemke, B., Lau, S., ... Krude, H. (2012). An Alu element-associated hypermethylation variant of the POMC gene is associated with childhood obesity. *PLoS Genetics, 8*(3), e1002543.

Kwako, L. E., Bickel, W. K., & Goldman, D. (2018). Addiction biomarkers: Dimensional approaches to understanding addiction. *Trends in Molecular Medicine, 24*, 121–128.

Lancaster, T. M., Ihssen, I., Brindley, L. M., & Linden, D. E. (2018). Preliminary evidence for genetic overlap between body mass index and striatal reward response. *Translational Psychiatry, 8*, 19.

Le Magnen, J. (1990). A role for opiates in food reward and food addiction. In E. D. Capaldi, & T. L. Powley (Eds.), *Taste, experience, and feeding* (pp. 241–252). Washington DC: American Psychological Association.

Lee, M. R., & Weerts, E. M. (2016). Oxytocin for the treatment of drug and alcohol use disorders. *Behavioural Pharmacology, 27*(8), 640–648.

Lee, K. W. K., Abrahamowicz, M., Leonard, G. T., Richer, L., Perron, M., Veillette, S., et al. (2015). Prenatal exposure to cigarette smoke interacts with *OPRM1* to modulate dietary preference for fat. *Journal of Psychiatry & Neuroscience, 40*, 38–45.

Leng, G., & Sabatier, N. (2017). Oxytocin – the sweet hormone? *Trends in Endocrinology and Metabolism, 28*, 365–376.

Lesseur, C., Armstrong, D. A., Paquette, A. G., Li, Z., Padbury, J. F., & Marsit, C. J. (2014). Maternal obesity and gestational diabetes are associated with placental leptin DNA methylation. *American Journal of Obstetrics and Gynecology, 211*(6), 654.e1-9.

Li, S., Zhao, J. H., Luan, J., Ekelund, U., Luben, R. N., Kha, K., & Loos, R. J. F. (2010). Physical Activity Attenuates the Genetic Predisposition to Obesity in 20,000 Men and Women from EPIC-Norfolk Prospective Population Study. *PLoS Med, 7*(8), e1000332.

Lilenfeld, L. R. R., Ringham, R., Kalarchian, M. A., & Marcus, M. D. (2008). A family history study of binge-eating disorder. *Comprehensive Psychiatry, 49*, 247–254.

Llewellyn, C. H., Trzaskowski, M., Plomin, R., & Wardle, J. (2013). Finding the missing heritability in pediatric obesity: The contribution of genome-wide complex trait analysis. *International Journal of Obesity, 37*(10), 1506–1509.

Llewellyn, C., & Wardle, J. (2015). Behavioral susceptibility to obesity: Gene-environment interplay in the development of weight. *Physiology and Behavior, 152*, 494–501.

Loos, R. J. F., & Yeo, G. S. H. (2014). The bigger picture of FTO – the first GWAS-identified obesity gene. *Nature Reviews Endocrinology, 10*(1), 51–61.

Loos, R. J. F. (2009). Recent progress in the genetics of common obesity. *British Journal of Clinical Pharmacology, 68*(6), 811–829.

Loos, R. J. F. (2018). The genetics of adiposity. *Current Opinion in Genetics & Development, 50*, 86–95.

Lopomo, A., Burgio, E., & Migliore, L. (2016). Epigenetics of obesity. *Progress in Molecular Biology and Translational Science, 140*, 151–184.

Love, T. M., Stohler, C. S., & Zubieta, J. K. (2009). Positron emission tomography measures of endogenous opioid neurotransmission and impulsiveness traits in humans. *Archives of General Psychiatry, 66*, 1124–1134.

Lucki, I. (1998). The spectrum of behaviors influenced by serotonin. *Biological Psychiatry, 44*, 151–162.

Lumey, L., Stein, A. D., Kahn, H. S., van der Pal-de Bruin, K. M., Blauw, G., Zybert, P. A., & Susser, E. S. (2007). Cohort profile: The Dutch hunger winter families study. *International Journal of Epidemiology, 36*(6), 1196–1204.

Lutter, M., Croghan, A. E., & Cui, H. (2016). Escaping the golden cage: Animal models of eating disorders in the post-Diagnostic and statistical manuel era. *Biological Psychiatry, 79*, 17–24.

Ma, Y., Yuan, W., Jiang, X., Cui, W.-Y., & Li, M. D. (2015). Updated findings of the association and functional studies of *DRD2/ANKK1* variants with addictions. *Molecular Neurobiology, 51*, 281–299.

Matthes, H. W. D., Maldonado, R., Simonin, F., Valverde, O., Slowe, S., Kitchen, I., et al. (1996). Loss of morphine-induced analgesia, reward effect and withdrawal symptoms in mice lacking the μ-opioid-receptor gene. *Nature, 383*, 819–823.

McCloy, J., & McCloy, R. F. (July 21 1979a). Enkephalins, hunger, and obesity. *The Lancet*, 156.

McCloy, J., & McCloy, R. F. (October 1979b). Enkephalins, hunger, and obesity. *The Lancet, 6*, 753.

McMurray, F., Church, C. D., Larder, R., Nicholson, G., Wells, S., Teboul, L., ... Cox, R. D. (2013). Adult onset global loss of the fto gene alters body composition and metabolism in the mouse. *PLoS Genetics, 9*(1), e1003166.

Melhorn, S. J., Askren, M. K., Chung, W. K., Kratz, M., Bosch, T. A., Tyagi, V., ... Schur, E. A. (2018). FTO genotype impacts food intake and corticolimbic activation. *American Journal of Clinical Nutrition, 107*(2), 145–154.

Miranda, G. G., Rodrigue, K. M., & Kennedy, K. M. (2019). Frontoparietal cortical thickness mediates the effect of COMT *Val^{158}Met* polymorphism on age-associated executive function. *Neurobiology of Aging, 73*, 104–114.

Moore, L. D., Le, T., & Fan, G. (2013). DNA methylation and its basic function. *Neuropsychopharmacology: Official Publication of the American College of Neuropsychopharmacology, 38*(1), 23–38.

Morton, L. M., Wang, S. S., Bergen, A. W., Chatterjee, N., Kvale, P., Welch, R., ... Caporaso, N. E. (2006). DRD2 genetic variation in relation to smoking and obesity in the prostate, lung, colorectal, and ovarian cancer screening trial. *Pharmacogenetics and Genomics, 12*, 901–910.

Muench, C., Wiers, C. E., Cortes, C. R., Momenan, R., & Lohoff, F. W. (2018). Dopamine transporter gene methylation is associated with nucleus accumbens activation during reward processing in healthy but not alcohol-dependent individuals. *Alcoholism: Clinical and Experimental Research, 42*, 21–31.

Mufford, M. S., Stein, D. J., Dalvie, S., Groenewold, N. A., Thompson, P. M., & Jahanshad, N. (2017). Neuroimaging genomics in psychiatry – a translational approach. *Genome Medicine, 9*, 102.

Munn-Chernoff, M. A., & Baker, J. H. (2016). A primer on the genetics of comorbid eating disorders and substance use disorders. *European Eating Disorders Review, 24*, 91–100.

Munn-Chernoff, M. A., McQueen, M. B., Stetler, G. L., Haberstick, B. C., Rhee, S. H., Sobik, L. E., et al. (2012). Examining associations between disordered eating and serotonin transporter gene polymorphisms. *International Journal of Eating Disorders, 45*, 556–561.

Muskiewicz, D. E., Uhl, G. R., & Hall, F. S. (2018). The role of cell adhesion molecule genes regulating neuroplasticity in addiction. *Neural Plasticity, 2018*. article ID 9803764.

Nikolova, Y. S., Ferrell, R. E., Manuck, S. B., & Harari, A. R. (2011). Multilocus genetic profile for dopamine signaling predicts ventral striatum reactivity. *Neuropsycho pharmacology, 36*, 1940–1947.

Nogueiras, R., Romaro-Pico, A., Vazquez, M. J., Novelle, M. G., Lopez, M., & Dieguez, C. (2012). The opioid system and food intake: Homeostatic and hedonic mechanisms. *Obesity Facts, 5*, 196–207.

Notaras, M., Hill, R., & van den Buuse, M. (2015). The BDNF gene Val66Met polymorphism as a modifier of psychiatric disorder susceptibility: Progress and controversy. *Molecular Psychiatry, 20*(8), 916–930.

Obermann-Borst, S. A., Eilers, P. H. C., Tobi, E. W., de Jong, F. H., Slagboom, P. E., Heijmans, B. T., & Steegers-Theunissen, R. P. M. (2013). Duration of breastfeeding and gender are associated with methylation of the LEPTIN gene in very young children. *Pediatric Research, 74*(3), 344–349.

Olivier, J. D. A., Van Der Hart, M. G. C., Van Swelm, R. P. L., Dederen, P. J., Homberg, J. R., Cremers, T., et al. (2008). A study in male and female 5-HT transporter knockout rats: An animal model for anxiety and depression disorders. *Neuroscience, 152*, 573–584.

Olszewski, P. K., & Levine, A. S. (2007). Central opioids and consumption of sweet tastants: When reward outweighs homeostasis. *Physiology & Behavior, 91*, 506–512.

Palmer, R. H., Young, S. E., Corley, R. P., Hopfer, C. J., Stallings, M. C., Hewitt, J. K., et al. (2013). Stability and change of genetic and environmental effects on the common liability to alcohol, tobacco, and cannabis DSM-IV dependence symptoms. *Behavior Genetics, 43*, 374–385.

Palmer, R. H. C., Brick, L., Nugent, N. R., Bidwell, L. C., McGeary, J. E., Knopik, V. S., et al. (2015). Examining the role of common genetic variants on alcohol, tobacco, cannabis and illicit drug dependence: Genetics of vulnerability to drug dependence. *Addiction, 110*, 530–537.

Park, J.-H., Kim, S.-G., Kim, J.-H., Lee, J.-S., Jung, W.-Y., & Kim, H.-K. (2017). Spicy food preference and risk for alcohol dependence in Korean. *Psychiatry Investigation, 14*, 825–829.

Pecina, S., & Berridge, K. C. (2005). Hedonic hot spots in nucleus accumbens shell: Where do mu-opioids cause increased hedonic impact of sweetness? *Journal of Neuroscience, 25*, 11777–11786.

Pecina, S., Smith, K. S., & Berridge, K. C. (2006). Hedonic hot spots in the brain. *The Neuroscientist, 12*, 500–511.

Pedram, P., Wadden, D., Amini, P., Gulliver, W., Randell, E., Cahill, F., ... Sun, G. (2013). Food addiction: Its prevalence and significant association with obesity in the general population. *PLoS One, 8*(9), e74832.

Pedram, P., Zhai, G., Gulliver, W., Zhang, H., & Sun, G. (2017). Two novel candidate genes identified in adults from the Newfoundland population with addictive tendencies towards food. *Appetite, 115*, 71–79.

Peters, U., North, K. E., Sethupathy, P., Buyske, S., Haessler, J., Jiao, S., ... Kooperberg, C. (2013). A systematic mapping approach of 16q12.2/FTO and BMI in more than 20,000 african Americans narrows in on the underlying functional variation: Results from the population architecture using genomics and epidemiology (PAGE) study. *PLoS Genetics, 9*(1), e1003171.

Pigeyre, M., Yazdi, F. T., Kaur, Y., Meyre, D., Fontaine, K. R., Redden, D. T., ... Gale, A. (2016). Recent progress in genetics, epigenetics and metagenomics unveils the pathophysiology of human obesity. *Clinical Science, 130*(12), 943–986.

Plagemann, A., Roepke, K., Harder, T., Brunn, M., Harder, A., Wittrock-Staar, M., ... Dudenhausen, J. W. (2010). Epigenetic malprogramming of the insulin receptor promoter due to developmental overfeeding. *Journal of Perinatal Medicine, 38*(4), 393–400.

Plomin, R., Haworth, C. M. A., & Davis, O. S. P. (2009). Common disorders are quantitative traits. *Nature Reviews Genetics, 10*, 872–878.

Pociot, F. (2017). Type 1 diabetes genome-wide association studies: Not to be lost in translation. *Clinical & Translational Immunology, 6*, e162.

Popkin, B. M., & Hawkes, C. (2016). Sweetening of the global diet, particularly beverages: Patterns, trends, and policy responses. *The Lancet. Diabetes & Endocrinology, 4*(2), 174–186.

Price, A. E., Anastasio, N. C., Stutz, S. J., Hommel, J. D., & Cunningham, K. A. (2018). Serotonin 5-HT_{2C} receptor activation suppresses binge intake and the reinforcing and motivational properties of high-fat food. *Frontiers in Pharmacology, 9*. article 821.

Qi, Q., Chu, A. Y., Kang, J. H., Jensen, M. K., Curhan, G. C., Pasquale, L. R., ... Qi, L. (2012). Sugar-sweetened beverages and genetic risk of obesity. *New England Journal of Medicine, 367*(15), 1387–1396.

Qi, Q., Kilpeläinen, T. O., Downer, M. K., Tanaka, T., Smith, C. E., Sluijs, I., ... Qi, L. (2014). FTO genetic variants, dietary intake and body mass index: Insights from 177 330 individuals. *Human Molecular Genetics, 23*(25), 6961–6972.

Reichborn-Kjennerud, T., Bulik, C. M., Tambs, K., & Harris, J. R. (2004). Genetic and environmental influences on binge eating in the absence of compensatory behaviors: A population-based twin study. *International Journal of Eating Disorders, 36*(3), 307–314.

Reichelt, A. C., Abbott, K. N., Westbrook, R. F., & Morris, M. J. (2016). Differential motivational profiles following adolescent sucrose access in male and female rats. *Physiology & Behavior, 157*, 13–19.

Resende, C. M. M., Durso, D. F., Borges, K. B. G., Pereira, R. M., Rodrigues, G. K. D., Rodrigues, K. F., ... Alvarez-Leite, J. I. (2018). The polymorphism rs17782313 near MC4R gene is related with anthropometric changes in women submitted to bariatric surgery over 60 months. *Clinical Nutrition, 37*(4), 1286–1292.

Rokholm, B., Silventoinen, K., Tynelius, P., Gamborg, M., Sørensen, T. I. A., & Rasmussen, F. (2011). Increasing genetic variance of body mass index during the Swedish obesity epidemic. *PLoS One, 6*(11), e27135.

Rönn, T., Volkov, P., Davegårdh, C., Dayeh, T., Hall, E., Olsson, A. H., ... Ling, C. (2013). A six months exercise intervention influences the genome-wide DNA methylation pattern in human adipose tissue. *PLoS Genetics, 9*(6), e1003572.

Sadri-Vakili, G. (2015). Cocaine triggers epigenetic alterations in the corticostriatal circuit. *Brain Research, 1628*, 50–59.

Sanchez, C., Khoury, A. E., Hassan, M., & Wegener, G. (2018). Sex-dependent behavior: Neuropeptide profile and antidepressant response in rat model of depression. *Behavioural Brain Research, 351*, 93–103.

Sayols-Baixeras, S., Subirana, I., Fernández-Sanlés, A., Sentí, M., Lluís-Ganella, C., Marrugat, J., & Elosua, R. (2017). DNA methylation and obesity traits: An epigenome-wide association study. The REGICOR study. *Epigenetics, 12*(10), 909–916.

Schwartz, M. W., Seeley, R. J., Zeltser, L. M., Drewnowski, A., Ravussin, E., Redman, L. M., & Leibel, R. L. (2017). Obesity pathogenesis: An endocrine society scientific statement. *Endocrine Reviews, 38*(4), 267–296.

Sharp, G. C., Lawlor, D. A., Richmond, R. C., Fraser, A., Simpkin, A., Suderman, M., ... Relton, C. L. (2015). Maternal pre-pregnancy BMI and gestational weight gain, offspring DNA methylation and later offspring adiposity: Findings from the avon longitudinal study of parents and children. *International Journal of Epidemiology, 44*(4), 1288–1304.

Shewchuk, B. M. (2017). Chapter 16 - epigenetics and obesity. In D. Yasui, J. Peedicayil, & D. Grayson (Eds.), *Neuropsychiatric disorders and epigentics* (pp. 309–334). Chicago, IL: Elsevier.

Shugart, Y. Y., Chen, L., Day, I. N. M., Lewis, S. J., Timpson, N. J., Yuan, W., ... Davey-Smith, G. (2009). Two British women studies replicated the association between the Val66Met polymorphism in the brain-derived neurotrophic factor (BDNF) and BMI. *European Journal of Human Genetics: European Journal of Human Genetics, 17*(8), 1050–1055.

Siegel, S. (2005). Drug tolerance, drug addiction, and drug anticipation. *Current Directions in Psychological Science, 14*(6), 296–300.

Silventoinen, K., Rokholm, B., Kaprio, J., & Sørensen, T. I. A. (2010). The genetic and environmental influences on childhood obesity: A systematic review of twin and adoption studies. *International Journal of Obesity, 34*(1), 29–40.

Singh, A., Babyak, M. A., Nolan, D. K., Brummett, B. H., Jiang, R., Siegler, I. C., ... Hauser, E. R. (2015). Gene by stress genome-wide interaction analysis and path analysis identify EBF1 as a cardiovascular and metabolic risk gene. *European Journal of Human Genetics: European Journal of Human Genetics, 23*(6), 854–862.

Song, L., Johnson, M. D., & Tamashiro, K. L. K. (2017). Chapter 8 – maternal and epigenetic factors that influence food intake and energy balance in offspring. Appetite and food intake: Central control. In R. B. S. Harris (Ed.), *Appetite and food intake: Central control* (2nd ed.). Boca Raton, FL: CRC Press/Taylor & Francis.

Sorli, J. V., Frances, F., Gonzalez, J. I., Guillen, M., Portoles, O., Sabater, A., et al. (2008). Impact of the -1438G>A polymorphism in the serotonin 2A receptor gene on anthropometric profile and obesity risk: A case-control study in a Spanish mediterranean population. *Appetite, 50*, 260–265.

Soubry, A., Schildkraut, J. M., Murtha, A., Wang, F., Huang, Z., Bernal, A., ... Hoyo, C. (2013). Paternal obesity is associated with IGF2hypomethylation in newborns: Results from a newborn epigenetics study (NEST) cohort. *BMC Medicine, 11*(1), 29.

Speaker, K. J., & Fleshner, M. (2012). Interleukin-1 beta: A potential link between stress and the development of visceral obesity. *BMC Physiology, 12*, 8.

Speakman, J. R., Rance, K. A., & Johnstone, A. M. (2008). Polymorphisms of the FTO gene are associated with variation in energy intake, but not energy expenditure. *Obesity, 16*(8), 1961–1965.

Stanfill, A. G., Conley, Y., Cashion, A., Thompson, C., Homayouni, R., Cowan, P., et al. (2015). Neurogenetic and neuroimaging evidence for a conceptual model dopaminergic contributions to obesity. *Biological Research For Nursing, 17*, 413–421.

Stevens, A., Begum, G., & White, A. (2011). Epigenetic changes in the hypothalamic proopiomelanocortin gene: A mechanism linking maternal undernutrition to obesity in the offspring? *European Journal of Pharmacology, 660*(1), 194–201.

Stunkard, A. J., Sørensen, T. I. A., Hanis, C., Teasdale, T. W., Chakraborty, R., Schull, W. J., & Schulsinger, F. (1986). An adoption study of human obesity. *New England Journal of Medicine, 314*(4), 193–198.

Stutzmann, F., Tan, K., Vatin, V., Dina, C., Jouret, B., Tichet, J., ... Meyre, D. (2008). Prevalence of melanocortin-4 receptor deficiency in Europeans and their age-dependent penetrance in multigenerational pedigrees. *Diabetes, 57*(9), 2511–2518.

Tan, L. J., Zhu, H., He, H., Wu, K.-H., Li, J., Chen, X.-D., ... Deng, H.-W. (2014). Replication of 6 obesity genes in a meta-analysis of genome-wide association studies from Diverse ancestries. *PLoS One, 9*(5), e96149.

Tanofsky-Kraff, M., Han, J. C., Anandalingam, K., Shomaker, L. B., Columbo, K. M., Wolkoff, L. E., ... Yanovski, J. A. (2009). The FTO gene rs9939609 obesity-risk allele and loss of control over eating 1-5. *American Journal of Clinical Nutrition, 90*, 1483–1491.

Taylor, A. E., Sandeep, M. N., Janipalli, C. S., Giambartolomei, C., Evans, D. M., Kranthi Kumar, M. V., ... Chandak, G. R. (2011). Associations of FTO and MC4R variants with obesity traits in Indians and the role of rural/urban environment as a possible effect modifier. *Journal of Obesity, 2011*, 307542.

Treur, J. L., Boosma, D. I., Ligthart, L., Willemsen, G., & Vink, J. M. (2016). Heritability of high sugar consumption through drinks and the genetic correlation with substance use. *American Journal of Clinical Nutrition, 104*, 1144–1150.

Tyrrell, J., Wood, A. R., Ames, R. M., Yaghootkar, H., Beaumont, R. N., Jones, S. E., ... Frayling, T. M. (2017). Gene–obesogenic environment interactions in the UK Biobank study. *International Journal of Epidemiology, 46*(2), 559–575.

Vaillancourt, K., Ernst, C., Mash, D., & Turecki, G. (2017). DNA methylation dynamics and cocaine in the brain: Progress and prospects. *Genes, 8*(5), 138.

Vasan, S. K., Fall, T., Neville, M. J., Antonisamy, B., Fall, C. H., Geethanjali, F. S., ... Karpe, F. (2012). Associations of variants in *FTO* and near *MC4R* with obesity traits in South asian Indians. *Obesity, 20*(11), 2268–2277.

Voigt, J. P., & Fink, H. (2015). Serotonin controlling feeding and satiety. *Behavioural Brain Research, 277*, 14–31.

Volkow, N. D., Wang, G.-W., Fowler, J. S., Tomasi, D., & Telang, F. (2011). Addition: Beyond dopamine reward circuitry. *Proceedings of the National Academy of Sciences, 108*, 15037–15042.

Volkow, N. D., Wang, G.-W., Telang, F., Fowler, J. S., Alexoff, D., Logan, J., et al. (2014). Decreased dopamine brain reactivity in marijuana abusers is associated with negative emotionality and addiction severity. *Proceedings of the National Academy of Sciences*, E3149–E3156.

Vucetic, Z., Kimmel, J., Totoki, K., Hollenbeck, E., & Reyes, T. M. (2010). Maternal high-fat diet alters methylation and gene expression of dopamine and opioid-related genes. *Endocrinology, 151*(10), 4756–4764.

Vucetic, Z., Kimmel, J., & Reyes, T. M. (2011). Chronic high-fat diet drives postnatal epigenetic regulation of μ-opioid receptor in the brain. *Neuropsychopharmacology: Official Publication of the American College of Neuropsychopharmacology, 36*(6), 1199–1206.

Walter, S., Mejía-Guevara, I., Estrada, K., Liu, S. Y., & Glymour, M. M. (2016). Association of a genetic risk score with body mass index across different birth cohorts. *Journal of the American Medical Association, 316*(1), 63.

Wang, G. J., Volkow, N. D., Logan, J., Pappas, N. R., Wong, C. T., Zhu, W., ... Fowler, J. S. (2001). Brain dopamine and obesity. *Lancet, 357*(9253), 354–357.

Wardle, J., Carnell, S., Haworth, C. M., & Plomin, R. (2008). Evidence for a strong genetic influence on childhood adiposity despite the force of the obesogenic environment. *American Journal of Clinical Nutrition, 87*(2), 398–404.

Whitaker, K. L., Jarvis, M. J., Beeken, R. J., Boniface, D., & Wardle, J. (2010). Comparing maternal and paternal intergenerational transmission of obesity risk in a large population-based sample. *American Journal of Clinical Nutrition, 91*(6), 1560–1567.

Wolff, G. L., Kodell, R. L., Moore, S. R., & Cooney, C. A. (1998). Maternal epigenetics and methyl supplements affect agouti gene expression in Avy/a mice. *The FASEB Journal: Official Publication of the Federation of American Societies for Experimental Biology, 12*(11), 949–957.

Yang, J., Manolio, T. A., Pasquale, L. R., Boerwinkle, E., Caporaso, N., Cunningham, J. M., ... Visscher, P. M. (2011). Genome-partitioning of genetic variation for complex traits using common SNPs. *Nature Genetics, 43*(6), 519–525.

Yang, J., Bakshi, A., Zhu, Z., Hemani, G., Vinkhuyzen, A. A., Hong Lee, S., ... Genet Author manuscript, N. (2015). Genetic variance estimation with imputed variants finds negligible missing heritability for human height and body mass index HHS Public Access Author manuscript. *Nature Genetics, 47*(10), 1114–1120.

Yazdi, F. T., Clee, S. M., & Meyre, D. (2015). Obesity genetics in mouse and human: Back and forth, and back again. *PeerJ, 3*, e856.

Yilmaz, Z., Davis, C., Loxton, N. J., Kaplan, A. S., Levitan, R. D., Carter, J. C., & Kennedy, J. L. (2015). Association between MC4R rs17782313 polymorphism and overeating behaviors. *International Journal of Obesity, 39*(1), 114–120.

Young, A. I., Wauthier, F., & Donnelly, P. (2016). Multiple novel gene-by-environment interactions modify the effect of FTO variants on body mass index. *Nature Communications, 7*(12724).

Zalbahar, N., Najman, J., McIntyre, H. D., & Mamun, A. (2017). Parental pre-pregnancy obesity and the risk of offspring weight and body mass index change from childhood to adulthood. *Clinical Obesity, 7*(4), 206–215.

Zheng, J., Xiao, X., Zhang, Q., Yu, M., Xu, J., Wang, Z., ... Wang, T. (2015). Maternal and post-weaning high-fat, high-sucrose diet modulates glucose homeostasis and hypothalamic POMC promoter methylation in mouse offspring. *Metabolic Brain Disease, 30*(5), 1129–1137.

Zoratto, F., Romano, E., Pascale, E., Pucci, M., Falconi, A., Dell'Osso, B., et al. (2017). Down-regulation of serotonin and dopamine transporter genes in individual rats expressing a gambling-prone profile: A possible role for epigenetic mechanisms. *Neuroscience, 340*, 101–116.

Further reading

Kessler, R. C., Berglund, P. A., Chiu, W. T., Deitz, A. C., Hudson, J. I., Shahly, V., et al. (2013). The prevalence and correlates of binge eating disorder in the world health organization world mental health surveys. *Biological Psychiatry, 73*, 904–914.

Papacostas-Quintanilla, H., Ortiz-Ortega, V. M., & Lopez-Rubalcava, C. (2017). Wistar-kyoto female rats are more susceptible to develop sugar binging: A comparison with wistar rats. *Frontiers in Nutrition, 4*. Article 15.

CHAPTER

11

Neuroimaging of compulsive disorders: similarities of food addiction with drug addiction

Sonja Yokum, Eric Stice

Oregon Research Institute, Eugene, Oregon, United States

The prevalence of obesity has increased dramatically over the past decades with more than 500 million people worldwide classified as obese. The overabundance of inexpensive foods that are high in sugar, fat, salt, and calories and the widespread marketing of these foods putatively contribute to overeating and subsequent unhealthy weight gain. Over the past few years, increased attention has been given to the neuronal substrates that underlie overeating of these foods. Research suggests that repeated exposure to tastes of palatable, high-calorie foods and cues signaling their availability increases responsivity of brain reward regions, resulting in overconsumption of these foods via learning/conditioning mechanisms in a manner similar to drugs of abuse (Alsio, Olszewski, Levine, & Schioth, 2012; Avena, Rada, & Hoebel, 2009). However, many individuals who consume these foods do not become obese or show persistent overeating despite negative health consequences, just as the majority of individuals who try an addictive drug such as cocaine do not progress to regular use with negative consequences. Both cross-sectional and prospective functional magnetic resonance imaging (fMRI) studies with humans have provided important advances in knowledge of individual differences in neural response to palatable, high-calorie foods and their cues, which increase risk for unhealthy weight gain, and regarding changes in neural responsivity that result from overeating, which may serve to maintain overeating. Findings from these studies suggest that some individuals have higher behavioral or physiological responses to palatable foods (because of either experience or genetic variation) and may be more sensitive to the rewarding effects of palatable, energy-dense foods (Devoto et al., 2018; Stice & Yokum, 2016).

It has been proposed that palatable, high-calorie foods can impact the reward circuitry of susceptible individuals in a manner similar to addictive drugs, resulting in the development of an addictive-like response (dependence) to certain foods (Gearhardt, Corbin, & Brownell, 2009a, 2009b), as they overconsume these foods despite experiencing negative health consequences (e.g., obesity, diabetes) and report the

Compulsive Eating Behavior and Food Addiction. https://doi.org/10.1016/B978-0-12-816207-1.00011-1
Copyright © 2019 Elsevier Inc. All rights reserved.

desire to cut down, paralleling behavioral features of substance-use disorders (American Psychiatric Association, 2013). Yet, the "food addiction" phenotype has been controversial (Fletcher & Kenny, 2018; Ziauddeen & Fletcher, 2013), and there is no agreed upon definition to classify who meets criteria for a "food addiction." Still, existing evidence suggests that there are some overlaps in neurobiological mechanisms, brain activation patterns, and behavioral outcomes in individuals reporting an addictive-like response to high-calorie foods and persons with substance-misuse disorders (Davis et al., 2013; Gearhardt, White, Masheb, & Grilo, 2013; Gearhardt et al., 2012; Gearhardt, Yokum, et al., 2011; Meule, Lutz, Vogele, & Kubler, 2012).

This chapter reviews the effects of intake of drugs of abuse and palatable, high-calorie food, as well as cues signaling their availability, on brain rewards circuitry. We also discuss shared dysfunctional neural systems and several neural vulnerability theories potentially explaining these dysfunctional neural systems as well as evidence consistent or inconsistent with these theories. Finally, we review differentiating factors of addictive-like eating and offer essential next steps in neuroimaging research for food addiction. It should be noted that previous studies investigating the neural mechanisms in overeating have often examined individuals with obesity or binge eating disorder (BED) (Kenny, 2011; Volkow, Wang, Fowler, & Telang, 2008; Volkow, Wang, Fowler, & Tomasi, 2012; Volkow, Wang, Tomasi, & Baler, 2013). Hence, we included these studies as well. However, while there exists overlap in individuals reporting an addictive-like response to certain foods and those with obesity (about 20%) (Gearhardt, Corbin, & Brownell, 2009b, Gearhardt, Corbin, & Brownell, 2016) and BED (about 50% (Davis et al., 2011), neither condition is necessary or sufficient for food addiction.

The role of dopaminergic and opioid pathways in drug and food reward

Dopaminergic and opioid pathways play an important role in both substance abuse (Morganstern, Liang, Ye, Kratayev, & Leibowitz, 2012; Negus et al., 1993) and consumption of palatable, high-calorie foods (Blasio, Steardo, Sabino, & Cottone, 2014; Giuliano, Robbins, Nathan, Bullmore, & Everitt, 2012). The opioid system (endogenous peptides and their receptors) has been strongly linked to the hedonic pleasure from drugs of abuse and palatable food intake. Opioid receptors are located in several brain regions related to reward processing and the regulation of energy homeostasis (Le Merrer, Becker, Befort, & Kieffer, 2009; Noguieiras et al., 2012). Among the different opioid receptor subtypes, μ-opioid receptors (MORs) are thought to be strongly implicated in reward (Fields & Margolis, 2015; Nogueiras et al., 2012). The mesolimbic dopaminergic pathway is a primary component of the brain's reward circuitry. This circuit is composed of dopamine (DA) cells that project primarily from the ventral tegmental area (VTA) to the nucleus accumbens (NAc) and into limbic and cortical regions, including the amygdala, orbitofrontal cortex (OFC), and anterior cingulate cortex (ACC) (Stice, Figlewicz, Gosnell, Levine, & Pratt, 2013). DA is a main catecholamine neurotransmitter implicated in reinforcement- and reward learning, such as motivation and craving. Brain reward regions contain both opioid and DA receptors (Ambrose, Unterwald, & van Bockstaele, 2004), and MOR and DA receptor availability in reward regions such as the striatum and VTA are highly correlated in humans (Tuominen et al., 2015). The two neurotransmitter systems also show crosstalk (Tuominen et al., 2015). For example, dopaminergic signals can be altered by other neurotransmitters, including endogenous opioids, and some opiate signal pathways need a functional DA D2 receptor to function (Davis et al., 2009).

Effects of drugs of abuse and food intake on opioid release

Many classes of addictive drugs release endogenous opioids or bind to opioid receptors, producing feelings of euphoria (Koob & Le Moal, 2001). For example, animal studies found that rats that self-administered heroin showed an increase in MOR binding in the striatum, hippocampus, and VTA (Fattore et al., 2007). Human positron emission tomography (PET) studies show that alcohol and cocaine dependence are associated with higher brain MOR availability in the reward circuitry (Gorelick et al., 2005; Heinz et al., 2005; Palpacuer et al., 2015), but that opiate use causes downregulation in MOR (Whistler, 2012). Furthermore, brain MOR availability in alcohol dependency and cocaine addiction remains elevated in abstinence (Gorelick et al., 2005; Heinz et al., 2005), while MORs availability in heroin-dependent individuals recovers after detoxification (Zubieta et al., 2000). Similar to opiate addictions, decreased MOR availability has been found in obesity, BED, and bulimia nervosa (Burghardt, Rothberg, Dykhuis, Burant, & Zubieta, 2015; Joutsa et al., 2018; Karlsson et al., 2015). Furthermore, bariatric surgery and the resulting weight loss normalized the initially downregulated opioid receptor availability (Karlsson et al., 2015). However, although several animal studies have established that palatable food consumption leads to endogenous opioid release in the hypothalamus, ACC, and striatum (Colantuoni et al., 2001; Dum, Gramsch, & Herz, 1983), the relation between endogenous opioid release and hedonic eating in humans is unclear. Preclinical and clinical studies that have upregulated (increasing hedonic properties of foods) or downregulated (decreasing hedonic properties of food) the MOR system pharmacologically showed an association between MOR activation and the hedonic response of feeding (Yeomans & Gray, 2002; Ziauddeen et al., 2013). In contrast, a recent study (Tuulari et al., 2017) using in vivo PET in healthy weight individuals found that both palatable (pizza and diet coke) and nonpalatable (nutritional drink) meal consumption led to MOR downregulation and endogenous opioid release in the ventral striatum, thalamus, and ACC, while only the palatable meal resulted in higher subjective hedonic ratings. Moreover, opioid release was significantly stronger following the nonpalatable meal than the palatable meal and was independent of subjective hedonic ratings (Tuulari et al., 2017). According to the authors, a possible explanation for the larger opioid release following the nonpalatable liquid meal could be that this meal was digested faster in both stomach and intestine, thus triggering more profound opioid release (Tuulari et al., 2017). These latter findings suggest that although the hedonic properties of palatable food consumption may contribute to opioid release, a mere change in the energy homeostasis following consumption is already enough to trigger endogenous opioid release in humans (Tuulari et al., 2017).

Acute effects of drugs of abuse and palatable food on dopamine signaling and neural activation in the mesolimbic circuitry

Both animal and human studies suggest that there are several parallels between the effects of intake of drugs of abuse and palatable food on mesolimbic dopaminergic signaling (Heinz et al., 2004; Volkow, Koob, & McLellan, 2016; Wang, Volkow, & Fowler, 2002). Drugs of abuse affect the NAc and VTA and enhance DA reward synaptic function in the NAc (Gardner, 2011; Volkow et al., 2016). This neurochemical effect makes most addictive substances acutely rewarding (Gardner, 2011). Intake of drugs of abuse causes DA release in the striatum and associated mesolimbic regions (Di Chiara, 2002; Heinz et al., 2004; Kalivas & O'Brien, 2008; Volkow, Fowler, Wang, & Goldstein, 2002; Volkow, Wang, Fowler, et al., 2008). In both humans and animals, administration of drugs of abuse leads to activation in the VTA and striatum (Kalivas & O'Brien, 2008; Koob & Bloom, 1988; Volkow et al., 2002; Volkow et al., 2016). Likewise, animal studies have shown that consumption of palatable foods results in DA release in the striatum (Avena et al., 2009) In humans, consumption of palatable foods resulted in striatal DA release with the amount released correlating with meal pleasantness ratings (Kringelbach, O'Doherty, Rolls, & Andrews, 2003; Small, Jones-Gotman, & Dagher, 2003) and energy density (Ferreira, Tellez, Ren, Yeckel, & de Araujo, 2012). Furthermore, fMRI studies have found that consumption of palatable, high-calorie food is associated with elevated activation in reward regions, including the midbrain, insula, ventral and dorsal striatum, subcallosal cingulate, and OFC (Kringelbach et al., 2003; Small, Zatorre, Dagher, Evans, & Jones-Gotman, 2001; Stice, Yokum, & Burger, 2013).

It has been hypothesized that palatable foods high in calories, fat, salt, and/or sugar (e.g., pizza, chocolate) may uniquely activate the reward system in a manner similar to drugs of abuse (Gearhardt, Davis, et al., 2011; Schulte, Avena, & Gearhardt, 2015). In support, animal studies have shown that exposure to a palatable, high-calorie diet results in increased synaptic density in the VTA (Liu et al., 2016). Administration of glucose in the livers of rats increased activity in regions associated with reward and homeostatic feeding, including the NAc, OFC, and hypothalamus (Delaere, Akaoka, De Vadder, Duchampt, & Mithieux, 2013). Furthermore, consumption of sucrose triggered opioid and DA release in the NAc, with downstream effects on other limbic and frontal regions (Pomonis et al., 2000). Sugar intake also indirectly modulates mesolimbic DA circuits by altering cholecystokinin, insulin, and ghrelin levels (Ochoa, Lalles, Malbert, & Val-Laillet, 2015). An fMRI study evaluating the effect of a high-fat or high-sugar equicaloric chocolate milkshake in healthy weight adolescents found that the high-sugar relative to high-fat milkshake resulted in greater activation in regions associated with reward and motivation (insula, putamen), oral somatosensation (Rolandic operculum), and gustatory stimulation (thalamus), whereas consumption of the high-fat milkshake resulted in greater activation in regions associated with reward learning (caudate, hippocampus) and somatosensory regions (postcentral gyrus) (Stice, Burger, & Yokum, 2013).

It is posited that the increase in DA release as a result of intake of drugs of abuse and palatable food elicits a reward signal that triggers associative learning or conditioning (Schultz, Dayan, & Montague, 1997). Although initially intake of drugs of abuse or palatable food results in DA signaling, after repeated pairings of the drugs of abuse or food with cues that precede them (e.g., mental state, environments in which reward has been taken), the DA cells start firing in an anticipatory response to the cues (Schultz et al., 1997; Tindell, Berridge, & Aldridge, 2004; Tobler, Fiorillo, & Schultz, 2005), motivating drug- and food-seeking behaviors when exposed to these cues (Volkow et al., 2016). Many fMRI studies have utilized cue reactivity paradigms to examine neural activation in response to drugs- and palatable food cues, finding similar patterns of activation in reward-related regions, including the amygdala, striatum, insula, and OFC (Pelchat, Johnson, Chan, Valdez, & Ragland, 2004; Tang, Fellows, Small, & Dagher, 2012). Moreover, several studies have found that elevated neural response in reward processing regions in response to food cues predict *ad libitum* food intake. For example, elevated activation in the NAc and amygdala during exposure to palatable food pictures (Lawrence, Hinton, Parkinson, & Lawrence, 2012; Mehta et al., 2012) and fast food commercials (relative to control images) (Gearhardt, Harris, Lumen, Epstein, & Yokum, 2019) predicted *ad libitum* high-calorie food intake. Another study found that elevated neural response in the medial prefrontal cortex in response to anticipating food compared to money is positively associated with subsequent ad lib healthy and high-calorie food intake (Adise, Geier, Roberts, White, & Keller, 2018). This latter study also found that elevated activation in the amygdala and OFC in response to winning food compared to money predicted greater intake of ad lib high-calorie food intake and healthy food intake, respectively (Adise et al., 2018).

Overall, these findings suggest that there is considerable overlap in the acute effects of drugs of abuse, palatable food, and their cues on DA signaling and neuronal activity in reward processing regions. In addition, palatable high-calorie foods relative to low-calorie foods seem to produce more powerful effects on the brain reward circuitry, prompting cravings for and increase consumption of these foods.

Effects of chronic substance misuse and overconsumption of high-calorie food intake on changes in the brain's reward circuitry

In addition to similarities in the acute effects of drugs of abuse and palatable high-calorie foods on brain reward structures and circuits, research suggests that there is considerable overlap between the effects of habitual drug use and overconsumption of high-calorie food on brain reward circuitry changes. Animal experiments have shown that chronic exposure to drugs of abuse reduces striatal D2 receptor availability (Nader et al., 2006; Porrino, Lyons, Smith, Daunais, & Nader, 2004) and sensitivity of the reward circuitry (Kenny, Chen, Kitamura, Markou, & Koob, 2006). Animal experiments also indicate that habitual psychostimulant and opiate use causes increased DR1 binding, decreased DR2 receptor sensitivity, increased μ-opioid receptor binding, decreased basal DA transmission, and enhanced NAc DA response (Unterwald, Kreek, & Cuntapay, 2001; Vanderschuren & Kalivas, 2000). Similarly, human neuroimaging studies have found that individuals with versus without alcohol, cocaine, heroin, or methamphetamine dependence show lower striatal D2 receptor availability and sensitivity (Volkow et al., 1997; Volkow et al., 1996; Wang et al., 1997). Furthermore, human cocaine abusers show lower DA release in response to stimulant drugs relative to controls (Martinez et al., 2005; Volkow et al., 2005) and tolerance to the euphoric effects of cocaine (O'Brien, Volkow, & Li, 2006). fMRI studies found that individuals with versus without various substance-use disorders showed greater activation of reward (e.g., amygdala, VTA, OFC) and attention regions (ACC) and reported greater craving in response to substance-use cues (Due, Huettel, Hall, & Rubin, 2002; George et al., 2001; Maas et al., 1998; Martinez et al., 2005; Myrick et al., 2004; Tapert et al., 2003; Volkow et al., 2006). Similarly, prolonged access to palatable, high-calorie foods in animals results in brain reward circuitry changes, including downregulation of DA (Alsio et al., 2012; Johnson & Kenny, 2010). Rats randomized to overeating conditions that result in weight gain versus control conditions show downregulation of postsynaptic D2 receptors, reduced D2 sensitivity, extracellular DA levels in the NAc and DA turnover, and lower sensitivity of DA reward circuitry to food intake, electrical stimulation, amphetamine administration, and potassium administration (Bello, Lucas, & Hajnal, 2002; Davis et al., 2008a, 2008b; Geiger et al., 2009; Johnson & Kenny, 2010; Thanos, Michaelides, Piyis, Wang, & Volkow, 2008). Rats that were given a glucose solution on a limited-access schedule showed greater DR1 and less DR2 binding in the striatum compared with chow-fed rats. Importantly, intake of palatable high-calorie food resulted in downregulation of striatal D1 and D2 receptors in rats relative to isocaloric intake of low-fat/sugar chow (Alsio et al., 2010), implying that it is intake of palatable, high-calorie foods versus a positive energy balance that causes changes in the reward circuitry. Several PET and single-photon emission computed tomography studies suggest that obesity is associated with lower D2/D3 receptor availability in the striatum (de Weijer et al., 2011; Volkow, Wang, Telang, et al., 2008; Wang et al., 2001), although some studies did not find an effect of obesity (Eisenstein et al., 2013; Karlsson et al., 2015). Obese versus lean humans showed greater responsivity of brain regions associated with reward (e.g., striatum, amygdala, OFC) and attention (e.g., ACC) to pictures of palatable, high-calorie foods (vs. control stimuli; Bruce et al., 2010; Martin et al., 2010; Rothemund et al., 2007; Stice, Yokum, Bohon, Marti, & Smolen, 2010; Stoeckel et al., 2008) and to pictorial cues that signal impending palatable food tastes (Ng, Stice, Yokum, & Bohon, 2011; Stice, Spoor, Bohon, & Small, 2008). Individuals with bulimia nervosa and BED similarly showed greater reward-related neural response (e.g., insula, medial OFC) to high-calorie food images (vs. control stimuli) compared with healthy weight and overweight nonbinge eaters (Geliebter et al., 2006; Schienle, Schafer, Hermann, & Vaitl, 2009; Simon et al., 2016; Uher et al., 2004). Interestingly, binge eating, but not body mass index (BMI), was associated with increased striatal DA release in response to food stimulation (Wang et al., 2011), suggesting a dissociation between BMI and DA transmission. In the only fMRI study of food addiction (Gearhardt, Yokum, et al., 2011), individuals scoring high versus low on indicators of food addiction (as measured with the Yale Food Addiction Scale [YFAS]) showed greater activation in reward regions (e.g., caudate) while anticipating a palatable, high-calorie food and diminished activation of inhibitory control regions (lateral OFC) during consumption of the food. A repeated-measures fMRI study found that young women who gained weight over a 6-month period showed a reduction in striatal responsivity to a palatable high-calorie food relative to women who remained weight stable (Stice, Yokum, Blum, & Bohon, 2010), suggesting that weight gain results in hyporesponsivity of reward regions to food intake. Another fMRI study found that adolescents who reported elevated intake of ice cream showed reduced striatal response during intake of a chocolate milkshake, independent of BMI (Burger & Stice, 2012).

Overall, these findings suggest that palatable, high-calorie food overconsumption and weight gain result in changes in components of the brain reward system known to be impacted by drugs of abuse.

Foods associated with addictive-like eating

Although the term "food addiction" does not differentiate which foods may be associated with addictive-like eating, the construct posits that palatable, processed foods have the greatest addictive potential, as these foods contain added fat, refined carbohydrates, and/or salt that uniquely activate the reward system in a manner similar to drugs of abuse (Gearhardt et al., 2009a; Gearhardt, Davis, et al., 2011). To date, only a few studies have tested the effects of specific foods or their macronutrients on addictive-like eating behavior. Animal studies found that rats exposed to cafeteria-style diets (a normal chow diet supplemented with high-fat, high-sugar human foods such as cake and meat pies) showed an increased motivation to obtain the food (Liu et al., 2016), overall feeding bout frequency (Martire, Holmes, Westbrook, & Morris, 2013), and consumption despite of negative consequences (footshock) (Johnson & Kenny, 2010). Other studies found that rats exposed to high-sugar diets showed binge consumption, craving, use despite of negative consequences, and withdrawal (anxiety, teeth chattering) (Avena, Rada, & Hoebel, 2008; Boggiano et al., 2007; Oswald, Murdaugh, King, & Boggiano, 2011). Rats did not demonstrate these behavioral indicators of addiction when they were exposed to nutritionally balanced chow, even when circumstances were implemented to increase risk for compulsive overeating (e.g., intermitted access to chow) (Johnson & Kenny, 2010; Oswald et al., 2011). However, two other studies did not find evidence of behavioral dependence in rats that were exposed to sweetened diets that were also high in fat (Avena et al., 2009; Bocarsly, Berner, Hoebel, & Avena, 2011). Studies in humans have shown that palatable, high-calorie foods relative to low-calorie foods are more likely to be craved (Gilhooly et al., 2007; Ifland et al., 2009; Yanovski, 2003) and consumed in greater quantities in response to negative affect (Epel, Lapidus, McEwen, & Brownell, 2001; Oliver, Wardle, & Gibson, 2000; Zellner et al., 2006) and during binge episodes (Vanderlinden, Dalle Grave, Vandereycken, & Noorduin, 2001; Yanovski et al., 1992). A neuroimaging study in humans (DiFeliceantonio et al., 2018) found that foods high in both fat and carbohydrate resulted in greater activity of brain areas involved in reward compared with the single nutrient foods and that individuals were willing to pay more for foods containing both fat and carbohydrate than foods containing only fat or carbohydrate. So far, only one study (Schulte et al., 2015) has systematically examined which foods may be associated with addictive-like eating. Consistent with animal research, this study found that foods high in calories, fat, salt, and/or sugar (e.g., pizza, chocolate, chips) were associated with greater loss of control, liking, pleasure, and craving based on participants self-report compared with healthier foods (Schulte et al., 2015). Importantly, these foods were consumed more frequently among individuals who met criteria for food addiction on the YFAS (Gearhardt et al., 2009b), relative to participants who did not meet criteria (Pursey, Collins, Stanwell, & Burrows, 2015).

Collectively, animal models suggest that palatable, high-calorie versus low-calorie foods elicit greater behavioral indicators of addictive-like eating. However, the existing evidence regarding which foods may be addictive in humans is limited. Although previous fMRI studies have provided an improved understanding of how certain foods or macronutrients activate the reward system (DiFeliceantonia et al., 2018; Green & Murphy, 2012; Rudenga & Small, 2012; Smith & Robbins, 2013; Stice, Burger, et al., 2013; Tryon et al., 2015), it remains unknown how certain foods or their macronutrients induce neuroplastic changes in the brain, resulting in addictive-like eating behavior.

Interaction between substance and behavioral indicators of addiction

The substance abuse literature suggests that certain circumstances or behavioral patterns of use may exacerbate the addictive potential of the substance. Intermittent access to addictive drugs, bingeing, and use in response to negative affect or stress are all behavioral components that enhance the addictive potential of a substance or process (Volkow & Morales, 2015). In a similar vein, the food addiction construct posits that certain behavioral patterns of engagement (e.g., intermittency, eating to cope with negative affect) exacerbate the addictive potential of certain foods. In support, animal studies have shown that restricting access to a palatable, high-calorie food induces binge eating of this food (Corwin, 2004, 2006; Dimitriou, Rice, & Corwin, 2000), as well as a reduction in chow consumption (Berner, Avena, & Hoebel, 2008; Cottone et al., 2009). Furthermore, animal models suggest that high-calorie foods need to be presented frequently over short periods of time for the bingeing to occur (Rada, Avena, & Hoebel, 2005; Wojnicki, Roberts, & Corwin, 2006). Deprivation of the palatable, high-calorie foods after being exposed to the food intermittently results in increased motivation to consume the food (Avena, Long, & Hoebel, 2005; Corwin et al., 1998; Grimm, Manaois, Osincup, Wells, & Buse, 2007) and accelerates habitual control of instrumental learning (Furlong, Jayaweera, Balleine, & Corbit, 2014). There is only very limited research in humans. Studies in binge-type eating disorders, such as bulimia nervosa and BED, have found that individuals exhibit more indicators of disordered eating with palatable, high-calorie foods relative to healthier foods (Hadigan, Kissileff, & Walsh, 1989; Yanovski et al., 1992) and report that their binge eating behavior would be intensified if they had access to specific high-calorie food (e.g., ice cream) (Yanovski et al., 1992). Additional research is needed to understand variability in food overconsumption under certain circumstances (e.g., caloric deprivation, instructed bingeing).

Neural vulnerability factors that increase risk for substance abuse and addictive-like eating

Scholars have proposed several neural vulnerability factors that could theoretically increase risk for substance abuse and addictive-like eating. Below, we will discuss these theories and evidence that is consistent or inconsistent with these theories.

Incentive sensitization theory

Scholars have posited that there are individual differences in the attribution of incentive salience to reward-related cues. Animal studies found that some animals exhibited greater motivation to engage with cues that predict a drug or food reward, an indication of increased incentive salience (so-called sign trackers), than to elements of reward receipt like the location of reward delivery (so-called goal trackers)

(Flagel, Akil, & Robinson, 2009). Healthy teens with versus without family substance-use disorders showed greater responsivity of reward and attention regions to pictures of alcohol (Tapert et al., 2003). Similarly, prospective fMRI studies found that elevated striatal response to palatable food images and food commercials and elevated OFC response to cues that signal impending presentation of palatable food images predicted future weight gain (Demos, Heatherton, & Kelley, 2012; Yokum, Gearhardt, Harris, Brownell, & Stice, 2014; Yokum, Ng, & Stice, 2011). In addition, youth who showed the greatest escalation in ventral pallidum responsivity to cues signaling impending receipt of a palatable food showed significantly larger increases in BMI over 2-year follow-up (Burger & Stice, 2014). However, because these samples included overweight individuals, it is possible that a period of overeating might have caused the elevated reward region responsivity to these food cues. To rule out the possibility that a history of overeating contributed to any aberrant neural responsivity at baseline, one study (Stice, Burger, & Yokum, 2015) recruited a sample of healthy weight adolescents to test whether elevated reward region response to receipt and anticipated receipt of a palatable high-calorie food precedes initial excessive weight gain. Replicating the previous studies, it was found that elevated OFC response to cues signaling palatable high-calorie food receipt predicted initial excessive body fat gain (Stice et al., 2015). However, another study (Stice & Yokum, 2018) did not find evidence of hyperresponsivity of the reward circuitry to cues signaling palatable food receipt predicting future weight gain in healthy weight adolescents. Healthy weight adolescents who were eating beyond objectively measured basal metabolic needs showed greater response to cues predicting impending palatable high-calorie food receipt in regions that encode salience (precuneus) and reward (striatum) (Burger & Stice, 2013). This latter finding may suggest that overeating palatable, high-calorie food, even if it has not yet resulted in excess weight gain, may be accompanied by elevated responsivity of reward-related regions to high-calorie food cues. Alternatively, it is possible that these individuals already have a history of eating palatable, high-calorie foods and so undergo a conditioning process in which the cues become strongly associated with reward from consuming these foods, prompting cravings for and increase consumption of these foods, resulting in weight gain. Overall, these results provide preliminary evidence that there are important individual differences in attribution of incentive salience to reward-related cues, which may render them more vulnerable to cue-induced cravings.

Reward surfeit theory

The reward surfeit model posits that elevated reward sensitivity may increase risk for a range of appetitive problems, including substance misuse and overeating (Davis, Strachan, & Berkson, 2004; Stice et al., 2008). Reward sensitivity is assumed to be a biologically based predisposition to seek out rewarding substances and pursue situations and stimuli with high reward potential (Gray & McNaughton, 2000). Consistent with this thesis, healthy adults with versus without family substance-use disorders report that benzodiazepines and alcohol are more euphoric (Ciraulo et al., 1996; Cowley et al., 1992; Streeter et al., 1998). Likewise, healthy adolescents with versus without family history of substance misuse showed greater activation of reward regions (e.g., putamen, midbrain) in response to anticipated monetary reward and tastes of high-calorie food (Stice & Yokum, 2014). Furthermore, elevated reward region responsivity (caudate, putamen) in response to monetary reward predicted future substance-use onset (Stice, Yokum, et al., 2013). Similarly, healthy weight adolescents at high-versus low risk for future weight gain based on parental obesity status showed greater activation of regions implicated in reward (caudate, putamen, OFC) in response to receipt of both high-calorie food and monetary reward in two separate studies (Shearrer, Stice, & Burger, 2018; Stice, Yokum, Burger, Epstein, & Small, 2011). Geha et al. (Geha, Aschenbrenner, Felsted, O'Malley, & Small, 2013) found that elevated response in the midbrain, thalamus, hypothalamus, ventral pallidum, and NAc to the taste of a high-calorie food predicted elevated weight gain over 1-year follow-up. In contrast, two other studies did not find a main effect of reward region response to food or money and future weight gain (Stice et al., 2008, 2015).

Interestingly, the latter two studies (Stice et al., 2008, 2015) did find a significant interaction wherein elevated caudate response to high-calorie food receipt predicted future weight gain for adolescents with a genetic propensity for greater DA signaling capacity by virtue of possessing the DRD2 receptor A2/A2 allele, but lower caudate response predicted weight gain for adolescents with a genetic propensity for lower DA signaling capacity by virtue of possessing one or more DRD2 receptor A1 allele (Stice et al., 2008, 2015). Another study (Sun et al., 2015) found that elevated amygdala response to high-calorie food receipt predicted future weight gain for adults with the DRD2 receptor A2/A2 allele and lower amygdala response predicted weight gain for adults with the DRD2 receptor A1 allele. However, this study did not replicate the significant interaction between DRD2 receptor allele status and caudate response to high-calorie food receipt in the prediction of future weight gain.

Additional genetic findings also seem to provide support for the reward surfeit model of overeating. Individuals with a greater number of genotypes putatively associated with greater DA signaling in the reward circuitry, as defined by the multilocus genetic composite risk score, showed elevated future weight gain in three samples, as well as significantly less weight loss in response to obesity treatment (Yokum, Marti, Smolen, & Stice, 2015). In addition, the multilocus genetic composite was higher in individuals with YFAS-diagnosed food addiction and correlated positively with binge eating, food cravings, and emotional overeating (Davis et al., 2013). In contrast, a genome-wide association study of 9314 females of European ancestry who were identified as having indicators of food addiction as measured with the YFAS did not find a significant association with any single-nucleotide polymorphisms or genes implicated in drug addiction (Cornelis et al., 2016). Overall, high-risk and prospective studies that investigated facets of the reward surfeit theory generated supportive findings as the results suggest that hyperresponsivity in the brain's reward circuitry to rewards in general increases risk for substance use and obesity.

Reward deficit model

It has been hypothesized that some individuals may be motivated to consume drugs of abuse or palatable, high-calorie foods to compensate for diminished DA receptor availability (Blum et al., 2000). Apparently, consistent with this reward deficiency theory, studies using PET and (^{11}C)-raclopride imaging on human subjects with alcohol, cocaine, heroin, and methamphetamine substance-use disorders showed less striatal D2-like receptor availability and sensitivity as well as lower in the DRD2 availability in the striatum (Martinez et al., 2005; Martinez et al., 2007; Volkow et al., 1997). Low striatal D2-like receptor availability in primates was also associated with increased future drug self-administration (Nader et al., 2006). As noted, some studies have found that obese versus healthy weight adults show lower striatal DA D2-like receptor availability (de Weijer et al., 2011; Volkow et al., 2008). Obese versus healthy weight adults also showed lower capacity of nigrostriatal neurons to synthesize DA (Wilcox, Braskie, Kluth, & Jagust, 2010) and less striatal responsivity to tastes of high-fat/sugar beverages (Babbs et al., 2013; Frank et al., 2012; Green, Jacobson, Haase, & Murphy, 2011; Stice, Spoor, Bohon, Veldhuizen, & Small, 2008). Obese individuals with BED exhibited less activation in limbic regions implicated in reward compared with obese individuals without BED when exposed to monetary rewards (Balodis et al., 2013). Animal studies found that obese versus lean rats have lower basal DA levels and D2-like receptor availability and less ex vivo DA release in response to electrical stimulation in the NAc and dorsal striatum tissue (Fetissov, Meguid, Sato, & Zhang, 2002; Geiger et al., 2008; Huang et al., 2006; Thanos et al., 2008).

However, other findings seem incompatible with the notion that a low sensitivity of DA reward circuitry leads to substance abuse and overeating. As noted, several prospective and experimental findings indicate that substance use and overeating contribute to reward region hyporesponsivity. Animal experiments found that regular substance use reduces striatal D2-like receptors and sensitivity of reward circuitry (Kenny et al., 2006; Nader et al., 2006). Chronic cocaine use has been associated with blunted DA release to stimulant drugs, tolerance to the euphoric effects of cocaine, and downregulated responses to both cocaine and food cues (O'Brien et al., 2006; Tomasi et al., 2015; Volkow et al., 2005). Likewise, increases in DA signaling lead to downregulated reward circuitry. Both cocaine-dependent and alcohol-dependent individuals showed lower DA release in the NAc in response to methylphenidate relative to healthy controls (Volkow, Wang, Fowler, Logan, Gatley, et al., 1997). Even adolescents with a relatively short history of substance use showed less caudate response to monetary reward relative to adolescents who had not initiated substance use (Stice et al., 2013). Similarly, mice in which reduced striatal DA signaling from food intake was experimentally induced through chronic intragastric infusion of fat worked *less* for acute intragastric infusion of fat and consumed *less* rat chow ad lib than control mice (Tellez et al., 2013). Experimentally induced DA depletion was associated with less ad lib food intake in humans (Hardman, Herbert, Brunstrom, Munafo, & Rogers, 2012). DA-deficient mice were unable to sustain appropriate levels of feeding and dysregulation of DA signaling in the dorsal striatum, in particular induced hypophagia (Sotak, Hnasko, Robinson, Kremer, & Palmiter, 2005; Zhou & Palmiter, 1995). As noted, animal studies have shown that rats randomized to overeating conditions that result in weight gain versus control conditions show downregulation of postsynaptic D2 receptors, reduced D2 sensitivity, and lower sensitivity of DA reward circuitry to food intake, electrical stimulation, amphetamine administration, and potassium administration (Bello et al., 2002; Davis et al., 2008a, 2008b; Geiger et al., 2009; Johnson & Kenny et al., 2010; Thanos et al., 2008). Also as noted, young women who gained weight over a 6-month period showed a reduction in striatal responsivity to palatable food receipt relative to women who remained weight stable (Stice et al., 2010). However, three studies in humans found significant interactions wherein a weaker striatal (Stice et al., 2008, 2015) and amygdala (Sun et al., 2015) response to receipt of palatable, high-calorie food predicted future weight gain for participants with the DRD2 receptor A1 allele. Furthermore, hyporesponsivity in reward-related regions (putamen, OFC) to palatable food images predicted future weight gain for adolescents possessing the A1 allele (Stice, Yokum, Bohon, et al., 2010). The presence of DRD2 receptor A1 allele has also found to be associated with compulsive overeating, which may suggest that reward hyposensitivity is a genetic risk factor for the development of problematic eating behavior (Blum et al., 2000; Davis et al., 2008a, 2008b). These latter findings suggest that the reward surfeit model may apply to individuals with a genetic propensity for greater DA signaling capacity and that the reward deficit model may apply to those with a genetic propensity for weaker DA signaling. That is, too much or too little DA signaling capacity and reward region responsivity may both increase risk for overeating, potentially because each perturbs homeostatic processes that maintain a balance between caloric intake and caloric expenditure.

Inhibitory control deficit theory

It has also been proposed that individuals with inhibitory control deficits, and by extension lower responsivity of brain regions implicated in inhibitory control, are more at risk for substance use and overeating, putatively because they are more likely to act on temptation when they are exposed to drug- and food cues and less likely to consider the negative consequences from their actions (Hester, Lubman, & Yucel, 2010; Nederkoorn, Braet, Van Eijs, Tanghe, & Jansen, 2006). Addicted individuals showed reduced activation in inhibitory control regions during decision-making tasks (Hester et al., 2010; MacKillop et al., 2011). Individuals with versus without various substance-use disorders had lower frontal inhibitory region volume (Franklin et al., 2002; Lyoo et al., 2006). Abstinent individuals with versus without family history of substance-use disorders showed less inhibitory control (Saunders et al., 2008) and less activation in frontal and parietal regions during a go/no-go task (Schweinsburg et al., 2004). Deficits in inhibitory control and immediate reward bias during childhood/adolescence increased risk for future substance-use onset (Ayduk et al., 2000; Ernst et al., 2006; McGue, Iacono, Legrand, Malone, & Elkins,

2001). Decreased inhibitory control activation was predictive of relapse (Paulus, Tapert, & Schuckit, 2005). Similarly, obese versus healthy weight adolescents showed less activation in inhibitory control regions (e.g., dorsolateral prefrontal cortex [dlPFC]) when trying to inhibit responses to high-calorie food images and also showed behavioral evidence of reduced inhibitory control (Batterink, Yokum, & Stice, 2010). Individuals who showed less recruitment of inhibitory control regions in tasks that require inhibition gained more weight over long-term follow-up (Kishinevsky et al., 2012), though this effect was not replicated in another study (Batterink et al., 2010). Individuals that showed less recruitment of the dlPFC during a delay discounting task showed significantly less weight loss in response to weight loss treatment (Weygandt et al., 2013) and less weight loss maintenance over a 1-year follow-up (Weygandt et al., 2015). Another study found that lower dlPFC response to high-calorie food images predicted greater food intake over the next 3 days (Cornier, Salzberg, Endly, Bessesen, & Tregellas, 2010). Less recruitment of an inhibitory control region (presupplemental motor area) in response to tastes of palatable, high-calorie foods was associated with greater weight gain over 3-year follow-up in initially healthy weight adolescents (Stice & Yokum, 2018). Overall, these findings suggest that deficits in executive control neural circuitry may contribute to impulsive decision-making in drug addiction and overeating.

Emotion dysregulation theory

Scholars have posited that stress and negative affect increase risk for both substance abuse and addictive-like eating. Negative affect has found to be associated with elevated drug use (Baker, Piper, McCarthy, Majeskie, & Fiore, 2004; Kenny et al., 2006; Sinha, Garcia, Paliwal, Kreek, & Rounsaville, 2006) and relapse (Miller, Westerberg, Harris, & Tonigan, 1996; Sinha et al., 2006). Individuals addicted to psychostimulants showed elevated activation in reward-related regions (e.g., striatum) and less activation in regions associated with processing of emotions (e.g., ACC) in response to stress (Sinha, 2018). Experimental studies have found that stress exposure increased drinking and nicotine smoking (Sinha, 2005). Emotional stress predicted future onset of substance use in adolescents (Tschann et al., 1994). Stress and adversity have also been associated with intake of fast food (Steptoe, Lipsey, & Wardle, 1998), snacks (Oliver & Wardle, 1999), high-calorie foods (Adam & Epel, 2007), compulsive food seeking of palatable foods (Lemmens, Rutters, Born, & Westerterp-Plantenga, 2011), and binge eating (Freeman & Gil, 2004). Elevated striatum activity in response to stress and food cues is related to stronger food cravings (Jastreboff et al., 2013). Chronic stress was associated with less activation of prefrontal inhibitory regions in response to high-calorie food cues, and this pattern of activation was related to greater consumption of high-calorie foods (Tryon, Carter, Decant, & Laugero, 2013). Acute stress was also associated with elevated amygdala activation in response to a palatable, high-calorie food (Rudenga, Sinha, & Small, 2013). Furthermore, prospective studies have found that high stress predicts longitudinal weight gain among adults (Block, He, Zaslavsky, Ding, &

Ayanian, 2009; Chao, Jastreboff, White, Grilo, & Sinha, 2017) and adolescents (Ruttle et al., 2013). Collectively, these findings suggest a positive association of stress and negative affect with both substance misuse and with overeating and weight gain.

Differences between addictive disorders and addictive-like eating

While existing evidence suggests that there are similarities in neurobiological mechanisms between addictive behaviors and addictive-like eating behavior, important differences exist. First, for drugs of abuse, an addictive agent has been defined (e.g., ethanol in alcohol and nicotine in cigarettes) and high concentrations of those ingredients are associated with an increased additive potential (Henningfield & Keenan, 1993). In contrast, the addictive agents in certain foods have not yet been isolated (Fletcher & Kenny, 2018; Ziauddeen & Fletcher, 2013). That is, although palatable high-calorie foods have been hypothesized to be most likely implicated in addictive-like eating because of the high levels of fat and/or refined carbohydrates (Gearhardt, Davis, et al., 2011), it is not yet clear which macronutrients or micronutrients (e.g., fat content) or which combination increases risk for addictive-like eating. However, sugar bingeing rat models suggest that there are parallels with addiction with regard to tolerance, withdrawal, cross-sensitization, and neurochemical changes (Avena, 2007). There are also preliminary data in humans suggesting that sugar taste more effectively recruits reward and gustatory regions compared with fat (Stice et al., 2013). Additionally, research has not tested whether a particular "dose" or quantity of certain food attributes would increase the abuse potential of an "addictive" food (Ziauddeen, Farooqi, & Fletcher, 2012). It is very likely that certain foods may have multiple addictive agents, such as sugar, fat, and salt, which may increase the food's addictive potential (DiFeliceantonio et al., 2018). Another difference between drug addiction and food addiction is the presence of withdrawal symptoms. Withdrawal symptoms are a necessary component of dependence (APA, 2013), and there is only limited evidence of physical or psychological withdrawal symptoms in food addiction in animal models. Future studies need to examine whether withdrawal symptoms may contribute to addictive-like eating behavior.

Conclusions and future directions

The above sections have examined similarities and differences between drug addiction and addictive-like eating. Several themes emerge from this chapter. First, both drugs of abuse and palatable, high-calorie food engage our reward circuitry in a similar fashion. Second, repeated intake of drugs of abuse and palatable food increases brain motivation circuits to drug cues and food cues, which appear to be critical for promoting escalated intake of both substances. Third, both humans and animal models suggest that certain individuals are more sensitive to the rewarding effects of drugs of abuse and food (due to either experience or genetic variation), resulting in higher psychological or physiological responses to drugs and food. Findings suggest that overlapping neural vulnerability factors (hyper- and hyporesponsivity of the reward circuitry, impulsivity, negative affect) and dysfunctional neural systems (inhibitory control, emotion regulation) appear to be implicated in both drug addiction and (compulsive) overeating.

There are several important gaps in the literature. First, it is still unclear which macronutrients or micronutrients have addictive potential in humans. The existing evidence for which foods may be addictive has been limited to self-report data (Schulte et al., 2015). Therefore, testing how foods that vary in sugar, fat, and salt content activate the brain reward circuitry and produce changes in brain function that lead to addictive or uncontrolled eating is essential to evaluating this perspective. Second, future research should also study individual differences that may increase risk of addictive-like eating behavior. Multiple risk factors have been implicated in drug addiction, including family history (Tapert et al., 2000), impulsivity (Dawe & Loxton, 2004), and genetic alleles associated with reward dysfunction (Dick & Foroud, 2003). These individual differences have also been implicated in obesity and eating disorders (Balodis et al., 2013; Davis et al., 2013; Gearhardt & Corbin, 2009;Yokum et al., 2015). Future research is needed to examine if these factors may increase the risk of addictive-like eating. Third, future research should explore whether mechanisms unique to addictive disorders, such as withdrawal and tolerance, may contribute to addictive-like consumption of certain foods in humans. Fourth, future research should assess the neural response to palatable, high-calorie foods and addictive substances among individuals who experience problems related to both food and drug of abuse (e.g., alcohol dependence). With this study design, it may be possible to compare the abuse liability of palatable, high-calorie foods relative to drugs of abuse in vulnerable individuals. Fifth, given the emerging evidence of overlap between addictive-like eating and binge-type eating disorders (e.g., BED, bulimia nervosa, anorexia binge/purge subtype) (Granero et al., 2014; Meule, Von Rezori, & Blecher, 2014), future research should explore the addiction-specific mechanism across various forms of binge eating. Finally, although it is important to explore similarities between substance misuse and addictive-like eating, the differences in overuse of drugs and food should be explored as well to provide a more complete picture of food addiction.

References

Adam, T. C., & Epel, E. S. (2007). Stress, eating and the reward system. *Physiology & Behavior, 91*(4), 449–458.

Adise, S., Geier, C., Roberts, N., White, C., & Keller, K. (2018). Is brain response to food reward related to overeating? A test of the reward surfeit model of overeating in children. *Appetite, 128*, 167–179.

Alsio, J., Olszewski, P., Norback, A., Gunarsson, Z., Levine, A., & Rickering, C. (2010). Dopamine D1 receptor gene expression decreases in the nucleus accumbens upon long-term exposure to palatable food and differs depending on diet-induced obesity phenotype in rats. *Neuroscience, 171*, 779–787.

Alsio, J., Olszewski, P. K., Levine, A. S., & Schioth, H. B. (2012). Feed-forward mechanisms: Addiction-like behavioral and molecular adaptations in overeating. *Frontiers in Neuroendocrinology, 33*(2), 127–139.

Ambrose, L., Unterwald, W., & van Bockstaele, E. (2004). Ultrastructural evidence for co-localization of dopamine D2 and micro-opioid receptors in the rat dorsolateral striatum. *The Anatomical Record Part A: Discoveries in Molecular, Cellular, and Evolutionary Biology, 279*, 583–591.

American Psychiatric Association. (2013). *Diagnostic and statistical manual of mental disorders: DSM-5*. Washington, DC: American Psychiatric Association.

Avena, N. M., Long, K. A., & Hoebel, B. G. (2005). Sugar-dependent rats show enhanced responding for sugar after abstinence: Evidence of a sugar deprivation effect. *Physiology & Behavior, 84*(3), 359–362.

Avena, N. M., Rada, P., & Hoebel, B. G. (2008). Evidence for sugar addiction: Behavioral and neurochemical effects of intermittent, excessive sugar intake. *Neuroscience & Biobehavioral Reviews, 32*(1), 20–39.

Avena, N. M., Rada, P., & Hoebel, B. G. (2009). Sugar and fat bingeing have notable differences in addictive-like behavior. *Journal of Nutrition, 139*(3), 623–628.

Avena, N. M. (2007). Examining the addictive-like properties of binge eating using an animal model of sugar dependence. *Experimental and Clinical Psychopharmacology, 15*(5), 481–491.

Ayduk, O., Mendoza-Denton, R., Mischel, W., Downey, G., Peake, P. K., & Rodriguez, M. (2000). Regulating the interpersonal self: Strategic self-regulation for coping with rejection sensitivity. *Journal of Personality and Social Psychology, 79*(5), 776–792.

Babbs, R. K., Sun, X., Felsted, J., Chouinard-Decorte, F., Veldhuizen, M. G., & Small, D. M. (2013). Decreased caudate response to milkshake is associated with higher body mass index and greater impulsivity. *Physiology & Behavior, 121*, 103–111.

Baker, T. B., Piper, M. E., McCarthy, D. E., Majeskie, M. R., & Fiore, M. C. (2004). Addiction motivation reformulated: An affective processing model of negative reinforcement. *Psychological Review, 111*(1), 33–51.

Balodis, I. M., Kober, H., Worhunsky, P. D., White, M. A., Stevens, M. C., Pearlson, G. D., … Potenza, M. N. (2013). Monetary reward processing in obese individuals with and without binge eating disorder. *Biological Psychiatry, 73*(9), 877–886.

Batterink, L., Yokum, S., & Stice, E. (2010). Body mass correlates inversely with inhibitory control in response to food among adolescent girls: An fMRI study. *NeuroImage, 52*(4), 1696–1703.

Bello, N. T., Lucas, L. R., & Hajnal, A. (2002). Repeated sucrose access influences dopamine D2 receptor density in the striatum. *NeuroReport, 13*(12), 1575–1578.

Berner, L. A., Avena, N. M., & Hoebel, B. G. (2008). Bingeing, self-restriction, and increased body weight in rats with limited access to a sweet-fat diet. *Obesity, 16*(9), 1998–2002.

Blasio, A., Steardo, L., Sabino, V., & Cottone, P. (2014). Opioid system in the medial prefrontal cortex mediates binge-like eating. *Addiction Biology, 19*(4), 652–662.

Block, J. P., He, Y., Zaslavsky, A. M., Ding, L., & Avanian, J. Z. (2009). Psychosocial stress and change in weight among US adults. *American Journal of Epidemiology, 170*(2), 181–192.

Blum, K., Braverman, E. R., Holder, J. M., Lubar, J. F., Monastra, V. J., Miller, D., … Comings, D. E. (2000). Reward deficiency syndrome: A biogenetic model for the diagnosis and treatment of impulsive, addictive, and compulsive behaviors. *Journal of Psychoactive Drugs, 32*(Suppl. i–iv), 1–112.

Bocarsly, M. E., Berner, L. A., Hoebel, B. G., & Avena, N. M. (2011). Rats that binge eat fat-rich food do not show somatic signs or anxiety associated with opiate-like withdrawal: Implications for nutrient-specific food addiction behaviors. *Physiology & Behavior, 104*, 865–872.

Boggiano, M. M., Artiga, A. I., Pritchett, C. E., Chandler-Laney, P. C., Smith, M. L., & Eldridge, A. J. (2007). High intake of palatable food predicts binge-eating independent of susceptibility to obesity: An animal model of lean vs obese binge-eating and obesity with and without binge-eating. *International Journal of Obesity, 31*(9), 1357–1367.

Bruce, A. S., Holsen, L. M., Chambers, R. J., Martin, L. E., Brooks, W. M., Zarcone, J. R., … Savage, C. R. (2010). Obese children show hyperactivation to food pictures in brain networks linked to motivation, reward and cognitive control. *International Journal of Obesity, 34*(10), 1494–1500.

Burger, K. S., & Stice, E. (2012). Frequent ice cream consumption is associated with reduced striatal response to receipt of an ice cream-based milkshake. *American Journal of Clinical Nutrition, 95*(4), 810–817.

Burger, K. S., & Stice, E. (2013). Elevated energy intake is correlated with hyperresponsivity in attentional, gustatory, and reward brain regions while anticipating palatable food receipt. *American Journal of Clinical Nutrition, 97*(6), 1188–1194.

Burger, K. S., & Stice, E. (2014). Greater striatopallidal adaptive coding during cue-reward learning and food reward habituation predict future weight gain. *NeuroImage, 99*, 122–128.

Burghardt, P. R., Rothberg, A. E., Dykhuis, K. E., Burant, C. F., & Zubieta, J. K. (2015). Endogenous opioid mechanisms are implicated in obesity and weight loss in humans. *The Journal of Cinical Endocrinology and Metabolism, 100*(8), 3193–3201.

Chao, A. M., Jastreboff, A. M., White, M. A., Grilo, C. M., & Sinha, R. (2017). Stress, cortisol, and other appetite-related hormones: Prospective prediction of 6-month changes in food cravings and weight. *Obesity, 25*(4), 713–720.

Ciraulo, D. A., Sarid-Segal, O., Knapp, C., Ciraulo, A. M., Greenblatt, D. J., & Shader, R. I. (1996). Liability to alprazolam abuse in daughters of alcoholics. *American Journal of Psychiatry, 153*(7), 956–958.

Colantuoni, C., Schwenker, J., McCarthy, J., Rada, P., Ladenheim, B., Cadet, J. L., … Hoebel, B. G. (2001). Excessive sugar intake alters binding to dopamine and mu-opioid receptors in the brain. *NeuroReport, 12*(16), 3549–3552.

Cornelis, M. C., Flint, A., Field, A. E., Kraft, P., Han, J., Rimm, E. B., & van Dam, R. M. (2016). A genome-wide investigation of food addiction. *Obesity, 24*(6), 1336–1341.

Cornier, M. A., Salzberg, A. K., Endly, D. C., Bessesen, D. H., & Tregellas, J. R. (2010). Sex-based differences in the behavioral and neuronal responses to food. *Physiology & Behavior, 99*(4), 538–543.

Corwin, R. L., Wojnicki, F. H., Fisher, J. O., Dimitriou, S. G., Rice, H. B., & Young, M. A. (1998). Limited access to a dietary fat option affects ingestive behavior but not body composition in male rats. *Physiology & Behavior, 65*(3), 545–553.

Corwin, R. L. (2004). Binge-type eating induced by limited access in rats does not require energy restriction on the previous day. *Appetite, 42*(2), 139–142.

Corwin, R. L. (2006). Bingeing rats: A model of intermittent excessive behavior? *Appetite, 46*(1), 11–15.

Cottone, P., Sabino, V., Roberto, M., Bajo, M., Pockros, L., Frihauf, J. B., … Zorrilla, E. P. (2009). CRF system recruitment mediates dark side of compulsive eating. *Proceedings of the National Academy of Sciences of the United States of America, 106*(47), 20016–20020.

Cowley, D. S., Roy-Byrne, P. P., Godon, C., Greenblatt, D. J., Ries, R., Walker, R. D., ... Hommer, D. W. (1992). Response to diazepam in sons of alcoholics. *Alcoholism: Clinical and Experimental Research, 16*(6), 1057–1063.

Davis, C., Strachan, S., & Berkson, M. (2004). Sensitivity to reward: Implications for overeating and overweight. *Appetite, 42*, 131–138.

Davis, C., Levitan, R. D., Kaplan, A. S., Carter, J., Reid, C., Curtis, C., ... Kennedy, J. L. (2008). Reward sensitivity and the D2 dopamine receptor gene: A case-control study of binge eating disorder. *Progress In Neuro-Psychopharmacology & Biological Psychiatry, 32*(3), 620–628.

Davis, J. F., Tracy, A. L., Schurdak, J. D., Tschop, M. H., Lipton, J. W., Clegg, D. J., & Benoit, S. C. (2008b). Exposure to elevated levels of dietary fat attenuates psychostimulant reward and mesolimbic dopamine turnover in the rat. *Behavioral Neuroscience, 122*(6), 1257–1263.

Davis, C. A., Levitan, R. D., Reid, C., Carter, J. C., Kaplan, A. S., Patte, K. A., ... Kennedy, J. L. (2009). Dopamine for "wanting" and opioids for "liking": A comparison of obese adults with and without binge eating. *Obesity, 17*(6), 1220–1225.

Davis, C., Curtis, C., Levitan, R. D., Carter, J. C., Kaplan, A. S., & Kennedy, J. L. (2011). Evidence that 'food addiction' is a valid phenotype of obesity. *Appetite, 57*(3), 711–717.

Davis, C., Loxton, N. J., Levitan, R. D., Kaplan, A. S., Carter, J. C., & Kennedy, J. L. (2013). 'Food addiction' and its association with a dopaminergic multilocus genetic profile. *Physiology & Behavior, 118*, 63–69.

Dawe, S., & Loxton, N. J. (2004). The role of impulsivity in the development of substance use and eating disorders. *Neuroscience & Biobehavioral Reviews, 28*(3), 343–351.

de Weijer, B. A., van de Giessen, E., van Amelsvoort, T. A., Boot, E., Braak, B., Janssen, I. M., ... Booij, J. (2011). Lower striatal dopamine D2/3 receptor availability in obese compared with non-obese subjects. *EJNMMI Research, 1*(1), 37.

Delaere, F., Akaoka, H., De Vadder, F., Duchampt, A., & Mithieux, G. (2013). Portal glucose influences the sensory, cortical and reward systems in rats. *European Journal of Neuroscience, 38*(10), 3476–3486.

Demos, K. E., Heatherton, T. F., & Kelley, W. M. (2012). Individual differences in nucleus accumbens activity to food and sexual images predict weight gain and sexual behavior. *Journal of Neuroscience, 32*, 5549–5552.

Devoto, F., Zapparoli, L., Bonandrini, R., Berlingeri, M., Ferrulli, A., Luzi, L., ... Paulesu, E. (2018). Hungry brains: A meta-analytical review of brain activation imaging studies on food perception and appetite in obese individuals. *Neuroscience & Biobehavioral Reviews, 94*, 271–285.

Di Chiara, G. (2002). Nucleus accumbens shell and core dopamine: Differential role in behavior and addiction. *Behavioural Brain Research, 137*(1–2), 75–114.

Dick, D. M., & Foroud, T. (2003). Candidate genes for alcohol dependence: A review of genetic evidence from human studies. *Alcoholism: Clinical and Experimental Research, 27*(5), 868–879.

DiFeliceantonio, A. G., Coppin, G., Rigoux, L., Thanarajah, S. E., Dagher, A., Tittgemeyer, et al. (2018). Supra-additive effects of combining fat and carbohydrate on food reward. *Cell Metabolism, 28*, 33–44.

Dimitriou, S. G., Rice, H. B., & Corwin, R. L. (2000). Effects of limited access to a fat option on food intake and body composition in female rats. *International Journal of Eating Disorders, 28*(4), 436–445.

Due, D. L., Huettel, S. A., Hall, W. G., & Rubin, D. C. (2002). Activation in mesolimbic and visuospatial neural circuits elicited by smoking cues: Evidence from functional magnetic resonance imaging. *American Journal of Psychiatry, 159*(6), 954–960.

Dum, J., Gramsch, C., & Herz, A. (1983). Activation of hypothalamic beta-endorphin pools by reward induced by highly palatable food. *Pharmacology Biochemistry and Behavior, 18*(3), 443–447.

Eisenstein, S. A., Antenor-Dorsey, J. A., Gredysa, D. M., Koller, J. M., Bihun, E. C., Ranck, S. A., ... Hershey, T. (2013). A comparison of D2 receptor specific binding in obese and normal-weight individuals using PET with (N-[(11)C]methyl)benperidol. *Synapse, 67*(11), 748–756.

Epel, E., Lapidus, R., McEwen, B., & Brownell, K. (2001). Stress may add bite to appetite in women: A laboratory study of stress-induced cortisol and eating behavior. *Psychoneuroendocrinology, 26*(1), 37–49.

Ernst, M., Luckenbaugh, D. A., Moolchan, E. T., Leff, M. K., Allen, R., Eshel, N., ... Kimes, A. (2006). Behavioral predictors of substance-use initiation in adolescents with and without attention-deficit/hyperactivity disorder. *Pediatrics, 117*(6), 2030–2039.

Fattore, L., Vigano, D., Fadda, P., Rubino, T., Fratta, W., & Parolaro, D. (2007). Bidirectional regulation of mu-opioid and CB1-cannabinoid receptor in rats self-administering heroin or WIN 55,212-2. *European Journal of Neuroscience, 25*(7), 2191–2200.

Ferreira, J. G., Tellez, L. A., Ren, X., Yeckel, C. W., & de Araujo, I. E. (2012). Regulation of fat intake in the absence of flavour signalling. *Journal of Physiology, 590*(4), 953–972.

Fetissov, S. O., Meguid, M. M., Sato, T., & Zhang, L. H. (2002). Expression of dopaminergic receptors in the hypothalamus of lean and obese Zucker rats and food intake. *American Journal of Physiology - Regulatory, Integrative and Comparative Physiology, 283*(4), R905–R910.

Fields, H. L., & Margolis, E. B. (2015). Understanding opioid reward. *Trends in Neurosciences, 38*(4), 217–225.

Flagel, S. B., Akil, H., & Robinson, T. E. (2009). Individual differences in the attribution of incentive salience to reward-related cues: Implications for addiction. *Neuropharmacology, 56*(Suppl. 1), 139–148.

Fletcher, P. C., & Kenny, P. J. (2018). Food addiction: A valid concept? *Neuropsychopharmacology, 43*, 2506–2513.

Frank, G. K., Reynolds, J. R., Shott, M. E., Jappe, L., Yang, T. T., Tregellas, J. R., & O'Reilly, R. C. (2012). Anorexia nervosa and obesity are associated with opposite brain reward response. *Neuropsychopharmacology, 307*, 2031–2046.

Franklin, T. R., Acton, P. D., Maldjian, J. A., Gray, J. D., Croft, J. R., Dackis, C. A., ... Childress, A. R. (2002). Decreased gray matter concentration in the insular, orbitofrontal, cingulate, and temporal cortices of cocaine patients. *Biological Psychology, 51*(2), 134–142.

Freeman, L. M., & Gil, K. M. (2004). Daily stress, coping, and dietary restraint in binge eating. *International Journal of Eating Disorders, 36*(2), 204–212.

Furlong, T. M., Jayaweera, H. K., Balleine, B. W., & Corbit, L. H. (2014). Binge-like consumption of a palatable food accelerates habitual control of behavior and is dependent on activation of the dorsolateral striatum. *Journal of Neuroscience, 34*(14), 5012–5022.

Gardner, E. L. (2011). Addiction and brain reward and antireward pathways. *Advances in Psychosomatic Medicine, 30*, 22–60.

Gearhardt, A. N., & Corbin, W. R. (2009). Body mass index and alcohol consumption: Family history of alcoholism as a moderator. *Psychology of Addictive Behaviors, 23*, 216–225.

Gearhardt, A. N., Corbin, W. R., & Brownell, K. D. (2009a). Food addiction: An examination of the diagnostic criteria for dependence. *Journal of Addiction Medicine, 3*(1), 1–7.

Gearhardt, A. N., Corbin, W. R., & Brownell, K. D. (2009b). Preliminary validation of the Yale food addiction Scale. *Appetite, 52*(2), 430–436.

Gearhardt, A. N., Davis, C., Kuschner, R., & Brownell, K. D. (2011). The addiction potential of hyperpalatable foods. *Current Drug Abuse Reviews, 4*, 140–145.

Gearhardt, A. N., Yokum, S., Orr, P. T., Stice, E., Corbin, W. R., & Brownell, K. D. (2011). Neural correlates of food addiction. *Archives of General Psychiatry, 68*(8), 808–816.

Gearhardt, A. N., White, M. A., Masheb, R. M., Morgan, P. T., Crosby, R. D., & Grilo, C. M. (2012). An examination of the food addiction construct in obese patients with binge eating disorder. *International Journal of Eating Disorders, 45*(5), 657–663.

Gearhardt, A. N., White, M. A., Masheb, R. M., & Grilo, C. M. (2013). An examination of food addiction in a racially diverse sample of obese patients with binge eating disorder in primary care settings. *Comprehensive Psychiatry, 54*(5), 500–505.

Gearhardt, A. N., Corbin, W. R., & Brownell, K. D. (2016). Development of the Yale food addiction Scale version 2.0. *Psychology of Addictive Behaviors, 30*(1), 113–121.

Gearhardt, A. N., Harris, J. L., Lumen, J. C., Epstein, L. H., & Yokum, S. (2019). *Neural response to unhealthy and healthier food commercials predicts food intake by adolescents in a simulated fast food restaurant* (submitted).

Geha, P. Y., Aschenbrenner, K., Felsted, J., O'Malley, S. S., & Small, D. M. (2013). Altered hypothalamic response to food in smokers. *American Journal of Clinical Nutrition, 97*(1), 15–22.

Geiger, B. M., Behr, G. G., Frank, L. E., Caldera-Siu, A. D., Beinfeld, M. C., Kokkotou, E. G., & Pothos, E. N. (2008). Evidence for defective mesolimbic dopamine exocytosis in obesity-prone rats. *The FASEB Journal, 22*(8), 2740–2746.

Geiger, B. M., Haburcak, M., Avena, N. M., Moyer, M. C., Hoebel, B. G., & Pothos, E. N. (2009). Deficits of mesolimbic dopamine neurotransmission in rat dietary obesity. *Neuroscience, 159*(4), 1193–1199.

Geliebter, A., Ladell, T., Logan, M., Schneider, T., Sharafi, M., & Hirsch, J. (2006). Responsivity to food stimuli in obese and lean binge eaters using functional MRI. *Appetite, 46*(1), 31–35.

George, M. S., Anton, R. F., Bloomer, C., Teneback, C., Drobes, D. J., Lorberbaum, J. P., … Vincent, D. J. (2001). Activation of prefrontal cortex and anterior thalamus in alcoholic subjects on exposure to alcohol-specific cues. *Archives of General Psychiatry, 58*(4), 345–352.

Gilhooly, C. H., Das, S. K., Golden, J. K., McCrory, M. A., Dallal, G. E., Saltzman, E., … Roberts, S. B. (2007). Food cravings and energy regulation: The characteristics of craved foods and their relationship with eating behaviors and weight change during 6 months of dietary energy restriction. *International Journal of Obesity, 31*(12), 1849–1858.

Giuliano, C., Robbins, T. W., Nathan, P. J., Bullmore, E. T., & Everitt, B. J. (2012). Inhibition of opioid transmission at the mu-opioid receptor prevents both food seeking and binge-like eating. *Neuropsychopharmacology, 37*(12), 2643–2652.

Gorelick, D. A., Kim, Y. K., Bencherif, B., Boyd, S. J., Nelson, R., Copersino, M., … Frost, J. J. (2005). Imaging brain mu-opioid receptors in abstinent cocaine users: Time course and relation to cocaine craving. *Biological Psychiatry, 57*(12), 1573–1582.

Granero, R., Hilker, I., Aguera, Z., Jimenez-Murcia, S., Sauchelli, S., Islam, M. A., ... Fernandez-Aranda, F. (2014). Food addiction in a Spanish sample of eating disorders: DSM-5 diagnostic subtype differentiation and validation data. *European Eating Disorders Review, 22*(6), 389–396.

Gray, J. A., & McNaughton, N. (2000). *Gray's neuropsychology of anxiety: An enquiry into the functions of septohippocampal theories* (2nd ed.).

Green, E., & Murphy, C. (2012). Altered processing of sweet taste in the brain of diet soda drinkers. *Physiology & Behavior, 107*(4), 560–567.

Green, E., Jacobson, A., Haase, L., & Murphy, C. (2011). Reduced nucleus accumbens and caudate nucleus activation to a pleasant taste is associated with obesity in older adults. *Brain Research, 1386*, 109–117.

Grimm, J. W., Manaois, M., Osincup, D., Wells, B., & Buse, C. (2007). Naloxone attenuates incubated sucrose craving in rats. *Psychopharmacology, 194*(4), 537–544.

Hadigan, C. M., Kissileff, H. R., & Walsh, B. T. (1989). Patterns of food selection during meals in women with bulimia. *American Journal of Clinical Nutrition, 50*(4), 759–766.

Hardman, C., Herbert, V., Brunstrom, J., Munafo, M., & Rogers, P. (2012). Dopamine and food reward: Effects of acute tyrosine/phenylalanine depletion on appetite. *Physiology & Behavior, 105*, 1202–1207.

Heinz, A., Siessmeier, T., Wrase, J., Hermann, D., Klein, S., Grusser, S. M., ... Bartenstein, P. (2004). Correlation between dopamine D(2) receptors in the ventral striatum and central processing of alcohol cues and craving. *American Journal of Psychiatry, 161*(10), 1783–1789.

Heinz, A., Reimold, M., Wrase, J., Hermann, D., Croissant, B., Mundle, G., ... Mann, K. (2005). Correlation of stable elevations in striatal mu-opioid receptor availability in detoxified alcoholic patients with alcohol craving: A positron emission tomography study using carbon 11-labeled carfentanil. *Archives of General Psychiatry, 62*(1), 57–64.

Henningfield, J. E., & Keenan, R. M. (1993). Nicotine delivery kinetics and abuse liability. *Journal of Consulting and Clinical Psychology, 61*(5), 743–750.

Hester, R., Lubman, D. I., & Yucel, M. (2010). The role of executive control in human drug addiction. *Current Topics in Behavioral Neurosciences, 3*, 301–318.

Huang, X. F., Zavitsanou, K., Huang, X., Yu, Y., Wang, H., Chen, F., ... Deng, C. (2006). Dopamine transporter and D2 receptor binding densities in mice prone or resistant to chronic high fat diet-induced obesity. *Behavioural Brain Research, 175*(2), 415–419.

Ifland, J. R., Preuss, H. G., Marcus, M. T., Rourke, K. M., Taylor, W. C., Burau, K., ... Manso, G. (2009). Refined food addiction: A classic substance use disorder. *Medical Hypotheses, 72*(5), 518–526.

Jastreboff, A. M., Sinha, R., Lacadie, C., Small, D. M., Sherwin, R. S., & Potenza, M. N. (2013). Neural correlates of stress- and food cue-induced food craving in obesity: Association with insulin levels. *Diabetes Care, 36*(2), 394–402.

Johnson, P. M., & Kenny, P. J. (2010). Dopamine D2 receptors in addiction-like reward dysfunction and compulsive eating in obese rats. *Nature Neuroscience, 13*(5), 635–641.

Joutsa, J., Karlsson, H. K., Majuri, J., Nuutila, P., Helin, S., Kaasinen, V., & Nummenmaa, L. (2018). Binge eating disorder and morbid obesity are associated with lowered mu-opioid receptor availability in the brain. *Psychiatry Research Neuroimaging, 276*, 41–45.

Kalivas, P. W., & O'Brien, C. (2008). Drug addiction as a pathology of staged neuroplasticity. *Neuropsychopharmacology, 33*(1), 166–180.

Karlsson, H. K., Tuominen, L., Tuulari, J. J., Hirvonen, J., Parkkola, R., Helin, S., ... Nummenmaa, L. (2015). Obesity is associated with decreased mu-opioid but unaltered dopamine D2 receptor availability in the brain. *Journal of Neuroscience, 35*(9), 3959–3965.

Kenny, P. J., Chen, S. A., Kitamura, O., Markou, A., & Koob, G. F. (2006). Conditioned withdrawal drives heroin consumption and decreases reward sensitivity. *Journal of Neuroscience, 26*(22), 5894–5900.

Kenny, P. J. (2011). Common cellular and molecular mechanisms in obesity and drug addiction. *Nature Reviews Neuroscience, 12*(11), 638–651.

Kishinevsky, F. I., Cox, J. E., Murdaugh, D. L., Stoeckel, L. E., Cook, E. W., 3rd, & Weller, R. E. (2012). fMRI reactivity on a delay discounting task predicts weight gain in obese women. *Appetite, 58*(2), 582–592.

Koob, G. F., & Bloom, F. E. (1988). Cellular and molecular mechanisms of drug dependence. *Science, 242*(4879), 715–723.

Koob, G. F., & Le Moal, M. (2001). Drug addiction, dysregulation of reward, and allostasis. *Neuropsychopharmacology, 24*, 97–129.

Kringelbach, M. L., O'Doherty, J., Rolls, E. T., & Andrews, C. (2003). Activation of the human orbitofrontal cortex to a liquid food stimulus is correlated with its subjective pleasantness. *Cerebral Cortex, 13*(10), 1064–1071.

Lawrence, N. S., Hinton, E. C., Parkinson, J. A., & Lawrence, A. D. (2012). Nucleus accumbens response to food cues predicts subsequent snack consumption in women and increased body mass index in those with reduced self-control. *NeuroImage, 63*(1), 415–422.

Le Merrer, J., Becker, J. A., Befort, K., & Kieffer, B. L. (2009). Reward processing by the opioid system in the brain. *Physiological Reviews, 89*(4), 1379–1412.

Lemmens, S. G., Rutters, F., Born, J. M., & Westerterp-Plantenga, M. S. (2011). Stress augments food 'wanting' and energy intake in visceral overweight subjects in the absence of hunger. *Physiology & Behavior, 103*(2), 157–163.

Liu, S., Globa, A. K., Mills, F., Naef, L., Qiao, M., Bamji, S. X., & Borgland, S. L. (2016). Consumption of palatable food primes food approach behavior by rapidly increasing synaptic density in the VTA. *Proceedings of the National Academy of Sciences of the United States of America, 113*(9), 2520–2525.

Lyoo, I. K., Pollack, M. H., Silveri, M. M., Ahn, K. H., Diaz, C. I., Hwang, J., ... Renshaw, P. F. (2006). Prefrontal and temporal gray matter density decreases in opiate dependence. *Psychopharmacology, 184*(2), 139–144.

Maas, L. C., Lukas, S. E., Kaufman, M. J., Weiss, R. D., Daniels, S. L., Rogers, V. W., ... Renshaw, P. F. (1998). Functional magnetic resonance imaging of human brain activation during cue-induced cocaine craving. *American Journal of Psychiatry, 155*(1), 124–126.

MacKillop, J., Amlung, M. T., Few, L. R., Ray, L. A., Sweet, L. H., & Munafo, M. R. (2011). Delayed reward discounting and addictive behavior: A meta-analysis. *Psychopharmacology, 216*(3), 305–321.

Martin, L. E., Holsen, L. M., Chambers, R. J., Bruce, A. S., Brooks, W. M., Zarcone, J. R., ... Savage, C. R. (2010). Neural mechanisms associated with food motivation in obese and healthy weight adults. *Obesity, 18*(2), 254–260.

Martinez, D., Gil, R., Slifstein, M., Hwang, D. R., Huang, Y., Perez, A., ... Abi-Dargham, A. (2005). Alcohol dependence is associated with blunted dopamine transmission in the ventral striatum. *Biological Psychiatry, 58*(10), 779–786.

Martinez, D., Narendran, R., Foltin, R. W., Slifstein, M., Hwang, D. R., Broft, A., ... Laruelle, M. (2007). Amphetamine-induced dopamine release: Markedly blunted in cocaine dependence and predictive of the choice to self-administer cocaine. *American Journal of Psychiatry, 164*(4), 622–629.

Martire, S. I., Holmes, N., Westbrook, R. F., & Morris, M. J. (2013). Altered feeding patterns in rats exposed to a palatable cafeteria diet: Increased snacking and its implications for development of obesity. *PLoS One, 8*(4), e60407.

McGue, M., Iacono, W. G., Legrand, L. N., Malone, S., & Elkins, I. (2001). Origins and consequences of age at first drink. I. Associations with substance-use disorders, disinhibitory behavior and psychopathology, and P3 amplitude. *Alcoholism: Clinical and Experimental Research, 25*(8), 1156–1165.

Mehta, S., Melhorn, S., Smeraglio, A., Tyagi, V., Grabowski, T., Schwartz, M., et al. (2012). Regional brain response to visual food cues is a marker of satiety that predicts food choice. *American Journal of Clinical Nutrition, 96*, 989–999.

Meule, A., Lutz, A., Vogele, C., & Kubler, A. (2012). Women with elevated food addiction symptoms show accelerated reactions, but no impaired inhibitory control, in response to pictures of high-calorie food-cues. *Eating Behaviors, 13*(4), 423–428.

Meule, A., von Rezori, V., & Blechert, J. (2014). Food addiction and bulimia nervosa. *European Eating Disorders Review, 22*(5), 331–337.

Miller, W. R., Westerberg, V. S., Harris, R. J., & Tonigan, J. S. (1996). What predicts relapse? Prospective testing of antecedent models. *Addiction, 91*(Suppl. l), S155–S172.

Morganstern, I., Liang, S., Ye, Z., Karatayev, O., & Leibowitz, S. F. (2012). Disturbances in behavior and cortical enkephalin gene expression during the anticipation of ethanol in rats characterized as high drinkers. *Alcohol, 46*(6), 559–568.

Myrick, H., Anton, R. F., Li, X., Henderson, S., Drobes, D., Voronin, K., & George, M. S. (2004). Differential brain activity in alcoholics and social drinkers to alcohol cues: Relationship to craving. *Neuropsychopharmacology, 29*(2), 393–402.

Nader, M. A., Morgan, D., Gage, H. D., Nader, S. H., Calhoun, T. L., Buchheimer, N., … Mach, R. H. (2006). PET imaging of dopamine D2 receptors during chronic cocaine self-administration in monkeys. *Nature Neuroscience, 9*(8), 1050–1056.

Nederkoorn, C., Braet, C., Van Eijs, Y., Tanghe, A., & Jansen, A. (2006). Why obese children cannot resist food: The role of impulsivity. *Eating Behaviors, 7*(4), 315–322.

Negus, S. S., Henriksen, S. J., Mattox, A., Pasternak, G. W., Portoghese, P. S., Takemori, A. E., … Koob, G. F. (1993). Effect of antagonists selective for mu, delta and kappa opioid receptors on the reinforcing effects of heroin in rats. *Journal of Pharmacology and Experimental Therapeutics, 265*(3), 1245–1252.

Ng, J., Stice, E., Yokum, S., & Bohon, C. (2011). An fMRI study of obesity, food reward, and perceived caloric density. *Does a low-fat label make food less appealing? Appetite, 57*, 65–72.

Nogueiras, R., Romero-Pico, A., Vazquez, M. J., Novelle, M. G., Lopez, M., & Dieguez, C. (2012). The opioid system and food intake: Homeostatic and hedonic mechanisms. *Obesity Facts, 5*(2), 196–207.

O'Brien, C. P., Volkow, N., & Li, T. K. (2006). What's in a word? Addiction versus dependence in DSM-V. *American Journal of Psychiatry, 163*(5), 764–765.

Ochoa, M., Lalles, J. P., Malbert, C. H., & Val-Laillet, D. (2015). Dietary sugars: Their detection by the gut-brain axis and their peripheral and central effects in health and diseases. *European Journal of Nutrition, 54*, 1–24.

Oliver, G., & Wardle, J. (1999). Perceived effects of stress on food choice. *Physiology & Behavior, 66*(3), 511–515.

Oliver, G., Wardle, J., & Gibson, E. L. (2000). Stress and food choice: A laboratory study. *Psychosomatic Medicine, 62*(6), 853–865.

Oswald, K. D., Murdaugh, D. L., King, V. L., & Boggiano, M. M. (2011). Motivation for palatable food despite consequences in an animal model of binge eating. *International Journal of Eating Disorders, 44*(3), 203–211.

Palpacuer, C., Laviolle, B., Boussageon, R., Reymann, J. M., Bellissant, E., & Naudet, F. (2015). Risks and benefits of nalmefene in the treatment of adult alcohol dependence: A systematic literature review and meta-analysis of published and unpublished double-blind randomized controlled trials. *PLoS Medicine, 12*(12), e1001924.

Paulus, M. P., Tapert, S. F., & Schuckit, M. A. (2005). Neural activation patterns of methamphetamine-dependent subjects during decision making predict relapse. *Archives of General Psychiatry, 62*(7), 761–768.

Pelchat, M. L., Johnson, A., Chan, R., Valdez, J., & Ragland, J. D. (2004). Images of desire: Food-craving activation during fMRI. *NeuroImage, 23*(4), 1486–1493.

Pomonis, J. D., Jewett, D. C., Kotz, C. M., Briggs, J. E., Billington, C. J., & Levine, A. S. (2000). Sucrose consumption increases naloxone-induced c-Fos immunoreactivity in limbic forebrain. *American Journal of Physiology - Regulatory, Integrative and Comparative Physiology, 278*, R712–R719.

Porrino, L. J., Lyons, D., Smith, H. R., Daunais, J. B., & Nader, M. A. (2004). Cocaine self-administration produces a progressive involvement of limbic, association, and sensorimotor striatal domains. *Journal of Neuroscience, 24*(14), 3554–3562.

Pursey, K. M., Collins, C. E., Stanwell, P., & Burrows, T. L. (2015). Foods and dietary profiles associated with 'food addiction' in young adults. *Addictive Behaviors Reports, 2*, 41–48.

Rada, P., Avena, N. M., & Hoebel, B. G. (2005). Daily bingeing on sugar repeatedly releases dopamine in the accumbens shell. *Neuroscience, 134*(3), 737–744.

Rothemund, Y., Preuschhof, C., Bohner, G., Bauknecht, H. C., Klingebiel, R., Flor, H., & Klapp, B. F. (2007). Differential activation of the dorsal striatum by high-calorie visual food stimuli in obese individuals. *NeuroImage, 37*, 410–421.

Rudenga, K. J., & Small, D. M. (2012). Amygdala response to sucrose consumption is inversely related to artificial sweetener use. *Appetite, 58*(2), 504–507.

Rudenga, K. J., Sinha, R., & Small, D. M. (2013). Acute stress potentiates brain response to milkshake as a function of body weight and chronic stress. *International Journal of Obesity, 37*(2), 309–316.

Ruttle, P. L., Javaras, K. N., Klein, M. H., Armstrong, J. M., Burk, L. R., & Essex, M. J. (2013). Concurrent and longitudinal associations between diurnal cortisol and body mass index across adolescence. *Journal of Adolescent Health, 52*(6), 731–737.

Saunders, B., Farag, N., Vincent, A. S., Collins, F. L., Jr., Sorocco, K. H., & Lovallo, W. R. (2008). Impulsive errors on a go-NoGo reaction time task: Disinhibitory traits in relation to a family history of alcoholism. *Alcoholism: Clinical and Experimental Research, 32*(5), 888–894.

Schienle, A., Schafer, A., Hermann, A., & Vaitl, D. (2009). Binge-eating disorder: Reward sensitivity and brain activation to images of food. *Biological Psychiatry, 65*(8), 654–661.

Schulte, E. M., Avena, N. M., & Gearhardt, A. N. (2015). Which foods may be addictive? The roles of processing, fat content, and glycemic load. *PLoS One, 10*(2), e0117959.

Schultz, W., Dayan, P., & Montague, P. R. (1997). A neural substrate of prediction and reward. *Science, 275*, 1593–1599.

Schweinsburg, A. D., Paulus, M. P., Barlett, V. C., Killeen, L. A., Caldwell, L. C., ... Tapert, S. F. (2004). An FMRI study of response inhibition in youths with a family history of alcoholism. *Annals of the New York Academy of Sciences, 1021*, 391–394.

Shearrer, G. E., Stice, E., & Burger, K. S. (2018). Adolescents at high risk of obesity show greater striatal response to increased sugar content in milkshakes. *American Journal of Clinical Nutrition, 107*(6), 859–866.

Simon, J. J., Skunde, M., Walther, S., Bendszus, M., Herzog, W., & Friederich, H. C. (2016). Neural signature of food reward processing in bulimic-type eating disorders. *Social Cognitive and Affective Neuroscience, 11*(9), 1393–1401.

Sinha, R., Garcia, M., Paliwal, P., Kreek, M. J., & Rounsaville, B. J. (2006). Stress-induced cocaine craving and hypothalamic-pituitary-adrenal responses are predictive of cocaine relapse outcomes. *Archives of General Psychiatry, 63*(3), 324–331.

Sinha, R. (2005). Stress and drug abuse. In N. H. K. T. Steckler, & J. M. H. M. Reul (Eds.), *Handbook of stress and the brain. Part 2 stress: Integrative and clinical aspects* (Vol. 15, pp. 333–356). Amsterdam: Elsevier.

Sinha, R. (2018). Role of addiction and stress neurobiology on food intake and obesity. *Biological Psychology, 131*, 5–13.

Small, D. M., Zatorre, R. J., Dagher, A., Evans, A. C., & Jones-Gotman, M. (2001). Changes in brain activity related to eating chocolate: From pleasure to aversion. *Brain, 124*(Pt 9), 1720–1733.

Small, D. M., Jones-Gotman, M., & Dagher, A. (2003). Feeding-induced dopamine release in dorsal striatum correlates with meal pleasantness ratings in healthy human volunteers. *NeuroImage, 19*(4), 1709–1715.

Smith, D. G., & Robbins, T. W. (2013). The neurobiological underpinnings of obesity and binge eating: A rationale for adopting the food addiction model. *Biological Psychiatry, 73*(9), 804–810.

Sotak, B., Hnasko, T., Robinson, S., Kremer, E., & Palmiter, R. (2005). Dysregulation of dopamine signaling in the dorsal striatum inhibits feeding. *Brain Research, 1061*, 88–96.

Steptoe, A., Lipsey, Z., & Wardle, J. (1998). Stress, hassles and variations in alcohol consumption, food choice and physical exercise: A diary study. *British Journal of Health Psychology, 3*, 51–63.

Stice, E., & Yokum, S. (2014). Brain reward region responsivity of adolescents with and without parental substance use disorders. *Psychology of Addictive Behaviors, 28*(3), 805–815.

Stice, E., & Yokum, S. (2016). Neural vulnerability factors that increase risk for future weight gain. *Psychological Bulletin, 142*, 447–471.

Stice, E., & Yokum, S. (2018). Relation of neural response to palatable food tastes and images to future weight gain: Using bootstrap sampling to enhance replicability of findings. *NeuroImage, 183*, 522–531.

Stice, E., Spoor, S., Bohon, C., & Small, D. M. (2008). Relation between obesity and blunted striatal response to food is moderated by TaqIA A1 allele. *Science, 322*(5900), 449–452.

Stice, E., Spoor, S., Bohon, C., Veldhuizen, M. G., & Small, D. M. (2008). Relation of reward from food intake and anticipated food intake to obesity: A functional magnetic resonance imaging study. *Journal of Abnormal Psychology, 117*, 924–935.

Stice, E., Yokum, S., Blum, K., & Bohon, C. (2010). Weight gain is associated with reduced striatal response to palatable food. *Journal of Neuroscience, 30*(39), 13105–13109.

Stice, E., Yokum, S., Burger, K. S., Epstein, L. H., & Small, D. M. (2011). Youth at risk for obesity show greater activation of striatal and somatosensory regions to food. *Journal of Neuroscience, 31*(12), 4360–4366.

Stice, E., Burger, K. S., & Yokum, S. (2013). Relative ability of fat and sugar tastes to activate reward, gustatory, and somatosensory regions. *American Journal of Clinical Nutrition, 98*(6), 1377–1384.

Stice, E., Figlewicz, D. P., Gosnell, B. A., Levine, A. S., & Pratt, W. E. (2013). The contribution of brain reward circuits to the obesity epidemic. *Neuroscience & Biobehavioral Reviews, 37*(9 Pt A), 2047–2058.

Stice, E., Yokum, S., & Burger, K. S. (2013). Elevated reward region responsivity predicts future substance use onset but not overweight/obesity onset. *Biological Psychiatry, 73*(9), 869–876.

Stice, E., Burger, K. S., & Yokum, S. (2015). Reward region responsivity predicts future weight gain and moderating effects of the TaqIA allele. *Journal of Neuroscience, 35*(28), 10316–10324.

Stoeckel, L. E., Weller, R. E., Cook, E. W., 3rd, Twieg, D. B., Knowlton, R. C., & Cox, J. E. (2008). Widespread reward-system activation in obese women in response to pictures of high-calorie foods. *NeuroImage, 41*, 636–647.

Streeter, C. C., Ciraulo, D. A., Harris, G. J., Kaufman, M. J., Lewis, R. F., Knapp, C. M., … Renshaw, P. F. (1998). Functional magnetic resonance imaging of alprazolam-induced changes in humans with familial alcoholism. *Psychiatry Research, 82*(2), 69–82.

Sun, X., Kroemer, N. B., Veldhuizen, M. G., Babbs, A. E., de Araujo, I. E., Gitelman, D. R., … Small, D. M. (2015). Basolateral amygdala response to food cues in the absence of hunger is associated with weight gain susceptibility. *Journal of Neuroscience, 35*(20), 7964–7976.

Tang, D. W., Fellows, L. K., Small, D. M., & Dagher, A. (2012). Food and drug cues activate similar brain regions: A meta-analysis of functional MRI studies. *Physiology & Behavior, 106*(3), 317–324.

Tapert, S. F., & Brown, S. A. (2000). Substance dependence, family history of alcohol dependence and neuropsychological functioning in adolescence. *Addiction, 95*(7), 1043–1053.

Tapert, S. F., Cheung, E. H., Brown, G. G., Frank, L. R., Paulus, M. P., Schweinsburg, A. D., … Brown, S. A. (2003). Neural response to alcohol stimuli in adolescents with alcohol use disorder. *Archives of General Psychiatry, 60*(7), 727–735.

Tellez, L. A., Ren, X., Han, W., Medina, S., Ferreira, J. G., Yeckel, C. W., & de Araujo, I. E. (2013). Glucose utilization rates regulate intake levels of artificial sweeteners. *Journal of Physiology, 591*(22), 5727–5744.

Thanos, P. K., Michaelides, M., Piyis, Y. K., Wang, G. J., & Volkow, N. D. (2008). Food restriction markedly increases dopamine D2 receptor (D2R) in a rat model of obesity as assessed with in-vivo muPET imaging ([11C] raclopride) and in-vitro ([3H] spiperone) autoradiography. *Synapse, 62*(1), 50–61.

Tindell, A. J., Berridge, K. C., & Aldridge, J. W. (2004). Ventral pallidal representation of pavlovian cues and reward: Population and rate codes. *Journal of Neuroscience, 24*, 1058–1069.

Tobler, P. N., Fiorillo, C. D., & Schultz, W. (2005). Adaptive coding of reward value by dopamine neurons. *Science, 307*, 1642–1645.

Tomasi, D., Wang, G. J., Wang, R., Caparelli, E. C., Logan, J., & Volkow, N. D. (2015). Overlapping patterns of brain activation to food and cocaine cues in cocaine abusers: Association to striatal D2/D3 receptors. *Human Brain Mapping, 36*(1), 120–136.

Tryon, M. S., Carter, C. S., Decant, R., & Laugero, K. D. (2013). Chronic stress exposure may affect the brain's response to high calorie food cues and predispose to obesogenic eating habits. *Physiology & Behavior, 120*, 233–242.

Tryon, M. S., Stanhope, K. L., Epel, E. S., Mason, A. E., Brown, R., Medici, V., ... Laugero, K. D. (2015). Excessive sugar consumption may Be a difficult habit to break: A view from the brain and body. *The Journal of Cinical Endocrinology and Metabolism, 100*(6), 2239–2247.

Tschann, J. M., Adler, N. E., Irwin, C. E., Jr., Millstein, S. G., Turner, R. A., & Kegeles, S. M. (1994). Initiation of substance use in early adolescence: The roles of pubertal timing and emotional distress. *Health Psychology, 13*(4), 326–333.

Tuominen, L., Tuulari, J., Karlsson, H., Hirvonen, J., Helin, S., Salminen, P., ... Nummenmaa, L. (2015). Aberrant mesolimbic dopamine-opiate interaction in obesity. *NeuroImage, 122*, 80–86.

Tuulari, J. J., Tuominen, L., de Boer, F. E., Hirvonen, J., Helin, S., Nuutila, P., & Nummenmaa, L. (2017). Feeding releases endogenous opioids in humans. *Journal of Neuroscience, 37*(34), 8284–8291.

Uher, R., Murphy, T., Brammer, M. J., Dalgleish, T., Phillips, M. L., Ng, V. W., ... Treasure, J. (2004). Medial prefrontal cortex activity associated with symptom provocation in eating disorders. *American Journal of Psychiatry, 161*(7), 1238–1246.

Unterwald, E. M., Kreek, M. J., & Cuntapay, M. (2001). The frequency of cocaine administration impacts cocaine-induced receptor alterations. *Brain Research, 900*(1), 103–109.

Vanderlinden, J., Dalle Grave, R., Vandereycken, W., & Noorduin, C. (2001). Which factors do provoke binge-eating? An exploratory study in female students. *Eating Behaviors, 2*(1), 79–83.

Vanderschuren, L. J., & Kalivas, P. W. (2000). Alterations in dopaminergic and glutamatergic transmission in the induction and expression of behavioral sensitization: A critical review of preclinical studies. *Psychopharmacology, 151*(2–3), 99–120.

Volkow, N. D., & Morales, M. (2015). The brain on drugs: From reward to addiction. *Cell, 162*(4), 712–725.

Volkow, N. D., Wang, G. J., Fowler, J. S., Logan, J., Gatley, S. J., MacGregor, R. R., ... Wolf, A. P. (1996). Measuring age-related changes in dopamine D2 receptors with 11C-raclopride and 18F-N-methylspiroperidol. *Psychiatry Research, 67*(1), 11–16.

Volkow, N. D., Wang, G. J., Fowler, J. S., Logan, J., Angrist, B., Hitzemann, R., ... Pappas, N. (1997). Effects of methylphenidate on regional brain glucose metabolism in humans: Relationship to dopamine D2 receptors. *American Journal of Psychiatry, 154*(1), 50–55.

Volkow, N. D., Fowler, J. S., Wang, G. J., & Goldstein, R. Z. (2002). Role of dopamine, the frontal cortex and memory circuits in drug addiction: Insight from imaging studies. *Neurobiology of Learning and Memory, 78*(3), 610–624.

Volkow, N. D., Wang, G. J., Ma, Y., Fowler, J. S., Wong, C., Ding, Y. S., ... Kalivas, P. (2005). Activation of orbital and medial prefrontal cortex by methylphenidate in cocaine-addicted subjects but not in controls: Relevance to addiction. *Journal of Neuroscience, 25*(15), 3932–3939.

Volkow, N. D., Wang, G. J., Telang, F., Fowler, J. S., Logan, J., Childress, A. R., ... Wong, C. (2006). Cocaine cues and dopamine in dorsal striatum: Mechanism of craving in cocaine addiction. *Journal of Neuroscience, 26*(24), 6583–6588.

Volkow, N. D., Wang, G. J., Fowler, J. S., & Telang, F. (2008). Overlapping neuronal circuits in addiction and obesity: Evidence of systems pathology. *Philosophical Transactions of the Royal Society of London B Biological Sciences, 363*(1507), 3191–3200.

Volkow, N. D., Wang, G. J., Telang, F., Fowler, J. S., Thanos, P. K., Logan, J., ... Pradhan, K. (2008). Low dopamine striatal D2 receptors are associated with prefrontal metabolism in obese subjects: Possible contributing factors. *NeuroImage, 42*(4), 1537–1543.

Volkow, N. D., Wang, G. J., Fowler, J. S., & Tomasi, D. (2012). Addiction circuitry in the human brain. *Annual Review of Pharmacology and Toxicology, 52*, 321–336.

Volkow, N. D., Wang, G. J., Tomasi, D., & Baler, R. D. (2013). Obesity and addiction: Neurobiological overlaps. *Obesity Reviews, 14*(1), 2–18.

Volkow, N. D., Koob, G. F., & McLellan, A. T. (2016). Neurobiologic advances from the brain disease model of addiction. *New England Journal of Medicine, 374*(4), 363–371.

Wang, G. J., Volkow, N. D., Fowler, J. S., Logan, J., Abumrad, N. N., Hitzemann, R. J., ... Pascani, K. (1997). Dopamine D2 receptor availability in opiate-dependent subjects before and after naloxone-precipitated withdrawal. *Neuropsychopharmacology, 16*(2), 174–182.

Wang, G. J., Volkow, N. D., Logan, J., Pappas, N. R., Wong, C. T., Zhu, W., ... Fowler, J. S. (2001). Brain dopamine and obesity. *Lancet, 357*(9253), 354–357.

Wang, G. J., Volkow, N. D., & Fowler, J. S. (2002). The role of dopamine in motivation for food in humans: Implications for obesity. *Expert Opinion on Therapeutic Targets, 6*(5), 601–609.

Wang, G. J., Geliebter, A., Volkow, N. D., Telang, F. W., Logan, J., Jayne, M. C., ... Fowler, J. S. (2011). Enhanced striatal dopamine release during food stimulation in binge eating disorder. *Obesity, 19*(8), 1601–1608.

Weygandt, M., Mai, K., Dommes, E., Leupelt, V., Hackmack, K., Kahnt, T., ... Haynes, J. D. (2013). The role of neural impulse control mechanisms for dietary success in obesity. *NeuroImage, 83*, 669–678.

Weygandt, M., Mai, K., Dommes, E., Ritter, K., Leupelt, V., Spranger, J., & Haynes, J. D. (2015). Impulse control in the dorsolateral prefrontal cortex counteracts post-diet weight regain in obesity. *NeuroImage, 109*, 318–327.

Whistler, J. L. (2012). Examining the role of mu opioid receptor endocytosis in the beneficial and side-effects of prolonged opioid use: From a symposium on new concepts in mu-opioid pharmacology. *Drug and Alcohol Dependence, 121*(3), 189–204.

Wilcox, C. E., Braskie, M. N., Kluth, J. T., & Jagust, W. J. (2010). Overeating behavior and striatal dopamine with 6-[F]-Fluoro-L-m-Tyrosine PET. *International Journal of Obesity, 2010.*

Wojnicki, F. H., Roberts, D. C., & Corwin, R. L. (2006). Effects of baclofen on operant performance for food pellets and vegetable shortening after a history of binge-type behavior in non-food deprived rats. *Pharmacology Biochemistry and Behavior, 84*(2), 197–206.

Yanovski, S. Z., Leet, M., Yanovski, J. A., Flood, M., Gold, P. W., Kissileff, H. R., & Walsh, B. T. (1992). Food selection and intake of obese women with binge-eating disorder. *American Journal of Clinical Nutrition, 56*(6), 975–980.

Yanovski, S. (2003). Sugar and fat: Cravings and aversions. *Journal of Nutrition, 133*(3), 835S–837S.

Yeomans, M. R., & Gray, R. W. (2002). Opioid peptides and the control of human ingestive behaviour. *Neuroscience & Biobehavioral Reviews, 26*(6), 713–728.

Yokum, S., Ng, J., & Stice, E. (2011). Attentional bias for food images associated with elevated weight and future weight gain: An fMRI study. *International Journal of Obesity, 19*, 1775–1783.

Yokum, S., Gearhardt, A., Harris, J., Brownell, K., & Stice, E. (2014). Individual differences in striatum activity to food commercials predicts weight gain in adolescence. *Obesity, 22*, 2544–2551.

Yokum, S., Marti, C. N., Smolen, A., & Stice, E. (2015). Relation of the multilocus genetic composite reflecting high dopamine signaling capacity to future increases in BMI. *Appetite, 87*, 38–45.

Zellner, D. A., Loaiza, S., Gonzalez, Z., Pita, J., Morales, J., Pecora, D., & Wolf, A. (2006). Food selection changes under stress. *Physiology & Behavior, 87*(4), 789–793.

Zhou, & Palmiter. (1995). Dopamine-deficient mice are severely hypoactive, adipsic, and aphagic. *Cell, 83*, 1197–1209.

Ziauddeen, H., & Fletcher, P. C. (2013). Is food addiction a valid and useful concept? *Obesity Reviews, 14*(1), 19–28.

Ziauddeen, H., Farooqi, I. S., & Fletcher, P. C. (2012). Obesity and the brain: How convincing is the addiction model? *Nature Reviews Neuroscience, 13*(4), 279–286.

Ziauddeen, H., Chamberlain, S. R., Nathan, P. J., Koch, A., Maltby, K., Bush, M., … Bullmore, E. T. (2013). Effects of the mu-opioid receptor antagonist GSK1521498 on hedonic and consummatory eating behaviour: A proof of mechanism study in binge-eating obese subjects. *Molecular Psychiatry, 18*(12), 1287–1293.

Zubieta, J., Greenwald, M. K., Lombardi, U., Woods, J. H., Kilbourn, M. R., Jewett, D. M., et al. (2000). Buprenorphine-induced changes in mu-opioid receptor availability in male heroin-dependent volunteers: A preliminary study. *Neuropsychopharmacology, 23*(3), 326–334.

CHAPTER

Modeling and testing compulsive eating behaviors in animals

12

Catherine F. Moore, Jonathan E. Cheng, Valentina Sabino, Pietro Cottone

Laboratory of Addictive Disorders, Departments of Pharmacology and Psychiatry, Boston University School of Medicine, Boston, MA, United States

Introduction

Compulsive eating is regarded as a strong, irresistible internal drive to overconsume food, typically contrary to one's will (American Psychiatric Association, 2013; Davis & Carter, 2009; Dalley, Everitt, & Robbins, 2011), similar to the construct of compulsive drug use that characterizes drug addiction. Compulsive eating is characteristic of certain forms of eating disorders and obesity (Davis & Carter, 2009; Moore, Sabino, Koob, & Cottone, 2017b) and has become a focus of the scientific and medical community in part because of increasing rates of overweight and obesity (Meule, 2015). The construct of compulsive eating has been proposed to consist largely of three main elements, which have been adapted from the drug addiction literature: (1) *habitual overeating,* (2) *overeating to alleviate a negative emotional state,* and (3) *overeating despite negative consequences (refer to Chapter 3: Dissecting compulsivity into three elements* (Moore et al., 2017b)).

The study of compulsive eating has only recently begun gaining attention in both preclinical and clinical research. At a preclinical level, the development of appropriate animal models and tests, which allow the study of complex behavioral constructs, has been a barrier to both investigating the neurobiological substrates of compulsive eating and refining pharmacological interventions for various forms of obesity and eating disorders. Within the last half-century of drug abuse research, a deeper understanding of the addiction process has been achieved through the development of animal models and tests, which mimic the multiple facets and stages of this chronic, relapsing disorder (Belin-Rauscent, Fouyssac, Bonci, & Belin, 2015; Olmstead, 2006).

Unfortunately, preclinical research in obesity and eating disorders has been classically hampered by a preferential focus on the energy-homeostatic and metabolic aspects of these disorders at the expense of their motivational, emotional, and cognitive characteristics. Animal models and tests of drug addiction have been refined across many decades of research and provide invaluable insights into which approach and methodology could be used to appropriately mimic other addiction-related behaviors, including compulsive eating. The purpose of this chapter is to describe the progress

Compulsive Eating Behavior and Food Addiction. https://doi.org/10.1016/B978-0-12-816207-1.00012-3
Copyright © 2019 Elsevier Inc. All rights reserved.

made in the development of animal models and tests to investigate compulsivity in the context of eating behavior. We will first describe the tests used to measure compulsive eating behavior in animal models and describe the different variables that are typically manipulated to produce a compulsive eating phenotype.

Compulsive eating in preclinical research

While often used interchangeably in neuropsychopharmacological preclinical research, the terms "model" and "test" represent two different aspects of the study of animal behavior (Charney, Sklar, Buxbaum, & Nestler, 2018). A model is a non-human subject that exhibits a specific phenotype, which is reminiscent of a physiological or pathological human condition, as a result of an experimental manipulation. A test is an experimental setup that enables the measurement of a specific behavior in the experimental subject (i.e., readout), relevant for a physiological or pathological human condition.

The development of preclinical animal models for psychiatric diseases, including addiction and compulsive eating, requires consideration of different types of validity (Geyer and Markou, 2000, 2002). Face validity, which refers to phenomenological similarities between the behavior of the model and the symptoms of the human condition, is most often used as a "starting point" in creation of preclinical models. However, preclinical models must ultimately display predictive validity (i.e., ability of a model to accurately predict the human phenomenon of interest) for utility in translational research (Geyer & Markou, 2000). This review will not be exhaustive in the description of the specific tests and model variables; thus, a discussion on the different validities (e.g., face, construct, predictive, convergent, etc.) is outside the scope of this chapter.

In the sections below, we will first focus on the tests used to measure compulsive eating as defined by the three aforementioned elements and then review the experimental manipulations that are most commonly used to make animals behave as compulsive eaters.

Tests of compulsive eating behavior

The elements of compulsive eating were first adapted from the drug addiction literature (Moore et al., 2017b), where compulsivity has been proposed to be a critical factor of disease manifestation (Belin-Rauscent et al., 2015; Everitt, 2014; Hopf & Lesscher, 2014; Koob, 2013; Piazza & Deroche-Gamonet, 2013). Compared with compulsive drug use, the field of compulsive eating behavior is still emerging, and little work has been done to systematically define and investigate compulsive eating behavior. Just as the elements have been adapted from the drug addiction literature, many of the behavioral tests used to study compulsive eating behavior in rodents are modified tests of compulsive drug use, using food as a primary reinforcer/reward instead of a drug of abuse. While discussing the tests used to measure compulsive eating, we will also acknowledge behaviors that have been exhibited by animal models of compulsive drug use in these same tests to draw certain parallels between these two compulsive behaviors. It is important to note that while we are able to differentiate between the elements of compulsive eating via different methods of testing, these elements are not mutually exclusive.

Habitual overeating

Compulsive, habitual eating behavior can emerge after a history of palatable food consumption and the higher the food palatability, the faster it appears. Habitual behavior refers to a shift from voluntary, goal-directed responding (action-outcome) to more stimulus-driven responding (stimulus-response) that is resistant to devaluation of the expected reward. Habitual eating behavior can be assessed through multiple methods of outcome devaluation.

Outcome devaluation

In an outcome devaluation test, instrumental responding for an outcome (i.e., seeking behavior) is evaluated after the value of that outcome has been reduced. If an animal is using a goal-directed strategy (i.e., relying on the action-outcome contingency), then responding for a devalued outcome should decrease. On the other hand, maintaining levels of responding for a devalued outcome suggests habitual, or stimulus-driven, behavior.

Currently, the three most popular methods of inducing outcome devaluation to test for habitual behavior in animal models of drug addiction are outcome-specific satiety, taste aversion learning, and contingency degradation (Ostlund & Balleine, 2008). Outcome-specific satiety lowers a palatable food's rewarding properties by allowing animals to become sated with that food during a period of ad libitum access before testing. An alternate method of outcome devaluation is taste aversion learning, which repeatedly pairs the reward with injections of lithium chloride, which induces malaise (Balleine, 1992; Loy & Hall, 2002). Contingency degradation refers to the devaluation of a reward through reducing the probability that it will be presented on the performance of a response and simultaneously increasing the probability that it will be presented noncontingently (Ostlund & Balleine, 2008). At this time and to the best of our knowledge, only outcome-specific satiety has been applied to animal studies of compulsive eating.

Outcome-specific satiety

In outcome-specific satiety, animals are trained to respond (e.g., lever press) for food rewards. On test days, the value of the reward is reduced by giving the animals ad libitum access to the specific reward before the operant sessions, therefore inducing satiety (Furlong, Jayaweera, Balleine, & Corbit, 2014; de Jong et al., 2013). In animal models of drug addiction, chronic long access to drugs results in resistance to devaluation (i.e., persistence in responding for the drug) (Corbit, Nie, & Janak, 2012; Miles, Everitt, & Dickinson, 2003).

Habitual responding for food has been shown to develop in animals with intermittent-short (< 12-hour) access to high-fat and high-sugar diet (HFSD) conditions (Furlong et al., 2014). In this experiment, three groups of animals (control diet, intermittent-short HFSD, continuous HFSD) were trained to lever press for palatable food. On the test day, animals were given 1-hour ad libitum access to that instrumental outcome/food before a test session where animals responded under extinction conditions (i.e., no food presentations) (Furlong et al., 2014). In this study, only the intermittent-short HFSD group displayed resistance to devaluation (i.e., their test performance remained high), while continuous HFSD and control rats maintained action–outcome responding (Furlong et al., 2014). However, a similar study found that both intermittent-short and continuous HFSD groups displayed habitual behavior, where control animal behavior remained goal-directed (Kosheleff et al., 2018). The major difference between these studies was the operant training and devaluation test, as Furlong et al. (2014) trained for one outcome, while Kosheleff et al. (2018) used two distinct palatable foods (chocolate pellets or sweetened condensed milk) and devalued just one before testing extinction responding on two levers previously reinforced by these food outcomes. Overall, these studies demonstrate that intermittent-short access to an HFSD increases habitual food-seeking behavior, while continuous HFSD may also increase habitual behavior in certain conditions.

In a similar procedure by Reichelt, Morris, and Westbrook (2014), continuous access to a HFSD resulted in habitual responding in two devaluation tasks. In a Pavlovian-conditioned approach procedure, rats were first trained to discriminate between two discrete audio cues that were paired with the delivery of two sucrose solutions from opposite magazines (e.g., noise → grape-maltodextrin, tone → -cherry-sucrose). On test days, animals were allowed to self-administer one of the two solutions for 20 min before a test session where cues were presented with no delivery of the solutions. The cue-outcome devaluation test measured approach behavior (i.e., magazine head entries) during the cue presentation that signaled availability of one of the two reward solutions. Rats with prior continuous access did not devalue or respond less for the solution they had previously consumed ad libitum (Reichelt et al., 2014). In another test of sensory-specific satiety, animals were given access to both solutions in the home cage. Preexposure to one solution *ad libitum* before testing did not cause a reduction in the consumption of the devalued outcome. A similar study using continuous access to a high-fat diet (HFD) also observed resistance to devaluation of a sucrose reward by specific satiety procedures (Tantot et al., 2017). Thus, continuous access can also result in similar compulsive and habitual eating behavior as seen under intermittent-short access conditions.

Contexts paired with palatable food eating have been shown to impair goal-directed responding, triggering increased habitual eating (Kendig, Cheung, Raymond, & Corbit, 2016). A study by Kendig et al. (2016) found that a single pairing of HFSD access with a specific context caused resistance to specific satiety devaluation in that context, but not in a context paired with standard chow. Interestingly, the habitual eating brought on by the palatable food-paired context could be ameliorated by the presentation of a chow-associated cue, suggesting the ability of cues to not only impair but also potentially restore decision-making processes (Kendig et al., 2016). This study demonstrates the potential influence of environments paired with palatable food on enhancing habitual overeating behavior.

Overeating to alleviate a negative emotional state

The negative emotional state hypothesized to drive overeating via negative reinforcement has been behaviorally well characterized in animal models of compulsive eating (Blasio et al., 2013; Cottone, Sabino, Roberto, et al., 2009; Cottone, Sabino, Steardo, & Zorrilla, 2009; Iemolo et al., 2012). This element of compulsive eating has two components: decreased reward and increased stress (Koob, 2015; Moore et al., 2017b). Results from the tests described below show reliable and reproducible evidence of decreased reward (increased anhedonia, depressive-like behavior, and reduced reward system functioning) and increased stress (increased anxiety-like behavior) in animal models of compulsive eating.

Decreased reward

The shift from positive to negative reinforcement mechanisms in part involves the emergence of a depressive-like state, characterized by decreased reward function and reduced motivation for ordinary life stimuli (i.e., hypohedonia/anhedonia), as well as behavioral despair. Sucrose consumption (preference for a natural reward), hypophagia of less palatable food alternatives, progressive ratio (PR) responding (motivation) for lesser rewards, and intracranial self-stimulation (ICSS) (reward system functioning) are common methods to assess hypohedonia in animals. Two common tests of depressive-like behavior, the forced swim test (rats and mice) and the tail suspension test (mice) rely on the principle of measuring behavioral despair when facing inescapable situations (Castagne, Moser, Roux, & Porsolt, 2011).

Sucrose consumption test

A classical test of hypohedonia is the sucrose consumption test, where sucrose consumption is used as a measure of responsiveness to natural rewards. Low consumption of and preference for sucrose is thought to represent a deficit in the reward system's response to a natural reward and is used as the operational definition of hypohedonia. Withdrawal from drugs of abuse, including amphetamine, cocaine, and heroin, has been shown to reduce motivation for natural rewards such as sucrose in rats (Der-Avakian & Markou, 2010). Intermittent-long access ($\geq$ 12-hour) to high-sugar diets (HSD) access resulted in reduced sucrose preference compared with control rats when tested during palatable food withdrawal (Iemolo et al., 2012). Renewed access to the HSD reduced this hypohedonic-like behavior (Iemolo et al., 2012). Models utilizing continuous HFD access consistently display decreased sucrose preference in obese experimental animals (Dutheil, Ota, Wohleb, Rasmussen, & Duman, 2016; Sharma, Fernandes, & Fulton, 2013; Yamada et al., 2011).

These results suggest that withdrawal from palatable food and/or an obese phenotype (resulting from continuous HFD access) can result in hypohedonic-like behaviors, likely indicating reward deficits.

Hypophagia of food with lower palatability

Interruption of palatable food access consistently results in hypophagia of the standard chow when this becomes the only food available (Blasio, Rice, Sabino, & Cottone, 2014; Colantuoni et al., 2002; Cottone, Sabino, Steardo, et al., 2009; Cottone, Sabino, Steardo, & Zorrilla, 2008a; Dore et al., 2014; Kreisler, Garcia, Spierling, Hui, & Zorrilla, 2017; Levin & Dunn-Meynell, 2002; Rossetti, Spena, Halfon, & Boutrel, 2014). While the decrease in food consumption occurs even after a single shift to a less palatable diet in intermittent access paradigms, hypophagia increases in magnitude with each additional palatable diet access cycle (Blasio et al., 2014; Cottone, Sabino, Steardo, et al., 2009, Cottone et al., 2008a).

Hypophagia was originally proposed as an energy-homeostatic compensatory response to previous hyperphagia and body weight gain induced by access to a highly palatable diet. When animals are shifted to a less palatable diet, hypophagia is accompanied by body weight loss (Levin & Dunn-Meynell, 2002). While this phenomenon could indeed represent a merely energy-homeostatic mechanism, hedonic mechanisms are also implicated in hypophagia of the less palatable food (Cottone et al., 2008a). The most striking evidence of this is when hypophagia was observed without overeating of a more preferred food (compared with controls) (Cottone et al., 2008a). In this instance, the alternating diets were similar in energy density and were both preferred over standard chow (Cottone et al., 2008a). However, following the removal of the more preferred food, rats significantly and progressively reduced their intake of the less preferred diet, suggesting that intermittent access to rewards, such as palatable food, can affect the reinforcing efficacy of otherwise satisfactory alternatives (Cottone et al., 2008a). Hypophagia is hypothesized to arise from a phenomenon called successive negative contrast (Flaherty, Coppotelli, Grigson, Mitchell, & Flaherty, 1995; Flaherty & Rowan, 1986), a reduction in reward consumption following recent experience with another reward of higher magnitude (Austen, Strickland, & Sanderson, 2016), implicating hedonic rather than energy-homeostatic mechanisms (Flaherty et al., 1995; Flaherty & Rowan, 1986).

Hypophagia may be also explained by the opponent process theory of motivation, which proposes progressively increasing habituation to a rewarding stimulus followed by a progressively increasing intensity of hedonic withdrawal with each exposure to that stimulus (Solomon & Corbit, 1974). Animals may become increasingly habituated to the palatable diet, leading to progressively increasing hyperphagia of this diet, and exhibit increasing hypophagia of the less palatable diet caused by an increasingly powerful withdrawal state. Hypophagia may also be reflected in models of drug addiction, where motivation for food and other nondrug reinforcers are reduced (discussed more in the section below).

Progressive ratio responding for food

Similar to hypophagia, during withdrawal from a palatable diet, animals also exhibit lower motivation to seek less palatable alternatives (e.g., standard chow) in PR operant task procedures. This procedure assesses animals' sensitivity to different food rewards by having them perform an exponentially increasing number of instrumental responses (e.g., button/lever press) to obtain a self-administered food reward in operant chambers (Cottone, Sabino, Steardo, & Zorrilla, 2008b). In these procedures, the breakpoint (i.e., last ratio completed) is used as a measure of motivation (Velazquez-Sanchez et al., 2014). In drug addiction studies, withdrawal from drugs of abuse resulted in lower breakpoints (i.e., decreased motivation) for sucrose (i.e., a reward of lower magnitude than the drugs of abuse) (Hoefer, Voskanian, Koob, & Pulvirenti, 2006; Zhang et al., 2007). During withdrawal from intermittent-long HSD access, rats exhibited a lower total number of PR responses and lower breakpoints for the less palatable food reward, compared with controls, showing lower motivation for the less palatable diet (Cottone et al., 2008b).

Intracranial self-stimulation

ICSS is a procedure used to assess brain reward function. Bipolar stimulating electrodes (implanted in specific brain areas, most commonly the medial forebrain bundle at the level of the lateral hypothalamus) deliver electrical stimulation to animals when they successfully perform an action (i.e., lever press or wheel turn) (Carlezon & Chartoff, 2007; Esposito & Kornetsky, 1977; Gallistel & Freyd, 1987; Iemolo et al., 2012; Johnson & Kenny, 2010; Markou & Koob, 1992). Brain reward function is measured as the animal's reward threshold, which is determined by calculations based on the minimum current intensity or minimum frequency of stimulation able to maintain self-stimulation behavior (Carlezon & Chartoff, 2007; Iemolo et al., 2012; Johnson & Kenny, 2010). Reward threshold increases and decreases throughout the trials are operationalized as lowered and elevated reward function, respectively (Iemolo et al., 2012; Johnson & Kenny, 2010). ICSS has been used to study the effects of drugs of abuse on reward system function; reward thresholds are increased in animal models of cocaine, amphetamine, and heroin addiction (Ahmed, Kenny, Koob, & Markou, 2002; Kenny, Chen, Kitamura, Markou, & Koob, 2006; Kokkinidis & McCarter, 1990; Markou & Koob, 1991).

Two main procedures are used to assess brain reward function: a rate-independent, discrete trial current intensity paradigm (Esposito & Kornetsky, 1977) and a response-rate frequency paradigm (Carlezon & Chartoff, 2007; Gallistel & Freyd, 1987). In the discrete trial current intensity protocol, trials begin with a noncontingent electrical stimulation (varying current), after which animals have a limited time to perform a response to self-stimulate at that same level of electrical current intensity (Iemolo et al., 2012; Johnson & Kenny, 2010). In the response-rate frequency paradigm, animals initially receive a noncontingent electrical stimulation (varying frequencies) and are then given a limited time to perform as many self-stimulating responses as desired (Carlezon & Chartoff, 2007).

A study by Johnson and Kenny (2010) investigated changes in brain reward thresholds using ICSS in rats with continuous HFSD access, intermittent-short HFSD access, and control animals (Johnson & Kenny, 2010). Rats that had continuous HFSD access gained significantly more body weight, which was coupled with decreases in brain reward responsiveness (i.e., increased ICSS thresholds) (Johnson & Kenny, 2010). Similarly, in a study of diet-induced obesity rats, continuous HFD access resulted in reduced reward system functioning, evidenced by decreased sensitivity to D-amphetamine's ability to decrease ICSS thresholds (Valenza, Steardo, Cottone, & Sabino, 2015). Intermittent-long HSD access did not result in changes in the brain reward threshold when tested during both palatable diet access and withdrawal conditions (Iemolo et al., 2012), mirroring what was found previously (Johnson & Kenny, 2010). Taken together, these results suggest that reward dysfunction may be more easily observed in an obese phenotype, which develops after continuous palatable diet access (Johnson & Kenny, 2010), but not in intermittent-long HSD (Iemolo et al., 2012) or intermittent-short HFSD access conditions (Johnson & Kenny, 2010). An alternative interpretation may be that pharmacological precipitation of withdrawal may be necessary to observe changes in the brain reward threshold in rats with intermittent access to palatable diets.

Forced swim test

Depressive-like behavior can be measured in rats and mice using the forced swim test, where depressive-like behavior is defined as increased immobility time compared with controls when placed in a pool of water. Animal models of cocaine (Hall, Pearson, & Buccafusco, 2010), opiate (Anraku, Ikegaya, Matsuki, & Nishiyama, 2001), amphetamine (Cryan, Hoyer, & Markou, 2003), and alcohol (Pang, Renoir, Du, Lawrence, & Hannan, 2013) addiction have displayed quicker onset of immobility and longer immobility times during withdrawal. Intermittent-long HSD access resulted in longer immobility times during a period of withdrawal from palatable food, indicating depressive-like behavior (Iemolo et al., 2012). Restoring access to the palatable diet reversed these depressive-like behaviors, with immobility time becoming similar to controls (Iemolo et al., 2012). Continuous HFD access resulted in similar overall depressive-like behavior during access to the HFD (Yamada et al., 2011) as well as after a withdrawal period (Sharma et al., 2013). Yamada et al. (2011) also tested animals after 3 weeks of withdrawal from the HFD, a period sufficient to decrease body weight to control levels and normalize markers of obesity (i.e., plasma levels of glucose, insulin, and leptin). At this time point, immobility time was no different from controls (Yamada et al., 2011), suggesting that the obese phenotype is associated with depressive-like behavior, or at the very least, that a 3-week withdrawal period was sufficient to ameliorate the depressive-like behavior.

Tail suspension test

Similar to the forced swim test, the tail suspension test can be used in mice to assess depressive-like behavior through analysis of time spent immobile when suspended upside down by the tail (Cryan, Markou, & Lucki, 2002). Mice that are in withdrawal from amphetamine exhibited increased immobility in this test (Cryan et al., 2003). Continuous HSD access resulted in increased immobility in the tail suspension test after 1 week of withdrawal from the palatable food (Kim, Shou, Abera, & Ziff, 2018). However, in a group of "reinstated" mice that were also given 1 week of withdrawal from the HSD (10% sucrose solution) but were then allowed access to a 2% sucrose solution for 2 days, there was no evidence of a depressive-like state (Kim et al., 2018).

Increased stress

Withdrawal from a palatable diet is associated with increased stress, also part of the emergence of a negative emotional state, and compulsive eating may reflect an attempt to reduce anxiety and improve mood (Moore et al., 2017b; Parylak, Koob, & Zorrilla, 2011). To assess the emergence of the increased stress response in animals, tests for anxiety-like behavior can be used and behavior can be compared across phases of palatable food access and withdrawal in intermittent access conditions. There are many iterations of unconditioned anxiety tests assessed in animal models of compulsive eating. The majority of anxiety tests described here are passive, exploration-based tests, as they simulate conditions that occur in nature by creating a conflict between rodents' innate desire to explore novel environments with their innate aversion to bright and open environments (Campos et al., 2013). The elevated plus maze (EPM), light/dark box, defensive withdrawal, and open field tests each assess anxiety through placing animals in an illuminated open space that usually contains a dark and/or enclosed area where animals can retreat to avoid the bright and open environment. Additionally, we describe the "active" anxiety tests marble burying (mice and rats) and defensive burying (rats), which measure active coping responses to increased anxiety (De Boer & Koolhaas, 2003).

Elevated plus maze

The EPM is a plus-shaped apparatus with two "open arms" (no walls and brightly lit) and two "closed arms" (walled and dimly lit). In the EPM test, increased anxiety-like behavior is operationalized as lower amounts of time spent in the open arms as a function of total arm time. In models of drug/alcohol addiction, rats with a history of chronic drug exposure spent less time in the open arms (i.e., higher anxiety-like behavior) while in withdrawal from alcohol (Bhattacharya, Chakrabarti, Sandler, & Glover, 1995; Knapp, Overstreet, Moy, & Breese, 2004), methamphetamine (Nawata, Kitaichi, & Yamamoto, 2012), cocaine (Hall et al., 2010; Sarnyai et al., 1995), opiates (Schulteis, Yackey, Risbrough, & Koob, 1998; Zhang & Schulteis, 2008), nicotine (Bhattacharya et al., 1995), and benzodiazepines (Bhattacharya et al., 1995; File, Baldwin, & Aranko, 1987). Rats in withdrawal from intermittent-long HSD access (Cottone, Sabino, Roberto, et al., 2009; Cottone et al., 2008b; Cottone, Sabino, Steardo, et al., 2009) or continuous HFSD access (Sharma et al., 2013) display decreased open arm time in the EPM (i.e., increased anxiety-like behavior). Furthermore, upon restored access to palatable food, open arm time increased to control levels, suggesting that palatable food can relieve the negative emotional state brought on by withdrawal (Iemolo et al., 2012; Sharma et al., 2013). There is evidence that after protracted withdrawal (4 days), rats with prior intermittent-long HSD access no longer show spontaneous anxiety-like behavior in the EPM (Blasio et al., 2013).

Other studies using intermittent-long HSD (Blasio et al., 2014) and HFD (Rossetti et al., 2014) access did not find increased anxiety-like behavior as assessed by the EPM. This could be explained by protocol differences, as in Blasio et al. (2014) a schedule of intermittent access that may not be sufficient to induce a spontaneous negative emotional state in female rats (cycles of 2 days of standard chow, 1 day of palatable food) was used. While Rossetti et al. (2014) attributed these negative results to a potential floor effect, in Blasio et al. (2014), anxiety-like behavior was able to be precipitated in intermittent-long HSD access animals by systemic administration of rimonabant, an antagonist/inverse agonist of cannabinoid type 1 (CB1) receptor.

An animal model of intermittent-long HSD access and food restriction (12-hour access, 12-hour food deprivation) resulted in a pharmacologically precipitated anxiety-like state, shown by decreased open arm time (Colantuoni et al., 2002). In another experiment with intermittent-short HSD access that occurred for only 10 min daily (Cottone et al., 2008b), rats showed decreased open arm time (increased anxiety-like behavior) that negatively correlated with the amount of palatable food eaten (Cottone et al., 2008b).

Light/dark box

The light/dark box test utilizes a two-compartment chamber divided into a "light" (illuminated) and a "dark" compartment. Longer latencies to first enter the light compartment and lower time spent in the light compartment reflect increased anxiety-like behavior. Withdrawal from chronic drug use in rodents has been shown to increase latency to enter the light compartment and decrease the time spent in the light compartment (Singh, Field, Vass, Hughes, & Woodruff, 1992; Timpl et al., 1998). Intermittent-long HSD access rats in withdrawal spent significantly less time in the light compartment than controls, providing further evidence that withdrawal from intermittent HSD access can induce anxiety (Iemolo et al., 2013).

Defensive withdrawal

Anxiety-like behavior in the defensive withdrawal test is defined by (1) longer latency to emerge from a withdrawal chamber located in an illuminated open field and (2) shorter durations of time spent in the open field, outside of the chamber (Cottone, Sabino, Steardo, et al., 2009). Rats in withdrawal from multiple drugs of abuse exhibit anxiety-like behavior in the defensive withdrawal test (Rodriguez de Fonseca, Carrera, Navarro, Koob, & Weiss, 1997; Skelton, Gutman, Thrivikraman, Nemeroff, & Owens, 2007). Withdrawal from intermittent-long HSD access resulted in longer latency and shorter time spent outside the chamber compared with controls, indicating greater anxiety-like behavior (Cottone, Sabino,

Steardo, et al., 2009). Another study using intermittent-long HSD access found no spontaneous anxiety-like behavior in protracted withdrawal (Blasio et al., 2013). However, in the same study, anxiety-like behavior in the defensive withdrawal test could be pharmacologically precipitated in rats protractedly withdrawn from palatable food by both systemic administration and site-specific administration into the central nucleus of the amygdala of rimonabant (Blasio et al., 2013).

Open field test

In an open field test, rodents are placed in the center of an illuminated box where decreased time spent in the center of the field and increased fecal boli production during the test are used as measures of anxiety-like behavior (Sharma & Fulton, 2013; Teegarden & Bale, 2007). Rats withdrawn from drugs of abuse (e.g., cocaine, alcohol) show reduced center time in an open field test (Craige, Lewandowski, Kirby, & Unterwald, 2015; Overstreet, Knapp, & Breese, 2004). Mice in withdrawal from continuous HFD or HFSD access exhibited anxiety-like behavior compared with groups fed only standard chow, which was displayed as less time spent in the center of the field (HFD) (Sharma & Fulton, 2013) or higher fecal boli production (HFSD) (Teegarden & Bale, 2007). A modified version of the open field test, in which a novel object is placed in the center following a habituation period, can also be used to assess anxiety-like behavior. Intermittent-long HSD access rats showed a reduced time in the open field and a reduced, but not significant, amount of time spent exploring the novel object (both considered measures of anxiety-like behavior) (Rossetti et al., 2014).

Defensive burying

Defensive burying procedures consist of animals being enclosed in a cage containing a noxious stimulus, most often a wall-mounted electric probe, which exerts a mild shock when contacted by the test animal. Active anxiety-like behavior is observed as increased frequency or duration of time spent burying the probe and/or shorter latency to begin burying, whereas immobility/freezing behavior is interpreted as passive anxiety (De Boer & Koolhaas, 2003). Animals in withdrawal from cocaine (Aujla, Martin-Fardon, & Weiss, 2008) or methamphetamine (Jang, Whitfield, Schulteis, Koob, & Wee, 2013) show increased defensive burying and greater time spent immobile. Literature investigating palatable diet effects on defensive burying behavior is sparse; however, it was observed that obese rats fed a continuous HSD exhibited a higher percentage of time spent burying the shock probe compared with control animals (Rebolledo-Solleiro et al., 2017).

Marble burying test

The marble burying test occurs in an enclosed arena with a layer of bedding covering the floor and a number of marbles evenly spread on top of this ground layer. After a test period, the number of marbles buried and frequency/duration of time spent burying is considered to be anxiety-like behavior (Njung'e & Handley, 1991). Increases in marbles buried have been observed in mice withdrawn from alcohol

(Umathe, Bhutada, Dixit, & Shende, 2008), morphine (Becker, Kieffer, & Le Merrer, 2017), and nicotine (Becker et al., 2017; Zhao-Shea et al., 2015). Rats with intermittent-short HFD access displayed increases in marbles buried (Satta et al., 2016). In a study of diet effects on marble burying behavior, mice fed continuous HFD buried more marbles than controls fed with a low-fat diet (Krishna et al., 2015). Another study investigated marble burying during HFD withdrawal compared with animals with current HFD access (Teegarden & Bale, 2007). Researchers observed an interaction of diet and withdrawal state, where number of marbles buried tended to be higher in withdrawal for HFD-fed mice, though in this case marbles buried was not different from controls at any point (Teegarden & Bale, 2007). Overall, there is evidence for the consumption of palatable diets to increase marble burying in rats and mice.

Importantly, there is some debate as to whether marble burying more appropriately assesses anxiety-versus compulsive-like behavior, as marbles are harmless, nonaversive stimuli (Joel, 2006). Furthermore, marble burying is reduced not only by anxiolytics (Njung'e & Handley, 1991) but also by drugs often used to treat obsessive-compulsive disorder (e.g., selective-serotonin reuptake inhibitors) (Broekkamp, Berendsen, Jenck, & Van Delft, 1989; Li, Morrow, & Witkin, 2006).

Overeating despite negative consequences

Animal models of compulsive eating, similar to models of drug addiction, show persistent and enduring palatable food seeking and eating in the face of aversive conditions or the threat of aversive conditions (Belin, Mar, Dalley, Robbins, & Everitt, 2008; Di Segni, Patrono, Patella, Puglisi-Allegra, & Ventura, 2014; Heyne et al., 2009; Oswald, Murdaugh, King, & Boggiano, 2011). Animals will continue to consume palatable food in the presence of either a conditioned or an unconditioned aversive stimulus. The light/dark conflict test, conditioned suppression of feeding procedure, punishment-induced suppression of intake procedure, and footshock maze require animals to experience an aversive physical environment or stimulus, or a cue that signals an aversive stimulus, to acquire the palatable food reward.

Light/dark conflict test

In the light/dark conflict test, animals are placed in the light compartment of a light/dark box that contains a bowl with palatable food (Cottone et al., 2012; Dore et al., 2014; Velazquez-Sanchez et al., 2014). Under normal circumstances, eating is suppressed in the brightly lit, aversive compartment; thus, continued palatable food consumption in this context is defined as compulsive-like eating behavior. This test was specifically developed to measure compulsive eating, rather than having been adapted from a test of compulsive drug use; thus, no light/dark conflict tests have been performed on models of drug addiction at this time.

In animal models with intermittent-long and intermittent-short HSD access, rats show increased compulsive-like eating in the light/dark conflict test (Cottone et al., 2012; Dore et al., 2014; Smith et al., 2015). Rats with intermittent-short HSD access that were more impulsive ate significantly more palatable food than low-impulsive rats with intermittent-short HSD access history, suggesting that palatable food can interact with underlying impulsive traits to potentiate compulsive-like eating (Velazquez-Sanchez et al., 2014). In a model of intermittent HSD access, Binge eating–prone (BEP) rats demonstrated significantly higher palatable food consumption in the light/dark conflict test than binge eating–resistant (BER) rats (Calvez & Timofeeva, 2016).

In an investigation into the *Cyfip2* gene's influence on compulsive-like eating, *Cyfip2* heterozygous knockout (one copy of the *Cyfip2* allele) and wild-type mice were tested in the light/dark conflict test (Kirkpatrick et al., 2017). Wild-type mice with prior access to an HSD ate more palatable food compared with wild-type controls with access to standard chow only; however, heterozygous *Cyfip2* knockout mice with a history of HSD access did not show compulsive-like eating (Kirkpatrick et al., 2017). This finding identified *Cyfip2* as a genetic factor in binge eating behavior and suggests that compulsive eating behaviors arise from interactions between genetics and palatable diet exposure.

Footshock maze

A novel paradigm developed by Oswald et al. (2011) assessed the motivation for palatable food despite punishment using a footshock maze. Similarly to the light/dark conflict test, the footshock maze was developed to investigate compulsive eating specifically; thus, it has not been used in any animal studies of drug addiction. In this procedure, a sated rat is placed in an apparatus with two arms, one containing standard chow without footshocks and the other arm containing palatable food with footshocks (Oswald et al., 2011). Palatable food consumption and tolerance of increasing footshock intensities were used as measures of compulsive-like eating (Oswald et al., 2011). BEP rats with intermittent-short HFSD access showed significantly higher HFSD consumption and tolerated higher intensities of footshock compared with BER rats with similar diet access history (Oswald et al., 2011). This suggests that individual differences in binge eating tendencies, rather than history of diet access, can affect compulsive-like eating (Oswald et al., 2011).

Punishment-induced suppression of intake

Punishment-induced suppression of intake tests pairs an aversive unconditioned stimulus (e.g., footshock) with a self-administered food reward (Rossetti et al., 2014). Food/drug intake is typically suppressed in the presence of the aversive stimulus compared with consumption without this punishment present. Cocaine intake remained unchanged despite punishment with footshocks in rats with a history of chronic cocaine self-administration (Belin & Everitt, 2008; Pelloux, Everitt, & Dickinson, 2007). Similarly, rats with prior intermittent-long HSD access continued food-seeking behavior (i.e., lever presses) in spite of footshocks, whereas food-seeking behavior was suppressed in controls (Rossetti et al., 2014). This persistent behavior demonstrates that intermittent exposure to a highly palatable diet can result in resistance to punishment-induced suppression of intake.

The addition of quinine to the palatable diet to adulterate its taste has also been used as a punishment to suppress intake (Heyne et al., 2009). Typically, animal studies of alcohol addiction use quinine adulteration to assess compulsive alcohol drinking (Hopf & Lesscher, 2014). A study by Heyne et al. (2009) investigated whether obese rats with 8 weeks of continuous HFSD access would continue to consume quinine-adulterated HFSD or consume an optional unadulterated standard chow diet. A considerable dichotomy was observed in the results, where a subset of the obese rats refused the bitter-tasting palatable diet, while the majority continued (Heyne et al., 2009). Moreover, the obese rats that refused the quinine adulterated HFSD did not compensate with the available standard chow, but instead exhibited profound hypophagia of both diets (Heyne et al., 2009). These findings suggest that compulsive-like eating, observed as continued overeating despite punishment, emerges after a history of palatable diet intake (Heyne et al., 2009).

Conditioned suppression of feeding behavior

Conditioned suppression of feeding behavior tests pairs an aversive/punishing unconditioned stimulus (e.g., footshock) with a harmless conditioned stimulus (e.g., light, noise) (Di Segni et al., 2014). Compulsive-like eating behavior is operationally defined as continued food seeking and/or consumption in the presence of the conditioned stimulus in this test (Di Segni et al., 2014; de Jong et al., 2013; Rossetti et al., 2014). Long, but not short, access to cocaine has been shown to cause resistance to the conditioned suppression of cocaine seeking (Vanderschuren & Everitt, 2004).

Johnson and Kenny (2010) found that rats with continuous HFSD access, but not intermittent-short HFSD access, displayed resistance to conditioned suppression of food intake. On the other hand, Velazquez-Sanchez et al. (2015) found that intermittent-short HSD access did result in compulsive-like responding in a conditioned suppression test. These discrepant findings are likely because of differences in type of diet used in the procedure (HSD vs. HFSD) and/or to the differences in procedures used for palatable food access (home cage vs. operant self-administration). Self-administration under a fixed-ratio schedule has been shown to promote habitual responding (Dickinson, Nicholas, & Adams, 1983), which could contribute to the development of other kinds of compulsive-like eating, including resistance to conditioned suppression of intake as measured by Velazquez-Sanchez et al. (2015).

Modeling compulsive eating

The scientific literature offers several strategies to develop models of aberrant forms of eating behavior in animals; however, in the specific context of compulsive eating, research is still in its infancy. Specifically, compulsive eating is considered as a strong, irresistible internal drive to consume food, and it represents a more nuanced and complex construct, defined most appropriately and accurately by its three main elements *(see Chapter 3: Dissecting compulsivity into three elements)*. Importantly, while compulsive eating can encompass overeating and binge eating, overeating and binge eating are not *necessarily* compulsive in their nature. Overeating is a general term, most often defined as consuming an excessive amount of food relative to energy expended, while binge eating is consuming an excess of what most people would eat in a discrete period of time (typically 1–2 h) (American Psychiatric Association, 2013). Although arguably maladaptive, overeating or binge eating alone is neither necessary nor sufficient for eating to be considered compulsive.

In the sections below, we will describe the different manipulations of experimental variables that reliably induce compulsive eating behavior in animals (as measured through the tests described above). These manipulated variables are based on hypotheses regarding the different genetic/environmental etiologies of compulsive eating behavior.

Food type

The increased availability of energy-dense food is proposed to be a risk factor for eating disorders and obesity (Hill, Wyatt, Reed, & Peters, 2003). Moreover, episodes of binge eating most often occur with foods high in sugar and/or fat (Yanovski et al., 1992). In this review, a palatable diet is defined as food that is uniformly preferred by animals over the standard laboratory chow. Notably, while preferred food is typically called "palatable" by researchers in the preclinical literature, this is loose and anthropomorphic terminology, with the more precise term being "preferred." Therefore, the majority of the existing models of compulsive eating have been developed by manipulating the food type provided to animals, where a more highly palatable diet is associated with greater propensity for compulsive eating.

Palatable diets can refer to HSD, HFD, or a combination of the two, referred to here as HFSD. HFSD can be a formulation of rodent chow that is high in both fat and sugar, or a so-called "cafeteria diet"/"junk food diet," typically consisting of multiple types of human food that is high in sugar and fat (e.g., Oreo's, lard, peanut butter, etc.) (Heyne et al., 2009; Martire, Westbrook, & Morris, 2015).

The highly palatable diet can be provided with or without concurrent standard chow access in a free-choice or forced-choice paradigm, respectively. However, a highly palatable diet is supposed to be preferred to a large extent over the standard chow diet. Under conditions of concurrent free-choice access, the palatable diet intake typically consists of $>90\%$ of the total caloric intake (Cottone et al., 2008a, 2008b; Johnson & Kenny, 2010). For this reason, the palatable diet is most often provided as the only option during scheduled access and can be assumed to be the case unless otherwise noted in this review. While food type is varied across models to different degrees, there does not seem to be any particular diet formulation (HSD vs. HFD vs. HFSD) that results in any more or less compulsive eating behavior apart from its level of preferredness/palatability.

Schedule of access to palatable food

Often individuals attempt to lose body weight by controlling the quality of foods consumed and by limiting their intake of energy-dense, highly palatable, "forbidden" foods and consuming only low-palatability "safe" foods (de Castro,

1995; Laessle, Tuschl, Kotthaus, & Pirke, 1989; Mela, 2001). Inevitably, restraining intake to only safe foods is ultimately associated with later disinhibited eating of forbidden foods (Herman & Polivy, 1990). Thus, in humans who display disordered eating behaviors, there is often a pattern of discrete alternations in intake of foods with different palatabilities. Similarly, in animal models, compulsive eating is often engendered by varying the schedule of access to palatable food.

In animals, palatable diet access is either given continuously (i.e., ad libitum) or intermittently. Intermittent access periods can range from daily 10-min periods (Cottone et al., 2008b) to multiple days (Cottone, Sabino, Roberto, et al., 2009). In this chapter, we have considered a short access period an intermittent access lasting less than 12 h and a long access as a period lasting 12 h or longer. Importantly, these designations are not equivalent to long and short access models of drug addiction, where short access conditions are meant to mimic recreational drug use, while long access conditions result in compulsive drug use (Ahmed & Koob, 1998; Ahmed, Walker, & Koob, 2000). Indeed, compulsive and binge eating behaviors are often readily observed in animal models that give as short as 10 min daily access to a palatable diet (Cottone et al., 2008b, 2012; Ferragud et al., 2016; Velazquez-Sanchez et al., 2015).

These variations of access to palatable food result in certain reproducible phenotypes related to food consumption patterns. Intermittent access models consistently show an *escalation of intake* over time (Dore et al., 2014; Parylak, Cottone, Sabino, Rice, & Zorrilla, 2012; Spierling et al., 2018; Velazquez-Sanchez et al., 2015), a feature that is considered a hallmark of addictive-like behaviors (Ahmed & Koob, 1998; Ahmed et al., 2000). Intermittency appears to be critical to this pattern of intake, likely because of a hyperevaluation of the palatable diet when unavailable, as well as underevaluation of the less preferred alternatives (i.e., standard laboratory chow diet). Additionally, overeating of the palatable food progressively shifts toward the beginning of the renewed access (Avena, Long, & Hoebel, 2005; Cottone et al., 2008a), a pattern resembling "deprivation effects" seen in models of intermittent drug and alcohol access (George et al., 2007; Rodd, Bell, Sable, Murphy, & McBride, 2004; Sabino et al., 2009; Wise, 1973).

In continuous access models, animals show overconsumption of the palatable diet, though this overeating occurs in a steadier and more continuous pattern, with no distinct binge periods. Intermittent palatable diet access enables binge-like consumption of palatable food, creating a phenotype distinct from continuous access (Corwin, 2006). However, continuous access is still able to produce compulsive eating behavior and brain reward deficits paralleling those seen in drug addiction (Johnson & Kenny, 2010). It is important to note that continuous palatable access results in increased body weight, making it often difficult to parse out effects of overweight from overeating behavior.

One commonly used intermittent-short access procedure allows animals to self-administer palatable food in 1-hour operant conditioning sessions; in this paradigm, animals display higher rates of responding for palatable food compared with controls who respond for standard chow (Cottone et al., 2012). Daily caloric intake escalates over time, resulting in up to four times higher consumption in animals responding for palatable food compared with chow; these animals also show multiple characteristics of compulsive eating behaviors (Smith et al., 2015; Velazquez-Sanchez et al., 2014).

An example of an intermittent-long access paradigm includes providing rats with standard chow for 5 days per week and a high-sucrose, palatable food for 2 days per week (Cottone, Sabino, Roberto, et al., 2009; Cottone, Sabino, Steardo, et al., 2009). Following repeated cycles of access to the palatable diet and chow, intake of palatable food on the first hour of renewed access escalates over time, and animals show compulsive-like eating behavior of the palatable diet, as well as withdrawal-dependent hypophagia of standard chow and anxiety-like behavior (Cottone, Sabino, Roberto, et al., 2009, Cottone, Sabino, Steardo, et al., 2009).

Food restriction/deprivation

Dieting, which typically consists of limiting daily caloric intake, promotes binge and compulsive eating behavior and is regarded as a risk factor for obesity (Dulloo & Montani, 2015; Herman & Polivy, 1990; Lowe, Doshi, Katterman, & Feig, 2013; Stice, Davis, Miller, & Marti, 2008). In animals, forced, experimental food restriction/deprivation is used to mimic the voluntary caloric restriction observed in human dieters and can be a useful experimental manipulation to generate a compulsive eating phenotype. Food restriction limits animals' daily food intake to a fixed amount that is significantly lower than that of ad libitum fed controls, whereas food deprivation prevents animals from accessing any food for a set period of time (Bi et al., 2003). Food restriction can be chronic (i.e., throughout the duration of the study) (Pankevich, Teegarden, Hedin, Jensen, & Bale, 2010) or given acutely in cycles of restriction and refeeding (Hagan & Moss, 1997).

It is important to note that, while caloric restriction and exposure to highly palatable food can both induce neuroadaptations promoting forms of compulsive-like behavior (Carr, 2016; Shalev, Yap, & Shaham, 2001), the underlying mechanisms are likely different (Cottone, Sabino, Steardo, et al., 2009, 2012; Smith et al., 2015). Caloric restriction causes deficits in energy intake, which leads to greater than normal consumption of food as an energy-homeostatic response. This homeostatic motivation behind hyperphagia of palatable food imposes a limitation on how accurately the hedonic mechanisms of palatable food overconsumption can be assessed.

One model of binge-like sucrose consumption involves 12-hour food deprivation followed by ad libitum access to chow and 25% glucose (Avena et al., 2005; Colantuoni et al., 2002; Rada, Avena, & Hoebel, 2005). In this animal model, compulsive eating may result from a negative emotional state, as emotional and physical withdrawal can be pharmacologically precipitated by treatment with the opioid receptor antagonist naloxone (Avena, Rada, & Hoebel, 2008).

Stress

While stress can affect feeding behavior bidirectionally, many individuals tend to overeat following stressful events (Adam & Epel, 2007). In people with eating disorders, stress is known to trigger binge and compulsive eating (Greeno & Wing, 1994; Heatherton, Herman, & Polivy, 1991). In animals, even though stress generally exerts an anorectic action (Hotta, Shibasaki, Arai, & Demura, 1999; Valles, Marti, Garcia, & Armario, 2000) in animals that have undergone experimental manipulations to induce aberrant feeding behavior, stress can instead induce binge-like and compulsive-like eating behavior (Calvez & Timofeeva, 2016; Hagan et al., 2002; Micioni Di Bonaventura et al., 2014). The consumption of palatable food has been shown to dampen the physiological stress response, compared with stressed animals with access to only standard chow (Pecoraro, Reyes, Gomez, Bhargava, & Dallman, 2004). Almost all of the animal models using stress couple palatable food consumption with periods of caloric restriction, as stress and caloric restriction have been shown to have a synergistic effect on food intake. For example, Hagan et al. (2002) found that a history of restriction (4 days of 66% of mean daily chow) and refeeding (6 days of ad libitum chow) followed by a footshock stressor induced greater overeating than no-stress and no-restriction. Other types of acute stress used include a forced swim stress (Consoli, Contarino, Tabarin, & Drago, 2009), chronic variable stress (e.g., restraint, predator odor, etc.) (Pankevich et al., 2010), and frustration stress induced by physical separation from the palatable food during a restriction period (Micioni Di Bonaventura et al., 2014).

Genes and phenotypes

Eating disorders and obesity are largely heritable (Bulik, Sullivan, & Kendler, 1998; Pigeyre, Yazdi, Kaur, & Meyre, 2016; Thornton, Mazzeo, & Bulik, 2011), and genetics can interact with environmental factors, such as the availability of highly palatable food, to determine compulsive eating behaviors (Davis, 2015). A large number of genes have been identified to contribute to hyperphagia. Historically, focus has been on genes regulating homeostatic food intake (e.g., polymorphisms in the fat mass and obesity-associated (*FTO*) gene, melanocortin-4 receptor (MC4R) gene, leptin gene Ob(lep)) that lead to hyperphagia and obesity (Chen et al., 1996; Church et al., 2009; Srisai et al., 2011). Recently, discovery-based genetic approaches, such as quantitative trait loci (QTL) mapping, have been applied to finding novel mechanisms of hedonic eating. One such study used QTL mapping to identify cytoplasmic FMR1-interacting protein 2 (*Cyfip2*) as a determinant in binge eating and compulsive eating behavior (Kirkpatrick et al., 2017).

Innate individual phenotypic differences may also be used to study compulsive eating behavior in animals (Calvez & Timofeeva, 2016; Velazquez-Sanchez et al., 2014). BEP and BER feeding phenotypes are commonly used to investigate innate individual differences in compulsive eating behavior. BEP rats model human eating behaviors that align with the DSM-V diagnostic criteria for BED, including high food consumption in short intervals of time and faster than normal consumption (Calvez & Timofeeva, 2016). BEP/BER studies classify test animals according to their general consumption of palatable food in a period of 1−4 h, under stressful or nonstressful conditions (Boggiano et al., 2007; Calvez & Timofeeva, 2016; Oswald et al., 2011). BEP animals display stable and behaviorally distinct phenotypes, consistently exhibiting greater consumption of intermittent palatable food, altered stress-like responses, and compulsive-like eating (Calvez & Timofeeva, 2016; Oswald et al., 2011). Another study classified animals by trait impulsive action and found that high trait impulsivity was associated with increased susceptibility to binge-like eating, high motivation, and compulsive-like eating (Velazquez-Sanchez et al., 2014).

Discussion

A systematic investigation of the construct of compulsivity in the context of feeding behavior has started only recently, leaving a gap in knowledge as well as missed opportunities to better understand the neurobiological mechanisms of aberrant feeding behaviors (Cottone, Sabino, Roberto, et al., 2009; Cottone et al., 2012; Moore, Sabino, Koob, & Cottone, 2017a, 2017b; Moore, Panciera, Sabino, & Cottone, 2018). At a preclinical level, this gap has been particularly profound, given that historically the motivational, emotional, and cognitive aspects related to eating behavior have been unappreciated in favor of a biased energy-homeostatic view of feeding-related disorders. The scientific community has already paid a high cost for this oversight. Only a decade ago, the antagonist of the cannabinoid receptor 1, rimonabant, was introduced in the market as a highly promising antiobesity drug because of its ability to reduce homeostatic feeding and food reward centrally, as well as metabolism and energy expenditure peripherally (Matias & Di Marzo, 2007). However, rimonabant was swiftly removed from the market following evidence of adverse psychiatric effects, including depression, anxiety, and suicidality (Blasio et al., 2013; Christensen, Kristensen, Bartels, Bliddal, & Astrup, 2007). After a decade, the lesson learned from rimonabant's failure is that dissociating energy-homeostatic feeding from higher-order emotional, motivational, and cognitive constructs can be risky. In this process, animal research can play a critical role. At the preclinical level, the battle against the epidemics of obesity and eating disorders needs to be brought to the next level through the integration of measures of complex behavioral constructs (e.g., negatively reinforced feeding, food craving, salience of food-related cues, food seeking, eating in spite of negative consequences, habitual/inflexible feeding/responding, stress/cues/primed food relapse). Retuning to such a construct and domain-driven methodological approach will be essential to understand the complex mechanisms underlying the maladaptive forms of food intake in obesity and eating disorders.

References

Adam, T. C., & Epel, E. S. (2007). Stress, eating and the reward system. *Physiology & Behavior, 91*(4), 449–458. https://doi.org/10.1016/j.physbeh.2007.04.011.

Ahmed, S. H., Kenny, P. J., Koob, G. F., & Markou, A. (2002). Neurobiological evidence for hedonic allostasis associated with escalating cocaine use. *Nature Neuroscience, 5*(7), 625–626. https://doi.org/10.1038/nn872.

Ahmed, S. H., & Koob, G. F. (1998). Transition from moderate to excessive drug intake: Change in hedonic set point. *Science, 282*(5387), 298–300.

Ahmed, S. H., Walker, J. R., & Koob, G. F. (2000). Persistent increase in the motivation to take heroin in rats with a history of drug escalation. *Neuropsychopharmacology, 22*(4), 413–421. https://doi.org/10.1016/S0893-133X(99)00133-5.

American Psychiatric Association. (2013). *Diagnostic and statistical manual of mental disorders*. Arlington, VA: American Psychiatric Publishing.

Anraku, T., Ikegaya, Y., Matsuki, N., & Nishiyama, N. (2001). Withdrawal from chronic morphine administration causes prolonged enhancement of immobility in rat forced swimming test. *Psychopharmacology, 157*(2), 217–220. https://doi.org/10.1007/s002130100793.

Aujla, H., Martin-Fardon, R., & Weiss, F. (2008). Rats with extended access to cocaine exhibit increased stress reactivity and sensitivity to the anxiolytic-like effects of the mGluR 2/3 agonist LY379268 during abstinence. *Neuropsychopharmacology, 33*(8), 1818–1826. https://doi.org/10.1038/sj.npp.1301588.

Austen, J. M., Strickland, J. A., & Sanderson, D. J. (2016). Memory-dependent effects on palatability in mice. *Physiology & Behavior, 167*, 92–99. https://doi.org/10.1016/j.physbeh.2016.09.001.

Avena, N. M., Long, K. A., & Hoebel, B. G. (2005). Sugar-dependent rats show enhanced responding for sugar after abstinence: Evidence of a sugar deprivation effect. *Physiology & Behavior, 84*(3), 359–362. https://doi.org/10.1016/j.physbeh.2004.12.016.

Avena, N. M., Rada, P., & Hoebel, B. G. (2008). Evidence for sugar addiction: Behavioral and neurochemical effects of intermittent, excessive sugar intake. *Neuroscience & Biobehavioral Reviews, 32*(1), 20–39. https://doi.org/10.1016/j.neubiorev.2007.04.019.

Balleine, B. (1992). Instrumental performance following a shift in primary motivation depends on incentive learning. *Journal of Experimental Psychology: Animal Behavior Processes, 18*(3), 236–250.

Becker, J. A. J., Kieffer, B. L., & Le Merrer, J. (2017). Differential behavioral and molecular alterations upon protracted abstinence from cocaine versus morphine, nicotine, THC and alcohol. *Addiction Biology, 22*(5), 1205–1217. https://doi.org/10.1111/adb.12405.

Belin-Rauscent, A., Fouyssac, M., Bonci, A., & Belin, D. (2015). How preclinical models evolved to resemble the diagnostic criteria of drug addiction. *Biological Psychiatry.* https://doi.org/10.1016/j.biopsych.2015.01.004.

Belin, D., & Everitt, B. J. (2008). Cocaine seeking habits depend upon dopamine-dependent serial connectivity linking the ventral with the dorsal striatum. *Neuron, 57*(3), 432–441. https://doi.org/10.1016/j.neuron.2007.12.019.

Belin, D., Mar, A. C., Dalley, J. W., Robbins, T. W., & Everitt, B. J. (2008). High impulsivity predicts the switch to compulsive cocaine-taking. *Science, 320*(5881), 1352–1355. https://doi.org/10.1126/science.1158136.

Bhattacharya, S. K., Chakrabarti, A., Sandler, M., & Glover, V. (1995). Rat brain monoamine oxidase A and B inhibitory (tribulin) activity during drug withdrawal anxiety. *Neuroscience Letters, 199*(2), 103–106.

Bi, S., Robinson, B. M., & Moran, T. H. (2003). Acute food deprivation and chronic food restriction differentially affect hypothalamic NPY mRNA expression. *American Journal of Physiology — Regulatory, Integrative and Comparative Physiology, 285*(5), R1030–R1036. https://doi.org/10.1152/ajpregu.00734.2002.

Blasio, A., Iemolo, A., Sabino, V., Petrosino, S., Steardo, L., Rice, K. C., et al. (2013). Rimonabant precipitates anxiety in rats withdrawn from palatable food: Role of the central amygdala. *Neuropsychopharmacology, 38*(12), 2498–2507. https://doi.org/10.1038/npp.2013.153.

Blasio, A., Rice, K. C., Sabino, V., & Cottone, P. (2014). Characterization of a shortened model of diet alternation in female rats: Effects of the CB1 receptor antagonist rimonabant on food intake and anxiety-like behavior. *Behavioral Pharmacology, 25*(7), 609–617. https://doi.org/10.1097/FBP.0000000000000059.

Boggiano, M. M., Artiga, A. I., Pritchett, C. E., Chandler-Laney, P. C., Smith, M. L., & Eldridge, A. J. (2007). High intake of palatable food predicts binge-eating independent of susceptibility to obesity: An animal model of lean vs obese binge-eating and obesity with and without binge-eating. *International Journal of Obesity, 31*(9), 1357–1367. https://doi.org/10.1038/sj.ijo.0803614.

Broekkamp, C. L., Berendsen, H. H., Jenck, F., & Van Delft, A. M. (1989). Animal models for anxiety and response to serotonergic drugs. *Psychopathology, 22*(Suppl. 1), 2–12. https://doi.org/10.1159/000284620.

Bulik, C. M., Sullivan, P. F., & Kendler, K. S. (1998). Heritability of binge-eating and broadly defined bulimia nervosa. *Biological Psychiatry, 44*(12), 1210–1218.

Calvez, J., & Timofeeva, E. (2016). Behavioral and hormonal responses to stress in binge-like eating prone female rats. *Physiology & Behavior, 157*, 28–38. https://doi.org/10.1016/j.physbeh.2016.01.029.

Campos, A. C., Fogaca, M. V., Aguiar, D. C., & Guimaraes, F. S. (2013). Animal models of anxiety disorders and stress. *Brazilian Journal of Psychiatry, 35*, S101–S111.

Carlezon, W. A., Jr., & Chartoff, E. H. (2007). Intracranial self-stimulation (ICSS) in rodents to study the neurobiology of motivation. *Nature Protocols, 2*(11), 2987–2995. https://doi.org/10.1038/nprot.2007.441.

Carr, K. D. (2016). Nucleus accumbens AMPA receptor trafficking upregulated by food restriction: An unintended target for drugs of abuse and forbidden foods. *Current Opinion in Behavioral Science, 9*, 32–39. https://doi.org/10.1016/j.cobeha.2015.11.019.

Castagne, V., Moser, P., Roux, S., & Porsolt, R. D. (2011). Rodent models of depression: Forced swim and tail suspension behavioral despair tests in rats and mice. *Current Protocols in Neuroscience*. https://doi.org/10.1002/0471142301.ns0810as55 (Chapter 8), Unit 8.10A.

de Castro, J. M. (1995). The relationship of cognitive restraint to the spontaneous food and fluid intake of free-living humans. *Physiology & Behavior, 57*(2), 287–295.

Charney, D. S., Sklar, P. B., Buxbaum, J. D., & Nestler, E. J. (2018). *Charney & Nestler's neurobiology of mental illness*.

Chen, H., Charlat, O., Tartaglia, L. A., Woolf, E. A., Weng, X., Ellis, S. J., et al. (1996). Evidence that the diabetes gene encodes the leptin receptor: Identification of a mutation in the leptin receptor gene in db/db mice. *Cell, 84*(3), 491–495.

Christensen, R., Kristensen, P. K., Bartels, E. M., Bliddal, H., & Astrup, A. (2007). Efficacy and safety of the weight-loss drug rimonabant: A meta-analysis of randomised trials. *Lancet, 370*(9600), 1706–1713. https://doi.org/10.1016/S0140-6736(07)61721-8.

Church, C., Lee, S., Bagg, E. A., McTaggart, J. S., Deacon, R., Gerken, T., et al. (2009). A mouse model for the metabolic effects of the human fat mass and obesity associated FTO gene. *PLoS Genetics, 5*(8), e1000599. https://doi.org/10.1371/journal.pgen.1000599.

Colantuoni, C., Rada, P., McCarthy, J., Patten, C., Avena, N. M., Chadeayne, A., et al. (2002). Evidence that intermittent, excessive sugar intake causes endogenous opioid dependence. *Obesity Research, 10*(6), 478–488. https://doi.org/10.1038/oby.2002.66.

Consoli, D., Contarino, A., Tabarin, A., & Drago, F. (2009). Binge-like eating in mice. *International Journal of Eating Disorders, 42*(5), 402–408. https://doi.org/10.1002/eat.20637.

Corbit, L. H., Nie, H., & Janak, P. H. (2012). Habitual alcohol seeking: Time course and the contribution of subregions of the dorsal striatum. *Biological Psychiatry, 72*(5), 389–395. https://doi.org/10.1016/j.biopsych.2012.02.024.

Corwin, R. L. (2006). Bingeing rats: A model of intermittent excessive behavior? *Appetite, 46*(1), 11–15.

Cottone, P., Sabino, V., Roberto, M., Bajo, M., Pockros, L., Frihauf, J. B., et al. (2009). CRF system recruitment mediates dark side of compulsive eating. *Proceedings of the National Academy of Sciences of the United States of America, 106*(47), 20016–20020. https://doi.org/10.1073/pnas.0908789106.

Cottone, P., Sabino, V., Steardo, L., & Zorrilla, E. P. (2008a). Intermittent access to preferred food reduces the reinforcing efficacy of chow in rats. *American Journal of Physiology – Regulatory, Integrative and Comparative Physiology, 295*(4), R1066–R1076. https://doi.org/10.1152/ajpregu.90309.2008.

Cottone, P., Sabino, V., Steardo, L., & Zorrilla, E. P. (2008b). Opioid-dependent anticipatory negative contrast and binge-like eating in rats with limited access to highly preferred food. *Neuropsychopharmacology, 33*(3), 524–535. https://doi.org/10.1038/sj.npp.1301430.

Cottone, P., Sabino, V., Steardo, L., & Zorrilla, E. P. (2009). Consummatory, anxiety-related and metabolic adaptations in female rats with alternating access to preferred food. *Psychoneuroendocrinology, 34*(1), 38–49. https://doi.org/10.1016/j.psyneuen.2008.08.010.

Cottone, P., Wang, X., Park, J. W., Valenza, M., Blasio, A., Kwak, J., et al. (2012). Antagonism of sigma-1 receptors blocks compulsive-like eating. *Neuropsychopharmacology, 37*(12), 2593–2604. https://doi.org/10.1038/npp.2012.89.

Craige, C. P., Lewandowski, S., Kirby, L. G., & Unterwald, E. M. (2015). Dorsal raphe 5-HT(2C) receptor and GABA networks regulate anxiety produced by cocaine withdrawal. *Neuropharmacology, 93*, 41–51. https://doi.org/10.1016/j.neuropharm.2015.01.021.

Cryan, J. F., Hoyer, D., & Markou, A. (2003). Withdrawal from chronic amphetamine induces depressive-like behavioral effects in rodents. *Biological Psychiatry, 54*(1), 49–58.

Cryan, J. F., Markou, A., & Lucki, I. (2002). Assessing antidepressant activity in rodents: Recent developments and future needs. *Trends in Pharmacological Sciences, 23*(5), 238–245.

Dalley, J. W., Everitt, B. J., & Robbins, T. W. (2011). Impulsivity, compulsivity, and top-down cognitive control. *Neuron, 69*(4), 680–694. https://doi.org/10.1016/j.neuron.2011.01.020.

Davis, C. (2015). The epidemiology and genetics of binge eating disorder (BED). *CNS Spectrums, 20*(6), 522–529. https://doi.org/10.1017/S1092852915000462.

Davis, C., & Carter, J. C. (2009). Compulsive overeating as an addiction disorder. A review of theory and evidence. *Appetite, 53*(1), 1–8. https://doi.org/10.1016/j.appet.2009.05.018.

De Boer, S. F., & Koolhaas, J. M. (2003). Defensive burying in rodents: Ethology, neurobiology and psychopharmacology. *European Journal of Pharmacology, 463*(1–3), 145–161.

Der-Avakian, A., & Markou, A. (2010). Withdrawal from chronic exposure to amphetamine, but not nicotine, leads to an immediate and enduring deficit in motivated behavior without affecting social interaction in rats. *Behavioral Pharmacology, 21*(4), 359–368. https://doi.org/10.1097/FBP.0b013e32833c7cc8.

Di Segni, M., Patrono, E., Patella, L., Puglisi-Allegra, S., & Ventura, R. (2014). Animal models of compulsive eating behavior. *Nutrients, 6*(10), 4591–4609. https://doi.org/10.3390/nu6104591.

Dickinson, A., Nicholas, D., & Adams, C. D. (1983). The effect of the instrumental training contingency on susceptibility to reinforcer devaluation. *The Quarterly Journal of Experimental Psychology Section B, 35*(1b), 35–51.

Dore, R., Valenza, M., Wang, X., Rice, K. C., Sabino, V., & Cottone, P. (2014). The inverse agonist of CB1 receptor SR141716 blocks compulsive eating of palatable food. *Addiction Biology, 19*(5), 849–861. https://doi.org/10.1111/adb.12056.

Dulloo, A. G., & Montani, J. P. (2015). Pathways from dieting to weight regain, to obesity and to the metabolic syndrome: An overview. *Obesity Reviews, 16*(Suppl. 1), 1–6. https://doi.org/10.1111/obr.12250.

Dutheil, S., Ota, K. T., Wohleb, E. S., Rasmussen, K., & Duman, R. S. (2016). High-fat diet induced anxiety and anhedonia: Impact on brain homeostasis and inflammation. *Neuropsychopharmacology, 41*(7), 1874–1887. https://doi.org/10.1038/npp.2015.357.

Esposito, R., & Kornetsky, C. (1977). Morphine lowering of self-stimulation thresholds: Lack of tolerance with long-term administration. *Science, 195*(4274), 189–191.

Everitt, B. J. (2014). Neural and psychological mechanisms underlying compulsive drug seeking habits and drug memories–indications for novel treatments of addiction. *European Journal of Neuroscience, 40*(1), 2163–2182. https://doi.org/10.1111/ejn.12644.

Ferragud, A., Howell, A. D., Moore, C. F., Ta, T. L., Hoener, M. C., Sabino, V., et al. (2016). The trace amine-associated receptor 1 agonist RO5256390 blocks compulsive, binge-like eating in rats. *Neuropsychopharmacology.* https://doi.org/10.1038/npp.2016.233.

File, S. E., Baldwin, H. A., & Aranko, K. (1987). Anxiogenic effects in benzodiazepine withdrawal are linked to the development of tolerance. *Brain Research Bulletin, 19*(5), 607–610.

Flaherty, C. F., Coppotelli, C., Grigson, P. S., Mitchell, C., & Flaherty, J. E. (1995). Investigation of the devaluation interpretation of anticipatory negative contrast. *Journal of Experimental Psychology: Animal Behavior Processes, 21*(3), 229–247.

Flaherty, C. F., & Rowan, G. A. (1986). Successive, simultaneous, and anticipatory contrast in the consumption of saccharin solutions. *Journal of Experimental Psychology: Animal Behavior Processes, 12*(4), 381–393.

Furlong, T. M., Jayaweera, H. K., Balleine, B. W., & Corbit, L. H. (2014). Binge-like consumption of a palatable food accelerates habitual control of behavior and is dependent on activation of the dorsolateral striatum. *Journal of Neuroscience, 34*(14), 5012–5022. https://doi.org/10.1523/JNEUROSCI.3707-13.2014.

Gallistel, C. R., & Freyd, G. (1987). Quantitative determination of the effects of catecholaminergic agonists and antagonists on the rewarding efficacy of brain stimulation. *Pharmacology Biochemistry and Behavior, 26*(4), 731–741.

George, O., Ghozland, S., Azar, M. R., Cottone, P., Zorrilla, E. P., Parsons, L. H., et al. (2007). CRF-CRF1 system activation mediates withdrawal-induced increases in nicotine self-administration in nicotine-dependent rats. *Proceedings of the National Academy of Sciences of the United States of America, 104*(43), 17198–17203. https://doi.org/10.1073/pnas.0707585104.

Geyer, M. A., & Markou, A. (2000). Animal models of psychiatric disorders. In F. E. Bloom, & D. J. Kupfer (Eds.), *Psychopharmacology: The fourth generation of progress* (pp. 787–798). New York, NY: Raven Press.

Geyer, M. A., & Markou, A. (2002). The role of preclinical models in the development of psychotropic drugs. In K. L. Davis, D. Charney, J. T. Coyle, & C. Nemeroff (Eds.), *Neuropsychopharmacology: The fifth generation of progress* (pp. 445–455). New York, NY: Lippincott Williams and Wilkins.

Greeno, C. G., & Wing, R. R. (1994). Stress-induced eating. *Psychological Bulletin, 115*(3), 444–464.

Hagan, M. M., & Moss, D. E. (1997). Persistence of binge-eating patterns after a history of restriction with intermittent bouts of refeeding on palatable food in rats: Implications for bulimia nervosa. *International Journal of Eating Disorders, 22*(4), 411–420.

Hagan, M. M., Wauford, P. K., Chandler, P. C., Jarrett, L. A., Rybak, R. J., & Blackburn, K. (2002). A new animal model of binge eating: Key synergistic role of past caloric restriction and stress. *Physiology & Behavior, 77*(1), 45–54.

Hall, B. J., Pearson, L. S., & Buccafusco, J. J. (2010). Effect of the use-dependent, nicotinic receptor antagonist BTMPS in the forced swim test and elevated plus maze after cocaine discontinuation in rats. *Neuroscience Letters, 474*(2), 84–87. https://doi.org/10.1016/j.neulet.2010.03.011.

Heatherton, T. F., Herman, C. P., & Polivy, J. (1991). Effects of physical threat and ego threat on eating behavior. *Journal of Personality and Social Psychology, 60*(1), 138–143.

Herman, C. P., & Polivy, J. (1990). From dietary restraint to binge eating: Attaching causes to effects. *Appetite, 14*(2), 123–125. discussion 142–123.

Heyne, A., Kiesselbach, C., Sahun, I., McDonald, J., Gaiffi, M., Dierssen, M., et al. (2009). An animal model of compulsive food-taking behaviour. *Addiction Biology, 14*(4), 373–383. https://doi.org/10.1111/j.1369-1600.2009.00175.x.

Hill, J. O., Wyatt, H. R., Reed, G. W., & Peters, J. C. (2003). Obesity and the environment: Where do we go from here? *Science, 299*(5608), 853–855. https://doi.org/10.1126/science.1079857.

Hoefer, M. E., Voskanian, S. J., Koob, G. F., & Pulvirenti, L. (2006). Effects of terguride, ropinirole, and acetyl-L-carnitine on methamphetamine withdrawal in the rat. *Pharmacology Biochemistry and Behavior, 83*(3), 403–409. https://doi.org/10.1016/j.pbb.2006.02.023.

Hopf, F. W., & Lesscher, H. M. (2014). Rodent models for compulsive alcohol intake. *Alcohol, 48*(3), 253–264. https://doi.org/10.1016/j.alcohol.2014.03.001.

Hotta, M., Shibasaki, T., Arai, K., & Demura, H. (1999). Corticotropin-releasing factor receptor type 1 mediates emotional stress-induced inhibition of food intake and behavioral changes in rats. *Brain Research, 823*(1–2), 221–225.

Iemolo, A., Blasio, A., St Cyr, S. A., Jiang, F., Rice, K. C., Sabino, V., et al. (2013). CRF-CRF1 receptor system in the central and basolateral nuclei of the amygdala differentially mediates excessive eating of palatable food. *Neuropsychopharmacology, 38*(12), 2456–2466. https://doi.org/10.1038/npp.2013.147.

Iemolo, A., Valenza, M., Tozier, L., Knapp, C. M., Kornetsky, C., Steardo, L., et al. (2012). Withdrawal from chronic, intermittent access to a highly palatable food induces depressive-like behavior in compulsive eating rats. *Behavioral Pharmacology, 23*(5–6), 593–602. https://doi.org/10.1097/FBP.0b013e328357697f.

Jang, C. G., Whitfield, T., Schulteis, G., Koob, G. F., & Wee, S. (2013). A dysphoric-like state during early withdrawal from extended access to methamphetamine self-administration in rats. *Psychopharmacology, 225*(3), 753–763. https://doi.org/10.1007/s00213-012-2864-0.

Joel, D. (2006). Current animal models of obsessive compulsive disorder: A critical review. *Progress in Neuro-Psychopharmacology & Biological Psychiatry, 30*(3), 374–388. https://doi.org/10.1016/j.pnpbp.2005.11.006.

Johnson, P. M., & Kenny, P. J. (2010). Dopamine D2 receptors in addiction-like reward dysfunction and compulsive eating in obese rats. *Nature Neuroscience, 13*(5), 635–641. https://doi.org/10.1038/nn.2519.

de Jong, J. W., Meijboom, K. E., Vanderschuren, L. J., & Adan, R. A. (2013). Low control over palatable food intake in rats is associated with habitual behavior and relapse vulnerability: Individual differences. *PLoS One, 8*(9), e74645. https://doi.org/10.1371/journal.pone.0074645.

Kendig, M. D., Cheung, A. M., Raymond, J. S., & Corbit, L. H. (2016). Contexts paired with junk food impair goal-directed behavior in rats: Implications for decision making in obesogenic environments. *Frontiers in Behavioral Neuroscience, 10*, 216. https://doi.org/10.3389/fnbeh.2016.00216.

Kenny, P. J., Chen, S. A., Kitamura, O., Markou, A., & Koob, G. F. (2006). Conditioned withdrawal drives heroin consumption and decreases reward sensitivity. *Journal of Neuroscience, 26*(22), 5894–5900. https://doi.org/10.1523/JNEUROSCI.0740-06.2006.

Kim, S., Shou, J., Abera, S., & Ziff, E. B. (2018). Sucrose withdrawal induces depression and anxiety-like behavior by Kir2.1 upregulation in the nucleus accumbens. *Neuropharmacology, 130*, 10–17. https://doi.org/10.1016/j.neuropharm.2017.11.041.

Kirkpatrick, S. L., Goldberg, L. R., Yazdani, N., Babbs, R. K., Wu, J., Reed, E. R., et al. (2017). Cytoplasmic FMR1-interacting protein 2 is a major genetic factor underlying binge eating. *Biological Psychiatry, 81*(9), 757–769. https://doi.org/10.1016/j.biopsych.2016.10.021.

Knapp, D. J., Overstreet, D. H., Moy, S. S., & Breese, G. R. (2004). SB242084, flumazenil, and CRA1000 block ethanol withdrawal-induced anxiety in rats. *Alcohol, 32*(2), 101–111. https://doi.org/10.1016/j.alcohol.2003.08.007.

Kokkinidis, L., & McCarter, B. D. (1990). Postcocaine depression and sensitization of brain-stimulation reward: Analysis of reinforcement and performance effects. *Pharmacology Biochemistry and Behavior, 36*(3), 463–471.

Koob, G. F. (2013). Addiction is a reward deficit and stress surfeit disorder. *Frontiers in Psychiatry, 4*, 72. https://doi.org/10.3389/fpsyt.2013.00072.

Koob, G. F. (2015). The dark side of emotion: The addiction perspective. *European Journal of Pharmacology, 753*, 73–87. https://doi.org/10.1016/j.ejphar.2014.11.044.

Kosheleff, A. R., Araki, J., Tsan, L., Chen, G., Murphy, N. P., Maidment, N. T., et al. (2018). Junk food exposure disrupts selection of food-seeking actions in rats. *Frontiers in Psychiatry, 9*, 350. https://doi.org/10.3389/fpsyt.2018.00350.

Kreisler, A. D., Garcia, M. G., Spierling, S. R., Hui, B. E., & Zorrilla, E. P. (2017). Extended vs. brief intermittent access to palatable food differently promote binge-like intake, rejection of less preferred food, and weight cycling in female rats. *Physiology & Behavior, 177*, 305–316. https://doi.org/10.1016/j.physbeh.2017.03.039.

Krishna, S., Keralapurath, M. M., Lin, Z., Wagner, J. J., de La Serre, C. B., Harn, D. A., et al. (2015). Neurochemical and electrophysiological deficits in the ventral hippocampus and selective behavioral alterations caused by high-fat diet in female C57BL/6 mice. *Neuroscience, 297*, 170–181. https://doi.org/10.1016/j.neuroscience.2015.03.068.

Laessle, R. G., Tuschl, R. J., Kotthaus, B. C., & Pirke, K. M. (1989). Behavioral and biological correlates of dietary restraint in normal life. *Appetite, 12*(2), 83–94.

Levin, B. E., & Dunn-Meynell, A. A. (2002). Defense of body weight depends on dietary composition and palatability in rats with diet-induced obesity. *American Journal of Physiology – Regulatory, Integrative and Comparative Physiology, 282*(1), R46–R54. https://doi.org/10.1152/ajpregu.2002.282.1.R46.

Li, X., Morrow, D., & Witkin, J. M. (2006). Decreases in nestlet shredding of mice by serotonin uptake inhibitors: Comparison with marble burying. *Life Sciences, 78*(17), 1933–1939. https://doi.org/10.1016/j.lfs.2005.08.002.

Lowe, M. R., Doshi, S. D., Katterman, S. N., & Feig, E. H. (2013). Dieting and restrained eating as prospective predictors of weight gain. *Frontiers in Psychology, 4*, 577. https://doi.org/10.3389/fpsyg.2013.00577.

Loy, I., & Hall, G. (2002). Taste aversion after ingestion of lithium chloride: An associative analysis. *Quarterly Journal of Experimental Psychology B, 55*(4), 365–380. https://doi.org/10.1080/02724990244000070.

Markou, A., & Koob, G. F. (1991). Postcocaine anhedonia. An animal model of cocaine withdrawal. *Neuropsychopharmacology, 4*(1), 17–26.

Markou, A., & Koob, G. F. (1992). Construct validity of a self-stimulation threshold paradigm: Effects of reward and performance manipulations. *Physiology & Behavior, 51*(1), 111–119.

Martire, S. I., Westbrook, R. F., & Morris, M. J. (2015). Effects of long-term cycling between palatable cafeteria diet and regular chow on intake, eating patterns, and response to saccharin and sucrose. *Physiology & Behavior, 139*, 80–88. https://doi.org/10.1016/j.physbeh.2014.11.006.

Matias, I., & Di Marzo, V. (2007). Endocannabinoids and the control of energy balance. *Trends in Endocrinology and Metabolism, 18*(1), 27–37. https://doi.org/10.1016/j.tem.2006.11.006.

Mela, D. J. (2001). Determinants of food choice: Relationships with obesity and weight control. *Obesity Research, 9*(Suppl. 4), 249S–255S. https://doi.org/10.1038/oby.2001.127.

Meule, A. (2015). Back by popular demand: A narrative review on the history of food addiction research. *Yale Journal of Biology and Medicine, 88*(3), 295–302.

Micioni Di Bonaventura, M. V., Ciccocioppo, R., Romano, A., Bossert, J. M., Rice, K. C., Ubaldi, M., et al. (2014). Role of bed nucleus of the stria terminalis corticotrophin-releasing factor receptors in frustration stress-induced binge-like palatable food consumption in female rats with a history of food restriction. *Journal of Neuroscience, 34*(34), 11316–11324. https://doi.org/10.1523/JNEUROSCI.1854-14.2014.

Miles, F. J., Everitt, B. J., & Dickinson, A. (2003). Oral cocaine seeking by rats: Action or habit? *Behavioral Neuroscience, 117*(5), 927–938. https://doi.org/10.1037/0735-7044.117.5.927.

Moore, C. F., Panciera, J. I., Sabino, V., & Cottone, P. (2018). Neuropharmacology of compulsive eating. *Philosophical Transactions of the Royal Society of London B Biological Sciences, 373*(1742). https://doi.org/10.1098/rstb.2017.0024.

Moore, C. F., Sabino, V., Koob, G. F., & Cottone, P. (2017a). Neuroscience of compulsive eating behavior. *Frontiers in Neuroscience, 11*, 469. https://doi.org/10.3389/fnins.2017.00469.

Moore, C. F., Sabino, V., Koob, G. F., & Cottone, P. (2017b). Pathological overeating: Emerging evidence for a compulsivity construct. *Neuropsychopharmacology, 42*(7), 1375–1389. https://doi.org/10.1038/npp.2016.269.

Nawata, Y., Kitaichi, K., & Yamamoto, T. (2012). Increases of CRF in the amygdala are responsible for reinstatement of methamphetamine-seeking behavior induced by footshock. *Pharmacology Biochemistry and Behavior, 101*(2), 297–302. https://doi.org/10.1016/j.pbb.2012.01.003.

Njung'e, K., & Handley, S. L. (1991). Evaluation of marble-burying behavior as a model of anxiety. *Pharmacology Biochemistry and Behavior, 38*(1), 63–67.

Olmstead, M. C. (2006). Animal models of drug addiction: Where do we go from here? *Quarterly Journal of Experimental Psychology (Hove), 59*(4), 625–653. https://doi.org/10.1080/17470210500356308.

Ostlund, S. B., & Balleine, B. W. (2008). On habits and addiction: An associative analysis of compulsive drug seeking. *Drug Discovery Today: Disease Models, 5*(4), 235–245. https://doi.org/10.1016/j.ddmod.2009.07.004.

Oswald, K. D., Murdaugh, D. L., King, V. L., & Boggiano, M. M. (2011). Motivation for palatable food despite consequences in an animal model of binge eating. *International Journal of Eating Disorders, 44*(3), 203–211. https://doi.org/10.1002/eat.20808.

Overstreet, D. H., Knapp, D. J., & Breese, G. R. (2004). Modulation of multiple ethanol withdrawal-induced anxiety-like behavior by CRF and CRF1 receptors. *Pharmacology Biochemistry and Behavior, 77*(2), 405–413.

Pang, T. Y., Renoir, T., Du, X., Lawrence, A. J., & Hannan, A. J. (2013). Depression-related behaviours displayed by female C57BL/6J mice during abstinence from chronic ethanol consumption are rescued by wheel-running. *European Journal of Neuroscience, 37*(11), 1803–1810. https://doi.org/10.1111/ejn.12195.

Pankevich, D. E., Teegarden, S. L., Hedin, A. D., Jensen, C. L., & Bale, T. L. (2010). Caloric restriction experience reprograms stress and orexigenic pathways and promotes binge eating. *Journal of Neuroscience, 30*(48), 16399–16407. https://doi.org/10.1523/JNEUROSCI.1955-10.2010.

Parylak, S. L., Cottone, P., Sabino, V., Rice, K. C., & Zorrilla, E. P. (2012). Effects of CB1 and CRF1 receptor antagonists on binge-like eating in rats with limited access to a sweet fat diet: Lack of withdrawal-like responses. *Physiology & Behavior, 107*(2), 231–242. https://doi.org/10.1016/j.physbeh.2012.06.017.

Parylak, S. L., Koob, G. F., & Zorrilla, E. P. (2011). The dark side of food addiction. *Physiology & Behavior, 104*(1), 149–156. https://doi.org/10.1016/j.physbeh.2011.04.063.

Pecoraro, N., Reyes, F., Gomez, F., Bhargava, A., & Dallman, M. F. (2004). Chronic stress promotes palatable feeding, which reduces signs of stress: Feedforward and feedback effects of chronic stress. *Endocrinology, 145*(8), 3754–3762. https://doi.org/10.1210/en.2004-0305.

Pelloux, Y., Everitt, B. J., & Dickinson, A. (2007). Compulsive drug seeking by rats under punishment: Effects of drug taking history. *Psychopharmacology, 194*(1), 127–137. https://doi.org/10.1007/s00213-007-0805-0.

Piazza, P. V., & Deroche-Gamonet, V. (2013). A multistep general theory of transition to addiction. *Psychopharmacology, 229*(3), 387–413. https://doi.org/10.1007/s00213-013-3224-4.

Pigeyre, M., Yazdi, F. T., Kaur, Y., & Meyre, D. (2016). Recent progress in genetics, epigenetics and metagenomics unveils the pathophysiology of human obesity. *Clinical Science, 130*(12), 943–986. https://doi.org/10.1042/CS20160136.

Rada, P., Avena, N. M., & Hoebel, B. G. (2005). Daily bingeing on sugar repeatedly releases dopamine in the accumbens shell. *Neuroscience, 134*(3), 737–744. https://doi.org/10.1016/j.neuroscience.2005.04.043.

Rebolledo-Solleiro, D., Roldán-Roldán, G., Díaz, D., Velasco, M., Larqué, C., Rico-Rosillo, G., et al. (2017). Increased anxiety-like behavior is associated with the metabolic syndrome in non-stressed rats. *PLoS One, 12*(5), e0176554. https://doi.org/10.1371/journal.pone.0176554.

Reichelt, A. C., Morris, M. J., & Westbrook, R. F. (2014). Cafeteria diet impairs expression of sensory-specific satiety and stimulus-outcome learning. *Frontiers in Psychology, 5*, 852. https://doi.org/10.3389/fpsyg.2014.00852.

Rodd, Z. A., Bell, R. L., Sable, H. J., Murphy, J. M., & McBride, W. J. (2004). Recent advances in animal models of alcohol craving and relapse. *Pharmacology Biochemistry and Behavior, 79*(3), 439–450. https://doi.org/10.1016/j.pbb.2004.08.018.

Rodriguez de Fonseca, F., Carrera, M. R., Navarro, M., Koob, G. F., & Weiss, F. (1997). Activation of corticotropin-releasing factor in the limbic system during cannabinoid withdrawal. *Science, 276*(5321), 2050–2054.

Rossetti, C., Spena, G., Halfon, O., & Boutrel, B. (2014). Evidence for a compulsive-like behavior in rats exposed to alternate access to highly preferred palatable food. *Addiction Biology, 19*(6), 975–985. https://doi.org/10.1111/adb.12065.

Sabino, V., Cottone, P., Zhao, Y., Steardo, L., Koob, G. F., & Zorrilla, E. P. (2009). Selective reduction of alcohol drinking in Sardinian alcohol-preferring rats by a sigma-1 receptor antagonist. *Psychopharmacology, 205*(2), 327–335. https://doi.org/10.1007/s00213-009-1548-x.

Sarnyai, Z., Biro, E., Gardi, J., Vecsernyes, M., Julesz, J., & Telegdy, G. (1995). Brain corticotropin-releasing factor mediates 'anxiety-like' behavior induced by cocaine withdrawal in rats. *Brain Research, 675*(1–2), 89–97.

Satta, V., Scherma, M., Giunti, E., Collu, R., Fattore, L., Fratta, W., et al. (2016). Emotional profile of female rats showing binge eating behavior. *Physiology & Behavior, 163*, 136–143. https://doi.org/10.1016/j.physbeh.2016.05.013.

Schulteis, G., Yackey, M., Risbrough, V., & Koob, G. F. (1998). Anxiogenic-like effects of spontaneous and naloxone-precipitated opiate withdrawal in the elevated plus-maze. *Pharmacology Biochemistry and Behavior, 60*(3), 727–731.

Shalev, U., Yap, J., & Shaham, Y. (2001). Leptin attenuates acute food deprivation-induced relapse to heroin seeking. *Journal of Neuroscience, 21*(4), RC129.

Sharma, S., Fernandes, M. F., & Fulton, S. (2013). Adaptations in brain reward circuitry underlie palatable food cravings and anxiety induced by high-fat diet withdrawal. *International Journal of Obesity, 37*(9), 1183–1191. https://doi.org/10.1038/ijo.2012.197.

Sharma, S., & Fulton, S. (2013). Diet-induced obesity promotes depressive-like behaviour that is associated with neural adaptations in brain reward circuitry. *International Journal of Obesity, 37*(3), 382–389. https://doi.org/10.1038/ijo.2012.48.

Singh, L., Field, M. J., Vass, C. A., Hughes, J., & Woodruff, G. N. (1992). The antagonism of benzodiazepine withdrawal effects by the selective cholecystokininB receptor antagonist CI-988. *British Journal of Pharmacology, 105*(1), 8–10.

Skelton, K. H., Gutman, D. A., Thrivikraman, K. V., Nemeroff, C. B., & Owens, M. J. (2007). The CRF1 receptor antagonist R121919 attenuates the neuroendocrine and behavioral effects of precipitated lorazepam withdrawal. *Psychopharmacology, 192*(3), 385–396. https://doi.org/10.1007/s00213-007-0713-3.

Smith, K. L., Rao, R. R., Velazquez-Sanchez, C., Valenza, M., Giuliano, C., Everitt, B. J., et al. (2015). The uncompetitive N-methyl-D-Aspartate antagonist memantine reduces binge-like eating, food-seeking behavior, and compulsive eating: Role of the nucleus accumbens shell. *Neuropsychopharmacology, 40*, 1163–1171. https://doi.org/10.1038/npp.2014.299.

Solomon, R. L., & Corbit, J. D. (1974). An opponent-process theory of motivation. I. Temporal dynamics of affect. *Psychological Reviews, 81*(2), 119–145.

Spierling, S. R., Kreisler, A. D., Williams, C. A., Fang, S. Y., Pucci, S. N., Kines, K. T., et al. (2018). Intermittent, extended access to preferred food leads to escalated food reinforcement and cyclic whole-body metabolism in rats: Sex differences and individual vulnerability. *Physiology & Behavior, 192*, 3–16. https://doi.org/10.1016/j.physbeh.2018.04.001.

Srisai, D., Gillum, M. P., Panaro, B. L., Zhang, X. M., Kotchabhakdi, N., Shulman, G. I., et al. (2011). Characterization of the hyperphagic response to dietary fat in the MC4R knockout mouse. *Endocrinology, 152*(3), 890–902. https://doi.org/10.1210/en.2010-0716.

Stice, E., Davis, K., Miller, N. P., & Marti, C. N. (2008). Fasting increases risk for onset of binge eating and bulimic pathology: A 5-year prospective study. *Journal of Abnormal Psychology, 117*(4), 941–946. https://doi.org/10.1037/a0013644.

Tantot, F., Parkes, S. L., Marchand, A. R., Boitard, C., Naneix, F., Laye, S., et al. (2017). The effect of high-fat diet consumption on appetitive instrumental behavior in rats. *Appetite, 108*, 203–211. https://doi.org/10.1016/j.appet.2016.10.001.

Teegarden, S. L., & Bale, T. L. (2007). Decreases in dietary preference produce increased emotionality and risk for dietary relapse. *Biological Psychiatry, 61*(9), 1021–1029. https://doi.org/10.1016/j.biopsych.2006.09.032.

Thornton, L. M., Mazzeo, S. E., & Bulik, C. M. (2011). The heritability of eating disorders: Methods and current findings. *Current Topics in Behavioral Neurosciences, 6*, 141–156. https://doi.org/10.1007/7854_2010_91.

Timpl, P., Spanagel, R., Sillaber, I., Kresse, A., Reul, J. M., Stalla, G. K., et al. (1998). Impaired stress response and reduced anxiety in mice lacking a functional corticotropin-releasing hormone receptor 1. *Nature Genetics, 19*(2), 162–166. https://doi.org/10.1038/520.

Umathe, S., Bhutada, P., Dixit, P., & Shende, V. (2008). Increased marble-burying behavior in ethanol-withdrawal state: Modulation by gonadotropin-releasing hormone agonist. *European Journal of Pharmacology, 587*(1–3), 175–180. https://doi.org/10.1016/j.ejphar.2008.03.035.

Valenza, M., Steardo, L., Cottone, P., & Sabino, V. (2015). Diet-induced obesity and diet-resistant rats: Differences in the rewarding and anorectic effects of D-amphetamine. *Psychopharmacology, 232*(17), 3215–3226. https://doi.org/10.1007/s00213-015-3981-3.

Valles, A., Marti, O., Garcia, A., & Armario, A. (2000). Single exposure to stressors causes long-lasting, stress-dependent reduction of food intake in rats. *American Journal of Physiology – Regulatory, Integrative and Comparative Physiology, 279*(3), R1138–R1144. https://doi.org/10.1152/ajpregu.2000.279.3.R1138.

Vanderschuren, L. J., & Everitt, B. J. (2004). Drug seeking becomes compulsive after prolonged cocaine self-administration. *Science, 305*(5686), 1017–1019. https://doi.org/10.1126/science.1098975.

Velazquez-Sanchez, C., Ferragud, A., Moore, C. F., Everitt, B. J., Sabino, V., & Cottone, P. (2014). High trait impulsivity predicts food addiction-like behavior in the rat. *Neuropsychopharmacology, 39*(10), 2463–2472. https://doi.org/10.1038/npp.2014.98.

Velazquez-Sanchez, C., Santos, J. W., Smith, K. L., Ferragud, A., Sabino, V., & Cottone, P. (2015). Seeking behavior, place conditioning, and resistance to conditioned suppression of feeding in rats intermittently exposed to palatable food. *Behavioral Neuroscience, 129*(2), 219–224. https://doi.org/10.1037/bne0000042.

Wise, R. A. (1973). Voluntary ethanol intake in rats following exposure to ethanol on various schedules. *Psychopharmacologia, 29*(3), 203–210.

Yamada, N., Katsuura, G., Ochi, Y., Ebihara, K., Kusakabe, T., Hosoda, K., et al. (2011). Impaired CNS leptin action is implicated in depression associated with obesity. *Endocrinology, 152*(7), 2634–2643. https://doi.org/10.1210/en.2011-0004.

Yanovski, S. Z., Leet, M., Yanovski, J. A., Flood, M., Gold, P. W., Kissileff, H. R., et al. (1992). Food selection and intake of obese women with binge-eating disorder. *American Journal of Clinical Nutrition, 56*(6), 975–980.

Zhang, Z., & Schulteis, G. (2008). Withdrawal from acute morphine dependence is accompanied by increased anxiety-like behavior in the elevated plus maze. *Pharmacology Biochemistry and Behavior, 89*(3), 392–403. https://doi.org/10.1016/j.pbb.2008.01.013.

Zhang, D., Zhou, X., Wang, X., Xiang, X., Chen, H., & Hao, W. (2007). Morphine withdrawal decreases responding reinforced by sucrose self-administration in progressive ratio. *Addiction Biology, 12*(2), 152–157. https://doi.org/10.1111/j.1369-1600.2007.00068.x.

Zhao-Shea, R., DeGroot, S. R., Liu, L., Vallaster, M., Pang, X., Su, Q., et al. (2015). Increased CRF signalling in a ventral tegmental area-interpeduncular nucleus-medial habenula circuit induces anxiety during nicotine withdrawal. *Nature Communications, 6*, 6770. https://doi.org/10.1038/ncomms7770.

CHAPTER

Sex and gender differences in compulsive overeating

13

Karen K. Saules, Kirstie M. Herb

Eastern Michigan University, Department of Psychology, Ypsilanti, MI, United States

Obesity

From the late 1970s through the late 1990s, the prevalence of obesity increased significantly among US adult men and women (Flegal, Carroll, Kuczmarski, & Johnson, 1998); rates remained high from 1999 to 2008, but escalation of obesity prevalence seemed to be leveling off, particularly for women (Flegal, Carroll, Ogden, & Curtin, 2010). However, this trend reversed from 2005 to 2014, when statistically significant increases in obesity rates for women were observed, particularly for Class 3 obesity (i.e., body mass index [BMI] $\geq$40). Overall, current age-adjusted prevalence estimates for obesity are 35.0% among men and 40.4% among women; overall age-adjusted prevalence of Class 3 obesity is 7.7% among men, but 9.9% among women (Flegal, Kruszon-Moran, Carroll, Fryar, & Ogden, 2016).

Overeating

Before embarking on a discussion of various forms of excessive and compulsive eating behavior, it should be noted that escalating rates of obesity are not simply because of pathology at the level of the individual. Rather, increased obesity prevalence has become a global pandemic that some have attributed to worldwide increases in food production, with associated decreases in time cost and increases in affordability and energy density of foods (Swinburn et al., 2011). Corresponding to these shifts in the food landscape, mean daily energy intake has increased in US adults over time since the 1970s, rather dramatically through the late 1980s and persisting since (Briefel & Johnson, 2004). It has been argued that increased dietary intake is sufficient to explain the escalation in obesity rates in the United States (Swinburn, Sacks, & Ravussin, 2009).

Dietary intake, however, varies by sex. Not surprisingly, dietary intake is higher for men (Leblanc, Bégin, Corneau, Dodin, & Lemieux, 2015), and much of this may be related to metabolic needs associated with their overall larger size. However, such increased dietary intake could be accomplished through steady but modest consumption; on the contrary, more men than women admit to eating episodes that involve intake of large quantities of food (Ivezaj et al., 2010; Striegel-Moore et al., 2009),

Compulsive Eating Behavior and Food Addiction. https://doi.org/10.1016/B978-0-12-816207-1.00013-5
Copyright © 2019 Elsevier Inc. All rights reserved.

but women are more likely to endorse a sense of loss of control (LOC) over how much they eat (Striegel-Moore et al., 2009). This disparity will be discussed in more depth under the section below on Binge Eating.

Across studies, rates of regular overeating by men range from 10% to 51% (Lynch, Everingham, Dubitzky, Hartman, & Kasser, 2000; Saules et al., 2009; Striegel-Moore et al., 2009). Interestingly, although overeating is more common among men, it may contribute more to change in weight status for women. In one study, mean total energy intake explained increases in body weight from 1986 to 2000 for a large sample of UK women, but not for men (for whom the combination of increased intake and reduced activity was presumed to be in play; Scarborough et al., 2011).

In addition to variation in overall energy intake, variation in macronutrient intake has been explored. For example, men with obesity tend to identify high-fat, high-protein foods as their favorites, while women with obesity prefer high-fat, high-carbohydrate items, particularly those high in sugar (Drewnowski, Kurth, Holden-Wiltse, & Saari, 1992). Consistent with this report, a more recent study that asked college students to define "binge" foods found that women were more likely to mention sweets, whereas men were more likely to reference pizza (Reslan & Saules, 2011). These preferences translate into actual food intake, whereby women with obesity have been observed to consume more high-sugar, high-fat foods than men with obesity (but no differences were observed between nonobese men and women; Macdiarmid, Vail, Cade, & Blundell, 1998). Whether these sex differences are a cause, consequence, or covariate of obesity remains to be determined. However, primate animal models support that this sweet preference is likely physiologically based (Knott, 1999).

Disordered eating

In contrast to the pattern observed for simple overeating, disordered eating (DE) is more common among females, with the female:male ratio in rates of anorexia nervosa (AN), bulimia nervosa (BN), and binge eating disorder (BED) ranging from 2:1 to 10:1 (American Psychiatric Association, 2013). The likelihood of occurrence of various eating disorder (ED) symptoms is typically lowest in younger boys and highest in older girls/young women (Mond et al., 2014; Smink, van Hoeken, Oldehinkel, & Hoek, 2014). Although women are more likely than men to endorse a wide range of ED features (binge eating [BE], strict dieting, purging, etc.), men nonetheless represent 39% of those who report at least one ED feature (Mitchison, Mond, Slewa-Younan, & Hay, 2013). In addition, men are not immune to weight concerns and dieting. For example, it is estimated that about 26% of college men and 43% of college women report dieting to lose weight in the last 30 days, and 1% of college men and 4% of college women report vomiting or laxative use to lose weight (Matthews-Ewald, Zullig, & Ward, 2014).

DE and weight concerns appear to have different trajectories for men versus women. In a 20-year longitudinal study, women's weight perception and dieting frequency decreased over time, whereas men's weight perception and dieting frequency increased. From adolescence into midlife, DE declined, but it declined more among women than men (Keel, Baxter, Heatherton, & Joiner, 2007). With respect to the narrower time frame of college years, concerns related to eating, weight, and shape appear to be stable for men (Cain, Epler, Steinley, & Sher, 2012). With respect to adolescence, such concerns are stable and persistent for both boys and girls (Allen, Byrne, Oddy, & Crosby, 2013; Colton et al., 2015).

Although determinants of overweight/obesity are complex and multifactorial, a major contributor is DE behavior, which can take many forms. Beyond simply overeating (i.e., eating large amounts and/or until uncomfortably full, discussed above), obesogenic phenotypes of DE include grazing/nibbling/picking, BE (eating large amounts of food with a sense of LOC, i.e., subthreshold BED), emotional eating (EE), food addiction (FA), and formal EDs, most notably night eating syndrome (NES), BN, and BED.

This chapter will review what is known about sex/gender differences with respect to each of these variants of excessive/compulsive eating behavior. We focus first on variants where less is known about sex differences; then we will get into more depth on the large body of research on sex differences in BE and BED.

Grazing

Although definitions and terminology tend to vary across studies, grazing (also called snacking, picking, and nibbling) refers to unplanned and repetitive eating of relatively small amounts of food over extended periods of time (Conceição et al., 2014; Lane & Szabó, 2013), commonly resulting in consuming larger amounts of food over time than initially intended. Grazing is prevalent among those with obesity, in general (Heriseanu, Hay, Corbit, & Touyz, 2017), as well as among those with other forms of DE, such as BN and BED (Masheb, Grilo, & White, 2011; Masheb, Roberto, & White, 2013). It is common among bariatric surgery candidates (Burgmer et al., 2005) and as an emergent form of DE after bariatric surgery (Nicolau et al., 2015; Saunders, 2004).

Data on sex differences in grazing behavior are scant, but at least one study suggests this phenotype is more common among female bariatric candidates (25%) relative to their male counterparts (9%) (Burgmer et al., 2005). However, in a study of postbariatric patients who were 18+ months postsurgery, no differences were observed in rates of grazing for men versus women (Nicolau et al., 2015). Likewise, no sex differences were observed in a sample of obese individuals seeking treatment for BED (Masheb et al., 2013), and a recent systematic review found no support for gender differences in grazing behavior (Heriseanu et al., 2017).

Nocturnal eating

Nocturnal eating or NES was introduced under "Other Specified Feeding or Eating Disorder" in the DSM-5 (American Psychiatric Association, 2013). NES is characterized by recurrent episodes of eating after awakening from sleep or by excessive food intake after the evening meal, accompanied by awareness of the episodes and associated distress or impairment. Although new to the DSM, this syndrome was first described over 60 years ago as involving nocturnal hyperphagia, insomnia, and morning anorexia (Stunkard, Grace, & Wolff, 1955). NES is associated with higher BMI (Kucukgoncu, Tek, Bestepe, Musket, & Guloksuz, 2014; Meule, Allison, & Platte, 2014), particularly in middle age (Meule, Allison, Brähler, & de Zwaan, 2014), and is prevalent among bariatric surgery candidates (Allison et al., 2008; Baldofski et al., 2014; De Zwaan, Marschollek, & Allison, 2015), those with other forms of DE (Lundgren et al., 2011; Tholin et al., 2009), those with depression (Kucukgoncu et al., 2014), and psychiatric outpatients, in general (Saraçli et al., 2015).

With respect to sex differences, the literature indicates only small and inconsistent differences in rates of NE for men versus women across studies (Kucukgoncu et al., 2014; Meule, Allison, Brähler, et al., 2014; Meule, Allison, & Platte, 2014; Saraçli et al., 2015; Tholin et al., 2009). The coherence of NES symptoms, however, varies by sex, with stronger co-occurrence of symptoms among women and less endorsement of distress by men; specifically, among men, endorsement of evening hyperphagia does not necessarily or commonly co-occur with other NES symptoms, nor is it commonly associated with distress (Allison et al., 2014). Notably, when nocturnal eating is assessed simply as a behavior, minus other symptoms and associated distress, men are more likely to screen positive for NE than are women (Root et al., 2011). Also, again conceptualizing NE as a behavior only, heritability estimates are moderate for both men and women, without significant disparity (Root et al., 2011).

In addition, there are sex differences with respect to which NES symptoms most strongly differentiate NES versus non-NES groups, with increased morning anorexia, evening/nighttime cravings, insomnia, needing to eat to sleep, distress, and impairment differentiating men with NES versus non-NES cases, whereas frequent morning anorexia, insomnia, needing to eat to sleep, distress, and lack of control of evening eating more strongly differentiated female cases (Allison et al., 2014).

Emotional eating

Research within the domain of EE is more plentiful than that related to other forms of compulsive eating, perhaps due in part to its comparatively higher prevalence rate. Studies investigating EE, or the tendency to eat (typically, overeat) in response to one's own emotional experiences (van Strien, Frijters, Bergers, & Defares, 1986), reveal that EE rates are high among the general population and particularly among women (Braden, Musher-Eizenman, Watford, & Emley, 2018; Camilleri et al., 2014;

Konttinen, Männistö, Sarlio-Lähteenkorva, Silventoinen, & Haukkala, 2010; Péneau, Ménard, Méjean, Bellisle, & Hercberg, 2013). For example, one study revealed 89% of female participants and 75% of male participants endorsed at least some level of EE (Camilleri et al., 2014).

The evidence is mixed with respect to the specific emotions that drive eating behavior for men versus women (Braden et al., 2018; Masheb & Grilo, 2006). A study of overweight/obese adults found female sex is related to elevations on the depression and boredom subscales of the Emotional Eating Scale, a scale assessing one's proclivity to eat in response to negative affect (Arnow, Kenardy, & Agras, 1995; Braden et al., 2018). In contrast, Masheb and Grilo (2006) found that, among overweight individuals with BED, men and women did not differ in EE in response to sadness. With respect to other emotions, there are converging findings to suggest no sex differences in EE in relation to anxiety, anger, or positive emotions (Braden et al., 2018; Masheb & Grilo, 2006). Although the data are not conclusive, it appears that there are no sex differences in EE in response to positive emotions; however, the relationship between sex and eating in response to various negative emotions is less clear.

Branching off of this work is that which has honed in specifically on EE's relationship to depression. It is well established that EE is associated with depressive symptoms in both men and women (Camilleri et al., 2014; Konttinen et al., 2010; Lazarevich, Irigoyen Camacho, Velázquez-Alva, & Zepeda Zepeda, 2016). Furthermore, EE appears to mediate the relationship between depression and BMI in both sexes (Lazarevich et al., 2016), providing some evidence for DE's contribution to overweight/obesity. In line with this, EE is associated with the consumption of energy-dense snack food in women, particularly among those with depressive symptoms (Camilleri et al., 2014). Intriguingly, for men, this relationship was found only in those without depressive symptoms.

As is the case with general overeating, it should be no surprise that there are differences in nutritional and caloric intake between men and women who engage in EE. Total daily caloric intake, nonsweet food intake, and sweet food intake have been found to be associated with EE in both sexes, yet EE was related to carbohydrate and fat intake only in males (Konttinen et al., 2010). Logically, one might think these differential food intake patterns may lead to divergent associations with weight. Yet, research to date has shown EE is associated with BMI regardless of sex (Konttinen et al., 2010; Péneau et al., 2013). At least one study has shown, however, that this association is stronger among women, particularly those who are overweight or obese (Péneau et al., 2013).

In addition to these non–sex-specific factors, research into EE has focused on a facet unique to females: ovarian hormones. This work has established that EE tends to peak during the midluteal phase, a period of time in the menstrual cycle where progesterone and estradiol levels are high (Klump et al., 2013a). Further research has shown that it is an interaction between these two hormones that predicts within-subject changes in EE (Klump et al., 2013b). A different but related study also suggests that EE, but not ovarian hormones, accounts for within-person changes in preoccupation with weight across the menstrual cycle (Hildebrandt et al., 2015). Hormonal influences on eating behavior will be discussed in more depth, later in this chapter.

Food addiction

Despite a growing presence in both popular culture and the scientific literature, the concept of FA is not new. Rather, it has a long and rich history, albeit with renewed interest in recent years (Meule, 2015). In the general population, the prevalence of FA appears to be about 15%–20% (Pursey, Stanwell, Gearhardt, Collins, & Burrows, 2014; Schulte & Gearhardt, 2017). FA is associated with BMI (Hauck, Weiβ, Schulte, Meule, & Ellrott, 2017; Pedram et al., 2013; Pursey et al., 2014; Schulte & Gearhardt, 2017) and unsurprisingly the prevalence rate increases to 24.9% in overweight/obese individuals (Pursey et al., 2014). In regard to sex differences, the literature suggests that FA is more prevalent among women (Hauck et al., 2017; Pedram et al., 2013; Pursey et al., 2014). In fact, there is evidence to suggest that prevalence rates of FA in women may be double that seen for men (Pedram et al., 2013; Pursey et al., 2014). Notably though, there has been at least one finding to the contrary (Schulte & Gearhardt, 2017).

Even with its growing body of research, little is known in regard to other sex differences in FA; study samples tend to be disproportionately female or tend not to report results classified by gender. It has been found, however, that the number of symptoms endorsed on the Yale Food Addiction Scale (YFAS), a diagnostic measure of FA, does not vary by sex (Schulte & Gearhardt, 2017). The same study also revealed elevated YFAS symptom count is associated with higher BMI in women, but not men. Given that FA is characterized by an affinity for foods high in fats and refined carbohydrates that drive weight gain (Schulte, Avena, & Gearhardt, 2015) and that both animal and human models support that females have a stronger preference for these foods than males (Asarian & Geary, 2013), this finding is perhaps unsurprising.

Bulimia nervosa

BN is characterized by recurrent episodes of BE and inappropriate compensatory behaviors such as self-induced vomiting, laxative and diuretic abuse, and excessive exercise (American Psychiatric Association, 2013). The disorder is by far more common among females with an estimated female to male ratio of 10:1. More specifically, estimates based on DSM-IV-TR diagnostic criteria suggest a lifetime prevalence of 0.9%–1.5% in women and 0.1%–0.7% in men (Nicdao, Hong, & Takeuchi, 2007; Smink, van Hoeken, & Hoek, 2012). This discrepancy is apparent not only in adults but also among adolescents (Smink et al., 2012). Studies using the more lenient DSM-5 diagnostic criteria have reported slightly higher lifetime prevalence rates (Lindvall Dahlgren, Wisting, & Rø, 2017; Smink et al., 2012, 2014), but none of these studies have reported on the nature of sex differences. Presumably, the ratio of sex differences is similar to that seen when using DSM-IV-TR diagnostic criteria, but this has not yet been verified.

Because BN is overwhelmingly more prevalent in females, research within the domain of sex and gender differences is lacking. The few studies that have been conducted provide some insight into sex differences in predictors of bulimic symptoms. However, they often use nonclinical samples and lack replication. Nonetheless, at least one study indicates men engage in bulimic behaviors to reduce states of anger, whereas women use these behaviors to decrease the chance of anger arising (Meyer et al., 2005). Women in this study also scored higher on the Bulimic Investigatory Scale, Edinburgh, indicating more bulimic eating pathology than men. In a similar vein, a study examining the role of coping responses to social stress found involuntary disengagement (emotional numbing, inaction) to predict bulimic symptoms in men (Kwan et al., 2014), after controlling for prior bulimic symptoms, depressive symptoms, and body dissatisfaction. This relationship was not present among women or for the two other coping responses examined.

Binge eating

Both with respect to animal models and studies of human behavior, BE and BED are the most widely studied phenotypic variants of excessive/compulsive eating behavior. Most research focuses on BE, in part because it is the core dysfunctional behavior across the range of EDs (aside from AN) and associated with significant medical, psychiatric (Mitchison & Mond, 2015; Smink et al., 2012), and weight problems (Barry, Grilo, & Masheb, 2002; Nicdao et al., 2007; Sonneville et al., 2013; Striegel, Bedrosian, Wang, & Schwartz, 2012). It is also associated with high levels of impairment among both men and women (albeit significantly higher among women; Striegel et al., 2012). As such, coverage of BE-related sex and gender differences will be more comprehensive than our review of the previously described forms of eating behavior. Because the research on BE and BED is tightly intertwined, findings on both subthreshold BE and full BED will be covered in this section and referred to as such to differentiate when studies were addressing a subthreshold variant of BE versus full BED syndrome.

The sex difference seen in prevalence rates of other forms of compulsive eating is much less pronounced when it comes to BE; this divergence appears again, however, in BED prevalence rates. Several studies have suggested that more females report BE than males (Ivezaj et al., 2010; Lynch et al., 2000; Nicdao et al., 2007; Saules et al., 2009; Striegel-Moore et al., 2009), though some have found the opposite (Forrester-Knauss & Zemp Stutz, 2012; Hudson, Hiripi, Pope, & Kessler, 2007). Importantly, these differences are small in most studies (4.71% in women and 3.94% in men, for example; Nicdao et al., 2007), suggesting a close to equal prevalence for both males and females. Conversely, studies have consistently revealed significant differences in BED prevalence rates between men and women (Hudson,

Coit, Lalonde, & Pope, 2012; Ivezaj et al., 2010; Nicdao et al., 2007; Saules et al., 2009). The overall lifetime prevalence rate for BED is 2.9% and rates based on DSM-5 criteria suggest that 3.6% of women and 2.1% of men will meet diagnostic criteria for the disorder at some point in their life (Hudson, Coit, Lalonde, & Pope, 2012). Furthermore, the overall point prevalence rate is 1.25% (1.7% in women and 0.8% in men; Hudson et al., 2012). This divergence is evident by adolescence (Kjelsås, Bjørnstrøm, & Götestam, 2004; Sonneville et al., 2013).

Although considerable BED research was fueled by its inclusion in the DSM-IV-TR (American Psychiatric Association, 2000) as a provisional diagnostic category and now the DSM-5 (American Psychiatric Association, 2013) as an official diagnosis, there are few studies that provide insight into the nature of gender differences. The majority of the work has focused on women; men are grossly underrepresented or completely absent in most samples, and the majority of studies that have included men have not reported gender comparisons. A qualitative study of men's experiences of BE, however, revealed that men regard overeating as consistent with the male gender role, whereas LOC over eating is not (Carey, Saules, & Carr, 2017); failure to endorse the LOC criterion would automatically negate the possibility of a BED diagnosis, thereby accounting, at least in part, for the lower prevalence rate seen among males and their subsequent underrepresentation in BED research. This work parallels previous studies that report women are more likely to endorse the LOC criterion than their male counterparts both at the subthreshold (Striegel-Moore et al., 2009) and clinical level (Reslan & Saules, 2011). This lack of representation may also be due partly to divergent interpretations of the "definitely large" criterion of BE episodes, with at least one study finding that men often report higher thresholds for "definitely large" amounts of food than women (Forney, Holland, Joiner, & Keel, 2015). This suggests that men may be less likely to self-identify BE episodes, which ultimately would lead to lower prevalence rates and research representation. Together, these studies suggest gender differences in what constitutes a "binge" may contribute to the disproportionate prevalence rates of BED seen among men and women.

Qualitative studies have revealed a number of other themes regarding BE that may be worthy of more exploration in relation to gender differences. For example, a tendency toward "mindless" overeating, overeating in response to unintentional dietary restriction, social encouragement to overeat, and concerns about overweight only when it begins to impact physical functioning and health were commonly mentioned among male participants in one study (Carey et al., 2017). Other qualitative work has revealed differences in the types of foods identified during a BE episode (Reslan & Saules, 2011). More specifically, females with BED referenced sweets more often while pizza was frequently mentioned by males.

An additional area that has been a large focus of empirical research is the association between BE, BED, and weight. Although several studies have established a link between BE/BED and overweight/obesity (with those meeting BE or BED criteria having higher rates compared with their nonclinical counterparts; Barry et al., 2002; Sonneville et al., 2013; Striegel et al., 2012; Tanofsky, Wilfley, Spurrell,

Welch, & Brownell, 1997), nuances of the relationship have yet to be parsed out. For example, the issue of causality is a complicated one often discussed in the literature. Given that, in many cases, it cannot be known whether BE is a cause or outcome of overweight/obesity, it is difficult to determine the typical temporal order of the two. On one hand, BE may lead to increases in BMI (a finding that is supported by cross-sectional data revealing binge size increases with BMI (Guss, Kissileff, Devlin, Zimmerli, & Walsh, 2002; Picot & Lilenfeld, 2003). On the other hand, an individual with BMI in the overweight or obese range may have an innately elevated level of food intake, thus inherently increasing the probability of BE occurring. The longitudinal data needed to test these hypotheses are lacking and even more so in relation to gender differences. Nonetheless, some data suggest that men with BED have significantly higher BMIs than females (Barry et al., 2002), whereas other studies have reported no gender differences in BMI (Striegel et al., 2012; Tanofsky et al., 1997). These discrepant findings, combined with the fact that studies used cross-sectional and not longitudinal data, make it difficult to draw conclusions about the nature of sex and gender differences in relation to BE/BED and weight.

Sex differences in the BED diagnostic criteria

Criterion A of the DSM-5 diagnostic criteria for BED requires that recurrent episodes of BE must be present. It elaborates further that the BE must be characterized by both eating an objectively large amount of food and an accompanying sense of lack of control. These two components are directly relevant to the present discussion, and as reviewed in the previous section, there are gender differences in the understanding and endorsement of both of these (Carey et al., 2017; Forney et al., 2015; Reslan & Saules, 2011; Striegel-Moore et al., 2009).

Much less is known about sex and gender differences related to the second diagnostic criterion for BED (criterion B), which describes associated features that must accompany BE episodes. To date, and to our knowledge, only one study has examined this criterion. Work by White and Grilo (2011) indicates that the most common feature of BE for men is eating more rapidly than usual, whereas for women it is feeling disgusted, depressed, or very guilty after BE. This study also calculated total predictive value scores (TPVs), which served as a measure of the indicator's utility for correctly predicting BED diagnosis. For men, the feature with the greatest TPV was feeling disgusted, depressed, or very guilty after BE. For women, eating large amounts of food when not physically hungry and eating alone because of embarrassment were the two features with the highest TPVs. Furthermore, TPVs for all associated features of BE were high across both genders. Together, the results suggest that although the strongest predictor of BED diagnosis varies by gender, the five associated features within criterion B generally have good diagnostic efficiency for both men and women.

Turning to criterion C (marked distress associated with BE), little is known about sex differences as only one study to date has reported results by sex. Specifically, results from an internet survey of 1075 community volunteers revealed that men were disproportionately likely to be in the BE—no distress group, relative to their distribution among BED, BN, and a no binge/purge obese control group (Grilo & White, 2011). Although clinical impairment is not explicitly mentioned within the BED diagnostic criteria, it closely intertwined with distress and often discussed together in the literature. Considerably more work has investigated sex and gender differences in BED-related impairment, and it has consistently revealed it to be higher among women (Hudson et al., 2007; Striegel et al., 2012; Striegel-Moore et al., 2009; Udo et al., 2013). Notably though, the effect sizes for these gender differences were small in the one study that reported them (Striegel et al., 2012).

Sex differences in psychological comorbidities

Compared with the general population, a disproportionately high number of individuals with DE meet criteria for other comorbid psychological disorders. In fact, one study revealed 78.9% of individuals with BED to have a lifetime history of at least one of the following comorbid conditions: mood, anxiety, impulse control, and substance use disorders (Hudson et al., 2007). Furthermore, nearly 50% of individuals had at least three or more comorbid disorders. More recent work has supported this finding by reporting similar rates of additional lifetime psychiatric disorders (67%; Grilo, White, Barnes, & Masheb, 2013). The results of this work also revealed the presence of current psychiatric comorbidity is associated with greater eating disordered psychopathology and clinical impairment.

Several studies have been interested in investigating gender-related differences in rates of specific DSM disorders; substance use and anxiety disorders and depression have been a large focus of such studies. In general, treatment-seeking males with BED have reported higher rates of what were formerly Axis I disorders, and substance abuse/dependence in particular, as compared with their female counterparts (Barry et al., 2002; Tanofsky et al., 1997; Wilfley et al., 2000). It has been proposed, however, that the sex difference seen in comorbid rates of substance abuse/dependence is because of the base rate of these disorders being higher among men in general. Although one study of obese individuals with BED revealed higher rates of anxiety disorders and posttraumatic stress disorder among men (Grilo et al., 2013), the majority of research has failed to find statistically significant gender differences in rates of anxiety disorders and depression (Barry et al., 2002; Tanofsky et al., 1997; Wilfley et al., 2000).

Sex differences in biopsychosocial/sociocultural factors

Myriad biopsychosocial and sociocultural variables are associated with DE, some only at a correlational level and others more established longitudinally as risk factors. In particular, internationalization of the thin ideal, shape/weight concerns, negative emotionality, perfectionism, and negative urgency are risk factors for dysregulated eating patterns, with higher rates of DE among girls and women aligning with higher levels on these risk factors (Culbert, Racine, & Klump, 2015).

Despite DE being less prevalent among men (Lundahl, Wahlstrom, Christ, & Stoltenberg, 2015), ED symptoms among men are associated with many similar behavioral, psychological, and sociocultural factors as those observed among women, including attentional and motor impulsivity, excessive exercise, body dissatisfaction, self-objectification, appearance-ideal internalization, dieting, and negative affectivity (Dakanalis, Pla-Sanjuanelo et al., 2016; Dakanalis, Timko, et al., 2016; Lundahl et al., 2015; Mitchison & Mond, 2015).

In the general population, body dissatisfaction affects between 23% and 73% of women and 15% and 61% of men (Fiske, Fallon, Blissmer, & Redding, 2014). A nationally representative sample found that when asked "How do you feel about your body size right now?" significant gender differences emerged, with 28.4% of women responding "very satisfied," 48.0% "somewhat satisfied," and 23.6% "not satisfied," while men responded with 38.0% "very satisfied," 48.2% "somewhat satisfied," and 12.5% "not satisfied" (Kruger, Lee, Ainsworth, & Macera, 2008). Data from experimental and prospective studies support that media exposure to the ideal physique and pressures to be thin has comparable effects on ED-related symptoms and body dissatisfaction for men and women (Hausenblas et al., 2013).

Gender differences do emerge, however, with respect to other covariates of excessive eating and weight concerns. For example, in terms of behavioral features, women are more likely than men to engage in body checking (Reas, White, & Grilo, 2006) and body avoidance (Reas et al., 2006; Striegel-Moore et al., 2009), although men are not immune to body checking: about 20% of women and 10% of men admit to checking their body size at least "very often" (Striegel-Moore et al., 2009). With respect to cognitive elements, women and girls are more likely to endorse body image dissatisfaction (Grilo & Masheb, 2005) and overvaluation of weight/shape (Bentley, Mond, & Rodgers, 2014; Mitchison et al., 2013; Mond et al., 2014), relative to their male counterparts. However, some evidence suggests that overvaluation of weight/shape may be increasing in men (Hay, Mond, Buttner, & Darby, 2008), and it may be prominent among men who binge eat only when they also desire to lose weight (De Young, Lavender, & Anderson, 2010). Overvaluation is not simply a reflection of weight concerns; rather, it is strongly associated with forms of dysregulated eating (Hrabosky, Masheb, White, & Grilo, 2007). For both men and women, role impairment is greater among those with ED behaviors who endorse overvaluation of weight/shape, relative to those with ED behaviors without overvaluation of weight/shape (Mond & Hay, 2007).

Drive for thinness is well-documented to be elevated among women (Anderson & Bulik, 2004); a related factor that is relatively common among boys/men, however, is drive for muscularity, which is extremely high among males, in general (Valls, Bonvin, & Chabrol, 2013), and also associated with DE (Kelley, Neufeld, & Musher-Eizenman, 2010). For example, a longitudinal study observed that body discontent motivates dieting among adolescent boys/young men who desire to lose subcutaneous body fat that hides muscle mass (Dakanalis et al., 2015). Related constructs of muscle dysmorphia and muscularity-oriented excessive exercise are also elevated among male samples (dos Santos Filho, Tirico, Stefano, Touyz, & Claudino, 2016) as well, but the research base for these is less well established (Mitchison & Mond, 2015). Symptoms of muscle dysmorphia include dangerous body change behaviors such as excessive exercise, high caloric intake, preoccupation with eating and exercise regimens, use of supplements, and use of anabolic steroids, all of which are higher among men relative to women (Cafri et al., 2005). That said, women are not immune to having a drive for muscularity. Women who report internalizing an athletic body ideal report having high drive for muscularity (Pritchard & Cramblitt, 2014), which may be a reflection of shifting cultural ideals for women to include having a toned, fit body in addition to being thin (Tiggemann & Zaccardo, 2015).

With respect to predicting ED onset and maintenance, self-objectification (i.e., the tendency to experience one's body from an observer's perspective, mainly as an object to be valued for its appearance) emerged as the strongest predictor for men (Dakanalis, Pla-Sanjuanelo et al., 2016). Longitudinal studies with adolescent girls have shown that body dissatisfaction and dieting increase risk for ED development (Culbert et al., 2015b). In a longitudinal study of both boys and girls, media-ideal internalization predicted later self-objectification, which predicted later negative emotional responses to one's body and appearance, which, in turn, predicted later dietary restraint and BE. In that study, these associations held up across genders (Dakanalis et al., 2015).

Genetics

Genetic effects on DE, particularly BE, have been reported. Specifically, heritability estimates are about 40%–70% for both men and women (Klump et al., 2012; Munn-chernoff et al., 2013; Reichborn-Kjennerud et al., 2003; Root et al., 2011), but there is only about 50% overlap in the genes accounting for BE among males versus females (Baker et al., 2009b; Reichborn-Kjennerud et al., 2003), and the association between candidate risk genes and BE may be stronger for women than men (Baker et al., 2009b; Micali, Field, Treasure, & Evans, 2015).

It has been proposed that gonadal hormones, which regulate gene transcription across a number of neural systems implicated in BE, may differentially regulate or activate genes in a way that organizes and activates sex differences in adulthood (Klump, Culbert, & Sisk, 2017). For example, ovarian hormones have been shown to increase genetic risk for eating pathology, both during puberty and across the menstrual cycle, such that higher levels of circulating estradiol were associated with increased genetic influences on DE among adolescent girls (Klump, Keel, Sisk, & Burt, 2010).

Influences of gonadal hormones

Biological sex differences in weight and eating behavior are well-established, particularly within animal models, with more recent work shedding light on the more complex relationships that impact human eating behavior. For example, it was observed over a century ago that removal of the ovaries results in accumulation of fat tissue in rats (Stotsenburg, 1913); variation in dietary intake across the ovarian cycle has also been firmly established (Slonaker, 1925). In consideration of the aforementioned higher rates of many forms of DE behavior among women, it is reasonable to implicate hormonal processes in eating and weight regulation. Indeed, abundant research, particularly from animal models, supports that both testosterone and ovarian hormones influence food intake in adulthood (Asarian & Geary, 2013). Specifically, removal of rat ovaries (ovariectomy) increases rats' food intake and body weight; estradiol treatment normalizes these effects; removal of the testicles (orchiectomy) decreases food intake and body weight, and testosterone replacement normalizes these effects (see Asarian & Geary, 2013 for a comprehensive review). In addition to hormonal influences on food intake, in general, there is strong evidence for changes in macronutrient intake across the ovarian cycle, most notably a cyclic change in the intake of sweets, with increased sweet preference during the luteal phase (Dye & Blundell, 1997).

With respect to DE per se, there are little data on hormonal influences for most of the specific variants discussed above, with the exception of "DE," defined broadly, and BE, for which abundant research support exists. Correlational data exist for other hormones, but it is only with gonadal hormones that methods have been used to establish etiologic influences (Culbert, Racine, & Klump, 2016). Furthermore, it is gonadal hormones that are of direct relevance to the issues of sex differences in eating behavior; hence, that work will be the focus here. In particular, through the Michigan State University Twin Registry (MSUTR) and related research, Klump, Culbert, et al. have pioneered our efforts to understand the contributions of hormonal influences on BE, specifically, concluding that "sex differences in binge eating in adulthood are partially the result of the protective effects of testosterone in males and the risky effects of estrogen and progesterone in females" (p. 187) (Klump et al., 2017). The work leading to this conclusion will be reviewed next.

Animal models

Animal models vary in their definitions of BE, but they share an assumption of overconsumption of highly palatable (i.e., high fat/sugar) foods that are characteristic of a binge (e.g., frosting, Crisco oil, peanut butter chips, etc.). As with all animal models, correspondence with human behavior is debatable, but all models (e.g., Babbs, Wojnicki, & Corwin, 2011; Klump, Racine, Hildebrandt, & Sisk, 2013) attempt to include core BE features: ingestion of highly palatable food, intermittent access to preferred foods, and consumption over a short time period. Some models also incorporate LOC (i.e., responding under adverse conditions) (Velázquez-Sánchez et al., 2015) and/or exposure to stressors that increase BE in humans (Balsevich et al., 2014; Bello, Walters, Verpeut, & Caverly, 2014; Oswald, Murdaugh, King, & Boggiano, 2012).

Animal models support higher prevalence of BE among female rats (Babbs et al., 2011; Freund, Thompson, Norman, Einhorn, & Andersen, 2015; Klump et al., 2013). Interestingly, sex differences only emerge in BE models that provide intermittent access to preferred foods; such differences are not evident when animals are given daily access to palatable food (Hardaway et al., 2016; Reichelt, Abbott, Westbrook, & Morris, 2016). Collectively, these studies support that sex differences are specific to BE rather than simply reflecting differences in overeating, more generally.

To better understand sex differences in BE behavior, some studies have compared binge-prone (BEP) and binge-resistant (BER) animals. In breeding such animals, it has been found that rates of BE proneness are 5–14 times higher (mean = six times) in females than males, and rates of BE resistance are 3–11 times higher (mean = five times) in males relative to female rats (Klump et al., 2013). In addition to differences in rates of binge proneness, sex differences in behavioral phenotypes are observed, with evidence that male rats decrease chow intake on feeding test days to compensate for increased palatable food intake and to achieve homeostasis. In contrast, female BEP rats do not behaviorally compensate for palatable food intake, instead consuming equal chow in comparable fashion to their BER counterparts on feeding test days (Klump et al., 2013).

Puberty seems to set the stage for behavior to become sexually differentiated in this regard. For example, studies find little difference in chow intake for male versus female rats before puberty, but escalation of palatable chow intake by females by midpuberty (Klump et al., 2017). Age of onset of exposure to a binge-type diet also seems to impact binge-like behavior, but only among females. That is, female mice, particularly those exposed to a high-fat diet starting in adolescence, demonstrated the emergence of binge-like behavior when given restricted access to a palatable food. These effects were not observed in male mice (Carlin et al., 2016). In another study of pubertal exposure to highly palatable food (intermittent sucrose), female rats were more motivated to seek out sucrose as adults, whereas similarly exposed male rats were not (Reichelt et al., 2016).

Human studies

Earlier work on hormonal influences on DE observed that low levels of estradiol and high levels of progesterone were associated with increased BE and EE in women with BN (Edler, Lipson, & Keel, 2007) and among women in the general population (Klump, Keel, Culbert, & Edler, 2008). In follow-up work, interactions between estrogen and progesterone were observed among non-ED women, with the highest levels of EE observed during the midluteal phase (i.e., when levels of both estradiol and progesterone are high) (Klump et al., 2013); these relationships are stronger among women with BE compared with their non-BE counterparts (Klump et al., 2014). With respect to testosterone, consistent with findings from animal models, higher levels of circulating testosterone are associated with lower levels of DE in adolescent boys, but only with the emergence of puberty (Culbert, Burt, Sisk, Nigg, & Klump, 2014).

In contrast to studies that have used direct assessment of circulating gonadal hormones, another line of research has used proxy measures of early hormonal influences among humans, including finger length ratios (index finger (2D)/ring finger (4D)) and opposite-sex twin pair studies. With respect to the former, as a function of prenatal testosterone exposure (Manning, Scutt, Wilson, & Lewis-Jones, 1998), males typically have a shorter index finger in relation to their ring finger, i.e., a lower ratio. With respect to the latter, females of opposite-sex twin pairs experience greater intrauterine testosterone exposure relative to same-sex female twins.

Both types of studies provide support for protective effects of early testosterone exposure against DE. For example, using the MSUTR, Culbert, Breedlove, Burt, and Klump (2008) concluded that prenatal exposure to gonadal hormones may contribute to the substantial gender difference in the prevalence of EDs. This team examined opposite-gender twin pairs where the female twin shared a prenatal environment with her male co-twin and, therefore, was presumably exposed to testosterone in utero. Levels of DE showed the expected linear trend as a function of testosterone exposure, with same-gender female twins exhibiting the highest levels followed by females from opposite-gender pairs, males from opposite-gender pairs, and finally same-gender male twin pairs exhibiting the lowest rates of DE (not limited to BE). The authors speculated that prenatal testosterone exposure may affect DE, in part, by decreasing CNS sensitivity to gonadal hormones in adulthood. Others, however, have attempted to replicate this finding using the Swedish Twin study of Child and Adolescent Development and similar ED measures but did not observe these effects (Baker, Lichtenstein, & Kendler, 2009). Nonetheless, MSUTR data consistently support that lower levels of DE in opposite-sex twins relative to nontwin females support the contribution of prenatal testosterone because of exposure effects from being raised in utero with a brother (Klump et al., 2017). Specifically, these twin studies lend strong support for sex differences in eating behavior being due, at least in part, to hormonal effects, as opposed to simply being due to socialization effects of being raised with/without a brother.

In addition, results from the MSUTR demonstrate that during midlate puberty, females from opposite-sex twin pairs exhibit more masculinized (i.e., lower) DE attitudes than females from same-sex twin pairs (i.e., females with a female co-twin), even after controlling for possible confounds (e.g., BMI, anxiety) (Culbert, Breedlove, Sisk, & Burt, 2013).

Finger length studies also support that higher prenatal testosterone exposure (lower 2D:4D) predicts lower DE (Klump et al., 2006; Oinonen & Bird, 2012), but it is also associated with anorexia (Quinton, Smith, & Joiner, 2011). In contrast, higher 2D:4D predicts BN among women (Quinton et al., 2011). Taken together, findings suggest that testosterone may protect against excessive/compulsive eating, but not against highly restrained eating such as is prototypic of anorexia.

In addition, more recently, it was observed that higher prenatal testosterone (assessed by both lower 2D:4D and as females from opposite-sex twin pairs vs. controls) predicted lower DE in early adolescence and young adulthood only, with no evidence during late adolescence, suggesting differential windows of expression over development (Culbert et al., 2015a). This finding of developmental windows of expression may shed light on why some previous studies have failed to observe relationships between indicators of testosterone exposure and DE status (e.g., Baker et al., 2009a).

Consistent with animal models, among humans, pubertal timing also seems to make a contribution to BE risk, with early maturing girls showing higher rates of BE and BN in adulthood, an effect not observed among their male counterparts (Klump, 2013). Relatedly, Baker, Thornton, Bulik, Kendler, and Lichtenstein (2012) have reported that the genetic factors that influence younger age at menarche are associated with increased liability for DE. Among males, recent work suggests that gonadal hormone effects on DE may emerge earlier, during adrenarche, when androgen levels are increasing, before the observable physical changes of puberty (Culbert, Burt, & Klump, 2017).

Treatment outcome

Given the wide array of gender and sex differences in excessive eating behavior discussed above, it is rather surprising that treatment outcome studies have given scant attention to how gender relates to treatment outcome or how treatments might be best adapted to address gender differences. Of course, many studies are limited in their capacity to explore gender differences because treatment samples tend to be predominantly or entirely female (Clark et al., 2003; Conason et al., 2013; Higa, Ho, Tercero, Yunus, & Boone, 2011; Kruseman, Leimgruber, Zumbach, & Golay, 2010; Lammers, Vroling, Ouwens, Engels, & van Strien, 2015; Maggard et al., 2005; Munsch, Meyer, & Biedert, 2012; Obeid et al., 2012; Zizza, Herring, Stevens, & Carey, 2003). Unfortunately, even when studies do include a sizable proportion of men, they tend to neglect to report gender differences in treatment outcome (e.g., Compare et al., 2013).

A notable exception is the Swedish Obesity Study, which has followed a large sample of men and women who participated in either conventional or surgical weight loss treatment, finding that long-term weight reduction was about the same for men and women (Karlsson, Taft, Rydén, Sjöström, & Sullivan, 2007). Studies with 1- and 2-year follow-ups of postgastric restrictive bariatric procedures also observed no differences between men and women on BMI changes or eating behavior (Burgmer et al., 2005, 2007). With respect to weight regain after achieving nadir postsurgical weight, again, no gender differences emerge (Magro et al., 2008).

Another study aggregated data from 11 randomized controlled psychosocial treatment studies (N = 1325: 208 male, 1117 female) (Shingleton, Thompson-brenner, Thompson, Pratt, & Franko, 2016). Findings revealed no main effects for gender after controlling for baseline gender differences in various dimensions of shape, weight, and eating concerns (which were all higher among women). A treatment outcome study examining the impact of cognitive behavioral therapy on the reduction of the BE among postbariatric surgery candidates similarly found no gender differences in the treatment's effectiveness (Ashton, Drerup, Windover, & Heinberg, 2009). However, perhaps the most important conclusion from this effort is that treatment research, overall, lacks inclusion of men and should improve in this regard so that our understanding of potential treatment by gender interactions can be expanded.

Results from a small qualitative study of men in ED treatment (three with anorexic and five with bulimic presentations) mirror the lack of representation of men in the ED literature, including research on treatment outcomes. A prominent theme was the sense that men with EDs are invisible (Robinson, Mountford, & Sperlinger, 2013). Another qualitative study, again with a mix of anorexic and bulimic presentations, highlighted men's experiences of delayed help seeking and discounting of ED symptoms by health professionals (Räisänen & Hunt, 2014).

Summary and conclusions

This chapter reviews gender and sex differences across a range of disordered and excessive eating behaviors, leading to a few broad conclusions. First, although men consume more food, overall, and consume food in larger quantities per episode than women, it is women who are more likely to experience overeating as distressing and out of their control. In addition, women are more likely to engage in several behavioral phenotypes of DE studied to date, including grazing (Burgmer et al., 2005), EE (Braden et al., 2018), and BN (Smink et al., 2012), with a more inconsistent picture emerging for rates of BE (Hudson et al., 2007; Striegel-Moore et al., 2009), nocturnal eating (Allison et al., 2014), and food addition (Schulte & Gearhardt, 2017). Sociocultural factors, most notably the "thin ideal," undoubtedly contribute to gender differences in DE, but gonadal hormones and sex differences in food preferences also appear to play complex roles.

Second, gender differences notwithstanding, men are not immune to DE and related weight concerns. A qualitative study of men's experiences of BE (Carey et al., 2017), however, suggests that although men experience low levels of overvaluation of weight and shape and do not vigorously attempt to control their eating, this may change as a function of weight status impairing functional capacity. That is, it is not weight or size, per se, that is of concern to men, but when weight begins to impact physical functioning, it becomes concerning and sometimes alarming. Given high rates of overweight and obesity among both men and women, efforts to intervene and promote healthier eating are warranted, regardless of level of distress. In fact, it may be men's lack of distress that contributes to eventual weight problems, whereas among women, excessive weight concerns may lead to DE and consequent weight problems. Differences in dieting and weight concern trajectories over the lifespan align with these differences, with women experiencing dieting and weight concerns earlier in life, whereas men show increases on both variables later in life (Keel et al., 2007), presumably as weight increases.

Third, framing DE as being outside of one's control may fail to capture what is in fact DE among men, who tend to conceptualize excessive eating as something they do without thinking, in "mindless" fashion, rather than as a violation of efforts to control their eating (Carey et al., 2017). Fourth, the biopsychosocial and sociocultural correlates of DE vary by gender, most notably with women more likely to endorse general body dissatisfaction (Kruger et al., 2008) and a drive for thinness (Anderson & Bulik, 2004), whereas men are more likely to strive for muscularity (dos Santos Filho et al., 2016). Finally, treatment outcomes studies that include men are sorely lacking, leaving considerable gaps in our understanding of their potentially unique treatment needs, resulting in a recent "call to action" for greater inclusion of men in ED research (Murray, Griffiths, & Nagata, 2018). Murray et al. (2018) specifically recommend that this call to action includes a shift in the nature of questions routinely included in community-based studies, which tend to focus on thinness-related DE.

In conclusion, attention to the unique and shared features of DE among men and women has been growing in recent years, with greater appreciation of the need to include men in both community-based and treatment research studies. Continued efforts in this regard are warranted to broaden our understanding of gender and sex

Table 13.1 Summary of sex differences covered in this report.

Females
• Higher prevalence rates of obesity • Prefer high-fat, high-carbohydrate/sugary foods • Higher prevalence of anorexia nervosa • More likely to endorse loss of control over eating • Higher prevalence of bulimia nervosa • Higher prevalence of binge eating disorder • More likely to diet • More likely to endorse body dissatisfaction and overvaluation of weight/shape • More likely to engage in body checking and body avoidance • More likely to experience "drive for thinness" • Nocturnal eating is more distressing and more strongly associated with other symptoms • Higher prevalence of emotional eating • Higher prevalence of food addiction • More likely to eat in response to depression and boredom • Associations between eating behavior and ovarian hormones • Stronger associations between candidate risk genes and binge eating

Table 13.1 Summary of sex differences covered in this report.—*cont'd*

Females
Males
• Higher rates of overeating • Greater overall dietary intake • Prefer high-fat, high-protein foods • Lower likelihood of eating disorder symptoms • Higher comorbidity among treatment-seeking males • More likely to experience "drive for muscularity," muscle dysmorphia, and muscularity-oriented excessive exercise • Protective effects of early testosterone exposure against disordered eating
Equivocal
• Rates of grazing/snacking/picking • Rates of nocturnal eating • Association between emotional eating and depression • Relationship between binge eating and body mass index • Similar behavioral, psychological, and sociocultural co-factors • Comparable treatment outcomes

differences in excessive and compulsive forms of eating behavior that likely contribute to the high prevalence of overweight and obesity in the United States and beyond (Table 13.1).

References

Allen, K. L., Byrne, S. M., Oddy, W. H., & Crosby, R. D. (2013). DSM-IV-TR and DSM-5 eating disorders in adolescents: Prevalence, stability, and psychosocial correlates in a population-based sample of male and female adolescents. *Journal of Abnormal Psychology, 122*(3), 720–732. https://doi.org/10.1037/a0034004.

Allison, K. C., Lundgren, J. D., O'Reardon, J. P., Martino, N. S., Sarwer, D. B., Wadden, T. A., … Stunkard, A. J. (2008). The night eating questionnaire (NEQ): Psychometric properties of a measure of severity of the night eating syndrome. *Eating Behaviors, 9*(1), 62–72. https://doi.org/10.1016/j.eatbeh.2007.03.007.

Allison, K. C., Lundgren, J. D., Stunkard, A. J., Bulik, C. M., Lindroos, A. K., Thornton, L. M., & Rasmussen, F. (2014). Validation of screening questions and symptom coherence of night eating in the Swedish twin registry. *Comprehensive Psychiatry, 55*(3), 579–587. https://doi.org/10.1016/j.comppsych.2013.01.006.

American Psychiatric Association. (2000). *Diagnostic and statistical manual of mental disorders (IV-TR)*. Washington, DC: American Psychiatric Publishing.

American Psychiatric Association. (2013). *Diagnostic and statistical manual of mental disorders* (5th ed.). Arlington, VA: American Psychiatric Publishing.

Anderson, C. B., & Bulik, C. M. (2004). Gender differences in compensatory behaviors, weight and shape salience, and drive for thinness. *Eating Behaviors, 5*(1), 1–11. https://doi.org/10.1016/j.eatbeh.2003.07.001.

Arnow, B., Kenardy, J., & Agras, W. S. (1995). The emotional eating scale: The development of a measure to assess coping with negative affect by eating. *International Journal of Eating Disorders, 18*(1), 79–90.

Asarian, L., & Geary, N. (2013). Sex differences in the physiology of eating. *The Australian Journal of Pharmacy: Regulatory, Integrative and Comparative Physiology, 305*(11), R1215–R1267. https://doi.org/10.1152/ajpregu.00446.2012.

Ashton, K., Drerup, M., Windover, A., & Heinberg, L. (2009). Brief, four-session group CBT reduces binge eating behaviors among bariatric surgery candidates. *Surgery for Obesity and Related Diseases, 5*(2), 257–262. https://doi.org/10.1016/j.soard.2009.01.005.

Babbs, R. K., Wojnicki, F. H. E., & Corwin, R. L. W. (2011). Sex differences in the effect of 2-hydroxyestradiol on binge intake in rats. *Physiology & Behavior, 103*(5), 508–512. https://doi.org/10.3174/ajnr.A1256.

Baker, J. H., Lichtenstein, P., & Kendler, K. S. (2009a). Intrauterine testosterone exposure and risk for disordered eating. *British Journal of Psychiatry, 194*(4), 375–376. https://doi.org/10.1192/bjp.bp.108.054692.

Baker, J. H., Maes, H. H., Lissner, L., Aggen, S. H., Lichtenstein, P., & Kendler, K. S. (2009b). Genetic risk factors for disordered eating in adolescent males and females. *Journal of Abnormal Psychology, 118*(3), 576–586. https://doi.org/10.1037/a0016314.Genetic.

Baker, J. H., Thornton, L. M., Bulik, C. M., Kendler, K. S., & Lichtenstein, P. (2012). Shared genetic effects between age at menarche and disordered eating. *Journal of Adolescent Health: Official Publication of the Society for Adolescent Medicine, 51*(5), 491. https://doi.org/10.1016/j.jadohealth.2012.02.013.

Baldofski, S., Rudolph, A., Tigges, W., Herbig, B., Jurowich, C., Kaiser, S., … Hilbert, A. (2014). Nonnormative eating behavior and psychopathology in prebariatric patients with binge-eating disorder and night eating syndrome. *Surgery for Obesity and Related Diseases, 11*(3), 621–626.

Balsevich, G., Uribe, A., Wagner, K. V., Hartmann, J., Santarelli, S., Labermaier, C., & Schmidt, M. V. (2014). Interplay between diet-induced obesity and chronic stress in mice: Potential role of FKBP51. *Journal of Endocrinology, 222*(1), 15–26. https://doi.org/10.1530/JOE-14-0129.

Barry, D. T., Grilo, C. M., & Masheb, R. M. (2002). Gender differences in patients with binge eating disorder. *International Journal of Eating Disorders, 31*(1), 63–70. https://doi.org/10.1002/eat.1112.

Bello, N. T., Walters, A. L., Verpeut, J. L., & Caverly, J. (2014). Dietary-induced binge eating increases prefrontal cortex neural activation to restraint stress and increases binge food consumption following chronic guanfacine. *Pharmacology Biochemistry and Behavior, 125*, 21–28. https://doi.org/10.1016/j.pbb.2014.08.003.

Bentley, C., Mond, J., & Rodgers, B. (2014). Sex differences in psychosocial impairment associated with eating-disordered behavior: What if there aren't any? *Eating Behaviors, 15*(4), 609–614. https://doi.org/10.1016/j.eatbeh.2014.08.015.

Braden, A., Musher-Eizenman, D., Watford, T., & Emley, E. (2018). Eating when depressed, anxious, bored, or happy: Are emotional eating types associated with unique psychological and physical health correlates? *Appetite*. https://doi.org/10.1016/J.APPET.2018.02.022.

Briefel, R. R., & Johnson, C. L. (2004). Secular trends in dietary intake in the United States. *Annual Review of Nutrition, 24*, 401–431. https://doi.org/10.1146/annurev.nutr.23.011702.073349.

Burgmer, R., Grigutsch, K., Zipfel, S., Wolf, A. M., de Zwaan, M., Husemann, B., ... Herpertz, S. (2005). The influence of eating behavior and eating pathology on weight loss after gastric restriction operations. *Obesity Surgery, 15*(5), 684–691. https://doi.org/10.1381/0960892053923798.

Burgmer, R., Petersen, I., Burgmer, M., de Zwaan, M., Wolf, A. M., & Herpertz, S. (2007). Psychological outcome two years after restrictive bariatric surgery. *Obesity Surgery, 17*(6), 785–791.

Cafri, G., Thompson, J., Ricciardelli, L., McCabe, M., Smolak, L., & Yesalis, C. (2005). Pursuit of the muscular ideal: Physical and psychological consequences and putative risk factors. *Clinical Psychology Review, 25*(2), 215–239. https://doi.org/10.1016/j.cpr.2004.09.003.

Cain, A., Epler, A., Steinley, D., & Sher, K. (2012). Concerns related to eating, weight, and shape: Typologies and transitions in men during the college years. *International Journal of Eating Disorders, 45*(6), 768–775. https://doi.org/10.1007/s10439-011-0452-9.

Camilleri, G. M., Mejean, C., Kesse-Guyot, E., Andreeva, V. A., Bellisle, F., Hercberg, S., & Peneau, S. (2014). The associations between emotional eating and consumption of energy-dense snack foods are modified by sex and depressive symptomatology. *Journal of Nutrition, 144*(8), 1264–1273. https://doi.org/10.3945/jn.114.193177.

Carey, J. B., Saules, K. K., & Carr, M. M. (2017). A qualitative analysis of men's experiences of binge eating. *Appetite, 116*, 184–195. https://doi.org/10.1016/j.appet.2017.04.030.

Carlin, J. L., McKee, S. E., Hill-Smith, T., Grissom, N. M., George, R., Lucki, I., & Reyes, T. M. (2016). Removal of high-fat diet after chronic exposure drives binge behavior and dopaminergic dysregulation in female mice. *Neuroscience, 326*, 170–179. https://doi.org/10.1016/j.neuroscience.2016.04.002.

Clark, M. M., Balsiger, B. M., Sletten, C. D., Dahlman, K. L., Ames, G., Williams, D. E., ... Sarr, M. G. (2003). Psychosocial factors and 2-year outcome following bariatric surgery for weight loss. *Obesity Surgery, 13*(5), 739–745. https://doi.org/10.1381/096089203322509318.

Colton, P. A., Olmsted, M. P., Daneman, D., Farquhar, J. C., Wong, H., Muskat, S., & Rodin, G. M. (2015). Eating disorders in girls and women with type 1 diabetes: A longitudinal study of prevalence, onset, remission, and recurrence. *Diabetes Care, 38*(7), 1212–1217. https://doi.org/10.2337/dc14-2646.

Compare, A., Calugi, S., Marchesini, G., Shonin, E., Grossi, E., Molinari, E., & Dalle Grave, R. (2013). Emotionally focused group therapy and dietary counseling in binge eating disorder: Effect on eating disorder psychopathology and quality of life. *Appetite, 71*, 361–368. https://doi.org/10.1016/j.appet.2013.09.007.

Conason, A., Teixeira, J., Hsu, C.-H., Puma, L., Knafo, D., & Geliebter, A. (2013). Substance use following bariatric weight loss surgery. *JAMA Surgery, 148*(2), 145–150. https://doi.org/10.1001/2013.jamasurg.265.

Conceição, E. M., Mitchell, J. E., Engel, S. G., Machado, P. P. P., Lancaster, K., & Wonderlich, S. A. (2014). What is "grazing"? Reviewing its definition, frequency, clinical characteristics, and impact on bariatric surgery outcomes, and proposing a standardized definition. *Surgery for Obesity and Related Diseases, 10*(5), 973–982. https://doi.org/10.1016/j.soard.2014.05.002.

Culbert, K., Breedlove, S., Burt, S., & Klump, K. (2008). Prenatal hormone exposure and risk for eating disorders: Comparison of opposite-sex and same-sex twins. *Archives of General Psychiatry, 65*(3), 329–336. https://doi.org/10.1001/archgenpsychiatry.2007.47.

Culbert, K. M., Breedlove, S. M., Sisk, C. L., & Burt, S. A. (2013). The emergence of sex differences in risk for disordered eating attitudes during puberty: A role for prenatal testosterone exposure. *Journal of Abnormal Psychology, 122*(2), 420–432. https://doi.org/10.1037/a0031791.

Culbert, K. M., Breedlove, S. M., Sisk, C. L., Keel, P. K., Neale, M. C., Boker, S. M., … Klump, K. L. (2015a). Age differences in prenatal testosterone's protective effects on disordered eating symptoms: Developmental windows of expression? *Behavioral Neuroscience, 129*(1), 18–36. https://doi.org/10.1037/bne0000034.Age.

Culbert, K. M., Racine, S. E., & Klump, K. L. (2015b). Research review: What we have learned about the causes of eating disorders - a synthesis of sociocultural, psychological, and biological research. *Journal of Child Psychology and Psychiatry, 56*(11), 1141–1164. https://doi.org/10.1111/jcpp.12441.

Culbert, K. M., Burt, S. A., & Klump, K. L. (2017). Expanding the developmental boundaries of etiologic effects: The role of adrenarche in genetic influences on disordered eating in males. *Journal of Abnormal Psychology, 126*(5), 593–606. https://doi.org/10.1037/abn0000226.

Çulbert, K. M., Burt, S. A., Sisk, C. L., Nigg, J. T., & Klump, K. L. (2014). The effects of circulating testosterone and pubertal maturation on risk for disordered eating symptoms in adolescent males. *Psychological Medicine, 44*(11), 2271–2286. https://doi.org/10.1126/science.1249098.Sleep.

Culbert, K., Racine, S., & Klump, K. (2016). Hormonal factors and disturbances in eating disorders. *Current Psychiatry Reports, 18*(7). https://doi.org/10.1007/s11920-016-0701-6.

Dakanalis, A., Carrà, G., Calogero, R., Fida, R., Clerici, M., Zanetti, M. A., & Riva, G. (2015). The developmental effects of media-ideal internalization and self-objectification processes on adolescents' negative body-feelings, dietary restraint, and binge eating. *European Child & Adolescent Psychiatry, 24*(8), 997–1010. https://doi.org/10.1007/s00787-014-0649-1.

Dakanalis, A., Pla-Sanjuanelo, J., Caslini, M., Volpato, C., Riva, G., Clerici, M., & Carrà, G. (2016). Predicting onset and maintenance of men's eating disorders. *International Journal of Clinical and Health Psychology, 16*(3), 247–255. https://doi.org/10.1016/j.ijchp.2016.05.002.

Dakanalis, A., Timko, A., Serino, S., Riva, G., Clerici, M., & Carrà, G. (2016). Prospective psychosocial predictors of onset and cessation of eating pathology amongst college women. *European Eating Disorders Review, 24*(3), 251–256. https://doi.org/10.1002/erv.2433.

De Young, K. P., Lavender, J. M., & Anderson, D. A. (2010). Binge eating is not associated with elevated eating, weight, or shape concerns in the absence of the desire to lose weight in men. *International Journal of Eating Disorders, 43*(8), 732–736. https://doi.org/10.1002/eat.20779.

De Zwaan, M., Marschollek, M., & Allison, K. C. (2015). The night eating syndrome (NES) in bariatric surgery patients. *European Eating Disorders Review, 23*(6), 426–434. https://doi.org/10.1002/erv.2405.

Drewnowski, A., Kurth, C., Holden-Wiltse, J., & Saari, J. (1992). Food preferences in human obesity: Carbohydrates versus fats. *Appetite, 18*(3), 207–221.

Dye, L., & Blundell, J. (1997). Menstrual cycle and appetite control: Implications for weight regulation. *Human Reproduction, 12*, 1142–1151.

Edler, C., Lipson, S. F., & Keel, P. K. (2007). Ovarian hormones and binge eating in bulimia nervosa. *Psychological Medicine, 37*(1), 131–141. https://doi.org/10.1017/S0033291706008956.

Fiske, L., Fallon, E., Blissmer, B., & Redding, C. (2014). Prevalence of body dissatisfaction among United States adults: Review and recommendations for future research. *Eating Behaviors, 15*, 357–359. https://doi.org/10.1016/j.eatbeh.2014.04.010.

Flegal, K., Carroll, M., Kuczmarski, R., & Johnson, C. (1998). Overweight and obesity in the United States: Prevalence and trends, 1960–1994. *International Journal of Obesity and Related Metabolic Disorders, 22*(1), 39–47.

Flegal, K. M., Carroll, M. D., Ogden, C. L., & Curtin, L. R. (2010). Prevalence and trends in obesity among US adults , 1999–2008. *Journal of American Medical Association, 303*(3), 235–241. https://doi.org/10.1001/jama.2009.2014.

Flegal, K. M., Kruszon-Moran, D., Carroll, M. D., Fryar, C. D., & Ogden, C. L. (2016). Trends in obesity among adults in the United States, 2005 to 2014. *Journal of the American Medical Association, 315*(21), 2284–2291. https://doi.org/10.1001/jama.2016.6458.

Forney, K. J., Holland, L. A., Joiner, T. E., & Keel, P. K. (2015). Determining empirical thresholds for "definitely large" amounts of food for defining binge-eating episodes. *Eating Disorders, 23*(1), 15–30. https://doi.org/10.1080/10640266.2014.931763.

Forrester-Knauss, C., & Zemp Stutz, E. (2012). Gender differences in disordered eating and weight dissatisfaction in Swiss adults: Which factors matter? *BMC Public Health, 12*(1). https://doi.org/10.1186/1471-2458-12-809.

Freund, N., Thompson, B. S., Norman, K. J., Einhorn, P., & Andersen, S. L. (2015). Developmental emergence of an obsessive-compulsive phenotype and binge behavior in rats. *Psychopharmacology, 232*(17), 3173–3181. https://doi.org/10.1007/s00213-015-3967-1.

Grilo, C., & Masheb, R. (2005). Correlates of body image dissatisfaction in treatment-seeking men and women with binge eating disorder. *International Journal of Eating Disorders, 38*(2), 162–166. https://doi.org/10.1002/eat.20162.

Grilo, C. M., & White, M. A. (2011). A controlled evaluation of the distress criterion for binge eating disorder. *Journal of Consulting and Clinical Psychology, 79*(4), 509–514. https://doi.org/10.1037/a0024259.A.

Grilo, C. M., White, M. a, Barnes, R. D., & Masheb, R. M. (2013). Psychiatric disorder comorbidity and correlates in an ethnically diverse sample of obese patients with binge eating disorder in primary care settings. *Comprehensive Psychiatry, 54*(3), 209–216. https://doi.org/10.1016/j.comppsych.2012.07.012.

Guss, J. L., Kissileff, H. R., Devlin, M. J., Zimmerli, E., & Walsh, B. T. (2002). Binge size increases with body mass index in women with binge-eating disorder. *Obesity Research, 10*(10), 1021–1029. https://doi.org/10.1038/oby.2002.139.

Hardaway, J. A., Jensen, J., Kim, M., Mazzone, C. M., Sugam, J. A., Diberto, J. F., … Kash, T. L. (2016). Nociceptin receptor antagonist SB 612111 decreases high fat diet binge eating. *Behavioural Brain Research, 307*, 25–34. https://doi.org/10.1016/j.bbr.2016.03.046.

Hauck, C., Weiß, A., Schulte, E. M., Meule, A., & Ellrott, T. (2017). Prevalence of "food addiction" as measured with the Yale Food Addiction Scale 2.0 in a representative German sample and its association with sex, age and weight categories. *Obesity Facts, 10*(1), 12–24. https://doi.org/10.1159/000456013.

Hausenblas, H. A., Campbell, A., Menzel, J. E., Doughty, J., Levine, M., & Thompson, J. K. (2013). Media effects of experimental presentation of the ideal physique on eating disorder symptoms: A meta-analysis of laboratory studies. *Clinical Psychology Review, 33*(1), 168–181. https://doi.org/10.1016/j.cpr.2012.10.011.

Hay, P. J., Mond, J., Buttner, P., & Darby, A. (2008). Eating disorder behaviors are increasing: Findings from two sequential community surveys in South Australia. *PLoS One, 3*(2), 1–5. https://doi.org/10.1371/journal.pone.0001541.

Heriseanu, A. I., Hay, P., Corbit, L., & Touyz, S. (2017). Grazing in adults with obesity and eating disorders: A systematic review of associated clinical features and meta-analysis of prevalence. *Clinical Psychology Review.* https://doi.org/10.1016/j.cpr.2017.09.004.

Higa, K., Ho, T., Tercero, F., Yunus, T., & Boone, K. (2011). Laparoscopic Roux-en-Y gastric bypass: 10-year follow-up. *Surgery for Obesity and Related Diseases, 7*, 516–525.

Hildebrandt, B. A., Racine, S. E., Keel, P. K., Burt, S. A., Neale, M., Boker, S., ... Klump, K. L. (2015). The effects of ovarian hormones and emotional eating on changes in weight preoccupation across the menstrual cycle. *International Journal of Eating Disorders, 48*(5), 477–486. https://doi.org/10.1002/eat.22326.

Hrabosky, J. I., Masheb, R. M., White, M. a, & Grilo, C. M. (2007). Overvaluation of shape and weight in binge eating disorder. *Journal of Consulting and Clinical Psychology, 75*(1), 175–180. https://doi.org/10.1037/0022-006X.75.1.175.

Hudson, J. I., Coit, C. E., Lalonde, J. K., & Pope, H. G. (2012). By how much will the proposed new DSM-5 criteria increase the prevalence of binge eating disorder? *International Journal of Eating Disorders, 45*(1), 139–141. https://doi.org/10.1002/eat.20890.

Hudson, J. I., Hiripi, E., Pope, H. G., & Kessler, R. C. (2007). The prevalence and correlates of eating disorders in the National Comorbidity Survey Replication. *Biological Psychiatry, 61*(3), 348–358. https://doi.org/10.1016/j.biopsych.2006.03.040.

Ivezaj, V., Saules, K. K., Hoodin, F., Alschuler, K., Angelella, N. E., Collings, A. S., ... Wiedemann, A. A. (2010). The relationship between binge eating and weight status on depression, anxiety, and body image among a diverse college sample: A focus on bi/multiracial women. *Eating Behaviors, 11*(1), 18–24. https://doi.org/10.1016/j.eatbeh.2009.08.003.

Karlsson, J., Taft, C., Rydén, a, Sjöström, L., & Sullivan, M. (2007). Ten-year trends in health-related quality of life after surgical and conventional treatment for severe obesity: The SOS intervention study. *International Journal of Obesity, 31*(8), 1248–1261. https://doi.org/10.1038/sj.ijo.0803573.

Keel, P. K., Baxter, M. G., Heatherton, T. F., & Joiner, T. E. (2007). A 20-year longitudinal study of body weight, dieting, and eating disorder symptoms. *Journal of Abnormal Psychology, 116*(2), 422–432. https://doi.org/10.1037/0021-843X.116.2.422.

Kelley, C. C. G., Neufeld, J. M., & Musher-Eizenman, D. R. (2010). Drive for thinness and drive for muscularity: Opposite ends of the continuum or separate constructs? *Body Image, 7*(1), 74–77.

Kjelsås, E., Bjørnstrøm, C., & Götestam, K. G. (2004). Prevalence of eating disorders in female and male adolescents (14–15 years). *Eating Behaviors, 5*, 13–25. https://doi.org/10.1016/S1471-0153(03)00057-6.

Klump, K. (2013). Puberty is a critical risk period for eating disorders: A review of human and animal studies. *Hormones and Behavior, 64*(2), 399–410.

Klump, K., Culbert, K., & Sisk, C. (2017). Sex differences in binge eating: Gonadal hormone effects across development. *Annual Review of Clinical Psychology, 13*, 183–207. https://doi.org/10.1146/annurev-clinpsy-032816.

Klump, K., Culbert, K., Slane, J., Burt, S., Sisk, C., & Nigg, J. (2012). The effects of puberty on genetic risk for disordered eating: Evidence for a sex difference. *Psychological Medicine, 42*(3), 627–637. https://doi.org/10.1017/S0033291711001541.

Klump, K., Gobrogge, K., Perkins, P., Thorne, D., Sisk, C., & Breedlove, S. (2006). Preliminary evidence that gonadal hormones organize and activate disordered eating. *Psychological Medicine, 36*(4), 539–546. https://doi.org/10.1017/S0033291705006653.

Klump, K. L., Keel, P. K., Culbert, K. M., & Edler, C. (2008). Ovarian hormones and binge eating: Exploring associations in community samples. *Psychological Medicine, 38*(12), 1749–1757. https://doi.org/10.1017/S0033291708002997.

Klump, K., Keel, P., Racine, S., Burt, S., … Hu, J. (2013a). The interactive effects of estrogen and progesterone on changes in emotional eating across the menstrual cycle. *Journal of Abnormal Psychology, 122*(1), 131–137. https://doi.org/10.1037/a0029524.

Klump, K. L., Racine, S., Hildebrandt, B., & Sisk, C. L. (2013b). Sex differences in binge eating patterns in male and female adult rats. *International Journal of Eating Disorders, 46*(7), 729–736. https://doi.org/10.1002/eat.22139.

Klump, K., Keel, P., Sisk, C., & Burt, S. (2010). Preliminary evidence that estradiol moderates genetic influences on disordered eating attitudes and behaviors during puberty. *Psychological Medicine, 40*(10), 1745–1753. https://doi.org/10.1017/S0033291709992236.

Klump, K., Racine, S., Hildebrandt, B., Burt, S., Neale, M., Sisk, C., … Keel, P. (2014). Ovarian hormone influences on dysregulated eating: A comparison of associations in women with versus women without binge episodes. *Clinical Psychological Science, 2*(4), 545–559. https://doi.org/10.1177/2167702614521794.

Knott, C. (1999). Changes in orangutan caloric intake, energy homeostasis and ketones in response to fluctuating fruit availability. *International Journal of Primatology, 19*, 1061–1079.

Konttinen, H., Männistö, S., Sarlio-Lähteenkorva, S., Silventoinen, K., & Haukkala, A. (2010). Emotional eating, depressive symptoms and self-reported food consumption. A population-based study. *Appetite, 54*(3), 473–479. https://doi.org/10.1016/j.appet.2010.01.014.

Kruger, J., Lee, C.-D., Ainsworth, B. E., & Macera, C. A. (2008). Body size satisfaction and physical activity levels among men and women. *Obesity, 16*(8), 1976–1979. https://doi.org/10.1038/oby.2008.311.

Kruseman, M., Leimgruber, A., Zumbach, F., & Golay, A. (2010). Dietary, weight, and psychological changes among patients with obesity, 8 years after gastric bypass. *Journal of the American Dietetic Association, 110*(4), 527–534. https://doi.org/10.1016/j.jada.2009.12.028.

Kucukgoncu, S., Tek, C., Bestepe, E., Musket, C., & Guloksuz, S. (2014). Clinical features of night eating syndrome among depressed patients. *European Eating Disorders Review, 22*(2), 102–108. https://doi.org/10.1002/erv.2280.

Kwan, M. Y., Gordon, K. H., Eddy, K. T., Thomas, J. J., Franko, D. L., & Troop-Gordon, W. (2014). Gender differences in coping responses and bulimic symptoms among undergraduate students. *Eating Behaviors, 15*(4), 632–637. https://doi.org/10.1016/j.eatbeh.2014.08.020.

Lammers, M. W., Vroling, M. S., Ouwens, M. A., Engels, R. C. M. E., & van Strien, T. (2015). Predictors of outcome for cognitive behaviour therapy in binge eating disorder. *European Eating Disorders Review, 23*(3), 219–228. https://doi.org/10.1002/erv.2356.

Lane, B., & Szabó, M. (2013). Uncontrolled, repetitive eating of small amounts of food or "grazing": Development and evaluation of a new measure of atypical eating. *Behaviour Change, 30*, 57–73.

Lazarevich, I., Irigoyen Camacho, M. E., Velázquez-Alva, M. D. C., & Zepeda Zepeda, M. (2016). Relationship among obesity, depression, and emotional eating in young adults. *Appetite, 107*, 639–644. https://doi.org/10.1016/j.appet.2016.09.011.

Leblanc, V., Bégin, C., Corneau, L., Dodin, S., & Lemieux, S. (2015). Gender differences in dietary intakes: What is the contribution of motivational variables? *Journal of Human Nutrition and Dietetics, 28*(1), 37–46. https://doi.org/10.1111/jhn.12213.

Lindvall Dahlgren, C., Wisting, L., & Rø, Ø. (2017). Feeding and eating disorders in the DSM-5 era: A systematic review of prevalence rates in non-clinical male and female samples. *Journal of Eating Disorders, 5*, 56. https://doi.org/10.1186/s40337-017-0186-7.

Lundahl, A., Wahlstrom, L. C., Christ, C. C., & Stoltenberg, S. F. (2015). Gender differences in the relationship between impulsivity and disordered eating behaviors and attitudes. *Eating Behaviors, 18*, 120–124. https://doi.org/10.1016/j.eatbeh.2015.05.004.

Lundgren, J. D., McCune, A., Spresser, C., Harkins, P., Zolton, L., & Mandal, K. (2011). Night eating patterns of individuals with eating disorders: Implications for conceptualizing the night eating syndrome. *Psychiatry Research, 186*(1), 103–108. https://doi.org/10.1016/j.psychres.2010.08.008.

Lynch, W. C., Everingham, A., Dubitzky, J., Hartman, M., & Kasser, T. (2000). Does binge eating play a role in the self-regulation of moods? *Integrative Physiological and Behavioral Science, 35*(4), 298–313.

Macdiarmid, J., Vail, A., Cade, J., & Blundell, J. (1998). The sugar-fat relationship revisited: Differences in consumption between men and women of varying BM. *International Journal of Obesity and Related Metabolic Disorders, 22*, 1053–1061.

Maggard, M. A., Shugarman, L. R., Suttorp, M., Maglione, M., Sugerman, H. J., Livingston, E. H., … others. (2005). Meta-analysis: Surgical treatment of obesity. *Annals of Internal Medicine, 142*(7), 547.

Magro, D. O., Geloneze, B., Delfini, R., Pareja, B. C., Callejas, F., & Pareja, J. C. (2008). Long-term weight regain after gastric bypass: A 5-year prospective study. *Obesity Surgery, 18*(6), 648–651. https://doi.org/10.1007/s11695-007-9265-1.

Manning, J., Scutt, D., Wilson, J., & Lewis-Jones, D. (1998). The ratio of 2nd to 4th digit length: A predictor of sperm numbers and concentrations of testosterone, luteinizing hormone and oestrogen. *Human Reproduction, 13*(11), 3000–3004.

Masheb, R. M., & Grilo, C. M. (2006). Emotional overeating and its associations with eating disorder psychopathology among overweight patients with binge eating disorder. *International Journal of Eating Disorders, 39*(2), 141–146. https://doi.org/10.1002/eat.20221.

Masheb, R. M., Grilo, C. M., & White, M. (2011). An examination of eating patterns in community women with bulimia nervosa and binge eating disorder. *International Journal of Eating Disorders, 44*(7), 618–624. https://doi.org/10.1002/eat.20853.

Masheb, R. M., Roberto, C. A., & White, M. A. (2013). Nibbling and picking in obese patients with binge eating disorder. *Eating Behaviors, 14*(4), 424–427. https://doi.org/10.1016/j.eatbeh.2013.07.001.

Matthews-Ewald, M. R., Zullig, K. J., & Ward, R. M. (2014). Sexual orientation and disordered eating behaviors among self-identified male and female college students. *Eating Behaviors, 15*(3), 441–444. https://doi.org/10.1016/j.eatbeh.2014.05.002.

Meule, A. (2015). Back by popular demand: A narrative review on the history of food addiction research. *Yale Journal of Biology & Medicine, 88*(3), 295–302.

Meule, A., Allison, K. C., Brähler, E., & de Zwaan, M. (2014). The association between night eating and body mass depends on age. *Eating Behaviors, 15*(4), 683–685. https://doi.org/10.1016/j.eatbeh.2014.10.003.

Meule, A., Allison, K. C., & Platte, P. (2014). Emotional eating moderates the relationship of night eating with binge eating and body mass. *European Eating Disorders Review, 22*(2), 147–151. https://doi.org/10.1002/erv.2272.

Meyer, C., Leung, N., Waller, G., Perkins, S., Paice, N., & Mitchell, J. (2005). Anger and bulimic psychopathology: Gender differences in a nonclinical group. *International Journal of Eating Disorders, 37*(1), 69–71. https://doi.org/10.1002/eat.20038.

Micali, N., Field, A. E., Treasure, J. L., & Evans, D. M. (2015). Are obesity risk genes associated with binge eating in adolescence? *Obesity, 23*(8), 1729–1736. https://doi.org/10.1002/oby.21147.

Mitchison, D., & Mond, J. (2015). Epidemiology of eating disorders, eating disordered behaviour, and body image disturbance in males: A narrative review. *Journal of Eating Disorders, 3*(1), 1–9. https://doi.org/10.1186/s40337-015-0058-y.

Mitchison, D., Mond, J., Slewa-Younan, S., & Hay, P. (2013). Sex differences in health-related quality of life impairment associated with eating disorder features: A general population study. *International Journal of Eating Disorders, 46*(4), 375–380. https://doi.org/10.1002/eat.22097.

Mond, J., Hall, A., Bentley, C., Harrison, C., Gratwick-Sarll, K., & Lewis, V. (2014). Eating-disordered behavior in adolescent boys: Eating disorder examination questionnaire norms. *International Journal of Eating Disorders, 47*(4), 335–341. https://doi.org/10.1002/eat.22237.

Mond, J. M., & Hay, P. J. (2007). Functional impairment associated with bulimic behaviors in a community sample of men and women. *International Journal of Eating Disorders, 40*(5), 391–398. https://doi.org/10.1002/eat.20380.

Munn-chernoff, M. A., Duncan, A. E., Grant, J. D., Wade, T. D., Agrawal, A., Bucholz, K. K., … Heath, A. C. (2013). A twin study of alcohol dependence, binge eating , and compensatory behaivors. *Journal of Studies on Alcohol and Drugs, 74*, 664–673.

Munsch, S., Meyer, A. H., & Biedert, E. (2012). Efficacy and predictors of long-term treatment success for cognitive-behavioral treatment and behavioral weight-loss-treatment in overweight individuals with binge eating disorder. *Behaviour Research and Therapy, 50*(12), 775–785. https://doi.org/10.1016/j.brat.2012.08.009.

Murray, S. B., Griffiths, S., & Nagata, J. M. (2018). Community-based eating disorder research in males: A call to action. *Journal of Adolescent Health, 62*(6), 649–650. https://doi.org/https://doi.org/10.1016/j.jadohealth.2018.03.008.

Nicdao, E., Hong, S., & Takeuchi, D. (2007). Prevalence and correlates of eating disorders among Asian Americans: Results from the National Latino and Asian American Study. *International Journal of Eating Disorders, 40*, S22–S26. https://doi.org/10.1002/eat.

Nicolau, J., Ayala, L., Rivera, R., Speranskaya, A., Sanchís, P., Julian, X., … Masmiquel, L. (2015). Postoperative grazing as a risk factor for negative outcomes after bariatric surgery. *Eating Behaviors, 18*, 147–150. https://doi.org/10.1016/j.eatbeh.2015.05.008.

Obeid, A., Long, J., Kakade, M., Clements, R. H., Stahl, R., & Grams, J. (2012). Laparoscopic Roux-en-Y gastric bypass: Long term clinical outcomes. *Surgical Endoscopy*, 3515–3520. https://doi.org/10.1007/s00464-012-2375-4.

Oinonen, K., & Bird, J. (2012). Age at menarche and digit ratio (2D:4D): Relationships with body dissatisfaction, drive for thinness, and bulimia symptoms in women. *Body Image, 9*, 302–306.

Oswald, K. D., Murdaugh, D. L., King, V. L., & Boggiano, M. M. (2012). Motivation for palatable food despite consequences in an animal model of binge-eating. *International Journal of Eating Disorders, 44*(3), 203–211. https://doi.org/10.1002/eat.20808.Motivation.

Pedram, P., Wadden, D., Amini, P., Gulliver, W., Randell, E., Cahill, F., ... Sun, G. (2013). Food addiction: Its prevalence and significant association with obesity in the general population. *PLoS One, 8*(9), e74832. https://doi.org/10.1371/journal.pone.0074832.

Péneau, S. P., Ménard, E. M., Méjean, C. M., Bellisle, F. B., & Hercberg, S. H. (2013). Gender and dieting modify the association between emotional eating and weight status. *American Journal of Clinical Nutrition, 97*(6), 1307–1313. https://doi.org/10.3945/ajcn.112.054916.

Picot, A. K., & Lilenfeld, L. R. R. (2003). The relationship among binge severity, personality psychopathology, and body mass index. *International Journal of Eating Disorders, 34*(1), 98–107. https://doi.org/10.1002/eat.10173.

Pritchard, M., & Cramblitt, B. (2014). Media influence on drive for thinness and drive for muscularity. *Sex Roles, 71*(5–8), 208–218. https://doi.org/10.1007/s11199-014-0397-1.

Pursey, K. M., Stanwell, P., Gearhardt, A. N., Collins, C. E., & Burrows, T. L. (2014). The prevalence of food addiction as assessed by the Yale Food Addiction Scale: A systematic review. *Nutrients, 6*(10), 4552–4590. https://doi.org/10.3390/nu6104552.

Quinton, S. J., Smith, A. R., & Joiner, T. (2011). The 2nd to 4th digit ratio (2D:4D) and eating disorder diagnosis in women. *Personality and Individual Differences, 51*(4), 402–405. https://doi.org/10.1016/j.paid.2010.07.024.

Räisänen, U., & Hunt, K. (2014). The role of gendered constructions of eating disorders in delayed help-seeking in men: A qualitative interview study. *BMJ Open, 4*(4), 1–8. https://doi.org/10.1136/bmjopen-2013-004342.

Reas, D. L., White, M. A., & Grilo, C. M. (2006). Body checking questionnaire : Psychometric properties and clinical correlates in obese men and women with binge eating disorder. *International Journal of Eating Disorders, 39*, 326–331. https://doi.org/10.1002/eat.

Reichborn-Kjennerud, T., Bulik, C. M., Kendler, K. S., Roysamb, E., Maes, H., Tambs, K., & Harris, J. R. (2003). Gender differences in binge-eating: A population-based twin study. *Acta Psychiatrica Scandinavica, 108*(3), 196–202. https://doi.org/10.1034/j.1600-0447.2003.00106.x.

Reichelt, A. C., Abbott, K. N., Westbrook, R. F., & Morris, M. J. (2016). Differential motivational profiles following adolescent sucrose access in male and female rats. *Physiology & Behavior, 157*, 13–19. https://doi.org/10.1016/j.physbeh.2016.01.038.

Reslan, S., & Saules, K. K. (2011). College students' definitions of an eating "binge" differ as a function of gender and binge eating disorder status. *Eating Behaviors, 12*(3), 225–227. https://doi.org/10.1016/j.eatbeh.2011.03.001.

Robinson, K. J., Mountford, V. A., & Sperlinger, D. J. (2013). Being men with eating disorders: Perspectives of male eating disorder service-users. *Journal of Health Psychology, 18*(2), 176–186. https://doi.org/10.1177/1359105312440298.

Root, T. L., Thornton, L., Lindroos, A. K., Stunkard, A. J., Lichtenstein, P., Pedersen, N. L., ... Bulik, C. M. (2011). Shared and unique genetic and environmental influences on binge eating and night eating: A Swedish twin study. *Eating Behaviors, 11*(2), 92–98. https://doi.org/10.1016/j.eatbeh.2009.10.004.Shared.

dos Santos Filho, C., Tirico, P., Stefano, S., Touyz, S., & Claudino, A. (2016). Systematic review of the diagnostic category muscle dysmorphia. *Australian and New Zealand Journal of Psychiatry, 50*(4), 322–333. https://doi.org/10.1177/0004867415614106.

Saraçli, Ö., Atasoy, N., Akdemir, A., Güriz, O., Konuk, N., Sevinçer, G. M., ... Atik, L. (2015). The prevalence and clinical features of the night eating syndrome in psychiatric out-patient population. *Comprehensive Psychiatry, 57*, 79–84. https://doi.org/10.1016/j.comppsych.2014.11.007.

Saules, K. K., Collings, A. S., Hoodin, F., Angelella, N. E., Alschuler, K., Ivezaj, V., ... Wiedemann, A. A. (2009). The contributions of weight problem perception, BMI, gender, mood, and smoking status to binge eating among college students. *Eating Behaviors, 10*(1), 1–9. https://doi.org/10.1016/j.eatbeh.2008.07.010.

Saunders, R. (2004). "Grazing": A high-risk behavior. *Obesity Surgery, 14*(1), 98–102. https://doi.org/10.1381/096089204772787374.

Scarborough, P., Burg, M. R., Foster, C., Swinburn, B., Sacks, G., Rayner, M., ... Allender, S. (2011). Increased energy intake entirely accounts for increase in body weight in women but not in men in the UK between 1986 and 2000. *British Journal of Nutrition, 105*(9), 1399–1404. https://doi.org/10.1017/S0007114510005076.

Schulte, E. M., Avena, N. M., & Gearhardt, A. N. (2015). Which foods may be addictive? The roles of processing, fat content, and glycemic load. *PLoS One, 10*(2), 1–18. https://doi.org/10.1371/journal.pone.0117959.

Schulte, E. M., & Gearhardt, A. N. (2017). Associations of food addiction in a sample recruited to be nationally representative of the United States. *European Eating Disorders Review, 26*(2), 112–119. https://doi.org/10.1002/erv.2575.

Shingleton, R., Thompson-brenner, H., Thompson, D. R., Pratt, E. M., & Franko, D. L. (2016). Gender differences in clinical trials of binge eating disorder: An analysis of aggregated data. *Journal of Consulting and Clinical Psychology, 83*(2), 382–386. https://doi.org/10.1037/a0038849.Gender.

Slonaker, J. (1925). The effect of copulation, pregnancy, pseudopregnancy, and lactation on the voluntary activity and food consumption of the albino rat. *American Journal of Physiology, 71*, 362–394.

Smink, F. R. E., van Hoeken, D., & Hoek, H. W. (2012). Epidemiology of eating disorders: Incidence, prevalence and mortality rates. *Current Psychiatry Reports, 14*(4), 406–414. https://doi.org/10.1007/s11920-012-0282-y.

Smink, F. R. E., van Hoeken, D., Oldehinkel, A. J., & Hoek, H. W. (2014). Prevalence and severity of DSM-5 eating disorders in a community cohort of adolescents. *International Journal of Eating Disorders, 47*(6), 610–619. https://doi.org/10.1002/eat.22316.

Sonneville, K. R., Horton, N. J., Micali, N., Crosby, R. D., Swanson, S. A., Solmi, F., & Field, A. E. (2013). Longitudinal associations between binge eating and overeating and adverse outcomes among adolescents and young adults. *JAMA Pediatrics, 167*(2), 149. https://doi.org/10.1001/2013.jamapediatrics.12.

Stotsenburg, J. (1913). The effect of spaying and semi-spaying young albino rats (Mus norvegicus albinus) on the growth in body weight and body length. *The Anatomical Record, 7*, 183–194.

Striegel-Moore, R. H., Rosselli, F., Perrin, N., Debar, L., Wilson, G. T., May, A., & Kraemer, H. C. (2009). Gender difference in the prevalence of eating disorder symptoms. *International Journal of Eating Disorders, 42*(5), 471–474. https://doi.org/10.1002/eat.20625.

Striegel, R. H., Bedrosian, R., Wang, C., & Schwartz, S. (2012). Why men should be included in research on binge eating: Results from a comparison of psychosocial impairment in men and women. *International Journal of Eating Disorders, 45*(2), 233–240. https://doi.org/10.1002/eat.20962.

van Strien, T., Frijters, J., Bergers, G., & Defares, P. (1986). The Dutch Eating Behavior Questionnaire (DEBQ) for assessment of restrained, emotional, and external eating behavior. *International Journal of Eating Disorders, 5*(2), 295–315.

Stunkard, A. J., Grace, W. J., & Wolff, H. G. (1955). The night-eating syndrome. *The American Journal of Medicine, 19*(1), 78–86. https://doi.org/10.1016/0002-9343(55)90276-X.

Swinburn, B., Sacks, G., Hall, K., McPherson, K., Finegood, D., Moodie, M., & Gortmaker, S. (2011). The global obesity pandemic: Shaped by global drivers and local environments. *The Lancet, 378*(9793), 804–814. https://doi.org/10.1016/S0140-6736(11)60813-1.

Swinburn, B., Sacks, G., & Ravussin, E. (2009). Increased food energy supply is more than sufficient to explain the US epidemic of obesity. *American Journal of Clinical Nutrition, 90*, 1453–1456. https://doi.org/10.3945/ajcn.2009.28595.1.

Tanofsky, M. B., Wilfley, D. E., Spurrell, E. B., Welch, R., & Brownell, K. D. (1997). Comparison of men and women with binge eating disorder. *International Journal of Eating Disorders, 21*(1), 49–54.

Tholin, S., Lindroos, A. K., Tynelius, P., Åkerstedt, T., Stunkard, A. J., Bulik, C. M., & Rasmussen, F. (2009). Prevalence of night eating in obese and nonobese twins. *Obesity, 17*(5), 1050–1055. https://doi.org/10.1038/oby.2008.676.

Tiggemann, M., & Zaccardo, M. (2015). "Exercise to be fit, not skinny": The effect of fitspiration imagery on women's body image. *Body Image, 15*, 61–67.

Udo, T., McKee, S. a., White, M. a., Masheb, R. M., Barnes, R. D., & Grilo, C. M. (2013). Sex differences in biopsychosocial correlates of binge eating disorder: A study of treatment-seeking obese adults in primary care setting. *General Hospital Psychiatry, 35*(6), 587–591. https://doi.org/10.1016/j.genhosppsych.2013.07.010.

Valls, M., Bonvin, P., & Chabrol, H. (2013). Association between muscularity dissatisfaction and body dissatisfaction among normal-weight French men. *Journal of Men's Health, 10*(4), 139–145.

Velázquez-Sánchez, C., Santos, J., Smith, K., Farragud, A., Sabino, V., & Cottone, P. (2015). Seeking behavior, place conditioning and resistance to conditioned suppression of feeding in rats intermittently exposed to palatable food. *Behavioral Neuroscience, 129*(2), 219–224. https://doi.org/10.1037/bne0000042.Seeking.

White, M. A., & Grilo, C. M. (2011). Diagnostic efficiency of DSM-IV indicators for binge eating episodes. *Journal of Consulting and Clinical Psychology, 79*(1), 75–83. https://doi.org/10.1037/a0022210.

Wilfley, D. E., Friedman, M. a, Dounchis, J. Z., Stein, R. I., Welch, R. R., & Ball, S. a. (2000). Comorbid psychopathology in binge eating disorder: Relation to eating disorder severity at baseline and following treatment. *Journal of Consulting and Clinical Psychology, 68*(4), 641–649.

Zizza, C. A., Herring, A. H., Stevens, J., & Carey, T. S. (2003). Bariatric surgeries in North Carolina, 1990 to 2001: A gender comparison. *Obesity Research, 11*(12), 1519–1525. https://doi.org/10.1038/oby.2003.203.

CHAPTER

Addressing controversies surrounding food addiction

14

Gemma Mestre-Bach[1,2], Susana Jiménez-Murcia[1,2,4], Fernando Fernández-Aranda[1,2,4], Marc N. Potenza[3,5,6,7]

Department of Psychiatry. Bellvitge University Hospital-IDIBELL, Barcelona, Spain[1]; Ciber Fisiopatología Obesidad y Nutrición (CIBERObn), Instituto de Salud Carlos III, Madrid, Spain[2]; Department of Psychiatry, Yale School of Medicine, New Haven, CT, United States[3]; Department of Clinical Sciences, School of Medicine, University of Barcelona, Barcelona, Spain[4]; Connecticut Council on Problem Gambling, Wethersfield, CT, United States[5]; Connecticut Mental Health Center, New Haven, CT, United States[6]; Department of Neuroscience and Child Study Center, Yale School of Medicine, New Haven, CT, United States[7]

Introduction

Since 1980, the prevalence of obesity worldwide has nearly doubled according to the World Health Organization (WHO), with 13% of the adult population in 2016 categorized as obese. Likewise, 39% of the adult population was reported to be overweight (body mass index (BMI) greater than or equal to 25 but less than 30) or obese (BMI greater than or equal to 30) (George, 2018; WHO, 2017). Obesity prevalence is higher for women, especially for those between 60 and 64 years old who live in countries with a high sociodemographic index (GBD 2015 Obesity Collaborators et al., 2017). Taking children into account, in 2015, approximately 107.7 million individuals were obese worldwide, with an overall prevalence of 5.0% (GBD 2015 Obesity Collaborators et al., 2017). Although increases in obesity seem to originate in high-income countries, middle-income and low-income countries also contribute to the global increase in obesity prevalence in adults and children (Swinburn et al., 2011). Another associated factor is the widespread increase in BMI, typical of obesity, which has been described as a risk factor associated with a wide range of chronic diseases, such as diabetes, cardiovascular disease, and musculoskeletal disorders (GBD 2015 Obesity Collaborators et al., 2017).

This considerable increase in obesity rates reflects a complex and remarkable change in society. Although some nonfood, noneating factors, including more sedentary lifestyles, may be contributing, some authors have contended that the increase is mainly because of the encouragement of industrial countries to purchase a wide range of consumer goods, including highly palatable foods (Blundell & Finlayson, 2011). Therefore, changes in eating behaviors and a reduction in healthy eating patterns need to be understood better, including environments in which hyperpalatable foods,

Compulsive Eating Behavior and Food Addiction. https://doi.org/10.1016/B978-0-12-816207-1.00014-7
Copyright © 2019 Elsevier Inc. All rights reserved.

usually high in fat and sugars with chemical additives to enhance flavor, are abundant (Brunstrom & Cheon, 2018). Such modifications in food environments have been very diverse, although a common denominator may be an increase in availability and accessibility to energy-dense foods, as well as the significant reduction in the costs to consumers of such foods. Likewise, the significant reduction of physical activity and the expansion of a sedentary lifestyle in recent years could also be considered a factor associated with increases in BMI (Swinburn et al., 2011).

Many groups do not agree on the origin of the rapid proliferation of obesogenic environments, which has sometimes been described as an epidemic. One sector claims that the sharp increase in obesity rates can exclusively be considered as a personal preference, and not as a disease, because individuals have the capacity to make their own decisions regarding food consumption. However, epidemiological studies have identified a considerable number of people who would like or are actively trying to lose weight, but do not achieve a sustained and significant weight loss because of, among other factors, the current lifestyle in modern societies (Brunstrom & Cheon, 2018).

Perhaps given the association between obesity and food addiction, interest in the food addiction framework by the scientific community has appeared to increase. An increased interest may relate to this proliferation of obesogenic environments and to the higher prevalence of obesity across the globe, although the relationship between food and addiction dates back to the 20th century (Meule, 2015). Theron Randolph proposed the food addiction concept in 1956 (Randolph, 1956). Food addiction was described as "a specific adaptation to one or more regularly consumed foods to which a person is highly sensitive which produces a common pattern of symptoms descriptively similar to those of other addictive processes." In that definition, he specified some of the most problematic foods, such as "corn, wheat, coffee, milk, eggs, potatoes, and other frequently eaten foods" (Meule, 2015; Randolph, 1956). From this perspective, experts are currently trying to decipher if there are addictive components in processed foods or if certain eating behaviors could be considered as addictive, which could explain, to some extent, the considerable increase in the prevalence of obesity in the last decades. Although empirical evidence supports the presence of addiction-like symptoms to food in different populations, the food addiction model is not without detractors. A lack of consensus regarding the theoretical framework of the food addiction construct, its assessment, treatment options, and possible overlap with other disorders has hindered food addiction becoming a more substantial part of public policy debate regarding foods. This chapter attempts, therefore, to provide a critical and comprehensive overview of the controversies surrounding food addiction.

Food addiction controversies

Controversies in food addiction diagnosis

Diagnostic criteria for food addiction were not included in the DSM-5 (American Psychiatric Association, 2013), and this situation may hinder research, prevention, and treatment efforts. In the absence of validated assessment instruments, initial empirical studies aimed to evaluate food addiction by self-report. However, this assessment can have considerable limitations and biases, especially in populations with a lack of insight regarding the problem (Gearhardt, Corbin, & Brownell, 2009b). Some investigators have used a structured clinical interview based on the DSM-IV criteria for addiction or developed an interview taking into account Gudman's criteria for addictive disorders to assess the presence of food addiction (Dimitrijević, Popović, Sabljak, Škodrić-Trifunović, & Dimitrijević, 2015). This important limitation in food addiction diagnosis highlights the need to develop valid diagnostic tools to assess food addiction (Gearhardt et al., 2009b). However, controversies relating to the conceptualization and classification of food addiction persist (Gordon, Ariel-Donges, Bauman, & Merlo, 2018).

On the one hand, the Yale Food Addiction Scale (YFAS) was designed based on the substance-based model (Gearhardt et al., 2009b) and currently has become the most accepted and widely used instrument for evaluating food addiction in adult populations (Dimitrijević et al., 2015). With the aim of exhaustively evaluating the consumption of hyperpalatable and processed food, this scale was based on the DSM-IV-TR substance-dependence criteria (American Psychiatric Association, 2000) as well as other diagnostic tools to evaluate behavioral addictions. It is a 25-item scale, in dichotomous and Likert-type format, which assesses food intake in larger amount, unsuccessful attempts to stop food consumption, time to obtain food, interference in vital domains, continued use despite physical and psychological problems, tolerance, and withdrawal symptoms. The presence of three or more criteria indicates dependency (Gearhardt et al., 2009b). The YFAS has shown discriminant validity with alcohol and inhibition and activation psychometrical instruments, as well as good internal reliability and good convergent validity with different eating pathology tools (Gearhardt et al., 2009b). It has been validated in different languages and populations, such as Italian (Ceccarini, Manzoni, Castelnuovo, & Molinari, 2015; Innamorati et al., 2015; Manzoni et al., 2018), Spanish (Granero et al., 2014), French (Brunault, Ballon, Gaillard, Réveillère, & Courtois, 2014), Malay (Swarna Nantha, Abd Patah, & Ponnusamy Pillai, 2016), Portuguese (Torres et al., 2017), and Chinese (Chen, Tang, Guo, Liu, & Xiao, 2015).

Starting with the original version of the YFAS, other versions have been developed to obtain more updated (Gearhardt, Corbin, & Brownell, 2016), shorter (Flint et al., 2014; Schulte & Gearhardt, 2017), and target-specific (Gearhardt, Roberto, Seamans, Corbin, & Brownell, 2013) diagnostic tools. To update the original psychometric scale, a second edition of the YFAS, the YFAS 2.0, was generated based on DSM-5 criteria for substance-use disorders (Gearhardt et al., 2016). The main modifications to the 11 new criteria involve the elimination of the distinction between abuse and dependence, as well as the inclusion of a new indicator, craving (American Psychiatric Association, 2013). The YFAS 2.0 is, therefore, composed of 35 items that evaluate these criteria, and it appears to show better internal consistency than the original (Gearhardt et al., 2016). The YFAS 2.0 has also been validated in different languages and is widely used in research (Aloi et al., 2017; Granero et al., 2018; Meule, Müller, Gearhardt, & Blechert, 2017).

With the aim of reducing the length of the original to facilitate the evaluation of the food addiction construct, the 9-item modified Yale Food Addiction Scale (mYFAS) was designed (Flint et al., 2014). This is a 9-item questionnaire evaluating the 7 diagnostic criteria for substance dependence and clinical factors. Like with the YFAS, food addiction status is met if three or more dependency criteria and clinical distress are present (Flint et al., 2014). The mYFAS has demonstrated good test–retest reliability, suggesting that it may be an adequate substitute for the YFAS (Lemeshow, Gearhardt, Genkinger, & Corbin, 2016). This short version was also adapted to DSM-5 criteria with the mYFAS 2.0, a 13-item measure with good reliability. It is, therefore, an appropriate option as a briefer screening instrument for food addiction (Schulte & Gearhardt, 2017).

Given the neuropsychological vulnerability that children may exhibit at this developmental stage, some authors have suggested the need to create tools to assess for food addiction in this population (Gearhardt et al., 2013). For this purpose, the YFAS-C was designed, adapting the criteria of the YFAS for adults to activities specific to children and the level of understanding for children. The scale consists of 25 items associated with the 7 diagnostic criteria of the DSM-IV-TR (American Psychiatric Association, 2000), and a diagnosis of food addiction is met when 3 or more symptoms are present and clinically significant impairment or distress is detected (Gearhardt et al., 2013).

Although some authors suggest the continued use of the original YFAS and its new versions to study the clinical relevance of food addiction (Gearhardt et al., 2016), the use of these instruments has not been exempt from criticism. There is a much debate on whether the YFAS really measures a substance-based addiction or rather an eating addiction, among other things because the instrument does not define the specific substance of abuse (Hebebrand et al., 2014). Moreover, some authors consider that addiction symptoms, such as tolerance, withdrawal syndrome, or interference in daily activities, are not appropriate within the context of food addiction (Ziauddeen, Farooqi, & Fletcher, 2012). This has led some authors to consider that the ecological validity of the YFAS is limited because the instrument is specifically focused on a substance-based model and that, when comparing substance addiction and food addiction, relevant differences arise between both constructs (Ruddock, Christiansen, Halford, & Hardman, 2017). Finally, the dichotomous diagnostic approach has also been questioned, with some authors arguing that a diagnosis should not be determined by a self-report tool, but rather by an experienced clinician (Ruddock et al., 2017). Additionally, some authors contend that the YFAS includes a measure related to food addiction "loss of control with respect to eating behavior" that overlaps with one of the diagnostic criteria for two of the eating disorders (binge eating disorder and bulimia nervosa), "loss of control over eating." This may explain some controversies regarding the overlap between both disorders and the difficulties when trying to assess the prevalence of both clinical conditions. They also suggest that other food addiction symptoms, especially "repeated attempts to cut down food," may be easily applicable to the general population, especially taking current dietary practices into account (Burrows et al., 2017a,b).

On the other hand, and from the food addiction as an eating addiction perspective, the Addiction-Like Eating Behavior Scale (AEBS) was validated with the aim of having a diagnostic instrument that may facilitate the operationalization of the eating addiction construct, furthering debate in the food addiction field (Ruddock et al., 2017). After identifying behaviors related to food addiction, the AEBS was developed to quantify these addiction-like eating behaviors in community samples, without providing a dichotomous diagnostic criterion for eating addiction, unlike the YFAS. Its 15 items are organized following a two-factor scale structure. Factor 1 refers to the appetitive drive/motivation and is comprised of nine items, such as "I find it difficult to limit what/how much I eat," "I eat until I feel sick," "I continue to eat despite feeling full or once I start eating certain foods," and "I can't stop until there's nothing left." Factor 2 assesses low dietary control through six items such as "Despite trying to eat healthily, I end up eating 'naughty' foods," "I don't eat a lot of high fat/sugar foods," or "I tend not to buy processed foods that are high in fat and/or sugar." Both subscales showed good internal consistency (Ruddock et al., 2017). This instrument has not been exempt from criticism either, given that some authors consider that it does not clearly distinguish between eating addiction and food addiction, and that although behaviors related to excessive eating are evaluated, a more thorough evaluation would be required to categorize the eating behavior as an addiction (Schulte et al., 2017b).

Other tools aim to assess features associated with food addiction. One example is the Food Craving Questionnaires (FCQs). The Food Craving Questionnaire-State (FCQ-S) evaluates, by means of 15 items, craving intensity, desires to ingest specific types of foods, in the current moment (Cepeda-Benito, Gleaves, Fernández, et al., 2000; Cepeda-Benito, Gleaves, Williams, & Erath, 2000). On the other hand, the Food Craving Questionnaire-Trait (FCQ-T) aims to assess, using 39 items, intensity and frequency of food craving at a general level (Cepeda-Benito, Gleaves, Fernández, et al., 2000; Cepeda-Benito, Gleaves, Williams, et al., 2000). Moreover, the FCQT-reduced (FCQ-T-r) has been designed to predict food cravings in daily life through 15 items and a 6-point scale (Meule, Hermann, & Kübler, 2014). Using this last instrument, along with the YFAS, Meule (2018) suggested a potential cutoff point of FCQ-T-r between 32 and 54 to differentiate between presence and absence of food addiction. Using a cutoff of 50, high sensitivity and specificity for discriminating presence and absence of food addiction were found.

From the same perspective, and to measure general and specific food craving in disturbed eating patterns, the Food-Craving Inventory (FCI) was designed. It is a 28-item questionnaire that operationalizes and assesses craving as intense desire to consume certain food, which is difficult to resist. The FCI uses a 5-point Likert scale (from 1 to 5) and it is divided into four subscales, measuring specific kinds of food craving: high-fat foods, sweets, complex carbohydrates/starches, and fast-food fats.

This instrument has demonstrated adequate internal reliability and concurrent validity in both clinical and community samples (White, Whisenhunt, Williamson, Greenway, & Netemeyer, 2002).

Another example of scales designed to assess features related to food addiction is the Brief Measure of Eating Compulsivity (MEC), oriented toward the evaluation of another component of addiction, compulsivity. The authors suggest that the YFAS only evaluates this factor as a part of a wider set of addiction features and that other instruments are needed to investigate more systematically compulsive aspects of food addiction. The MEC10 is, therefore, a brief 10-item instrument focused on the evaluation of compulsive eating. Validation of the tool showed excellent test–retest reliability and high internal consistency. Furthermore, scores on the MEC10 have been found to be statistically predictive of being diagnosed with food addiction based on YFAS criteria (Schroder, Sellman, & Adamson, 2017).

The studies included in this chapter have used different evaluation tools for food addiction, although the YFAS has been employed in most studies.

Controversies surrounding the existence of food addiction and its categorization

Although research has increased significantly in recent years, the existence of food addiction and classification of the construct remain controversial (Onaolapo & Onaolapo, 2018). Although other possibilities exist, three main positions can be found in the scientific literature: that food addiction should be considered as a behavioral/eating addiction; that it should be categorized as a substance addiction; and a third approach questioning the previous two (i.e., whether "food addiction" is a real entity). The first two positions agree that addictive-like eating behavior is possible, although both differ in the specific role that food as a substance may have in perpetuating compulsive eating (Schulte, Potenza, & Gearhardt, 2017a, 2017b).

On the one hand, it can be argued that food addiction could be conceptualized as a behavioral/eating addiction. What differentiates this perspective from the view of considering food addiction as a substance addiction would be to consider that foods do not directly contribute to the characteristic phenotype of addiction (Schulte et al., 2017a, 2017b). Likewise, some authors suggest that there are significant differences between food and addictive substances, especially in the intensity of their effects on neurobiological processes, with drugs argued to having longer-lasting and impactful effects (Ruddock et al., 2017). This model focuses on overeating behavior per se and shifts attention away from the importance of the properties of food itself (Hebebrand et al., 2014). As with other behaviors, eating may become an addiction in predisposed individuals under specific environmental circumstances (Hebebrand et al., 2014). Overeating often occurs independently from feelings of hunger and may be triggered by negative emotional states such as boredom or high levels of anxiety, suggesting that eating behavior may be separated from the nutritional features of food (Hebebrand et al., 2014). Some authors argue that eating behavior is intrinsically rewarding and that the processes involved in eating behavior, such as appetite or satiation, among others, are already associated with the brain reward pathways (Bellisle, Drewnowski, Anderson, Westerterp-Plantenga, & Martin, 2012). Other theories argue that appetitive reward systems are inherently associated with regulatory systems (Wiers et al., 2007) and that an evident risk factor for the development of a behavioral addiction would constitute being highly sensitive to reward-related stimuli (Bechara, 2005).

On the other hand, a substance-based model has been proposed. It asserts that certain foods, most likely those including large amounts of processed fats and sugars, are capable of promoting overeating and addictive-like behaviors by activating brain reward systems, especially in certain risk population groups (Schulte, Avena, & Gearhardt, 2015). Such processed foods may be made more palatable by varying the fat content or adding substances such as biopolymers, soy protein sauce, carrageenan mix, flavor enhancers, or food sweeteners (Onaolapo & Onaolapo, 2018). From an evolutionary point of view, historically food was not processed, justifying the biologic propensity to seek energy-dense foods to increase energy stores as a preventative mechanism (Gearhardt, Corbin, & Brownell, 2009a). However, the relatively rapid changes in the food environment in which hyperpalatable, energy-dense foods are readily available and highly affordable may have changed advantageous phenotypes or characteristics to disadvantageous ones.

Examining this perspective in greater depth, similar neurobiological and behavioral changes have been observed in animal models of food and substance addictions (Schulte, Joyner, Potenza, Grilo, & Gearhardt, 2015). Specifically, animal models have found that animals exposed to hyperpalatable foods presented a pattern of behavior characterized by compulsive food seeking, binge eating, tolerance, and withdrawal (Carter et al., 2016). When sugar-enhanced food is removed, animals show a behavioral pattern similar to drug withdrawal, characterized by aggression, high anxiety levels, teeth chattering, and head shaking, suggesting that sugar intake could cross-sensitize to substances of abuse (Davis et al., 2011).

From a neurobiological perspective, repeated exposure to food or drugs has been associated with changes in opioid and dopaminergic systems, specifically with extracellular dopamine levels that are released in the nucleus accumbens (Gordon et al., 2018). Moreover, lower striatal availability of dopamine D2-like receptors have been found in some substance-use disorders and food addiction or obesity studies (Gearhardt et al., 2009a). Downregulation of these circuits may reduce sensitivity to rewards and may be associated with anhedonic states, which may lead individual to try to activate these circuits through drug-seeking or food-seeking behaviors (Carter et al., 2016). An incentive salience indicator for food or drugs may involve reactivity in these brain areas related to reward processing, which may lead to a propensity to gain weight and difficulty in obtaining food-intake moderation (Schulte, Yokum, Potenza, & Gearhardt, 2016). This model also suggests that food addiction would be associated with neurocognitive deficits such as impairments in executive cognitive control that may include maladaptive decision-making and poor emotional regulation (Carter et al., 2016).

Some authors claim that these findings, as well as data suggesting that certain individuals experience difficulties in reducing food intake despite negative health consequences, are consistent with a conceptualization of food addiction akin to a substance-use disorder (Gearhardt et al., 2009a). The DSM-IV-TR (American Psychiatric Association, 2000) defined substance dependence as a "cluster of cognitive, behavioral, and physiological symptoms associated with the continued use of the substance despite significant substance-related problems." If three or more of the seven criteria for substance dependence were met, generating clinically significant impairment, a diagnosis of substance dependence was given. These diagnostic criteria for dependence included tolerance, withdrawal, taking the substance often in larger amounts or over a longer period than was intended, a persistent desire or unsuccessful effort to cut down or control substance use, spending a great deal of time in activities necessary to obtain or use the substance or to recover from its effects, giving up social, occupational, or recreational activities because of substance use, and continuing the substance use with the knowledge that it is causing or exacerbating a persistent or recurrent physical or psychological problem. However, more research is needed to understand overeating and food addiction, taking these criteria into account (Gearhardt et al., 2009a).

Finally, some people have questioned how food or eating behavior may be considered as addictive since eating food constitutes a primary need for human survival (Schulte, Joyner, et al., 2015). This position reopens the debate on how best to define what degrees of involvement in a behavior may be considered normal versus excessive or problematic (Maraz, Király, & Demetrovics, 2015). This debate is valid not only in the field of food addiction but also in the conceptualization of other putative behavioral addictions such as hypersexual disorder or compulsive sexual behavior disorder (Kraus et al., 2018; Kraus, Voon, Kor, & Potenza, 2016; Kraus, Voon, & Potenza, 2016). In this particular case, there are also controversies about how to define "normal" sexual behavior because sex, like food, is a behavior that is essential for survival, at least from a species perspective. It has been suggested that two key factors to consider may involve assessing if control over behavior is impaired and if the behavior persists despite associated negative consequences (Kor, Fogel, Reid, & Potenza, 2013). Some authors have noted that many people with weight problems may not present a neurobiological profile that suggests addiction, and these authors argue that such a view undermines support for a proposed addiction model (Ziauddeen et al., 2012). Relatedly, other studies highlight that there is no consensus regarding the concept of "addiction" and that there is, therefore, not enough empirical evidence regarding the possible addictive potentials of certain foods (Gordon et al., 2018). Although the debate is still open, most agree that more research is needed (Gordon et al., 2018).

Controversies related to prevalence, comorbidities, and vulnerability factors

A recent systematic review found that prevalence estimates of food addiction in general populations have ranged from 5.4% to 56.8% (Pursey, Stanwell, Gearhardt, Collins, & Burrows, 2014). This considerable variation may in part be explained by differences in how the disorder is defined and the thresholds employed across studies. This finding is similar to the wide variation in prevalence estimates for internet gaming disorder before diagnostic criteria having been introduced in Section III of the DSM-5 (Petry & O'Brien, 2013). Other factors, such as differences in age, sex, BMI, or other measures, may also influence prevalence estimates. Taking sociodemographic and clinical features into account, some at-risk groups have been identified, with food addiction being more prevalent in older (Flint et al., 2014; Mies et al., 2017), female (Fattore, Melis, Fadda, & Fratta, 2014), and overweight (Lee, Hall, Lucke, Forlini, & Carter, 2014) populations. In addition, close relationships between food addiction and obesity (Davis et al., 2011; Fernández-Aranda, Steward, Mestre-Bach, & Jiménez-Murcia, Gearhardt, 2019), eating disorders (Hilker, Sánchez, et al., 2016; Wolz et al., 2016), addictions, both substances (Mies et al., 2017) and behavioral (Jiménez-Murcia et al., 2017), and other mental disorders (Goluza, Borchard, Kiarie, Mullan, & Pai, 2017) have been described.

Food addiction, obesity, and eating disorders

Although food addiction may occur in nonclinical populations and in people with a healthy BMI, an overlap of food addiction with eating disorders and obesity has been observed, with prevalence estimates between 34% and 40% in individuals with obesity (Ceccarini et al., 2015; Meule, Heckel, Jurowich, Vögele, & Kübler, 2014). Obesity is a heterogeneous phenotype, in which physiologic, genetic, psychological, and environmental features may contribute (De Ridder et al., 2016). Neurobiological parallels have been suggested between obesity and substance addiction, with lower activation in reward-related regions in relation to consumption and greater activation in motivation-related regions in response to cues (Gearhardt, Boswell, & White, 2014). De Ridder et al. (2016) found similar altered brain activity in a group with alcohol abuse and in a group with high levels of food addiction, involving the anterior cingulate cortex/dorsal medial prefrontal cortex, pregenual anterior cingulate cortex extending into the medial orbitofrontal cortex, parahippocampal area, and precuneus. These results suggest that both pathologies may be considered as addictions as they share neurophysiological substrates and parallels in genetic and behavioral characteristics. The authors also propose that there are distinguishable neurobiological mechanisms in people with obesity with or without food addiction.

An association between food addiction and eating disorders has also been described, reaching up to 50% the prevalence of food addiction in people with binge eating disorder (Burrows, Kay-Lambkin, Pursey, Skinner, & Dayas, 2018). Granero et al. (2014) found that those eating disorder groups that showed the highest food addiction prevalence were those with binge eating behaviors (bulimia nervosa, binge eating disorder, and anorexia nervosa binge eating–purging subtype). In other studies, high levels of food addiction were associated with eating disorder symptomatology and higher psychopathology (Gearhardt et al., 2013). People with co-occurring food addiction and binge eating disorder as compared with those with binge eating disorder alone experienced greater food craving, higher impulsivity, and more depressive symptomatology (Burrows, Skinner, McKenna, & Rollo, 2017a,b). As in the case of obesity, some authors suggest that the compulsive overeating characteristic of binge eating disorder has parallels with impaired control observed in individuals with substance-use disorders (Davis et al., 2011), which may explain why many women with binge eating disorder also meet criteria for drug dependence (Cassin & von Ranson, 2007).

Similarities between food addiction, obesity, and binge eating disorder may contribute to misunderstanding or debates regarding the food addiction construct (Gordon et al., 2018; Imperatori et al., 2016). Food addiction, like binge eating disorder, may contribute to weight gain, leading, in some cases, to obesity because of recurrent episodes of overeating, associated with a poor self-control (Fernández-Aranda, Steward, Mestre-Bach, Jiménez-Murcia, & Gearhardt, 2019). For this reason, some authors warn of the need to take into consideration differences in diagnosis, mainly between binge eating disorder and food addiction (Gearhardt, White, & Potenza, 2011b), and specifically the function that food fulfills for the individual, the circumstances in which eating behavior occurs, reactions when deprived of eating food, and awareness of the disorder (Bąk-Sosnowska, 2017). Other differences between both clinical conditions have been debated, suggesting that the function of eating in binge eating disorder may be to reduce distress, usually associated with cognitive distortions related to food, whereas in the case of food addiction, food may be more frequently used to induce hedonistic satisfaction and a sense of pleasure. Moreover, in binge eating disorder, there are episodes of disturbed eating behavior in which large amounts of food are consumed, even though the food is not perceived as a source of pleasure, whereas in food addiction, disturbing eating behavior may be more continuous and characterized by intense cravings for particular food, usually perceived as extremely palatable (Bąk-Sosnowska, 2017).

Moreover, although Davis et al. (2011) found a considerable overlap between food addiction and binge eating disorder in populations with high BMIs, they detected differences between both disorders, which suggested to the authors that food addiction and binge eating disorder are two independent clinical conditions. Furthermore, the data suggest that bingeing is not the only possible consumption pattern, but it is one of the several possible forms of excessive food intake. Other authors, however, suggest that food addiction could be considered as a subtype of binge eating disorder because of the observation, among other things, such that both disorders can occur in populations with a healthy weight, although they are more frequently observed in overweight or obese people (Davis, 2013).

Another clinical entity to consider is night eating syndrome. It is characterized by changes in appetite at different moments of the day, especially loss of appetite in the morning and hyperphagia, insomnia, and depressed mood in the evening associated with nocturnal ingestions. This syndrome is associated with emotional eating and binge eating disorder, although both are independent entities (Nolan & Geliebter, 2017). Nolan and Geliebter (2016) founded a positive relationship between food addiction, and night eating syndrome in both student and older community samples. The symptoms most associated with night eating syndrome were craving and food tolerance. Other studies have also found a positive association between measures of food addiction and scores on the Night Eating Diagnostic Questionnaire (NEDQ) (Nolan & Geliebter, 2017).

Food addiction and other addictions

Although the comorbidity between food addiction and behavioral addictions has not been examined in depth, an association with gambling disorder has been described, with a prevalence of 9.7% among individuals with a diagnosis of gambling disorder. When considering sex, the prevalence of food addiction was significantly greater in females (30.5%) than in males (6.02%), and all symptoms of food addiction symptoms were more prevalent in women (Jiménez-Murcia et al., 2017). This same study also found, through a predictive model for food addiction, that factors statistically predicting a food addiction diagnosis include being female, being younger, and having higher level of self-transcendence and harm avoidance. The authors suggested that patients with comorbidity between food addiction and gambling disorder showed a specific phenotype in comparison with those with only a gambling disorder diagnosis, highlighting the need for food addiction to be considered as an independent clinical condition.

On the other hand, and regarding substance-use disorders, Canan, Karaca, Sogucak, Gecici, and Kuloglu (2017) reported that patients with opioid-use disorder to heroin were likely to experience food addiction, and the co-occurrence of both disorders was related to higher craving levels and a higher likelihood of suicide attempts. Chao, White, Grilo, and Sinha (2017) evaluated substance cravings toward hyperpalatable foods, a food addiction symptom, in this population. Current smokers showed more total food craving, in comparison with former smokers and never smokers. Moreover, the consumption of fast-food fats was higher in the group of current smokers compared with the other two groups. However, the consumption of sweets and carbohydrates/starches did not differ by smoking status. Current smokers also showed higher depressive symptomatology and perceived stress in comparison with former smokers. Finally, some studies have examined the relationship between food addiction and substance use in patients having undergone bariatric surgery, obtaining mixed results (Ivezaj, Wiedemann, & Grilo, 2017). For example, one study found that those patients with substance misuse at the postoperative stage after bariatric surgery showed higher food addiction scores at the presurgical stage (Reslan, Saules, Greenwald, & Schuh, 2014), while others have not observed differences after bariatric surgery in terms of substance use according to food addiction (Clark & Saules, 2013).

Food addiction, other mental disorders, and associated factors

Research, to date, has not systematically examined the prevalence of many comorbid mental disorders among individuals with food addiction, and different assessment instruments have been used across existing studies (Burrows et al., 2018). However, the presence of higher psychopathology, such as depression and anxiety, has been described in patients meeting criteria for food addiction (Burrows et al., 2018).

The relationship between emotional eating and food addiction has yet to be examined in depth. Frayn, Sears, and von Ranson (2016) examined attention to food in women with food addiction, before and after a negative mood induction. Participants with food addiction showed higher attention to unhealthy food images and decreased attention to healthy food images after negative mood induction. The authors suggested that individuals with food addiction may be using unhealthy food images to improve negative moods. In the case of healthy control group, the pattern was different. Sad mood induction had no effect on total fixation times for either unhealthy or healthy food images, suggesting specific cognitive processes involved in food addiction.

In another study assessing comorbidity, a significant relation between food addiction, impulsivity, depression, anxiety, and stress in relation to BMI in people with type-two diabetes was found, with food addiction and nonplanning impulsivity being the two factors showing the strongest relationships with BMI (Raymond & Lovell, 2015). By considering impulsivity traits with the UPPS-P framework, patients with eating disorders that met criteria for food addiction have shown increased negative urgency, namely the tendency to engage in risky behaviors when experiencing negative emotions. This would imply that individuals with food addiction may use food intake when attempting to regulate emotional distress (Wolz, Granero, & Fernández-Aranda, 2017).

The association between the presence of posttraumatic stress disorder (PTSD) and food addiction has also been studied (Brewerton, 2017). A strong association between childhood physical and sexual abuse and food addiction in a female population has been described (Mason, Flint, Field, Austin, & Rich-Edwards, 2013). Similarly, in a sample of female and male trauma-exposed veterans, PTSD was positively associated with food addiction and eating disorder symptomatology (Mitchell & Wolf, 2016).

With respect to schizophrenia, Goluza et al. (2017) reported that the prevalence of food addiction was 26.9%, a higher figure than in the general population, although not as high as in other disorders, such as eating disorders. Moreover, the food addiction symptom that was most commonly reported was a persistent desire or repeated attempts to stop consumption.

Food addiction in children and adolescents

Recently, the availability of unhealthy food has increased significantly in environments in which children and adolescents have access, such as schools and supermarkets. Furthermore, there have been increases in food advertisements that may be very attractive for youth. So-called obesogenic environments may influence the nutritional choices of children and adolescents, promoting health problems such as obesity (Keser et al., 2015). However, very little is currently known about food addiction and caloric intake in children and adolescents. Addictive substances may interfere in the psychological and neuromaturational development of children and adolescents, making them more vulnerable to their negative effects (Lubman, Cheetham, & Yücel, 2015; Lydon, Wilson, Child, & Geier, 2014). Although food may have lower addictive potentials in comparison with some other substances, food consumption begins earlier, thereby promoting repeated consumption, which may raise the risk for disruptive eating behaviors and related negative health consequences, such as obesity and binge eating disorder, later in life (Burrows et al., 2017a,b).

An association between addictive-like eating patterns and BMI has been found in recent studies of children (Gearhardt et al., 2013). Higher tendencies to engage in emotional eating was found in this at-risk population, suggesting the presence of an addictive process related to problematic eating behaviors (Gearhardt et al., 2013). Other epidemiological research in children had found an association between food addiction scores and caloric consumption. This association has been observed particularly in younger children, and a possible explanation could be that, during this development stage, children present major difficulties in inhibiting their eating behavior. The authors suggest that older children may possibly be better able to inhibit overeating behavior in public eating contexts (Richmond, Roberto, & Gearhardt, 2017). Burrows et al. (2017a,b) found that 22.7% of children in the general population met food addiction criteria, a higher percentage than those described in previous studies (4%–7.2%) (Gearhardt et al., 2013; Laurent & Sibold, 2016). They also found a moderate association between the diagnosis of food addiction in children and the food addiction symptomatology of their parents. Keser et al. (2015) found that the most "addictive" foods included chocolate (70%), ice cream (58%), and carbonated beverages (59%), among others.

Adolescence is characterized by high impulsivity and low inhibitory control. As such, adolescents may be more vulnerable to addictive behaviors including substance use (Chuang et al., 2017). Mies et al. (2017) found in a representative sample of adolescents between 14 and 21 years that 2.6% met criteria for food addiction, a lower percentage in relation to children and adults. However, authors suggested that this difference could be because of the lack of studies in at-risk populations. In the case of adolescents with obesity, the percentage increases to 10%, although it is still lower than that of adults with obesity, with percentages around 20% (Schulte, Jacques-Tiura, Gearhardt, & Naar, 2017). These findings suggest that food addiction may follow different developmental trajectories than other addictions and that more research is needed in adolescents.

Controversies in treatment options for food addiction

Obesity is associated with high health care–related costs, reaching $215 billion annually in countries such as the United States (Hammond & Levine, 2010).

However, current public health interventions have not appeared to be effective in reducing obesity rates, highlighting an essential need to implement new treatment and prevention strategies (De Ridder et al., 2016).

Although there is an association between obesity and food addiction, few studies have examined relationships between food addiction and treatment outcomes, and the few studies conducted report seemingly contradictory results. There is a scarce knowledge about the effect of food addiction on weight loss after bariatric surgery, although some studies have reported that greater food addiction levels in the presurgical stage are associated with poorer weight loss at the follow-up, 1 year postsurgery (Miller-Matero et al., 2018). In other studies regarding weight loss treatments, food addiction symptomatology negatively correlated with weight loss, and it may interfere with weight loss programs, specifically with lower attendance and higher dropout levels, especially in adolescents with obesity (Tompkins, Laurent, & Brock, 2017). In this vein, after 7 weeks of a behavioral weight loss program, Burmeister, Hinman, Koball, Hoffmann, and Carels (2013) found that patients who showed higher food addiction symptomatology at baseline experienced relatively minor weight losses. The relationship between food addiction and treatment outcomes could, therefore, justify the use of certain therapeutic options that address food addiction symptomatology. However, there is no clear consensus on the best approach to food addiction or if every therapeutic intervention fits equally well with all types of patients presenting with food addiction symptomatology.

Pharmacological treatments have been suggested for targeting food craving (Potenza & Grilo, 2014). However, it is not clear if approaches oriented to the opioid system are effective (Vella & Pai, 2017). To address relapses frequent in addictions, some authors have proposed pharmacological treatments aimed at the dopaminergic system (Grosshans, Loeber, & Kiefer, 2011), particularly given dopaminergic involvement in eating disorders such as binge eating disorder that shares features with food addiction (Kessler, Hutson, Herman, & Potenza, 2016).

On the other hand, the use of psychological treatments, especially cognitive behavioral therapy, has been suggested to improve abstinence from certain types of food to reduce impulsivity and compulsivity and to develop healthy coping strategies (An, He, Zheng, & Tao, 2017). This therapy usually includes consumption monitoring, identification of triggers for problematic eating behaviors, identification of automatic thoughts, and implementation of adaptive coping strategies. However, although its application has been demonstrated to be effective in the reduction of binge eating episodes, especially in patients with binge eating disorder, it has not proven effective in reducing excess body weight (Bąk-Sosnowska, 2017).

Hilker et al. (2016) sought to assess if a brief psychoeducational treatment for patients with a diagnosis of bulimia nervosa reduced food addiction symptomatology. Patients received a brief outpatient group psychoeducational intervention focused on finding a pattern of regular and healthy eating and reducing eating psychopathology, especially bingeing and purging behaviors. The authors found that food addiction was a statistical predictor of short-term treatment outcome and that patients with greater food addiction severity when starting the intervention were less likely to achieve abstinence from bingeing/purging episodes after the brief intervention. Moreover, food addiction severity was significantly reduced after the psychoeducational treatment (from 90.6% to 72.9%) in individuals with bulimia nervosa. However, although symptomatology related to the eating disorder was significantly reduced, the prevalence of food addiction symptomatology remained relatively high, suggesting the need to strengthen interventions focused on food addiction.

Relatedly, some authors suggest similarities between treatments for food addiction and eating disorders, especially for binge eating disorder, particularly with respect to limiting access to hyperpalatable and processed food. Moreover, treatments for both conditions may identify risky situations associated with overeating behavior and automatic thoughts related to food and eating to prevent relapses and to develop new coping strategies in the setting of negative emotions and stress (Bąk-Sosnowska, 2017). However, discrepancies between treatments for traditional eating disorders and substance-use—related approaches in the food addiction field should be considered (Gearhardt et al., 2011a). Regarding the goals of treatment, a primary concept for treating eating disorders may suggest that there are no "harmful" or "bad" foods, and all should be consumed in moderation to obtain a healthy balance. In this approach, an objective may involve combating food-related irrational thoughts, and when some foods are perceived as fattening or harmful, normalizing eating though healthy meals. Addiction treatment suggests that cutting down on or eliminating specific problematic foods that could have addictive potential is important (Brewerton, 2017). A main goal of this addiction-based intervention may involve correcting behavior through altering dysfunctional brain circuitry and reversing bodily alterations related to the overconsumption of food (for example, at hormonal levels) (Wiss & Brewerton, 2017). An example from this framework is Overeaters Anonymous, founded in Los Angeles in 1960 to fight against compulsive overeating, bingeing behaviors, and being overweight or obese. A main conception was to consider processed food, such as sugar and white flour, as addictive (Brewerton, 2017). Moreover, while eating disorder treatment includes factors exploring one's emotional relationship with one's own body and weight, in the treatment of food addiction, clinical features related to addictive mechanisms should to be addressed (Bąk-Sosnowska, 2017). Therefore, the different approaches, both for eating disorders and food addiction, should take into account the complexity of both disorders, mainly in relation to their biological and psychological mechanisms, and their similarities, differences, and possible co-occurrence. Moreover, although a proposal similar to the existing treatment for addictions is suggested for food addiction, there is still no evidence to support the effectiveness of this approach. Hence, it does not mean that a 12-step program for addictions is exclusively valid to address the disorders or that total abstinence is the only option. Moreover, to enhance its effectiveness, components of addiction therapy could be adapted, including a diversity of approaches, such as cognitive behavioral therapy, mindfulness, dialectical behavior therapy, motivational interviewing, and motivational enhancement therapy, among others (Brewerton, 2017).

Finally, other proposals in this field refer to dietary or macronutrient intervention, combined with exercise, to obtain metabolic changes and nutritional balance in patients with food addiction (Wiss & Brewerton, 2017; Xiao & Tao, 2017). Moreover, some authors suggest the use of new therapies, for example, neurofeedback, based on targeting the neurobehavioral correlates of self-regulation (Imperatori et al., 2016), or transcranial direct current stimulation, applied to reduce food craving (Sauvaget et al., 2015) and food addiction levels.

Controversies in food addiction and implications for policy and regulatory efforts

In most industrialized countries, there has been a considerable increase in the consumption of processed foods, and this has coincided with increases in rates of obesity and being overweight (Gearhardt, Grilo, Dileone, Brownell, & Potenza, 2011a). The increase in obesogenic environments may in part reflect efforts of the food industry to maximize the immediate rewarding properties of certain foods by increasing mainly sugar, fat, and palatability (Carter et al., 2016). Changes in food consumption patterns and lifestyle can have negative consequences on the health of significant portions of the population, which may then lead to higher health care costs (Gearhardt et al., 2011a,b).

Considering the possible addictive potential of certain foods may be important in motivating behavioral changes and generating policy interventions (Swinburn et al., 2011). With respect to the latter, possible prevention initiatives and regulatory policies have been proposed, with the objective of reducing the consumption of potentially addictive foods. In these efforts, governments may have essential roles in promoting the healthiness of food environments through different policies and regulatory efforts (Mahesh, Vandevijvere, Dominick, & Swinburn, 2018). However, although the 2004 global strategy from the World Health Organization on diet, physical activity, and health includes a thorough guide to implement societal changes, governments of most countries have moved only slowly toward implementation of the proposed policies (Swinburn et al., 2011).

To improve and to hasten efforts, some authors have suggested applying efforts successful in other domains (e.g., regulating tobacco) as data suggest potential overlaps in strategies across industries with respect to marketing techniques and strategies to promote continuity of consumption (Capewell & Lloyd-Williams, 2018). To control tobacco use, different strategies have focused on affordability, acceptability, and availability, with affordability possibly involving increases in taxation rates and prices, acceptability possibly involving laws limiting consumption, advertising bans and packaging warnings, and availability possibly involving age verification and the elimination of vending machines. With those strategies, reductions in tobacco consumption, lowering smoking initiation rates, and greater cessation rates have been achieved (Capewell & Lloyd-Williams, 2018). Although some approaches may translate from tobacco control, some authors consider that it is essential to highlight that food regulation may be more complex because of, among other things, the fact that tobacco may be eliminated but food is essential to survival; furthermore, numerous tons of highly processed food are produced annually (Capewell & Lloyd-Williams, 2018). Moreover, pressures for market liberalization may hinder implementation of effective regulations to limit marketing of obesogenic foods (Swinburn et al., 2011).

Some authors highlight that it is essential to assess the complexity of obesogenic environments and that multifactorial interventions may be required at different levels to achieve effects (Capewell & Lloyd-Williams, 2018). On the one hand, the Health Select Committee recommended 12 evidence-based interventions: restrictions on advertising to children, reformulation, 20% tax on sugared soft drinks, price promotions, placement of food and drink in retail environments, labeling, portion sizes and caps, nutrition standards in all schools, supporting local authorities and wider public sectors, early interventions, calorie reductions (not just from sugar), and promotion of physical activity (Capewell & Lloyd-Williams, 2018). On the other hand, INFORMAS (the International Network for Food and Obesity/non-communicable diseases Research, Monitoring and Action Support) created the Healthy Food Environment Policy Index (Food-EPI) to evaluate the extent of implementation of regulations by national governments, focused on promoting healthy food environments. It is divided into two components relating to "policy" and "infrastructure support," and both include policy domains and good practice indicators (Mahesh et al., 2018). Although there is an evident growing international interest in preventing the seemingly exponential increase of obesogenic environments, there is little empirical literature focused on evaluating the effects of different food policies and taking into account that not all of the recommended food regulations contribute at the same level to improving healthy eating behaviors patterns (Mahesh et al., 2018).

Food taxes and food availability

Currently, highly processed foods are often available at very low prices, which may substantially increase their consumption (Gearhardt et al., 2009a). To limit access to these products, one suggested alternative involves bans and economic regulations for specific foods (Fletcher, Frisvold, & Tefft, 2010). Tax policy strategies are focused on addressing unhealthy food consumption, decreasing the purchase of these products, and, consequently, empowering the food industry to rethink the production of this highly processed food (George, 2018). This price regulation strategy also generates revenue for governments, and such funds could be reinvested in new health policies (George, 2018). This approach uses the successful regulations imposed on the tobacco industry as a model (Gearhardt et al., 2011a,b). In this vein, this policy is supported by different entities, such as the Society of Behavioral Medicine, which considers that it can have a positive impact on health, improving diet-related behaviors and reducing the likelihood of suffering from chronic diseases associated with the excessive consumption of sugar and other harmful substances (Taber et al., 2018).

However, this initiative is controversial and has not received the support of some sectors, especially the food and beverage industry which has voiced intentions for legal actions due to what they consider to be unlawful taxes (Carter et al., 2016). In some countries, such as Denmark, this measure has even been abolished because of aspects such as inadequately conceived design of the tax, negative media coverage, and insufficient involvement from the health sector (George, 2018). There is no consensus identifying the impact that food taxes may have on health and well-being (Mahesh et al., 2018). Some authors suggest, therefore, exploring this field more fully to identify conditions that favor tax proposals with respect to feasibility, acceptability, and successful implementation (Le Bodo & De Wals, 2017). Moreover, cooperation between public health and finance policy-makers may be particularly important in efforts to make food and beverage taxation feasible and successful (Le Bodo & De Wals, 2017). In addition, other potential difficulties may involve guiding the selection process of specific food products whose prices should be increased, as well as the potential need to increase the availability of alternative products in the market (Schroeter, Lusk, & Tyner, 2008). However, increasing the availability of healthy foods is sometimes difficult, especially in "food deserts" (i.e., generally urban areas where large groups of people often have limited access to healthy food). These may be generated in part because of the investment costs of having properties where healthy foods may be sold, as well as the often shorter shelf lives of such foods (White, 2018).

Another initiative aims to promote the availability of healthy foods at schools. After the implementation of government policies targeting changes in school food environments, the consumption of highly processed food decreased. School-based food policies are important, especially when considering the hours that children spend in schools and the potential impact of changing food preferences and habits in this developing population (Mahesh et al., 2018).

Food advertising restrictions

Eating behaviors are sensitive to external cues, such as advertising, and thus marketing may contribute importantly to obesogenic environments (Smithers, Lynch, & Merlin, 2014). Data suggest that food advertising may have strong influences on children's eating preferences and, consequently, on their food purchases and food consumption (Galbraith-Emami & Lobstein, 2013). Some authors suggest that children should be considered as the primary at-risk population because of the fact that incomplete development of cognitive abilities may make them vulnerable to food marketing efforts (Carter et al., 2016; Whalen, Harrold, Child, Halford, & Boyland, 2018). Studies examining the impact of "junk food" advertising during sport events on alcohol consumption and gambling behavior in youth suggest that an association with decrease decreased risk perception of these unhealthy behaviors (Lopez-Gonzalez & Griffiths, 2018). People with low socioeconomic status may also be more susceptible to effects of food advertising, and such a vulnerability may explain concentrations of food advertising in specific areas and in part account for socioeconomic disparities relating to unhealthy food consumption (Zimmerman & Shimoga, 2014).

Given such data, some have proposed the inclusion of regulatory measures on advertising, such as advertising bans, and sanctions for violating advertisement restrictions to protect the population, especially youth (Bacardí-Gascón & Jiménez-Cruz, 2015). These policies may include restrictions on advertising unhealthy foods that do not meet government standards, especially during children's programs (Smithers et al., 2014). In addition, the promotion of media messages related to balanced diets and physical activity is recommended by institutions such as the International Food and Beverage Alliance and the Food and Drink Federation (Whalen et al., 2018).

Restrictions on advertising seem to be effective in some countries, although they may have lower impacts on English-speaking children (Galbraith-Emami & Lobstein, 2013). These measures have been controversial, and the lack of sanctions for breaches of such regulation and the possibly low effectiveness in the cooperation between governments and the food industry has been criticized in some countries (Brown et al., 2018). Moreover, there are often discrepancies between conclusions in industry-sponsored and independent scientific reports (Galbraith-Emami & Lobstein, 2013).

As food marketing has spread to diverse media (e.g., on the internet), regulating advertising has become complicated and may require international cooperation (Gearhardt et al., 2011a,b). Currently, several countries are experiencing substantial growth of online advertisement, mainly through social networking sites, which younger children may often access easily. This new development raises concerns about the need to have new regulatory policies to control children's exposure to unhealthy food marketing (Galbraith-Emami & Lobstein, 2013).

Labeling and health and nutritional education

Another factor to consider is the value of policy initiatives that promote health and nutritional education programs that provide warnings of the characteristics of foods. Some authors contend that consumers have the right to access to food information, as well as to know the consequences of their choices on their own health and well-being (Dragone, Manaresi, & Savorelli, 2016). As such, some individuals advocate for the implementation of strategies such as the use of media to educate the population about the main characteristics and the potentially addictive properties of specific foods (Carter et al., 2016).

Stricter regulation of specific foods with possible addictive potential represents another possible initiative (Gearhardt, Roberts, & Ashe, 2012). Front-of-pack traffic-light nutritional labeling has received support from some groups as a means for improving nutrition. This approach consists of highlighting color-coded nutritional information, such as the total fat, saturated fat, sugar, and salt content, on the front panel of food packages (Sacks, Rayner, & Swinburn, 2009). Although, the effectiveness of this initiative has not been systematically investigated, some studies suggest that this approach may be highly cost-effective and may impact health, including in at-risk populations, and that this policy should be considered for implementation (Sacks, Veerman, Moodie, & Swinburn, 2011).

However, some policies oriented to personal decision and education through the provision of information may be ineffective, and some authors suggest that for this reason they have received partial support from the industry (Capewell & Lloyd-Williams, 2018). As such, other environment-based initiatives may be more effective than education-based initiatives in terms of long-term changes (Schulte, Joyner, et al., 2015), and some countries, such as the United Kingdom, are examining the strengths and weaknesses of these initiatives to develop more effective interventions (Capewell & Lloyd-Williams, 2018).

Food accessibility and other environmental interventions

One factor that is part of the complex nature of obesity involves organizational and community issues (Karacabeyli, Allender, Pinkney, & Amed, 2018). Exposure to processed food takes place from early developmental stages, which is why prevention strategies based on policies that limit the accessibility of certain foods are important (Gearhardt et al., 2011a,b). Furthermore, the implementation of urban environments that promote physical activity and, consequently, that allow for the maintenance of a healthy lifestyle, such as the use of bicycles or spaces for walking, would be another possible intervention on an environmental level (Gearhardt et al., 2009a; Koohsari et al., 2015). Although Karacabeyli et al. (2018) examined policies and practices in community environments designed to cope with unhealthy behaviors, findings are difficult to interpret because of multiple interventions being used simultaneously.

With regard to school environments, some authors have suggested classifying schools based on the level of involvement in health promotion activities. In this classification, some prevention strategies that have been considered include training school professionals regarding health promotion issues, having spaces for students to conduct physical activity, integrating school-based interventions in the school curriculum, and implementing other new prevention policies (Fair, Solari Williams, Warren, McKyer, & Ory, 2018).

Policies focused on environments may be more efficacious in comparison with health education programs because they may affect larger populations. However, such environmental policies may be more difficult to implement because of, among other things, reluctance of the general public toward modification of environments to which they are accustomed, such as traveling by car through cities instead of using alternative means that promote improvements in health (Swinburn et al., 2011).

Cultural and socioeconomic considerations related to food addiction

The consumption patterns of highly processed foods may differ between Hispanic, white, and black populations. On the one hand, Hispanic populations have shown greater rates of addictive-like eating behavior in comparison with both white and black groups (Schulte & Gearhardt, 2018). Different results have been found when comparing different racial/ethnic groups on aspects of eating disorders. Bodell et al. (2018) found that white as compared with black females were more likely to exhibit increasing trajectories of eating disorder symptomatology. Taking the association between food addiction and BMI into account, differences between Hispanic, white, and black populations were also identified, finding a positive association between food addiction and BMI in white individuals and the opposite association in the Hispanic group (Schulte & Gearhardt, 2018). Knowledge related to racial/ethnic differences could help to identify potentially vulnerable populations and to implement more effective policies to reduce health disparities.

People with lower incomes show greater tendencies to consume highly processed foods, and this suggests that low economic costs associated with these foods, as well as difficulties in accessing healthier food, may contribute to food addiction (Schulte & Gearhardt, 2018). The application of public policies that consider vulnerabilities of populations with low socioeconomic resources could help reduce health disparities (Gerhard & Monsey, 2014). Such policies could target the availability and affordability of healthier foods through, for example, subsidies for certain foods (Schulte & Gearhardt, 2018).

Conclusions

Interest in the food addiction construct has been growing in the setting of obesogenic environments in developed countries and widespread access to highly processed foods. Some populations, especially children and people of lower socioeconomic status, have been considered as at-risk because of the proliferation of these environments that may promote substantial changes in eating patterns and which may have detrimental health consequences.

Although the core goal in research has been to illuminate characteristics of food addiction, food addiction remains a controversial topic that fails to obtain consensus from either clinicians or researchers. There exist numerous controversies surrounding food addiction, and these relate to its definition, categorization, diagnosis, and management.

Two different positions may be defined in relation to the food addiction construct. On the one hand, some individuals contend that food addiction presents numerous similarities with substance-use disorders, and that the construct may best be considered within an addiction framework. From this perspective, the possible addictive potential of certain foods (especially those containing sugars and refined ingredients) is taken into account and repeated exposure to these types of food could generate patterns of behavior in which clinical aspects such as tolerance, craving, and withdrawal emerge. However, another proposed perspective suggests the existence of an eating addiction. This model focuses mainly on the behavior of eating, arguing that overeating may occur, for example, under negative emotions regardless of the type of food that is eaten.

The debates in reference to the conceptualization of food addiction may hinder its evaluation. Food addiction is not universally accepted as a mental disorder; thus, there do not exist formal diagnostic criteria. Nonetheless, diagnostic instruments from both food and eating addiction perspectives have been developed. The concurrent use of available psychometric tools may ultimately help resolve current controversies.

Currently, however, the existing controversies both in conceptualization and diagnosis of food addiction may limit effective treatments. There is a growing interest in the development of new approaches, both psychological and pharmacological, for food addiction, and these should be tested in empirical studies. Some suggested therapies utilize cognitive behavioral approaches used in the treatment of substance-use disorders.

Furthermore, the influence of the food industry in food consumption habits may pose an additional challenge. Numerous regulatory policies have been proposed to reduce obesogenic environments and promote healthy eating habits. These have been included policies on food taxation, food availability, advertising restrictions, labeling, health and nutritional education, food accessibility, and other environmental interventions. However, because governments, food industry, and entities that promote health may have conflicting interests, the implementation of these regulations, as well as the evaluation of their effectiveness in different populations, may be slow in developing.

In conclusion, multiple controversies currently exist surrounding food addiction. Resolving these controversies through additional research may help address public health concerns and promote the development of more effective prevention, treatment, and policy efforts.

References

Aloi, M., Rania, M., Rodríguez Muñoz, R. C., Jiménez Murcia, S., Fernández-Aranda, F., De Fazio, P., & Segura-Garcia, C. (2017). Validation of the Italian version of the Yale Food Addiction Scale 2.0 (I-YFAS 2.0) in a sample of undergraduate students. *Eating and Weight Disorders, 22*(3), 527–533.

American Psychiatric Association. (2000). Diagnostic and statistical manual of mental disorders. In *Text Revision (DSM-IV-TR)* (4th ed.). Washington, DC: American Psychiatric Association.

American Psychiatric Association. (2013). *Diagnostic and statistical manual of mental disorders* (5th ed.). Washington, DC: American Psychiatric Association.

An, H., He, R.-H., Zheng, Y.-R., & Tao, R. (2017). Cognitive-behavioral therapy. In *Substance and non-substance addiction* (pp. 321–329). Singapore: Springer.

Bacardí-Gascón, M., & Jiménez-Cruz, A. (2015). Revisión TV food advertising geared to children in Latin-American countries and Hispanics in the USA: A review. *Nutricion Hospitalaria, 3131*(5), 1928–1935.

Bechara, A. (2005). Decision making, impulse control and loss of willpower to resist drugs: A neurocognitive perspective. *Nature Neuroscience, 8*(11), 1458.

Bellisle, F., Drewnowski, A., Anderson, G. H., Westerterp-Plantenga, M., & Martin, C. K. (2012). Sweetness, satiation, and satiety. *Journal of Nutrition, 142*(6), 1149S–1154S.

Blundell, J. E., & Finlayson, G. (2011). Food addiction not helpful: The hedonic component — implicit wanting — is important. *Addiction, 106*(7), 1216–1218.

Bodell, L. P., Wildes, J. E., Cheng, Y., Goldschmidt, A. B., Keenan, K., Hipwell, A. E., & Stepp, S. D. (2018). Associations between race and eating disorder symptom trajectories in black and white girls. *Journal of Abnormal Child Psychology, 46*(3), 625–638.

Brewerton, T. D. (2017). Food addiction as a proxy for eating disorder and obesity severity, trauma history, PTSD symptoms, and comorbidity. *Eating and Weight Disorders, 22*(2), 241–247.

Brown, V., Ananthapavan, J., Veerman, L., Sacks, G., Lal, A., Peeters, A., ... Moodie, M. (2018). The potential cost-effectiveness and equity impacts of restricting television advertising of unhealthy food and beverages to Australian children. *Nutrients, 10*(5), 622.

Brunault, P., Ballon, N., Gaillard, P., Réveillère, C., & Courtois, R. (2014). Validation of the French version of the Yale Food Addiction Scale: An examination of its factor structure, reliability, and construct validity in a nonclinical sample. *Canadian Journal of Psychiatry, 59*(5), 276–284.

Brunstrom, J. M., & Cheon, B. K. (2018). Do humans still forage in an obesogenic environment? Mechanisms and implications for weight maintenance. *Physiology & Behavior.* https://doi.org/10.1016/j.physbeh.2018.02.038 [Epub ahead of print].

Burmeister, J. M., Hinman, N., Koball, A., Hoffmann, D. A., & Carels, R. A. (2013). Food addiction in adults seeking weight loss treatment. Implications for psychosocial health and weight loss. *Appetite, 60*(1), 103–110.

Burrows, T., Kay-Lambkin, F., Pursey, K., Skinner, J., & Dayas, C. (2018). Food addiction and associations with mental health symptoms: A systematic review with meta-analysis. *Journal of Human Nutrition and Dietetics.* https://doi.org/10.1111/jhn.12532 [Epub ahead of print].

Burrows, T., Skinner, J., Joyner, M. A., Palmieri, J., Vaughan, K., & Gearhardt, A. N. (2017a). Food addiction in children: Associations with obesity, parental food addiction and feeding practices. *Eating Behaviors, 26*, 114–120.

Burrows, T., Skinner, J., McKenna, R., & Rollo, M. (2017b). Food addiction, binge eating disorder, and obesity: Is there a relationship? *Behavioral Sciences, 7*(3), 54.

Bąk-Sosnowska, M. (2017). Differential criteria for binge eating disorder and food addiction in the context of causes and treatment of obesity. *Psychiatria Polska, 51*(2), 247–259.

Canan, F., Karaca, S., Sogucak, S., Gecici, O., & Kuloglu, M. (2017). Eating disorders and food addiction in men with heroin use disorder: A controlled study. *Eating and Weight Disorders, 22*(2), 249–257.

Capewell, S., & Lloyd-Williams, F. (2018). The role of the food industry in health: Lessons from tobacco? *British Medical Bulletin, 125*(1), 131–143.

Carter, A., Hendrikse, J., Lee, N., Yücel, M., Verdejo-Garcia, A., Andrews, Z., & Hall, W. (2016). The neurobiology of "food addiction" and its implications for obesity treatment and policy. *Annual Review of Nutrition, 36*(1), 105–128.

Cassin, S. E., & von Ranson, K. M. (2007). Is binge eating experienced as an addiction? *Appetite, 49*(3), 687–690.

Ceccarini, M., Manzoni, G. M., Castelnuovo, G., & Molinari, E. (2015). An evaluation of the Italian version of the Yale Food Addiction Scale in obese adult inpatients engaged in a 1-month-weight-loss treatment. *Journal of Medicinal Food, 18*(11), 1281–1287.

Cepeda-Benito, A., Gleaves, D. H., Fernández, M. C., Vila, J., Williams, T. L., & Reynoso, J. (2000). The development and validation of Spanish versions of the state and trait food cravings questionnaires. *Behaviour Research and Therapy, 38*(11), 1125–1138.

Cepeda-Benito, A., Gleaves, D. H., Williams, T. L., & Erath, S. A. (2000). The development and validation of the state and trait food-cravings questionnaires. *Behavior Therapy, 31*(1), 151–173.

Chao, A. M., White, M. A., Grilo, C. M., & Sinha, R. (2017). Examining the effects of cigarette smoking on food cravings and intake, depressive symptoms, and stress. *Eating Behaviors, 24*, 61–65.

Chen, G., Tang, Z., Guo, G., Liu, X., & Xiao, S. (2015). The Chinese version of the Yale Food Addiction Scale: An examination of its validation in a sample of female adolescents. *Eating Behaviors, 18*, 97–102.

Chuang, C. W. I., Sussman, S., Stone, M. D., Pang, R. D., Chou, C. P., Leventhal, A. M., & Kirkpatrick, M. G. (2017). Impulsivity and history of behavioral addictions are associated with drug use in adolescents. *Addictive Behaviors, 74*, 41–47.

Clark, S. M., & Saules, K. K. (2013). Validation of the Yale Food Addiction Scale among a weight-loss surgery population. *Eating Behaviors, 14*(2), 216–219.

Davis, C. (2013). From passive overeating to "food addiction": A spectrum of compulsion and severity. *ISRN Obesity, 2013*, 1–20.

Davis, C., Curtis, C., Levitan, R. D., Carter, J. C., Kaplan, A. S., & Kennedy, J. L. (2011). Evidence that "food addiction" is a valid phenotype of obesity. *Appetite, 57*(3), 711–717.

De Ridder, D., Manning, P., Leong, S. L., Ross, S., Sutherland, W., Horwath, C., & Vanneste, S. (2016). The brain, obesity and addiction: An EEG neuroimaging study. *Scientific Reports, 6*, 1–13.

Dimitrijević, I., Popović, N., Sabljak, V., Škodrić-Trifunović, V., & Dimitrijević, N. (2015). Food addiction-diagnosis and treatment. *Psychiatria Danubina, 27*(1), 101–106.

Dragone, D., Manaresi, F., & Savorelli, L. (2016). Obesity and smoking: Can we kill two birds with one tax? *Health Economics, 25*(11), 1464–1482.

Fair, K. N., Solari Williams, K. D., Warren, J., McKyer, E. L. J., & Ory, M. G. (2018). The influence of organizational culture on school-based obesity prevention interventions: A systematic review of the literature. *Journal of School Health, 88*(6), 462–473.

Fattore, L., Melis, M., Fadda, P., & Fratta, W. (2014). Sex differences in addictive disorders. *Frontiers in Neuroendocrinology, 35*(3), 272–284.

Fernández-Aranda, F., Steward, T., Mestre-Bach, G., Jiménez-Murcia, S., & Gearhardt, A. (2019). Obesity and food addiction. In *Encyclopedia of endocrine diseases* (2nd ed.). *1* pp. 414–419. Elsevier.

Fletcher, J. M., Frisvold, D., & Tefft, N. (2010). Taxing soft drinks and restricting access to vending machines to curb child obesity. *Health Affairs, 29*(5), 1059–1066.

Flint, A. J., Gearhardt, A. N., Corbin, W. R., Brownell, K. D., Field, A. E., & Rimm, E. B. (2014). Food addiction scale measurement in 2 cohorts of middle-aged and older women. *American Journal of Clinical Nutrition, 99*(3), 578–586.

Frayn, M., Sears, C. R., & von Ranson, K. M. (2016). A sad mood increases attention to unhealthy food images in women with food addiction. *Appetite, 100*, 55–63.

Galbraith-Emami, S., & Lobstein, T. (2013). The impact of initiatives to limit the advertising of food and beverage products to children: A systematic review. *Obesity Reviews, 14*(12), 960–974.

GBD 2015 Obesity Collaborators, Afshin, A., Forouzanfar, M. H., Reitsma, M. B., Sur, P., Estep, K., … Murray, C. J. L. (2017). Health effects of overweight and obesity in 195 countries over 25 years. *New England Journal of Medicine, 377*(1), 13–27.

Gearhardt, A. N., Boswell, R. G., & White, M. A. (2014). The association of "food addiction" with disordered eating and body mass index. *Eating Behaviors, 15*(3), 427–433.

Gearhardt, A. N., Corbin, W. R., & Brownell, K. D. (2009a). Food addiction: An examination of the diagnostic criteria for dependence. *Journal of Addiction Medicine, 3*(1), 1–7.

Gearhardt, A. N., Corbin, W. R., & Brownell, K. D. (2009b). Preliminary validation of the Yale Food Addiction Scale. *Appetite, 52*(2), 430–436.

Gearhardt, A. N., Corbin, W. R., & Brownell, K. D. (2016). Development of the Yale Food Addiction Scale version 2.0. *Psychology of Addictive Behaviors, 30*(1), 113–121.

Gearhardt, A. N., Grilo, C. M., Dileone, R. J., Brownell, K. D., & Potenza, M. N. (2011a). Can food be addictive? Public health and policy implications. *Addiction, 106*(7), 1208–1212.

Gearhardt, A. N., Roberto, C. A., Seamans, M. J., Corbin, W. R., & Brownell, K. D. (2013). Preliminary validation of the Yale Food Addiction Scale for children. *Eating Behaviors, 14*(4), 508–512.

Gearhardt, A., Roberts, M., & Ashe, M. (2012). If sugar is addictive what does it mean for the law? *Journal of Law Medicine & Ethics*, 46–49.

Gearhardt, A. N., White, M. A., & Potenza, M. N. (2011b). Binge eating disorder and food addiction. *Current Drug Abuse Reviews, 4*(3), 201–207.

George, A. (2018). Not so sweet refrain: Sugar-sweetened beverage taxes, industry opposition and harnessing the lessons learned from tobacco control legal challenges. *Health Economics, Policy and Law*, 1–27.

Gerhard, D. M., & Monsey, M. S. (2014). Obesity: From public health to public policy: An interview with Marlene Schwartz, PhD. *Yale Journal of Biology & Medicine, 87*(2), 167.

Goluza, I., Borchard, J., Kiarie, E., Mullan, J., & Pai, N. (2017). Exploration of food addiction in people living with schizophrenia. *Asian Journal of Psychiatry, 27*, 81–84.

Gordon, E., Ariel-Donges, A., Bauman, V., & Merlo, L. (2018). What is the evidence for "food addiction?" A systematic review. *Nutrients, 10*(4), 477.

Granero, R., Hilker, I., Agüera, Z., Jiménez-Murcia, S., Sauchelli, S., Islam, M. A., … Fernández-Aranda, F. (2014). Food addiction in a Spanish sample of eating disorders: DSM-5 diagnostic subtype differentiation and validation data. *European Eating Disorders Review, 22*(6), 389–396.

Granero, R., Jiménez-Murcia, S., Gerhardt, A., Agüera, Z., Aymamí, N., Gómez-Peña, M., … Fernández-Aranda, F. (2018). Clinical correlates and validation of the Spanish version of the Yale Food Addiction Scale 2.0 (YFAS 2.0) and in a sample of eating disorder, gambling disorder and healthy control participants. *Frontiers in Psychiatry, 9*, 208.

Grosshans, M., Loeber, S., & Kiefer, F. (2011). Implications from addiction research towards the understanding and treatment of obesity. *Addiction Biology, 16*(2), 189–198.

Hammond, R., & Levine, R. (2010). The economic impact of obesity in the United States. *Diabetes, Metabolic Syndrome and Obesity: Targets and Therapy, 3*, 285.

Hebebrand, J., Albayrak, Ö., Adan, R., Ante, J., Dieguez, C., De Jong, J., … Dickson, S. L. (2014). "Eating addiction", rather than "food addiction", better captures addictive-like eating behavior. *Neuroscience & Biobehavioral Reviews, 47*, 295–306.

Hilker, I., Sánchez, I., Steward, T., Jiménez-Murcia, S., Granero, R., Gearhardt, A. N., … Fernández-Aranda, F. (2016). Food addiction in bulimia nervosa: Clinical correlates and association with response to a brief psychoeducational intervention. *European Eating Disorders Review, 24*(6), 482–488. https://doi.org/10.1002/erv.2473.

Imperatori, C., Fabbricatore, M., Vumbaca, V., Innamorati, M., Contardi, A., & Farina, B. (2016). Food addiction: Definition, measurement and prevalence in healthy subjects and in patients with eating disorders. *Rivista di Psichiatria, 51*(2), 60–65.

Innamorati, M., Imperatori, C., Manzoni, G. M., Lamis, D. A., Castelnuovo, G., Tamburello, A., … Fabbricatore, M. (2015). Psychometric properties of the Italian Yale

Food Addiction Scale in overweight and obese patients. *Eating and Weight Disorders, 20*(1), 119–127.

Ivezaj, V., Wiedemann, A. A., & Grilo, C. M. (2017). Food addiction and bariatric surgery: A systematic review of the literature. *Obesity Reviews, 18*(12), 1386–1397.

Jiménez-Murcia, S., Granero, R., Wolz, I., Baño, M., Mestre-Bach, G., Steward, T., ... Fernández-Aranda, F. (2017). Food addiction in gambling disorder: Frequency and clinical outcomes. *Frontiers in Psychology, 8*, 473.

Karacabeyli, D., Allender, S., Pinkney, S., & Amed, S. (2018). Evaluation of complex community-based childhood obesity prevention interventions. *Obesity Reviews*. https://doi.org/10.1111/obr.12689 [Epub ahead of print].

Keser, A., Yüksel, A., Yeşiltepe-Mutlu, G., Bayhan, A., Özsu, E., & Hatun, Ş. (2015). A new insight into food addiction in childhood obesity. *Turkish Journal of Pediatrics, 57*(3), 219–224.

Kessler, R. M., Hutson, P. H., Herman, B. K., & Potenza, M. N. (2016). The neurobiological basis of binge-eating disorder. *Neuroscience & Biobehavioral Reviews, 63*, 223–238.

Koohsari, M. J., Sugiyama, T., Sahlqvist, S., Mavoa, S., Hadgraft, N., & Owen, N. (2015). Neighborhood environmental attributes and adults' sedentary behaviors: Review and research agenda. *Preventive Medicine, 77*, 141–149.

Kor, A., Fogel, Y., Reid, R. C., & Potenza, M. N. (2013). Should hypersexual disorder be classified as an addiction? *Sexual Addiction & Compulsivity, 20*(1–2), 27–47.

Kraus, S. W., Krueger, R. B., Briken, P., First, M. B., Stein, D. J., Kaplan, M. S., ... Reed, G. M. (2018). Compulsive sexual behaviour disorder in the ICD-11. *World Psychiatry, 17*(1), 109–110.

Kraus, S. W., Voon, V., Kor, A., & Potenza, M. N. (2016). Searching for clarity in muddy water: Future considerations for classifying compulsive sexual behavior as an addiction. *Addiction, 111*(12), 2113–2114.

Kraus, S. W., Voon, V., & Potenza, M. N. (2016). Should compulsive sexual behavior be considered an addiction? *Addiction, 111*(12), 2097–2106.

Laurent, J. S., & Sibold, J. (2016). Addictive-like eating, body mass index, and psychological correlates in a community sample of preadolescents. *Journal of Pediatric Health Care, 30*(3), 216–223.

Le Bodo, Y., & De Wals, P. (2017). Soda taxes: The importance of analysing policy processes comment on the untapped power of soda taxes: Incentivising consumers, generating revenue, and altering corporate behaviours. *International Journal of Health Policy and Management, 7*(5), 470–473.

Lee, N. M., Hall, W. D., Lucke, J., Forlini, C., & Carter, A. (2014). Food addiction and its impact on weight-based stigma and the treatment of obese individuals in the U.S. and Australia. *Nutrients, 6*(11), 5312–5326.

Lemeshow, A. R., Gearhardt, A. N., Genkinger, J. M., & Corbin, W. R. (2016). Assessing the psychometric properties of two food addiction scales. *Eating Behaviors, 23*, 110–114.

Lopez-Gonzalez, H., & Griffiths, M. D. (2018). Betting, forex trading, and fantasy gaming sponsorships—a responsible marketing inquiry into the 'gamblification' of English football. *International Journal of Mental Health and Addiction, 16*(2), 404–419.

Lubman, D. I., Cheetham, A., & Yücel, M. (2015). Cannabis and adolescent brain development. *Pharmacology & Therapeutics, 148*, 1–16.

Lydon, D. M., Wilson, S. J., Child, A., & Geier, C. F. (2014). Adolescent brain maturation and smoking: What we know and where we're headed. *Neuroscience & Biobehavioral Reviews, 45*, 323–342.

Mahesh, R., Vandevijvere, S., Dominick, C., & Swinburn, B. (2018). Relative contributions of recommended food environment policies to improve population nutrition: Results from a Delphi study with international food policy experts. *Public Health Nutrition*, 1–7.

Manzoni, G. M., Rossi, A., Pietrabissa, G., Varallo, G., Molinari, E., Poggiogalle, E., ... Castelnuovo, G. (2018). Validation of the Italian Yale Food Addiction Scale in postgraduate university students. *Eating and Weight Disorders, 23*(2), 167–176.

Maraz, A., Király, O., & Demetrovics, Z. (2015). Commentary on: Are we overpathologizing everyday life? A tenable blueprint for behavioral addiction research. *Journal of Behavioral Addictions, 4*(3), 151–154.

Mason, S. M., Flint, A. J., Field, A. E., Austin, S. B., & Rich-Edwards, J. W. (2013). Abuse victimization in childhood or adolescence and risk of food addiction in adult women. *Obesity, 21*(12), E775–E781.

Meule, A. (2015). Back by popular demand: A narrative review on the history of food addiction research. *Yale Journal of Biology & Medicine, 88*(3), 295–302.

Meule, A. (2018). Food cravings in food addiction: Exploring a potential cut-off value of the food cravings questionnaire-trait-reduced. *Eating and weight disorders - Studies on anorexia. Bulimia and Obesity, 23*(1), 39–43.

Meule, A., Heckel, D., Jurowich, C. F., Vögele, C., & Kübler, A. (2014). Correlates of food addiction in obese individuals seeking bariatric surgery. *Clinical Obesity, 4*(4), 228–236.

Meule, A., Hermann, T., & Kübler, A. (2014). A short version of the food cravings questionnaire-trait: The FCQ-T-reduced. *Frontiers in Psychology, 5*, 190.

Meule, A., Müller, A., Gearhardt, A. N., & Blechert, J. (2017). German version of the Yale Food Addiction Scale 2.0: Prevalence and correlates of 'food addiction' in students and obese individuals. *Appetite, 115*, 54–61.

Mies, G. W., Treur, J. L., Larsen, J. K., Halberstadt, J., Pasman, J. A., & Vink, J. M. (2017). The prevalence of food addiction in a large sample of adolescents and its association with addictive substances. *Appetite, 118*, 97–105.

Miller-Matero, L. R., Bryce, K., Saulino, C. K., Dykhuis, K. E., Genaw, J., & Carlin, A. M. (2018). Problematic eating behaviors predict outcomes after bariatric surgery. *Obesity Surgery*, 1–6.

Mitchell, K. S., & Wolf, E. J. (2016). PTSD, food addiction, and disordered eating in a sample of primarily older veterans: The mediating role of emotion regulation. *Psychiatry Research, 243*, 23–29.

Nolan, L. J., & Geliebter, A. (2016). "Food addiction" is associated with night eating severity. *Appetite, 98*, 89–94.

Nolan, L. J., & Geliebter, A. (2017). Validation of the night eating diagnostic questionnaire (NEDQ) and its relationship with depression, sleep quality, "food addiction", and body mass index. *Appetite, 111*, 86–95.

Onaolapo, A. Y., & Onaolapo, O. J. (2018). Food additives, food and the concept of 'food addiction': Is stimulation of the brain reward circuit by food sufficient to trigger addiction? *Pathophysiology*. https://doi.org/10.1016/j.pathophys.2018.04.002 [Epub ahead of print].

Petry, N. M., & O'Brien, C. P. (2013). Internet gaming disorder and the DSM-5. *Addiction, 108*(7), 1186–1187.

Potenza, M. N., & Grilo, C. M. (2014). How relevant is food craving to obesity and its treatment? *Frontiers in Psychiatry, 5*, 164.

Pursey, K. M., Stanwell, P., Gearhardt, A. N., Collins, C. E., & Burrows, T. L. (2014). The prevalence of food addiction as assessed by the Yale Food Addiction Scale: A systematic review. *Nutrients, 6*(10), 4552–4590.

Randolph, T. G. (1956). The descriptive features of food addiction; addictive eating and drinking. *Quarterly Journal of Studies on Alcohol, 17*, 198–224.

Raymond, K. L., & Lovell, G. P. (2015). Food addiction symptomology, impulsivity, mood, and body mass index in people with type two diabetes. *Appetite, 95*, 383–389.

Reslan, S., Saules, K. K., Greenwald, M. K., & Schuh, L. M. (2014). Substance misuse following Roux-en-Y gastric bypass surgery. *Substance Use & Misuse, 49*(4), 405–417.

Richmond, R. L., Roberto, C. A., & Gearhardt, A. N. (2017). The association of addictive-like eating with food intake in children. *Appetite, 117*, 82–90.

Ruddock, H. K., Christiansen, P., Halford, J. C. G., & Hardman, C. A. (2017). The development and validation of the addiction-like eating behaviour scale. *International Journal of Obesity, 41*(11), 1710–1717.

Sacks, G., Rayner, M., & Swinburn, B. (2009). Impact of front-of-pack 'traffic-light' nutrition labelling on consumer food purchases in the UK. *Health Promotion International, 24*(4), 344–352.

Sacks, G., Veerman, J. L., Moodie, M., & Swinburn, B. (2011). Traffic-light nutrition labelling and junk-food tax: A modelled comparison of cost-effectiveness for obesity prevention. *International Journal of Obesity, 35*(7), 1001–1009.

Sauvaget, A., Trojak, B., Bulteau, S., Jiménez-Murcia, S., Fernández-Aranda, F., Wolz, I., … Grall-Bronnec, M. (2015). Transcranial direct current stimulation (tDCS) in behavioral and food addiction: A systematic review of efficacy, technical, and methodological issues. *Frontiers in Neuroscience, 9*, 349.

Schroder, R., Sellman, J. D., & Adamson, S. (2017). Development and validation of a brief measure of eating compulsivity (MEC). *Substance Use & Misuse, 52*(14), 1918–1924.

Schroeter, C., Lusk, J., & Tyner, W. (2008). Determining the impact of food price and income changes on body weight. *Journal of Health Economics, 27*(1), 45–68.

Schulte, E. M., Avena, N. M., & Gearhardt, A. N. (2015). Which foods may be addictive? The roles of processing, fat content, and glycemic load. *PLoS One, 10*(2), e0117959.

Schulte, E. M., & Gearhardt, A. N. (2017). Development of the modified Yale Food Addiction Scale version 2.0. *European Eating Disorders Review, 25*(4), 302–308.

Schulte, E. M., & Gearhardt, A. N. (2018). Associations of food addiction in a sample recruited to be nationally representative of the United States. *European Eating Disorders Review, 26*(2), 112–119.

Schulte, E. M., Jacques-Tiura, A. J., Gearhardt, A. N., & Naar, S. (2017). Food addiction prevalence and concurrent validity in African American adolescents with obesity. *Psychology of Addictive Behaviors, 32*(2), 187.

Schulte, E. M., Joyner, M. A., Potenza, M. N., Grilo, C. M., & Gearhardt, A. N. (2015). Current considerations regarding food addiction. *Current Psychiatry Reports, 17*(4), 19.

Schulte, E. M., Potenza, M. N., & Gearhardt, A. N. (2017a). A commentary on the "eating addiction" versus "food addiction" perspectives on addictive-like food consumption. *Appetite, 115*, 9–15.

Schulte, E. M., Potenza, M. N., & Gearhardt, A. N. (2017b). How much does the addiction-like eating behavior Scale add to the debate regarding food versus eating addictions? *International Journal of Obesity, 42*(4), 946.

Schulte, E. M., Yokum, S., Potenza, M. N., & Gearhardt, A. N. (2016). Neural systems implicated in obesity as an addictive disorder: From biological to behavioral mechanisms. In *Progress in brain research, 223* pp. 329–346). Elsevier.

Smithers, L. G., Lynch, J. W., & Merlin, T. (2014). Industry self-regulation and TV advertising of foods to Australian children. *Journal of Paediatrics and Child Health, 50*(5), 386–392.

Swarna Nantha, Y., Abd Patah, N. A., & Ponnusamy Pillai, M. (2016). Preliminary validation of the Malay Yale Food Addiction Scale: Factor structure and item analysis in an obese population. *Clinical Nutrition ESPEN, 16*, 42–47.

Swinburn, B. A., Sacks, G., Hall, K. D., McPherson, K., Finegood, D. T., Moodie, M. L., & Gortmaker, S. L. (2011). The global obesity pandemic: Shaped by global drivers and local environments. *The Lancet, 378*(9793), 804–814.

Taber, D. R., Dulin-Keita, A., Fallon, M., Chaloupka, F. J., Andreyeva, T., Schwartz, M. B., & Harris, J. L. (2018). Society of Behavioral Medicine (SBM) position statement: Enact taxes on sugar sweetened beverages to prevent chronic disease. *Translational Behavioral Medicine*. https://doi.org/10.1093/tbm/iby035 [Epub ahead of print].

Tompkins, C. L., Laurent, J., & Brock, D. W. (2017). Food addiction: A barrier for effective weight management for obese adolescents. *Childhood Obesity, 13*(6), 462–469.

Torres, S., Camacho, M., Costa, P., Ribeiro, G., Santos, O., Vieira, F. M., ... Oliveira-Maia, A. J. (2017). Psychometric properties of the Portuguese version of the Yale Food Addiction Scale. *Eating and Weight Disorders, 22*(2), 259–267.

Vella, S.-L. C., & Pai, N. B. (2017). A narrative review of potential treatment strategies for food addiction. *Eating and Weight Disorders, 22*(3), 387–393.

Whalen, R., Harrold, J., Child, S., Halford, J., & Boyland, E. (2018). The health Halo Trend in UK Television food advertising viewed by children: The rise of implicit and explicit health messaging in the promotion of unhealthy foods. *International Journal of Environmental Research and Public Health, 15*(3), 560.

White, E. J. (2018). The problem of obesity and dietary nudges. *Politics and the Life Sciences, 37*(01), 120–125.

White, M. A., Whisenhunt, B. L., Williamson, D. A., Greenway, F. L., & Netemeyer, R. G. (2002). Development and validation of the food-craving inventory. *Obesity Research, 10*(2), 107–114.

Wiers, R. W., Bartholow, B. D., van den Wildenberg, E., Thush, C., Engels, R. C. M. E., Sher, K. J., ... Stacy, A. W. (2007). Automatic and controlled processes and the development of addictive behaviors in adolescents: A review and a model. *Pharmacology Biochemistry and Behavior, 86*(2), 263–283.

Wiss, D. A., & Brewerton, T. D. (2017). Incorporating food addiction into disordered eating: The disordered eating food addiction nutrition guide (DEFANG). *Eating and Weight Disorders – Studies on Anorexia, Bulimia and Obesity, 22*(1), 49–59.

Wolz, I., Granero, R., & Fernández-Aranda, F. (2017). A comprehensive model of food addiction in patients with binge-eating symptomatology: The essential role of negative urgency. *Comprehensive Psychiatry, 74*, 118–124.

Wolz, I., Hilker, I., Granero, R., Jiménez-Murcia, S., Gearhardt, A. N., Dieguez, C., ... Fernandez-Aranda, F. (2016). "Food addiction" in patients with eating disorders is associated with negative urgency and difficulties to focus on long-term goals. *Frontiers in Psychology, 7*, 1–10.

World Health Organization (WHO). (2017). *Obesity and overweight.*

Xiao, L.-J., & Tao, R. (2017). Nutrition support therapy. *Advances in Experimental Medicine & Biology, 1010*, 281–293.

Ziauddeen, H., Farooqi, I. S., & Fletcher, P. C. (2012). Obesity and the brain: How convincing is the addiction model? *Nature Reviews Neuroscience, 13*(4), 279.

Zimmerman, F. J., & Shimoga, S. V. (2014). The effects of food advertising and cognitive load on food choices. *BMC Public Health, 14*(1), 342.

CHAPTER

Food addiction and its associations to trauma, severity of illness, and comorbidity

15

Timothy D. Brewerton

Department of Psychiatry and Behavioral Sciences, Medical University of South Carolina, Charleston, SC, United States

The concept of food addiction

Randolph coined the term "food addiction" (FA) in 1956, and he associated it with "addictive drinking" (Randolph, 1956). In this initial report, FA was defined as "a specific adaptation to one or more regularly consumed foods to which a person is highly sensitive" and which "produces a common pattern of symptoms descriptively similar to those of other addictive processes." However, FA went relatively unexplored for many decades until this millennium when considerable scientific focus began in earnest and has increasingly progressed ever since (Brewerton, 2017). As this research base has expanded with this systematic inquiry, FA has increasingly become a valuable and arguably valid clinical syndrome that has significant implications for health care practice, research, and policy (Gearhardt, Grilo, DiLeone, Brownell, & Potenza, 2011; Lee et al., 2013; The National Center on Addiction and Substance Abuse, 2016). Highly palatable foods, largely consisting of ultra-processed simple carbohydrates and saturated fats, are postulated to act via similar mechanisms as both licit and illicit drugs of abuse in the brain (Avena, Gold, Kroll, & Gold, 2012; Avena, Rada, & Hoebel, 2009; Avena, Wang, & Gold, 2011; Benton, 2010; Fortuna, 2012; Gearhardt, Corbin, & Brownell, 2009; Gearhardt, Davis, Kuschner, & Brownell, 2011; Gearhardt et al., 2012; Gearhardt, White, & Potenza, 2011; Gold, Graham, Cocores, & Nixon, 2009; Hoebel, Avena, Bocarsly, & Rada, 2009; Joranby, Pineda, & Gold, 2005; Liu, von Deneen, Kobeissy, & Gold, 2010; Lustig, 2010; Wiss, Criscitelli, Gold, & Avena, 2017). Most of the clinical research in humans has been expedited by the publication of the Yale Food Addiction Scale (YFAS), which was authored by Ashley Gearhardt et al. (Gearhardt et al., 2009; Gearhardt, Roberto, Seamans, Corbin, & Brownell, 2013; Pursey, Stanwell, Gearhardt, Collins, & Burrows, 2014). The YFAS has exhibited excellent validity and test–retest reliability. Gearhart et al. have also published a version of the YFAS

Compulsive Eating Behavior and Food Addiction. https://doi.org/10.1016/B978-0-12-816207-1.00015-9
Copyright © 2019 Elsevier Inc. All rights reserved.

for use in children (YFAS-C) (Gearhardt, Roberto, et al., 2013), as well as an updated version of the original YFAS, which is constructed around the new DSM-5 criteria for substance-use disorder (SUD) (YFAS-2) (Meule & Gearhardt, 2014).

Despite the advances in understanding and validating the concept of FA, other investigators remain unconvinced of its existence, consider FA to be a "misnomer," and have proposed an alternative concept termed "eating addiction," which is based on a behavioral addiction model rather than a substance addiction one (Hebebrand et al., 2014). In response to this, Schulte et al. outlined behavioral components characteristic of all SUDs, including FA, the fact that not all foods are equally addictive, and the major differences between these disparate perspectives (Schulte, Avena, & Gearhardt, 2015; Schulte, Potenza, & Gearhardt, 2017; Schulte, Smeal, & Gearhardt, 2017). A recent study by Lemeshow et al. using the Nurses' Health Study I and II examined FA and food consumption and demonstrated that FA was positively associated with consumption of foods that include combinations of carbohydrates and fats rather than just carbohydrates alone (Lemeshow et al., 2018).

Food addiction and its links to eating disorder and obesity severity

Gearhardt et al. were the first to demonstrate high rates of FA in individuals with binge eating disorder (BED) (Gearhardt, White, Masheb, & Grilo, 2013; Gearhardt et al., 2012; Gearhardt, White, et al., 2011). In one study of 81 obese, treatment-seeking patients with current DSM-IV—defined BED, 57% met criteria for current FA. This subcategory of participants with BED plus FA had significantly higher measures of eating disorder (ED) psychopathology, significantly higher degrees of depression, negative affect, and emotional dysregulation, and significantly lower degrees of self-esteem. Notably, the authors reported that participants' YFAS scores were meaningful predictors of binge eating (BE) frequency over and above other psychometric instruments. Collectively, the BED-FA faction represented a "more disturbed variant" of BED (Gearhardt et al., 2012). In a related study of a racially varied group of 96 participants with BED, 41.5% of them met the criteria for FA. Yet again the BED-FA subcategory of individuals had significantly greater measures of ED psychopathology, emotional dysregulation, negative affect, and low self-esteem (Gearhardt, White, et al., 2013). Furthermore, higher YFAS scores were once more the greatest predictors of BE frequency compared with all other measures. YFAS scores were also linked to an earlier age of initially becoming overweight as well as dieting. Interestingly, an earlier age of onset of BE has been linked with higher rates of any ED, greater severity of bulimia nervosa (BN), earlier onset of depression and first trauma, as well as higher rates of victimization, posttraumatic stress disorder (PTSD), substance abuse, and dependence (Brewerton, Rance, Dansky, O'Neil, & Kilpatrick, 2014).

The construct of FA as defined by the YFAS has been investigated in other EDs besides BED. In one study involving 26 women with current BN, 20 women with BN in remission, and 63 healthy control women, 100% of those women with active BN met FA criteria, while 30% of the group with remitted BN and 0% of the controls met YFAS criteria for FA (Meule, von Rezori, & Blechert, 2014). The investigators determined that eating behavior within the context of active BN can be designated as "addiction-like." In addition, it seems likely that a diagnosis of FA declines when signs and symptoms of BN, and perhaps BED, abate, although this remains to be demonstrated with prospective, long-term studies.

In another study, 125 patients with various DSM-5–defined EDs were compared with 82 healthy controls using the YFAS Spanish version (YFAS-S). The YFAS-S exhibited suitable discriminative capacity to differentiate ED patients and healthy controls, as well as a suitable sensitivity to screen for distinctive ED subtypes (Granero et al., 2014). The greatest prevalence estimates of FA were observed in binge-type ED patient subgroups in the following rank order: (1) anorexia nervosa (AN) binge–purge subtype (AN-BP) (85.7%); (2) BN (81.5%); and (3) BED (76.9%). The lowest prevalence estimate was found in the AN restrictive subtype (AN-R) (50%). Once again, it was demonstrated that high rates of FA were associated with higher levels of eating psychopathology, negative affect, depression, general psychopathology, and body mass index (BMI). Taken together, greater FA scores were connected with binge-type ED patient subgroups and with greater eating severity and psychopathology. Moreover, the construct of FA distinguished between ED and controls, although the investigators appealed for additional research on this topic. Given the high rates of FA in AN-R in this study, the issue of possible false positives of FA in ED populations was illustrated, particularly those without BE as a feature. Therefore, it is important for clinicians not to solely rely on the YFAS for clinical assessment, but to rather use it in the context of a comprehensive psychiatric evaluation. Nevertheless, the experience of being addicted to food appears to be common in all forms of EDs.

In a study of 112 patients who were requesting weight loss treatment, the authors found that 33.9% of these individuals met YFAS criteria for FA (Imperatori et al., 2014). The severity of FA was strongly correlated with the severity of BE as measured by the Binge Eating Scale. The combination of FA and BE was moderately associated with psychopathology, which was measured with the Symptom Checklist 90-Revised (SCL-90-R). These researchers also found that the relationship between FA and psychopathology was entirely mediated by BE severity.

Chao et al. studied 178 participants who were seeking weight loss treatment and measured FA using the YFAS (Chao et al., 2017). In contrast to other studies, only 6.7% of participants met FA criteria, but 33.3% of these participants met criteria for co-occurring BED. This study further illustrates the significant overlap between FA and binge-type EDs, even in an otherwise obese population. As others have noted, obesity per se does not constitute FA (Corsica & Pelchat, 2010).

Food addiction and its links to trauma and posttraumatic stress disorder

Binge-related behaviors, including BE and purging, and especially binge-type EDs, including AN-BP, BN, and BED, have been highly linked with prior traumatic events and subsequent PTSD and partial PTSD (Brewerton, 2004, 2007; Brewerton & Brady, 2014; Brewerton, Dansky, Kilpatrick, & O'Neil, 1999; Brewerton, Dansky, O'Neil, & Kilpatrick, 2015; Brewerton et al., 2014; Dansky, Brewerton, O'Neil, & Kilpatrick, 1997; Mitchell, Mazzeo, Schlesinger, Brewerton, & Smith, 2012; Wonderlich, Brewerton, Jocic, Dansky, & Abbott, 1997). Likewise, obesity has also been associated with victimization histories and diagnoses of PTSD (Brewerton, O'Neil, Dansky, & Kilpatrick, 2015; Noll, Zeller, Trickett, & Putnam, 2007; Pagoto et al., 2012; Vieweg et al., 2006; Vieweg et al., 2007; Whitaker, Phillips, Orzol, & Burdette, 2007). Similar to how other addictive substances or behaviors are used, excessive indulgence in hedonic foods and drinks may consequently become an inexpensive, legal, and easily accessible way that distressed individuals may adopt to numb and distract themselves from disturbing trauma-related thoughts, feelings, and memories and to reduce emotional arousal (Brewerton, 2004, 2007, 2011, 2015; Brewerton & Brady, 2014; Brewerton et al., 1999, 2014). Given the relative, short-lived relief or escape that may be experienced, BE with or without purging often become highly reinforcing behaviors that promote increased severity and chronicity of illness over time. The contribution of negative reinforcement mechanisms in compulsive eating behaviors has been discussed in detail elsewhere (see Chapter 6, The Dark Side of Compulsive Eating and Food Addiction: Affective Dysregulation, Negative Reinforcement, and Negative Urgency) (Moore, Sabino, Koob, & Cottone, 2017).

In an investigation of this model, Hirth et al. surveyed over 3000 adult women patients at five different public health clinics in Texas in regard to (1) their fast food and sugary soda consumption, (2) their ED behaviors (e.g., dieting, BE, purging), and (3) their current PTSD symptoms. The researchers detected a statistically significant association between symptoms of PTSD and (1) their frequency of fast food and sugary soda intake and (2) their ED symptoms, including severe dieting, purging, and compulsive exercising, but not with BMI (Hirth, Rahman, & Berenson, 2011). This contribution is unique because it is the first study to establish a link between PTSD symptoms (and thus trauma history) and eating highly palatable foods known to be comparatively unhealthy and related with the concept of FA, i.e., foods with elevated concentrations of processed sugars and saturated fats.

Research findings published by Mason and her collaborators have further established significant associations between prior traumatic exposure, diagnosis of PTSD, and the presence of FA. Assessments of lifetime childhood physical and sexual abuse were obtained in over 57,000 adult participants in the National Nurses' Health

Study II (NNHSII) in 2001, which was followed by assessment of a current diagnosis of FA in 2009 (Mason, Flint, Field, Austin, & Rich-Edwards, 2013). Notably, 80% of those with histories of childhood maltreatment met criteria for FA. Moreover, severe physical and sexual abuse histories were linked with roughly a 90% increase in the risk for FA. The authors calculated the relative risk (RR) of FA for participants with histories of abuse and found the following statistics: physical abuse 1.92 (95% CI: 1.76–2.09) and sexual abuse 1.87 (95% CI: 1.69–2.05). For those with the combination of severe physical and sexual abuse, the RR was 2.40 (95% CI: 2.16–2.67). The investigators took note of the robust relationship between histories of childhood physical and sexual abuse and FA in this large sample of registered nurses. It is critical to mention that other types of childhood trauma, e.g., emotional abuse, emotional neglect, physical neglect, accidents, disasters, etc., were not described in this study. In addition, symptoms or diagnoses of PTSD were also not reported.

However, Mason et al. published another study using the NNHSII database in which 49,408 participants completed a customized version of the YFAS as well as assessments of lifetime traumatic events and PTSD symptoms (Mason et al., 2014). This study overlapped somewhat with the previously described study, but in this report multiple types of trauma were examined in light of PTSD and FA. Again, approximately 80% of the sample participants reported some type of traumatic event, and 66% of those who were trauma-exposed endorsed one or more lifetime PTSD symptom. The criteria for FA were met by 8% of the total group of participants, and most notably, as the number of lifetime PTSD symptoms increased, so did the prevalence estimates of FA. The nurses with the highest number of PTSD symptoms, i.e., six to seven symptoms, had more than double the prevalence estimate of FA as nurses with neither trauma histories nor PTSD symptoms. PTSD symptoms were more convincingly associated with FA when the onset of FA symptoms occurred at a younger age. Importantly, the PTSD–FA connection did not differ significantly by the type of trauma exposure endorsed. The researchers noted that strategies to diminish obesity concomitant with PTSD may necessitate psychological and behavioral interventions that focus on the use of food to cope with distress and/or addiction to food. Although there is a clear relationship between severity of PTSD and the presence of FA, the direction of causation was not clearly established.

In a more recent study, Rainey et al. investigated the rates of FA as defined by the YFAS in sexual minorities, who are previously known to have an increased risk of developing EDs and SUDs. In their study of 356 participants, they reported that high rates of "heterosexist harassment" were associated with FA in the group of sexual minorities compared with the group of heterosexuals. They also found that compassion for self appeared to play a protective role against developing FA (Rainey, Furman, & Gearhardt, 2018).

Food addiction and its links to trauma and obesity

The associations between child maltreatment, PTSD, and obesity have been well-documented (Anda et al., 2009; Felitti, 1991, 1993; Mason et al., 2016; Mason, MacLehose, et al., 2015). Using data generated from the National Women's Study, Brewerton et al. demonstrated statistically significant associations between extreme obesity (defined as a lifetime BMI >40) and endorsements of rape, childhood sexual abuse, any childhood maltreatment, and current and lifetime PTSD (Brewerton et al., 2015). Furthermore, women with extreme obesity were significantly more likely to endorse BE and purging behaviors to meet DSM-IV criteria for BN, any binge-type disorder (BN or BED), and lifetime major depressive disorder.

Other publications also describe critical links between trauma histories, FA, and obesity. In a study mentioned earlier using the NNHSII, Mason et al. reported that FA was not only strongly associated with childhood abuse but also with significantly higher weights and BMIs. Those who met criteria for FA had BMIs on average 6 units higher than women without FA (Mason et al., 2013). In addition, the authors found that nearly two out of three women with FA had BMIs greater than 30 and thereby met criteria for obesity.

Three hundred one overweight or obese women seeking weight loss treatment completed self-report instruments assessing childhood trauma histories, BE, and FA (Imperatori et al., 2016). The investigators discovered that childhood trauma severity was moderately and independently correlated with both FA and BE severity. This relationship persisted even after controlling for possible confounders. Importantly, the co-occurrence of FA and BE was linked with greater BMI and more severe depression and anxiety. In another study by the same group of investigators including 112 men and women seeking weight loss treatment, they found that 33.9% of these patients met YFAS criteria for FA (Imperatori et al., 2014). The severity of FA was significantly correlated with BE, and the combination of FA and BE was moderately related to the presence of psychopathology.

Food addiction and psychiatric comorbidity

As previously noted in this chapter, there have been important links established between FA, greater histories of trauma exposure, higher ED and obesity severity, as well as PTSD and its symptoms. It has been reported in the literature that lifetime diagnoses of PTSD are predictive of greater psychiatric comorbidity (Brady, Killeen, Brewerton, & Lucerini, 2000; Brewerton, 2004, 2007; Brewerton & Brady, 2014), including binge-type EDs, mood disorders, SUDs, and other psychopathologies.

In a recent study of trauma-exposed male ($n = 642$) and female ($n = 55$) veterans, the associations between FA and PTSD were assessed (Mitchell & Wolf, 2016).

PTSD was discovered to be significantly and positively linked with ED symptoms, FA, expressive suppression of emotions, and cognitive reappraisal. Specifically, the expressive suppression of emotions, which is a feature of PTSD and other trauma-related disorders, was significantly related to ED symptoms and was found to mediate the PTSD–ED link. These findings underscore the importance of further examining PTSD as a risk factor for FA and binge-type ED symptoms. Also, the likely mediating role of emotion regulation in the development of eating and related disorders and PTSD to ascertain targets for therapies was an important observation.

As discussed earlier, the combination of FA and BE was linked to more general psychopathology, particularly significantly higher levels of depression and anxiety (Imperatori et al., 2014, 2016). This FA–depression link has been noted in other studies as well (Burmeister, Hinman, Koball, Hoffmann, & Carels, 2013; Davis et al., 2011; Gearhardt et al., 2012; Granero et al., 2018; Koball et al., 2016). Women with obesity and self-identified carbohydrate craving were found to ingest significantly more carbohydrates after a dysphoric mood induction (Corsica & Spring 2008). Further research would do well to identify whether FA is indeed linked to DSM-5–defined mood and/or anxiety disorders and not just depressive and/or anxiety symptoms.

In an interesting human experiment, Frayn et al. studied attention to food images using an eye tracking device in women with and without FA and before and after the induction of sad mood (Frayn, Sears, & von Ranson, 2016). Those with FA increased their attention toward "unhealthy food images" after the sad mood induction, whereas there was no change in the non-FA group. Secondly, the FA group decreased their attention to unhealthy food images after the sad mood induction was over. The authors interpreted these data as evidence that women with FA display increased emotional reactivity and poorer emotional regulation than those without, a phenomenon that is seen transdiagnostically across various forms of psychopathologies.

In a study by Koball et al. in 923 adult outpatients seeking bariatric surgery, only 14% of patients met YFAS criteria for FA (Koball et al., 2016). However, like in other studies of obese individuals, those with FA were more likely to show higher levels of depression, anxiety, BE episodes, and low eating self-efficacy. In addition, those with FA were more likely to meet criteria for night eating syndrome (NES), but not current substance use. Notably, FA was not predictive of bariatric surgery outcome at 6 or 12 months. Another study has shown that FA appeared to be associated with NES in older individuals (Nolan & Geliebter, 2016). NES is now classified as an "other specified feeding and eating disorder" in DSM-5. In another study of 178 patients who had bariatric surgery, 57.8% met criteria for FA before surgery, but at 6- and 12-month postop, the rates of FA reduced to 7.2% and 13.7%, respectively (Sevincer, Konuk, Bozkurt, & Coskun, 2016).

Another important comorbid psychiatric disorder that has been shown in the last few years to be linked to FA, BE, and obesity is attention-deficit hyperactivity disorder (ADHD) (Davis et al., 2011; Davis, Levitan, Smith, Tweed, & Curtis, 2006). In one study using the National Comorbidity Survey Replication (NCS-R) database, both women and men with a lifetime and current history of BE and BED had significantly higher prevalence estimates of ADHD than those without BE and BED. In addition, women with lifetime and current BN and lifetime AN had significantly higher prevalence estimates of ADHD compared with those women without an ED (Brewerton & Duncan, 2016). In a larger follow-up study using data from both the NCS-R and the National Survey of American Life, the investigators examined the associations between ADHD and EDs after controlling for (1) demographic variables (race, gender, and age), and (2) comorbid psychiatric disorders, including any mood disorder, any anxiety disorder, any SUD, and PTSD (Ziobrowski, Brewerton, & Duncan, 2017). Before controlling for comorbid disorders, significantly higher ORs were found for BN, BED, any ED, and subthreshold BED, although these relationships were attenuated after controlling for comorbidity. Nevertheless, significant relationships between ADHD and BN and any ED remained significant. Although these studies did not measure FA, Davis et al. went a step further and showed that FA was linked not only to BED, depression, addictive traits, hedonic eating, and emotional eating but also to ADHD symptoms in obese adults (Davis et al., 2011). Furthermore, the FA group displayed greater impulsivity and greater emotional reactivity than obese controls. Interestingly, ADHD has also been linked to addictive disorders, anxiety, PTSD, and obesity (Adler, Kunz, Chua, Rotrosen, & Resnick, 2004; Antshel et al., 2013; Biederman et al., 2013; Cortese et al., 2008; Cortese & Vincenzi, 2012; Daviss & Diler, 2012; Khalife et al., 2014; Pagoto et al., 2009; Reimherr, Marchant, Gift, & Steans, 2017; Wilson, 2007; de Zwaan et al., 2011). A large twin study determined that the genetic correlation between ADHD symptoms and BE behavior was 0.35 (95% CI: 0.25−0.46) (Capusan et al., 2017). In addition, the covariance between ADHD and BE behavior was predominantly explained by genetic factors (91%). Furthermore, overlapping neurobehavioral circuits have been reported from neuroimaging studies for ADHD, obesity, and BE (Seymour, Reinblatt, Benson, & Carnell, 2015). Finally, the involvement of dopamine circuits in ADHD, obesity, and BE has also been described (Kaplan, Howlett, & Levitan, 2009).

Hardy et al. assessed comorbid PTSD, depression, childhood and adult trauma, and emotional dysregulation in 229 women who met criteria for FA only, SUD criteria only, or neither (Hardy, Fani, Jovanovic, & Michopoulos, 2018). Prevalence estimates of FA and SUD were 18.3% and 30.6%, respectively. Not surprisingly, in comparison with women without FA or SUD, women with FA and SUD endorsed (1) more PTSD and depression symptoms and (2) higher total emotion dysregulation scores, which included difficulties in goal-directed behaviors, nonacceptance of emotional responses, impulse control, limited access to emotion regulation strategies, and lack of emotional clarity. These results demonstrated how women with FA and SUD share very similar comorbidities and psychological characteristics. However, those women with FA did endorse higher scores on the Childhood Trauma Questionnaire than women with SUD, although this was not a representative sample.

Using the YFAS-C, Mies and coinvestigators examined the prevalence of FA and its association with addictive substances in a large group of Dutch adolescents (n = 2653) (Mies et al., 2017). Symptoms of FA were positively related with alcohol, cannabis, and nicotine use, as well as sugar intake. The authors speculated on possible shared genetic and neurobiological mechanisms that might underlie these links. Others have discussed the neural responses to visual food cues associated with weight status (Brooks, Cedernaes, & Schioth, 2013; Pursey, Stanwell, Callister, et al., 2014), as well as the neural correlates of FA (Gearhardt, Yokum, et al., 2011), which share commonalities with other addictive disorders (Schulte, Grilo, & Gearhardt, 2016).

Interestingly, the concept of FA has been explored in a group of 100 outpatients with schizophrenia attending a clozapine clinic (Goluza, Borchard, Kiarie, Mullan, & Pai, 2017). The prevalence of YFAS-defined FA in this group was 26.7%, higher than numbers found in the general population and lower than those found in populations of EDs or disordered eating. Among those not meeting FA criteria, 77.4% did endorse three or more symptoms of FA but did not endorse distress or impairment, which is a requirement for the diagnosis of FA. Whether schizophrenia itself, or treatment with clozapine, predisposes to FA was unable to be determined in the study.

Food addiction, obesity, and medical morbidity and mortality

FA has been clearly linked to overweight and obesity. Together, they have become the rule rather than the exception in the United States, and this trend has been noted in other Westernized and developing countries and nations (Bhurosy & Jeewon, 2014; Zukiewicz-Sobczak et al., 2014). In addition, the literature is replete with evidence documenting obesity as a major risk factor for several medical conditions and causes of death (Must et al., 1999). In parallel to this line of research, the adverse childhood experiences research studies have identified childhood adversity as a risk factor for obesity and all of the foremost causes of death (Anda et al., 2009; Brown et al., 2009; Felitti, 1991; Felitti et al., 1998). Since the initiation of these studies, other investigators have replicated these results and have searched for possible mechanisms. Studies of stress in animals and PTSD in humans have established major changes in several biological systems, including the cardiovascular system, central and autonomic nervous systems, and endocrine and immunological systems (Baker, Nievergelt, & O'Connor, 2012; Bick et al., 2015; Cohen et al., 2012; Danese et al., 2009, 2008; Menard, Pfau, Hodes, & Russo, 2016; Michopoulos, Norrholm, & Jovanovic, 2015; Pace & Heim, 2011; Rooks, Veledar, Goldberg, Bremner, & Vaccarino, 2012). Other researchers have discovered evidence of telomere shortening as a consequence of severe stress, and this finding has been thought to serve as a marker for augmented aging and early death (Epel et al., 2004; Mason, Prescott, Tworoger, DeVivo, & Rich-Edwards, 2015). An extensive discussion of this topic is beyond the scope of this article. However, when taken together, evidence strongly suggests that the interrelated conditions of obesity, severe binge-type EDs, FA, childhood trauma, PTSD, and major depression have all been associated with significantly increased morbidity and mortality. This perspective forces the astute, responsible clinician to carry out complete psychiatric and medical evaluations, to recognize any and all medical conditions, and to introduce protective measures against further deterioration in health.

Implications for treatment

The emergence and validation of the concept of FA has important implications for assessment and treatment. Several inherent conflicts have been identified between conventional concepts of ED treatment and SUD treatment, including FA (Gearhardt, White, et al., 2011). One of the foundations within the ED treatment community has been the premise that there is no such thing as "bad foods." All foods should be eaten in moderation and in balance with each other, unless a food allergy or intolerance has been medically documented. This has been a particularly essential principle when treating AN, a disorder depicted by unfounded fears about weight gain. Consequently, it is perceived that eating "fattening" or "bad" foods is to be avoided at all costs.

In contrast to this view, the addiction treatment community is much more accepting of the elimination of and abstinence from certain substances, including explicit foods that are experienced as addicting. The typical example of this viewpoint is represented by the philosophy of Overeaters Anonymous (OA). OA was founded in 1960 in California under the premise that those besieged with compulsive overeating, BE, overweight, and/or obesity are addicted to processed foods, including white flour and white sugar.

Given the nascence of this field, evidence-based treatments for FA remain in their infancy. Nevertheless, it can be stated somewhat assuredly what the concept of FA *does not mean* for treatment. Very importantly, it *does not mean* that a 12-step approach should be the exclusive treatment for FA, nor does it mean it should be the treatment of choice. Nevertheless, many addiction treatment professionals erroneously assume that the only useful treatment for any addiction, regardless of type, is a 12-step approach. Furthermore, the FA concept *does not mean* that the OA canon of abolishing white flour and sugar from one's diet is legitimate. This premise used within the principles of OA has never been satisfactorily investigated scientifically. Only anecdotes pro and con this strategy have been published. The concept of FA also *does not mean* that complete abstinence from the addictive substance or behavior is required for improvement. Since the founding of OA nearly 60 years ago, the addiction treatment community has incorporated the "middle road" concept of harm reduction (HR) instead of an "all-or-nothing" position where there is no room for slips or "failure" (Logan & Marlatt, 2010).

In contrast, FA *may mean* that evidence-based therapies used for addictive and related disorders could be potentially useful. This, of course, necessitates further research starting with previously determined effective treatments for addictive disorders that may prove applicable to the treatment of FA. Approaches that have shown efficacy include motivational interviewing and motivational enhancement therapy, cognitive behavioral therapy, dialectical behavior therapy, and other mindfulness-based therapies, such as mindfulness-based relapse prevention, family therapies, and various substance-specific psychopharmacotherapies (Brewerton & Dennis, 2014). Twelve-step facilitation has been found to be effective for alcohol and cocaine dependence and may be helpful for FA. Traditionally, abstinence from addictive substances and behaviors has been a major goal of substance-use treatment, although it is certainly not always achieved. Evidence suggests that a significant reduction of substance use or addictive behaviors may be now seen as successful outcomes. This HR strategy has evolved for addictions to licit and illicit drugs and alcohol to replace the extremes of a rigid "all-or-nothing" attitude or "zero tolerance approach." This HR philosophy emerged to more effectively meet the needs of patients struggling with chronic conditions, and it may be especially useful for FA given how pervasive and available processed and ultra-processed foods have become. Similarly, significant reductions in binge frequency—not just achievement of full remission of symptoms—have been defined as successes in the ED field. HR stresses practical instead of idealized goals, and it is founded in the continuously growing public health and advocacy movements. In fact, HR has been embraced by Canada as well as other major countries as officially accepted policy given its proven efficacy shown (Logan & Marlatt, 2010). A guiding feature of HR is acceptance of the fact that some drug users for multiple reasons cannot be expected to cease their drug use at the present time.

Summary

This chapter reviews the existing literature on FA, as defined by the YFAS, and its associations with EDs, obesity, trauma, and comorbid disorders (see Table 15.1). An argument is made that FA, as defined by the YFAS, may serve as a marker for (1) the presence and severity of binge-type EDs, (2) the presence and severity of obesity, (3) the presence of significant trauma histories, (4) the presence and severity of PTSD and its symptoms, (5) the presence of psychiatric and medical comorbidities, and (6) the combination of the above features. Implications for the treatment of FA have also been reviewed in terms of what FA implies and does not imply. Finally, the justification for a comprehensive approach that necessitates thorough attentiveness to both psychiatric and medical assessment and integrated care that encompasses trauma-informed care is supported (Brewerton, 2018).

Table 15.1 Comorbid disorders reported to be linked to food addiction.

I. Binge-type eating disorders
A. Bulimia nervosa
B. Binge eating disorder
C. Anorexia nervosa binge–purge type
D. Night eating syndrome (other specified feeding and eating disorder)
II. Other psychiatric disorders
A. Mood disorders
B. Posttraumatic stress disorder
C. Substance-use disorders
D. Anxiety disorders
E. Attention-deficit hyperactivity disorder
F. Schizophrenia
III. Medical morbidity and mortality
A. Obesity
B. Cardiovascular disease
C. Central and autonomic nervous system disease
D. Endocrine disease
E. Immunological disease

References

Adler, L. A., Kunz, M., Chua, H. C., Rotrosen, J., & Resnick, S. G. (2004). Attention-deficit/hyperactivity disorder in adult patients with posttraumatic stress disorder (PTSD): Is ADHD a vulnerability factor? *Journal of Attention Disorders, 8*(1), 11–16.

Anda, R. F., Dong, M., Brown, D. W., Felitti, V. J., Giles, W. H., Perry, G. S., … Dube, S. R. (2009). The relationship of adverse childhood experiences to a history of premature death of family members. *BMC Public Health, 9*, 106. https://doi.org/10.1186/1471-2458-9-106.

Antshel, K. M., Kaul, P., Biederman, J., Spencer, T. J., Hier, B. O., Hendricks, K., & Faraone, S. V. (2013). Posttraumatic stress disorder in adult attention-deficit/hyperactivity disorder: Clinical features and familial transmission. *Journal of Clinical Psychiatry, 74*(3), e197–204. https://doi.org/10.4088/JCP.12m07698.

Avena, N. M., Gold, J. A., Kroll, C., & Gold, M. S. (2012). Further developments in the neurobiology of food and addiction: Update on the state of the science. *Nutrition, 28*(4), 341–343. https://doi.org/10.1016/j.nut.2011.11.002.

Avena, N. M., Rada, P., & Hoebel, B. G. (2009). Sugar and fat bingeing have notable differences in addictive-like behavior. *Journal of Nutrition, 139*(3), 623–628. https://doi.org/10.3945/jn.108.097584.

Avena, N. M., Wang, M., & Gold, M. S. (2011). Implications of food addiction and drug use in obesity. *Psychiatric Annals, 41*(10), 478–482. https://doi.org/10.3928/00485713-20110921-06.

Baker, D. G., Nievergelt, C. M., & O'Connor, D. T. (2012). Biomarkers of PTSD: Neuropeptides and immune signaling. *Neuropharmacology, 62*(2), 663–673. https://doi.org/10.1016/j.neuropharm.2011.02.027.

Benton, D. (2010). The plausibility of sugar addiction and its role in obesity and eating disorders. *Clinical Nutrition, 29*(3), 288–303. https://doi.org/10.1016/j.clnu.2009.12.001.

Bhurosy, T., & Jeewon, R. (2014). Overweight and obesity epidemic in developing countries: A problem with diet, physical activity, or socioeconomic status? *ScientificWorldJournal, 2014*, 964236. https://doi.org/10.1155/2014/964236.

Bick, J., Nguyen, V., Leng, L., Piecychna, M., Crowley, M. J., Bucala, R., ... Grigorenko, E. L. (2015). Preliminary associations between childhood neglect, MIF, and cortisol: Potential pathways to long-term disease risk. *Developmental Psychobiology, 57*(1), 131–139. https://doi.org/10.1002/dev.21265.

Biederman, J., Petty, C. R., Spencer, T. J., Woodworth, K. Y., Bhide, P., Zhu, J., & Faraone, S. V. (2013). Examining the nature of the comorbidity between pediatric attention deficit/hyperactivity disorder and post-traumatic stress disorder. *Acta Psychiatrica Scandinavica, 128*(1), 78–87. https://doi.org/10.1111/acps.12011.

Brady, K. T., Killeen, T. K., Brewerton, T., & Lucerini, S. (2000). Comorbidity of psychiatric disorders and posttraumatic stress disorder. *Journal of Clinical Psychiatry, 61*(Suppl. 7), 22–32.

Brewerton, T. D. (2004). Eating disorders, victimization, and comorbidity: Principles of treatment. In T. D. Brewerton (Ed.), *Clinical handbook of eating disorders: An integrated approach* (pp. 509–545). New York: Marcel Decker.

Brewerton, T. D. (2007). Eating disorders, trauma, and comorbidity: Focus on PTSD. *Eating Disorders, 15*(4), 285–304. https://doi.org/10.1080/10640260701454311.

Brewerton, T. D. (2011). Posttraumatic stress disorder and disordered eating: Food addiction as self-medication. *Journal of Women's Health (Larchmt), 20*(8), 1133–1134. https://doi.org/10.1089/jwh.2011.3050.

Brewerton, T. D. (2015). Stress, trauma, and adversity as risk factors in the development of eating disorders. In L. Smolak, & M. Levine (Eds.), *Wiley handbook of eating disorders* (pp. 445–460). New York: Guilford.

Brewerton, T. D. (2017). Food addiction as a proxy for eating disorder and obesity severity, trauma history, PTSD symptoms, and comorbidity. *Eating and Weight Disorders, 22*(2), 241–247. https://doi.org/10.1007/s40519-016-0355-8.

Brewerton, T. D. (2018). An overview of trauma-informed care and practice for eating disorders. *Journal of Aggression, Maltreatment & Trauma*, 1–18. https://doi.org/10.1080/10926771.2018.1532940.

Brewerton, T. D., & Brady, K. (2014). The role of stress, trauma, and PTSD in the etiology and treatment of eating disorders, addictions, and substance use disorders. In T. D. Brewerton, & A. B. Dennis (Eds.), *Eating disorders, addictions, and substance use disorders: Research, clinical and treatment perspectives* (pp. 379–404). Berlin: Springer.

Brewerton, T. D., Dansky, B. S., Kilpatrick, D. G., & O'Neil, P. M. (1999). Bulimia nervosa, PTSD, and forgetting: Results from the national women's study. In L. M. Williams, & V. L. Banyard (Eds.), *Trauma and memory* (pp. 127–138). Durham: Sage.

Brewerton, T. D., Dansky, B. S., O'Neil, P. M., & Kilpatrick, D. G. (2015). The number of divergent purging behaviors is associated with histories of trauma, PTSD, and comorbidity in a national sample of women. *Eating Disorders*, 1–8. https://doi.org/10.1080/10640266.2015.1013394.

Brewerton, T. D., & Dennis, A. B. (2014). In T. D. Brewerton, & A. B. Dennis (Eds.), *Eating disorders, addictions and substance use disorders*. Berlin: Springer.

Brewerton, T. D., & Duncan, A. E. (2016). Associations between attention deficit hyperactivity disorder and eating disorders by gender: Results from the National Comorbidity

Survey Replication. *European Eating Disorders Review, 24*(6), 536–540. https://doi.org/10.1002/erv.2468.

Brewerton, T. D., O'Neil, P. M., Dansky, B. S., & Kilpatrick, D. G. (2015). Extreme obesity and its associations with victimization, PTSD, major depression and eating disorders in a national sample of women. *Journal of Obesity & Eating Disorders, 1*(2), 1–9. https://doi.org/10.4172/2471-8203.100010.

Brewerton, T. D., Rance, S. J., Dansky, B. S., O'Neil, P. M., & Kilpatrick, D. G. (2014). A comparison of women with child-adolescent versus adult onset binge eating: Results from the national women's study. *International Journal of Eating Disorders, 47*(7), 836–843. https://doi.org/10.1002/eat.22309.

Brooks, S. J., Cedernaes, J., & Schioth, H. B. (2013). Increased prefrontal and parahippocampal activation with reduced dorsolateral prefrontal and insular cortex activation to food images in obesity: A meta-analysis of fMRI studies. *PLoS One, 8*(4), e60393. https://doi.org/10.1371/journal.pone.0060393.

Brown, D. W., Anda, R. F., Tiemeier, H., Felitti, V. J., Edwards, V. J., Croft, J. B., & Giles, W. H. (2009). Adverse childhood experiences and the risk of premature mortality. *American Journal of Preventive Medicine, 37*(5), 389–396. https://doi.org/10.1016/j.amepre.2009.06.021.

Burmeister, J. M., Hinman, N., Koball, A., Hoffmann, D. A., & Carels, R. A. (2013). Food addiction in adults seeking weight loss treatment. Implications for psychosocial health and weight loss. *Appetite, 60*(1), 103–110. https://doi.org/10.1016/j.appet.2012.09.013.

Capusan, A. J., Yao, S., Kuja-Halkola, R., Bulik, C. M., Thornton, L. M., Bendtsen, P., … Larsson, H. (2017). Genetic and environmental aspects in the association between attention-deficit hyperactivity disorder symptoms and binge-eating behavior in adults: A twin study. *Psychological Medicine*, 1–13. https://doi.org/10.1017/S0033291717001416.

Chao, A. M., Shaw, J. A., Pearl, R. L., Alamuddin, N., Hopkins, C. M., Bakizada, Z. M., … Wadden, T. A. (2017). Prevalence and psychosocial correlates of food addiction in persons with obesity seeking weight reduction. *Comprehensive Psychiatry, 73*, 97–104. https://doi.org/10.1016/j.comppsych.2016.11.009.

Cohen, S., Janicki-Deverts, D., Doyle, W. J., Miller, G. E., Frank, E., Rabin, B. S., & Turner, R. B. (2012). Chronic stress, glucocorticoid receptor resistance, inflammation, and disease risk. *Proceedings of the National Academy of Sciences of the United States of America, 109*(16), 5995–5999. https://doi.org/10.1073/pnas.1118355109.

Corsica, J. A., & Pelchat, M. L. (2010). Food addiction: True or false? *Current Opinion in Gastroenterology, 26*(2), 165–169. https://doi.org/10.1097/MOG.0b013e328336528d.

Corsica, J. A., & Spring, B. J. (2008). Carbohydrate craving: A double-blind, placebo-controlled test of the self-medication hypothesis. *Eating Behaviors, 9*(4), 447–454. https://doi.org/10.1016/j.eatbeh.2008.07.004.

Cortese, S., Angriman, M., Maffeis, C., Isnard, P., Konofal, E., Lecendreux, M., … Mouren, M. C. (2008). Attention-deficit/hyperactivity disorder (ADHD) and obesity: A systematic review of the literature. *Critical Reviews in Food Science and Nutrition, 48*(6), 524–537. https://doi.org/10.1080/10408390701540124.

Cortese, S., & Vincenzi, B. (2012). Obesity and ADHD: Clinical and neurobiological implications. *Current Topics in Behavioral Neurosciences, 9*, 199–218. https://doi.org/10.1007/7854_2011_154.

Danese, A., Moffitt, T. E., Harrington, H., Milne, B. J., Polanczyk, G., Pariante, C. M., … Caspi, A. (2009). Adverse childhood experiences and adult risk factors for age-related disease: Depression, inflammation, and clustering of metabolic risk markers. *Archives of Pediatrics and Adolescent Medicine, 163*(12), 1135–1143. https://doi.org/10.1001/archpediatrics.2009.214.

Danese, A., Moffitt, T. E., Pariante, C. M., Ambler, A., Poulton, R., & Caspi, A. (2008). Elevated inflammation levels in depressed adults with a history of childhood maltreatment. *Archives of General Psychiatry, 65*(4), 409–415. https://doi.org/10.1001/archpsyc.65.4.409.

Dansky, B. S., Brewerton, T. D., O'Neil, P. M., & Kilpatrick, D. G. (1997). The National Womens Study: Relationship of victimization and posttraumatic stress disorder to bulimia nervosa. *International Journal of Eating Disorders, 21*, 213–228.

Davis, C., Curtis, C., Levitan, R. D., Carter, J. C., Kaplan, A. S., & Kennedy, J. L. (2011). Evidence that 'food addiction' is a valid phenotype of obesity. *Appetite, 57*(3), 711–717. https://doi.org/10.1016/j.appet.2011.08.017.

Davis, C., Levitan, R. D., Smith, M., Tweed, S., & Curtis, C. (2006). Associations among overeating, overweight, and attention deficit/hyperactivity disorder: A structural equation modelling approach. *Eating Behaviors, 7*(3), 266–274. https://doi.org/10.1016/j.eatbeh.2005.09.006.

Daviss, W. B., & Diler, R. (2012). Does comorbid depression predict subsequent adverse life events in youth with attention-deficit/hyperactivity disorders? *Journal of Child and Adolescent Psychopharmacology, 22*(1), 65–71. https://doi.org/10.1089/cap.2011.0046.

Epel, E. S., Blackburn, E. H., Lin, J., Dhabhar, F. S., Adler, N. E., Morrow, J. D., & Cawthon, R. M. (2004). Accelerated telomere shortening in response to life stress. *Proceedings of the National Academy of Sciences of the United States of America, 101*(49), 17312–17315. https://doi.org/10.1073/pnas.0407162101.

Felitti, V. J. (1991). Long-term medical consequences of incest, rape, and molestation. *Southern Medical Journal, 84*, 328–331.

Felitti, V. J. (1993). Childhood sexual abuse, depression and family dysfunction in adult obese patients: A case control study. *Southern Medical Journal, 86*, 732–736.

Felitti, V. J., Anda, R. F., Nordenberg, D., Williamson, D. F., Spitz, A. M., Edwards, V., … Marks, J. S. (1998). Relationship of childhood abuse and household dysfunction to many of the leading causes of death in adults. *American Journal of Preventive Medicine, 14*, 245–258.

Fortuna, J. L. (2012). The obesity epidemic and food addiction: Clinical similarities to drug dependence. *Journal of Psychoactive Drugs, 44*(1), 56–63. https://doi.org/10.1080/02791072.2012.662092.

Frayn, M., Sears, C. R., & von Ranson, K. M. (2016). A sad mood increases attention to unhealthy food images in women with food addiction. *Appetite, 100*, 55–63. https://doi.org/10.1016/j.appet.2016.02.008.

Gearhardt, A. N., Corbin, W. R., & Brownell, K. D. (2009). Preliminary validation of the Yale food addiction Scale. *Appetite, 52*(2), 430–436. https://doi.org/10.1016/j.appet.2008.12.003.

Gearhardt, A. N., Davis, C., Kuschner, R., & Brownell, K. D. (2011a). The addiction potential of hyperpalatable foods. *Current Drug Abuse Reviews, 4*(3), 140–145.

Gearhardt, A. N., Grilo, C. M., DiLeone, R. J., Brownell, K. D., & Potenza, M. N. (2011b). Can food be addictive? Public health and policy implications. *Addiction, 106*(7), 1208–1212. https://doi.org/10.1111/j.1360-0443.2010.03301.x.

Gearhardt, A. N., Roberto, C. A., Seamans, M. J., Corbin, W. R., & Brownell, K. D. (2013a). Preliminary validation of the Yale food addiction Scale for children. *Eating Behaviors, 14*(4), 508–512. https://doi.org/10.1016/j.eatbeh.2013.07.002.

Gearhardt, A. N., White, M. A., Masheb, R. M., & Grilo, C. M. (2013b). An examination of food addiction in a racially diverse sample of obese patients with binge eating disorder in primary care settings. *Comprehensive Psychiatry, 54*(5), 500–505. https://doi.org/10.1016/j.comppsych.2012.12.009.

Gearhardt, A. N., White, M. A., Masheb, R. M., Morgan, P. T., Crosby, R. D., & Grilo, C. M. (2012). An examination of the food addiction construct in obese patients with binge eating disorder. *International Journal of Eating Disorders, 45*(5), 657–663. https://doi.org/10.1002/eat.20957.

Gearhardt, A. N., White, M. A., & Potenza, M. N. (2011c). Binge eating disorder and food addiction. *Current Drug Abuse Reviews, 4*, 201–207.

Gearhardt, A. N., Yokum, S., Orr, P. T., Stice, E., Corbin, W. R., & Brownell, K. D. (2011d). Neural correlates of food addiction. *Archives of General Psychiatry, 68*(8), 808–816. https://doi.org/10.1001/archgenpsychiatry.2011.32.

Gold, M. S., Graham, N. A., Cocores, J. A., & Nixon, S. J. (2009). Food addiction? *Journal of Addiction Medicine, 3*(1), 42–45. https://doi.org/10.1097/ADM.0b013e318199cd20.

Goluza, I., Borchard, J., Kiarie, E., Mullan, J., & Pai, N. (2017). Exploration of food addiction in people living with schizophrenia. *Asian Journal of Psychiatry, 27*, 81–84. https://doi.org/10.1016/j.ajp.2017.02.022.

Granero, R., Hilker, I., Aguera, Z., Jimenez-Murcia, S., Sauchelli, S., Islam, M. A., … Fernandez-Aranda, F. (2014). Food addiction in a Spanish sample of eating disorders: DSM-5 diagnostic subtype differentiation and validation data. *European Eating Disorders Review, 22*(6), 389–396. https://doi.org/10.1002/erv.2311.

Granero, R., Jiménez-Murcia, S., Gerhardt, A. N., Agüera, Z., Aymamí, N., Gómez-Peña, M., … Fernández-Aranda, F. (2018). Validation of the Spanish version of the Yale food addiction Scale 2.0 (YFAS 2.0) and clinical correlates in a sample of eating disorder, gambling disorder, and healthy control participants. *Frontiers in Psychiatry, 9*. https://doi.org/10.3389/fpsyt.2018.00208.

Hardy, R., Fani, N., Jovanovic, T., & Michopoulos, V. (2018). Food addiction and substance addiction in women: Common clinical characteristics. *Appetite, 120*, 367–373. https://doi.org/10.1016/j.appet.2017.09.026.

Hebebrand, J., Albayrak, O., Adan, R., Antel, J., Dieguez, C., de Jong, J., … Dickson, S. L. (2014). "Eating addiction", rather than "food addiction", better captures addictive-like eating behavior. *Neuroscience & Biobehavioral Reviews, 47*, 295–306. https://doi.org/10.1016/j.neubiorev.2014.08.016.

Hirth, J. M., Rahman, M., & Berenson, A. B. (2011). The association of posttraumatic stress disorder with fast food and soda consumption and unhealthy weight loss behaviors among young women. *Journal of Women's Health (Larchmt), 20*(8), 1141–1149. https://doi.org/10.1089/jwh.2010.2675.

Hoebel, B. G., Avena, N. M., Bocarsly, M. E., & Rada, P. (2009). Natural addiction: A behavioral and circuit model based on sugar addiction in rats. *Journal of Addiction Medicine, 3*(1), 33–41. https://doi.org/10.1097/ADM.0b013e31819aa621.

Imperatori, C., Innamorati, M., Contardi, A., Continisio, M., Tamburello, S., Lamis, D. A., … Fabbricatore, M. (2014). The association among food addiction, binge eating severity and psychopathology in obese and overweight patients attending low-energy-diet therapy.

Comprehensive Psychiatry, 55(6), 1358–1362. https://doi.org/10.1016/j.comppsych.2014.04.023.

Imperatori, C., Innamorati, M., Lamis, D. A., Farina, B., Pompili, M., Contardi, A., & Fabbricatore, M. (2016). Childhood trauma in obese and overweight women with food addiction and clinical-level of binge eating. *Child Abuse & Neglect, 58*, 180–190. https://doi.org/10.1016/j.chiabu.2016.06.023.

Joranby, L., Pineda, K. F., & Gold, M. S. (2005). Addiction to food and brain reward systems. *Sexual Addiction & Compulsivity, 12*(2–3), 201–217. https://doi.org/10.1080/10720160500203765.

Kaplan, A. S., Howlett, A. L., & Levitan, R. D. (2009). Attention deficit hyperactivity disorder and binge eating shared phenomenology, genetics and response to treatment. *International Journal of Child Health and Adolescent Health, 2*(2), 165–174.

Khalife, N., Kantomaa, M., Glover, V., Tammelin, T., Laitinen, J., Ebeling, H., … Rodriguez, A. (2014). Childhood attention-deficit/hyperactivity disorder symptoms are risk factors for obesity and physical inactivity in adolescence. *Journal of the American Academy of Child & Adolescent Psychiatry, 53*(4), 425–436. https://doi.org/10.1016/j.jaac.2014.01.009.

Koball, A. M., Clark, M. M., Collazo-Clavell, M., Kellogg, T., Ames, G., Ebbert, J., & Grothe, K. B. (2016). The relationship among food addiction, negative mood, and eating-disordered behaviors in patients seeking to have bariatric surgery. *Surgery for Obesity and Related Diseases, 12*(1), 165–170. https://doi.org/10.1016/j.soard.2015.04.009.

Lee, N. M., Lucke, J., Hall, W. D., Meurk, C., Boyle, F. M., & Carter, A. (2013). Public views on food addiction and obesity: Implications for policy and treatment. *PLoS One, 8*(9), e74836. https://doi.org/10.1371/journal.pone.0074836.

Lemeshow, A. R., Rimm, E. B., Hasin, D. S., Gearhardt, A. N., Flint, A. J., Field, A. E., & Genkinger, J. M. (2018). Food and beverage consumption and food addiction among women in the nurses' health studies. *Appetite, 121*, 186–197. https://doi.org/10.1016/j.appet.2017.10.038.

Liu, Y., von Deneen, K. M., Kobeissy, F. H., & Gold, M. S. (2010). Food addiction and obesity: Evidence from bench to bedside. *Journal of Psychoactive Drugs, 42*(2), 133–145. https://doi.org/10.1080/02791072.2010.10400686.

Logan, D. E., & Marlatt, G. A. (2010). Harm reduction therapy: A practice-friendly review of research. *Journal of Clinical Psychology, 66*(2), 201–214. https://doi.org/10.1002/jclp.20669.

Lustig, R. H. (2010). Fructose: Metabolic, hedonic, and societal parallels with ethanol. *Journal of the American Dietetic Association, 110*(9), 1307–1321. https://doi.org/10.1016/j.jada.2010.06.008.

Mason, S. M., Bryn Austin, S., Bakalar, J. L., Boynton-Jarrett, R., Field, A. E., Gooding, H. C., … Rich-Edwards, J. W. (2016). Child maltreatment's heavy toll: The need for trauma-informed obesity prevention. *American Journal of Preventive Medicine, 50*(5), 646–649. https://doi.org/10.1016/j.amepre.2015.11.004.

Mason, S. M., Flint, A. J., Field, A. E., Austin, S. B., & Rich-Edwards, J. W. (2013). Abuse victimization in childhood or adolescence and risk of food addiction in adult women. *Obesity, 21*(12), E775–E781. https://doi.org/10.1002/oby.20500.

Mason, S. M., Flint, A. J., Roberts, A. L., Agnew-Blais, J., Koenen, K. C., & Rich-Edwards, J. W. (2014). Posttraumatic stress disorder symptoms and food addiction in women by timing and type of trauma exposure. *JAMA Psychiatry, 71*(11), 1271–1278. https://doi.org/10.1001/jamapsychiatry.2014.1208.

Mason, S. M., MacLehose, R. F., Katz-Wise, S. L., Austin, S. B., Neumark-Sztainer, D., Harlow, B. L., & Rich-Edwards, J. W. (2015). Childhood abuse victimization, stress-related eating, and weight status in young women. *Annals of Epidemiology, 25*(10), 760–766. https://doi.org/10.1016/j.annepidem.2015.06.081. e762.

Mason, S. M., Prescott, J., Tworoger, S. S., DeVivo, I., & Rich-Edwards, J. W. (2015). Childhood physical and sexual abuse history and leukocyte telomere length among women in middle adulthood. *PLoS One, 10*(6), e0124493. https://doi.org/10.1371/journal.pone.0124493.

Menard, C., Pfau, M. L., Hodes, G. E., & Russo, S. J. (2016). Immune and neuroendocrine mechanisms of stress vulnerability and resilience. *Neuropsychopharmacology*, 1–19. https://doi.org/10.1038/npp.2016.90.

Meule, A., & Gearhardt, A. N. (2014). Food addiction in the light of DSM-5. *Nutrients, 6*(9), 3653–3671. https://doi.org/10.3390/nu6093653.

Meule, A., von Rezori, V., & Blechert, J. (2014). Food addiction and bulimia nervosa. *European Eating Disorders Review, 22*(5), 331–337. https://doi.org/10.1002/erv.2306.

Michopoulos, V., Norrholm, S. D., & Jovanovic, T. (2015). Diagnostic biomarkers for posttraumatic stress disorder: Promising horizons from translational neuroscience research. *Biological Psychiatry, 78*(5), 344–353. https://doi.org/10.1016/j.biopsych.2015.01.005.

Mies, G. W., Treur, J. L., Larsen, J. K., Halberstadt, J., Pasman, J. A., & Vink, J. M. (2017). The prevalence of food addiction in a large sample of adolescents and its association with addictive substances. *Appetite, 118*, 97–105. https://doi.org/10.1016/j.appet.2017.08.002.

Mitchell, K. S., Mazzeo, S. E., Schlesinger, M. R., Brewerton, T. D., & Smith, B. N. (2012). Comorbidity of partial and subthreshold PTSD among men and women with eating disorders in the National Comorbidity Survey-Replication Study. *International Journal of Eating Disorders, 45*(3), 307–315. https://doi.org/10.1002/eat.20965.

Mitchell, K. S., & Wolf, E. J. (2016). PTSD, food addiction, and disordered eating in a sample of primarily older veterans: The mediating role of emotion regulation. *Psychiatry Research, 243*, 23–29. https://doi.org/10.1016/j.psychres.2016.06.013.

Moore, C. F., Sabino, V., Koob, G. F., & Cottone, P. (2017). Pathological overeating: Emerging evidence for a compulsivity construct. *Neuropsychopharmacology, 42*(7), 1375–1389. https://doi.org/10.1038/npp.2016.269.

Must, A., Spadano, J., Coakley, E. H., Field, A. E., Golditz, G., & Dietz, W. H. (1999). The disease burden associated with overweight and obesity. *Journal of the American Medical Association, 282*(16), 1523–1529.

Nolan, L. J., & Geliebter, A. (2016). "Food addiction" is associated with night eating severity. *Appetite, 98*, 89–94.

Noll, J. G., Zeller, M. H., Trickett, P. K., & Putnam, F. W. (2007). Obesity risk for female victims of childhood sexual abuse: A prospective study. *Pediatrics, 120*(1), e61–67. https://doi.org/10.1542/peds.2006-3058.

Pace, T. W., & Heim, C. M. (2011). A short review on the psychoneuroimmunology of posttraumatic stress disorder: From risk factors to medical comorbidities. *Brain, Behavior, and Immunity, 25*(1), 6–13. https://doi.org/10.1016/j.bbi.2010.10.003.

Pagoto, S. L., Curtin, C., Lemon, S. C., Bandini, L. G., Schneider, K. L., Bodenlos, J. S., & Ma, Y. (2009). Association between adult attention deficit/hyperactivity disorder and obesity in the US population. *Obesity, 17*(3), 539–544. https://doi.org/10.1038/oby.2008.587.

Pagoto, S. L., Schneider, K. L., Bodenlos, J. S., Appelhans, B. M., Whited, M. C., Ma, Y., & Lemon, S. C. (2012). Association of post-traumatic stress disorder and obesity in a nationally representative sample. *Obesity, 20*(1), 200–205. https://doi.org/10.1038/oby.2011.318.

Pursey, K. M., Stanwell, P., Callister, R. J., Brain, K., Collins, C. E., & Burrows, T. L. (2014). Neural responses to visual food cues according to weight status: A systematic review of functional magnetic resonance imaging studies. *Frontiers in Nutrition, 1*, 7. https://doi.org/10.3389/fnut.2014.00007.

Pursey, K. M., Stanwell, P., Gearhardt, A. N., Collins, C. E., & Burrows, T. L. (2014). The prevalence of food addiction as assessed by the Yale food addiction Scale: A systematic review. *Nutrients, 6*(10), 4552–4590. https://doi.org/10.3390/nu6104552.

Rainey, J. C., Furman, C. R., & Gearhardt, A. N. (2018). Food addiction among sexual minorities. *Appetite, 120*, 16–22. https://doi.org/10.1016/j.appet.2017.08.019.

Randolph, T. G. (1956). The descriptive features of food addiction; addictive eating and drinking. *Quarterly Journal of Studies on Alcohol – Part A, 17*(2), 198–224.

Reimherr, F. W., Marchant, B. K., Gift, T. E., & Steans, T. A. (2017). ADHD and anxiety: Clinical significance and treatment implications. *Current Psychiatry Reports, 19*(12), 109. https://doi.org/10.1007/s11920-017-0859-6.

Rooks, C., Veledar, E., Goldberg, J., Bremner, J. D., & Vaccarino, V. (2012). Early trauma and inflammation: Role of familial factors in a study of twins. *Psychosomatic Medicine, 74*(2), 146–152. https://doi.org/10.1097/PSY.0b013e318240a7d8.

Schulte, E. M., Avena, N. M., & Gearhardt, A. N. (2015). Which foods may be addictive? The roles of processing, fat content, and glycemic load. *PLoS One, 10*(2), e0117959. https://doi.org/10.1371/journal.pone.0117959.

Schulte, E. M., Grilo, C. M., & Gearhardt, A. N. (2016). Shared and unique mechanisms underlying binge eating disorder and addictive disorders. *Clinical Psychology Review.*

Schulte, E. M., Potenza, M. N., & Gearhardt, A. N. (2017). A commentary on the "eating addiction" versus "food addiction" perspectives on addictive-like food consumption. *Appetite, 115*, 9–15. https://doi.org/10.1016/j.appet.2016.10.033.

Schulte, E. M., Smeal, J. K., & Gearhardt, A. N. (2017). Foods are differentially associated with subjective effect report questions of abuse liability. *PLoS One, 12*(8), e0184220. https://doi.org/10.1371/journal.pone.0184220.

Sevincer, G. M., Konuk, N., Bozkurt, S., & Coskun, H. (2016). Food addiction and the outcome of bariatric surgery at 1-year: Prospective observational study. *Psychiatry Research, 244*, 159–164. https://doi.org/10.1016/j.psychres.2016.07.022.

Seymour, K. E., Reinblatt, S. P., Benson, L., & Carnell, S. (2015). Overlapping neurobehavioral circuits in ADHD, obesity, and binge eating: Evidence from neuroimaging research. *CNS Spectrums, 20*(4), 401–411. https://doi.org/10.1017/S1092852915000383.

The_National_Center_on_Addiction_and_Substance_Abuse. (2016). *Understanding and addressing food addiction: A science-based approach to policy, practice and research.*

Vieweg, W. V., Fernandez, A., Julius, D. A., Satterwhite, L., Benesek, J., Feuer, S. J., … Pandurangi, A. K. (2006). Body mass index relates to males with posttraumatic stress disorder. *Journal of the National Medical Association, 98*(4), 580–586.

Vieweg, W. V., Julius, D. A., Bates, J., Quinn, J. F., 3rd, Fernandez, A., Hasnain, M., & Pandurangi, A. K. (2007). Posttraumatic stress disorder as a risk factor for obesity among male military veterans. *Acta Psychiatrica Scandinavica, 116*(6), 483–487. https://doi.org/10.1111/j.1600-0447.2007.01071.x.

Whitaker, R. C., Phillips, S. M., Orzol, S. M., & Burdette, H. L. (2007). The association between maltreatment and obesity among preschool children. *Child Abuse & Neglect, 31*(11–12), 1187–1199. https://doi.org/10.1016/j.chiabu.2007.04.008.

Wilson, J. J. (2007). ADHD and substance use disorders: Developmental aspects and the impact of stimulant treatment. *American Journal on Addictions, 16*(Suppl. 1), 5–11. https://doi.org/10.1080/10550490601082734. quiz 12–13.

Wiss, D. A., Criscitelli, K., Gold, M., & Avena, N. (2017). Preclinical evidence for the addiction potential of highly palatable foods: Current developments related to maternal influence. *Appetite, 115*, 19–27. https://doi.org/10.1016/j.appet.2016.12.019.

Wonderlich, S. A., Brewerton, T. D., Jocic, Z., Dansky, B. S., & Abbott, D. W. (1997). Relationship of childhood sexual abuse and eating disorders. *Journal of the American Academy of Child & Adolescent Psychiatry, 36*(8), 1107–1115.

Ziobrowski, H., Brewerton, T. D., & Duncan, A. E. (2017). Associations between ADHD and eating disorders in relation to comorbid psychiatric disorders in a nationally representative sample. *Psychiatry Research, 260*, 53–59. https://doi.org/10.1016/j.psychres.2017.11.026.

Zukiewicz-Sobczak, W., Wroblewska, P., Zwolinski, J., Chmielewska-Badora, J., Adamczuk, P., Krasowska, E., … Silny, W. (2014). Obesity and poverty paradox in developed countries. *Annals of Agricultural and Environmental Medicine, 21*(3), 590–594. https://doi.org/10.5604/12321966.1120608.

de Zwaan, M., Gruss, B., Muller, A., Philipsen, A., Graap, H., Martin, A., … Hilbert, A. (2011). Association between obesity and adult attention-deficit/hyperactivity disorder in a German community-based sample. *Obesity Facts, 4*(3), 204–211. https://doi.org/10.1159/000329565.

Index

'*Note*: Page numbers followed by "f" indicate figures and "t" indicate tables.'

D

E

F

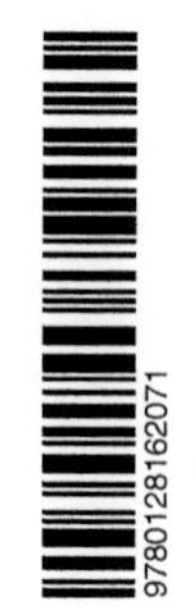